# Food Analogues: Emerging Methods and Challenges

Özlem Pelin CAN
Meryem GÖKSEL SARAÇ
Duygu ASLAN TÜRKER
Editors

# Food Analogues: Emerging Methods and Challenges

Springer

*Editors*
Özlem Pelin CAN
Department of Food Hygiene and Technology
Sivas Cumhuriyet University
Sivas, Türkiye

Meryem GÖKSEL SARAÇ
Department of Food Technology
Sivas Cumhuriyet University
Sivas, Türkiye

Duygu ASLAN TÜRKER
Department of Food Engineering
Erciyes University
Kayseri, Türkiye

ISBN 978-3-031-69871-2 ISBN 978-3-031-69872-9 (eBook)
https://doi.org/10.1007/978-3-031-69872-9



This Springer imprint is published by the registered company Springer Nature Switzerland AG
The registered company address is: Gewerbestrasse 11, 6330 Cham, Switzerland

# Preface

The world population is increasing exponentially, and with this increase, the demand for food is also rising significantly. According to the United Nations (UN) 2050 estimates, the world population is expected to reach approximately 10 billion people. This rapid population growth puts significant pressure on food production and natural resources. As a result, there is a significantly increasing interest in food analogues, which are alternatives that imitate the properties of products that humans have traditionally produced and consumed since their existence. This growing importance is due to consumers' increasing health awareness and environmental concerns. Factors affecting this transformation include changing nutritional preferences, decreasing fat and calorie intake, reducing greenhouse gas emissions, and protecting natural resources.

Food analogues are defined as natural or synthetic substances used in place of traditional food products or ingredients. These alternatives go by a variety of names, such as food alternatives, substitutes, imitations, or even fake food.

This book is a collection of literature on food analogue groups, raw materials used in their production, production techniques, product types and their characterization, packaging techniques, and sensory characterization.

It contains very useful information for businesses producing analogues, consumers who prefer such products, and researchers working in this field.

It is evident in both the retail and food service sectors that the food analogue industry is growing rapidly. Therefore, the main purpose of writing this book is to provide comprehensive information on the properties of food analogue products for food manufacturers, dietitians, nutritionists, regulators, and most importantly consumers. This book can be used successfully in their decision-making efforts.

Additionally, this book aims to be a reference source for assessing the future growth and development of the food analogues industry. The editors express their sincere gratitude to all authors who contributed to the realization of this important mission.

We would like to thank all the chapter authors for their cooperation and valuable contributions to the preparation of this book. We would also like to thank Julie Goldberg, Food Science Associate Editor at Springer Nature, and Vyshnavi Kalanatha, Production Editor at Springer Nature, for their valuable suggestions and guidance.

Sivas, Türkiye — Meryem GÖKSEL SARAÇ
Kayseri, Türkiye — Duygu ASLAN TÜRKER
Sivas, Türkiye — Özlem Pelin CAN

# Contents

# About the Editors

**Özlem Pelin CAN** received her undergraduate degree from Fırat University Faculty of Veterinary Medicine and her doctorate degree from Fırat University Faculty of Veterinary Medicine, Department of Food Hygiene and Technology. She entered academic life in 2001. She worked as a research assistant at Fırat University, Faculty of Veterinary Medicine, Department of Food Hygiene and Technology, and then as a lecturer at Sivas Cumhuriyet University, Department of Food Engineering. Since 2022, she has been a faculty member at Sivas Cumhuriyet University, Faculty of Veterinary Medicine, Department of Food Hygiene and Technology. She teaches courses on food hygiene, food microbiology, food additives, meat technology, and dairy technology. Dr. Can published articles in internationally cited journals and presented papers at national and international congresses and conferences. She works on food microbiology, product development, and animal foods and carries out projects with industrial organizations on the development and improvement of production systems. She worked as the Director of the Science Institute of Sivas Cumhuriyet University between 2017 and 2022 and has been working as the Director of the Institute of Health Sciences of Sivas Cumhuriyet University since 2022.

**Meryem GÖKSEL SARAÇ** graduated from Erciyes University with a bachelor's degree in food engineering and completed her master's and doctoral studies at the same university. Since 2012, she has worked in the Food Technology program at Sivas Cumhuriyet University and continues to work in the Department of Cookery. She has been working on food science, especially in the fields of food rheology and texture, food additives, food chemistry, and research and development. She is also working on sensory evaluations, interfacial rheology, and food waste utilization and recovery. She has articles, book chapters, and projects in the related field. She also serves as the deputy director of Sivas Cumhuriyet University Food Studies Application and Research Center. Göksel Saraç continues her R&D studies at the research center.

**Duygu ASLAN TÜRKER** is an accomplished food engineer and research assistant at the Department of Food Engineering, Erciyes University, Kayseri, Turkey. She holds a BSc (2012) and an MSc (2015) in Food Engineering from Erciyes University, where she also completed her PhD. Her research focuses on food rheology, chemistry, and the development of milk and dairy products, contributing significantly to the field through her extensive studies on the rheological and textural properties of food. Since 2014, Aslan Türker has been a member of her department, engaging in various research projects. She has authored numerous publications in prestigious journals, including works on gluten-free cake formulations and innovative extraction techniques for anthocyanins from black carrots. Her research extends to the interfacial properties of alternative proteins, and the stability and sensory attributes of dairy products, showcasing her diverse expertise. Aslan Türker's contributions to the field have been recognized with several awards and scholarships, including the TÜBİTAK scholarships for her PhD and postgraduate studies. She has also received certificates of honor and appreciation from Erciyes University.

# Chapter 1
# Meat Analogs: Prospects and Challenges

**Pavan Kumar, Neelesh Sharma, Muhammad Nizam Hayat, Akhilesh K. Verma, Pramila Umaraw, Nitin Mehta, Mohammad Rashedi Ismail-Fitry, and Awis Qurni Sazili**

## 1.1 Introduction

Meat has been an integral part of the human diet since time immemorial and has played an important role in fulfilling the nutritional requirement for growth and development in addition to social and cultural roles. Meat is an excellent source of animal proteins having higher bioavailability and digestibility, essential amino acids, lipids, minerals, vitamins, and bioactive peptides. Compared to plant proteins, meat proteins have higher nutritive value due to the presence of essential

P. Kumar (✉)
Institute of Tropical Agriculture and Food Security, Universiti Putra Malaysia, Serdang, Selangor, Malaysia

Department of Livestock Products Technology, College of Veterinary Science, Guru Angad Dev Veterinary and Animal Sciences University, Ludhiana, Punjab, India
e-mail: pavankumar@gadvasu.in

N. Sharma
Division of Veterinary Medicine, Faculty of Veterinary Sciences and Animal Husbandry, Sher-e-Kashmir University of Agricultural Sciences and Technology of Jammu, Jammu, Jammu and Kashmir, India

M. N. Hayat
Department of Animal Science, Faculty of Agriculture, Universiti Putra Malaysia, Serdang, Selangor, Malaysia

A. K. Verma · P. Umaraw
Department of Livestock Products Technology, College of Veterinary Science, and Animal Science, Sardar Vallabhbhai Patel University of Agriculture & Technology, Meerut, UP, India

N. Mehta
Department of Livestock Products Technology, College of Veterinary Science, Guru Angad Dev Veterinary and Animal Sciences University, Ludhiana, Punjab, India



Ö. P. CAN et al. (eds.), *Food Analogues: Emerging Methods and Challenges*,
https://doi.org/10.1007/978-3-031-69872-9_1

amino acids and higher digestibility of these proteins in human digestive tracts (Kaur et al. 2022; Kumar et al. 2022a).

The meat production system is basically an inefficient conversion of vegetable proteins into meat proteins by rearing animals, followed by their slaughter. Thus, compared to production of plant proteins, meat production requires higher inputs and has high carbon, water, and land footprints (Kumar et al. 2017). In addition to this, the issue of animal welfare due to the involvement of animals in meat production, food safety risks (antibiotic resistance, food borne-diseases, illness, metabolic diseases), risk of zoonotic diseases (due to agriculture intensification, meat packaging plants, etc.), also pose significant challenges in the meat production. Meat and meat products also need to have an efficient cold chain facility in place to prevent their spoilage and to extend the shelf life (Kumar et al. 2020a, b). In addition to animal welfare issues (during intensive rearing, slaughtering, and habitat destruction), consumption of meat and meat products, mainly red and processed meat products, has been associated with an increasing risk of cardiovascular diseases, hypertension, and colorectal cancers (Bouvard et al. 2015). Thus, it is widely recommended to reduce meat consumption to improve the health of humans and the environment (Kumar et al. 2022b).

The world's population has recorded rapid growth in the last century, especially in Asia and Africa, and touched the 8 billion marks on 15th Nov 2022 (United Nations 2022). It is expected to reach 9.8 billion in the year 2050 and 11.32 billion in the year 2100. A 60% higher agriculture production and 76% more meat production will be required to meet the food requirement for the population in the year 2050 as compared to agriculture and meat production levels in the year 2005–2006 (Alexandratos and Bruinsma 2012). Further, in the last five decades, a four-fold growth in food consumption was observed with a two-fold increase in the population, thus, on average, twice the increase in food consumption (Weis 2015). Ensuring proper food availability and food safety for this rapidly increasing population will be an upheaval task, considering that all the natural resources are under severely strained conditions and have already been used to their full potential; thus, this sector needs to be shifted from a more sustainable production system.

M. R. Ismail-Fitry
Department of Food Technology, Faculty of Food Science and Technology, Universiti Putra Malaysia, Serdang, Selangor, Malaysia

Halal Products Research Institute, Putra Inforport, Universiti Putra Malaysia, Serdang, Selangor, Malaysia
e-mail: ismailfitry@upm.edu.my

A. Q. Sazili
Department of Animal Science, Faculty of Agriculture, Universiti Putra Malaysia, Serdang, Selangor, Malaysia

Halal Products Research Institute, Putra Inforport, Universiti Putra Malaysia, Serdang, Selangor, Malaysia
e-mail: awis@upm.edu.my

As per one estimate, about 83% of total farmland is used for animal production activities, contributing to 56–58% of total emissions but only 37% of total protein supply and 18% of total energy supply (Poore and Nemecek 2018). The environmental footprint of vegetable proteins is markedly lower than those of meat proteins as vegetable proteins have 2–6 g $CO_2$e/kg protein carbon footprint, 38–205 blue water footprint, and 6–8 $m^2$ year/kg protein land use footprint as compared to their corresponding values for the beef production viz., 174–184, 1607 and 1310–1311, respectively (Thrane et al. 2017). Bluewater refers to the total surface water and groundwater. Bluewater footprints indicate the volume of freshwater evaporated from the global blue water resources, viz., groundwater, lakes, reservoirs, ponds, glaciers and rivers (Mehta et al. 2021).

With the advancement of processing technologies, plants and other non-meat proteins can be processed into meat analogs by imitating the meat products in their structure, chemical nature, and sensory attributes (Kumar et al. 2022c). These plant-based meat analogs provide sustainable and healthy alternatives to meat and have noticed increasing acceptance among consumers who prefer a sustainable and healthy diet (Kaur et al. 2022). Similarly, the EAT-Lancet Commission also recommended reducing meat and dairy consumption and increasing plant-based diets comprising more legumes, beans, and nuts (Willett et al. 2019). Thus, meat analogs could play a vital role in ensuring the sustainability of global food security (Fig. 1.1).

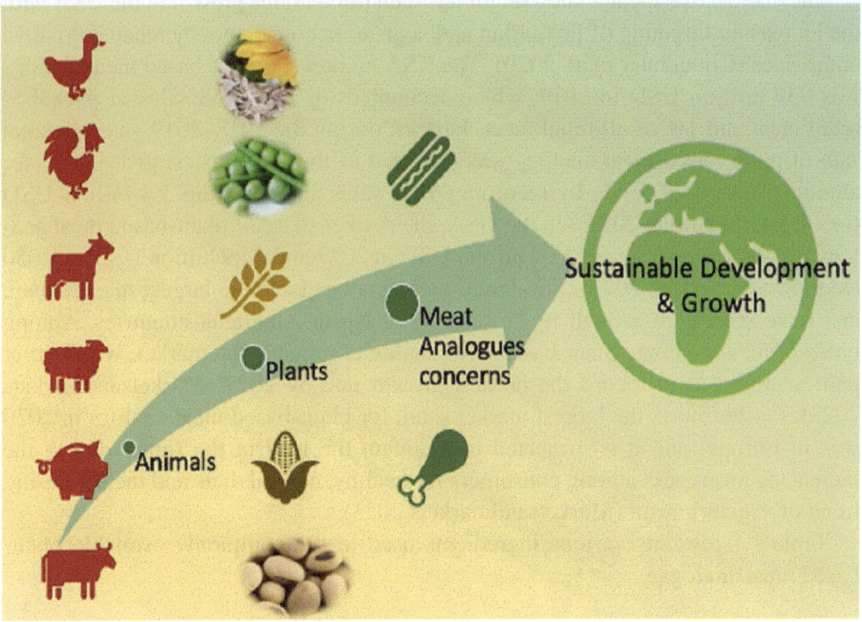


**Fig. 1.1** Role of plant-based meat analogs in sustainable development and growth

## 1.2 Current Status

The global meat analog market has grown tremendously over the last two decades. According to a Bloomberg Intelligence report on the plant-based food market, the market size of plant-based meat analogs is expected to touch 74 billion USD in 2030 from 4.2 billion USD in 2020 (Bloomberg Intelligence 2021). The report attributed the higher growth in this sector to the increasing awareness among consumers towards health consciousness, sustainability merits, increasing availability, and price advantages of plant-based meat analogs. Further, if the current trends continue, the sale of plant-based meat analogs could reach 118 billion USD in 2030 (Bloomberg Intelligence 2021). In 2023, the total sale of meat alternatives was projected to touch 10.15 billion USD by following a CAGR (compound annual growth rate 2022–2028) of 10.58%, with China ranked first in the total growth (Statista 2023). Globally, Europe has the largest market for plant-based meat analogs with 51.5% of the market share, followed by 26.8% share of North America, 11.8% share of Asia Pacific, 6.3% share of Latin America and 3.6% share of Middle East and Africa (OECD 2021).

The main drivers for this growth are increasing numbers of start-ups and food processing companies that make the availability of meat analogs at lower prices, higher diversity, and varied tastes as per local preferences, advanced packaging, improving the nutritive value and functionality of plant proteins, application of a more sustainable, cheaply available, and unconventional source of plant and insect proteins in the preparation of meat analogs (Sharma et al. 2022).

The plant-based meat analog sector has seen tremendous growth in the USA with the increasing targeting of flexitarian and vegetarian consumers by meat alternative companies (Choudhury et al. 2020). The USA market for plant-based meat analogs was 939 million USD in 2019, which accounted for 2% of all sales of packaged retail meat and 1% of all retail meat. Further, during the 2018–2019 year, the total sale of plant-based meat analogs was recorded as the third fastest growing sector among plant-based foods by recording 19% sales and touching 1.4 billion USD (Good Food Institute 2023). In the USA, the market share of plant-based meat analogs was projected to reach 15.7 billion USD in 2027 and 27.9 billion USD by 2050 (Marketsandmarkets 2023). Soy-based meat analogs have the largest market share and have seen high growth in European and North American countries. Among types of meat analogs, plant-based beef analog dominates the market, with burger patties projected to record the highest growth rate by 2027 (Marketsandmarkets 2023). Furthermore, the largest market share for plant-based meat analogs in 2020 was in Europe, and it is projected to maintain the lead in the future due to the increasing awareness among consumers for healthy, natural diets and the increasing trend of vegetarianism (Marketsandmarkets 2023).

Table 1.1 presents various ingredients used in the commonly available plant-based meat analogs.

**Table 1.1** Ingredient list of some commonly available plant-based meat analogs

| Analog product | Manufacturer | Main ingredients | | | |
|---|---|---|---|---|---|
| | | Protein | Lipid | Carbohydrate | Additives |
| Beyond burger | Beyond Foods, USA | Pea protein isolates, rice protein, mung protein | Canola oil, sunflower oil, coconut oil | Potato starch, methylcellulose, | Apple extract, beet juice, vinegar, lecithin, salt, pomegranate fruit powder |
| Impossible burger | Impossible Foods, USA | SPC, SPI, potato protein | Coconut oil, sunflower oil | Modified starch, methylcellulose, cultured dextrose | Soy leghemoglobin, salt, yeast extract, zinc, vitamins $B_1$, E, C, $B_2$, and $B_{12}$ |
| MorningStar farms grillers burger | MorningStar Farms, Kellogg's, USA | Wheat gluten, soy flour, SPI, egg whites | Corn oil, sunflower oil, canola oil | Corn starch, potato starch, xanthan gum, sugar, gum acacia, methylcellulose | Salt, yeast extract, onion powder, soy sauce, carrot juice, garlic powder, whey, tomato paste, onion juice, calcium caseinate |
| Boca all American veggie burger | Boca Food Company, Kraft Foods, USA | SPC, wheat gluten, SPI, cheese powder, low-fat cheddar cheese | Corn oil, sesame oil | Methylcellulose | Additives: spice, salt, herbs, dried garlic, HSP, mannitol, yeast extract |
| Gardein meatless meat ball | Garden Protein International, Pinnacle Foods, Canada | SPC, wheat gluten, SPI, barley malt extract, pea protein | Canola oil | Wheat flour, sugar, cane sugar | Salt, yeast extract, sea salt, spice, dried garlic, dried onion, vinegar, vitamin $B_2$, $B_5$, $B_1$, folic acid |
| Tofurky ham roast with glaze | Tofurky, Turtle Island Foods, USA | Wheat gluten, tofu, | Canola oil | Oat fiber, cane sugar, konjac fiber, carrageenan, xanthan gum, dextrose | Spice, lycopene, carrot juice, salt, garlic, KCl |
| Quorn brand chik'n nuggets | Quorn Foods, Inc., UK | Mycoprotein, egg white, wheat gluten, whole egg | Canola oil | Wheat flour, sugar, cane sugar, turbinado sugar, dextrose, pea fiber | Yeast extract, salt, calcium acetate, black pepper, yeast, onion powder, calcium chloride |

Adapted and modified from Bohrer (2017)
*SPC* soy protein concentrate, *SPI* soy protein isolates, *HSP* hydrolyzed soy protein, *KCl* potassium chloride

## 1.3 Major Ingredients

In the development of meat analogs, protein, lipids, and moisture form the major components, followed by carbohydrates, minerals, vitamins, and other food additives (colorants, flavorings, taste enhancers, etc.). In plant-based meat analogs, plant protein is the important component that affects the product's texture, appearance, and nutritional quality. The meat analogs with the desirable nutritive and sensory attributes could be prepared with a suitable selection of ingredients and processing methods. Plant-based meat analogs contain about 50–80% moisture, 4–25% protein, 2–30% carbohydrate, 0–15% lipids, and 0–15% additives comprising seasonings, salt, spices, pigments, colorants, flavorings, and binding agents (Chen et al. 2022; Huang et al. 2022). Protein plays an important role in nutritive value and structural attributes. Carbohydrates make cross-linking with proteins and act as fillers, binders, and thickeners (Chen et al. 2021; Huang et al. 2022). The incorporation of lipids improves nutritive quality, sensory attributes, and storage stability of meat analogs (McClements and Grossmann 2021).

### *1.3.1 Plant Protein Sources*

Vegetable proteins, due to their high nutritive value and relatively cheap availability, are commonly used in the development of canned meat analogs and pet foods (Featherstone 2015). Soybean and wheat gluten remain the primary source of vegetable proteins in the development of plant-based meat analogs in addition to groundnut, zein, cereal proteins, pulses/legumes, nuts, fungi, and yeast, also commonly used in the preparation of meat analogs (Kumar et al. 2022c; Sharma et al. 2022).

The easy availability, low cost, high nutritive value, and textural attributes make soy protein the most popular and common ingredient in developing plant-based meat analogs (Krintiras et al. 2015; Kumar et al. 2023). Soy in different forms, such as soy flour, spray-dried soy milk, texturized soy protein (TSP), soy protein concentrate (SPC), and soy protein isolates (SPI), are commonly used due to their nutritional, functional attributes (texturizer, emulsifier, stabilizers, water holding capacity), satiating properties, sensory attributes, wide availability, and inexpensive/ low cost (Kumar et al. 2022c). The biological value of SPI is reported to be comparable to meat, with both having equivalent PDCAAS (protein digested corrected amino acid score) of 1.0 (Kumar et al. 2023). In addition, the intake of soy-based meat analogs leads to a long-lasting satiating effect and causes lower food intake after consuming these products (Williamson et al. 2006).

The legumes (peas, lupin, lentils, chickpea, and mung bean) have better amino acid profiles than cereals and have lysine, leucine, and phenylalanine content equivalent to soy protein. Legume seeds have high protein content (ranging from 20–35%), dietary fiber, saturated fat, vitamins, and minerals; hence their inclusion in meat analogs results in improving the nutritive quality of meat analogs (Doss et al. 2022). The digestibility of legume protein is lower as compared to soy protein; however, by suitable processing technologies, legumes are increasingly used in the development of functional meat analogs such as improved textural and fibrousness, gluten-free, lower allergen ingredients and organoleptic attributes (Vatansever et al. 2020).

Wheat gluten is the most commonly used traditional cereal protein in the development of meat analogs. The visco-elastic properties of wheat gluten form the fibrous texture and impart strength to the product, thereby improving binding, solubility, baking properties, swelling, and leavening properties (Kumar et al. 2012). However, it should be noted that cereals are a rich source of carbohydrates, and their digestibility and protein content are lower than those of soy (Joshi and Kumar 2015; Mota et al. 2016). Thus, it needs to be compensated with other suitable ingredients to get the desired nutritional profile of the meat analogs. In addition, wheat gluten is a rich source of glutamine amino acid but deficient in threonine and lysine amino acids. Although used in meat analogs, zein protein available in maize is not widespread due to its inability to form fiber structure or sheet-like texture at ambient temperature (Jeong et al. 2017). It can be successfully incorporated into meat analog preparations with suitable technological interventions such as hydrolysis, cross-linking, deamination, or mixing with other protein sources (Glusac and Fishman 2021).

Rice protein is used in the preparation of gluten-free meat analogs for people with gluten intolerance or celiac disease. The incorporation of rice bran in meat analogs improves the nutritive and functional quality due to its high dietary fiber content (27.6–33.3%) (Xiao et al. 2022). The rice protein does not have a strong flavor like soy protein in addition, it has a hypoallergic and hypocholesterolemic effect (Detchewa et al. 2022). Further, the nutritive value of rice protein in terms of digestibility and biological values are similar to SPI (Han et al. 2015), thus a suitable nutrition source for infants and elderly persons (Jeong et al. 2017). Rice protein has methionine, arginine, valine, histidine, and cysteine amino acids in high concentrations, and its lysine content (3.8%) and leucine content (8.2%) are higher than wheat protein (Kumar Singh et al. 2016).

Further with the advancement in processing, extraction and purification of plant proteins, various other non-conventional sources of plant proteins such as algae, fungi, single-cell-protein (SCP)/microbial proteins/edible unicellular micrograms, leafy vegetables, agriculture byproducts, seaweeds, oilseeds proteins, jackfruits, pseudocereals (such as Quinoa and Amaranthus, etc.) are gaining popularity due to their sustainability potential and nutritive value (high protein content, presence of essential amino acids) and price advantage in the meat analogs (Delshadi et al. 2020; Kurek et al. 2022; Yesuraj et al. 2022). Quinoa seed protein has excellent binding, moisture retention, emulsifying, and foaming properties equivalent to soy proteins. Further, soy quinoa seed protein has good gelling properties at low pH (Kaspchak et al. 2017).

The proteins obtained from potato, chia seed, hemp, and pumpkin seeds have good solubility in alkaline medium and other function attributes such as gelling, foaming, and moisture retention. They could be harvested for their potential use in the formulation of meat analogs (López et al. 2018). Konjac fiber obtained from konjac plant root is gaining popularity as an ingredient in meat analogs due to associated health benefits. In addition, the application of konjac fiber in Tempeh (a fermented soy product) was observed to improve functional attributes such as viscoelasticity, fibrousness, increased water retention, and overall texture of the products (Yuliarti et al. 2023). Further, protein harvested from leaves from grass, tobacco, alfalfa, sugarcane, etc., could also be utilized as a cheaper source of proteins in meat analogs (Singh et al. 2021).

Mycoprotein (protein obtained from fungi) has a comparatively lower environmental impact as it requires 10–20 times less land and has a ten times lower carbon footprint as compared to beef production, whereas four times lower carbon footprint as compared to poultry (Hashempour-Baltork et al. 2020). These are rich sources of nutrients and have a good quantity of essential amino acids, minerals, carotene, and vitamins. The mycoprotein from *Fusarium graminearum* (a filamentous edible fungi) has a nutritive quality similar to that of meat, such as meaty flavor and taste, fibers, fibrous texture, and high digestibility. On average, mycoprotein contains 45% protein and 25% fiber, mainly contains 75% β-glucan and chitin, 13% fat, and 10% carbohydrates (Finnigan et al. 2019). Mycoprotein forms the fibrous chitin-glucan matrix and is recognized as 'high in fiber food' by the European Union (Commission 2020).

Mushroom incorporation in the meat analogs imparts a specific flavor, light grey color, and nutritive value. The texture and binding of such products (mushroom-based) could be further improved by adding various gelling thickening agents and binding, such as carrageenan and xanthan gum. A good quality plant-based meat analog was developed by adding mushrooms in texturized soy protein, wheat gluten, egg albumen, and food additives (Kumar et al. 2011, 2012).

### *1.3.2 Carbohydrate Sources*

During the preparation of meat analogs, carbohydrates are incorporated to modify texture, structure, and consistency (Yao et al. 2004). The primary sources of carbohydrates in meat analogs are either as binders/fillers such as starches from corn, potato, rice, wheat, peas, and cassava or in minor quantities of gums, methylcellulose, and carrageenan. In meat analogs, carbohydrates play an important role in imparting functional properties such as gelation, reducing power, gelling, dehydration, and degradation (Huang et al. 2022). Carbohydrates in their native or refined form are classified as food additives and used in food to improve taste, appearance, and mouthfeel (Huang et al. 2022). As such, meat contains very low levels of carbohydrates; thus, while preparing analogs for fresh meat cuts, the carbohydrate content is kept very low, similar to their meat counterparts, and has a role in the Maillard reaction affecting color and flavor similar to fresh meat cuts (Sohail et al. 2022).

However, in the development of analogs for processed meat products, the level of carbohydrates could be high to maintain the structural and textural properties.

The starch in water forms a gel-like structure, which gives smoothness and fat-like properties, thus used to prepare low-fat meat analogs (Joshi and Kumar 2015). The application of sugar balances the hardening effect of salt and improves the flavor and taste of the product. To get the maximum health benefits from carbohydrates, they should be utilized in minimally processed form as refined or processed carbohydrates could alter their nutritive value, especially by decreasing the fiber content (Viuda-Martos et al. 2010). Methylcellulose, a dietary fiber obtained from cellulose, is used as a binder in the development of meat analogs (Schuh et al. 2013).

### *1.3.3 Lipid Source*

Lipid forms an essential component of the meat analog as it affects the sensory, textural, and nutritive quality of the meat analogs. Fat improves the mouthfeel, juiciness, appearance, tenderness, and flavor of the product by interacting with the other macronutrients in food products (Chen et al. 2023a, b). There is a significant difference in the structure and composition of lipids derived from animal and protein sources, such as isomerization, degree of unsaturation, physical state (solid state of animal fat vs liquid state of vegetable oil), fatty acid profiles, and physicochemical properties. These differences cause variations in the sensorial attributes of the meat analogs, such as mouthfeel, aroma, spreadability, juiciness, and texture (McClements and Grossmann 2021).

Thus, proper selection of lipids is required during the preparation of meat analogs by carefully considering various factors such as sensory attributes, nutritive value, functionality, cost, availability, chemical stability, and traditions/ customs (Chen et al. 2023a, b). Based on the origin, the lipids used in the meat analog manufacturing can be divided into two categories viz., (i) exogenous lipid source obtained from oilseeds and tropical sources such as coconut oil, canola oil, sunflower oil, flaxseed oil, and soybean oil; and (ii) endogenous lipid already present in the less-refined or partially fractionated protein ingredients (Bohrer 2019). Further, with increasing focus on healthier food and sustainable food production, the lipids obtained from nuts, fruit seeds, fruit peels, grain embryos, and agro-byproducts are increasingly used to prepare meat analogs (Chen et al. 2023a, b).

At present, the primary sources of lipids in plant-based meat analogs are canola oil, sunflower oil, palm oil, coconut oil, sesame oil, cocoa butter, and corn oil (Bohrer 2019). The lipid content of commonly available meat analogs is equivalent to the meat products. Furthermore, vegetable oils are considered a healthier option due to the absence of cholesterol (Dekkers et al. 2018; Pietsch et al. 2017, 2019). Oleogels (having properties of solid fat with >90% fat w/w, reduced amount of trans and saturated fat) are used to replace detrimental fat in food products (Manzoor et al. 2022). These oleogels are nowadays used as replacements for fat in meat analogs (Martins et al. 2019). Various fat replacers also used in meat analogs, such as oat soluble fiber in low-fat meat analogs, improve the water-binding properties and structural attributes (Piñero et al. 2008; Summo et al. 2020).

### 1.3.4 Miscellaneous Ingredients

In addition to protein, fat, and carbohydrates, meat analogs and other ingredients play a major role in affecting sensory attributes and the nutritive quality of meat analogs. Flavor is an important parameter that determines the consumer acceptance and marketability of food products. Meat is a rich source of zinc, vitamin $B_1$, $B_{12}$, $B_3$, $B_6$, and sodium ascorbate, and to improve the nutritive value similar to meat, these compounds are added during the development of meat analogs (Damayanti et al. 2018).

#### 1.3.4.1 Flavorings

Developing the meat-like flavor in plant-based meat analogs remains a challenge as the flavor of the meat is formed by a complex mechanism involving more than 1000 compounds (Wang et al. 2018). The free amino acids, short peptides, inorganic salts, and metabolic products of nucleic acids such as ribose inosine play important roles in the flavor development of meat. Furthermore, the aroma developed from volatile compounds during heating, such as sulfur-containing compounds, ketones, aldehydes, and heterocyclic compounds, also have major role in the flavor development of meat (Robbins et al. 2003). The flavor of meat products is the result of very complex mechanisms involving several products by Maillard reaction, oxidation of lipids, and thiamine degradation (Robbins et al. 2003). The vegetable protein most commonly used in the preparation of plant-based meat analogs does not have these critical compounds needed for flavor development.

Soy has a beany, grassy, astringent, and chalky odor, which gives an undesirable off-flavor to soy protein-based meat analogs (Damodaran and Arora 2013; Zhu and Damodaran 2018). The origin of this off-flavor is credited to the oxidation of unsaturated fatty acids by lipoxygenase enzyme released due to breaking cell wall (Kumari et al. 2016), consequently leading to the formation of non-volatile substances such as esters and long-chain alcohols (Zhu and Damodaran 2018). The lipid forms up to 20–33% of raw soybean, and it contains mainly monounsaturated fatty acids (as oleic acid) and polyunsaturated fatty acids (as linoleic and linolenic acid) (Damodaran and Arora 2013). This problem of off-flavor by soy protein usage in meat analogs could be alleviated using various processing methods such as enzymatic decomposition, acid treatment, genetic engineering, and solvent extraction (Li and Li 2020).

To get a flavor similar to meat, various flavorings/flavor enhancers are added to meat analogs, such as aromatic plants or herbs, spices, hydrolyzed vegetable proteins, yeast extracts, vegetable oils, seasonings, etc. Hydrolyzed vegetable protein

(HVP) imparts flavor to meat analogs in addition to increasing the protein content of the meat analogs (Aaslyng et al. 1998; Kumar et al. 2017). HVPs are prepared by enzymatic hydrolysis under mild acidic and temperature conditions of plant proteins such as wheat, soy, corn, mushroom, etc., into short peptides and free amino acids. HVP contains several volatile compounds such as furans, pyridines, furanones, organic acids, pyrrole, pyrazines, alcohols, aldehydes, esters, ketones, sulfur-containing compounds, and phenols (Aaslyng et al. 1998). These compounds produce a meat-like flavor upon heating with reducing sugars and yeast autolysis (Aaslyng et al. 1998). Similarly, the hydrolysis of rapeseed protein with trypsin and neutral protease obtained from *Bacillus subtilis* was observed to facilitate Maillard's reaction and meat-like favor (Kale et al. 2022).

Yeast extract (YE) prepared from food-grade yeast (brewer's yeast and baker's yeast) is a rich source of protein, volatile compounds, reducing sugars, amino acids, peptides, thiamine, lipids, peptides, and nucleic acid content (Aaslyng et al. 1998). The volatile flavor compounds such as thiophene and pyrazine impart a sweet and meaty aroma to roasted products (Lin et al. 2014). Spice and herbs have their typical flavor and aroma and are used to prepare food products to develop aroma, flavor, and taste. Due to their essential oils and phenolic compounds, these exert antioxidant and antimicrobial effects, leading to improving shelf life and masking undesirable flavors (Awad et al. 2021, 2022). Allicin present in garlic degrades to alkyl sulfides, leading to the development of a meat-like flavor (Lanzotti 2006).

#### 1.3.4.2 Colorants

The appearance and color of meat analogs should match the color and appearance of their meat counterparts to increase consumer acceptance. Further, it is also desirable to have raw and cooked meat analogs whose color and appearance are similar to that of meat products (Mancini and Hunt 2005). Color is an intrinsic attribute of the food product that consumers notice and affects the consumer perception of taste and overall acceptability of a food product. Altering the hue and chroma, along with the taste of a food product, was observed to have a dramatic effect on consumer expectations and experience (Spence 2015).

Meat color is mainly affected by the oxidative state of myoglobin and, to some extent, other components such as lipids, protein status, etc. The bright red color of fresh red meat is attributed to oxymyoglobin (bloom). In contrast, the brown color of cooked meat is attributed to metmyoglobin formation during cooking (Suman and Joseph 2013). The color of raw plant-based meat analogs is yellow or beige due to various pigments, and it has been a challenge to mimic the typical brown color of cooked red meat by applying suitable colorants (Sakai et al. 2022). To imitate the meat-like color of plant-based meat analogs, the industry commonly uses beet red pigment and soy leghemoglobin (Bohrer 2019).

The beet red pigment consists of betanidin/betanin pigment extracted from the beet plant (*Beta vulagaris*). It has a high margin of safety and a low production cost; thus, it is widely used in food products to impart red color (Strack et al. 2003). In addition to imparting color, betanin has a high antioxidant effect, thus positively affecting consumers' health by protecting cells against oxidative stress (Fernández-López et al. 2020; Vulić et al. 2013). However, unlike myoglobin, betanin is thermostable and photostable, thus maintaining its natural red color even after cooking and overheating, it gives a yellow color, which has poor consumer acceptance (Bohrer 2019; Sakai et al. 2022; Strack et al. 2003).

Leghemoglobin is a genetically modified protein expressed in *Pichia pastoris* yeast. It resembles myoglobin chemically and structurally and imparts a brown color to cooked meat analogs similar to the meat (Goldstein et al. 2017). Although it is safe to consume, some consumers still have concerns about its consumption. In addition to developing the desired level of Maillard reaction (browning) for color and flavor development, various precursors such as caramel color, reducing sugars, malt extracts, and amino acids are added to mask the red color and impart the brown color of cooked meat analog (Dennis et al. 2015).

To stabilize the color of meat analogs during cooking, various ingredients such as citrus fruit extract, ascorbic acid, and apple extract are added in combination with natural color pigments such as lycopene, and annatto (Mattice and Marangoni 2020). Apple extracts for color and citrus fibers to provide a meat-like texture were used by Herbafood Inc. to produce vegan burger patties. In addition to the color, the application of extract also improved the antioxidant potential of the meat analogs. Beet juice is used in Beyond Burgers and tomato paste by MorningStar to get the desired color of the meat analogs (Fraser et al. 2018). Recently, the application of freshwater unicellular green algae *Haematococcus Pluvialis* was used for imparting color as well as improving oxidative stability to the meat analogs due to its high content of carotenoids as 69.7% astaxanthin, 3% beta-carotene, and 27% lutein (Huang et al. 2023). Further, the incorporation of astaxanthin was observed to provide a meat-like color to meat analogs (Fu et al. 2021). Similarly, application of natural pigments viz., oleoresin paprika, red beet (0.4–1.5 mg/g), sorghum, cacao pigment (1.1–1.3 mg/g), and monascus red, were used to develop color to a plant-based meat analog product similar to cooked meat color (Ryu et al. 2023).

Table 1.2 Summarize the ingredients used and their role in the development of plant-based meat.

## 1.4 Processing of Plant Protein

Plant proteins are processed using various technologies such as extrusion, spinning, and cross-linking to impart fibrous texture. The functional attributes of plant proteins, such as gelling ability, solubility, water binding ability, viscosity, gelation, flavor, texturization, dough bindings, etc., are affected by composition (amino acid content) and processing conditions. These functional attributes play a crucial role in imparting meat-like properties to meat analogs, and these could be modified by using suitable processing such as denaturation induced by heat, alkali or acid treatment, extraction methodology, enzymatic action, and hydrolysis. The enzymatic

**Table 1.2** Various ingredients in plant-based meat analog and their role

| Ingredients | Amount (range) | Sources | Role |
|---|---|---|---|
| Moisture | 40–75% | Potable water | Carrier medium for other ingredients, plasticizers, juiciness, texture, palatability |
| Protein | 3–28% | Conventional: Soy, wheat gluten, rice, zein, peas, pulses, legumes<br>Unconventional: leafy protein, grass, potato, oilseed proteins, algal, mycoprotein, SCP | Texture, structure, functional attributes, nutritive value |
| Carbohydrates | 1–30% | Starch and sugar from<br>Major components: corn, potato, rice, wheat, peas, cassava<br>Minor components: gums, methylcellulose, carrageenan | Filler, binder, gelling, hydration, dehydration, flavor, color by Maillard reaction, taste, |
| Lipids | 2–30% | Exogenous: soybean oil, sunflower oil, cocoa, flaxseed<br>Endogenous: less refined or partially fractionated legume proteins, maize, cereals | Flavor, mouthfeel, juiciness, tenderness, nutritive value, texture |
| Additives | 0–12% | Salt: sodium chloride | Taste, flavor, preservation |
| | | Colorants, pigments: apple fruit extract, betanin, leghemoglobin, tomato juice, lycopene, carotene, green algae | Color, antioxidant, and antimicrobial |
| | | Flavorings, seasoning: spices, herbs, HVP, yeast extracts | Flavor, aroma, taste, nutritive value, preservation |
| | | Vitamins: vitamin $B_1$, $B_{12}$, $B_3$, $B_6$, sodium ascorbate | Nutritive value |
| | | Minerals: Zinc, iron, calcium, selenium | Nutritive value |

*SCP* single-cell protein, *HVP* hydrolyzed vegetable proteins

treatment of plant proteins could result in improving solubility, heat stability, and reduced allergenicity (Butré et al. 2012; Jeewanthi et al. 2015).

The texturization and structuring of plant proteins to a meat-like texture significantly affect the consumer's acceptance of meat analogs (Sun et al. 2021). The separation of protein-rich and water-rich fractions is the first step in fiber formation during the processing of proteins (Tolstoguzov 2006), and this process could be further sped up by adding some polymeric compounds (hydrocolloids or carbohydrates by increasing mutual exclusion among various polymers (Tolstoguzov 2003).

Extrusion cooking is the most commonly used technology for texturizing and structuring plant proteins while preparing meat analogs by passing them between dies at suitable pressure, moisture, and pressure conditions. Food products of various shapes and varieties are manufactured by modifying the size and shape of dies (Kumar et al. 2019). Hot extrusion is performed at 100 °C temperature, thus leading to a cooking effect that causes various chemical and structural changes in proteins.

These changes depend on various factors such as moisture content, barrel temperature, composition and quality of raw ingredients, protein content, particle size, barrel diameter, barrel temperature, extrusion rate, and pressure (Ryu 2020; Wild 2016). The high moisture extrusion is characterized by high heat transfer and lower mechanical dissipation due to a larger barrel surface (Wild 2016). It facilitates a high degree of texturization, elasticity, nutrient retention at lower temperatures, and elasticity but increases storage cost and, at the same time, reduces aroma and taste. At high temperatures, plant protein gets denatured, and screw rotation causes shear force leading to the unfolding of peptides and destruction of the three-dimensional structure of proteins. This leads to the formation of disulfide bonds and cross-linking between denatured proteins (Ryu 2020). However, the high energy usage, high cost of extrusion, and size constraints are challenges faced by the extrusion industry (Kumar et al. 2022d).

Other technologies used in the processing of plant proteins are spinning, 3D/4D printing, Couette shear cells, and freeze structuring. Under spinning, ultrathin fibers are produced from vegetable proteins to develop meat analogs. Further, the fiber characteristics vary with the spinneret's parameters (temperature, humidity, and distance between spinnerets to collectors), viscosity, surface tension, and conductivity (Moomand and Lim 2015). However, the high amount of waste generated during the process, the requirement of low pH, the application of additives, and high salt concentration make this process need to be addressed by the food industry.

Figure 1.2 represents the various steps in the processing of plant proteins into plant-based meat analogs.

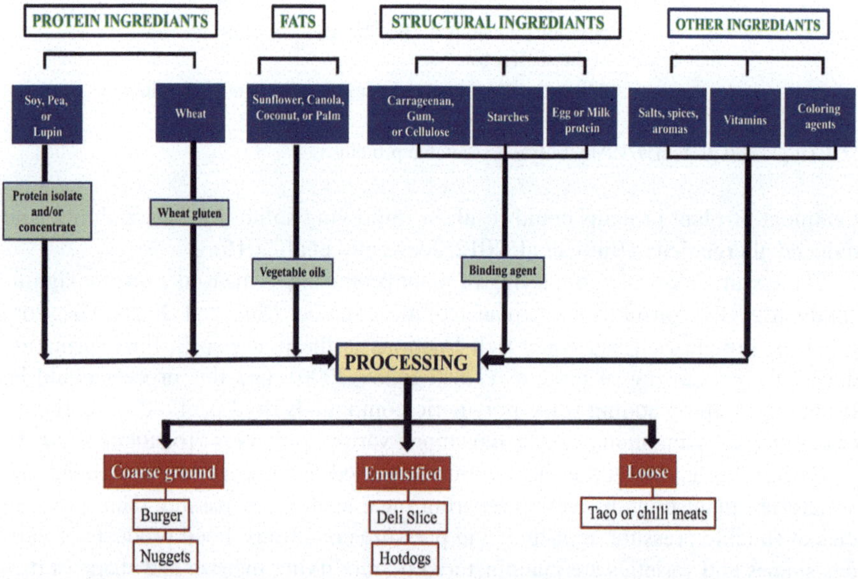


**Fig. 1.2** Formulation of plant-based meat analogs. (Adapted from Ishaq et al. (2022). This article was published in Ishaq et al. (2022), Copyright Elsevier 2022)

## 1.5 Prospects and Challenges

The plant-based meat analogs have bright prospects due to their sustainability potential, lack of animal welfare and ethical issues, and ability to supply food for a burgeoning population with already strained resources. But to harness their potential, these products should be accepted among the broader population as meat substitutes.

### *1.5.1 Economics and Marketing*

The price of meat analogs is comparatively lower than meat due to the use of cheaply available proteins, among which soy proteins still form the most commonly used source of protein in plant-based meat analogs (Kumar et al. 2023; Rubio et al. 2020). At the standardized cost based on 2009 data, (Lusk and Norwood 2009) noted a significantly lower per-gram price for plant proteins as compared to meat proteins, e.g., $0.01/g for soybeans, $0.03 for wheat) whereas $0.22/g for pigs and $0.12/g for chickens. However, this price depends upon several factors, and premium meat analogs usually cost more and maybe equivalent or higher to meat counterparts.

Unlike traditional meat, the post-processing cost is significantly higher for meat analogs (up to 94.3% of the total retail cost of crop products) as compared to 50% of the retail cost for beef (Lusk and Norwood 2009). In addition, to impart flavor, texture, and nutritional value to meat analogs, a range of additives, such as colorants, vitamins, minerals, lipids, etc., are added, consequently increasing the overall cost of production for these products (Tziva et al. 2020). However, we should note that with increasing incomes and lifestyles, premium meat analogs are also accepted by health-conscious and affluent consumers.

### *1.5.2 Consumer Acceptance*

Consumer acceptance of food products is essential to get it popularized among consumers and its marketing. The consumer acceptance of meat analogs, similar to other food products, is largely affected by their organoleptic qualities and also price compared to their meat counterparts. The packaging also plays an important role in attracting consumer attention and creating interest in food products. Thus, using a proper packaging system enlisting its nutritive value, sustainable merits, and highlighting animal welfare compliance could result in higher consumer acceptance of the meat analogs. Similarly, consumer acceptance varies from region to region due to food habits, availability of dispensable income, and availability/exposure to typical food products. Thus, to improve consumer acceptance on a large scale, the imitation of common meat products resembling them in taste and other attributes could be a proper strategy.

The consumer acceptance of plant-based meat analogs was observed to be very high for China (95.6%) and India (94.5%), in comparison to the USA, where it was reported to be 74.7% (Bryant et al. 2019). The significant roadblocks in consumer acceptance among EU consumers were observed in the lack of familiarity and lower sensory attractiveness; thus, in these countries, a very close imitation of meat analogs to their meat counterparts could improve the marketing of these products (Hoek et al. 2011). The factors affecting the acceptance of meat analogs vary from country to countries such as in Germany and France, the issue of animal welfare and its positive impact on the environment and health are the main driving force, whereas, in the Netherlands, animal welfare, health, and lower-quality meat were identified as the main driving force for the adoption of meat analogs (Weinrich 2018). However, in all these above-mentioned three countries viz., In Germany, Netherlands, and France, the taste of meat analog is the key factor in determining consumer acceptance and other factors such as consumer habits, price competitiveness, and convenience (Schouteten et al. 2016). However, under emotional and sensory profiling under blind testing conditions, the plant burgers were awarded with 'distrust, disappointed and discontented' whereas meat burgers with 'happy, contented, and pleasant' by the 97 young adults, with an insect-based burger and plant-based burgers were having comparable liking (Schouteten et al. 2016).

Among various sensory attributes, taste is the most important criterion affecting consumer acceptance of meat analogs. As per the Mintel Store report on 'US plant-based protein report 2023', more than half (51%) of consumers surveyed were on the view that plant-based meat analogs should taste indistinguishable from their meat counterparts, followed by price (25%) concerns (Mintel Store 2023). Further, in the same report, consumers tried the meat analogs once or twice but did not follow them due to their taste, 46% followed by those disappointed due to these products did not meet the expectations of 35% of consumers.

Some meat products are highly processed to mimic the sensory, chemical, and physical attributes of meat. These novel meat analogs are considered by consumers to be 'highly processed' as compared to traditional burgers and thus considered an unnatural method of food production (Asioli et al. 2017). The negative marketing approaches such as preservative-free and antibiotic-free also have an impact on consumer acceptance. Thus, those consumers who have a liking for the 'clean label' have some concern for these products due to their unnaturalness (Asioli et al. 2017).

The consumer acceptance of insect-based meat analogs is challenging in societies where entomophagy is not generally practiced, such as Europe, North America, and India. This is due to entomophobia (fear and disgust towards insects), attachment to insects eating to low social strata with consideration of food of the poor, and neophobia (Deroy et al. 2015; Shelomi 2016; Van Huis et al. 2013).

### 1.5.3 Sensory Attributes

Appearance, flavor, and texture are the primary sensory attributes that determine the overall quality of the meat analogs. The flavor of meat is imparted by a unique combination of a range of compounds formed by the interaction of meat lipids, proteins, and minerals, and the muscle fibers in the meat form texture. The presence of heme pigments contributes to the color of the meat and meat products. An ideal meat analog product should mimic its meat counterpart in shape, convenience, texture, appearance, flavor, aroma, and nutritive value.

For imparting the red color of fresh meat to meat analogs, various plant extracts and juices such as apple extract, tomato juice, carrot juice, beet juice, carotene pigments, or recombinant heme protein sourced from soy expressed in *Pichia pastoris* as soy-leghemoglobin are used. The merit of leghemoglobin is that it forms a brown color upon cooking the product, similar to meat products (Jin et al. 2018). Further, it also improves the flavor profile of meat analogs (Fraser et al. 2018; Kyriakopoulou et al. 2021). Vegetable proteins are texturized to give texture and functional properties similar to meat by applying mechanical, thermal, or shear pressure, such as various extrusion, shear cell technology, 3D/4D printing, and other technologies (Kumar et al. 2022c). Plant lipids such as coconut oil and cocoa butter are used in meat analogs to give smoothness, taste, texture, and flavor to meat products (Rubio et al. 2020).

Various additives (3–10%) are added to the meat analogs preparation to mimic the flavor as well as to mask or modify the flavor (Asgar et al. 2010). Several plant proteins have inherent bitterness and astringent taste, which further make the selection of ingredients for the preparation of meat analog very crucial. Alternatively, by adopting advanced processing technologies, this bitterness and astringency could be modified. Soy protein has beany and grassy flavor due to the presence of isoflavone, lipoxygenase, and saponin compounds, which can be mitigated by soaking or heating (Kumar et al. 2012, 2017, 2023). However, the increasing processing levels decrease the nutritive value and, thus, the need to compensate for these nutrients by adding higher amounts. The application of filamentous fungi provides meat-like microstructure to mycoprotein-based meat analogs such as Quorn™ (Joshi and Kumar 2015; Kumar et al. 2011; Wiebe 2002).

Texture and flavor are the two main factors that determine the consumer acceptance of meat analogs. The application of high-moisture extrusion is widely used to develop fibrous and meat-like textures to plant proteins (Palanisamy et al. 2018; Ryu 2020). These texturized proteins are used as meat analogs to impart texture and appearance similar to meat. The structure and composition of the meat analogs have an effect on flavor either by retention by entrapment of volatile compounds or by the specific chemical or non-chemical interactions/bindings between the ions, water molecules, and large food molecules (Wang and Arntfield 2017). During high thermoplastic extrusion, the volatile compounds are lost in steam at the time of expansion of materials immediately after coming out from the die, thus leading to poor flavor scores (Yuliani et al. 2004). Various factors that affect the flavor of meat analogs are raw ingredients used, extrusion conditions such as time, temperature, pressure, and moisture content, size of the end product, vapor loss during expansion, and nature/diffusability of volatile compounds (Bhandari et al. 2001; Reifsteck and Jeon 2000).

As such, food proteins do not have any prominent flavor, but at high temperatures, proteins form bonds with flavor compounds (Wang et al. 2017). Thus, upon altering the protein's ability to bind with the aromatic compounds, the flavor of the food products is also altered (Guichard 2006). Interestingly, this binding of protein to aromatic compounds is reversible and comprises hydrogen and hydrophobic bindings (Farrell et al. 2002), which have a direct impact on the desirable flavors (Wang and Arntfield 2017). Further, meat analogs developed by using various combinations of soy proteins having wheat gluten from 10% to 40% and moisture content varies from 50% to 80%; the highest flavor retention was measured with the product having the highest wheat gluten and minimum moisture (Guo et al. 2020). Further, the authors (Guo et al. 2020) the alteration of microstructure upon increasing wheat gluten content in product with the most common compounds affecting flavor as alcohols, esters, phenols, alkanes, and alkenes (Guo et al. 2020).

There is a need for a cluster-specific/region-specific approach while considering the consumer acceptance of meat analogs, as a study by Lemken et al. (2019) reported the consumer acceptance of legume-based meat analogs in New Zealand, whereas in another cluster, a substitution of legume in meat products was preferred. Similarly, a higher acceptance for plant-based meat analogs was reported in North Belgium, especially among youths and women, but highlighted the need for more varieties and better imitation of meat products (Bryant and Sanctorum 2021). Further, to increase consumer acceptance of plant-based meat analogs, there is a need to develop, market, and promote plant-based meals and plant-forward living based on consumption orientation rather than opposing meat consumption (Graça et al. 2019). Among Danish consumers, negative attitudes toward high meat consumption towards satiety effect, taste, protein content, and environmental and health effects were reported as the main limiting factors in adopting a plant-based diet among these consumers. Whereas consumers having low-meat intake are more attracted to plant-based diets (Reipurth et al. 2019).

The consumer acceptance of meat analogs could be further increased by appropriate marketing strategies highlighting the nutritive role, environmental impact, and positive health effects of the consumption of meat analogs. Giving more front space in supermarkets and placing them next to equivalent meat products could also increase consumer interest and facilitate the transition from meat to plant-based diets. An increase of 67% in meat alternatives sales was recorded by this marketing intervention (Coucke et al. 2022).

Various prospects of plant-based meat analogs are presented in Fig. 1.3.

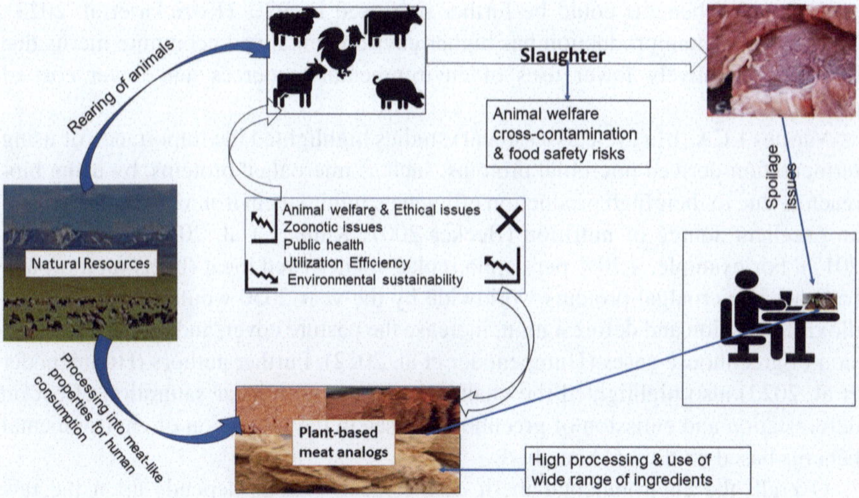


**Fig. 1.3** Prospects of development of plant-based meat analogs

## 1.6 Impact of Meat Analogs

The popularization and adoption of plant-based meat analogs in diets would have a major effect on the environment, public health, animal welfare, and food security.

### *1.6.1 Environmental Impact*

As the environmental impact of meat analogs (plant and insect-based meat alternatives) is markedly lower than that of conventional meat, thus, a reduction in meat production and consumption could have a positive effect on the environment. Even the environmental impact of the animal product with the lowest environmental impact among animal products is higher than the meat analogs (Poore and Nemecek 2018). Increasing red meat production has created newer challenges in the form of rapid conversion of forests into agriculture to produce more feed and grains (deforestation), emission of greenhouse gases leading to climate changes, water pollution, and eutrophication (Poore and Nemecek 2018). This highlighted the importance of the immediate dietary shift towards more environmentally friendly and healthy plant protein-based substitutes from meat and meat products.

A 50% substitution of meat and dairy by 2050 with plant-based meat analogs could stop the deforestation of land and reduce greenhouse gas emissions by 31% as compared to 2020 levels (Kozicka et al. 2023). Among all animal products, replacement of beef was reported to have the highest environmental impact. Interestingly authors pointed out that by restoring the agricultural land to forest, the environmental benefits could be further improved by 92% (Kozicka et al. 2023). Thus, plant protein production has higher environmental and economic merits due to its comparatively lower uses of environmental resources and lower cost of production.

Various LCA (life cycle assessment) studies highlighted the importance of using fermentation-derived microbial proteins, such as microalgal proteins, by using bioreactors due to their high production efficiency, minimal environmental impact, and an excellent source of nutrition (Becker 2007; Kumar et al. 2022d; Wells et al. 2017). For example, a 20% per capita replacement of red meat (beef) on a protein basis with microalgal proteins worldwide by the year 2050 would half the carbon dioxide emission and deforestation, increase the pasture cover, and reduce the emission of greenhouse gases (Humpenöder et al. 2022). Further authors (Humpenöder et al. 2022) also highlighted the challenges of the non-linear saturation effect on deforestation and emission of greenhouse gases in the prediction of environmental benefits based on the LCA method.

Overall, the environmental merit of the meat analogs depends upon the raw ingredients used and the level of processing used during their manufacture. Based on LCA assessment, plant-based beef analogs produced by Beyond Meat® and Impossible™ Foods have a lower environmental impact, such as significantly lower land usage and water eutrophication as compared to their beef, pork, and chicken counterparts, whereas, in terms of energy usage, the meat analog exceeds the energy usage of pork and chicken. Further, the greenhouse gas emission was calculated as lower than pork and beef but higher than chicken (Heller and Keoleian 2018; Khan et al. 2019).

The Beyond burger (1/4 pound) was calculated to emit 90% less greenhouse gas emission, 93% less land usage, 46% lower energy requirement, and more than 99% less water requirement as compared to its beef counterpart having the same weight (Heller and Keoleian 2018). Further, in this burger, the major ingredients used (pea protein, coconut oil, and canola oil) are mainly responsible for greenhouse gas emissions and land and energy usage. The polypropylene tray used for packing these products also has an environmental effect, and a 100% recycling of this tray has the potential to further reduce greenhouse gas emissions by 2% and energy usage by 10% (Heller and Keoleian 2018). Similarly, the Impossible Burger® has 87–96% lower environmental impact (87% less water usage, 96% less land usage, 89% lower greenhouse gas emission, and 92% less eutrophication) as compared to conventional ground beef burgers (Khan et al. 2019). However, the ingredients have a major environmental impact, while the packaging of the product was reported to have a negligible impact on eutrophication, water and land usage, and emission of greenhouse gases (Khan et al. 2019).

In terms of energy and land footprint, mycoproteins (Quoron™) are more efficient than Beyond Meat® and Impossible™ burgers. The water footprint of the plant-based meat analogs mainly depends upon the major plant protein present in these products. Based on the LCA, mycoprotein was observed to have a water footprint of 40 kg/kg of mycoprotein which stands very high as compared to other common sources of plant proteins such as wheat gluten with 0.954 kg/kg and soy protein as 0.73 kg/kg (Smetana et al. 2015). Further, Fresán et al. (2019) calculated the water footprint of 39 meat analog products and reported 3800 $m^3$ of water consumption for the production of 1 ton of plant-based meat analog products, with a major chunk of this water is used during the harvesting of protein sources.

Thus, there is an urgent need to mitigate the environmental impact of the producers by following a multi-pronged or holistic way. Even the communication for the lower environmental impact to the consumers is equally important to increase consumer acceptance, thus making it more economically viable.

### *1.6.2 Impact on Employment*

The development and popularization of meat analogs/alternatives might have an effect on the agriculture sector and, more specifically, on meat production practices, consequently impacting social and cultural aspects. As this sector is still in the growing phase and still remains to tap the actual potential, the exact effect on agriculture, social, and cultural aspects is yet to be clear. The job profile of conventional meat production is different than the production of meat alternatives. As in livestock rearing and meat production practices, with job profile can be suited for semi-skilled or trained personnel. However, the production of meat analogs warrants the application of advanced technologies and the application of a number of additives in a specific manner to get the desired attributes in the developed meat analogs. Thus, the meat analogs are expected to create new employment for skilled and highly skilled jobs. However, the actual impact may vary with the geographical locations and place to place due to differences in the production systems. In a study conducted among experts in Brazil, the USA, and Europe, (Morais-da-Silva et al. 2022) reported pressure on animal farmers due to the fast transition, with the Brazilian farmers being more positive/optimistic about the meat alternatives in creating new job opportunities.

### *1.6.3 Impact on Public Health*

Although meat is an important source of high-quality nutrients its overconsumption of meat, especially red and processed meat, has been correlated to various health issues. Overconsumption of cured and processed meat products has been linked to colorectal cancer and cardiovascular diseases (Badar et al. 2021; Bouvard et al. 2015; Domingo and Nadal 2017). This could be due to high salt intake and high levels of processing, risking the formation of carcinogenic compounds. Meat is deficient in dietary fiber and antioxidant potential, which are abundantly present in vegetables (Kumar et al. 2020b).

The vegetarian diet was reported to exert higher health benefits and provide protection against such as from cardiovascular diseases, metabolic conditions, cancers, and mortality. In addition, it also offers protection against obesity and diabetes (type-2), with higher positive effects recorded in males rather than females (Le and Sabaté 2014). However, plant foods also lack some important nutrients (vitamin $B_{12}$, long-chain $n-3$ fatty acids, retinol) and thus need a careful approach while formulating plant-based diets (Sun 2021). The clinical trials conducted by the Stanford School of Medicine also demonstrated that lower serum trimethylamine-N-oxide levels in the blood lead to reduced risk of cardiovascular diseases (Crimarco et al. 2020). Thus, the consumption of plant-based foods and the reduction in red and processed meat consumption are included in most dietary guidelines for hypertension, obesity, and reducing cholesterol (Craig and Mangels 2009).

Meat is an excellent medium for the growth of food spoilage and pathogenic microorganisms. Further, various steps during the production of meat and post-slaughter processing by coming in contact with contaminated water, packaging, contaminated surfaces, etc., also increase the risk of contamination of pathogenic microorganisms such as *E. coli*, *Campylobacter*, *Salmonella*, *Staphylococcus aureus*, *Shigella*, *Listeria monocytogenes*, *Yersinia enterocolitica*, etc. In addition, the oxidation of meat lipids and proteins also reduces the storage life and spoilage in huge quantities, leading to substantial environmental costs during production, and environmental pollution. The non-judicious use of antibiotics in animal treatment and as feed additives also leads to the entry of antibiotics into the food chain and increases the risk of antimicrobial resistance (AMR). Interestingly, the issue of spoilage, growth of pathogenic microorganisms, and application of antibiotics are comparatively lower than these incidences in meat production. Antibiotic usage in agriculture in the USA was reported to be minimal with only 0.12% of animal agriculture (Stockwell and Duffy 2012).

### *1.6.4 Impact on Animal Welfare*

Due to their production nature, the meat analogs production process does not involve animal rearing and is thus free from any ethical and animal welfare issues. These products are free from animal byproducts except the dairy and egg-based meat analogs or additives. The treatment meted out to animals during the rearing and meat production remain major concerns due to the intensive rearing practices, managemental practices (such as tail docking, dehorning, beak trimming, early weaning, etc.), stress, pain, and distress during transportation and slaughtering operations. The production of plant-based meat analogs would have a direct impact on animal welfare by avoiding these issues. Further, there is an indirect benefit on animal welfare by the substitution of meat with meat analogs, that is by preventing degradation of forest lands and reforestation leading to better animal welfare of wild animals and preservation of biodiversity.

However, this should be noted that in the case of preparation of insect-based meat analogs, there is a concern for animal welfare during their harvesting. There should be proper handling during their harvesting and it should be specific for species types and insect growth stages (Adámková et al. 2017). Some of the humane methods for killing insects are fast-freezing and boiling water (Adámková et al. 2017). Ensuring proper management practices during insect rearing and harvesting would have a positive impact on their growth and productivity in addition to increasing the acceptance of insects as food (De Goede et al. 2013).

### *1.6.5 Nutritive Quality and Food Safety Issues*

At present, the main component of meat analogs remains protein, lipid, and moisture in addition to the addition of several vitamins, trace minerals, colorants, flavorings, and additives. At present, soy and wheat gluten remain the main source of protein in plant-based meat analogs, with an increasing trend in the utilization of pulses/legumes, non-conventional protein sources, fungi, and SCP (Kumar et al. 2017; Kyriakopoulou et al. 2019). The nutritive value of meat analogs is affected by the formulations and composition of these products.

Plant-based meat analogs could be prepared as per the desired nutritive quality by proper selection of raw ingredients and processing technologies. Further, a high level of processing is performed to get the meaty texture of plant proteins. The exposure to high temperature and pressure applied during processing may cause loss of nutrients and nutritive quality of some ingredients. Thus, plant-based meat analogs may have lower nutritive quality than raw ingredients, which needs to compensate for these losses.

The increased consumption of these ultra-processed food products in place of unprocessed/whole food could cause nutritional deficiencies (Hall et al. 2019; Moodie et al. 2013). Further, in an effort to closely mimic the nutritive and organoleptic attributes of meat, some synthetic or animal-origin additives are used during the preparation of plant-based meat analogs. This warrants proper labeling information on the product packages to develop consumer confidence (Choudhury et al. 2020). The food safety concerns are more with respect to the use of single-cell-proteins (SCP) and insect-based meat analogs. These two sources are considered highly efficient in production and environmental sustainability.

Insects, mainly due to their growing ability on wastes and hazardous materials and as a survival strategy under hostile conditions, accumulate allergens, toxins, and anti-nutrient substances (Meyer-Rochow et al. 2021), in addition, being a carrier/vector of pathogens. A higher amount of these toxins, allergens, and anti-nutrient factors are noted when insects are reared on poor quality food sources such as solid wastes, and pollutants (bioaccumulation), and these substances in insects could be controlled by rearing on food sources free from these compounds (Kumar et al. 2022d). Phytates, oxalates, hydrocyanides, tannins, heavy metals, and cyanogenic glycosides are some common anti-nutrients present in insects. These compounds, however, are well below the toxic levels in insects, interfere with the absorption of nutrients in our gut, and could adversely affect human health (Ekop et al. 2010; Omotoso 2006). Further chitin, forming the exoskeleton of the insects, could only be digested in countries (mostly tropical countries) where there is a tradition of entomophagy (the practice of eating edible insects) due to the presence of chitinolytic enzymes secreted by gut microbiota (Finke 2007). Similarly, the consumption of African silkworms causes thiamine deficiency due to the presence of thiaminase enzyme in these insects, thus requiring proper heat treatment before consumption to neutralize it (Nishimune et al. 2000).

Some SCP sources are also having issues of accumulation of heavy metals, pesticides, toxins and allergens (Putten et al. 2011). The endotoxin released by fungi and bacteria also poses health risks e.g., aflatoxins by *Aspergillus flavus*. The presence of a high amount of nucleic acid (upto 16% of dry weight) in bacterial SCP could pose a risk of kidney stones and gout due to the conversion of nucleic acid into uric acid (Bux and Chisti 2016). Some microalgae also pose neurotoxins and hepatotoxins, such as microcystin hepatotoxin by Cyanobacteria may risk liver cancer (Testai et al. 2016). Similarly, the presence of some neurotoxins, such as anatoxins and saxitoxins in *Anabaena*, *Aphanizominon*, *Oscillatoria*, and *Trichodesmium*, may risk food poisoning (Mulvenna et al. 2012).

## 1.7 Conclusion

At present, plant-based meat analogs for beef, pork, and chicken are becoming popular, and various types of these products are widely available in the market. The plant-based analogs for seafood are still in the developing phase. To increase the consumer acceptance and economics of the analogs, these products should be formulated and processed to remain indistinguishable from their meat counterparts in taste, color, texture, and nutritive value. There is a need to have uniform regulatory frameworks to facilitate easier global growth and trade among countries. The high level of processing for texturizing and structuring plant proteins similar to meat proteins and the addition of various additives to improve flavor, color, and nutritive value lead to various issues of food safety, lack of consumer confidence, increasing cost, clean labeling, sustainability, loss of nutritive value, and permissible limits in the development of plant-based meat analogs.

To reduce the cost and improve the sustainability of plant-based meat analogs, it is of utmost importance to use non-conventional sources of vegetable proteins such as mycoprotein, single-cell proteins (SCP), grass, leaf proteins, agriculture industry byproducts that are yet to be harnessed and have high environmental and nutritive benefits. There is a need to follow a cluster/region-specific approach and proper packaging and marketing strategies to make this product popular among the masses.

## References

Aaslyng MD, Martens M, Poll L, Nielsen PM, Flyge H, Larsen LM (1998) Chemical and sensory characterization of hydrolyzed vegetable protein, a savory flavoring. J Agric Food Chem 46(2):481–489. https://doi.org/10.1021/jf970556e

Adámková A, Adámek M, Mlček J, Borkovcová M, Bednávrová M, Kouvrimská L, Skácel J, Vitová E (2017) Welfare of the mealworm (*Tenebrio molitor*) breeding with regard to nutrition value and food safety. Potravinarstvo Slovak J Food Sci 11(1):460–465

Alexandratos N, Bruinsma J (2012) World agriculture towards 2030/2050: the 2012 revision (p. ESA Working paper No. 12-03). https://ageconsearch.umn.edu/record/288998/

Asgar MA, Fazilah A, Huda N, Bhat R, Karim AA (2010) Nonmeat protein alternatives as meat extenders and meat analogs. Compr Rev Food Sci Food Saf 9(5):513–529. https://doi.org/10.1111/j.1541-4337.2010.00124.x

Asioli D, Aschemann-Witzel J, Caputo V, Vecchio R, Annunziata A, Næs T, Varela P (2017) Making sense of the "clean label" trends: a review of consumer food choice behavior and discussion of industry implications. Food Res Int 99:58–71. https://doi.org/10.1016/j.foodres.2017.07.022

Awad AM, Kumar P, Ismail-Fitry MR, Jusoh S, Aziz MFA, Sazili AQ (2021) Green extraction of bioactive compounds from plant biomass and their application in meat as natural antioxidant. Antioxidants 10(9):1465. https://doi.org/10.3390/antiox10091465

Awad AM, Kumar P, Ismail-Fitry MR, Jusoh S, Ab Aziz MF, Sazili AQ (2022) Overview of plant extracts as natural preservatives in meat. J Food Process Preserv 46(8):e16796. https://doi.org/10.1111/jfpp.16796

Badar IH, Liu H, Chen Q, Xia X, Kong B (2021) Future trends of processed meat products concerning perceived healthiness: a review. Compr Rev Food Sci Food Saf 20(5):4739–4778. https://doi.org/10.1111/1541-4337.12813

Becker EW (2007) Micro-algae as a source of protein. Biotechnol Adv 25(2):207–210. https://doi.org/10.1016/j.biotechadv.2006.11.002

Bhandari B, D'Arcy B, Young G (2001) Flavour retention during high temperature short time extrusion cooking process: a review. Int J Food Sci Technol 36(5):453–461. https://doi.org/10.1046/j.1365-2621.2001.00495.x

Bloomberg Intelligence (2021) Plant-based foods market to hit $162 billion in next decade. Projects Bloomberg Intelligence. https://www.bloomberg.com/company/press/plant-based-foods-market-to-hit-162-billion-in-next-decade-projects-bloomberg-intelligence/

Bohrer BM (2017) Nutrient density and nutritional value of meat products and non-meat foods high in protein. Trends Food Sci Technol 65:103–112

Bohrer BM (2019) An investigation of the formulation and nutritional composition of modern meat analogue products. Food Sci Human Wellness 8(4):320–329. https://doi.org/10.1016/j.fshw.2019.11.006

Bouvard V, Loomis D, Guyton KZ, Grosse Y, Ghissassi F, Benbrahim-Tallaa L, Guha N, Mattock H, Straif K (2015) Carcinogenicity of consumption of red and processed meat. Lancet Oncol 16(16):1599–1600. https://doi.org/10.1016/S1470-2045(15)00444-1

Bryant C, Sanctorum H (2021) Alternative proteins, evolving attitudes: comparing consumer attitudes to plant-based and cultured meat in Belgium in two consecutive years. Appetite 161:105161. https://doi.org/10.1016/j.appet.2021.105161

Bryant C, Szejda K, Parekh N, Desphande V, Tse B (2019) A survey of consumer perceptions of plant-based and clean meat in the USA, India, and China. Front Sustain Food Syst 3. https://doi.org/10.3389/fsufs.2019.00011

Butré CI, Wierenga PA, Gruppen H (2012) Effects of ionic strength on the enzymatic hydrolysis of diluted and concentrated whey protein isolate. J Agric Food Chem 60(22):5644–5651. https://doi.org/10.1021/jf301409n

Bux F, Chisti Y (2016) Algae biotechnology: products and processes, 1st edn. Springer. https://doi.org/10.1007/978-3-319-12334-9

Chen Y, Liang Y, Jia F, Chen D, Zhang X, Wang Q, Wang J (2021) Effect of extrusion temperature on the protein aggregation of wheat gluten with the addition of peanut oil during extrusion. Int J Biol Macromol 166:1377–1386. https://doi.org/10.1016/j.ijbiomac.2020.11.017
Chen YP, Feng X, Blank I, Liu Y (2022) Strategies to improve meat-like properties of meat analogs meeting consumers' expectations. Biomaterials 287:121648. https://doi.org/10.1016/j.biomaterials.2022.121648
Chen Q, Chen Z, Zhang J, Wang Q, Wang Y (2023a) Application of lipids and their potential replacers in plant-based meat analogs. Trends Food Sci Technol 138:645–654. https://doi.org/10.1016/j.tifs.2023.07.007
Chen Q, Zhang J, Liu H, Li T, Wang Q (2023b) Mechanism of high-moisture extruded protein fibrous structure formation based on the interactions among pea protein, amylopectin, and stearic acid. Food Hydrocoll 136:108254. https://doi.org/10.1016/j.foodhyd.2022.108254
Choudhury D, Singh S, Seah JSH, Yeo DCL, Tan LP (2020) Commercialization of plant-based meat alternatives. Trends Plant Sci 25(11):1055–1058. https://doi.org/10.1016/j.tplants.2020.08.006
Commission E (2020) A farm to fork strategy-for a fair, healthy and environmentally-friendly food system (p. 381). https://www.eufic.org/en/food-production/article/the-eu-farm-to-fork-strategy-can-we-make-the-european-food-system-healthier-and-sustainable
Coucke N, Vermeir I, Slabbinck H, Geuens M, Choueiki Z (2022) How to reduce agri-environmental impacts on ecosystem services: the role of nudging techniques to increase purchase of plant-based meat substitutes. Ecosyst Serv 56:101444. https://doi.org/10.1016/j.ecoser.2022.101444
Craig WJ, Mangels AR (2009) Position of the American dietetic association: vegetarian diets. J Am Diet Assoc 109(7):1266–1282. https://doi.org/10.1016/j.jada.2009.05.027
Crimarco A, Springfield S, Petlura C, Streaty T, Cunanan K, Lee J, Fielding-Singh P, Carter MM, Topf MA, Wastyk HC, Sonnenburg ED, Sonnenburg JL, Gardner CD (2020) A randomized crossover trial on the effect of plant-based compared with animal-based meat on trimethylamine-N-oxide and cardiovascular disease risk factors in generally healthy adults: study with appetizing plantfood—Meat eating alternative trial (SWAP-MEAT). Am J Clin Nutr 112(5):1188–1199. https://doi.org/10.1093/ajcn/nqaa203
Damayanti D, Jaceldo-Siegl K, Beeson W, Fraser G, Oda K, Haddad E (2018) Foods and supplements associated with vitamin B12 biomarkers among vegetarian and non-vegetarian participants of the adventist health study-2 (AHS-2) Calibration Study. Nutrients 10(6):722. https://doi.org/10.3390/nu10060722
Damodaran S, Arora A (2013) Off-flavor precursors in soy protein isolate and novel strategies for their removal. Annu Rev Food Sci Technol 4(1):327–346. https://doi.org/10.1146/annurev-food-030212-182650
De Goede DM, Erens J, Kapsomenou E, Peters M (2013) In: Röcklinsberg H, Sandin P (eds) Large scale insect rearing and animal welfare BT - The ethics of consumption: the citizen, the market and the law. Wageningen Academic Publishers, pp 236–242. https://doi.org/10.3920/978-90-8686-784-4_38
Dekkers BL, Boom RM, van der Goot AJ (2018) Structuring processes for meat analogues. Trends Food Sci Technol 81:25–36. https://doi.org/10.1016/J.TIFS.2018.08.011
Delshadi R, Bahrami A, Tafti AG, Barba FJ, Williams LL (2020) Micro and nano-encapsulation of vegetable and essential oils to develop functional food products with improved nutritional profiles. Trends Food Sci Technol 104:72–83. https://doi.org/10.1016/j.tifs.2020.07.004
Dennis C, Karim F, Smith JS (2015) Evaluation of Maillard reaction variables and their effect on heterocyclic amine formation in chemical model systems. J Food Sci 80(2):T472–T478. https://doi.org/10.1111/1750-3841.12737
Deroy O, Reade B, Spence C (2015) The insectivore's dilemma, and how to take the West out of it. Food Qual Prefer 44:44–55. https://doi.org/10.1016/j.foodqual.2015.02.007
Detchewa P, Prasajak P, Phungamngoen C, Sriwichai W, Naivikul O, Moongngarm A (2022) Substitution of rice flour with rice protein improved quality of gluten-free rice spaghetti processed using single screw extrusion. LWT-Food Sci Technol 153:112512. https://doi.org/10.1016/j.lwt.2021.112512

Domingo JL, Nadal M (2017) Carcinogenicity of consumption of red meat and processed meat: a review of scientific news since the IARC decision. Food Chem Toxicol 105:256–261. https://doi.org/10.1016/J.FCT.2017.04.028
Doss A, Esther A, Rajalakshmi R (2022) Influence of UV-B treatment on the accumulation of free phenols and tannins in the legumes of *Abrus precatorius* L. and *Vigna mungo* (L.) Hepper. Phytomed Plus 2(1):100189. https://doi.org/10.1016/J.PHYPLU.2021.100189
Ekop EA, Udoh AI, Akpan PE (2010) Proximate and anti-nutrient composition of four edible insects in Akwa Ibom State, Nigeria. World J Appl Sci Technol 2(2):224–231
Farrell HM, Qi PX, Brown EM, Cooke PH, Tunick MH, Wickham ED, Unruh JJ (2002) Molten globule structures in milk proteins: implications for potential new structure-function relationships. J Dairy Sci 85(3):459–471. https://doi.org/10.3168/jds.S0022-0302(02)74096-4
Featherstone S (2015) Ingredients used in the preparation of canned foods. In: A complete course in canning and related processes. Woodhead Publishing, pp 147–211
Fernández-López JA, Fernández-Lledó V, Angosto JM (2020) New insights into red plant pigments: more than just natural colorants. RSC Adv 10(41):24669–24682. https://doi.org/10.1039/d0ra03514a
Finke MD (2007) Estimate of chitin in raw whole insects. Zoo Biol 26(2):105–115. https://doi.org/10.1002/zoo.20123
Finnigan TJA, Wall BT, Wilde PJ, Stephens FB, Taylor SL, Freedman MR (2019) Mycoprotein: the future of nutritious nonmeat protein, a symposium review. Curr Develop Nutr 3(6). https://doi.org/10.1093/cdn/nzz021
Fraser RZ, Shitut M, Agrawal P, Mendes O, Klapholz S (2018) Safety evaluation of soy leghemoglobin protein preparation derived from *Pichia pastoris*, intended for use as a flavor catalyst in plant-based meat. Int J Toxicol 37(3):241–262. https://doi.org/10.1177/1091581818766318
Fresán U, Marrin D, Mejia M, Sabaté J (2019) Water footprint of meat analogs: selected indicators according to life cycle assessment. Water 11(4):728. https://doi.org/10.3390/w11040728
Fu Y, Chen T, Chen SHY, Liu B, Sun P, Sun H, Chen F (2021) The potentials and challenges of using microalgae as an ingredient to produce meat analogues. Trends Food Sci Technol 112:188–200. https://doi.org/10.1016/j.tifs.2021.03.050
Glusac J, Fishman A (2021) Enzymatic and chemical modification of zein for food application. Trends Food Sci Technol 112:507–517. https://doi.org/10.1016/j.tifs.2021.04.024
Goldstein B, Moses R, Sammons N, Birkved M (2017) Potential to curb the environmental burdens of American beef consumption using a novel plant-based beef substitute. PLoS One 12(12):e0189029. https://doi.org/10.1371/journal.pone.0189029
Good Food Institute (2023) U.S. retail market insights for the plant-based industry. https://gfi.org/marketresearch/?utm_source=linkedin&utm_medium=social&utm_campaign=SPINS2019&utm_content=staff
Graça J, Truninger M, Junqueira L, Schmidt L (2019) Consumption orientations may support (or hinder) transitions to more plant-based diets. Appetite 140:19–26. https://doi.org/10.1016/j.appet.2019.04.027
Guichard E (2006) Flavour retention and release from protein solutions. Biotechnol Adv 24(2):226–229. https://doi.org/10.1016/j.biotechadv.2005.11.003
Guo Z, Teng F, Huang Z, Lv B, Lv X, Babich O, Yu W, Li Y, Wang Z, Jiang L (2020) Effects of material characteristics on the structural characteristics and flavor substances retention of meat analogs. Food Hydrocoll 105:105752. https://doi.org/10.1016/j.foodhyd.2020.105752
Hall KD, Ayuketah A, Brychta R, Cai H, Cassimatis T, Chen KY, Chung ST, Costa E, Courville A, Darcey V, Fletcher LA, Forde CG, Gharib AM, Guo J, Howard R, Joseph P, McGehee S, Ouwerkerk R, Raisinger K et al (2019) Ultra-processed diets cause excess calorie intake and weight gain: an inpatient randomized controlled trial of ad libitum food intake. Cell Metab 30(1):67–77.e3. https://doi.org/10.1016/j.cmet.2019.05.008
Han S-W, Chee K-M, Cho S-J (2015) Nutritional quality of rice bran protein in comparison to animal and vegetable protein. Food Chem 172:766–769. https://doi.org/10.1016/j.foodchem.2014.09.127

Hashempour-Baltork F, Khosravi-Darani K, Hosseini H, Farshi P, Reihani SFS (2020) Mycoproteins as safe meat substitutes. J Clean Prod 253:119958. https://doi.org/10.1016/j.jclepro.2020.119958

Heller MC, Keoleian GA (2018) Beyond meat's beyond burger life cycle assessment: a detailed comparison between a plant-based and an animal-based protein source. CSS Report, University of Michigan, Ann Arbor, pp 1–38. https://css.umich.edu/publications/research-publications/beyond-meats-beyond-burger-life-cycle-assessment-detailed

Hoek AC, Luning PA, Weijzen P, Engels W, Kok FJ, de Graaf C (2011) Replacement of meat by meat substitutes. A survey on person- and product-related factors in consumer acceptance. Appetite 56(3):662–673. https://doi.org/10.1016/j.appet.2011.02.001

Huang M, Mehany T, Xie W, Liu X, Guo S, Peng X (2022) Use of food carbohydrates towards the innovation of plant-based meat analogs. Trends Food Sci Technol 129:155–163. https://doi.org/10.1016/j.tifs.2022.09.021

Huang Z, Liu Y, An H, Kovacs Z, Abddollahi M, Sun Z, Zhang G, Li C (2023) Utilizing haematococcus pluvialis to simulate animal meat color in high-moisture meat analogues: texture quality and color stability. Food Res Int 113685. https://doi.org/10.1016/j.foodres.2023.113685

Humpenöder F, Bodirsky BL, Weindl I, Lotze-Campen H, Linder T, Popp A (2022) Projected environmental benefits of replacing beef with microbial protein. Nature 605(7908):90–96. https://doi.org/10.1038/s41586-022-04629-w

Ishaq A, Irfan S, Sameen A, Khalid N (2022) Plant-based meat analogs: a review with reference to formulation and gastrointestinal fate. Curr Res Food Sci 5:973–983. https://doi.org/10.1016/j.crfs.2022.06.001

Jeewanthi RKC, Lee N-K, Paik H-D (2015) Improved functional characteristics of whey protein hydrolysates in food industry. Korean J Food Sci Anim Resour 35(3):350–359. https://doi.org/10.5851/kosfa.2015.35.3.350

Jeong S, Kim HW, Lee S (2017) Rheological and secondary structural characterization of rice flour-zein composites for noodles slit from gluten-free sheeted dough. Food Chem 221:1539–1545. https://doi.org/10.1016/J.FOODCHEM.2016.10.139

Jin Y, He X, Andoh-Kumi K, Fraser RZ, Lu M, Goodman RE (2018) Evaluating potential risks of food allergy and toxicity of soy leghemoglobin expressed in *Pichia pastoris*. Mol Nutr Food Res 62(1):1700297. https://doi.org/10.1002/mnfr.201700297

Joshi VK, Kumar S (2015) Meat analogues: plant based alternatives to meat products-a review. Int J Food Ferment Technol 5(2):107–119. https://doi.org/10.5958/2277-9396.2016.00001.5

Kale P, Mishra A, Annapure US (2022) Development of vegan meat flavour: a review on sources and techniques. Future Foods 5:100149. https://doi.org/10.1016/j.fufo.2022.100149

Kaspchak E, de Oliveira MAS, Simas FF, Franco CRC, Silveira JLM, Mafra MR, Igarashi-Mafra L (2017) Determination of heat-set gelation capacity of a quinoa protein isolate (Chenopodium quinoa) by dynamic oscillatory rheological analysis. Food Chem 232:263–271. https://doi.org/10.1016/J.FOODCHEM.2017.04.014

Kaur L, Mao B, Beniwal AS, Kaur R, Chian FM, Singh J (2022) Alternative proteins vs animal proteins: the influence of structure and processing on their gastro-small intestinal digestion. Trends Food Sci Technol 122:275–286. https://doi.org/10.1016/j.tifs.2022.02.021

Khan S, Dettling J, Loyola C, Hester J, Moses R (2019) Environmental life cycle analysis: impossible Burger 2.0. (Quantis, 2019). https://impossiblefoods.com/sustainable-food/burger-life-cycle-assessment-2019

Kozicka M, Havlík P, Valin H, Wollenberg E, Deppermann A, Leclère D, Lauri P, Moses R, Boere E, Frank S, Davis C, Park E, Gurwick N (2023) Feeding climate and biodiversity goals with novel plant-based meat and milk alternatives. Nat Commun 14(1):5316. https://doi.org/10.1038/s41467-023-40899-2

Krintiras GA, Göbel J, van der Goot AJ, Stefanidis GD (2015) Production of structured soy-based meat analogues using simple shear and heat in a Couette Cell. J Food Eng 160:34–41. https://doi.org/10.1016/J.JFOODENG.2015.02.015

Kumar Singh A, Singh R, Subramani R, Kumar R, Wankhede PD (2016) Molecular approaches to understand nutritional potential of coarse cereals. Curr Genomics 17(3):177–192. https://doi.org/10.2174/1389202917666160202215308

Kumar P, Sharma BD, Kumar RR (2011) Optimization of the level of mushroom in analogue meat nuggets. J Meat Sci 7(1):53–55

Kumar P, Sharma BD, Kumar RR, Kumar A (2012) Optimization of the level of wheat gluten in analogue meat nuggets. Indian J Vet Res (The) 21(1):54–59

Kumar P, Chatli MK, Mehta N, Singh P, Malav OP, Verma AK (2017) Meat analogues: health promising sustainable meat substitutes. Crit Rev Food Sci Nutr 57(5). https://doi.org/10.1080/10408398.2014.939739

Kumar P, Verma AK, Kumar D, Umaraw P, Mehta N, Malav OP (2019) Chapter 11 - Meat snacks: a novel technological perspective. In: Innovations in traditional foods. https://doi.org/10.1016/B978-0-12-814887-7.00011-3

Kumar P, Verma AK, Umaraw P, Kumar D, Mehta N (2020a) Superchilling preserves freshness: the novel technology has emerged to improve long term storage life. Fleischwirtschaft Int 2:45–49

Kumar P, Verma AK, Umaraw P, Mehta N, Malav OP (2020b) Plant phenolics as natural preservatives in food system. In: Plant phenolics in sustainable agriculture. Springer Singapore, pp 367–406. https://doi.org/10.1007/978-981-15-4890-1_16

Kumar M, Tomar M, Punia S, Dhakane-Lad J, Dhumal S, Changan S, Senapathy M, Berwal MK, Sampathrajan V, Sayed AAS, Chandran D, Pandiselvam R, Rais N, Mahato DK, Udikeri SS, Satankar V, Anitha T, Reetu R et al (2022a) Plant-based proteins and their multifaceted industrial applications. LWTi Food Sci Technol 154:112620. https://doi.org/10.1016/J.LWT.2021.112620

Kumar P, Abubakar AA, Verma AK, Umaraw P, Nizam MH, Mehta N, Ahmed MA, Kaka U, Sazili AQ (2022b) New insights in improving sustainability in meat production: opportunities and challenges. Crit Rev Food Sci Nutr. https://doi.org/10.1080/10408398.2022.2096562

Kumar P, Mehta N, Abubakar AA, Verma AK, Kaka U, Sharma N, Sazili AQ, Pateiro M, Kumar M, Lorenzo JM (2022c) Potential alternatives of animal proteins for sustainability in the food sector. Food Rev Intl 39:1–26. https://doi.org/10.1080/87559129.2022.2094403

Kumar P, Sharma N, Ahmed MA, Verma AK, Umaraw P, Mehta N, Abubakar AA, Hayat MN, Kaka U, Lee S-J, Sazili AQ (2022d) Technological interventions in improving the functionality of proteins during processing of meat analogs. Front Nutr 9. https://doi.org/10.3389/fnut.2022.1044024

Kumar P, Sharma M, Abubakar AA, Nizam bin Hayat M, Ahmed MA, Kaka U, Sazili AQ (2023) Soybean: sustainability issues. In: Reference module in food science. Elsevier. https://doi.org/10.1016/B978-0-12-823960-5.00021-4

Kumari S, Memba LJ, Dahuja A, Vinutha T, Saha S, Sachdev A (2016) Elucidation of the role of oleosin in off-flavour generation in soymeal through supercritical CO2 and biotic elicitor treatments. Food Chem 205:264–271. https://doi.org/10.1016/j.foodchem.2016.03.028

Kurek MA, Onopiuk A, Pogorzelska-Nowicka E, Szpicer A, Zalewska M, Półtorak A (2022) Novel protein sources for applications in meat-alternative products—insight and challenges. Food Secur 11(7):957. https://doi.org/10.3390/foods11070957

Kyriakopoulou K, Dekkers B, van der Goot AJ (2019) Plant-based meat analogues. In: Sustainable meat production and processing. Academic Press, pp 103–126. https://doi.org/10.1016/B978-0-12-814874-7.00006-7

Kyriakopoulou K, Keppler JK, van der Goot AJ, Boom RM (2021) Alternatives to meat and dairy. Annu Rev Food Sci Technol 12(1):29–50. https://doi.org/10.1146/annurev-food-062520-101850

Lanzotti V (2006) The analysis of onion and garlic. J Chromatogr A 1112(1–2):3–22. https://doi.org/10.1016/j.chroma.2005.12.016

Le L, Sabaté J (2014) Beyond meatless, the health effects of vegan diets: findings from the adventist cohorts. Nutrients 6(6):2131–2147. https://doi.org/10.3390/nu6062131

Lemken D, Spiller A, Schulze-Ehlers B (2019) More room for legume—Consumer acceptance of meat substitution with classic, processed and meat-resembling legume products. Appetite 143:104412. https://doi.org/10.1016/j.appet.2019.104412

Li X, Li J (2020) The flavor of plant-based meat analogues. Cereal Foods World 65(4). https://doi.org/10.1094/CFW-65-4-0040

Lin M, Liu X, Xu Q, Song H, Li P, Yao J (2014) Aroma-active components of yeast extract pastes with a basic and characteristic meaty flavour. J Sci Food Agric 94(5):882–889. https://doi.org/10.1002/jsfa.6330

López DN, Ingrassia R, Busti P, Wagner J, Boeris V, Spelzini D (2018) Effects of extraction pH of chia protein isolates on functional properties. LWT-Food Sci Technol 97:523–529. https://doi.org/10.1016/J.LWT.2018.07.036

Lusk JL, Norwood FB (2009) Some economic benefits and costs of vegetarianism. Agric Res Econ Rev 38(2):109–124. https://doi.org/10.1017/S1068280500003142

Mancini RA, Hunt MC (2005) Current research in meat color. Meat Sci 71(1):100–121. https://doi.org/10.1016/j.meatsci.2005.03.003

Manzoor S, Masoodi FA, Naqash F, Rashid R (2022) Oleogels: promising alternatives to solid fats for food applications. Food Hydrocoll Health 2:100058. https://doi.org/10.1016/j.fhfh.2022.100058

Marketsandmarkets (2023) Plant-based meat market worth $15.7 billion by 20. https://www.marketsandmarkets.com/PressReleases/plant-based-meat.asp

Martins AJ, Lorenzo JM, Franco D, Vicente AA, Cunha RL, Pastrana LM, Quiñones J, Cerqueira MA (2019) Omega-3 and polyunsaturated fatty acids-enriched hamburgers using sterol-based oleogels. Eur J Lipid Sci Technol 121(11):1900111. https://doi.org/10.1002/ejlt.201900111

Mattice KD, Marangoni AG (2020) Evaluating the use of zein in structuring plant-based products. Curr Res Food Sci 3:59–66. https://doi.org/10.1016/J.CRFS.2020.03.004

McClements DJ, Grossmann L (2021) The science of plant-based foods: constructing next-generation meat, fish, milk, and egg analogs. Compr Rev Food Sci Food Saf 20(4):4049–4100. https://doi.org/10.1111/1541-4337.12771

Mehta N, Kumar P, Verma AK, Umaraw P, Ranjan R (2021) Missing the real culprit? Insight on the water footprint of the meat Industry. Fleishwirtschaft Int 2:54–58

Meyer-Rochow VB, Gahukar RT, Ghosh S, Jung C (2021) Chemical composition, nutrient quality and acceptability of edible insects are affected by species, developmental stage, gender, diet, and processing method. Food Secur 10(5). https://doi.org/10.3390/foods10051036

Mintel Store (2023) US pant-based protein report 2023. https://store.mintel.com/us/report/us-plant-based-proteins-market-report/

Moodie R, Stuckler D, Monteiro C, Sheron N, Neal B, Thamarangsi T, Lincoln P, Casswell S (2013) Profits and pandemics: prevention of harmful effects of tobacco, alcohol, and ultra-processed food and drink industries. Lancet 381(9867):670–679. https://doi.org/10.1016/S0140-6736(12)62089-3

Moomand K, Lim LT (2015) Effects of solvent and n-3 rich fish oil on physicochemical properties of electrospun zein fibres. Food Hydrocoll 46:191–200. https://doi.org/10.1016/J.FOODHYD.2014.12.014

Morais-da-Silva RL, Villar EG, Reis GG, Sanctorum H, Molento CFM (2022) The expected impact of cultivated and plant-based meats on jobs: the views of experts from Brazil, the United States and Europe. Humanit Soc Sci Commun 9(1):297. https://doi.org/10.1057/s41599-022-01316-z

Mota C, Santos M, Mauro R, Samman N, Matos AS, Torres D, Castanheira I (2016) Protein content and amino acids profile of pseudocereals. Food Chem 193:55–61. https://doi.org/10.1016/j.foodchem.2014.11.043

Mulvenna V, Dale K, Priestly B, Mueller U, Humpage A, Shaw G, Allinson G, Falconer I (2012) Health risk assessment for cyanobacterial toxins in seafood. Int J Environ Res Public Health 9(3):807–820. https://doi.org/10.3390/ijerph9030807

Nishimune T, Watanabe Y, Okazaki H, Akai H (2000) Thiamin is decomposed due to Anaphe spp. entomophagy in seasonal ataxia patients in Nigeria. J Nutr 130(6):1625–1628. https://doi.org/10.1093/jn/130.6.1625

OECD, Agriculture Organization FAO, F., & of the United Nations, A. O (2021) OECD-FAO agricultural outlook 2021-2030. Organisation for Economic Co-operation and Development. https://doi.org/10.1787/19428846-en

Omotoso OT (2006) Nutritional quality, functional properties and anti-nutrient compositions of the larva of *Cirina forda* (Westwood) (Lepidoptera: Saturniidae). J Zhejiang Univ Sci B 7(1):51–55. https://doi.org/10.1631/jzus.2006.B0051

Palanisamy M, Töpfl S, Aganovic K, Berger RG (2018) Influence of iota carrageenan addition on the properties of soya protein meat analogues. LWT-Food Sci Nutr 87:546–552. https://doi.org/10.1016/j.lwt.2017.09.029

Pietsch VL, Emin MA, Schuchmann HP (2017) Process conditions influencing wheat gluten polymerization during high moisture extrusion of meat analog products. J Food Eng 198:28–35. https://doi.org/10.1016/j.jfoodeng.2016.10.027

Pietsch VL, Werner R, Karbstein HP, Emin MA (2019) High moisture extrusion of wheat gluten: relationship between process parameters, protein polymerization, and final product characteristics. J Food Eng 259:3–11. https://doi.org/10.1016/J.JFOODENG.2019.04.006

Piñero MP, Parra K, Huerta-Leidenz N, Arenas de Moreno L, Ferrer M, Araujo S, Barboza Y (2008) Effect of oat's soluble fibre (β-glucan) as a fat replacer on physical, chemical, microbiological and sensory properties of low-fat beef patties. Meat Sci 80(3):675–680. https://doi.org/10.1016/j.meatsci.2008.03.006

Poore J, Nemecek T (2018) Reducing food's environmental impacts through producers and consumers. Science 360(6392):987–992. https://doi.org/10.1126/science.aaq0216

Reifsteck BM, Jeon IJ (2000) Retention of volatile flavors in confections by extrusion processing. Food Rev Intl 16(4):435–452. https://doi.org/10.1081/FRI-100102318

Reipurth MFS, Hørby L, Gregersen CG, Bonke A, Perez Cueto FJA (2019) Barriers and facilitators towards adopting a more plant-based diet in a sample of Danish consumers. Food Qual Prefer 73:288–292. https://doi.org/10.1016/j.foodqual.2018.10.012

Robbins K, Jensen J, Ryan KJ, Homco-Ryan C, McKeith FK, Brewer MS (2003) Effect of dietary vitamin E supplementation on textural and aroma attributes of enhanced beef clod roasts in a cook/hot-hold situation. Meat Sci 64(3):317–322. https://doi.org/10.1016/S0309-1740(02)00203-6

Rubio NR, Xiang N, Kaplan DL (2020) Plant-based and cell-based approaches to meat production. Nat Commun 11(1):6276. https://doi.org/10.1038/s41467-020-20061-y

Ryu G-H (2020) Extrusion cooking of high-moisture meat analogues. In: Extrusion cooking. Elsevier, pp 205–224. https://doi.org/10.1016/B978-0-12-815360-4.00007-9

Ryu KK, Kang YK, Jeong EW, Baek Y, Lee KY, Lee HG (2023) Applications of various natural pigments to a plant-based meat analog. LWT-Food Sci Technol 174:114431. https://doi.org/10.1016/j.lwt.2023.114431

Sakai K, Sato Y, Okada M, Yamaguchi S (2022) Synergistic effects of laccase and pectin on the color changes and functional properties of meat analogs containing beet red pigment. Sci Rep 12(1):1168. https://doi.org/10.1038/s41598-022-05091-4

Schouteten JJ, De Steur H, De Pelsmaeker S, Lagast S, Juvinal JG, De Bourdeaudhuij I, Verbeke W, Gellynck X (2016) Emotional and sensory profiling of insect-, plant- and meat-based burgers under blind, expected and informed conditions. Food Qual Prefer 52:27–31. https://doi.org/10.1016/j.foodqual.2016.03.011

Schuh V, Allard K, Herrmann K, Gibis M, Kohlus R, Weiss J (2013) Impact of carboxymethyl cellulose (CMC) and microcrystalline cellulose (MCC) on functional characteristics of emulsified sausages. Meat Sci 93(2):240–247. https://doi.org/10.1016/j.meatsci.2012.08.025

Sharma M, Kaur S, Kumar P, Mehta N, Umaraw P, Ghosh S (2022) Development, prospects, and challenges of meat analogs with plant-based alternatives. In: Recent advances in food biotechnology. Springer Nature Singapore, pp 275–299. https://doi.org/10.1007/978-981-16-8125-7_14

Shelomi M (2016) The meat of affliction: insects and the future of food as seen in Expo 2015. Trends Food Sci Technol 56:175–179. https://doi.org/10.1016/J.TIFS.2016.08.004

Singh M, Trivedi N, Enamala MK, Kuppam C, Parikh P, Nikolova MP, Chavali M (2021) Plant-based meat analogue (PBMA) as a sustainable food: a concise review. Eur Food Res Technol 247(10):2499–2526. https://doi.org/10.1007/s00217-021-03810-1
Smetana S, Mathys A, Knoch A, Heinz V (2015) Meat alternatives: life cycle assessment of most known meat substitutes. Int J Life Cycle Assess 20(9):1254–1267. https://doi.org/10.1007/s11367-015-0931-6
Sohail A, Al-Dalali S, Wang J, Xie J, Shakoor A, Asimi S, Shah H, Patil P (2022) Aroma compounds identified in cooked meat: a review. Food Res Int 157:111385. https://doi.org/10.1016/j.foodres.2022.111385
Spence C (2015) On the psychological impact of food colour. Flavour 4(1):21. https://doi.org/10.1186/s13411-015-0031-3
Statista (2023) Meat substitutes. https://www.statista.com/outlook/cmo/food/meat/meat-substitutes/worldwide
Stockwell VO, Duffy B (2012) Use of antibiotics in plant agriculture. Revue Scientifique et Technique de l'OIE 31(1):199–210. https://doi.org/10.20506/rst.31.1.2104
Strack D, Vogt T, Schliemann W (2003) Recent advances in betalain research. Phytochemistry 62(3):247–269. https://doi.org/10.1016/S0031-9422(02)00564-2
Suman SP, Joseph P (2013) Myoglobin chemistry and meat color. Annu Rev Food Sci Technol 4(1):79–99. https://doi.org/10.1146/annurev-food-030212-182623
Summo C, De Angelis D, Difonzo G, Caponio F, Pasqualone A (2020) Effectiveness of oat-hull-based ingredient as fat replacer to produce low fat burger with high beta-glucans content. Food Secur 9(8):1057. https://doi.org/10.3390/foods9081057
Sun W (2021) Vegetarian diet: why is it beneficial?. In: IOP conference series: earth and environmental science, vol 714(2). IOP Publishing, p 022004. https://doi.org/10.1088/1755-1315/714/2/022004
Sun C, Ge J, He J, Gan R, Fang Y (2021) Processing, quality, safety, and acceptance of meat analogue products. Engineering 7(5):674–678. https://doi.org/10.1016/j.eng.2020.10.011
Testai E, Scardala S, Vichi S, Buratti FM, Funari E (2016) Risk to human health associated with the environmental occurrence of cyanobacterial neurotoxic alkaloids anatoxins and saxitoxins. Crit Rev Toxicol 46(5):385–419. https://doi.org/10.3109/10408444.2015.1137865
Thrane M, Paulsen PV, Orcutt MW, Krieger TM (2017) Soy protein: impacts, production, and applications. In: Sustainable protein sources. Academic Press, pp 23–45. https://doi.org/10.1016/B978-0-12-802778-3.00002-0
Tolstoguzov V (2003) Some thermodynamic considerations in food formulation. Food Hydrocoll 17(1):1–23. https://doi.org/10.1016/S0268-005X(01)00111-4
Tolstoguzov V (2006) Texturising by phase separation. Biotechnol Adv 24(6):626–628. https://doi.org/10.1016/J.BIOTECHADV.2006.07.001
Tziva M, Negro SO, Kalfagianni A, Hekkert MP (2020) Understanding the protein transition: the rise of plant-based meat substitutes. Environ Innov Soc Trans 35:217–231. https://doi.org/10.1016/j.eist.2019.09.004
United Nations (2022) Day of 8 Billion. https://www.un.org/en/dayof8billion#:~:text=Day%20of%20Eight%20Billion,a%20milestone%20in%20human%20development
Van Huis A, Van Itterbeeck J, Klunder H, Mertens E, Halloran A, Muir G, Vantomme P (2013) Edible insects: future prospects for food and feed security (Issue 171). Food and Agriculture Organization of the United Nations
van Putten MC, Kleter GA, Gilissen LJ, Gremmen B, Wichers HJ, Frewer LJ (2011) Novel foods and allergy: regulations and risk-benefit assessment. Food Control 22(2):143–157. https://doi.org/10.1016/J.FOODCONT.2010.08.002
Vatansever S, Tulbek CM, Riaz NM (2020) Low- and high-moisture extrusion of pulse proteins as plant-based meat ingredients: a review. Cereal Foods World 65(4). https://doi.org/10.1094/CFW-65-4-0038
Viuda-Martos M, López-Marcos MC, Fernández-López J, Sendra E, López-Vargas JH, Pérez-Álvarez JA (2010) Role of fiber in cardiovascular diseases: a review. Compr Rev Food Sci Food Saf 9(2):240–258. https://doi.org/10.1111/j.1541-4337.2009.00102.x

Vulić JJ, Ćebović TN, Čanadanović VM, Ćetković GS, Djilas SM, Čanadanović-Brunet JM, Velićanski AS, Cvetković DD, Tumbas VT (2013) Antiradical, antimicrobial and cytotoxic activities of commercial beetroot pomace. Food Funct 4(5):713. https://doi.org/10.1039/c3fo30315b

Wang K, Arntfield SD (2017) Effect of protein-flavour binding on flavour delivery and protein functional properties: a special emphasis on plant-based proteins. Flavour Fragr J 32(2):92–101. https://doi.org/10.1002/ffj.3365

Wang K, Li C, Wang B, Yang W, Luo S, Zhao Y, Jiang S, Mu D, Zheng Z (2017) Formation of macromolecules in wheat gluten/starch mixtures during twin-screw extrusion: effect of different additives. J Sci Food Agric 97(15):5131–5138. https://doi.org/10.1002/jsfa.8392

Wang X, Zhu L, Han Y, Xu L, Jin J, Cai Y, Wang H (2018) Analysis of volatile compounds between raw and cooked beef by HS-SPME–GC–MS. J Food Process Preserv 42(2):e13503. https://doi.org/10.1111/jfpp.13503

Weinrich R (2018) Cross-cultural comparison between German, French and Dutch Consumer preferences for meat substitutes. Sustain For 10(6):1819. https://doi.org/10.3390/su10061819

Weis T (2015) SFSGEC - Meatification and the madness of the doubling narrative. Canadian Food Studies/La Revue Canadienne Des Études Sur l'alimentation 2(2):296–303. https://doi.org/10.15353/cfs-rcea.v2i2.105

Wells ML, Potin P, Craigie JS, Raven JA, Merchant SS, Helliwell KE, Smith AG, Camire ME, Brawley SH (2017) Algae as nutritional and functional food sources: revisiting our understanding. J Appl Phycol 29(2):949–982. https://doi.org/10.1007/s10811-016-0974-5

Wiebe M (2002) Myco-protein from *Fusarium venenatum*: a well-established product for human consumption. Appl Microbiol Biotechnol 58(4):421–427. https://doi.org/10.1007/s00253-002-0931-x

Wild F (2016) Manufacture of meat analogues through high moisture extrusion. In: Reference module in food science. Elsevier. https://doi.org/10.1016/B978-0-08-100596-5.03281-9

Willett W, Rockström J, Loken B, Springmann M, Lang T, Vermeulen S, Garnett T, Tilman D, DeClerck F, Wood A, Jonell M, Clark M, Gordon LJ, Fanzo J, Hawkes C, Zurayk R, Rivera JA, De Vries W, Majele Sibanda L et al (2019) Food in the anthropocene: the EAT–Lancet Commission on healthy diets from sustainable food systems. Lancet 393(10170):447–492. https://doi.org/10.1016/S0140-6736(18)31788-4

Williamson DA, Geiselman PJ, Lovejoy J, Greenway F, Volaufova J, Martin CK, Arnett C, Ortego L (2006) Effects of consuming mycoprotein, tofu or chicken upon subsequent eating behaviour, hunger and safety. Appetite 46(1):41–48. https://doi.org/10.1016/j.appet.2005.10.007

Xiao Z, Jiang R, Huo J, Wang H, Li H, Su S, Gao Y, Duan Y (2022) Rice bran meat analogs: relationship between extrusion parameters, apparent properties and secondary structures. LWT 163:113535. https://doi.org/10.1016/j.lwt.2022.113535

Yao G, Liu KS, Hsieh F (2004) A new method for characterizing fiber formation in meat analogs during high-moisture extrusion. J Food Sci 69(7):303–307. https://doi.org/10.1111/j.1365-2621.2004.tb13634.x

Yesuraj D, Deepika C, Ravishankar GA, Ranga Rao A (2022) Seaweed-based recipes for food, health-food applications, and innovative products including meat and meat analogs. In: Sustainable Global Resources of Seaweeds, vol 2. Springer International Publishing, pp 267–292. https://doi.org/10.1007/978-3-030-92174-3_14

Yuliani S, Bhandari B, Rutgers R, D'Arcy B (2004) Application of microencapsulated flavor to extrusion product. Food Rev Intl 20(2):163–185. https://doi.org/10.1081/FRI-120037159

Yuliarti O, Ng L, Koh WM, Abdullah Tan MFBMF, Dwi Sentana A (2023) Structural properties of meat analogue with added konjac gels. Food Hydrocoll 142:108716. https://doi.org/10.1016/j.foodhyd.2023.108716

Zhu D, Damodaran S (2018) Removal of off-flavour-causing precursors in soy protein by concurrent treatment with phospholipase A2 and cyclodextrins. Food Chem 264:319–325. https://doi.org/10.1016/j.foodchem.2018.05.045

# Chapter 2
# Milk and Dairy Products Analogues

**Athira Syamala, Khushi Dubey, and Prafulla Salunke**

## 2.1 Introduction

For centuries, milk and its derivatives have been integral to human diets, providing vital nutrients such as protein, fats, carbohydrates, and essential elements crucial for growth and development. The duration of milk feeding varies among mammals, marked by distinct weaning periods contingent on the species (Chalupa-Krebzdak et al. 2018; StuartMacadam 2017; Pereira 2014). Unlike other mammals, humans continue the consumption of milk from various species throughout their lives as liquid milk for drinking or in the form of milk products. And hence, over the years, milk has become an integral part of the human diet. However, evolving human life-styles have given rise to health concerns over milk consumption, including issues like lactose intolerance, milk allergies, saturated fat, worries about calorie intake, a high prevalence of cholesterol issues, concerns about growth hormones or antibiotic residues in cow's milk. and more recently, environmental challenges linked to milk production. In addition, the modern consumer is increasingly emphasizing sustain-ability, environmental considerations, animal welfare, and, consequently, there is an evolving trend to shift away from traditional milk and its products towards other alternatives.

These alternatives, known by various names such as analogs, imitation or non-dairy substitute, have evolved to offer alternatives made from non-dairy ingredients aimed to replicate the moisture, fat, protein, carbohydrates, and minerals, such as calcium, while closely mimicking the flavor and texture of traditional milk and milk products. Other misnomers used are analog, non-dairy, substitute, filled, mimic, simulated, alternate, plant-based, or dairy deceptions, etc. Categorically, these

A. Syamala · K. Dubey · P. Salunke (✉)
Department of Dairy and Food Science, South Dakota State University, Brookings, SD, USA
e-mail: athira.syamala@sdstate.edu; Khushi.dubey@sdstate.edu;
prafulla.salunke@sdstate.edu



Ö. P. CAN et al. (eds.), *Food Analogues: Emerging Methods and Challenges*,
https://doi.org/10.1007/978-3-031-69872-9_2

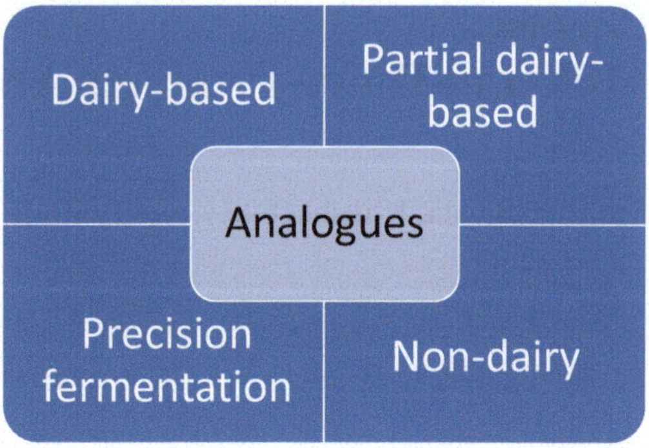


**Fig. 2.1** The base of analogue products

analogues fall into four major types based on their formulation or manufacturing ingredients. These include dairy-based, partial dairy-based, non-dairy, and emerging proteins from precision fermentation or single-cell proteins (Fig. 2.1). Non-dairy alternatives are mostly plant-based dairy substitutes that can be broadly classified into dairy alternatives and dairy analogues. Dairy alternatives merely replace carriers for nutrients and probiotics with plant-based ingredients, while dairy analogues transform plant-based materials to recreate the flavor, texture, appearance, and often nutrition of real dairy products. Examples range from fortified aqueous extracts to probiotic-enriched juices as dairy alternatives, and milks, yoghurts, and cheeses as popular dairy analogues.

Dairy analogues, including plant-based milk and milk product substitutes, are experiencing rapid growth due to perceived positive health effects and the expanding vegan market. The food industry views these analogues as economically advantageous due to lower production costs. Responding to consumer demand, the industry developed a diverse range of dairy analogues, including yogurt and cheese, using a variety of plant materials, such as almonds and potatoes (Pua et al. 2022). While these analogues aim to mirror the composition, nutrition, functionality, sensory attributes, health benefits, dietary aspects, and sustainability of the products they emulate, many of these analogues were initially nutritionally unbalanced and had unacceptable flavors (Tangyu et al. 2019; Tan and McClements 2021). These products often fall short in quality, organoleptic qualities, and texture compared to dairy products. In addition, many substitutes have high sugar and sodium levels, low protein, and unwanted stabilizers and emulsifiers added to match the functionality of dairy products. Careless substitution in dairy analogues can result in deficiencies in key nutrients such as protein, calcium, vitamin B12, vitamin D, omega-3 fatty acids, iron, and zinc (Carvalho et al. 2001; Tan and McClements 2021). Many dairy analogues are formulated products where manufacturers, to compete and reduce costs, resort to cheaper substitutions, resulting in unwanted ingredients. This contradicts the current trend where consumers demand clean-label products. Considering the aforementioned reasons, a detailed examination and improvement of the formulation processes for dairy analogues are imperative to address nutritional shortcomings, enhance product quality, and align with consumer preferences for clean and transparent labeling.

## 2.2 Legal Status

Milk has been defined legally in many countries, and the International Dairy Federation's position paper and Codex Committee on milk and milk products have developed definitions to apply to dairy products accurately under The Codex General Standard for Use of Dairy Terms (GSUDT). Milk has been defined as "normal mammary secretion of milking animals obtained from one or more milkings," the term dairy in milk and milk products should be limited to use in the products conforming to the definition of milk. The definition excludes precision fermentation ingredients or lab-grown ingredients as they do not meet the definition. Similarly, the European Commission regulates the term milk clearly to be exclusively used for the normal mammary secretion obtained from one or more milkings without either addition thereto or extraction therefrom (Council of the European Union 2007). Further designations related to dairy products are also reserved exclusively for milk products, i.e., whey, cream, butter, buttermilk, butteroil, caseins, anhydrous milk fat, cheese, yoghurt, kefir, koumiss, (Council of the European Union 2007; Jeske et al. 2018). According to Regulation (EU) No. 1308/2013 of the European Parliament and the Council of the European Union, the important aspect to be mentioned with milk is the animal species "as regards milk, the animal species from which the milk originates shall be stated, if it is not bovine" (European Union regulations 2013; Plamada et al. 2023)

However, the interpretation and use of the definition vary as not all milk species are covered, or other milk analogues have been allowed by certain countries. The guide uses correct terms that are universally accepted for dairy products. In the late 1990s, the Codex Committee on Milk and Milk Products developed the guidance to ensure dairy terms are accurately applied to milk and milk products to avoid potential misuse of the term dairy if they do not contain milk or milk ingredients. This definition applies only to products consumed by humans and not used for non-human food uses like animal feeds or medical uses. Similarly, the fermented, cultured, or lab-grown ingredients will not be labeled in dairy terms. European Commission (2010) allowed coconut milk and almond milk to be labeled as milk in the European Union (Jeske et al. 2018). Soy milk is labeled as soy milk worldwide except in the European Union, which does not meet their standards (Sethi et al. 2016).

Labeling of products is a huge problem and any undue advantage to gain access to milk and milk product goodness by misleading the consumers should be avoided. Milk and cheese are traditional dairy terms, and consumers relate such products to being dairy-derived. However, PBMSs are often labeled with the same descriptors to achieve familiar expectations (Jeske et al. 2018). Any wrong, false, misleading, deceptive, confusing, or erroneous impression should be avoided. The terms explained in the GSUDT include milk, milk products, reconstituted milk products, recombined milk products, and composite milk products. The labeling becomes essential, and consumers need to be accurately informed on species origin, compositional changes (lactose-free, change in fat or water content, etc.), nutritional properties modifications, changes in physical conditions and shelf life, etc. The common dairy terms (e.g., milk, cream, cheese, specific cheese names, butter, yogurt, etc.) are reserved for products made from milk and milk products because applying these terms to other products fundamentally misleads consumers. However, for certain products, cocoa butter, canned coconut milk, or coconut cream, the term is allowed because of the historical perspective, traditions, and practice, as these products are not used as substitutes for dairy terms.

## 2.3 Ingredient Sourcing for Dairy Analogues

### *2.3.1 Lipid Sources*

Lipids or fat sources utilized in analogues contribute to flavor and texture, serving as carriers for various fat-soluble components such as vitamins and flavonoids. For dairy-based or partially dairy-based analogue products, milk fat is commonly employed as a lipid source. Milk fat sources encompass high-fat milk, cream (18–22% fat), cream (40–50% fat), plastic cream (>70% fat), salted or unsalted butter (80% fat), concentrated milk fat (78–82% fat), and anhydrous milk fat or butter oil (99.5–99.9% fat). In contrast, non-dairy analogues or fat-substituted analogues typically incorporate vegetable oils, such as palm oil, sunflower oil, and canola oil. However, it is crucial to avoid trans fatty acids due to well-documented health concerns. Most vegetable oils have a fatty acid core, where triacylglycerol is surrounded by a phospholipid monolayer, and oleosins protein provides structure to the oil. Additionally, alternatives like margarine, shortenings, and modified oils, including hydrogenated oils, are available, along with animal fat sources like lard or tallow.

Each oil possesses distinct characteristics and varies in chemical and physical parameters such as solid fat index, melting and crystallization points, viscosity, and sensory attributes. The properties of various oils and fats used in analogue product formulations are given in Table 2.1. These parameters significantly influence the flavor and texture of analogue products. Plant-derived lipids contain varying levels of phospholipids, which contribute to emulsion stability by acting as emulsifiers with both hydrophilic and lipophilic ends (Tan and McClements 2021). The essential criterion for plant-based fats in analogue products is their ability to function similarly to fats from animal sources.

The primary distinction between animal and plant lipids lies in the composition of saturated and unsaturated fats, along with the solid fat index they provide. Plant oils and fats are desirable for their high levels of polyunsaturated fatty acids (PUFA), offering positive health benefits. However, increased unsaturation in fats makes them more susceptible to oxidation and rancidity, leading to flavor issues. Since

**Table 2.1** Properties of various oils and fats used in analogue product formulation

| Properties | Specific gravity at 25 °C | Refractive Index | Iodine value | Saponification value | Unsaponifiable matter, % | Melting point, °C |
|---|---|---|---|---|---|---|
| Soyabean oil | 0.917 to 0.921 | 1.470 to 1.476 | 123.0–139.0 | 185 to 195 | 1.5 max. | −23.0 to −20.0 |
| Cottonseed oil | 0.916 to 0.918 | 1.468 to 1.472 | 99.0 to 113.0 | 189 to 198 | 1.5 max. | 30 to 37 |
| Peanut Oil | 0.910 to 0.915 | 1.467 to 1.470 | 84.0 to 100.0 | 188 to 195 | 0.2 to 1.0 | −2 |
| Corn Oil | 0.915 to 0.920 | 1.470 to 1.474 | 103 to 128 | 187 to 193 | 2.0 max. | −12.0 to −10.0 |
| Sunflower Oil | 0.915 | 1.472 to 1.474 | 125.0–136.0 | 188 to 194 | 1.5 max. | −18.0 to −16.0 |
| Safflower Oil | 0.919 to 0.924 | 1.748 to 1.4752 | 140.0 to 150 | 186 to 197 | 1.5 max. | −18.0 to −16.0 |
| Canola Oil | 0.914 to 0.920 | 1.470 to 1.474 | 110.0–126.0 | 182 to 193 | 15 g/kg max. | |
| Palm Oil | 0.8919 to 0.8932 | 1.457 to 1.459 | 46.0 to 56.0 | 196 to 202 | 0.2 to 0.5 | 36.0 to 45.0 |
| Olive oil | 0.909 to 0.915 | 1.468 to 1.471 | 80.0 to 88.0 | 188 to 196 | 1.8 max. | 0 |
| Coconut oil | 0.917 to 0.919 | 1.448 to 1.450 | 7.5 to 10.5 | 250 to 264 | 0.5 max. | 23.0 to 26.0 |
| Palm kernel oil | 0.899 to 0.914 | 1.448 to 1.452 | 13.0 to 23.0 | 230 to 254 | 0.8 max. | 24.0 to 26.0 |
| Lard | 0.858 to 0.864 | 45 to 52 | 46.0 to 70.0 | 195 to 202 | 1.0 max. | |
| Tallow | 0.860 to 0.870 | 46.0 to 49.0 | 38.0 to 48.0 | 193 to 202 | 0.8 max. | |
| Milk fat | 0.907 to 0.912 | 1.4465 | 25.0 to 42.0 | 210 to 250 | 0.4 max. | 38 to 42 |

Adapted from O'Brien (1998)

many plant fats and oils are liquid at room temperature due to their high unsaturation, stability problems arise in most analogues. Techniques such as hydrogenation, homogenization, emulsification, oleogel formations, or choosing lipids rich in saturated fats (like coconut oil or cocoa butter) can potentially address this issue.

Plant-based analogous drinks typically consist of long-chain fatty acids, whereas those made from coconut fat or milk fat contain short-chain fatty acids and relatively high amounts of medium-chain fatty acids (Walther et al. 2022). The crystallization behavior and solid fat index of lipids significantly influence the structure, texture, and functionality of the foods in which they are used. Therefore, simulating and restructuring the crystallization behavior of fats is crucial to match the characteristics of the finished product, such as spreadability (butter), foamability (whipped cream), meltability (cheese), and hardness (ice cream), among others (McClements and Grossmann 2021).

In short, fat plays a crucial role in determining the flavor of products and can pose flavor-related challenges if not adequately managed throughout the supply chain. Therefore, maintaining specific conditions, such as regulating temperature, oxygen levels, lighting, incorporating antioxidants, preventing contamination by pro-oxidation compounds, utilizing chelating agents, or structuring oils, can mitigate the development of off-flavors in fats and oils (Jacobsen et al. 2013; Jacobsen 2015; McClements and Decker 2018; McClements and Grossmann 2021). It is important to highlight that once the auto-oxidation of fats and oils initiates, slowing down or halting the process becomes exceedingly difficult. Given the inherent susceptibility of most lipids to oxidation, they are typically packaged under nitrogen and stored below 20 °C, preferably below 10 °C, away from direct sunlight or artificial lights. At these temperatures, many lipids assume a solid or semi-solid state, necessitating warming or tempering in a heated room where the temperature is maintained above the melting point of the fat, which poses significant challenges.

In recent years, there has been ongoing interest in the development of texturized or structured oils, such as oleogels or a combination of hydrogels and oleogels. Although these innovations, as highlighted by Chaudhary et al. (2023), are gaining attention, their widespread adoption may require some time. Many have started to use oleogels in products as they not only contribute to the texture but also to the functionality of the end product in which it is used (Chaudhary et al. 2023).

### 2.3.2 *Protein Sources*

Milk proteins consist of caseins and whey or serum proteins. Caseins constitute about 80% of the total protein in bovine milk and form approximately 2.5% of its weight. 95% of the casein exists in milk as large colloidal particles, known as Micelles, which hold 10% of the volume due to their high hydration behavior (Dalgleish and Corredig 2012). The white color of milk is due largely to light scattering by the casein micelles (de Kruif et al. 2012).

Casein is by far the most important and valuable component of milk from a product, technological, and industry perspective. The characteristics and existence of numerous dairy products depend on milk protein casein, while fat, lactose, and notably salts, also exert considerable influence. The intense heat applied to many dairy products is feasible due to the exceptionally high heat stability of the casein. Caseins easily bind to air-water and oil-water interfaces owing to their open structure, high polar amino acid content, and uneven amino acid distribution. This grants caseins exceptional emulsification and foaming properties (Fox et al. 2015).

According to FAO, replacing the currently accepted Protein Digestibility Corrected Amino Acid Score (PDCAAS) with the Digestible Indispensable Amino Acid Score (DIAAS) for an improved assessment of the protein quality found in human foods and ingredients (FAO 2013). According to this method, all types of milk proteins have greater DIAAS i.e., more than 100 compared to plant proteins (Mathai et al. 2017). Therefore, milk protein products are the first ideal protein choice of the food industry when it comes to developing high-protein foods. Besides nutritional value, milk protein products have desirable functional properties like emulsifying, foaming, gelling, water binding, and viscosity development (Agarwal et al. 2015). The functional properties of milk proteins in food products are the result of the intrinsic properties of native proteins (amino acid composition and sequence, molecular weight, genetic variation, charge, hydrophobicity, and protein configuration) and extrinsic factors (temperature, pH, salts, and concentration) (Kinsella 1984; Henning et al. 2006).

As a result of understanding how various factors influence milk proteins, we have been able to tailor ingredients for specific purposes. Developmental approaches in manufacturing milk protein-enriched ingredients encompass three main categories: (1) Ingredients with normal ratio of milk constituents (skimmed milk powder, non-fat dry milk, whole milk powder). (2) Ingredients with whey proteins (whey protein concentrate (34–80% protein), whey protein isolates). (3) Ingredients focusing on casein (rennet casein, acid casein, milk protein concentrate, micellar casein concentrate) (Salunke et al. 2023a). Additional categories include fat and lactose-only products, along with modified versions like hydrolysates and calcium- or sodium casinates (Salunke et al. 2023a). The manufacturing principles primarily involve separation, concentration, and drying processes. Drying ensures ingredient stability, functionality preservation, and increased shelf life. These specialized ingredients offer advantages such as concentrated form and enhanced functionality in formulations with smaller quantities. Enzymatic, acidification, chromatographic, and ion exchange techniques are employed for milk protein fractionation. Commercially available high-protein powders are casein-based, whey-based, or a blend of both. While skimmed milk powder (SMP), nonfat dry milk (NFDM), and whole milk powder (WMP) contain high levels of caseins and whey proteins, but their other constituents like lactose and minerals might limit certain applications. Caseins and whey proteins can be isolated together through a method called co-precipitation by utilizing heat and calcium salts. The main attraction of these products is not only high yield but also unique functional properties (Fox et al. 2015).

Rennet casein is manufactured from milk by treating with a selected proteinase (rennet). Rennet casein powder is the preferred choice in imitation cheese or process cheese product applications due to its superior textural and functional properties. It consists of 100% intact casein, offering improved emulsification, strong structural support, and the highest viscosity per gram of protein (Salunke and Metzger 2022a, b; Salunke et al. 2022a, b). However, for rennet casein powder to effectively hydrate and emulsify the oil phase in cheese analogues, it requires the disruption of colloidal calcium mediated protein interactions by calcium-sequestering salts (Caric and Kalab 1993; Aimutis 1995; Ennis and Mulvihill 1997). So, in dairy-based imitation cheese formulations, introducing other proteins may lead to textural issues. But crosslinking enzymes like transglutaminase have shown potential in enhancing functionality (Salunke and Metzger 2023a) and improving the texture and melting properties of imitation mozzarella or process cheese products (Salunke et al. 2022a, b; Salunke and Metzger 2022b; Salunke et al. 2023b).

Milk proteins play a vital role in dairy products, contributing essential texture and functional properties. They are responsible for many physicochemical properties relevant to dairy products such as water-holding capacity, gelation, and gel strength, as well as the generation of flavor precursors and/or compounds (Pua et al. 2022). However, there has been a recent surge in the popularity of plant proteins among manufacturers due to their associated beneficial properties. These include stability, sustainability, non-allergenicity, non-toxicity, low cost, biodegradability, functional attributes, and consumer acceptance stemming from their clean label status (Yin et al. 2012; Doublier et al. 2000). These plant proteins can be derived from various sources, including soybeans, chickpeas, cowpeas, peas, faba beans, mung beans, lentils, wheat gluten, rice glutelin, flaxseed protein, algae, and microalgae, each possessing unique characteristics (Karaca et al. 2011; Tang 2017; Jeske et al. 2018; Burger and Zhang 2019). Milk proteins exhibit functionalities over a broad pH range whereas plant proteins exhibit functionalities in a narrow range. The composition of various dairy protein powders is shown in Table 2.2. These powders offer various functional properties based on the constituents present (Table 2.3). The

**Table 2.2** Composition (%) of dairy protein powders

| Parameters | Moisture | Fat | Protein | Lactose | Ash |
|---|---|---|---|---|---|
| SMP | 4.3 | 1.0 | 35.0 | 51.9 | 7.8 |
| WMP | 3.5 | 26.2 | 25.2 | 38.1 | 7.0 |
| MPC 42 | 3.5 | 1.0 | 42.0 | 46.0 | 7.5 |
| MPC 56 | 3.8 | 1.2 | 56.0 | 31.0 | 8.0 |
| MPC 70 | 4.2 | 1.4 | 70.0 | 16.2 | 8.2 |
| MPC 75 | 5.0 | 1.5 | 75.0 | 10.9 | 7.6 |
| MPC 80 | 3.9 | 1.8 | 80.0 | 4.1 | 7.4 |
| MPC 85 | 4.9 | 1.6 | 85.0 | 1.0 | 7.1 |
| MPI | 5.0 | 1.5 | 87.1 | 3.5 | 6.9 |
| MCC | 3.5 | 1.1 | 86.3 | 4.2 | 8.4 |
| Sweet Whey | 4.5 | 1.0 | 12.0 | 73.0 | 8.0 |
| Acid Whey | 4.5 | 0.5 | 12.0 | 68.0 | 11.0 |
| WPC34 | 4.0 | 3.0 | 35.0 | 51.0 | 6.0 |
| WPC50 | 4.0 | 4.0 | 50.0 | 35.0 | 7.0 |
| WPC80 | 4.0 | 5.0 | 80.0 | 4.0 | 4.0 |
| WPI | 4.5 | 1.0 | 92.0 | 0.5 | 2.0 |
| Co- precipitate | 5.0 | 1.5 | 89–94 | 1.5 | 4.5 |
| Acid Casein | 9.6 | 1.2 | 87.3 | 0.1 | 1.8 |
| Rennet Casein | 11.0 | 0.5 | 80.6 | 0.1 | 7.8 |
| Sodium Caseinate | 4.6 | 1.1 | 90.4 | 0.1 | 3.8 |
| Calcium Caseinate | 4.6 | 1.1 | 90.5 | 0.1 | 3.7 |
| Calcium Casein Caseinate | 4.6 | 0.1 | 89.3 | 0.1 | 4.9 |

Adapted from Huffman and Harper (1999), Caric (1994)
*SMP* Skimmed milk powder, *WMP* Whole milk powder, *MPC* Milk protein concentrate, *MPI* Milk protein isolate, *MCC* Micellar casein concentrate, *WPC* Whey protein concentrate, *WPI* Whey protein isolate

**Table 2.3** Functional properties of high protein powders

| Application | Caseinates | MPC | WPC | Hydrolysates | TMP |
|---|---|---|---|---|---|
| Emulsion stability | + | + | + | | + |
| Heat stability | + | + | | + | + |
| Opacity | + | + | + | | + |
| Flavor | | + | | | + |
| Water binding | + | | + | | + |
| Protein fortification | + | + | + | + | + |
| Heat gelation | | | + | | |
| Acid stability | | | + | + | |
| Whipping | + | | + | + | + |

Adapted from Baldwin and Pearce (2005)
*TMP* total milk protein

**Table 2.4** Use of high protein powders in various food ingredients

| Application | Caseinates | MPC | WPC | Hydrolysates | TMP |
|---|---|---|---|---|---|
| Bakery | + | + | + | | + |
| Dressings | + | | + | | |
| Sauce/soup | + | + | + | | + |
| Ham | + | | + | | + |
| Reformed poultry | + | | + | | + |
| Sausages | + | | + | | + |
| Coffee whitener | + | | | | + |
| Low-fat spreads | + | | | | |
| Egg white replacer | | | + | | |
| Processed cheese | + | + | + | | + |
| Cultured dairy | + | + | + | | + |
| Cheese milk extension | + | + | | | + |
| Nutritional bars | + | + | + | + | + |
| Thermal processed beverages | + | + | + | + | + |
| Powdered beverages | + | + | + | + | + |
| Pasta/noodles | | | + | | |
| Vegetable analogs | | | + | | |
| Surimi | | | + | | |

Adapted from Baldwin and Pearce (2005)
*TMP* total milk protein

manufacturers have a wide range of ingredients and constituents available through these dairy powders to be used in dairy or partial dairy products (Table 2.4).

The functionality of these proteins is influenced by their biological origin and any changes in their association and native states during isolation and purification. Most plant proteins are globular with ordered structures, consisting of different types of proteins held together by physical and/or chemical bonds. As they adsorb at interfaces, they undergo entropy changes through structural rearrangement in secondary and tertiary changes (Euston 2004). The globular nature of plant proteins, with exposed hydrophobic groups, leads to their adsorption to nonpolar groups of oil droplets in non-dairy alternatives. This results in strong and long-range hydrophobic attractive forces, overcoming repulsive forces and causing particle aggregation. Proteins with inherently disordered structures exhibit better surface activity compared to ordered proteins. For instance, β-lactoglobulin in milk generally has an unordered structure, whereas α-lactalbumin has a helical structure. Unstructured proteins like casein, with an open random coil structure, undergo conformational rearrangement as emulsifiers, causing rapid changes compared to globular proteins. Studies have shown that the competitive adsorption of proteins takes place at the oil–water interface in non-dairy milk, and among the mixture of proteins, some proteins adsorb more effectively compared to others based on their structure and the partitioning of hydrophobic and hydrophilic residues.

A significant challenge in the plant-based food sector is the lack of plant proteins with consistent functional attributes. Table 2.5 depicts the protein content and total amino acid content (essential and non-essential) present in various protein isolates. It is very difficult to match the quality of dairy proteins and hence fortification becomes necessary. Proteins exhibit variability in emulsion formation and stability under external conditions like temperature, pH, salt concentrations, and sugars. Plant proteins (peas, legumes, faba bean) lack stability at pH close to their isoelectric point, high temperatures, and high salt concentrations (Gumus et al. 2017). To achieve stability in plant-based milks, specific protein modification strategies, such as denaturation, emulsion stability, surface activity, and solubility, should be considered when formulating these products.

**Table 2.5** Protein content (% of raw material) and amino acid (essential and non-essential) content of various dietary protein sources

| | | Amino acids | |
|---|---|---|---|
| Protein isolate | Protein % | Essential | Non-essential |
| Oat | 64 | 13.7 | 24.7 |
| Lupin | 61 | 13.1 | 28.2 |
| Wheat | 81 | 18 | 48.9 |
| Hemp | 51 | 11.6 | 22.4 |
| Soy | 61–91 | 19.9 | 31.9 |
| Brown rice | 79 | 22.1 | 36.8 |
| Pea | 80 | 23.6 | 34.4 |
| Corn | 65 | 21 | 32.3 |
| Potato | 80 | 29.3 | 27.8 |
| Whey | 72–84 | 34.1 | 34.9 |
| Milk | 73–85 | 30.3 | 38.6 |
| Caseinate | 86 | 32.8 | 40.4 |
| Casein | 67–78 | 24.8 | 32.5 |

Adapted from Gorissen et al. (2018)

### 2.3.3 *Carbohydrate Sources*

Milk primarily contains lactose as its main carbohydrate. The prevalence of lactose intolerance, especially among newer generations in various countries, has prompted many consumers to either reduce their milk intake or switch to analogue products. Analogues often incorporate additional carbohydrates to enhance body, texture, and mouthfeel in the final products. Sugars and various carbohydrates, including oligosaccharides or polysaccharides, serve as functional ingredients (McClements and Grossmann 2021). According to Walther et al. (2022), milk analogues may contain sucrose, fructose, glucose, lactose, starch, and fiber carbohydrates. In unsweetened beverages, naturally occurring carbohydrates contribute to the sugar content, potentially resulting in lower sweetness. However, in some analogues, sweeteners such as sugar or sugar equivalents are added, raising potential health concerns (Bleich and Vercammen 2018; Malik et al. 2010; Thorning et al. 2016).

Plant-derived carbohydrates can exhibit diverse properties, including molecular, physicochemical, functional, and biological aspects, depending on their sources and extraction methods (McClements and Grossmann 2021). The total carbohydrate content in analogues can vary significantly. In contrast to milk, where lactose is the predominant carbohydrate, analogues may contain sucrose and glucose as the main carbohydrates (Walther et al. 2022).

Other polysaccharides commonly used are starch (potato, corn, tapioca, etc.), cellulose, alginate, carrageenan, pectin, agar, gellan gum, xanthan gum, gum Arabic, and their derivatives (Tan and McClements 2021). Besides sweetness, they can help provide varying levels of solubility, thickening, gelling, foaming, emulsifying, structure formation, stabilization, and water-holding properties and are used in varying concentrations (McClements and Grossmann 2021; Tan and McClements 2021). For example, the concentration of starch varied from nil to 32 g/kg in a cashew drink (Walther et al. 2022).

Dietary fibers can be used as binding agents, texturizers, and emulsifiers in analogues (Kyriakopoulou et al. 2021a, b; McClements et al. 2019). Some of these fibers are digestible or indigestible and may be fermentable (McClements and Grossmann 2021), and these may have a positive impact on health and nutrition. Selecting the appropriate carbohydrate based on the food matrix and intended functionality is crucial. Polysaccharides can be combined with proteins to achieve desired textural and sensory properties in plant-based foods through phase separation and interactive mechanisms (McClements and Grossmann 2021).

Notwithstanding the increased demand for carbohydrate analogues, it is essential to note that numerous dairy products are naturally low in lactose or entirely lactose-free. This includes items like cheese, lactose-free milk, and ice cream. High-fat milk products, such as butter oil, anhydrous milk fat, or ghee, typically do not contain lactose. Additionally, protein powders derived from membrane systems often have low lactose content. Thus, consumers may find suitable dairy options that align with their dietary needs without entirely forgoing milk-based products.

### 2.3.4 Other Functional Additives and Attributes

In the formulation and production of high-quality analogues, it is common to incorporate fortification and enrichment along with enhanced functionality additives. These additives typically contribute to various aspects such as color, flavor, body, and texture with an aim to improve the overall appearance and taste of the product. Additionally, they potentially serve as buffers, preservatives, and crosslinking agents, with an aim to enhance specific health-related attributes, provide stability, and potentially extend the shelf life of the product. However, several drawbacks with the current functional attributes exist.

***Flavors*** Analogous products often lack distinctive flavors or may carry flavors from their constituent ingredients. Consequently, these products frequently rely on the use of masking and overpowering flavors to mitigate the original taste and align with the intended flavor profile of the product they aim to emulate.

***Color*** Despite the vibrant colors naturally present in plant-based ingredients, many analogue products face criticism for their dull appearance. To address this, various colors, whether of natural or synthetic origin, are incorporated to mask and enhance the product's visual appeal. Therefore, the process of masking or enhancing specific colors plays a crucial role.

***Other Healthy Additives*** Due to the perception that analogs are less healthy, manufacturers often include a variety of healthy ingredients to counteract this notion. For instance, plant-based foods are fortified with various micronutrients, polyphenols, and nutraceuticals or phytochemicals. Some of them include natural carotenoids; precursors of vitamin A (natural pigments in colored fruits and vegetables) (Walther et al. 2022), curcuminoids, anthocyanins, catechins, and tannins (McClements et al. 2019), bioactive substances like tocopherols, tocotrienols, coenzyme Q10, and bioactive peptides (Sha and Xiong 2020) or preservatives (e.g., essential oils or antimicrobial peptides) (McClements and Grossmann 2021). Both macro- and micro-nutrients, including protein additives, minerals like calcium, vitamins, and antioxidants, may be incorporated. In terms of micronutrients, it's worth noting that plant-based milk substitutes generally have lower mineral and vitamin content, and their absorption is less consistent compared to milk. Astolfi et al. (2020) reported that soy and coconut milk were good sources of magnesium, hazelnut milk had high amounts of sodium whereas only cow and goat milk provided a significant amount of calcium, potassium, magnesium, sodium, and phosphorus. Nevertheless, the inclusion of selected and concentrated forms of these healthy ingredients may contribute to an overall increase in the cost of the products in which they are utilized.

***Antinutrients*** Despite analogues being promoted as a healthy alternative to dairy, it's important to note that those made from plant-based ingredients may contain anti-nutrients. These are compounds that interfere with mineral absorption and reduce protein digestibility. Common antinutrients found in these products include phytate, phytin or phytic acid, amylase and trypsin inhibitors, tannins, oxalate, phenols and polyphenols, saponins, agglutinin, alkaloids, cyanide and inositol phosphates (Plamada et al. 2023), however, they vary a lot depending on the sources.

## 2.4 Milk and Dairy Analogues

### *2.4.1 Milk Analogues*

Milk is a complex fluid that has several components, which are distributed in three different phases. A major part of milk is a true solution of lactose, organic and inorganic salts, vitamins, enzymes, and other small molecules in water. Proteins are also dispersed in true solution, where whey proteins are at the molecular level and caseins exist as larger colloidal aggregates. Along with that, fat is present in the form of emulsified globules. Each component has a major impact on the nutritional, technological, and sensory properties of milk and dairy products. Milk is a complex mixture of a diverse array of components, including fat, protein, lactose, minerals, vitamins, and various macro and micro-nutrients, existing in either emulsion, colloidal or soluble form. These constituents maintain a delicate balance, influencing the color, appearance, body, texture, and, notably, the organoleptic qualities of milk (McClements 2020; Schiano et al. 2017).

Substitutes for animal milk are commonly referred to as milk analogues, plant-based or vegetable milk, or non-dairy milk alternatives. These analogues are essentially water-soluble extracts derived from plant-based raw materials, including cereals, pseudocereals, nuts, legumes, and oilseeds (Dhankhar 2023). The production process involves reducing the size of the raw material (Plamada et al. 2023) and creating emulsions through the homogenization of plant components and water (Mäkinen et al. 2016; Röös et al. 2016; Sethi et al. 2016). The objective is to replicate the consistency and appearance of cow milk (Haas et al. 2019) or specifically imitate homogenized and pasteurized milk (McClements et al. 2019), to mirror the sensory and functional attributes of dairy milk (Reyes-Jurado et al. 2021). The extraction process and processing techniques used are depicted in Fig. 2.2.

Milk analogues typically comprise ingredients from plant sources, water, emulsifiers, and additives (Reyes-Jurado et al. 2021). The nutritional composition of the final product is contingent upon the chosen plant, as well as the methods employed in processing and fortification. Legumes, owing to their rich nutritional profile, are often favored in this regard (Mäkinen et al. 2016; Paul et al. 2020; Reyes-Jurado et al. 2021). However, the objective of closely resembling cow milk, both in terms of nutrition and organoleptic qualities, introduces challenges during the manufacturing process (Tangyu et al. 2019). As natural emulsions, plant-based beverages must ensure optimal homogenization and possess stable emulsion properties to meet consumer expectations (Silva and Smetana 2022).

Given that milk analogues typically manifest as oil-in-water emulsions in complex multi-component systems, comprising fats, proteins, polysaccharides,

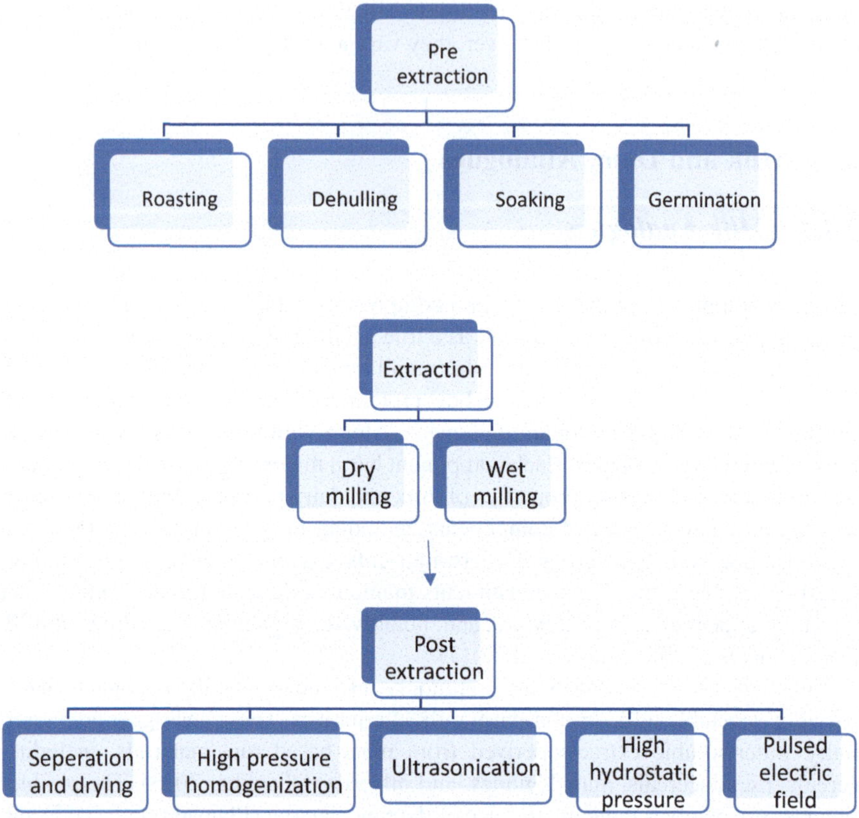


**Fig. 2.2** Overview of possible processing techniques for plant-based ingredient extraction for dairy analogue production. (Adapted from Pua et al. 2022)

additives, water, sugars, flavors, and other small molecular-weight compounds, inherent instability issues arise, leading to phenomena such as aggregation, creaming, coalescence, and sedimentation. Effective formulation is imperative, necessitating the judicious selection of emulsion stabilizers (surfactants, emulsifiers, and hydrocolloids) to enhance the stability of milk substitutes. The manufacturing process encompasses various steps, including size reduction of raw materials, extraction of raw material in water, homogenization, and aseptic processing (McClements et al. 2019; Reyes-Jurado et al. 2021; Dhankhar 2023).

Soy milk stands as the most widely consumed milk analogue, with its initial product launch occurring in Hong Kong in 1940 (Chen 1989; Yadav et al. 2017). The soymilk market experienced significant growth in Asian countries during the late 1970s with the development of industrial technology (Yadav et al. 2017). Over the years, various new milk analogues from diverse plant sources have been introduced, categorized broadly based on the origin of the plant. This classification results in five general categories: cereal-based (e.g., oat milk, rice milk, corn milk, spelt milk), legume-based (e.g., soy milk, peanut milk, lupin milk, cowpea milk), nut-based (e.g., almond milk, cashew nut milk, coconut milk, hazelnut milk, pistachio milk, walnut milk, tiger nut milk), seed-based (e.g., sesame milk, flax milk, hemp milk, sunflower milk), and pseudocereal-based (e.g., quinoa milk, teff milk, amaranth milk) (Sethi et al. 2016). The nutritional quality available from these plant-based ingredients is given in Table 2.6, which varies a lot hence proper ingredient selection is important. Milk analogues available in the market show wide variation in major and minor constituents (Table 2.7) and only a few were close to milk nutrients. Milk analogues, characterized as water extracts of plant-based materials, aim to emulate animal milk in both appearance and consistency (Yadav et al. 2017). The growth of milk analogues in recent times is predominantly attributed to health-related components and the benefits they offer, including antibacterial effects, antioxidant properties with free radical scavenging capabilities, cardioprotective effects, improvements in gut health, and the rising consumer interest in vegan diets (Dhankhar 2023). These products have gained popularity due to their rich antioxidant activity and fatty acid profiles, contributing to the reduction of various diseases such as cardiovascular diseases, cancer, atherosclerosis, and diabetes (Zujko and Witkowska 2014). Additionally, they serve as suitable carriers for probiotics, maintaining adequate probiotic levels throughout the product shelf life, and imparting various health-related benefits (Rasika et al. 2021).

While milk analogues are often lauded for their perceived health benefits, they do have drawbacks in terms of health and nutrition. These downsides encompass lower protein levels, inferior protein quality, reduced bioavailability of crucial minerals and vitamins, and potential oral health concerns. However, there are strategies available to mitigate some of these negative effects. Techniques like fermentation (Mäkinen et al. 2016), the reduction of additives like sweeteners (Aydar et al. 2020), and the incorporation of additives that boost the nutritional value of these alternatives can address certain shortcomings. Fermentation proves effective in neutralizing anti-nutrients and polyphenols, which might hinder the body's absorption of nutrients (Dubey and Patel 2018; Rekha and Vijayalakshmi 2010; Rosado et al. 2005; Aydar et al. 2020; Tangyu et al. 2019). It's always advisable to compare the energy and nutritional content of milk analogues against that of bovine milk to make informed dietary choices.

Milk analogs are colloidal mixtures of a variety of particles such as oil bodies, fat droplets, protein aggregates, plant tissue fragments, and insoluble calcium carbonate particles. These particles are dispersed within an aqueous medium containing soluble proteins, polysaccharides, sugars, and salts. Preparation of these analogs typically involves two main methods: disrupting plant tissues or using homogenization. It involves a detailed process aimed at breaking down plant materials like soybeans, flaxseeds, almonds, or coconut flesh into smaller particles. This process employs various steps including soaking, mechanical disruption, enzymatic hydrolysis, separation, formulation, homogenization, and thermal treatment. Beginning with raw material preprocessing—dehulling, soaking, grinding, sprouting, and blanching are conducted to improve extraction efficiency, nutritional value, and

**Table 2.6** Nutritional profile of dietary sources (g/100 g) available for analogues

| | | | | | | | | Total fatty acids | | |
|---|---|---|---|---|---|---|---|---|---|---|
| Source | Water | Energy | Protein | Lipid | Ash | Carbohydrate | Total dietary fibre | Saturated | Monosaturated | Polyunsaturated |
| Oat | 8.22 | 389 | 16.9 | 6.9 | 1.72 | 66.3 | 10.6 | 1.22 | 2.18 | 2.54 |
| Rice | 12.9 | 360 | 6.61 | 0.58 | 0.58 | 79.3 | – | 0.16 | 0.18 | 0.15 |
| Corn | 10.4 | 365 | 9.42 | 4.74 | 1.2 | 74.3 | 7.3 | 0.67 | 1.25 | 2.16 |
| Quinoa | 13.3 | 368 | 14.1 | 6.07 | 2.28 | 64.2 | 7 | 0.71 | 1.61 | 3.29 |
| Teff | 8.82 | 367 | 13.3 | 2.38 | 2.37 | 73.1 | 8 | 0.45 | 0.59 | 1.07 |
| Amaranth | 11.3 | 371 | 13.6 | 7.02 | 2.88 | 65.2 | 6.7 | 1.46 | 1.68 | 2.73 |
| Buckwheat | 9.75 | 343 | 13.2 | 3.4 | 2.1 | 71.5 | 10 | 0.74 | 1.04 | 1.04 |
| Soy | 7.25 | 327 | 51.46 | 1.22 | 1.3 | 33.92 | 17.5 | 0.14 | 0.21 | 0.53 |
| Peanut | 6.5 | 567 | 25.8 | 49.2 | 2.3 | 16.1 | 8.5 | 6.3 | 24.4 | 15.6 |
| Lupin | 5.52 | 371 | 92.6 | – | 5.3 | 11 | <0.10 | 0.48 | 0.25 | 0.2 |
| Pea | 78.9 | 81 | 5.42 | 0.4 | 0.87 | 14.4 | 5.7 | 0.07 | 0.03 | 0.19 |
| Chickpea | 60.2 | 164 | 6.05 | 2.59 | 0.92 | 62.95 | 7.6 | 0.6 | 1.38 | 2.73 |
| Almond | 4.7 | 575 | 21.22 | 49.42 | 2.99 | 21.67 | 12.2 | 3.73 | 30.89 | 12.07 |
| Hazelnut | 5.31 | 628 | 15 | 60.7 | 2.29 | 16.7 | 9.7 | 4.46 | 45.7 | 7.91 |
| Walnut | 3.62 | 630 | 16.66 | 66.9 | 1.81 | 13.7 | 6.7 | 6.13 | 8.93 | 47.2 |
| Pistachio | 4.37 | 560 | 26.16 | 45.4 | 2.99 | 29 | 10.6 | 5.6 | 24.5 | 13.3 |
| Coconut | 47 | 354 | 3.33 | 33.5 | 0.97 | 15.2 | 9 | 29.7 | 1.42 | 0.387 |
| Sesame | 4.69 | 573 | 17.6 | 49.7 | 4.48 | 9.85 | 14.9 | 7.09 | 18.8 | 21.8 |
| Flaxseed | – | 536 | 20 | 42.9 | – | 29 | 28 | 3.57 | 7.02 | 24.3 |
| Hempseed | 4.96 | 553 | 31.6 | 48.8 | 6.06 | 8.67 | 4 | 4.6 | 5.4 | 38.1 |
| Sunflower | 4.73 | 584 | 20.8 | 51.5 | 3.02 | 20 | 8.6 | 4.46 | 18.5 | 23.1 |

From Plamada et al. (2023)

**Table 2.7** Milk analogues available in the market

| Product | Moisture % | protein % | Total solids % | Total fat% | PUFA, % | Ash % | Carbohydrate % | Fiber % | pH |
|---|---|---|---|---|---|---|---|---|---|
| Almond | 72–93 | 0.8–1.7 | 3.36–8.11 | 1.06–8.89 | 0.81 | 0.09–3.04 | 0.08–4.5 | 0.58–1.25 | 5.72–6.92 |
| Cashew | 87.12 | 2.05 | 13 | 3.3 | 1.06 | 2.63 | 4.38 | 1.15 | |
| Coconut | 50–54 | 3.28–3.6 | 14 | 30.13–34.7 | 0 | 1.2 | 6.6 | 0 | 6.608 |
| Hazelnut | 94.1 | 0.65–4.17 | 5.3–9.6 | 1.55–4.02 | 6.16 | 0.2 | 3.23–3.99 | 0.4 | 6.35–7.15 |
| Peanut | 82–87 | 2.05–2.33 | | 2.40–3.48 | 3.66 | | 5.50–5.60 | | 6.82–6.85 |
| Soy | 88–91 | 2.36–8.71 | 8.11 | 2.35–4.35 | 0.98 | 0.84–0.81 | 4.78 ± 5% | 0.7 | 6.9–7.39 |
| Oat | | 0.78 | | 0.28 | 4.21 | 0.48 | 2.75 | 0.8 | 7.16 |
| Rice | 87 | 0.07–0.6 | 12.9 | 0.87–2.33 | 0.94 | 0.13–0.42 | 25.28 | 0 | 5.21–6.10 |
| Hemp | 91.6 | 2.6–4 | | 4.61–5 | 4.17 | 0.47 | 0.3 | 0.4 | |
| Sesame | | 2.48–2.97 | 10.19–13.91 | 6.43–8.02 | | 0.22–0.48 | 4.04 | 0–0.2 | 6.81–8.23 |
| Tiger nut | 76–81 | 0.47–7.7 | 12.99–23.2 | 1.88–5.5 | | 0.13–0.7 | 2.31–11 | 0.53–0.65 | 4.4–7.02 |
| Walnut | 86 | 1.3–1.7 | | 7.37–18 | | 0.2–0.41 | 4.76 | 0.4 | 5.08 |
| Quinoa | | 0.57 | 5.63 | 0.11 | | 0.04 | | 0.43 | |
| Chickpea extract | 94 | 2.1 | | 0.39 | | 0.37 | 3.39 | | |
| Pistachio | | 5.20–5.80 | 9.50–10.30 | | | | | | 6.50–8.50 |
| Melon seed | | 2.03–3.67 | | 7.97–13.00 | | 0.40–1.10 | 3.4 | | 5.86–6.45 |
| Amaranth | | 3.42 | | | | 0.43 | 3.07 | 1.9 | 10 |
| Millet | | 0.55–9.17 | | 9.17 | | 0.35–1.69 | 78.73 | 0.9 | 4.5–6.47 |

From Aydar et al. (2020)

sensory qualities (Pingali et al. 2023). Two extraction methods, wet and dry, are utilized in this process. In the wet method, the plant-based material is soaked and ground with water to form a slurry, while in the dry method, the material is ground into flour and then extracted using water. This helps extract protein, fat, and soluble solids, followed by filtration to remove insoluble or larger particles, resulting in an aqueous phase. After this separation, any remaining coarse or insoluble components might undergo hydrolysis (McClements et al. 2019; Reyes-Jurado et al. 2021). Subsequent steps involve incorporating additional ingredients like oil, sugar, salts, colors, flavors, and stabilizers. Homogenization and thermal processing are then employed to produce non-dairy milk alternatives with specific desired qualities. Fortification of heat-sensitive ingredients can occur post-heat treatment but before packaging (McClements et al. 2019; Reyes-Jurado et al. 2021). This sequence of processing steps ensures the development of milk analogues with targeted nutritional compositions and sensory attributes. Many times, milk analogues are formulated by reconstituting protein isolates or concentrates in water and adding oils, sugars, salts, and stabilizers, and then processing it. Figure 2.3 depicts the general flow chart for producing milk analogue.

Producing high-quality milk analogs involves several key steps. To prevent particle clumping and separation, it's crucial to maintain small particle sizes (<500 nm) using chemical, enzymatic, or mechanical methods. Stabilizers like plant-based emulsifiers (soy, pea, fava bean, lentil proteins), thickening agents (modified starches, pectin, locust bean gum, gellan gum, starch, methylcellulose, carrageenan, alginate), phospholipids (soy, sunflower lecithin), or surfactants (quillaja, tea saponins) help adjust texture and prevent particle separation, significantly enhancing overall stability. However, plant-based milk often lacks essential nutrients found in real bovine milk. To bridge this nutritional gap, fortification with key micronutrients like vitamin A, vitamin D, vitamin B12, and calcium becomes crucial for supplementing plant-based diets. Advanced encapsulation technologies are employed to fortify these analogs with highly absorbable forms of these nutrients (McClements and Grossmann 2021). By incorporating stabilizers and fortification methods, producers can improve the nutritional profile and stability of milk analogs, attempting to mimick these alternatives more closely with real milk while maintaining consistent quality and functionality.

Plant-based milks couldn't match with the sensory and functional traits of bovine milk, which can be a concern for consumers. Appearance, the first thing consumers notice, matters significantly (Sethi et al. 2016). To create a creamy look in milk analogs, it's essential to control the concentration and size of particles like oil bodies, fat droplets, and tissue fragments they contain. The perceived lightness of these alternatives increases when particle concentration aligns with light wave dimensions (380–780 nm). The natural color of milk analogs depends on the types and amounts of pigments present (Jeske et al. 2019). To achieve an appealing appearance, adjustments to these natural pigments may be necessary. Adding or removing specific components helps fine-tune the visual appeal of plant-based milk alternatives.

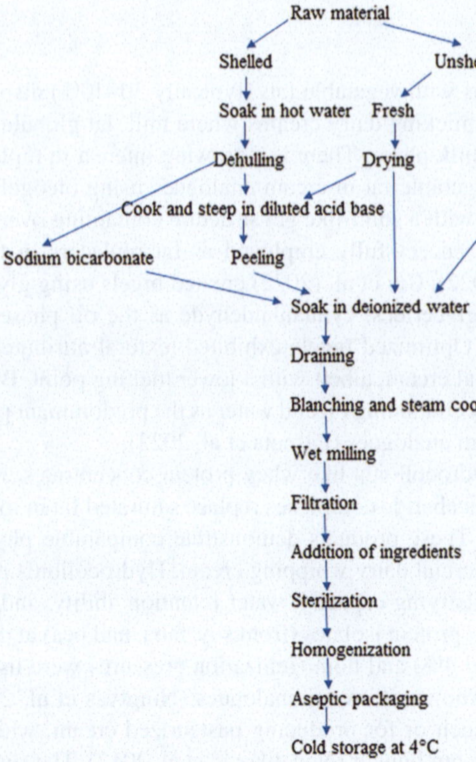


**Fig. 2.3** Flow chart for production of plant-based milk substitute. (Adapted from Aydar et al. 2020; Mäkinen et al. 2016)

Recreating the sensory qualities of cow's milk poses a challenge due to its distinct mild flavor profile, typically comprising over 100 volatile compounds (Drake 2007; Bendall 2001). In contrast, milk analogs derive their flavors from the raw materials of plants and are further influenced by processing and storage. For example, soymilk often carries a beany taste, while hazelnut milk offers a nutty flavor (Sethi et al. 2016). Additionally, phytochemicals like phenols and glucosinolates present in these analogs might introduce undesirable tastes such as bitterness, earthiness, or vegetative notes (Watson 2019). Methods like blanching and fermentation (Tangyu et al. 2019) are being developed to minimize or eliminate these off-flavors, enhancing the overall taste profile of these alternatives. The flavor profile of plant-based milks can also be improved by adding sweeteners. Although the viscosity of plant-based milk often resembles that of dairy milk, the texture can occasionally become gritty or sandy due to insoluble particles, affecting its mouthfeel (Sethi et al. 2016).

### 2.4.2 Cream Analogues

Cream alternatives are emulsions with vegetable fats (typically 30–40%) dispersed in a continuous water phase, mimicking dairy creams where milk fat globules disperse in the aqueous skimmed milk phase. There is a growing interest in replacing milk fat in dairy cream with vegetable fat in cream analogues using oleogels and bigels as alternatives. Oleogels, with a solid-like gel structure containing over 90% liquid vegetable oil, have been successfully employed as fat replacers in filling cream production (Kim et al. 2022). Cui et al. (2022) created bigels using glycerol monolaurate, medium-chain triglycerides, cynnamaldehyde as the oil phase, and chitosan as the aqueous phase. Optimized bigels exhibited textural attributes and appearance similar to commercial cream, albeit with a lower melting point. Bigels, polysaccharide-based edible gels containing oil and water as the predominant phase, hold immense potential for cream analogues (Pascuta et al. 2022).

In dairy-based analogues, hydrocolloids like whey protein concentrates, modified starches, and protein–polysaccharide complexes replace saturated fat in low-fat whipping cream (20–30% fat). These products demonstrate comparable physical and textural properties to commercial dairy whipping cream. Hydrocolloids play a crucial role in maintaining emulsifying capacity, water retention ability, and high viscosity (Jiang et al. 2023. Plant protein isolates (from soy, faba, and pea) at different concentrations (2%, 3%, and 4%) and homogenization pressures were used to create stable cream emulsions known as cream analogues (Ningtyas et al. 2021). This cream can be utilized as such or for producing pasteurized cream, whipped cream, cheese, ice cream, and cream liqueur (Nedeljkovic et al. 2017). The surge in demand for vegan products encourages manufacturers to seek innovative solutions, leading to the development of new products like cream analogues (Plamada et al. 2023).

On the other hand, the market boasts numerous plant-based analogues of fermented cream products, including butter, sour creams, and sour cream-based dips. Cashew, almond, coconut, and oats often serve as base ingredients, along with plant-based oils. Starches, gums, and plant-derived lecithin are common textural aids. Although most products do not disclose the specific cultures used for fermentation, some, like wild brine and Forager Project, indicate the use of lactobacilli and lactic acid bacteria (LAB) commonly found in yogurt fermentation (Pua et al. 2022).

### 2.4.3 Butter Analogues

Butter is a semi-solid fat created through a mechanical process that involves converting the dairy cream phase, an oil-in-water emulsion, into a water-in-oil emulsion (Gutierrez-Luna et al. 2022). According to the Codex Alimentarius (2011), butter is defined as a fatty product exclusively derived from milk and/or milk-derived products, primarily existing as an emulsion of the water-in-oil type. The development of butter-like products has been driven not only by the aim to reduce production costs but also to enhance certain properties of butter, such as its storage time. This innovation is particularly motivated by the susceptibility of essential fatty acids (EFAs) to autoxidation. Traditional dairy butter, whether salted or unsalted, contains 55.2 g/100 g of saturated fat and 222.5 mg/100 g of cholesterol (Md. Asif et al. 2022). However, the technical advantages notwithstanding, the presence of saturated and trans fatty acids in butter has been associated with various health concerns, including cardiovascular disease, hypercholesterolemia, diabetes, and an increased risk of cancer (Lichtenstein 2014; Lu et al. 2016).

Over the years, the consumption of butter analogues, an option for dairy butter, has increased considerably (Gorrepati et al. 2015). Alternative plant-derived components and formulations have been proposed and tested to eliminate animal fat, introduce alternate fatty acid arrangements, and include other beneficial compounds in various dietary matrices (Gutierrez-Luna et al. 2022; Lu et al. 2017). According to U.S. statistics, peanut butter stands out as the most favored vegan butter substitute, with 299.34 million people consuming it in 2020, and an anticipated increase to 307.7 million by 2024. Recent developments have introduced new butter analogues to address peanut allergies, including alternatives made from soy, almond, cashew, pistachio, and sesame (Abd-Elsattar and Abdel-Haleem 2016; Mandalari et al. 2018; Agrahar-Murugkar et al. 2018). Among these, peanut butter has the highest lipid content (58.7 g/100 g) and calorie count (2768 KJ/100 g), while soybean butter leads in moisture (3.7–5.4 g/100 g) and protein (25.8–30 g/100 g). However, maintaining the health and nutritional value of these butter analogues necessitates careful consideration of roasting, grinding, and storage temperatures during the manufacturing process (Gutierrez-Luna et al. 2022). Mimicking the peculiar diacetyl flavor of butter will be the most important challenge in butter analogue production.

To address potential allergens and enhance nutritional value, Sanders et al. (2014) explored consumer acceptability of peanut butter fortified with ground peanut skins, finding that this addition maintained overall acceptability, irrespective of heat treatment. Additionally, investigations into the texture and milling properties of sesame butter under different conditions revealed that oven-roasted samples exhibited more pronounced aroma, characterized by a robust nut-like odor, compared to those heated in a microwave oven. Thus, oven roasting of sesame butter holds the potential to enhance volatile content and flavor (Zhang et al. 2019). Further exploration into plant-based nutritionally enriched butter formulations, incorporating chia seed, sesame seed, watermelon seed, pumpkin seed, and olive oil in various ratios, indicated that the chia + sesame + olive oil formulation exhibited the highest total phenolic content (68.73 ± 0.01 µg GAE/mL), while the chia, watermelon and olive oil formulation showed the highest antioxidant activity (52.30 ± 0.01%) (Ghosh et al. 2021). Consequently, butter analogues serve as viable substitutes for dairy butter, given that nuts and seeds are rich sources of protein, fiber, essential fatty acids, and other essential nutrients (Sanders et al. 2014; Zhang et al. 2019; Gutierrez-Luna et al. 2022; Md. Asif et al. 2022; Plamada et al. 2023).

Rege et al. (2012) reported on the use of kalonji seeds ethanol extract (KEE) and curcuminoids to stabilize essential fatty acids (EFAs) in butter-like products, revealing synergistic antioxidant activity. KEE demonstrated effective antioxidant properties in these products, enhancing physical attributes like spreadability. Furthermore, the incorporation of EFAs elevated the nutritional profile of the end product, with added antioxidants providing medicinal benefits. Mickle et al. (1963) conducted rearrangements of milk fat under controlled times, temperatures, and catalyst concentrations, observing that the hardness of butter-like spreads increased with prolonged time but decreased with higher catalyst concentrations due to the formation of diglycerides.

### 2.4.4 Cheese Analogues

Cheese analogues are products derived from a blend of dairy, non-dairy, or partially dairy ingredients, designed to mirror the appearance, texture, and flavor profile of traditional cow milk-based cheese. Different types of substitutes available are shown in Fig. 2.4. The primary objective is to closely replicate the physicochemical, organoleptic, and functional characteristics of conventionally marketed cheese. The rising popularity of cheese analogues can be attributed to their ease of production, the potential substitution of dairy with vegetable ingredients, and lower production costs compared to ripened and processed cheeses (Cunha et al. 2010; Mohamed et al. 2013). The global cheese analogue market is projected to reach US$ 2232.4 million by 2023, with an anticipated 6.5% Compound Annual Growth Rate (CAGR) leading to a market value of US$ 4052.6 million by 2033, fueled by the increasing demand for non-dairy cheese alternatives (Future Market Insights 2023).

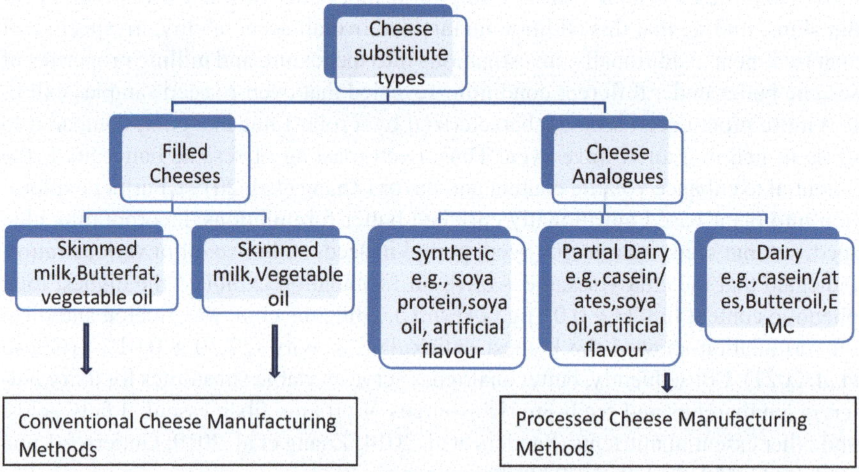


**Fig. 2.4** Different types of substitutes. (Adapted from Shaw 1984; Bachmann 2001)

**Table 2.8** Typical formulations used for cheese analogues

| Ingredients | A | B | C | D | E | F | G | H | I |
|---|---|---|---|---|---|---|---|---|---|
| Acid casein | – | 21.00 | – | 28.41 | – | – | – | – | – |
| Acid curd | – | – | – | – | – | – | 49.77 | – | – |
| Rennet casein | – | – | – | – | – | 21.05 | – | – | – |
| Ca/Na caseinate | 23.00 | – | 14.39 | – | 26.00 | – | – | – | – |
| MCC Powder | – | – | – | – | – | – | 7.19 | 25.63 | – |
| MPC Powder | – | – | – | – | – | – | – | – | 26.8 |
| Peanut protein isolate | – | – | 14.39 | – | – | – | – | – | – |
| Vegetable oil | 25.00 | 12.50 | 20.17 | 16.65 | 25.00 | 19.98 | 19.89 | 20.04 | 20.21 |
| Starch | 2.00 | 5.00 | – | – | – | – | – | – | – |
| Emulsifying salts | 2.00 | 2.50 | 0.75 | 1.49 | – | – | – | – | – |
| Sodium chloride | – | 1.50 | 1.83 | 1.60 | 1.50 | 1.50 | 1.50 | 1.00 | 1.00 |
| Emulsifier | – | 0.15 | – | – | – | 1.4 | – | 3.00 | 3.00 |
| Stabilizer | – | 0.42 | – | – | 1.00 | – | – | – | – |
| Flavor | 2.00 | 0.30 | – | – | 1.50 | – | – | – | – |
| Flavor enhancer | 1.00 | – | – | – | – | – | – | – | – |
| Acidifying agent | 0.40 | 0.27 | 1.38 | 0.58 | 1.00 | – | – | – | – |
| Color | 0.04 | – | – | – | – | – | – | – | – |
| Preservative | 0.10 | – | – | 0.09 | – | – | – | – | – |
| Calcium chloride | – | 0.36 | – | – | – | – | – | – | – |
| Water | 38.50 | 56.40 | 46.81 | 42.99 | 34.00 | 46.44 | 11.56 | 46.55 | 46.43 |
| Condensate | 7.00 | – | – | – | – | – | – | – | – |
| Milk permeate powder | – | – | – | – | – | 8.73 | 10.09 | – | – |
| Cheddar cheese | – | – | – | – | 10.00 | – | – | – | – |
| Deproteinized whey powder | – | – | – | – | – | – | – | 2.48 | 1.26 |
| Citric acid | – | – | – | – | – | 0.90 | – | 0.50 | 0.50 |
| Acidulant | – | – | – | – | – | – | – | 0.80 | 0.80 |

Various formulations from Fox et al. (2000), Jana (1998), Krishnaswamy et al. (1971), Montesinos-Herrero et al. (2006), Bachmann (2001), Ahmed (2019), Salunke et al. (2022a, b)

Traditional cheese, numbering over 2000 varieties, are predominantly derived from a singular raw material—milk. The production process involves the use of chymosin enzyme, starter culture, or acid and heat. While some cheeses are consumed fresh, others, especially rennet cheese varieties, undergo varying periods of ripening at specific time-temperature and relative humidity conditions to achieve the desired flavor profile. Processed cheese analogues are manufactured by blending dairy ingredients (various sources of protein, fat, carbohydrates, etc.) and non-dairy ingredients (salt, water, mold inhibitor, preservatives, emulsifying salts, color, flavor, additives, etc.) followed by continuous mixing, heating and cooling to extend shelf life (Guinee 2007; Kapoor and Metzger 2008; Kammerlehner 2009; Salunke et al. 2022a, b). Typical ingredient blends used in cheese analogue manufacturing are shown in Table 2.8, and this indicates there are a lot of ingredient choices available to the manufacturers. The manufacturing process relies on several key principles including calcium sequestration, pH adjustment and stabilization, para-CN

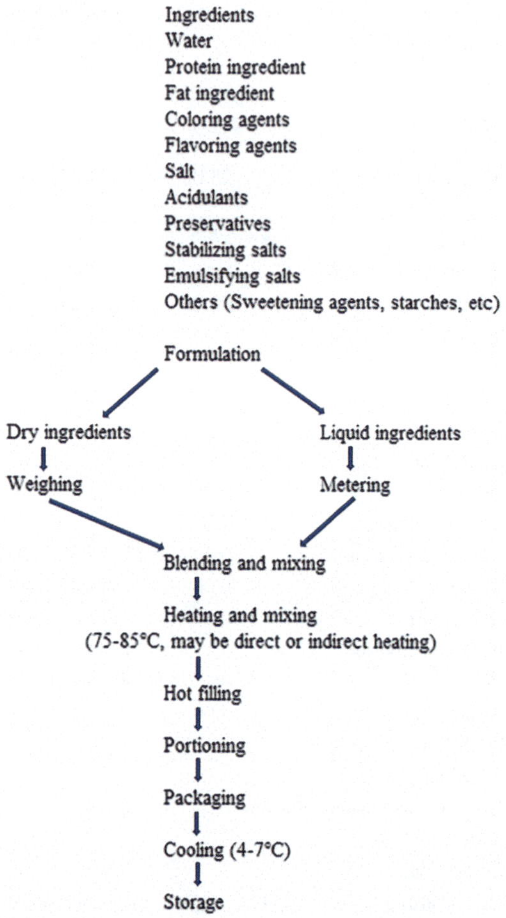


**Fig. 2.5** Typical imitation cheese manufacturing process

dispersion, water binding, and emulsification (Henning et al. 2006, Guinee 2007; Kapoor and Metzger 2008; Kammerlehner 2009; Salunke et al. 2022a, b). The typical manufacturing process of imitation cheeses is shown in Fig. 2.5. Composition and functionality are balanced by selecting correct ingredients, emulsifiers, and stabilizers. Table 2.9 shows the comparison between cheese and cheese analogues. The composition of cheese analogues available in the market in the block, shred, slice, or wedge format is given in Table 2.10.

There's a different kind of cheese product known as pasteurized process cheese products (PCP). These resemble various Processed cheese (PC) categories but might include ingredients not permitted in standard cheeses or fail to meet their compositional standards (Lu et al. 2007; Kapoor and Metzger 2008). PCP falls into two groups: substitute or imitation cheese (Chandan and Kapoor 2011). They use ingredients like milk protein concentrate or micellar casein concentrate, which aren't allowed under CFR regulations, to mimic the texture and functionality of other

**Table 2.9** Typical composition of cheese and cheese analogues

| Constituents | Moisture% | Protein% | Fat % | Fat on dry matter % | Salt % | Salt in moisture % | Ash % | Ca, mg/100 g | pH |
|---|---|---|---|---|---|---|---|---|---|
| Low-moisture mozzarella cheese | 46.4 | 26 | 23.2 | 44.6 | – | 3.1 | 3.9 | 27.5 | 5.5 |
| Analogue pizza cheese | 48.8 | 18.5 | 25 | 49 | – | 3.5 | 4.2 | 34.4 | 6.1 |
| natural mozzarella cheese | 47.53 | 23.06 | 24.6 | 46.81 | 1.29 | – | 3.18 | 0.62 | 5.4 |
| Analogue pizza cheese | 53.35 | 20.13 | 15.2 | 32.66 | 0.9 | – | 3.91 | 0.39 | 5.7 |
| Imitation mozzarella cheese | 47.9 | 19.9 | 21 | 40.4 | 1.2 | – | – | – | 5.8 |

From Fox et al. (2000), Jana (1998), Salunke et al. (2022a, b)

**Table 2.10** Composition of cheese analogues available in the market

| Commercial plant-based cheese sample | Moisture % | TS | Fat | Salt | Protein | Ash | Starch test | pH |
|---|---|---|---|---|---|---|---|---|
| Block cheese | | | | | | | | |
| Mozzarella block | 67.5 | 32.5 | 15.5 | 1.2 | 4.1 | 1.7 | + | 4.4 |
| Almond-style mozzarella block | 54.3 | 45.7 | 13.0 | 1.7 | 18.1 | 3.8 | + | 5.6 |
| Medium cheddar block | 48.8 | 51.2 | 22.0 | 1.8 | 3.1 | 3.2 | + | 4.1 |
| Shred cheese | | | | | | | | |
| Parmesan shred | 35.4 | 64.6 | 26.5 | 3.8 | 0.6 | 2.3 | + | 4.5 |
| Mozzarella shred | 51.4 | 48.6 | 24.5 | 2.0 | 1.2 | 1.8 | + | 3.5 |
| Mozzarella shred | 48.2 | 51.8 | 22.5 | 2.2 | 4.4 | 3.5 | + | 4.1 |
| Almond-style mozzarella shred | 51.1 | 48.9 | 12.5 | 1.6 | 17.0 | 3.5 | + | 5.3 |
| Cheddar shreds | 49.8 | 50.2 | 26.5 | 2.1 | 1.0 | 1.9 | + | 4.0 |
| Slice | | | | | | | | |
| Cheese Slice | 52.4 | 47.6 | 23.5 | 2.4 | 0.3 | 2.4 | + | 3.7 |
| Gouda slices | 52.6 | 47.4 | 23.5 | 2.5 | 0.2 | 2.4 | + | 4.1 |
| Wedge | | | | | | | | |
| Parmesan wedge | 44.9 | 55.1 | 15.3 | 3.4 | 2.2 | 4.0 | + | 4.0 |

process cheese types (Kapoor and Metzger 2008). PCP is more affordable as it incorporates non-cheese dairy ingredients, and it can be tailored to suit specific functional needs (Lu et al. 2007; Kammerlehner 2009).

During the manufacturing of PC, extensive protein-based interactions occur that lead to a strong protein network with restricted flow properties (Garimella Purna et al. 2006). In PCP applications, Gelling ability of casein is exploited (Augustin et al. 2011), and CN provides a desirable un-melted firm texture and a stringy melted texture, whereas whey protein forms a heat-induced irreversible gel and produces a PCP that has restricted melt characteristics (Garimella Purna et al. 2006; Brickley et al. 2007; Kapoor and Metzger 2008; Kammerlehner 2009; Chandan and Kapoor 2011; Salunke et al. 2022a, b; Salunke and Metzger 2022a, b, 2023a, b; Salunke et al. 2023b).

Dairy Analogues are also known as Imitation Cheeses which are produced with partial or complete substitution of natural cheese preparations by proteins and fats derived from milk or other sources, with the addition of emulsifying salts, under heat and agitation (Chavan and Jana 2007). The protein source may be dairy (cheese, high protein powders) or non-dairy such as vegetable proteins, legumes e.g., pea, soy, lupin or, potato, nuts, and seeds, zein, etc. The fat source can be a mix of solid (e.g., cocoa, coconut, palm oil) and liquid (avocado, canola, sesame, rapeseed, soybean, palm, sunflower oil) fats whereas carbohydrate sources such as polysaccharides (especially from starches of different plants) (Mohamed et al. 2013). The use of plant proteins in cheese analogues is associated with their poor functional properties like low elasticity, insufficient hardness, high adhesiveness, poor stretchability, and often unacceptable taste (Chavan and Jana 2007). Replicating the authentic flavor of dairy-based cheeses is another challenge. Recent studies underscore the significant impact of extraction and processing parameters on the texture of cheese analogues, even with the inclusion of fillers and stabilizers (Foegeding 2004; Solowiej et al. 2008: Mefieh et al. 2022). The traditional process of manufacturing soy cheese is shown in Fig. 2.6.

Achieving an authentic dairy-like flavor is a distinct challenge, considering the substantial differences in proteins and fats between plant-based and animal milk ingredients (Jeewanthi and Park 2018; McClements et al. 2019; Ben-Harb et al. 2020). To address this, cultures with robust enzymatic activities, particularly in proteolytic and lipolytic pathways, are being explored to enhance the textural qualities of plant-based cheese analogues. Notably, the fermentation characteristics of cultures may vary in different matrices, emphasizing the need for tailored evaluations (Jeske et al. 2018). While sensory assessments have been conducted to evaluate product likability and rate textural attributes, specific studies on the reduction of off-flavors and the evolution of cheese and dairy-like notes in plant-based cheese analogues remain limited. It is noteworthy that many fermented, plant-based cheese analogues currently available in the market are crafted from cashews and almonds (Pua et al. 2022). The incorporation of plant proteins into cheese analogues has been linked to several challenges, including inadequate elasticity, insufficient hardness, high adhesiveness, poor stretchability, and sometimes unfavorable taste (Chavan and Jana 2007). Consequently, the utilization of isolated milk proteins, such as casein, or a combination of casein and whey proteins, emerges as a more suitable option (Mleko and Foegeding 2001; Solowiej et al. 2008).

Fig. 2.6 The traditional process of "sufu" (soya cheese) (Jeske et al. 2018)

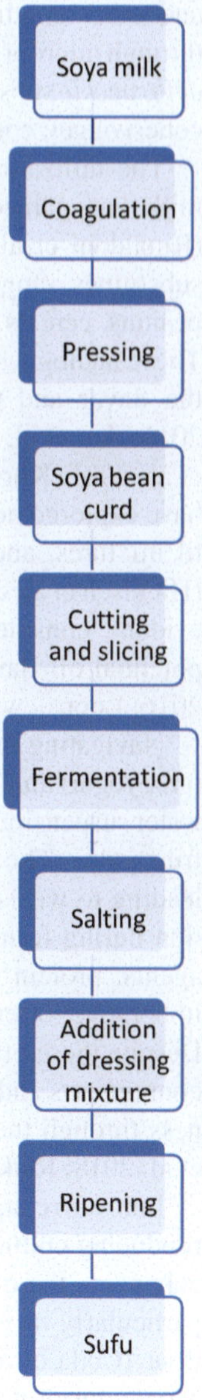


### 2.4.5 *Fermented Dairy Analogues (Yoghurt-like Products)*

Yogurt, encompassing popular variations like the Greek-style rendition, stands as a prevalent choice among fermented milk products, available in a diverse array of set, stirred, flavored, and drinkable types. Beyond the traditional parameters, it often incorporates enrichments such as nuts, fruits, or vegetables, along with sweeteners and other condiments. The classical milk-based yogurt undergoes fermentation through microorganisms, particularly *Streptococcus thermophilus* and *Lactobacillus delbrueckii* subsp. *bulgaricus*, resulting in a distinct texture marked by a firm and cohesive gel, coupled with the characteristic acidic flavor.

The utilization of plant-based alternatives, such as soy, almond, and coconut milk has gained considerable traction in the production of yogurt analogues (Rinaldoni et al. 2012; Rahmautuzzaman et al. 2021). The diverse landscape of substitutes, ranging from soybeans and African yam bean to cowpeas, mung beans, peanuts, cereals, oat, almond, cashew, and coconut, has expanded the possibilities. These analogues often incorporate yogurt cultures or probiotic cultures to emulate the flavor and texture of conventional milk-based yogurt (Pandey and Mishra 2015; Ani et al. 2018; Bruckner-Guhmann et al. 2019; Ogundipe et al. 2021; Yang et al. 2021; Khrundin et al. 2021; Aydar et al. 2021). Some innovative approaches have explored non-yogurt strains, utilizing lupin protein isolate, rice–chickpea–lentil mixtures, and other non-yogurt cultures to achieve yogurt-like characteristics (Hickisch et al. 2016; Pontonio et al. 2020). The need for tailored solutions becomes evident, considering variations in protein structure, coagulation properties, and potential off-flavors inherent in plant matrices (Boeck et al. 2021; Hickisch et al. 2016; Lopusiewicz et al. 2020).

Navigating the challenges associated with achieving a comparable protein structure, yogurt analogues are typically crafted by fermenting aqueous extracts or flour–water suspensions of cereals, pseudocereals, legumes, and nut flours, or homogenized fruit pulps. The absence of casein in plant matrices presents a significant hurdle, leading to weaker gels and heightened susceptibility to syneresis or water separation during fermentation. To address these challenges, the addition of structuring agents, protein extracts, inulin, thickeners, stabilizers, and emulsifiers becomes imperative, often incorporating starch, gums, and/or pectins in the ingredient list. Despite these efforts, yogurt analogues tend to exhibit weaker gels than their dairy counterparts and are highly prone to syneresis, necessitating improvements in firmness through the addition of thickeners and emulsifiers (Wang et al. 2018; Ermis et al. 2018; Raikos et al. 2020; Pachekrepapol et al. 2021).

Innovative approaches, such as the utilization of exopolysaccharide (EPS)-producing cultures and starch gelatinization by heating, have been instrumental in enhancing the quality of yogurt analogues. These techniques have shown efficacy, particularly in yogurt analogues made from soy, lupin protein isolate, and quinoa flour (Coda et al. 2012; Hickisch et al. 2016; Ripari 2019; Pontonio et al. 2020; Montemurro et al. 2021). Critiques of the organoleptic properties of yogurt analogues often revolve around the presence of antinutritional ingredients, coupled with distinct flavors like bitter, beany, astringent, and herbaceous tastes. An unusual odor, perceived by regular consumers of dairy yogurt, poses an additional challenge (Betsche et al. 2005; Aminigo et al. 2009; Montemurro et al. 2021).

Beyond their culinary aspects, the lactic acid fermentation of legume-based fermented milk extends shelf life and enhances the nutritional value of the final product (Beuchat and Nail 1978; Rao et al. 1988). Meanwhile, soy-based fermented foods offer diverse functional properties, including antiatherogenic and anticholesterolemic effects, hypolipidemic benefits, and reduced allergenicity (Mishra and Mishra 2013). The exploration of yogurt analogues unfolds as researchers and manufacturers navigate the intricacies of plant-based formulations to align with the evolving preferences of consumers.

### 2.4.6 Dairy Dessert Analogues (Dairy Dessert-like Products)

Ice cream stands out as an ideal candidate for analog conversion, especially towards partial dairy analogs, mainly due to its high fat and sugar content, and crucially, economic considerations. The formulated nature of ice cream simplifies the substitution of ingredients with more affordable alternatives, such as vegetable oils and sugar replacements. However, due to the specific legal definition of ice cream, most analogs are termed frozen desserts. Dairy dessert analogs have gained popularity among lactose-intolerant individuals, vegetarians, and vegans.

Formulating an ice cream analog is straightforward, but the real challenge lies in maintaining rheological properties, particularly viscosity during mix preparation and structural integrity post-freezing. Various compounds, such as inulin, maltodextrin, starches, gums, polydextrose, psyllium, and pectin, or enzymatic hydrolysis of protopectin in vegetable purées, are employed to enhance rheological properties by binding water and creating a proper structure (Plamada et al. 2023).

From a technological standpoint, achieving a specific consistency and favorable sensory characteristics in ice cream analogs often involves a maximum of 6% added and 8% additional pectin fibers, especially when used with soy protein isolate, almond, and hemp milk, combined with dietary fibers like psyllium (6%) (Plamada et al. 2023). Commercially available ice cream may lack certain beneficial nutrients, prompting the addition of compounds like soy extract, soy kefir, and dehydrated jaboticaba peel (Mendonça et al. 2022).

Almond milk-based ice cream analogs, known for their sweet aroma, have been favored over hemp milk-based counterparts with superior rheological properties (Reyes-Jurado et al. 2021). The incorporation of fermented ingredients has been suggested to enhance overall product quality (Reyes-Jurado et al. 2021).

The introduction of coconut milk, fruits, or fruit juices as substitutes for or additions to other milk further improves product quality. Functional food ingredients, such as red beetroot pomace, have been suggested to enhance antioxidant activity, as demonstrated in the traditional Indian frozen dessert, kulfi. Integrating fruits and vegetables during processing emerges as an effective method for producing functional dairy products with elevated nutritional content (Plamada et al. 2023).

Advancements in technology enable the adjustment of structural properties in matrices through controlled modifications of food ingredients and the substitution of milk with vegetable by-products. In the realm of dessert-like products, those based on soy or soy protein isolate are prevalent. Rheological properties, or texture, can be enhanced using fats and fat substitutes like starches and gums (Spada et al. 2015). Water-soluble soy extract and guava pulp have been reported to enhance properties (Granato et al. 2010). Despite soy extract's astringent flavor and chalkiness, combining it with pink guava overcomes these challenges. Pink guava, rich in soluble fibers, carotenoids, bioactive flavonoids, and ascorbic acid, provides a favorable solution for improving the sensory qualities of soy-based desserts enriched with fruit pulp, exhibiting similar properties to commercial samples (Spada et al. 2015; Mendonça et al. 2022).

### *2.4.7 Infant Formula*

While breast milk remains the optimal nutrient source for infants, certain circumstances, such as health considerations, allergenicity, and lifestyle choices, may lead parents to rely on infant formulas. These formulas often serve as either the sole or partial source of nutrition for infants, especially within the first year of their lives. Therefore, it becomes crucial for these formulas to not only meet the nutritional needs of infants but also contribute to their long-term health benefits.

Over the past few decades, there has been a significant improvement in the nutritional quality of infant formulas. Ongoing research and development have led to the creation of new ingredients that not only fulfill the essential nutritional requirements of infants but also offer additional advantages, such as immunity support and enhanced cognitive development. Examining the fatty acid (FA) composition of mature human milk from various continents reveals that breastfed infants benefit from preformed components like arachidonic acid (ARA), stearidonic acid (SDA, 18:4n-3), eicosapentaenoic acid (EPA), docosapentaenoic acid (DPA), and docosahexaenoic acid (DHA). However, the safety and long-term health effects of incorporating such ingredients into infant formulas require thorough evaluation before implementation.

In the formulation of infant formulas, it is not only the quantity of nutrients that matters but also their physical and chemical forms. Fats play a crucial role in the growth and development of infants. Achieving a physical size and structure in infant formula fat analogs similar to that of milk fat globules becomes essential for optimal absorption and metabolism. Additionally, the introduction of long-chain polyunsaturated fatty acids (LCPUFAs) may pose challenges related to oxidative deterioration. Thus, exploring appropriate antioxidants on a case-by-case basis is imperative to stabilize each product.

With the increasing emphasis on clean labels and natural ingredients, there is a growing need to determine the efficacy and safety of natural antioxidants. As the field of infant formula evolves, a holistic approach considering nutritional content, physical attributes, and safety measures will contribute to the continued enhancement of infant formulas (Zou et al. 2016).

## 2.5 Targeted and Precision Fermentation

Targeted and precision fermentation is also known as fermentation-derived dairy (Mendly-Zambo et al. 2021), animal-free dairy (Broad et al. 2022), and sometimes cellular agriculture (Glaros et al. 2023).

Targeted and precision fermentation holds promise for the production of ingredients with specific qualities and functionalities, particularly within the realm of dairy analogs. Although the technology is in its early stages, some startups are at the forefront of its exploration, successfully creating whey proteins with equivalent constituents, notably β-lactoglobulin. However, the emulation of casein micelle remains a significant challenge, highlighting the technology's current limitations. The broader acceptance of precision fermented ingredients faces hurdles related to labeling, legal considerations, and consumer acceptance, which is expected to evolve over time. A more significant issue is the legal standing in many countries which can be challenging.

Furthermore, precision fermentation processes incorporate genetically modified microorganisms, such as bacteria, yeast, or fungi, to synthesize dairy proteins and fats. These synthesized components serve as a foundation for producing conventional dairy products, adding a novel dimension to dairy analog creation (Pua et al. 2022). Targeted fermentation, a subset of precision fermentation, is utilized to either eliminate or generate specific flavoring compounds in raw materials. These compounds can then be employed in analogs or extracted to obtain desired flavor profiles. For instance, a study demonstrated that fermenting pea protein isolates using lactic acid bacteria (LAB) and yeast reduced off-flavors, such as bitter taste (Youssef et al. 2020; Shi et al. 2021; Arteaga et al. 2021). Diacetyl and acetaldehyde, crucial compounds contributing to a dairy-like aroma, are commonly reported in LAB fermentation of cereal- and soy-based matrices (Houackova et al. 2015; Peyer et al. 2016; Tangyu et al. 2019).

Despite its potential, the application of genetic modification (GM) technology in precision fermentation faces significant resistance from consumers. The associated high cost of microbial engineering poses an additional challenge. Substantial investments and extensive research are anticipated to bridge the gap, making GM-produced dairy commercially competitive against conventional dairy products or plant-based dairy analogs (Pua et al. 2022).

Fermentation-derived dairy ingredients will be available and compete with both Animal-derived and plant-derived dairy ingredients because of the relative ease with which it is made (Mendly-Zambo et al. 2021) and it provides economic opportunities (Glaros et al. 2023). Challenges facing fermentation-derived dairy include environmental and sustainability questions (land and water use), acceptance of technology, policy, and regulation, and consumer acceptance as an alternative to culturally accepted dairy products (Mendly-Zambo et al. 2021), safety, its potential contributions to individual health and climate change mitigation, its organoleptic qualities, nutritional components, animal welfare implications, price to consumers (Broad et al. 2022) and feed the world (Glaros et al. 2023).

## 2.6 Conclusion

In conclusion, the historical significance of milk and its derivatives in human diets is undeniable, serving as a fundamental source of essential nutrients for centuries. However, the evolution of human lifestyles has given rise to health concerns and environmental challenges associated with traditional milk consumption. As a response to these concerns and changing consumer preferences, non-dairy alternatives, including dairy analogues, have gained prominence. The research interest in dairy analogues and restructured dairy products is burgeoning, as evidenced by numerous publications. This dynamic landscape aligns with the increasing demands of consumers for products that promote health and well-being. Restructured dairy products, facilitated by innovative technologies, have the potential to retain health-promoting compounds such as whey proteins, active peptides, and exogenous amino acids. Meanwhile, the production of dairy analogues can be streamlined and accelerated through certain technologies, offering cost-effectiveness and new product development opportunities on a global scale.

While the rapid growth of the dairy analogue market is driven by perceived health benefits and the expanding vegan market, it is crucial to recognize that these alternatives often fall short in terms of quality, flavor, and nutritional balance. Addressing these challenges requires a critical examination and improvement of the formulation processes. The increase in research not only opens avenues for addressing the needs of consumers with conditions like lactose intolerance and food allergies but also presents opportunities for continued innovation in the food industry.

In essence, the future perspectives for the field of dairy analogues and restructured dairy products are promising, with ongoing research paving the way for customizable products and diverse end-use applications. As the industry strives to meet the evolving demands of consumers, a concerted effort towards enhancing product quality, nutritional balance, and transparency in labeling is imperative for the sustained success of dairy analogues in the ever-changing landscape of food consumption.

The biggest problem with the milk and dairy analogues is that most of the products are formulated, and slight modifications or any changes may lead to destabilization and bad product quality. Also, the major problem with analogue milk is that it cannot be processed further to make new products as with milk. Other things like the perceived health and safety of plant products and environmental and sustainability of dairy products concerns may be exaggerated and need more research, especially improving protein quality and getting rid of anti-nutritional factors in plant products. Many of the products are trying to mimic dairy products, leading to wrong positioning in the market; the example of mayonnaise, peanut butter, or soy products are successful and should be investigated as they have created their own space and may not be competing with dairy products. Many unknown benefits of plant products need to be explored, and more effort is needed to improve formulation, texture, stability, and sensory issues, leading to new usage and market. The milk and dairy analogues being formulated product is more suited for 3-D food printing-type applications. On the other hand, many lactose-free products are available in dairy products, and consumers need to be educated on these issues so that they don't reduce their intake of dairy which is needed for their overall health.

## References

Abd-Elsattar HH, Abdel-Haleem AMH (2016) Production of soybean butter using different technological treatments. Lwt-Food Sci Technol 69:40–46

Agarwal S, Beausire RL, Patel S, Patel H (2015) Innovative uses of milk protein concentrates in product development. J Food Sci 80(S1):A23–A29

Agrahar-Murugkar D, Dwivedi S, Dixit-Bajpai P, Kumar M (2018) Effect of natural fortification with calcium and protein rich ingredients on texture, nutritional quality and sensory acceptance of cookies. Nutr Food Sci 48:807–818

Ahmed H (2019) Production and storage stability of high concentrated micellar casein and its effect on the functional properties of process cheese products. South Dakota State University

Aimutis WR (1995) Dairy protein usage in processed and imitation cheeses. Food Technol Europe 2:30–34

Aminigo ER, Metzger L, Lehtola PS (2009) Biochemical composition and storage stability of a yogurt-like product from African yam bean (Sphenostylis stenocarpa). Int J Food Sci Technol 44:560–566

Ani E, Amove J, Igbabul B (2018) Physicochemical, microbiological, sensory properties and storage stability of plant-based yoghurt produced from bambaranut, soybean and *Moringa oleifera* seed milks. Am J Food Nutr 6:115–125. https://doi.org/10.12691/ajfn-6-4-4

Arteaga VG, Leffler S, Muranyi I, Eisner P, Schweiggert-Weisz U (2021) Sensory profile, functional properties and molecular weight distribution of fermented pea protein isolate. Curr Res Food Sci 4:1–10

Astolfi ML, Marconi E, Protano C, Canepari S (2020) Comparative elemental analysis of dairy milk and plant-based milk alternatives. Food Control 107327. https://doi.org/10.1016/j.foodcont.2020.107327

Augustin MA, Oliver CM, Hemar Y (2011) Casein, caseinates and milk protein concentrates Chapter 7. In: Chandan RC, Kilara A (eds) Dairy ingredients for food processing. Wiley-Blackwell, A John Wiley & Sons, Inc., Publication, Iowa, USA, pp 161–178

Aydar EF, Tutuncu S, Ozcelik B (2020) Plant-based milk substitutes: bioactive compounds, conventional and novel processes, bioavailability studies, and health effects. J Funct Foods 70:103975

Aydar AY, Mataracı CE, Sağlam TB (2021) Development and modeling of a novel plant-based yoghurt produced by Jerusalem artichoke and almond milk using l-optimal mixture design. J Food Meas Charact 15:3079–3087. https://doi.org/10.1007/s11694-021-00884-z

Bachmann HP (2001) Cheese analogues: a review. Int Dairy J 11:505–515

Baldwin A, Pearce D (2005) Milk powder. In: Onwulata C (ed) Encapsulated and powdered foods. CRC Press, Taylor & Francis Group, Boca Raton, pp 387–434

Bendall JG (2001) Aroma compounds of fresh milk from New Zealand cows fed different diets. J Agric Food Chem 49:4825–4832

Ben-Harb S, Irlinger F, Saint-Eve A, Panouillé M, Souchon I, Bonnarme P (2020) Versatility of microbial consortia and sensory properties induced by the composition of different milk and pea protein-based gels. LWT 118:108720

Betsche T, Azeke M, Buening-Pfaue H, Fretzdorff B (2005) Food safety and security: fermentation as a tool to improve the nutritional value of African yam bean. In: Proceedings of conference on international agricultural research development, Stuttgart-Hohenheim, pp 1–5

Beuchat LR, Nail BJ (1978) Fermentation of peanut milk with Lactobacillus bulgaricus and L. acidophilus. J Food Sci 43:1109–1112

Bleich SN, Vercammen KA (2018) The negative impact of sugar-sweetened beverages on children's health: an update of the literature. BMC Obes 5:6. https://doi.org/10.1186/s40608-017-0178-9

Boeck T, Sahin AW, Zannini E, Arendt EK (2021) Nutritional properties and health aspects of pulses and their use in plant-based yogurt alternatives. Compr Rev Food Sci Food Saf 20:3858–3880. https://doi.org/10.1111/1541-4337.12778

Brickley CA, Auty MAE, Piraino P, McSweeney PLH (2007) The effect of natural cheddar cheese ripening on the functional and textural properties of the processed cheese manufactured therefrom. J Food Sci 72:483–490

Broad GM, Thomas OZ, Dillard C, Bowman D, Le Roy B (2022) Framing the futures of animal-free dairy: using focus groups to explore early-adopter perceptions of the precision fermentation process. Front Nutr 9:997632. https://doi.org/10.3389/fnut.2022.997632

Brückner-Gühmann M, Banovic M, Drusch S (2019) Towards an increased plant protein intake: rheological properties, sensory perception and consumer acceptability of lactic acid fermented, oat-based gels. Food Hydrocoll 96:201–208. https://doi.org/10.1016/j.foodhyd.2019.05.016

Burger TG, Zhang Y (2019) Recent progress in the utilization of pea protein as an emulsifier for food applications. Trends Food Sci Technol 86:25–33

Caric M (1994) Concentrated and dried dairy products. VCH Publisher Inc, New York

Caric M, Kalab M (1993) Processed cheese products. In: Fox PF (ed) Cheese: chemistry, physics and microbiology II, 2nd edn. Elsevier Science Publishers, London, pp 467–505

Carvalho NF, Kenney RD, Carrington PH, Hall DE (2001) Severe nutritional deficiencies in toddlers resulting from health food milk alternatives. Pediatrics 107(4):e46–e46. https://doi.org/10.1542/peds.107.4.e46

Chalupa-Krebzdak S, Long CJ, Bohrer BM (2018) Nutrient density and nutritional value of milk and plant-based milk alternatives. Int Dairy J 87:84–92. https://doi.org/10.1016/j.idairyj.2018.07.018

Chandan RC, Kapoor R (2011) Manufacturing outlines and applications of selected cheese varieties. Chapter 11. In: Chandan RC, Kilara A (eds) Dairy ingredients for food processing. Wiley-Blackwell, A John Wiley & Sons, Inc., Publication, Iowa, pp 267–316

Chavan RS, Jana A (2007) Cheese substitutes: an alternative to natural cheese—a review. Int J Food Sci Technol Nutr 2:25–39

Chen S (1989) Preparation of fluid soymilk. In: Proceedings of the World Congress on Vegetable Protein Utilization in Human Foods and Animal Feedstuffs, pp 341–351

Chowdhury B, Sharma A, Akshit FNU, Mohan MS, Salunke P, Anand S (2023) A review of oleogels applications in dairy foods. Crit Rev Food Sci Nutr. https://doi.org/10.1080/10408398.2023.2215871

Coda R, Lanera A, Trani A, Gobbetti M, Di Cagno R (2012) Yogurt-like beverages made of a mixture of cereals, soy and grape must: microbiology, texture, nutritional and sensory properties. Int J Food Microbiol 155:120–127

Codex Alimentarius (2011) Codex Alimentarius Commission. International Food Standards. Standard for Butter. FAO and WHO, Rome, pp 4

Council of the European Union (2007) Regulation, C. No 1234/2007 of 22 October 2007 establishing a common organization of agricultural markets and specific provisions for certain agricultural products (Single CMO Regulation). Official Journal of the European Union, 16

Cui H, Tang C, Wu S, Julian McClements D, Liu S, Li B, Li Y (2022) Fabrication of chitosan-cinnamaldehyde-glycerol monolaurate bigels with dual gelling effects and application as cream analogs. Food Chem 384:132589

Cunha CR, Dias AI, Viotto WH (2010) Microstructure, texture, colour and sensory evaluation of a spreadable processed cheese analogue made with vegetable fat. Food Res Int 43(3):723–729

de Kruif CG, Huppertz T, Urban VS, Petukhov AV (2012) Casein micelles and their internal structure. Adv Colloid Interface Sci 171–172:36–52

Dalgleish DG, Corredig M (2012) The structure of the casein micelle of milk and its changes during processing. Annu Rev Food Sci Technol 3:449–467

Dhankhar J (2023) A perspective on the pros and cons, manufacturing aspects, and recent advances in non-dairy milk alternatives. J Microbiol Biotechnol Food Sci 12(5):e9543

Doublier JL, Garnier C, Renard D, Sanchez C (2000) Protein–polysaccharide interactions. Curr Opin Colloid Interface Sci 5(3–4):202–214

Drake MA (2007) Invited review: sensory analysis of dairy foods. J Dairy Sci 90:4925–4937

Dubey MR, Patel VP (2018) Probiotics: a promising tool for calcium absorption. Open Nutr J 12(1). https://doi.org/10.2174/1874288201812010059

Ennis MP, Mulvihill DM (1997) Cheese analogues. In: Cogan TM, Fox PF, Ross RP (eds) Proceedings of the fifth cheese symposium. Teagasc, Dublin, pp 1–14

Ermiş E, Güneş R, Zent I, Çağlar MY, Yılmaz MT (2018) Characterization of hazelnut milk fermented by *Lactobacillus delbrueckii* subsp. *bulgaricus* and *Streptococcus thermophilus*. Gida/J Food 43:677–686. https://doi.org/10.15237/gida.GD18022

EU (2013) REGULATION (EU) No 1308/2013 OF THE EUROPEAN PARLIAMENT AND OF THE COUNCIL of 17 December 2013, Establishing a Common Organization of the Markets in Agricultural Products and Repealing Council Regulations (EEC) No 922/72, (EEC) No 234/79, (EC) No 1037/2001 and (EC) No 1234/2007; EU, Maastricht

European Commission (2010) Commission decision of 20 December 2010 listing the products referred to in the second subparagraph of point III (1) of Annex XII to Council regulation (EC) No 1234/2007

Euston SR (2004) Computer simulation of proteins: adsorption, gelation and self-association. Curr Opin Colloid Interface Sci 9(5):321–327

FAO (2013) Dietary protein quality evaluation in human nutrition. Food and Agriculture Organization of the United Nations, Rome

Foegeding EA (2004) The viscosity, texture and other rheological properties of dairy products. In: Proceedings of 4th international symposium on recombined milk and milk products, Cancun, Mexico, p 28

Fox PF, Guinee TP, Cogan TM, McSweeney PLH (2000) Processed cheese and substitute or imitation cheese products. In: Fundamentals of cheese science. Aspen, Gaithersburg, pp 429–451

Fox PF, Uniacke-Lowe T, McSweeney PLH, O'Mahony JA (2015) Milk proteins. In: Dairy chemistry and biochemistry, 2nd edn, pp 145–183

Future Market Insights (2023) https://www.futuremarketinsights.com/reports/cheese-analogue-market. Downloaded in August 2023

Garimella Purna SK, Pollard A, Metzger LE (2006) Effect of formulation and manufacturing parameters on process cheese food functionality—I. Trisodium Citrate. J Dairy Sci 89:2386–2396

Ghosh N, Singha S, Ghosh M (2021) Formulation and characterization of chia (Salvia hispanica) seed spread with incorporation of sesame (Sesamum indicum) seed, watermelon (Citrullus lanatus) seed, and pumpkin (Cucurbita pepo) seed. Appl Biochem Biotechnol 193:1898–1908

Glaros A, Newell R, Fraser E, Newman LL (2023) Socio-economic futures for cellular agriculture: the development of a novel framework. Front Sustain Food Syst 7:970369. https://doi.org/10.3389/fsufs.2023.970369

Gorissen SHM, Crombag JJR, Senden JMG, Huub Waterval WA, Bierau J, Verdijk LB, van Loon LJC (2018) Protein content and amino acid composition of commercially available plant-based protein isolates. Amino Acids 50(12):1685–1695. https://doi.org/10.1007/s00726-018-2640-5

Gorrepati K, Balasubramanian S, Chandra P (2015) Plant based butters. J Food Sci Technol 52:3965–3976

Granato D, Ribeiro JCB, Castro IA, Masson ML (2010) Sensory evaluation and physicochemical optimisation of soy-based desserts using response surface methodology. Food Chem 121:899–906

Guinee TP (2007) In: McSweeney PLH (ed) Processed cheese. Cheese problems solved. CRC Press, Woodhead Publishing Ltd., Cambridge, pp 365–383

Gumus CE, Decker EA, McClements DJ (2017) Formation and stability of ω-3 oil emulsion-based delivery systems using plant proteins as emulsifiers: lentil, pea, and faba bean proteins. Food Biophys 12(2):186–197

Gutiérrez-luna K, Ansorena D, Astiasarán I (2022) Use of hydrocolloids and vegetable oils for the formulation of a butter replacer: optimization and oxidative stability. Lwt 153:112538

Haas R, Schnepps A, Pichler A, Meixner O (2019) Cow milk versus plant-based milk substitutes: a comparison of product image and motivational structure of consumption. Sustain For 11:5046

Henning DR, Baer RJ, Hassan AN, Dave R (2006) Major advances in concentrated and dry milk products, cheese, and milk fat-based spreads. J Dairy Sci 89:1179–1188

Hickisch A, Beer R, Vogel RF, Toelstede S (2016) Influence of lupin-based milk alternative heat treatment and exopolysaccharide-producing lactic acid bacteria on the physical characteristics of lupin-based yogurt alternatives. Food Res Int 84:180–188. https://doi.org/10.1016/j.foodres.2016.03.037

Horáˇcková Š, Mühlhansová A, Sluková M, Schulzová V, Plocková M (2015) Fermentation of soymilk by yoghurt and bifidobacteria strains. Czech J Food Sci 33:313–319

Huffman LM, Harper WJ (1999) Maximizing the value of milk through separation technologies. Symposium: marketing dairy value through technolgy. J Dairy Sci 82:2238–2244

Jacobsen C (2015) Some strategies for the stabilization of long chain n-3 PUFA-enriched foods: a review. Eur J Lipid Sci Technol 117:1853–1866

Jacobsen C, Sorensen ADM, Nielsen NS (2013) Food enrichment with Omega-3 fatty acids. In: Woodhead Publishing Series in Food Science Technology and Nutrition, vol 252, pp 130–149

Jana AH (1998) Process standardization for manufacture of Mozzarella cheese analogue Ph.D thesis, submitted to Gujarat Agricultural University, Anand, Gujarat State

Jeewanthi RKC, Paik HD (2018) Modifications of nutritional, structural, and sensory characteristics of non-dairy soy cheese analogs to improve their quality attributes. J Food Sci Technol 55:4384–4394

Jeske S, Zannini E, Arendt EK (2018) Past, present and future: the strength of plant-based dairy substitutes based on gluten-free raw materials. Food Res Int 110:42–51

Jeske S, Bez J, Arendt EK, Zannini E (2019) Formation, stability, and sensory characteristics of a lentil-based milk substitute as affected by homogenisation and pasteurisation. Eur Food Res Technol 245:1519–1531

Jiang X, Shekarforoush E, Muhammed MK, Whitehead KA, Arneborg N, Risbo J (2023) Lactic acid bacteria as structural building blocks in non-fat whipping cream analogues. Food Hydrocoll 135:108137

Kammerlehner J (2009) Processed cheese-processed cheese food. In: Cheese technology. Published by Kammerlehner, J., Freising, Germany. pp 742–798

Kapoor R, Metzger LE (2008) Process cheese: scientific and technological aspects—a review. Compr Rev Food Sci Food Saf 7:194–214

Karaca AC, Low N, Nickerson M (2011) Emulsifying properties of chickpea, faba bean, lentil and pea proteins produced by isoelectric precipitation and salt extraction. Food Res Int 44(9):2742–2750

Khrundin D, Ponomarev V, Yunusov E, Ezhkova G (2021) The use of plant proteins in the technology of fermented dairy-free products. IOP Conf Ser Earth Environ Sci 715:012066. https://doi.org/10.1088/1755-1315/715/1/012066

Kim M, Hwang HS, Jeong S, Lee S (2022) Utilization of oleogels with binary oleogelator blends for filling creams low in saturated fat. LWT 155:112972

Kinsella JE (1984) Milk proteins: physicochemical and functional properties. CRC Crit Rev Food Sci Nutr 21:197–262

Krishnaswamy MA, Patel JD, Dhanraj S, Govindraj VS (1971) Curd from miltone. J Food Sci Technol 8:41–45

Kyriakopoulou K, Dekkers B, van der Goot AJ (2021a) Plant-based meat analogues. In: Williams PA, Phillips GO (eds) Handbook of hydrocolloids, 3rd edn. (Woodhead Publishing, 2021) (Academic Press, 2019)

Kyriakopoulou K, Keppler JK, van der Goot AJ (2021b) Functionality of ingredients and additives in plant-based meat analogues. Food Secur 10(3):600

Lichtenstein AH (2014) Dietary trans fatty acids and cardiovascular disease risk: past and present. Curr Atheroscler Rep 16:433

Lu Y, Shirashoji N, Lucey JA (2007) Rheological, textural and melting properties of commercial samples of some of the different types of pasteurized processed cheese. Int J Dairy Technol 60:74–80

Lu W, Chen H, Niu Y, Wu H, Xia D, Wu Y (2016) Dairy products intake and cancer mortality risk: a meta-analysis of 11 population-based cohort studies. Nutr J 15:91

Lu F, Kuhnle GK, Cheng Q (2017) Vegetable oil as fat replacer inhibits formation of heterocyclic amines and polycyclic aromatic hydrocarbons in reduced fat pork patties. Food Control 81:113–125

Łopusiewicz Ł, Drozłowska E, Siedlecka P, Mężyńska M, Bartkowiak A (2020) Preparation and characterization of novel flaxseed oil cake yogurt-like plant milk fortified with inulin. J Food Nutr Res 59:61–70

Mäkinen OE, Wanhalinna V, Zannini E, Arendt EK (2016) Foods for special dietary needs: non-dairy plant- based milk substitutes and fermented dairy-type products. Food Sci Nutr 56(3):339–349

Malik VS, Popkin BM, Bray GA, Després J-P, Hu FB (2010) Sugar-sweetened beverages, obesity, type 2 diabetes mellitus, and cardiovascular disease risk. Circulation 121:1356–1364

Mandalari G, Parker ML, Grundy MM, Grassby T, Smeriglio A, Bisignano C, Raciti R, Trombetta D, Baer DJ, Wilde PJ (2018) Understanding the effect of particle size and processing on almond lipid bioaccessibility through microstructural analysis: from mastication to faecal collection. Nutrients 10:213

Mathai JK, Liu Y, Stein HH (2017) Values for digestible indispensable amino acid scores (DIAAS) for some dairy and plant proteins may better describe protein quality than values calculated using the concept for protein digestibility-corrected amino acid scores (PDCAAS). Br J Nutr 117(4):490–499

McClements DJ (2020) Development of next-generation nutritionally fortified plant-based milk substitutes: structural design principles. Food Secur 9(4). https://doi.org/10.3390/foods9040421

McClements DJ, Decker E (2018) Interfacial antioxidants: a review of natural and synthetic emulsifiers and coemulsifiers that can inhibit lipid oxidation. J Agric Food Chem 66:20–35

McClements DJ, Grossmann L (2021) The science of plant-based foods: constructing next-generation meat, fish, milk, and egg analogs. Compr Rev Food Sci Food Saf 20:4049–4100

McClements DJ, Newman E, McClements IF (2019) Plant-based milks: a review of the science underpinning their design, fabrication, and performance. Compr Rev Food Sci Food Saf 18:2047–2067

Md Asif AH, Sarker MAH, Deb GK, Habib MR, Arefin S, Bari MS, Islam MZ, Harun-Ur-Rashid M, Siddiki MSR, Shahjadee UF et al (2022) Fatty acid and amino acid profiles of cheese, butter, and ghee made from buffalo milk. J Adv Vet Anim Res 9:144–154

Mefleh M, Pasqualone A, Caponio F, Faccia M (2022) Legumes as basic ingredients in the production of dairy-free cheese alternatives: a review. J Sci Food Agric 102:8–18

Mendly-Zambo Z, Powell LJ, Newman LL (2021) Dairy 3.0: cellular agriculture and the future of milk. Food, Cult Soc 24(5):675–693. https://doi.org/10.1080/15528014.2021.1888411

Mendonça GMN, Oliveira EMD, Rios AO, Pagno CH, Cavallini DCU (2022) Vegan ice cream made from soy extract, soy kefir and jaboticaba peel: antioxidant capacity and sensory profile. Food Secur 11:3148

Mickle JB, Von Gunten RL, Morrison RD (1963) Rearrangement of milk fat as a means for adjusting hardness of butterlike products. J Dairy Sci 46:1357–1361

Mishra S, Mishra HN (2013) Effect of synbiotic interaction of fructooligosaccharide and probiotics on the acidification profile, textural and rheological characteristics of fermented soy milk. Food Bioprocess Technol 6:3166–3176

Mleko S, Foegeding EA (2001) Incorporation of polymerized whey proteins into processed cheese. Milchwissenschaft 56:612–615

Mohamed AG, Abo-El-Khair BE, Shalaby SM (2013) Quality of novel healthy processed cheese analogue enhanced with marine microalgae Chlorella vulgaris biomass. Word Appl Sci J 23(7):914–925

Montemurro M, Pontonio E, Coda R, Rizzello CG (2021) Plant-based alternatives to yogurt: state-of-the-art and perspectives of new biotechnological challenges. Food Secur 10(2):316. https://doi.org/10.3390/foods10020316

Montesinos-Herrero C, Cottell C, O'Riordan ED, O'Sullivan M (2006) Partial replacement of fat by functional fibre in imitation cheese: effects on rheology and microstructure. Int Dairy J 16:910–919

Nedeljkovic A, Tomasevic I, Miocinovic J, Pudja P (2017) Feasibility of discrimination of dairy creams and cream-like analogues using Raman spectroscopy and chemometric analysis. Food Chem 232:487–492

Ningtyas DW, Bhandari B, Prakash S (2021) Modulating the fat globules of plant-based cream emulsion: influence of the source of plant proteins. Innov Food Sci Emerg Technol 74:102852

O'Brien RD (1998) Fats and oils. Formulating and processing for applications. Technomic Publishing Co. Inc, Lancaster, pp 1–46

Ogundipe OO, Fasogbon BM, Ogundipe FO, Oredope O, Amaezenanbu RU (2021) Nutritional composition of non-dairy yogurt from sprouted tigernut tubers. J Food Process Preserv 45:e15884. https://doi.org/10.1111/jfpp.15884

Pachekrepapol U, Kokhuenkhan Y, Ongsawat J (2021) Formulation of yogurt-like product from coconut milk and evaluation of physicochemical, rheological, and sensory properties. Int J Gastron Food Sci 2021(25):100393. https://doi.org/10.1016/j.ijgfs.2021.100393

Pandey SM, Mishra HN (2015) Optimization of the prebiotic & probiotic concentration and incubation temperature for the preparation of synbiotic soy yoghurt using response surface methodology. LWT 62:458–467. https://doi.org/10.1016/j.lwt.2014.12.003

Pascuta MS, Varvara RA, Teleky BE, Szabo K, Plamada D, Nemeş SA, Mitrea L, Martău LGA, Ciont C, Călinoiu LF et al (2022) Polysaccharide-based edible gels as functional ingredients: characterization, applicability, and human health benefits. Gels 8:524

Paul AA, Kumar S, Kumar V, Sharma R (2020) Milk analog: plant based alternatives to conventional milk, production, potential and health concerns. Crit Rev Food Sci Nutr 60(18):3005–3023

Pereira PC (2014) Milk nutritional composition and its role in human health. Nutrition 30(6):619–627. https://doi.org/10.1016/j.nut.2013.10.011

Peyer LC, Zannini E, Arendt EK (2016) Lactic acid bacteria as sensory biomodulators for fermented cereal-based beverages. Trends Food Sci Technol 54:17–25

Pingali P, Boiteau J, Choudhry A, Hall A (2023) Making meat and milk from plants: a review of plant-based food for human and planetary health. World Dev 170:106316

Plamada D, Teleky BE, Nemes SA, Mitrea L, Szabo K, Calinoiu LF, Pascuta MS, Varvara RA, Ciont C, Martau GA, Simon E, Barta G, Dulf FV, Vodnar DC, Nitescu M (2023) Plant-based dairy alternatives—a future direction to the milky way. Food Secur 12:1883. https://doi.org/10.3390/foods12091883

Pontonio E, Raho S, Dingeo C, Centrone D, Carofiglio VE, Rizzello CG (2020) Nutritional, functional, and technological characterization of a novel gluten- and lactose-free yogurt-style snack produced with selected lactic acid bacteria and *Leguminosae* flours. Front Microbiol 11:1664. https://doi.org/10.3389/fmicb.2020.01664

Pua A, Tang VCY, Goh RMV, Sun J, Lassabliere B, Liu SQ (2022) Ingredients, processing, and fermentation: addressing the organoleptic boundaries of plant-based dairy analogues. Food Secur 11:875

Rahmatuzzaman Rana M, Babor M, Sabuz AA (2021) Traceability of sweeteners in soy yogurt using linear discriminant analysis of physicochemical and sensory parameters. J Agric Food Res 5:100155

Raikos V, Juskaite L, Vas F, Hayes HE (2020) Physicochemical properties, texture, and probiotic survivability of oat-based yogurt using aquafaba as a gelling agent. Food Sci Nutr 8:6426–6432. https://doi.org/10.1002/fsn3.1932

Rao DR, Pulusani SR, Chawan CB (1988) Preparation of a yoghurt-like product from cowpeas and mung beans. Int J Food Sci Technol 23:195–198

Rasika DMD, Vidanarachchi JK, Rocha RS, Balthazar CF, Cruz AG, Sant'Ana AS, Ranadheera CS (2021) Plant-based milk substitutes as emerging probiotic carriers. Curr Opin Food Sci 38:8–20

Rege SA, Momin SA, Bhowmick DN, Pratap AA (2012) Stabilization of emulsion and butter like products containing essential fatty acids using kalonji seeds extract and curcuminoids. J Oleo Sci 61(1):11–16

Rekha CR, Vijayalakshmi G (2010) Bioconversion of isoflavone glycosides to aglycones, mineral bioavailability and vitamin B complex in fermented soymilk by probiotic bacteria and yeast. J Appl Microbiol I 109:1198–1208

Reyes-Jurado F, Soto-Reyes N, Dávila-Rodríguez M, Lorenzo- Leal AC, Jiménez-Munguía MT, Mani-López E, López-Malo A (2021) Plant-based milk alternatives: types, processes, benefits, and characteristics. Food Rev Intl:1–32

Rinaldoni AN, Campderrós ME, Pérez Padilla A (2012) Physico-chemical and sensory properties of yogurt from ultrafiltreted soy milk concentrate added with inulin. LWT—Food Sci Technol 45:142–147

Ripari V (2019) Techno-functional role of exopolysaccharides in cereal-based, yogurt-like beverages. Beverages 5:16

Röös E, Patel M, Spångberg J (2016) Producing oat drink or cow's milk on a Swedish farm — Environmental impacts considering the service of grazing, the opportunity cost of land and the demand for beef and protein. Agric Syst 142:23–32

Rosado JL, Díaz M, González K, Griffin I, Abrams SA, Preciado R (2005) The addition of milk or yogurt to a plant-based diet increases zinc bioavailability but does not affect iron bioavailability in women. J Nutr 135(3):465–468

Salunke P, Metzger LE (2022a) Functional characteristics of process cheese product as affected by milk protein concentrate and micellar casein concentrate at different usage levels. Int Dairy J 128:105324

Salunke P, Metzger LE (2022b) Impact of transglutaminase treatment given to the skim milk before or after microfiltration on the functionality of micellar casein concentrate used in process cheese product and comparison with rennet casein. Int Dairy J 128:105317. https://doi.org/10.1016/j.idairyj.2022.105317

Salunke P, Metzger LE (2023a) Functional properties of micellar casein concentrate and milk protein concentrate as affected by transglutaminase treatment. Food Hydrocoll 137:108367

Salunke P, Metzger LE (2023b) Impact of transglutaminase treated micellar casein concentrate and milk protein concentrate on the functionality of process cheese product slice formulations. Food Research International 3:31. https://doi.org/10.48130/FMR-2023-0031

Salunke P, Marella C, Amamcharla JK, Muthukumarappan K, Metzger LE (2022a) Use of micellar casein concentrate and milk protein concentrate treated with transglutaminase in imitation cheese products—melt and stretch properties. J Dairy Sci 105:7904–7916. https://doi.org/10.3168/jds.2022-22253

Salunke P, Marella C, Amamcharla JK, Muthukumarappan K, Metzger LE (2022b) Use of micellar casein concentrate and milk protein concentrate treated with transglutaminase in imitation cheese products—unmelted texture. J Dairy Sci 105:7891–7903. https://doi.org/10.3168/jds.2022-21852

Salunke P, Syamala A, Metzger LE (2023a) Microstructural characterization of high-protein dairy powders. Dairy 4:462–481. https://doi.org/10.3390/dairy4030031

Salunke P, Biswas AC, Metzger LE (2023b) Manufacture of loaf-type processed cheese products using transglutaminase crosslinked micellar casein concentrate and milk protein concentrate. Int J Dairy Technol. https://doi.org/10.1111/1471-0307.12977

Sanders CT, DeMasie CL, Kerr WL, Hargrove JL, Pegg RB, Swanson RB (2014) Peanut skins-fortified peanut butters: effects on consumer acceptability and quality characteristics. LWT—Food Sci Technol 59:222–228

Schiano AN, Harwood WS, Drake MA (2017) A 100-year review: sensory analysis of milk. J Dairy Sci 100(12):9966–9986. https://doi.org/10.3168/jds.2017-13031

Sethi S, Tyagi S, Anurag R (2016) Plant-based milk alternatives an emerging segment of functional beverages: a review. J Food Sci Technol 53(9):3408–3423

Sha L, Xiong YLL (2020) Plant protein-based alternatives of reconstructed meat: science, technology, and challenges. Trends Food Sci Technol 102:51–61

Shaw M (1984) Cheese substitutes: threat or opportunity? J Soc Dairy Technol 37:27–31

Shi Y, Singh A, Kitts DD, Pratap-Singh A (2021) Lactic acid fermentation: a novel approach to eliminate unpleasant aroma in pea protein isolates. LWT 150:111927

Silva BQ, Smetana S (2022) Review on milk substitutes from an environmental and nutritional point of view. Appl Food Res 2:100105

Solowiej B, Mleko S, Gustaw W (2008) Physicochemical properties of acid casein processed cheese analogs obtained with different whey products. Milchwissenschaft Milk Sci Int 63(3):299–302

Spada JC, Marczak LDF, Tessaro IC, Flores SH, Cardozo NSM (2015) Rheological modelling, microstructure and physical stability of custard-like soy-based desserts enriched with guava pulp. CyTA J Food 13(3):373–384

Stuart-Macadam P (2017) In: Stuart-Macadam P, Dettwyler KA (eds) Breastfeeding. Routledge. https://doi.org/10.4324/9781315081984

Tan Y, McClements DJ (2021) Improving the bioavailability of oil-soluble vitamins by optimizing food matrix effects: a review. Food Chem 348:129148

Tang CH (2017) Emulsifying properties of soy proteins: a critical review with emphasis on the role of conformational flexibility. Crit Rev Food Sci Nutr 57(12):2636–2679

Tangyu M, Muller J, Bolten CJ, Wittmann C (2019) Fermentation of plant-based milk alternatives for improved flavour and nutritional value. Appl Microbiol Biotechnol 103:9263–9275. https://doi.org/10.1007/s00253-019-10175-9

Thorning TK, Raben A, Tholstrup T, Soedamah-Muthu SS, Givens I, Astrup A (2016) Milk and dairy products: good or bad for human health? An assessment of the totality of scientific evidence. Food Nutr Res 60. https://doi.org/10.3402/fnr.v60.32527

Walther B, Guggisberg D, Badertscher R, Egger L, Portmann R, Dubois S, Haldimann M, Kopf-Bolanz K, Rhyn P, Zoller O, Veraguth R, Rezzi S (2022) Comparison of nutritional composition between plant-based drinks and cow's milk. Front Nutr 9:988707

Wang S, Chelikani V, Serventi L (2018) Evaluation of chickpea as alternative to soy in plant-based beverages, fresh and fermented. LWT 97:570–572. https://doi.org/10.1016/j.lwt.2018.07.067

Watson E (2019) Food navigator. https://www.foodnavigator-usa.com/Article/2019/2008/2022/In-conversation-with-Givaudan-How-do-you-create-meaty-flavors-inplant-based-meat

Yadav D, Bansal S, Jaiswal AK, Singh R (2017) Plant based dairy analogues: an emerging food. Agric Res Technol, Open Access Journal ISSN: 2471-6774

Yang M, Li N, Tong L, Fan B, Wang L, Wang F, Liu L (2021) Comparison of physicochemical properties and volatile flavor compounds of pea protein and mung bean protein-based yogurt. LWT 152:112390. https://doi.org/10.1016/j.lwt.2021.112390

Yin B, Deng W, Xu K, Huang L, Yao P (2012) Stable nano-sized emulsions produced from soy protein and soy polysaccharide complexes. J Colloid Interface Sci 380(1):51–59

Youssef CE, Bonnarme P, Fraud S, Péron AC, Helinck S, Landaud S (2020) Sensory improvement of a pea protein-based product using microbial co-cultures of lactic acid bacteria and yeasts. Food Secur 9:349

Zhang W, Xu T, Yang R (2019) Effect of roasting and grinding on the processing characteristics and organoleptic properties of sesame butter. Eur J Lipid Sci Technol 121:1800401

Zou L, Pande G, Akoh CC (2016) Infant formula fat analogs and human milk fat: new focus on infant developmental needs. Annu Rev Food Sci Technol is online at https://www.food.annualreviews.org. https://doi.org/10.1146/annurev-food-041715-033120

Zujko ME, Witkowska AM (2014) Antioxidant potential and polyphenol content of beverages, chocolates, nuts, and seeds. Int J Food Prop 17:8

# Chapter 3
# Egg Analogues

**Muhammed Yüceer**

## 3.1 Introduction to Novel Egg Analogues

In the global food system, hen's eggs are widely consumed by people of all ages. Nonetheless, there has been debate over certain health (egg allergies, avian influenza, cholesterol, phenylketonuria, and vegetarians or vegans), allergic, religious belief, cost reduction and environmental issues with the public's perception occasionally lagging behind evolving scientific findings in recent years (Myers and Ruxton 2023; Yazici and Ozer 2021; Yüceer 2023). The global market for vegan foods, such as substitutes/alternatives for dairy, eggs and meat has increased significantly worldwide in recent years (Beate and Klaus 2020; Boukid and Gagaoua 2022; McClements 2020). It might be difficult to find suitable substitutes for animal products when many producers and consumers desire to switch to plant-based food items entirely or partially (Yüceer 2019, 2020b; Yüceer et al. 2012). Novel egg analogs refer to plant-based or alternate ingredients (vegetarian/vegan egg, eggless egg) that aim to mimic the properties and functionality of eggs in various food applications (Guillaume et al. 2022; Yazici and Ozer 2021).

Egg analogs are products, usually plant-based, designed to replace real eggs. They are intended to provide alternatives for vegan diets or for people with egg allergies, but today they are designed as natural product substitutes. Egg analogs are formulated to behave similarly to the various functions provided by real eggs. Today, egg substitutes are used in factories that use egg products (industrial cake manufacturing, etc.) and are preferred because they offer a significant advantage to the factory in terms of low price. This can lead to significant cost advantages, especially in mayonnaise, meringue and cake manufacturing where eggs are used in

M. Yüceer (✉)
Canakkale Onsekiz Mart University, Department of Food Processing, Terzioglu Campus, Canakkale, Turkiye
e-mail: myuceer@comu.edu.tr



Ö. P. CAN et al. (eds.), *Food Analogues: Emerging Methods and Challenges*,
https://doi.org/10.1007/978-3-031-69872-9_3

bulk. It is also known that the use of egg substitutes increased during the spread of Avian Influenza and Newcastle disease (Yüceer 2023). Egg analogues come in various forms, including plant-based and laboratory-grown options. Plant-based alternatives, such as tofu, chia seed, banana, pea, hemp, soy flour, zucchini, aquafaba (the liquid from canned chickpeas), and flaxseed gel, replicate in recipes the binding and leavening qualities of eggs (Boukid and Gagaoua 2022; He et al. 2021; Niccum 2021; Raikos et al. 2020; Saleh 2022). These options are particularly appealing to vegans and individuals with egg allergies or dietary restrictions. However, developments in cellular agriculture have produced lab-grown proteins that, in terms of flavor, texture, and usefulness, nearly mimic the real thing (Santo et al. 2020). Mung bean protein is usually the component in egg analogues that gives the cooked "egg" its semi-solid texture. Carotenoids found in carrots and curcumin found in turmeric give these goods their yellowish hue. Since these additives are very hydrophobic compounds, unique delivery strategies must be created in order to include them into plant-based foods like emulsions or nanoemulsions (McClements and Grossmann 2021b). Eggless cake formulations containing isolated pea protein, 0.1% xanthan gum, and 1% soy lecithin were shown to be nearly identical to the values of traditional cake (control) when made with emulsifier mixes, xanthan gum, and isolated pea protein (Lin et al. 2017a).

The development of egg analogues is not only driven by dietary considerations but also by environmental concerns. Traditional egg production can be resource-intensive and have a significant ecological footprint (Guillaume et al. 2022). By opting for more sustainable alternatives, consumers and food manufacturers can reduce the environmental impact associated with animal-based product such as egg production (Leinonen et al. 2014; Prag and Henriksen 2020). Moreover, egg substitutes are essential to the development of the food sector. They enable the creation of innovative products that cater to evolving consumer preferences. From plant-based egg patties for breakfast sandwiches to vegan meringues that replicate the texture of their egg-based counterparts, these analogues offer versatility and open new avenues for culinary experimentation (Hefferon et al. 2023; Runte et al. 2024; Xiao et al. 2023). The discovery of a mixture of plant proteins that unfold and assemble similarly to egg proteins in a temperature range of roughly 63–93 °C, provide a similar texture-temperature profile, and When heated above their thermal denaturation temperature, many plant proteins, including those found in peas, chickpeas, beans, soybeans, and sunflower seeds, can gel; however, the texture varies depending on the protein structure, protein concentration, pH, salt, and temperature. In order to replicate some of the characteristics of actual eggs, egg analogs also need a lot of other parts. You can add an emulsified plant-based oil to provide the required mouthfeel, flavor, texture, and appearance. Some of the functional characteristics that lipoproteins in real eggs ordinarily supply are simulated by these oil droplets. Curcumin and carotenoids are examples of natural pigments that are frequently added to eggs to replicate the desired yellowish color. Add a variety of flavorings, including sugars, salts, herbs, or spices, to create a fragrance and taste similar to eggs (Lima et al. 2022). In a survey of 326 American food processing plants conducted by the American Egg Board (AEB), 59% of the food industry used only egg products (liquid and powder), 35% used both egg products and egg analogues together in the product recipe, and 6% preferred only egg analogues, and 75% preferred only egg products and 16% preferred only egg analogues in the recipe of formulations. In the same survey, it was found that the product category group in which eggs and/or egg analogues are mostly used is bakery products and cake manufacturing with 59%. In addition, egg products are preferred for coagulation and crystallization control while egg analogues are preferred for foaming (aeration) and crystallization control. In the study, it was observed that the utilization rate in terms of texture-structure was close to each other in terms of egg products and analogues. In 2017, it was reported in the report that the consumption of egg products increased in 38% of the industry users, while it remained unchanged in 54% and decreased in 7% due to the decrease in the supply of eggs in the market and allergenicity, as the impact of the bird flu case decreased and egg prices decreased. In the case of egg analogues, 39% of food industry users increased their use of egg analogues due to food safety and lack of allergen risk, 36% remained unchanged and 15% switched to egg analogues due to consumer demand for a declaration of the use of egg products on product packaging. Accordingly, the report found that egg prices and especially the availability of egg supply in the market can increase interest in egg analogues (Yüceer 2021, 2023). Many food ingredients are currently used as egg analogues, including chickpea flour/canned chickpea juice (Ali et al. 2018; Aslan and Ertaş 2020; Mustafa et al. 2018; Raikos et al. 2020), Chia seeds/gel (*Salvia hispanica L.*) (Borneo et al. 2010; Gallo et al. 2018), whey powder (Arunepanlop et al. 1996), lentil protein (Jarpa-Parra et al. 2017), bean protein (Arozarena et al. 2014), soy milk/flour (Rahmati and Mazaheri Tehrani 2015), milk and polysaccharides (Lin et al. 2017b; Mahdi et al. 2015; Shamooshaki et al. 2015), guar gum, starch, xanthan gum (Axner and Lagerson 2016; Kohrs et al. 2010).

In this era of changing dietary habits and heightened environmental consciousness, the introduction of novel egg analogues is a significant milestone in the culinary world. As research and development continue to expand the possibilities, we can expect to see even more creative and sustainable solutions emerge, further reshaping the way we cook and consume eggs in the future. The egg analogues are designed to imitate the texture, binding properties, and taste of eggs, providing a suitable alternative for individuals who follow a vegan or vegetarian diet, have allergies or dietary restrictions, or simply choose to reduce their consumption of animal products (Boukid and Gagaoua 2022; Lima et al. 2022). The potential of innovative egg mimics to mimic the binding, foaming, emulsifying, and texturizing capabilities of eggs is what makes them functional. The egg analogues can effectively bind ingredients together in recipes, acting as a binder in plant-based substitutes or as an emulsifier in mayonnaise alternatives (Kyriakopoulou et al. 2021).

Analogs can improve food products' mouthfeel and texture, giving consumers a dining experience akin to that of dishes made with eggs. The new egg substitutes can be used in baked products, sauces, dressings, and other recipes that call for eggs. They can also be utilized as plant-based meat alternatives. These analogues are particularly useful in vegan or vegetarian recipes, where they can provide the desired functionality and texture without the use of animal-derived ingredients. Additionally, novel egg analogues can also be used in products that aim to reduce cholesterol or fat content, appealing to individuals seeking healthier alternatives (Kurek et al. 2022; Zhang et al. 2021).

Novel egg analogues serve as sustainable and functional alternatives to traditional eggs, replicating their unique characteristics such as whipping, binding, emulsifying, and texturizing. They can be utilized in many different ways, including as plant-based meat alternatives, baked items, dressings, sauces, and dishes that are vegan or vegetarian. These analogues are carefully formulated to ensure a similar texture and appearance to real eggs, providing a seamless transition for consumers looking for plant-based or healthier options (McClements and Grossmann 2021a). The unique combination of plant proteins in egg analogues allows them to unfold and aggregate over the same temperature range as real eggs, resulting in similar texture and functionality (Lima et al. 2022). Plant-based egg substitutes have the potential to enhance the sustainability of the food sector by harnessing underutilized co-product streams to extract viable functional substitutes (Bohrer 2019; Langyan et al. 2021). Some consumers may still prefer traditional eggs due to their taste and nutritional profile. However, for those who choose to embrace plant-based or healthier options, novel egg analogues offer a viable alternative that can replicate the functionalities of traditional eggs in a wide range of culinary applications. Novel egg analogues have a range of functionalities and applications that make them highly useful in various food products (Langyan et al. 2021; McClements and Grossmann 2021a).

Whole hen's eggs are a flexible component that can be used in cuisine for a variety of purposes, including gelation, emulsification, foaming, and water retention (Yüceer 2020a). These desired characteristics, such as the capacity to go through a sol-gel transition when cooked in an environment similar to that of real eggs, should be replicated in plant-based egg analogs. The type of protein, protein concentration, and environmental factors all affect the sort of gel that forms. Egg analogs made from plant-based components should also have high emulsifying qualities, like mayonnaise or dressings. One of the most popular egg analogs is JUST Egg™, which offers two products: frozen egg slices for breakfast sandwiches and fluid eggs for omelets or scrambled eggs. The primary ingredients are emulsified canola oil and mung bean protein; canola oil droplets add to the mouthfeel, texture, flavor profile, and opaque appearance. Additional useful components are included to more precisely mimic the characteristics of actual eggs (McClements and Grossmann 2021a; Yüceer 2021).

## 3.2 Exploring the Functionality of Egg Analogues

Innovative alternatives to conventional eggs known as "egg analogues" have drawn a lot of attention recently because of their potential to solve a number of dietary, ethical, and environmental issues. Beyond their ability to replace eggs in recipes, it is crucial to delve into the functionality of these analogues and understand how they perform in cooking and baking applications. Some examples of the functionality and applications of novel egg analogues include:

- Binding and Emulsification: Eggs are well-known for their binding and emulsifying properties in culinary preparations. They act as a glue, holding ingredients together in dishes like meatloaf and meatballs. Additionally, they serve as excellent emulsifiers in dressings, sauces, and mayonnaise, helping oil and water-based ingredients stay mixed. Egg analogues must replicate these functions effectively. Some plant-based alternatives like flaxseed gel and chia seed gel exhibit binding properties, making them suitable for vegan baking. Moreover, aquafaba, the viscous liquid from canned chickpeas, excels at emulsification, enabling it to replace egg whites in recipes like meringues (Buhl et al. 2019; He et al. 2021; Langyan et al. 2021; Raikos et al. 2020; Starmer et al. 2018).
- Leavening and Structure: Eggs also contribute to the leavening and structural integrity of baked goods. In baking, they can add volume, moisture, and tenderness to cakes, muffins, and pancakes. For egg analogues to be successful in these applications, they must mimic these characteristics. Plant-based options like silken tofu and mashed bananas can serve as leavening agents and provide moisture in vegan recipes. In some cases, commercial egg replacers specifically designed for baking offer the desired rise and structure (Langyan et al. 2021; Yazici and Ozer 2021).
- Texture and Mouthfeel: The texture and mouthfeel of dishes often depend on eggs. In custards, quiches, and scrambled eggs, eggs create a creamy, smooth texture. To achieve similar results with egg analogues, manufacturers have developed products like silken tofu and dairy-free yogurt. These alternatives offer creaminess and texture that closely resemble real eggs. Flavor and Aroma: Eggs contribute to the flavor and aroma of many dishes. Their unique taste can be challenging to replicate. While some egg analogues have a neutral flavor and do not significantly impact the final taste of a dish, others, like nutritional yeast or black salt (kala namak), can add an eggy flavor to vegan recipes.
- Challenges and Advancements: Despite the progress in egg analogue development, challenges persist. Achieving the exact functionality of eggs in all culinary applications remains a complex task. Moreover, some egg analogues may lack the nutritional profile of real eggs, including essential vitamins and minerals.

## 3.3 Key Applications of Egg Analogues

Key Applications of Egg Analogues include:

- Plant-based meat substitutes: Egg analogues can serve as binders in plant-based meat substitutes, providing structure and texture similar to real eggs. They can also contribute to the moisture and fat retention of these meat substitutes, improving their overall texture and appearance.

- Baked goods: In order to give baked goods like cakes, cookies, muffins, and other baked goods structure and air retention during baking, egg analogs can be utilized. This will result in a light and fluffy texture.
- Emulsifying agent: Egg analogues can act as emulsifiers in various food products, helping to stabilize oil and water mixtures and create a smooth and creamy texture.
- Sauces and dressings: Egg analogues can be used as emulsifiers in sauces and dressings, helping to bind together oil and water molecules and create a smooth and creamy consistency.
- Vegan and allergy-friendly products: Novel egg analogues offer a plant-based alternative to traditional eggs, making them suitable for vegans and individuals with egg allergies or dietary restrictions. Novel egg alternatives can be employed as emulsifiers in sauces and dressings, as foaming agents in sweets and confectioneries, as binders in plant-based meat substitutes, and to give structure and texture to baked goods. Innovative egg substitutes present a plethora of opportunities for the production of vegan and allergy-friendly goods, and their application can aid in the advancement of more ethical and sustainable food options for consumers (Langyan et al. 2021; McClements and Grossmann 2021a).
- Alternative protein sources: Egg substitutes can be included into a variety of food products as an additional source of protein, giving customers a sustainable and wholesome choice. The innovative egg analogs can be utilized in a variety of food products, such as baked foods, sauces, dressings, desserts, and plant-based meat alternatives. They are particularly valuable for creating vegan and allergy-friendly options, as well as offering alternative protein sources. They can also contribute to the reduction of animal agriculture's environmental impact by offering more sustainable and ethically produced food options. The functionality and applications of novel egg analogues are diverse and extensive. They can function as emulsifiers in sauces and dressings, binders in plant-based meat substitutes, provide baked goods structure and texture, and foaming agents in desserts and confections. Their use can help create more sustainable and ethical food options for customers, and they provide a wealth of options for making vegan and allergy-friendly goods. Egg analogs can also be a source of protein for people on vegetarian or vegan diets and utilized as binders in plant-based meat replacements. They can also be utilized as a component for producing allergen-free products, catering to individuals with specific dietary restrictions or allergies (Boukid and Gagaoua 2022; Langyan et al. 2021).

Egg analogues, innovative substitutes for traditional eggs, have found diverse applications across the culinary landscape. Their versatility makes them valuable in various contexts, catering to a wide range of dietary preferences and restrictions. Some key applications of egg analogues:

- Vegan and Vegetarian Cooking: Egg analogues are a boon for those following vegan or vegetarian diets. They can replace eggs in recipes for scrambled eggs, omelets, and quiches, allowing individuals to enjoy familiar dishes without animal-derived ingredients.

- Baking: In baking, eggs serve multiple purposes, from providing structure and leavening to adding moisture and richness (Yüceer 2019). Egg analogues such as applesauce, mashed bananas, and commercial egg replacers are commonly used in vegan baking to create tender cakes, muffins, and cookies (Boukid and Gagaoua 2022).
- Binding Agent: Eggs are known for their binding properties in dishes like meatloaf, burgers, and meatballs. Plant-based alternatives like flaxseed gel and chia seed gel work well as binding agents, holding ingredients together in meatless recipes.
- Emulsification: Egg yolks are excellent emulsifiers, helping oil and water-based ingredients blend seamlessly in salad dressings, sauces, and mayonnaise. Aquafaba, the liquid from canned chickpeas, is a popular egg white substitute with remarkable emulsifying properties.
- Desserts: Egg analogues play a crucial role in creating creamy desserts like custards, puddings, and ice creams. Ingredients like silken tofu and dairy-free yogurt can replicate the smooth texture and creaminess of traditional egg-based desserts.
- Allergen-Friendly Alternatives: Egg allergies are common, prompting the development of egg analogues that offer allergen-friendly options. These alternatives allow individuals with egg allergies to enjoy a diverse range of foods without compromising on safety.
- Environmental Sustainability: As concerns about the environmental impact of conventional egg production grow, egg analogues offer a more sustainable alternative. Raising plant-based egg substitutes is less resource-intensive and emits less greenhouse gas than raising chickens for eggs.
- Food Product Enhancement: The versatility of egg analogues contributes to the improvement of existing food products. These analogues enhance the nutritional profile, texture, and stability of various foods, fostering innovation within the food industry. Egg analogs are essential because they provide a foundation for the development of novel goods that depend on them for texture, binding, and emulsification, such as plant-based meat substitutes and dairy-free cheeses.
- Culinary Research and Innovation: Egg analogues serve as indispensable tools in food research and development. They provide crucial functionality for texture, binding, and emulsification, which makes it easier to create innovative plant-based food products like dairy and meat substitutes.
- Functional Ingredient Replacement: Egg analogues are employed as substitutes in culinary and baking applications. In baked goods, they replicate the leavening, binding, and emulsifying properties of eggs, ensuring the desired texture and structure of various products.
- Vegan and Vegetarian Cuisine: Egg analogues are pivotal in vegan and vegetarian cooking, enabling the creation of egg-free versions of classic dishes like omelets, scrambles, and quiches. Their utilization accommodates dietary restrictions and preferences.

## 3.4 Benefits and Drawbacks of Egg Analogues

The benefits of egg analogues include their ability to provide a plant-based alternative for consumers with dietary restrictions or preferences. Other benefits include their sustainability and potential for utilizing food industry by-products. Egg analogues offer a cost-effective solution as they can be produced at a lower price compared to traditional eggs. They also have a longer shelf life and can be easily stored, transported, and incorporated into various food products. On the other hand, drawbacks of egg analogues may include a different taste compared to traditional eggs and potential allergenicity for individuals with specific plant protein allergies (Boukid and Gagaoua 2022; Bryant 2022; Langyan et al. 2021). One potential drawback of egg analogues is that they may not provide the exact same nutritional profile as traditional eggs, although efforts are being made to enhance their nutritional content through fortification. The functionality and applications of novel egg analogues are vast and diverse. They can be used to omelettes, scrambles, and frittatas, among other delectable dishes, and utilized as binders in plant-based meat alternatives and to give baked goods texture and structure (Boukid and Gagaoua 2022; Kyriakopoulou et al. 2021; Langyan et al. 2021; Setser and Racette 1992). The functionality and application of novel egg analogues extend beyond the realm of meat substitutes and include their incorporation into a wide range of culinary delights such as baked goods, pasta, custards, and even beverages like smoothies and milkshakes. Furthermore, novel egg analogues can also be used as a coating for fried foods, providing a crispy and golden exterior. The novel egg analogues are extensive, offering versatility in various food products while catering to the needs of individuals with dietary restrictions or preferences and contributing to the development of sustainable and innovative food solutions (Boukid and Gagaoua 2022; Kyriakopoulou et al. 2021). The functionality and applications of novel egg analogues are extensive, offering versatility in various food products while catering to the needs of individuals with dietary restrictions or preferences and contributing to the development of sustainable and innovative food solutions. The innovative egg analogs can be added to a variety of culinary treats, coated for fried foods, employed as binders in plant-based meat alternatives, and even utilized in drinks like milkshakes and smoothies. They can also give texture and structure to baked items (Langyan et al. 2021; McClements and Grossmann 2021b; Setser and Racette 1992; Yang 2014).

## 3.5 Understanding the Science Behind Egg Analogues

Understanding the science behind egg analogues involves studying their composition, structure, and functionality. By studying the composition of egg analogues, researchers can determine the ingredients and proportions necessary to mimic the nutritional profile and sensory characteristics of traditional eggs. Additionally, they can investigate the structural characteristics of egg substitutes to find out how well they bind, emulsify, foam/whip, and provide texture to a variety of products, including sauces, dressings, and baked items. The functionality of egg analogues in baked goods lies in their ability to act as emulsifiers, providing stability and texture to the products. The goal of creating plant-based meat substitutes is to replicate the flavor and texture of actual meat (Kyriakopoulou et al. 2021). This involves understanding the rheological properties of egg analogues and their ability to retain moisture, bind ingredients together, and provide a desirable mouthfeel. In terms of applications, the functionality of novel egg analogues allows for their wide use in various food products. The versatility of novel egg analogues makes them suitable for use in both vegan and vegetarian diets, providing a convenient alternative for individuals who choose to avoid or reduce their consumption of traditional eggs for various reasons such as ethical, health, or dietary preferences. Novel egg analogues have a broad range of applications and functionality that make it possible to include them into a variety of food products, such as baked goods, sauces, dressings, and plant-based meat alternatives (Boukid and Gagaoua 2022; Langyan et al. 2021).

The capacity of new egg mimics to replicate the sensory qualities and nutritional makeup of regular eggs is crucial for supplying consistency, solidity, and moisture retention in a range of food products. They facilitate the creation of plant-based substitutes, satisfying the dietary requirements of vegans and vegetarians, and providing accessible options for those with diverse tastes. The creation of plant-based meat alternatives, baked goods, sauces, dressings, and a variety of culinary specialties all benefit from the use of innovative egg analogues. They give food makers a plethora of options for producing goods that both satisfy different dietary requirements and preferences and replicate the qualities and sensory experience of actual eggs (Andreani et al. 2023; Boukid and Gagaoua 2022; Kyriakopoulou et al. 2021).

## 3.6 Innovation in the Field of Egg Analogues

Researchers and food producers have been working hard in recent years to create creative plant-based and sustainable egg substitutes in order to meet the growing demand for them. Researchers have been experimenting with several plant-based components, including legumes, algae, and starches, to produce textures and features reminiscent of eggs (Söderberg 2013). The binding, leavening, and emulsifying qualities of real eggs are intended to be replicated in these novel egg analogs. Moreover, developments in food technology have made it possible to produce egg substitutes with enhanced sensory attributes including flavor and texture. These developments have created new opportunities for the use of egg substitutes in a variety of culinary applications, such as plant-based protein products, vegan baking, and even cake, omelets, scrambles, frittatas, salad dressings, and mayonnaise without eggs (Boukid and Gagaoua 2022; Langyan et al. 2021).

## 3.7 Future Prospects of Egg Analogues

The future prospects of egg analogs include exploring the use of novel ingredients and technologies to further improve their functionality and sensory properties. Scientists are always looking for novel plant-based components that can improve the flavor, texture, and nutritional profile of egg analogs while closely emulating the special qualities of eggs. The creation of environmentally friendly and sustainable egg substitutes is gaining popularity. This involves utilizing components that have a smaller negative influence on the environment, including microalgae or insect proteins, and utilizing cutting-edge processing techniques to use less energy and produce less waste. The progress made in egg analogs serves not just the needs of those with certain dietary requirements or allergies, but also advances the creation of more ethical and ecological food solutions. These solutions can lessen the negative effects of food production on the environment and satisfy the growing demand for plant-based substitutes in the international market (Boukid and Gagaoua 2022; Joshi and Kumar 2015).

Egg analogs offer potential benefits for individuals with specific dietary restrictions or allergies. For instance, people who are vegan or vegetarian or have egg allergies can still enjoy the flavor and texture of eggs without sacrificing their dietary choices or concerns. Novel egg analogs provide opportunities for food manufacturers to create products with similar properties and sensory experiences as real eggs, catering to different dietary preferences and needs. These innovative egg analogs can also be applied in non-food applications such as cosmetic formulations, where components derived from eggs are commonly used they have the power to completely transform a number of sectors by offering plant-based, sustainable substitutes that are just as functional and have comparable qualities to conventional eggs. These novel egg analogs offer numerous opportunities for food manufacturers to create plant-based products with similar taste, texture, and culinary uses as real eggs. Furthermore, the use of egg analogs can contribute to the development of more sustainable and environmentally friendly food production practices (Boukid and Gagaoua 2022; Langyan et al. 2021).

Egg analogs have the potential to be used in pharmaceutical formulations, such as vaccines and drug delivery systems, as eggs are commonly used in vaccine production. This demonstrates the broad range of possibilities and opportunities that novel egg analogs present in various sectors, making them a versatile and valuable ingredient in today's evolving food industry. Therefore, the functionality and applications of novel egg analogs have the potential to revolutionize not only the food industry but also the pharmaceutical industry by providing sustainable alternatives that cater to reducing environmental impact (Boukid and Gagaoua 2022; Geels and Ye 2010; Hegde 2015).

## 3.8 Environmental Impact of Egg Analogues

In terms of the environmental impact, novel egg analogues have the potential to reduce the carbon footprint associated with traditional egg production. In addition, compared to conventional egg production techniques, egg analogues need less natural resources, such as land and water, because they are primarily produced from plant-based sources. As a result, the use of novel egg analogues can contribute to sustainable food systems and help reduce the environmental burden caused by conventional egg production. The functionality and applications of novel egg analogues are vast and diverse. Their ability to offer sustainable, allergen-free, plant-based substitutes for eggs might completely transform the food business. In addition, the versatility and functionality of egg analogues make them an ideal ingredient for various food applications. These applications include vegan baking, plant-based protein products, egg-free mayo or salad dressings, and even culinary dishes. By exploring the amino acid sequences and properties of different plant proteins, researchers can design and modify protein-rich ingredients to suit specific functional requirements in various food applications, thereby expanding the possibilities for using egg analogues in different products and processes. The development and use of novel egg analogues offer numerous opportunities for the food industry. These advancements not only cater to the needs of individuals with dietary restrictions but also contribute to the development of more sustainable and plant-based food options (Alcorta et al. 2021). The functionality of novel egg analogues are extensive, ranging from catering to dietary restrictions and allergies to addressing sustainability and ethical concerns in food production. Because the new egg analogues are plant-based, sustainable, and allergy-free substitutes for conventional eggs, they have the potential to completely transform the food sector. Furthermore, because innovative egg mimics have the same texture, flavor, and functioning as traditional eggs, they can help meet the growing demand for plant-based and vegan products. By using less animal husbandry, they can also help lessen the impact that food production has on the environment. In conclusion, there are many uses for innovative egg analogs and a wide range of functionalities. They have the potential to meet the needs of individuals with dietary restrictions, contribute to sustainability efforts, and provide innovative options for food manufacturers and consumers alike (Boukid and Gagaoua 2022; Polyak et al. 2023).

## 3.9 Comparative Study: Egg Analogues Versus Traditional Eggs

A comparative study between egg analogues and traditional eggs can provide insights into their nutritional, environmental, and food safety characteristics. This study can help identify the potential benefits and drawbacks of using egg analogues, as well as highlight areas for improvement. By analyzing factors such as nutritional composition, protein content, cholesterol levels, and allergenicity, researchers can determine how egg analogues stack up against traditional eggs in terms of their health impacts and suitability for different dietary needs. By contrasting the environmental effects of traditional eggs with egg mimics, they may also assess the environmental impact of egg production, including greenhouse gas emissions, trash generation, and the use of water and land. Additionally, conducting food safety assessments on egg analogues can ensure that they meet the necessary quality standards and regulations for consumer protection. By understanding the preferences and perceptions of different age groups towards upcycled foods, researchers can develop strategies to improve the acceptability of novel egg analogues and increase their adoption among consumers. In the food business, using innovative egg substitutes can help reduce food waste and promote sustainability (Boukid and Gagaoua 2022).

## 3.10 Consumer Acceptance of Egg Analogues

Utilizing new egg analogs as binders in plant-based meat replacements is one of their main uses. These egg substitutes can offer the binding qualities required to bind plant proteins together, creating a cohesive texture that resembles actual meat. This is particularly important in minced products, where the texture and structure are crucial in replicating the mouthfeel of meat. Another important application of novel egg analogues is their incorporation into baked goods. They can act as a leavening agent, providing structure and promoting the rise of baked goods such as cakes and muffins. In addition to these applications, novel egg analogues can also act as emulsifiers, stabilizers, and thickeners in various food products. Because of their functionality, innovative egg mimics can be used in a variety of culinary applications to mimic the functions of traditional eggs. The binding qualities required to keep plant proteins together and create a cohesive texture resembling actual flesh can be found in egg analogs. This is especially significant for minced products since the structure and texture play a critical role in emulating the mouthfeel of meat. The utilization of new egg substitutes in vegan baking is a significant additional application (Hefferon et al. 2023; Yüceer 2021).

Novel egg analogues can provide texture and structure to baked goods, acting as leavening agents and promoting the rise of cakes and muffins. The novel egg analogues can also function as emulsifiers, stabilizers, and thickeners in various food products (Kyriakopoulou et al. 2021). Because of their uses, they are valued ingredients in the food industry and can be used to manufacture vegan and plant-based products that mimic conventional egg-containing products. Novel egg mimics have a wide range of uses due to their functionality and adaptability. Some potential applications of novel egg analogues include: creating vegan omelets, preparing egg-free mayonnaise, developing plant-based custards and puddings, making vegetarian quiches and frittatas, and even producing vegan meringues (Akcicek et al. 2022; He et al. 2021).

The functionality of novel egg analogues allows for their application in various food systems as binders in plant-based meat substitutes, leavening agents in baked goods, emulsifiers, stabilizers, and thickeners in food products, and the creation of vegan alternatives to traditional egg-containing products (Appiani et al. 2023; Benkovic et al. 2023).

## 3.11 Conclusion

In conclusion, egg analogues offer a promising solution to various culinary scenarios, addressing dietary, ethical, and environmental concerns. Their ability to replicate the functionalities of traditional eggs opens up a world of possibilities for chefs, home cooks, and food manufacturers, making it easier for people to embrace alternative dietary choices and sustainable food practices. Egg analogues can effectively replicate many of the essential functions of eggs in cooking and baking, but their success depends on the specific application and the chosen analogue. As research and development in this field continue, we can anticipate even more sophisticated egg analogues that offer improved functionality, further expanding the possibilities for egg-free cooking and baking. Exploring these alternatives not only opens doors to innovative culinary creations but also contributes to more sustainable and inclusive food practices. The new egg substitutes can be used in many other culinary preparations, such as plant-based protein products and vegan baking. For those who have dietary limitations or allergies, they provide a good alternative that lets them still enjoy the flavor and texture of eggs without sacrificing their preferences. Because they lessen the reliance on conventional animal agriculture and the completion of the sentence will be based on the health or ethical considerations, these egg analogues also have the potential to support the promotion of environmentally friendly and sustainable food choices. The increasing demand for plant-based and sustainable food options in the global market provides a promising future for the growth and acceptance of novel egg analogues in various industries, including food service, retail, and food manufacturing.

Eggs are considered one of the best foods for "good health" from a nutritional perspective, and they have been a part of our everyday lives for thousands of years. Eggs are still a highly nutritious food that is consumed by people all over the world, including youngsters and the elderly. In order to add softness, rise, flavor, color, and texture to a variety of meals, including baked products, dairy desserts, cakes, ice cream, mayonnaise, noodles, and pasta, eggs, pasteurized liquid eggs, and egg powder are utilized as functional ingredients. Due to many factors such as cholesterol intake, dietary preferences (such as veganism and vegetarianism), avian influenza, egg allergies, and environmental concerns, a growing number of consumers and food manufacturers are restricting their intake of eggs and substituting them with egg alternatives. Egg analogs can be found on the market under different names, such as vegan eggs, plant-based eggs, egg substitutes, egg alternatives, vegetable eggs, artificial eggs, etc. A variety of components, including soy flour, wheat flour, starch, casein, rye flour, whey, and blood plasma, can be found in egg analogs. Egg substitutes can be preferred in the food industry all over the world where eggs and their products are used instead of eggs. Some factors affect the use of egg analogues, the most important of which is product cost. There are many products in the world as egg analogues. Many food components that can be used as analogues for eggs are known. All of the functional qualities that the egg produces alone, such as binding, foaming, and flavor, cannot be replicated, though. Generally, the egg substitute products developed have one or two functionalities but do not meet the multiple techno-functional characterizations provided by the egg.

In conclusion, new egg analogs have a wide range of uses and functionalities. They might be anything from enhancing the sensory qualities of food items to offering substitutes for people following certain dietary plans or taking ethical issues into account. Egg analogs can be added to a variety of culinary applications, utilized as binders in plant-based egg analogues, and to provide baked goods structure and texture. To enhance the nutritional profiles and functional versatility of egg analogs, more study is required.

## References

Akcicek A, Karasu S, Bozkurt F, Kayacan S (2022) Egg yolk-free vegan mayonnaise preparation from pickering emulsion stabilized by gum nanoparticles with or without loading olive pomace extracts. ACS Omega 7:26316–26327

Alcorta A, Porta A, Tarrega A, Alvarez MD, Vaquero MP (2021) Foods for plant-based diets: challenges and innovations. Food Secur 10(2):293

Ali R, Saeed SMG, Ali SA, Sayed SA, Ahmed R, Mobin L (2018) Effect of black gram flour as egg replacer on microstructure of biscuit dough and its impact on edible qualities. J Food Meas Charact 12:1641–1647

Andreani G, Sogari G, Marti A, Froldi F, Dagevos H, Martini D (2023) Plant-based meat alternatives: technological, nutritional, environmental, market, and social challenges and opportunities. Nutrients 15(2):452

Appiani M, Cattaneo C, Laureati M (2023) Sensory properties and consumer acceptance of plant-based meat, dairy, fish and eggs analogs: a systematic review. Front Sustain Food Syst 7:1268068

Arozarena I, Bertholo H, Empis J, Bunger A, Sousa I (2014) Study of the total replacement of egg by white lupine protein, emulsifiers and xanthan gum in yellow cakes. Eur Food Res Technol 213:312–316

Arunepanlop B, Morr CV, Karleskind D, Laye I (1996) Partial replacement of egg white proteins with whey proteins in angel food cakes. J Food Sci 61:1085–1093

Aslan M, Ertaş N (2020) Possibility of using 'chickpea aquafaba' as egg replacer in traditional cake formulation. Harran Tarım ve Gıda Bilimleri Dergisi 24:1–8

Axner E, Lagerson E (2016) Cryopreservation of dog semen in a tris extender with 1% or 2% soya bean lecithin as a replacement of egg yolk. Reprod Domest Anim 51:262–268

Beate G, Klaus H (2020) Plant-based foods for future: results of consumer and professional expert interviews in five European countries, EIT-Food Project "The V-Place", Hohenheimer Agrarökonomische Arbeitsberichte, No. 29, Universität Hohenheim, Institut für Agrarpolitik und Landwirtschaftliche Marktlehre, Stuttgart

Benkovic M, Jurinjak Tusek A, Sokac Cvetnic T, Jurina T, Valinger D, Gajdos Kljusuric J (2023) An overview of ingredients used for plant-based meat analogue production and their influence on structural and textural properties of the final product. Gels 9(12):921

Bohrer BM (2019) An investigation of the formulation and nutritional composition of modern meat analogue products. Food Sci Human Wellness 8:320–329

Borneo R, Aguirre A, Leon AE (2010) Chia (Salvia hispanica L) gel can be used as egg or oil replacer in cake formulations. J Am Diet Assoc 110:946–949

Boukid F, Gagaoua M (2022) Vegan egg: a future-proof food ingredient? Food Secur 11(2):161

Bryant CJ (2022) Plant-based animal product alternatives are healthier and more environmentally sustainable than animal products. Future Foods 6:100174

Buhl TF, Christensen CH, Hammershøj M (2019) Aquafaba as an egg white substitute in food foams and emulsions: protein composition and functional behavior. Food Hydrocoll 96:354–364

Gallo LRDR, Assunção Botelho RB, Ginani VC, de Oliveira LD, Riquette RFR, Leandro EDS (2018) Chia (Salvia hispanica L.) gel as egg replacer in chocolate cakes: applicability and microbial and sensory qualities after storage. J Culin Sci Technol 18:29–39

Geels M, Ye K (2010) Developments in high-yield system expressed vaccines and immunotherapy. Recent Pat Biotechnol 4:189–197

Guillaume A, Hubatová-Vacková A, Kočí V (2022) Environmental impacts of egg production from a life cycle perspective. Agriculture 12(3):355

He Y, Purdy SK, Tse TJ, Tar'an B, Meda V, Reaney MJT, Mustafa R (2021) Standardization of aquafaba production and application in vegan mayonnaise analogs. Food Secur 10(9):1978

Hefferon KL, De Steur H, Perez-Cueto FJA, Herring R (2023) Alternative protein innovations and challenges for industry and consumer: an initial overview. Front Sustain Food Syst 7:1038286

Hegde NR (2015) Cell culture-based influenza vaccines: a necessary and indispensable investment for the future. Hum Vaccin Immunother 11:1223–1234

Jarpa-Parra M, Wong L, Wismer W, Temelli F, Han J, Huang W, Eckhart E, Tian Z, Shi K, Sun T, Chen L (2017) Quality characteristics of angel food cake and muffin using lentil protein as egg/milk replacer. Int J Food Sci Technol 52:1604–1613

Joshi VK, Kumar S (2015) Meat analogues: plant based alternatives to meat products- a review. Int J Food Ferment Technol 5:107–119

Kohrs D, Herald TJ, Aramouni FM, Abughoush M (2010) Evaluation of egg replacers in a yellow cake system. Emir J Food Agric 22:340–352

Kurek MA, Onopiuk A, Pogorzelska-Nowicka E, Szpicer A, Zalewska M, Poltorak A (2022) Novel protein sources for applications in meat-alternative products-insight and challenges. Food Secur 11(7):957

Kyriakopoulou K, Keppler JK, van der Goot AJ (2021) Functionality of ingredients and additives in plant-based meat analogues. Food Secur 10(3):600

Langyan S, Yadava P, Khan FN, Dar ZA, Singh R, Kumar A (2021) Sustaining protein nutrition through plant-based foods. Front Nutr 8:772573

Leinonen I, Williams AG, Kyriazakis I (2014) The effects of welfare-enhancing system changes on the environmental impacts of broiler and egg production. Poult Sci 93:256–266

Lima M, Costa R, Rodrigues I, Lameiras J, Botelho G (2022) A narrative review of alternative protein sources: highlights on meat, fish, egg and dairy analogues. Food Secur 11(14):2053

Lin M, Tay SH, Yang H, Yang B, Li H (2017a) Development of eggless cakes suitable for lacto-vegetarians using isolated pea proteins. Food Hydrocoll 69:440–449

Lin M, Tay SH, Yang H, Yang B, Li H (2017b) Replacement of eggs with soybean protein isolates and polysaccharides to prepare yellow cakes suitable for vegetarians. Food Chem 229:663–673

Mahdi GS, Alireza SM, Mohammas G, Yaya M, Aman MZ (2015) Effect of milk and Xanthan as egg replacement on the chemical and microbial safety properties of Mayonnaise. Scholarly J Agric Sci 5:256–261

McClements DJ (2020) Development of next-generation nutritionally fortified plant-based milk substitutes: structural design principles. Food Secur 9(4):421

McClements DJ, Grossmann L (2021a) A brief review of the science behind the design of healthy and sustainable plant-based foods. NPJ Sci Food 5:17

McClements DJ, Grossmann L (2021b) The science of plant-based foods: constructing next-generation meat, fish, milk, and egg analogs. Compr Rev Food Sci Food Saf 20:4049–4100

Mustafa R, He Y, Shim YY, Reaney MJT (2018) Aquafaba, wastewater from chickpea canning, functions as an egg replacer in sponge cake. Int J Food Sci Technol 53(10):2247–2255
Myers M, Ruxton CHS (2023) Eggs: healthy or risky? a review of evidence from high quality studies on hen's eggs. Nutrients 15(12):2657
Niccum C (2021) Can pea and white hemp proteins be used as a more sustainable protein source to improve nutritional and sensory results in gluten-free baked products? London South Bank University, Ph.D. dissertation, London
Polyak E, Breitenbach Z, Frank E, Mate O, Figler M, Zsalig D, Simon K, Szijarto M, Szabo Z (2023) Food and sustainability: is it a matter of choice? Sustain For 15(9):7191
Prag A, Henriksen C (2020) Transition from animal-based to plant-based food production to reduce greenhouse gas emissions from agriculture—the case of Denmark. Sustain For 12(19):8228
Rahmati NF, Mazaheri Tehrani M (2015) Replacement of egg in cake: effect of soy milk on quality and sensory characteristics. J Food Process Preserv 39:574–582
Raikos V, Hayes H, Ni H (2020) Aquafaba from commercially canned chickpeas as potential egg replacer for the development of vegan mayonnaise: recipe optimisation and storage stability. Int J Food Sci Technol 55(5):1935–1942
Runte M, Guth JN, Ammann J (2024) Consumers' perception of plant-based alternatives and changes over time. A linguistic analysis across three countries and ten years. Food Qual Prefer 113:105057
Saleh S (2022) Zucchini puree as a novel egg substitute in cake: comparing with other substitutes. Alex Sci Exch J 43:219–228
Santo RE, Kim BF, Goldman SE, Dutkiewicz J, Biehl EMB, Bloem MW, Neff RA, Nachman KE (2020) Considering plant-based meat substitutes and cell-based meats: a public health and food systems perspective. Front Sustain Food Syst 4:134
Setser CS, Racette WL (1992) Macromolecule replacers in food products. Crit Rev Food Sci Nutr 32:275–297
Shamooshaki MG, Mahounak AS, Ghorbani M, Maghsouldloo Y, Ziaeifar AM (2015) Effect of milk and xanthan as egg replacement on the physical properties of mayonnaise. Int Lett Nat Sci 49:24–34
Söderberg J (2013) Functional properties of legume proteins compared to egg proteins and their potential as egg replacers in vegan food. Swedish University of Agricultural Sciences, Second cycle, A2E. SLU, Dept. of Food Science, Uppsala, p 378
Starmer D, Coate K, Terry P (2018) The effects of creating a vegan alternative to hard meringues by substituting aquafaba for egg whites. J Acad Nutr Diet 118:A51
Xiao X, Zou PR, Hu F, Zhu W, Wei ZJ (2023) Updates on plant-based protein products as an alternative to animal protein: technology, properties, and their health benefits. Molecules 28(10):4016
Yang X (2014) Optimizing textural properties of soft solid foods. In: Food texture design and optimization, pp 45–73
Yazici GN, Ozer MS (2021) A review of egg replacement in cake production: effects on batter and cake properties. Trends Food Sci Technol 111:346–359
Yüceer M (2019) Yumurta ve Yumurta Ürünleri İşleme Teknolojisi ve Uygulamaları Sidas Medya Ltd. Şti. İzmir, Turkiye
Yüceer M (2020a) Evaluation of physicochemical properties on meringue prepared from phospholipase A2 enzyme-hydrolyzed liquid egg albumen. Eur Food Res Technol 246:1847–1856
Yüceer M (2020b) Structural and rheological characterization of liquid egg white modified with phospholipase A2 enzyme. J Food Process Preserv 44:e14450
Yüceer M (2021) Yumurta Alternatifleri ve Yumurta İkamesi olarak Kullanım İmkanları. Yumurta 56:8–10
Yüceer M (2023) Gıda Analoglarının Sağlıklı ve Helal Olma Şartları Yumurta Örneği. Helal Yaşam Rehberi Dergisi 17:22–25
Yüceer M, Temizkan R, Caner C (2012) Fonksiyonel Gıda Olarak Yumurta: BileŞenleri ve Fonksiyonel Özellikleri. Akademik Gıda 10:70–76
Zhang T, Dou W, Zhang X, Zhao Y, Zhang Y, Jiang L, Sui X (2021) The development history and recent updates on soy protein-based meat alternatives. Trends Food Sci Technol 109:702–710

# Chapter 4
# The Production, Ingredients and Resulting Sensory Properties of Plant-Based Chicken Analogues

Matthew B. McSweeney

## 4.1 Introduction

Sustainability has various definitions and explanations however, one of the most used is that of the United Nations (UN). Sustainability is defined by the UN as a "development that meets the needs of the present without compromising the ability of future generations to meet their own needs" (United Nations 1987). To achieve sustainability, different elements are needed including social responsibilities, environmental concerns, and economic development. The UN developed 17 Sustainable Development Goals (SDGs), and one of the primary focuses of the SDGs is to "protect the planet from degradation, including through sustainable consumption and production, sustainably managing its natural resources and taking urgent action on climate change" (Rosa 2017). Meat production is significant contributor to greenhouse gases and in turn, alternatives to meat products with increased sustainability need to be investigated (Meybeck and Gitz 2017).

Over the last decade, consumers have started to change their diet to incorporate more plant-based options. Additionally, more consumers have started to follow flexitarian or plant-based diets. Flexitarians work to consciously decrease their meat consumption and these consumers worry about the ethical implications of meat consumption (Derbyshire 2017). Consumers are more aware of the environmental impact of the food they are consuming and they consider the meat industry to have a significant carbon footprint (Sanchez-Sabate and Sabaté 2019). This knowledge of the carbon emissions caused by the meat industry can also lead some consumers to shift their diet to plant-based or flexitarian (Arora and Mishra 2019). As more consumers begin to make these dietary shifts, food companies have also started to tailor products to these new customers.

M. B. McSweeney (✉)
School of Nutrition and Dietetics, Acadia University, Wolfville, NS, Canada
e-mail: matthew.mcsweeney@acadiau.ca



Ö. P. CAN et al. (eds.), *Food Analogues: Emerging Methods and Challenges*,
https://doi.org/10.1007/978-3-031-69872-9_4

Furthermore, more energy, water and land resources are needed to produce animal-protein-based diets, when compared to a flexitarian or plant-based diet (Pimentel and Pimentel 2003). A diet incorporating further plant-based foods is considered to be sustainable and although both plant and animal diets indeed require significant amounts of non-renewable fossil fuels during production, the current Western meat-consuming diets require more resources than plant-based diets (Pimentel and Pimentel 2003; Sanchez-Sabate and Sabaté 2019).

Researchers have identified nutritional benefits to plant-based and flexitarian diets (Fehér et al. 2020; Green et al. 2022; Himics et al. 2022; Kahleova et al. 2017). Plant-based foods include whole grains, fruits, vegetables, nuts, and pulses. Pulses and nuts are good sources of fibre, antioxidants and vitamins (Satija and Hu 2018) and are usually low in cholesterol and saturated fat (Gardner et al. 2005). These properties have been associated with positive cardiovascular impacts (Satija and Hu 2018). Furthermore, plant-based diets have been found to reduce the risk of certain cancers (Lea et al. 2006; Radnitz et al. 2015). In addition to health concerns, consumers also identify animal rights, environmental concerns, and ethical considerations as reasons why they choose to consume a plant-based foods (Hopwood et al. 2020).

Other factors including gender, age, socioeconomic status and education level impact food purchasing and consumption, as well as attitudes towards sustainability and sustainable food (Lea et al. 2006). Women that possess a higher education level are more likely to consume plant-based foods (Lea et al. 2006; Modlinska et al. 2020). Similarly, more individuals identifying as females consider plant-based foods are sufficient in nutrients and that it is beneficial to the environment (Lea et al. 2006). Food costs, level of education, and consumer information influence food choices (Meybeck and Gitz 2017). In addition, although consumers may want to change their food behaviour, it can be difficult and consumers who choose to reduce their animal consumption are a minority compared to conventional consumers (Radnitz et al. 2015; Sanchez-Sabate and Sabaté 2019).

Plant-based alternatives (PBAs) strive to have similar sensory characteristics and nutritional components of animal-based foods are currently being marketed toward consumers, especially those who do not want to change their eating behaviour. These products offer the same functionality as their animal protein-based counterparts. However, the ingredients used to produce these PBAs need to be carefully selected as they need to provide appropriate texture and flavours, as well as ensure similar colour, functionality and nutritional components as their conventional counterparts (Kyriakopoulou et al. 2021a, b). This is challenging to the food industry as plant-based ingredients can lead to differences in appearance, texture, flavour, nutritional composition, and functionality than animal-based products. Additionally, there are physiochemical and molecular differences between plant proteins and animal proteins (Xing et al. 2022). All these aspects make it challenging for the food industry to create PBAs that successfully mimic their conventional animal-based products.

Chicken presents a unique challenge for product developers as it difficult to replicate the taste and texture using plant-based ingredients (Kamani et al. 2019). The juiciness and tenderness of chicken-based products are highly valued by consumers, and they need to be present in PBAs to chicken. Consumers are interested in chicken and the chicken market continues to grow (Izar-Tenorio et al. 2020). In the United States, consumers have increased their purchases of chicken by 40% from 2000 to 2021 (Tonsor and Lusk 2022). There is a demand and growing need for PBAs to chicken-based products as consumers are interested in sustainability but they are not willing to sacrifice the quality, functionality or sensory properties of conventional chicken products (Lang 2020; McClements 2020). Especially as consumers are concerned with the sustainability of large scale farming of chickens (Pinto da Rosa et al. 2021). The time is now to produce high-quality PBAs to chicken-based products. This chapter will explore the variety of PBAs that are manufactured to replicate that of chicken-based products. Furthermore, it will discuss the different ingredients used in PBAs to chicken, different production methods and the resulting sensory properties.

## 4.2 Ingredients

PBAs to chicken usually consist of a base ingredient or ingredients that are considered the main ingredients, and they also contain hydrocolloids, lipids, and fibre, as well as flavouring and colourants. Hydrocolloids or gums are used as stabilizers, gelling agents, and thickeners in many different food products, including PBAs. The most commonly used hydrocolloids are gelatin, agar, pectin and carrageenan (Dickinson and Pawlowsky 1997). Soy protein-based PBAs usually contain pectin and carrageenan to improve the functional aspects of the food item (Sun et al. 2022). Gums are also able to improve the water-holding capacity of zein and soybean-based PBAs to chicken (Mattice and Marangoni 2020).

Lipids are added to enhance the tenderness and flavours of the PBAs, as well as acting as plasticizers throughout the manufacturing procedure (Offiah et al. 2019; Younis et al. 2023). Fibre is usually added to PBAs to add nutritional value, but their water-holding capacity and textural properties are important during the PBA manufacturing process (Almeida et al. 2014). Flavours and colourants are very important to the consumer acceptance and purchase of PBAs. Seasoning and spices are added to PBAs to increase their consumer acceptance. Also, reducing sugars, amino acid, and thiamine that are present in the manufacturing process impact the flavours of the PBAs (Mottram et al. 1995). Enzyme-hydrolyzed soy protein can be used to produce chicken-like aromas (Wu et al. 2000; Wu and Cadwallader 2002).

Colour is another important attribute that impacts consumers' willingness to purchase (Ardeshiri and Rose 2018). Caramel colours and malt extracts, as well as reducing sugars which react with proteins during cooking or processing to produce a brown colour can be used in PBAs to mimic conventional chicken products (Kyriakopoulou et al. 2019). Titanium dioxide is used in PBAs to chicken to brighten (or whiten) the colour if they are considered to be too brown after production (Liao et al. 2020).

**Table 4.1** Examples of different ingredients that have been used to create chicken analogues

| Ingredients | Type of plant-based alternative (or conventional product used for comparison) | Reference |
|---|---|---|
| Soy protein isolate and wheat gluten | Chicken breast | Krintiras et al. (2016) |
| Wheat gluten and pea protein isolate | Chicken breast | Schreuders et al. (2019) |
| Zein and soy protein isolate | Chicken strips | Mattice and Marangoni (2020) |
| Zein and pea protein | Chicken nuggets | Salgado et al. (2023) |
| Wheat gluten and soy protein concentrate | Chicken breast | Chiang et al. (2019) |
| Soy protein isolate and soy flour | Chicken breast | Mateen et al. (2023) |
| Wheat gluten and soy protein isolate | Chicken breast | Chiang et al. (2021) |
| Pea protein and wheat proteins | Chicken nugget | Zhu et al. (2021) |
| Soy protein isolate, corn starch and wheat gluten | Chicken nugget | Jung et al. (2022) |
| Soy protein isolate, wheat gluten and chickpea flour | Chicken sausages | Kamani et al. (2019) |

To manufacture products with appropriate sensory attributes similar to that of chicken, a variety of different plant proteins have been used including legumes, cereal grains, oilseeds and soy proteins (including isolate, textured, concentrate and flour) (Malav et al. 2015). These ingredients have been found to have appropriate functional properties including emulsification, oil absorption and water absorption capacities. They have also been found to possess some of the nutritional benefits of chicken including amino acids, essential fatty acids and vitamins (Asgar et al. 2010; Malav et al. 2015). A study evaluating commercially available PBAs to chicken found that the most common base ingredients were soy protein, cereal grains (mainly wheat) and pulses (Ettinger et al. 2022). These ingredients will be further discussed. Table 4.1 presents examples of the different base ingredients that have been used to produce PBAs to chicken products.

### *4.2.1 Soy*

Soybeans were traditionally cultivated in Asia and are a source of high-quality protein (Qin et al. 2022). Currently, USA is the largest producer of soy (45%), followed by Brazil (20%) and China (12%) (Cheng and Rosentrater 2017). Soy proteins are the most prevalent ingredient in PBAs due to their gelation properties, nutritional components and low costs (Zhang et al. 2021). Tofu, soybean-based, can be considered the original PBA and it has a long history of consumption as well as being consumed today (Kyriakopoulou et al. 2019). Soy proteins can create a fibrous anisotropic structure, which is important to the textural attributes of PBAs (Elzerman et al. 2011). Heat processing treatments, including autoclaving, irradiation and cooking, are regularly used processing methods on soybeans (Sá et al. 2020).

Soy protein is an important component of PBAs due to its' gelation properties, but also because it possesses nine essential amino acids, however it is lacking in sulphur amino acids (Urade 2011). Commonly, textured vegetable protein (TVP) is manufactured using soy flour and soybean oil. In this process, the protein is spun to create products that aim to mimic meat (Elzerman et al. 2013). The TVP process not only makes it possible to produce PBAs that have similar sensory properties to meat, but it also can create products with a similar nutritional composition (Elzerman et al. 2013; Kyriakopoulou et al. 2021a, b). Originally used as extenders in meat products, TVP is now commonly mixed with other plant ingredients to manufacture PBAs. TVP is commonly combined with wheat gluten and the ratio of these two ingredients can directly impact the textural properties of the PBA (Cornet et al. 2020). For instance, soy protein isolate and wheat gluten have been mixed to create PBAs that possessed the textural properties of chicken breast (Krintiras et al. 2016). As stated by Zhang et al. (2021), TVP can be used in the production of PBAs but it is encouraged to include other plant-based proteins in the formulation.

### 4.2.2 *Pulses*

Pulses can also be used in PBAs as they are a significant source of protein, fibre, and minerals (Singh 2017). Pulses refers to the dried seed of the legume (Capurso et al. 2018). The pulses family includes navy beans, red kidney beans, soybeans, as well as dried peas, lentils, and chickpeas (Capurso et al. 2018). Pulses have gained popularity among manufacturers of PBAs as they add many benefits and functional properties. In addition to their nutritional benefits (fibre, protein, and various micronutrients) pulses can create PBAs with a texture and mouthfeel that replicates that of animal proteins.

Some PBAs have a fibrous texture and have been described as being elastic, rubbery and hard, which is unacceptable to consumers (Kumar et al. 2017). Although pulses are sustainable and may improve the textural properties of PBAs, they also possess off-flavours that negatively impact consumer acceptability (Singh 2017). The Maillard reaction during the heating of the amino acids and sugars present in the pulses causes this off-flavours to be present in the food item. This reaction causes nuttiness, bitterness and astringency to be present in the PBAs and they can directly impact consumer liking (Barker and McSweeney 2022). Pulses have shown the potential to be used to create PBAs to chicken. For instance, pea protein isolate and wheat gluten were to manufacture PBAs to chicken breast and were found to have a similar strength as chicken breast (Schreuders et al. 2019).

### 4.2.3 Cereal Grains

Wheat gluten and extrusion is used to create PBAs because of its' viscoelastic properties which allow for a gel to form (Singh and MacRitchie 2001). Specifically, it can be used as a texturing agent in PBAs, as well as a thickener (Zhang et al. 2021). However, due to its deficiency in lysine, it is usually paired with soy protein. Wheat gluten has been found to improve the sensory qualities of PBAs, and improve its' functionality (Fiorentini et al. 2020). Also, it can be mixed with other plant proteins to create PBAs with optimized moisture contents, without directly impacting the sensory properties (Patil et al. 2016).

Starch is an important ingredient in the production of PBAs as well and it helps form the fibrous structure (Zhang et al. 2016). Consumers want PBAs to mimic those of conventional chicken as such this fibrous structure is important for the success of the product. The most commonly used starches are wheat starch and corn starch, as they can improve the textural aspects without drastically impacting the flavour of the product (Asgar et al. 2010) due to their high water absorption capacity. For instance, Lin et al. (2002) found that wheat starch mixed with soy protein isolate was able to create a fibrous structure after extrusion. During extrusion, the starch can become embedded in the protein phase and increase the stability of the hydrophobic interactions, also the starch can lead to proper colour and taste development from the Maillard reaction (Zhang et al. 2019; Zhu et al. 2010).

Furthermore, zein fibres have been created that can be incorporated into soy protein isolate-based PBAs that can mimic the texture of chicken (Mattice and Marangoni 2020). Also, zein and pea protein formulations have been found to have similar textural properties to commercially produced chicken nuggets (Salgado et al. 2023). Millet is another cereal grain that has been mixed with soy protein isolate to create PBAs that can mimic chicken breast (Mateen and Singh 2023).

### 4.2.4 Mycoprotein

Mycoproteins are considered to be single-cell proteins and they are usually produced from filamentous fungi (Nasseri et al. 2011). The first commercial PBAs made with mycelia or mycoprotein were produced using edible filamentous fungi, *Fusarium graminearu,* and they are usually referred to by their commercial name, Quorn™ (Trinci 1994). Mycoprotein usually contains approximately 13 grams of fats, 10 grams of carbohydrates, 45 grams of protein and 25 grams of fibres in 100 g of dry matter (Finnigan et al. 2017). Mycoproteins have a fibrous texture and they also possess a meaty flavour due to the presence of glutamic acid, as well as sulphur-containing amino acids. Due to their alignment of protein filaments that have a flaky and fibrous texture, as well as possessing few off-flavours. Also, colours and flavourings can be readily added to mycoprotein to help it imitate the flavour of a conventional meat product.

PBAs made with mycoprotein have been widely used in Europe as an alternative to chicken (Kumar et al. 2017). Mycoproteins have been found to have a more favourable flavour property than TVP (Smetana et al. 2021) and they can be mixed with other seasonings and flavourings to improve their flavour profile. Mycoprotein has been used successfully to create a nugget that mimics the sensory properties of chicken nuggets (Hashempour-Baltork et al. 2023) and received higher acceptance than other commercially available nuggets. However, Hashempour-Baltork et al. (2023) identified that mycoproteins have not been used often to create PBAs due to a lack of proper production technologies and their market may be limited due to possible allergic effects. More studies are needed to investigate the use of mycoproteins to create PBAs to chicken.

### 4.2.5 *Other Base Ingredients*

While soy and wheat are the main ingredients in a significant number of PBAs, nuts and seeds can be used to create new PBAs. Certain nuts (ex: cashews, almonds, and walnuts) are good sources of protein, fat and can be manufactured to possess sensory properties similar to meat products (Alae-Carew et al. 2020). These ingredients can also create PBAs that are nutrient-dense, have a long shelf life and are considered to be sustainable options (Alae-Carew et al. 2020; Banovic et al. 2018). Many more ingredients need to be explored as currently, PBAs do not possess the same textural, flavour and appearance of conventional chicken products.

## 4.3 Variety of Products

Chicken products on the market usually include whole chickens and their corresponding cut pieces (e.g. breast, wings), but also strips, tenders, chunks, patties, burgers, blocks, ground products, nuggets, steaks and sausages (Kyriakopoulou et al. 2021a, b). Each products has different corresponding textures and flavours, which need to be found in the PBAs as well. The most popular type of PBAs are emulsion-type products, and they are a mixture of water, fats, proteins, carbohydrates, salt, and spices. These products aim to have sensory properties like conventional meat products (Dekkers et al. 2018).

When designing a PBA to chicken it is important to envision the final product, its' functionality, and its use. PBAs can also be separated into categories based on which meat products that are mimicking and that can include ground, comminuted and whole muscle meat products. Although most studies have compared PBAs to chicken breast (outlined in Tables 4.1, 4.2 and 4.3), most commercial products are comminuted products that include nuggets, tenders (or strips) pieces (sometimes referred to as popcorn chicken or burgers that mimic conventional chicken products (*Impossible™ Chicken Made From Plants* n.d.; *Plant-Based Meat Products* |

**Table 4.2** Examples of processing methods that have created plant-based alternatives to chicken

| Processing Method | Type of Product (or conventional product compared to) | Reference |
|---|---|---|
| Low moisture extrusion | Chicken breast | Kaleda et al. (2021) |
| High moisture extrusion | Chicken breast | Mateen and Singh (2023) |
| High moisture extrusion | Chicken breast | Mateen et al. (2023) |
| High moisture extrusion | Chicken breast | Samard and Ryu (2019) |
| Freeze structuring | Chicken nugget | Yuliarti et al. (2021) |
| Freeze structuring | Chicken breast and tenderloin | Chantanuson et al. (2022) |
| Electrospinning | Chicken strips | Mattice and Marangoni (2020) |
| 3D printing | Fried chicken breast | Chen et al. (2021) |
| 3D printing | Ground chicken | Wang et al. (2023) |
| 3D printing | Chicken surimi | Chao et al. (2022) |

**Table 4.3** Examples of sensory evaluation studies involving plant-based alternatives to chicken

| Type of product | Sensory evaluation method | Reference |
|---|---|---|
| Nuggets | Hedonic scales, check-all-that-apply (n = 105 untrained participants) | Ettinger et al. (2022) |
| Nuggets and strips | Hedonic scales and visual analog scales (n = 71 untrained participants) | Godschalk-Broers et al. (2022) |
| PBA to chicken breast | Intensity scales (n = 46 untrained participants) | Chiang et al. (2019) |
| Nuggets | Ranking scale (n = 42 untrained participants) | Yuliarti et al. (2021) |
| Sausage | Hedonic scales (n = 8 trained panelists) | Kamani et al. (2019) |
| Minced | Hedonic scales (n = 55 untrained participants) | Chiang et al. (2020) |
| Strips | Descriptive analysis (n = 14 trained panelists) | Katayama and Wilson (2008) |
| Nuggets | Hedonic scales (n = 110 untrained participants) | Sharima-Abdullah (2018) |
| Nuggets | Hedonic scales (n = 11 trained panelists) | Hashempour-Baltork et al. (2023) |
| Fillet pieces | Hedonic scales, appropriateness, and intention-to-use scales (n = 93 untrained participants) | Elzerman et al. (2011) |

*Beyond Meat* n.d.). Furthermore, fast food companies are also selling these nuggets or burgers (Jones 2022). These products usually have an intermediate moisture content and fibrous structure that mimics that of cut, shredded, ground and minced meat. Further studies need to be completed to improve comminuted PBAs. Furthermore, studies are needed to investigate new PBAs that are manufactured to mimic whole-muscle meat products.

### 4.3.1 Nutritional Composition

PBAs aim to replicate the nutritional composition of animal-based foods (Bakhsh et al. 2021). Also, consumers believe that PBAs are healthier than their conventional animal-based products (Ngapo 2022), although these products are sometimes classified as ultra-processed foods (Bohrer 2019). PBAs have been found to possess the nutritional composition of the conventional meat products (Bohrer 2019), however, there are some compositional differences.

Nuggets created with TVP, mushrooms and wheat gluten were compared to conventional chicken nuggets (Kumar et al. 2011). The conventional chicken nuggets' protein and lipid content was higher than the PBA. However, the PBA's sodium content and cholesterol were significantly lower than the conventional chicken nugget. Furthermore, the PBA contained higher amounts of potassium, zinc, copper, and iron. Also, Bohrer (2019) compared the nutritional composition of commercially available products (one a PBA and the other a conventional chicken nugget). The PBA had a lower calorie content, as well as lower protein, fat, cholesterol content and a higher total carbohydrates and dietary fibre content. Also, the PBA had a similar sodium content to the sodium and iron content to the conventional chicken nugget. Nuggets made of chickpeas TVP were found to have higher amounts of ash, protein and carbohydrates than a commercially available chicken nugget and lower amounts of fat (Sharima-Abdullah 2018).

Overall, the nutritional composition of PBAs to chickens varies but they usually have a more fibre and less saturated fat than commercially available chicken products (Bryngelsson et al. 2022). Salt content is usually lower as well, but very few PBAs would be able to be labelled with a health-related claim of low in salt (Bryngelsson et al. 2022). Researchers evaluated a total of 132 PBAs (that aimed to mimic meat products) and 658 meat products on the Australian market that were part of the Health Star Rating (energy, protein, saturated fat, sodium, total sugars and fibre content) and concluded that generally, PBAs had a higher health star rating than their conventional meat counterparts (Melville et al. 2023). However, the PBAs were also classified as ultra-processed and did not contain key micronutrients found in meat products. For instance, the bioavailability of iron in PBAs needs to be further investigated, especially as PBAs may contain antinutrients like phytic acid (Schlemmer et al. 2009). Also, processed food items or food that consumers perceive to be processed are usually disliked by consumers (Cardello et al. 2007) and this could impact their acceptance on the market.

Most importantly for PBAs, they need to contain adequate amounts of protein. Consumers are interested in their foods' protein content (Estell et al. 2021) and consumers want PBAs to contain a similar amount of protein as meat products as this allows for easy substitution in meals (as well as functionality). Specifically, PBAs need to consider protein quality as many PBAs have similar protein content to conventional meat products but are not of the same quality (animal proteins containing all essential amino acids are high quality) (Adhikari et al. 2022). Also, animal proteins usually have a higher digestibility than plant proteins, but digestibility can be affected by the processing method used to produce the PBA (Pasqualone et al. 2020).

## 4.4 Processing Methods

When developing PBAs, the processing methods need to be especially concerned about the textural properties of the finished product. There are a variety of different processing methods that can be used to produce PBAs including extrusion, freeze structuring, electrospinning and 3D printing. Extrusion is the most common manufacturing process used to create PBAs to chicken, but other methods including 3D food printing (an emerging method) will be discussed. Table 4.2 lists examples of studies that have investigated different processing methods to create PBAs to chicken.

### *4.4.1 Extrusion*

The PBAs are produced using different processing methods, but currently, they are mainly produced via extrusion. After extrusion, the resulting items can be further processed either by cooking, freezing, curing, or marinating. These processing steps create products that have not only the structure of the chicken product but also the appearance (colour), aroma and taste. Extrusion technology is extremely effective at creating meat-like textures in PBAs using plant proteins (Zhang et al. 2022) and also allows PBAs to have a moisture content that resembles that of meat products (Roland et al. 2017). Extrusion uses pressure, temperature and mechanical shear to process the plant-based protein to create an ingredient or product that possesses the desired physical properties (Elzerman et al. 2013). For instance, extrusion has been used to create PBAs from pulses that have an acceptable nutritional component as well as the sensory properties (more specifically the textural properties), allowing them to be used in the production of PBAs. Additionally, extrusion may prevent thermal degradation and oxidation from occurring, which leads to off flavours in the PBAs. Extrusion processing methods are usually separated into two categories, low and high moisture extrusion based on the feed's moisture content.

Low moisture extrusion produces items that expand and absorb water quickly. These products are used as additives to processed meat products as they improve the water retention capacity and prevent shrinkage of the food item during cooking (Gnanasambandam and Zayas 1994). It can be used to produce PBAs that have structural and textural similarities to meat, however, there is appearance is quite different from that of meat. Furthermore, low moisture extruded products are usually rehydrated (water or other liquids) when being used to produce PBAs (Zhang et al. 2021). Researchers used oat and pea protein blends with low moisture extrusion to create PBAs (Kaleda et al. 2021). They found that some of the formulations could create PBAs that possessed chicken-like fibres and neutral flavour, but they identified that the moisture content of the PBAs, as well as the water holding capacity were sub-optimal. Overall, low moisture extrusion seems better suited to produce ingredients that can be used in processed meat products.

High moisture extrusion (wet extrusion) is the most popular method used to create PBAs (Wang et al. 2022). For successful high moisture extrusion, a twin screw intermeshed and corotating extruder is necessary. The main component of the extruder is the screw in the barrel and the cooling die (Zhang et al. 2022). There are usually five steps involved in extrusion, which include (1) raw material feeding, (2) mixing with water, (3) melting, (4) die shaping and (5) a cooling step for texturization. All steps impact the final product. This manufacturing process occurs at high moisture (above 50%) and high temperature (usually between 140 °C and 180 °C). This process causes the plant proteins to unfold, unalign, and cross-link. Resulting in the formation of different fibre and layer structures, which although not the same as the muscle fibres of animal proteins, can be perceived by consumers as similar (Sha and Xiong 2020). Also, the cooling die design can impact the properties of the PBAs as a short and thick cooling zone creates a rough surface with a cross-section fibrous structure that can mimic that of chicken (Sun et al. 2022). However, many PBAs produced by high moisture extrusion are perceived as being dry and not very juicy, and these properties are important to consumers (Birke Rune et al. 2022; Ettinger et al. 2022). Furthermore, off-flavours can also be created by high moisture extrusion (Kaleda et al. 2021). Many studies have investigated the high moisture extrusion method to create PBAs to chicken products.

Mateen et al. (2023) investigated the effect of different manufacturing parameters during high moisture extrusion on the textural properties of soy-based PBAs. The resulting products were compared to meat cuts of chicken, beef, and lamb. They identified that the hardness, chewiness and cutting strength of the meat were like that of PBAs, However the springiness, cohesiveness, and resilience with different from the cuts of meat. These results lead them to conclude that extrusion of soy may not be able to attain textural equivalence to meat. Similarly, Samard and Ryu (2019) extruded texturized vegetable protein and compared its' textural properties to that of chicken. They concluded that the cutting strength and chewiness of the PBAs were like chicken, but they differed in other textural properties. Another study compared the properties of extruded PBAs, made from soy protein concentrate and wheat gluten to chicken breast (Chiang et al. 2019). They found that the PBAs made with 20% and 30% wheat gluten were the most like chicken breast's structural properties, but the other sensory properties of the PBA and chicken breast were different. Soy protein isolate was also mixed with three different grain types (sorghum, pearl millet and finger millet) (Mateen and Singh 2023). This study found very promising results as the PBAs made with soy protein isolate and millet were found to have textural properties more like chicken breast than PBAs made with soy protein isolate, soy protein concentrate and wheat gluten. Furthermore, different grains allowed for different fibre patterns to be produced and colours, which could allow for a wide array of PBAs to be developed. Although high moisture extrusion has produced PBAs with similar characteristics to chicken products, many properties still need to be improved.

### 4.4.2 Freeze Structuring

The freeze structuring method creates a unique fibrous structure by freezing a protein emulsion (Dekkers et al. 2018). Once frozen, removal of the ice crystals results in a fibrous and porous structure which is like that of animal proteins. The processing methods changes the appearance of the protein as the freeze-structured sample and the original ingredient. This is due to a change in the proteins' level of organization (Singh and Sit 2022). After the freezing process, the protein has two levels of orientation which include parallel-oriented protein sheets and randomly arranged entangled fibres within the protein sheets. These two distinct levels of orientation differentiate freeze structuring from other protein texturizing methods (Consolacion and Jelen 1986). This structure is highly dependent on the protein source because different plant proteins have different gelation, hydration, texture formation and solubility functions. These all impact the final structure of the PBA (Meade et al. 2005).

As explained by Sha and Xiong (2020), there are five steps to protein texturization by freeze structuring:

1. The protein is frozen to produce ice crystals. These ice crystals are aligned perpendicular to the cooling surface.
2. This formation of ice crystals creates parallel zones, which entrap the protein molecules.
3. The moisture and protein content is not affected during the process and this generates an elongated fibrous structure.
4. Freeze drying removes the water and a mass of fibrous protein remains after freeze drying.
5. A solution of stabilizers, fats, flavours, and colours stabilizers, can be used to rehydrate it the mass of fibrous protein.

Freeze structuring has been used with a mixture of different plant proteins to create nuggets (Yuliarti et al. 2021). These PBAs were compared to chicken-based products. The mixture of plant proteins included pea protein and wheat protein. The researchers concluded that freeze structuring process created a fibrous and layered structure in the plant-based nuggets. Their results also reinforced that the protein source has a direct impact on the final textural properties and identified that wheat and pea protein in a 3:1 ratio had a similar textural characteristic to chicken-based nuggets. Freeze structuring was also used with different soy protein concentrations and compared to chicken breast and tenderloin (Chantanuson et al. 2022). They found that final textural properties were highly reliant on the type of soy protein (soy flour and two different soy protein isolates) included in the freeze structuring process. Also, to manufacture a meat-like textures using the freezing process, then the soy protein's gelling ability needs to be lowered. Freeze structuring allows for a PBA with adequate hardness and elasticity to be created, however, optimal freezing conditions for different proteins and protein mixtures need to be determined (Sha and Xiong 2020). Based on these studies, freeze structuring is a promising technique for producing PBAs to chicken and can produce textural properties similar to that of chicken-based products.

### 4.4.3 Electrospinning

During the electrospinning process, a hollow needle with electric potential connected to a ground electrode is needed and a biopolymer is pushed through the hollow needle (Dekkers et al. 2018). This reaction causes very thin fibres to be attracted to the ground electrode (Schiffman and Schauer 2008). Electrospinning has been usually used by the food industry to create delivery systems for probiotics and polyphenols (Librán et al. 2017), but has also been used to create fibres that can be used in PBAs. Zein has been successfully spun into fibres (Ghorani and Tucker 2015), as well as being used as a carrier for soy protein (Nieuwland et al. 2014).

The study by Mattice and Marangoni (2020) evaluated different processing methods, including electrospinning, and investigated how they could process zein to create fibrous material. A 3 mL plastic syringe with a 16-gauge blunt-end stainless steel needle spinneret was connected to the positive electrode of a direct current power supply. Inside the syringe was placed a zein solution of 25% (w/w) polymer in 70:30 ethanol in water. A voltage of 20 kV was applied and the spinneret was 20 cm from the collector plate. The fibres were collected and soy-zein composite gels were created with the electrospun fibres. The electrospun fibres were compared to zein that had undergone mechanical elongation and antisolvent precipitation. The zein that had been mechanically elongated had a texture that was most alike chicken meat and the researchers concluded that more studies are required to improve the efficiency and throughput of electrospinning zein.

### 4.4.4 3D Food Printing

3D food printing (additive manufacturing), creates three-dimensional food items by printing the product one layer at a time (Fahmy et al. 2020). There are five basic steps for 3D food printing including designing the shape (food item) using software, pre-treatment of ingredients to ensure they are suitable for printing (usually semi-solid pastes), filling the cartridges, slicing processing and then extruding the ingredients from the nozzle on to the printing platform (Chuanxing et al. 2018). These pre-treatment steps for ingredient preparation can include mixing dried powders with water, heating water-containing foods to make a fine paste, as well as grinding products to create a puree (Aregawi et al. 2015; Fahmy et al. 2020). Ko et al. (2021) stated that the advantages of 3D printing for creation of PBAs are that it can create structural diversification of products, as well as create personalized shapes for different consumers. Furthermore 3D printing methods have been created that are able to manufacture food items with multi-materials (Kim et al. 2021). By using different materials microfibers and anisotropic structures similar to animal proteins can be created (Wen et al. 2023). Secondly, 3D printing can personalize the textural and nutritional composition of food items and in turn, hopefully reduce food waste (Wen et al. 2023). Lastly, 3D food printing only requires a small production space (Lipton et al. 2015).

3D printing was used to manufacture an analogue to commercially available fried chicken breast (Chen et al. 2021). The study used texturized soy protein and drawing soy protein mixed with hydrocolloids to create 3D-printed analogues and then fried them. The researchers identified that the texturized soy protein mixed with xanthan gum had preferable printing characteristics and it was able to maintain its integrity after frying. They also determined that an infill rate of 60% while using a triangle infilling pattern was able to create a PBA with similar textural properties (chewiness, gumminess, and hardness) to the commercially available fried chicken breast. 3D printing has also used pea protein isolate to create PBAs to ground chicken (chicken mince) (Wang et al. 2023). Furthermore, Chao et al. (2022) used mealworm isolate and 3D printing to create PBA to chicken surimi for older adults. 3D printing has been found to produce PBAs with similar textural properties to chicken products, however, more studies need to be conducted on the use of 3D printing to create PBAs to chicken.

#### 4.4.4.1 Other Processing Methods

Shear cell technology can also be used to manufacture fibrous products from plant proteins that can mimic the textural properties of animal proteins (Manski et al. 2007). Shear cells can apply an intensive shear within a cone-in-cone or couette geometry (Krintiras et al. 2016). Shear cell technology and soy protein concentrate have been used to create a PBA (Grabowska et al. 2016), as well as a mixture of soy protein isolate and wheat gluten (Grabowska et al. 2014). Shear cell technology also can process a mixture of pea protein isolate and wheat gluten to create PBAs that possessed a tensile strength similar to chicken filets (Schreuders et al. 2019).

Another processing method that has shown potential for the production of PBAs to chicken is ohmic cooking (Jung et al. 2022). Ohmic cooking consists of a very rapid and uniform increase of the temperature of food due to internal heat generation. This heat generation is caused by an alternating current (Varghese et al. 2014). It has many applications in the food industry including cooking processes as well as sterilization, pasteurization, and thawing (Varghese et al. 2014). A mixture of soy protein isolate and wheat gluten was processed using ohmic cooking and it was identified that it produced a product with a similar hardness to commercial chicken nuggets.

Lastly, most processing methods are currently used to create acceptable textural and colour properties of the PBA, but the flavour needs to be considered as well by manufacturers. Besides adding spices and seasonings to the PBA (before or after processing), fermentation could be used to incorporate flavours that mimic that of chicken products. Fermentation can reduce the off-flavours that are present in PBAs produced from soybeans (Kaczmarska et al. 2018). Also, the fermentation of different plant proteins may be able to help achieve different colours in PBAs (Agboyibor et al. 2018). Furthermore, fermentation can help with the plant proteins' digestibility (Sá et al. 2020), as well as produce ingredients like mycoprotein that can be used to create PBAs (Dekkers et al. 2018). Many different processing methods could be used to produce PBAs to chicken; however, they all still need to be investigated as there is still room for improvement.

## 4.5 Sensory Properties

PBAs have suffered from poor consumer acceptability and consumer perception. For instance, sausages created from textured soy protein and soy protein isolate were found to have unacceptable textural properties (Majzoobi et al. 2017). Table 4.3 lists examples of the different sensory evaluation studies that have been conducted on PBAs to chicken.

The sensory studies on PBAs to chicken, have mainly focused on comminuted products, that include nuggets, strips, sausage, and fillet pieces. Also, the studies have mainly included untrained participants (consumers) evaluating the PBAs. Some of the studies have used trained panelists and descriptive analysis, however, studies should not use trained panelists and hedonic scales to evaluate products (rather trained panelists and intensity scales). Sensory properties have a significant impact on consumer purchase and acceptance and therefore need to be evaluated. Ettinger et al. (2022) evaluated conventional chicken nuggets and five plant-based nuggets using hedonic scales (1 = Dislike Extremely and 9 = Like Extremely) and check-all-that-apply (CATA). The researchers found that the consumers could identify the conventional chicken nugget and identified that the PBAs were associated with aftertaste. Furthermore, the consumers also associated PBAs with fibrous, beany, and chewy. The consumers also found that the crunchiness, juiciness, saltiness, and crispiness increased their liking of the nuggets, however, they associated these attributes with the conventional chicken nuggets and not the PBAs. The results indicated that improvement of the texture, flavour and appearance of PBAs is necessary if PBAs wish to successfully mimic chicken nuggets.

Godschalk-Broers et al. (2022) also conducted asking consumers to evaluate PBAs, however before they consumed the product, the participants were asked to evaluate the attractiveness of the uncooked samples. This is an important evaluation as the appearance and the colour of the PBA can impact consumer purchase (Mancini and Hunt 2005). The PBAs were separated into precooked and raw appearance categories, and the consumers found the raw appearance more appealing. Once, the consumers ate the PBAs and evaluated it compared to conventional chicken, the PBAs generally scored higher for fattiness and colour, while they scored lower for meaty flavour as well as the textural properties including hardness, cohesiveness, and fibrousness. Also, the juiciness, chewiness and flavour intensity fluctuated higher and lower than the conventional chicken based on which PBA was being evaluated. However, textural attributes of PBAs that resembled chicken did not lead to similar liking scores, but the researchers that juiciness and meaty flavour related directly to consumer liking.

Other than commercially available products, researchers have compared extruded soy protein isolate and wheat gluten PBAs to cooked chicken breast (Chiang et al. 2019). The different formulations were compared using untrained participants and intensity scales. The researchers identified that increasing wheat gluten and decreasing soy protein isolate increased the fibrous structure scores but increased the chewiness and hardness scores. The grilled chicken had the lowest hardness scores in the study. Yuliarti et al. (2021) also investigated the sensory attributes of PBAs and compared them to chicken. These nuggets were produced using pea protein isolate and freeze structuring and then compared the different formulations using a ranking test. The results found there were no significant differences in the taste of the different samples, but did identify textural differences, however, a commercial chicken nugget was not included in the sensory trial. Therefore, they concluded the sensory properties are directly related to the base ingredient and processing method used to manufacture the PBA to chicken.

Tenderness and juiciness are the main determinants of consumer liking of PBAs (Ettinger et al. 2022; Godschalk-Broers et al. 2022; Hoek et al. 2004). Juiciness is one of the main sensory properties that can be used to improve PBAs that are supposed to resemble chicken. In the study by Godschalk-Broers et al. (2022), juiciness was not found to be related to cooking loss, moisture content, or expressible moisture in a study on PBAs. Therefore, new methods to evaluate juiciness of PBAs need to be developed. Also, consumers' opinion of juiciness in PBAs and meat products needs to be better defined. Many studies have evaluated PBAs that hoped to mimic beef burgers, but not as many studies have been conducted on PBAs to chicken. More studies need to investigate PBAs to chicken as consumers are continuing to purchase chicken and the demand for chicken is growing, especially in the Western World.

Most studies conducted on PBAs to chicken have been conducted with untrained participants and hedonic scales. More studies need to use descriptive analysis and trained panelists to evaluate the different sensory properties of PBAs. This will allow for differences in products based on different processing methods, as well as based ingredients to be identified. Also, more studies are needed on products that mimic whole meat products. Studies that do use untrained participants and hedonic scales should consider using attribute diagnostic questions (e.g., just-about-right) or CATA to provide more information about the PBAs and identify areas of improvement for product developers. Furthermore, sensory studies may want to include extrinsic cues like health claims and sustainability claims to develop the acceptability of PBAs as consumers are concerned about the environment (de Oliveira Padilha et al. 2022) and nutritional benefits of PBAs (Michel et al. 2021). Lastly, textural and flavour properties are important to consumer acceptance of PBAs and studies should consider how these characteristics change during product breakdown or mastication using dynamic sensory methodologies like temporal check-all-that-apply (Ares et al. 2015) or temporal dominance of sensations (Meyners and Pineau 2010). These dynamic sensory methodologies can identify how PBAs' sensory properties change throughout the mastication process (Barker and McSweeney 2022).

## 4.6 Conclusion

New processing methods and improvements in processing methods need to continue to be investigated to optimize the sensory properties of PBAs to chicken, as they are currently not well-liked by consumers. PBAs need to better mimic their conventional chicken counterparts. The processing method cannot negatively impact the nutritional benefits of PBAs as consumers consider them to be healthy. New ingredients should be explored as TVP may lead to beneficial textural properties but is associated with off-flavours. Mycoproteins as well as zein may be ingredients that can overcome some of these off-flavours if they can be produced consistently, efficiently, and cost-effectively. The future is bright for PBAs to chicken as consumer demand for plant-based options and chicken continues to expand, but improvements need to be made to increase consumer purchase and acceptance. But for acceptable PBAs to chicken to be created comprehensive studies evaluating the nutritional composition, sensory properties and structural components need to be conducted.

## References

Adhikari S, Schop M, de Boer IJM, Huppertz T (2022) Protein quality in perspective: a review of protein quality metrics and their applications. Nutrients 14(5):5. https://doi.org/10.3390/nu14050947

Agboyibor C, Kong W-B, Chen D, Zhang A-M, Niu S-Q (2018) Monascus pigments production, composition, bioactivity and its application: a review. Biocatal Agric Biotechnol 16:433–447. https://doi.org/10.1016/j.bcab.2018.09.012

Alae-Carew C, Nicoleau S, Bird FA, Hawkins P, Tuomisto HL, Haines A, Dangour AD, Scheelbeek PFD (2020) The impact of environmental changes on the yield and nutritional quality of fruits, nuts and seeds: a systematic review. Environ Res Lett 15(2):023002. https://doi.org/10.1088/1748-9326/ab5cc0

Almeida CM, Wagner R, Mascarin LG, Zepka LQ, Campagnol PCB (2014) Production of low-fat emulsified cooked sausages using amorphous cellulose gel. J Food Qual 37(6):437–443. https://doi.org/10.1111/jfq.12104

Ardeshiri A, Rose JM (2018) How Australian consumers value intrinsic and extrinsic attributes of beef products. Food Qual Prefer 65:146–163. https://doi.org/10.1016/j.foodqual.2017.10.018

Aregawi WA, Verboven P, Vancauwenberghe V, Bongaers E, Van den Eijnden E, Vanbommel K, Diaz J, Nicolai B (2015, November) Structural-mechanical analysis of cookies produced by conventional and 3D printing techniques. Food science research and innovation: delivering sustainable solutions to the global economy and society. EFFoST International Conference, 2015/11/10 - 2015/11/12, Athens, Greece. https://lirias.kuleuven.be/1715948

Ares G, Jaeger SR, Antúnez L, Vidal L, Giménez A, Coste B, Picallo A, Castura JC (2015) Comparison of TCATA and TDS for dynamic sensory characterization of food products. Food Res Int 78:148–158. https://doi.org/10.1016/j.foodres.2015.10.023
Arora NK, Mishra I (2019) United Nations sustainable development goals 2030 and environmental sustainability: race against time. Environ Sustain 2(4):339–342. https://doi.org/10.1007/s42398-019-00092-y
Asgar MA, Fazilah A, Huda N, Bhat R, Karim AA (2010) Nonmeat protein alternatives as meat extenders and meat analogs. Compr Rev Food Sci Food Saf 9(5):513–529. https://doi.org/10.1111/j.1541-4337.2010.00124.x
Bakhsh A, Lee S-J, Lee E-Y, Hwang Y-H, Joo S-T (2021) Evaluation of rheological and sensory characteristics of plant-based meat analog with comparison to beef and pork. Food Sci Animal Res 41(6):983–996. https://doi.org/10.5851/kosfa.2021.e50
Banovic M, Arvola A, Pennanen K, Duta DE, Brückner-Gühmann M, Lähteenmäki L, Grunert KG (2018) Foods with increased protein content: a qualitative study on European consumer preferences and perceptions. Appetite 125:233–243. https://doi.org/10.1016/j.appet.2018.01.034
Barker S, McSweeney MB (2022) Sensory characterization of yellow pea and ground chicken hybrid meat burgers using static and dynamic methodologies. J Food Sci 87(12):5390–5401. https://doi.org/10.1111/1750-3841.16380
Birke Rune CJ, Song Q, Clausen MP, Giacalone D (2022) Consumer perception of plant-based burger recipes studied by projective mapping. Future Foods 6:100168. https://doi.org/10.1016/j.fufo.2022.100168
Bohrer BM (2019) An investigation of the formulation and nutritional composition of modern meat analogue products. Food Sci Human Wellness 8(4):320–329. https://doi.org/10.1016/j.fshw.2019.11.006
Bryngelsson S, Moshtaghian H, Bianchi M, Hallström E (2022) Nutritional assessment of plant-based meat analogues on the Swedish market. Int J Food Sci Nutr 73(7):889–901. https://doi.org/10.1080/09637486.2022.2078286
Capurso A, Crepaldi G, Capurso C (2018) Legumes and pulses. In: Benefits of the mediterranean diet in the elderly patient. Springer International Publishing, pp 285–324. https://doi.org/10.1007/978-3-319-78084-9_12
Cardello AV, Schutz HG, Lesher LL (2007) Consumer perceptions of foods processed by innovative and emerging technologies: a conjoint analytic study. Innovative Food Sci Emerg Technol 8(1):73–83. https://doi.org/10.1016/j.ifset.2006.07.002
Chantanuson R, Nagamine S, Kobayashi T, Nakagawa K (2022) Preparation of soy protein-based food gels and control of fibrous structure and rheological property by freezing. Food Struct 32:100258. https://doi.org/10.1016/j.foostr.2022.100258
Chao C, Hwang JS, Kim IW, Choi RY, Kim HW, Park HJ (2022) Coaxial 3D printing of chicken surimi incorporated with mealworm protein isolate as texture-modified food for the elderly. J Food Eng 333:111151. https://doi.org/10.1016/j.jfoodeng.2022.111151
Chen Y, Zhang M, Bhandari B (2021) 3D printing of steak-like foods based on textured soybean protein. Foods 10(9):9. https://doi.org/10.3390/foods10092011
Cheng M-H, Rosentrater KA (2017) Profitability analysis of soybean oil processes. Bioengineering 4(4):4. https://doi.org/10.3390/bioengineering4040083
Chiang JH, Loveday SM, Hardacre AK, Parker ME (2019) Effects of soy protein to wheat gluten ratio on the physicochemical properties of extruded meat analogues. Food Struct 19:100102. https://doi.org/10.1016/j.foostr.2018.11.002
Chiang JH, Hardacre AK, Parker ME (2020) Effects of Maillard-reacted beef bone hydrolysate on the physicochemical properties of extruded meat alternatives. J Food Sci 85(3):567–575. https://doi.org/10.1111/1750-3841.14960
Chiang JH, Tay W, Ong DSM, Liebl D, Ng CP, Henry CJ (2021) Physicochemical, textural and structural characteristics of wheat gluten-soy protein composited meat analogues prepared with the mechanical elongation method. Food Struct 28:100183. https://doi.org/10.1016/j.foostr.2021.100183

Chuanxing F, Qi W, Hui L, Quancheng Z, Wang M (2018) Effects of pea protein on the properties of potato starch-based 3D printing materials. Int J Food Eng 14(3). https://doi.org/10.1515/ijfe-2017-0297

Consolacion F, Jelen P (1986) Freeze texturization of proteins effect of the alkali acid and freezing treatments on texture formation. Food Microstructure 5(1):33–40

Cornet SHV, van der Goot AJ, van der Sman RGM (2020) Effect of mechanical interaction on the hydration of mixed soy protein and gluten gels. Curr Res Food Sci 3:134–145. https://doi.org/10.1016/j.crfs.2020.03.007

de Oliveira Padilha LG, Malek L, Umberger WJ (2022) Consumers' attitudes towards lab-grown meat, conventionally raised meat and plant-based protein alternatives. Food Qual Prefer 99:104573. https://doi.org/10.1016/j.foodqual.2022.104573

Dekkers BL, Boom RM, van der Goot AJ (2018) Structuring processes for meat analogues. Trends Food Sci Technol 81:25–36. https://doi.org/10.1016/j.tifs.2018.08.011

Derbyshire EJ (2017) Flexitarian diets and health: a review of the evidence-based literature. Front Nutr 3:55. https://doi.org/10.3389/fnut.2016.00055

Dickinson E, Pawlowsky K (1997) Effect of ι-Carrageenan on flocculation, creaming, and rheology of a protein-stabilized emulsion. J Agric Food Chem 45(10):3799–3806. https://doi.org/10.1021/jf970304d

Elzerman JE, Hoek AC, van Boekel MAJS, Luning PA (2011) Consumer acceptance and appropriateness of meat substitutes in a meal context. Food Qual Prefer 22(3):233–240. https://doi.org/10.1016/j.foodqual.2010.10.006

Elzerman JE, van Boekel MAJS, Luning PA (2013) Exploring meat substitutes: consumer experiences and contextual factors. Br Food J 115(5):700–710. https://doi.org/10.1108/00070701311331490

Estell M, Hughes J, Grafenauer S (2021) Plant protein and plant-based meat alternatives: consumer and nutrition professional attitudes and perceptions. Sustainability 13(3):3. https://doi.org/10.3390/su13031478

Ettinger L, Falkeisen A, Knowles S, Gorman M, Barker S, Moss R, McSweeney MB (2022) Consumer perception and acceptability of plant-based alternatives to chicken. Foods 11(15):15. https://doi.org/10.3390/foods11152271

Fahmy AR, Becker T, Jekle M (2020) 3D printing and additive manufacturing of cereal-based materials: quality analysis of starch-based systems using a camera-based morphological approach. Innovative Food Sci Emerg Technol 63:102384. https://doi.org/10.1016/j.ifset.2020.102384

Fehér A, Gazdecki M, Véha M, Szakály M, Szakály Z (2020) A comprehensive review of the benefits of and the barriers to the switch to a plant-based diet. Sustainability 12(10):10. https://doi.org/10.3390/su12104136

Finnigan T, Needham L, Abbott C (2017) Chapter 19 - Mycoprotein: a healthy new protein with a low environmental impact. In: Nadathur SR, Wanasundara JPD, Scanlin L (eds) Sustainable protein sources. Academic, pp 305–325. https://doi.org/10.1016/B978-0-12-802778-3.00019-6

Fiorentini M, Kinchla AJ, Nolden AA (2020) Role of sensory evaluation in consumer acceptance of plant-based meat analogs and meat extenders: a scoping review. Foods 9(9):9. https://doi.org/10.3390/foods9091334

Gardner CD, Coulston A, Chatterjee L, Rigby A, Spiller G, Farquhar JW (2005) The effect of a plant-based diet on plasma lipids in hypercholesterolemic adults. Ann Intern Med 142(9):725–733. https://doi.org/10.7326/0003-4819-142-9-200505030-00007

Ghorani B, Tucker N (2015) Fundamentals of electrospinning as a novel delivery vehicle for bioactive compounds in food nanotechnology. Food Hydrocoll 51:227–240. https://doi.org/10.1016/j.foodhyd.2015.05.024

Gnanasambandam R, Zayas JF (1994) Chemical and bacteriological stability of Frankfurters extended with wheat germ, corn germ and soy proteins. J Food Process Preserv 18(2):159–171. https://doi.org/10.1111/j.1745-4549.1994.tb00250.x

Godschalk-Broers L, Sala G, Scholten E (2022) Meat analogues: relating structure to texture and sensory perception. Food Secur 11(15):2227. https://doi.org/10.3390/foods11152227

Grabowska KJ, Tekidou S, Boom RM, van der Goot AJ (2014) Shear structuring as a new method to make anisotropic structures from soy–gluten blends. Food Res Int 64:743–751. https://doi.org/10.1016/j.foodres.2014.08.010

Grabowska KJ, Zhu S, Dekkers BL, de Ruijter NCA, Gieteling J, van der Goot AJ (2016) Shear-induced structuring as a tool to make anisotropic materials using soy protein concentrate. J Food Eng 188:77–86. https://doi.org/10.1016/j.jfoodeng.2016.05.010

Green A, Blattmann C, Chen C, Mathys A (2022) The role of alternative proteins and future foods in sustainable and contextually-adapted flexitarian diets. Trends Food Sci Technol 124:250–258. https://doi.org/10.1016/j.tifs.2022.03.026

Hashempour-Baltork F, Jannat B, Dadgarnejad M, Mirza Alizadeh A, Khosravi-Darani K, Hosseini H (2023) Mycoprotein as chicken meat substitute in nugget formulation: physicochemical and sensorial characterization. Food Sci Nutr 11(7):4289–4295. https://doi.org/10.1002/fsn3.3354

Himics M, Giannakis E, Kushta J, Hristov J, Sahoo A, Perez-Dominguez I (2022) Co-benefits of a flexitarian diet for air quality and human health in Europe. Ecol Econ 191:107232. https://doi.org/10.1016/j.ecolecon.2021.107232

Hoek AC, Luning PA, Stafleu A, de Graaf C (2004) Food-related lifestyle and health attitudes of Dutch vegetarians, non-vegetarian consumers of meat substitutes, and meat consumers. Appetite 42(3):265–272. https://doi.org/10.1016/j.appet.2003.12.003

Hopwood CJ, Bleidorn W, Schwaba T, Chen S (2020) Health, environmental, and animal rights motives for vegetarian eating. PLoS One 15(4):e0230609. https://doi.org/10.1371/journal.pone.0230609

Impossible™ Chicken Made From Plants (n.d.) Retrieved 17 Oct 2023, from https://impossiblefoods.com/ca/products/chicken

Izar-Tenorio J, Jaramillo P, Griffin WM, Small M (2020) Impacts of projected climate change scenarios on heating and cooling demand for industrial broiler chicken farming in the Eastern U.S. J Clean Prod 255:120306. https://doi.org/10.1016/j.jclepro.2020.120306

Jones P (2022) Plant-based food in the hospitality industry: an exploratory case study of leading fast food outlets. Athens J Tourism 9. https://www.athensjournals.gr/ajt/v9i2

Jung AH, Hwang JH, Jun S, Park SH (2022) Application of ohmic cooking to produce a soy protein-based meat analogue. LWT 160:113271. https://doi.org/10.1016/j.lwt.2022.113271

Kaczmarska KT, Chandra-Hioe MV, Frank D, Arcot J (2018) Aroma characteristics of lupin and soybean after germination and effect of fermentation on lupin aroma. LWT 87:225–233. https://doi.org/10.1016/j.lwt.2017.08.080

Kahleova H, Levin S, Barnard N (2017) Cardio-metabolic benefits of plant-based diets. Nutrients 9(8):8. https://doi.org/10.3390/nu9080848

Kaleda A, Talvistu K, Vaikma H, Tammik M-L, Rosenvald S, Vilu R (2021) Physicochemical, textural, and sensorial properties of fibrous meat analogs from oat-pea protein blends extruded at different moistures, temperatures, and screw speeds. Future Foods 4:100092. https://doi.org/10.1016/j.fufo.2021.100092

Kamani MH, Meera MS, Bhaskar N, Modi VK (2019) Partial and total replacement of meat by plant-based proteins in chicken sausage: evaluation of mechanical, physico-chemical and sensory characteristics. J Food Sci Technol 56(5):2660–2669. https://doi.org/10.1007/s13197-019-03754-1

Katayama M, Wilson LA (2008) Utilization of soybeans and their components through the development of textured soy protein foods. J Food Sci 73(3):S158–S164. https://doi.org/10.1111/j.1750-3841.2008.00663.x

Kim SM, Kim HW, Park HJ (2021) Preparation and characterization of surimi-based imitation crab meat using coaxial extrusion three-dimensional food printing. Innovative Food Sci Emerg Technol 71:102711. https://doi.org/10.1016/j.ifset.2021.102711

Ko HJ, Wen Y, Choi JH, Park BR, Kim HW, Park HJ (2021) Meat analog production through artificial muscle fiber insertion using coaxial nozzle-assisted three-dimensional food printing. Food Hydrocoll 120:106898. https://doi.org/10.1016/j.foodhyd.2021.106898

Krintiras GA, Gadea Diaz J, van der Goot AJ, Stankiewicz AI, Stefanidis GD (2016) On the use of the Couette Cell technology for large scale production of textured soy-based meat replacers. J Food Eng 169:205–213. https://doi.org/10.1016/j.jfoodeng.2015.08.021

Kumar P, Sharma B, Kumar R (2011) Optimization of the level of mushroom in analogue meat nuggets. Journal Meat Sci 7:53–55

Kumar P, Chatli MK, Mehta N, Singh P, Malav OP, Verma AK (2017) Meat analogues: health promising sustainable meat substitutes. Crit Rev Food Sci Nutr 57(5):923–932. https://doi.org/10.1080/10408398.2014.939739

Kyriakopoulou K, Dekkers B, van der Goot AJ (2019) Chapter 6—Plant-based meat analogues. In: Galanakis CM (ed) Sustainable meat production and processing. Academic, pp 103–126. https://doi.org/10.1016/B978-0-12-814874-7.00006-7

Kyriakopoulou K, Keppler JK, van der Goot AJ (2021a) Functionality of ingredients and additives in plant-based meat analogues. Food Secur 10(3):600. https://doi.org/10.3390/foods10030600

Kyriakopoulou K, Keppler JK, van der Goot AJ, Boom RM (2021b) Alternatives to meat and dairy. Annu Rev Food Sci Technol 12(1):29–50. https://doi.org/10.1146/annurev-food-062520-101850

Lang M (2020) Consumer acceptance of blending plant-based ingredients into traditional meat-based foods: evidence from the meat-mushroom blend. Food Qual Prefer 79:103758. https://doi.org/10.1016/j.foodqual.2019.103758

Lea EJ, Crawford D, Worsley A (2006) Public views of the benefits and barriers to the consumption of a plant-based diet. Eur J Clin Nutr 60(7):828–837. https://doi.org/10.1038/sj.ejcn.1602387

Liao C, Li Y, Tjong SC (2020) Visible-light active titanium dioxide nanomaterials with bactericidal properties. Nanomaterials 10(1):1. https://doi.org/10.3390/nano10010124

Librán CM, Castro S, Lagaron JM (2017) Encapsulation by electrospray coating atomization of probiotic strains. Innovative Food Sci Emerg Technol 39:216–222. https://doi.org/10.1016/j.ifset.2016.12.013

Lin S, Huff HE, Hsieh F (2002) Extrusion process parameters, sensory characteristics, and structural properties of a high moisture soy protein meat analog. J Food Sci 67(3):1066–1072. https://doi.org/10.1111/j.1365-2621.2002.tb09454.x

Lipton JI, Cutler M, Nigl F, Cohen D, Lipson H (2015) Additive manufacturing for the food industry. Trends Food Sci Technol 43(1):114–123. https://doi.org/10.1016/j.tifs.2015.02.004

Majzoobi M, Talebanfar S, Eskandari MH, Farahnaky A (2017) Improving the quality of meat-free sausages using κ-carrageenan, konjac mannan and xanthan gum. Int J Food Sci Technol 52(5):1269–1275. https://doi.org/10.1111/ijfs.13394

Malav OP, Talukder S, Gokulakrishnan P, Chand S (2015) Meat analog: a review. Crit Rev Food Sci Nutr 55(9):1241–1245. https://doi.org/10.1080/10408398.2012.689381

Mancini RA, Hunt MC (2005) Current research in meat color. Meat Sci 71(1):100–121. https://doi.org/10.1016/j.meatsci.2005.03.003

Manski JM, van der Goot AJ, Boom RM (2007) Formation of fibrous materials from dense calcium caseinate dispersions. Biomacromolecules 8(4):1271–1279. https://doi.org/10.1021/bm061008p

Mateen A, Singh G (2023) Evaluating the potential of millets as blend components with soy protein isolate in a high moisture extrusion system for improved texture, structure, and colour properties of meat analogues. Food Res Int 173:113395. https://doi.org/10.1016/j.foodres.2023.113395

Mateen A, Mathpati M, Singh G (2023) A study on high moisture extrusion for making whole cut meat analogue: characterization of system, process and product parameters. Innovative Food Sci Emerg Technol 85:103315. https://doi.org/10.1016/j.ifset.2023.103315

Mattice KD, Marangoni AG (2020) Comparing methods to produce fibrous material from zein. Food Res Int 128:108804. https://doi.org/10.1016/j.foodres.2019.108804

McClements DJ (2020) Development of next-generation nutritionally fortified plant-based milk substitutes: structural design principles. Foods 9(4):4. https://doi.org/10.3390/foods9040421

Meade SJ, Reid EA, Gerrard JA (2005) The impact of processing on the nutritional quality of food proteins. J AOAC Int 88(3):904–922. https://doi.org/10.1093/jaoac/88.3.904

Melville H, Shahid M, Gaines A, McKenzie BL, Alessandrini R, Trieu K, Wu JHY, Rosewarne E, Coyle DH (2023) The nutritional profile of plant-based meat analogues available for sale in Australia. Nutr Diet 80(2):211–222. https://doi.org/10.1111/1747-0080.12793

Meybeck A, Gitz V (2017) Sustainable diets within sustainable food systems. Proc Nutr Soc 76(1):1–11. https://doi.org/10.1017/S0029665116000653

Meyners M, Pineau N (2010) Statistical inference for temporal dominance of sensations data using randomization tests. Food Qual Prefer 21(7):805–814. https://doi.org/10.1016/j.foodqual.2010.04.004

Michel F, Hartmann C, Siegrist M (2021) Consumers' associations, perceptions and acceptance of meat and plant-based meat alternatives. Food Qual Prefer 87:104063. https://doi.org/10.1016/j.foodqual.2020.104063

Modlinska K, Adamczyk D, Maison D, Pisula W (2020) Gender differences in attitudes to vegans/vegetarians and their food preferences, and their implications for promoting sustainable dietary patterns–a systematic review. Sustainability 12(16):16. https://doi.org/10.3390/su12166292

Mottram DS, Madruga MS, Whitfield FB (1995) Some novel meatlike aroma compounds from the reactions of alkanediones with hydrogen sulfide and furanthiols. J Agric Food Chem 43(1):189–193. https://doi.org/10.1021/jf00049a035

Nasseri AT, Rasoul-Amini S, Morowvat MH, Younes G (2011) Single cell protein: production and process. Am J Food Technol 6. https://doi.org/10.3923/ajft.2011.103.116

Ngapo TM (2022) Meat analogues, the Canadian Meat Industry and the Canadian consumer. Meat Sci 191:108846. https://doi.org/10.1016/j.meatsci.2022.108846

Nieuwland M, Geerdink P, Brier P, van den Eijnden P, Henket JTMM, Langelaan MLP, Stroeks N, van Deventer HC, Martin AH (2014) Reprint of "Food-grade electrospinning of proteins". Innovative Food Sci Emerg Technol 24:138–144. https://doi.org/10.1016/j.ifset.2014.07.006

Offiah V, Kontogiorgos V, Falade KO (2019) Extrusion processing of raw food materials and by-products: a review. Crit Rev Food Sci Nutr 59(18):2979–2998. https://doi.org/10.1080/10408398.2018.1480007

Pasqualone A, Costantini M, Coldea TE, Summo C (2020) Use of legumes in extrusion cooking: a review. Foods 9(7):7. https://doi.org/10.3390/foods9070958

Patil SS, Brennan MA, Mason SL, Brennan CS (2016) The effects of fortification of legumes and extrusion on the protein digestibility of wheat based snack. Foods 5(2):2. https://doi.org/10.3390/foods5020026

Pimentel D, Pimentel M (2003) Sustainability of meat-based and plant-based diets and the environment. Am J Clin Nutr 78(3):660S–663S. https://doi.org/10.1093/ajcn/78.3.660S

Pinto da Rosa P, Pio Ávila B, Damé Veber Angelo I, Garavaglia Chesini R, Albandes Fernandes T, da Silva Camacho J, Bugoni M, Roll VFB, Gularte MA (2021) Impact of different chicken meat production systems on consumers' purchase perception. Br Poult Sci 62(3):387–395. https://doi.org/10.1080/00071668.2020.1857335

Plant-Based Meat Products | Beyond Meat (n.d.) Retrieved 17 Oct 2023, from https://www.beyondmeat.com/en-US/products/

Qin P, Wang T, Luo Y (2022) A review on plant-based proteins from soybean: health benefits and soy product development. J Agric Food Res 7:100265. https://doi.org/10.1016/j.jafr.2021.100265

Radnitz C, Beezhold B, DiMatteo J (2015) Investigation of lifestyle choices of individuals following a vegan diet for health and ethical reasons. Appetite 90:31–36. https://doi.org/10.1016/j.appet.2015.02.026

Roland WSU, Pouvreau L, Curran J, van de Velde F, de Kok PMT (2017) Flavor aspects of pulse ingredients. Cereal Chem 94(1):58–65. https://doi.org/10.1094/CCHEM-06-16-0161-FI

Rosa W (2017) Transforming our world: the 2030 agenda for sustainable development. In: A new era in global health. Springer Publishing Company. https://doi.org/10.1891/9780826190123.ap02

Sá AGA, Moreno YMF, Carciofi BAM (2020) Food processing for the improvement of plant proteins digestibility. Crit Rev Food Sci Nutr 60(20):3367–3386. https://doi.org/10.1080/10408398.2019.1688249

Salgado AM, Ozturk OK, Hamaker BR, Campanella OH (2023) Matching textural properties of commercial meat and cheese products using zein as the viscoelastic agent and calcium hydroxide as the textural modifier in plant-based formulations. Food Hydrocoll 137:108323. https://doi.org/10.1016/j.foodhyd.2022.108323

Samard S, Ryu G-H (2019) A comparison of physicochemical characteristics, texture, and structure of meat analogue and meats. J Sci Food Agric 99(6):2708–2715. https://doi.org/10.1002/jsfa.9438

Sanchez-Sabate R, Sabaté J (2019) Consumer attitudes towards environmental concerns of meat consumption: a systematic review. Int J Environ Res Public Health 16(7):1220. https://doi.org/10.3390/ijerph16071220

Satija A, Hu FB (2018) Plant-based diets and cardiovascular health. Trends Cardiovasc Med 28(7):437–441. http://dx.doi.org.ezproxy.acadiau.ca:2048/10.1016/j.tcm.2018.02.004

Schiffman JD, Schauer CL (2008) A review: electrospinning of biopolymer nanofibers and their applications. Polym Rev 48(2):317–352. https://doi.org/10.1080/15583720802022182

Schlemmer U, Frølich W, Prieto RM, Grases F (2009) Phytate in foods and significance for humans: food sources, intake, processing, bioavailability, protective role and analysis. Mol Nutr Food Res 53(S2):S330–S375. https://doi.org/10.1002/mnfr.200900099

Schreuders FKG, Dekkers BL, Bodnár I, Erni P, Boom RM, van der Goot AJ (2019) Comparing structuring potential of pea and soy protein with gluten for meat analogue preparation. J Food Eng 261:32–39. https://doi.org/10.1016/j.jfoodeng.2019.04.022

Sha L, Xiong YL (2020) Plant protein-based alternatives of reconstructed meat: science, technology, and challenges. Trends Food Sci Technol 102:51–61. https://doi.org/10.1016/j.tifs.2020.05.022

Sharima-Abdullah N, Hassan CZ, Arfin N, Huda-Faujan N (2018) Physicochemical properties and consumer preference of imitation chicken nuggets produced from chickpea flour and textured vegetable protein. Int Food Res J 25(3):1016–1025

Singh N (2017) Pulses: an overview. J Food Sci Technol 54(4):853–857. https://doi.org/10.1007/s13197-017-2537-4

Singh H, MacRitchie F (2001) Application of polymer science to properties of gluten. J Cereal Sci 33(3):231–243. https://doi.org/10.1006/jcrs.2000.0360

Singh A, Sit N (2022) Meat analogues: types, methods of production and their effect on attributes of developed meat analogues. Food Bioprocess Technol 15(12):2664–2682. https://doi.org/10.1007/s11947-022-02859-4

Smetana S, Profeta A, Voigt R, Kircher C, Heinz V (2021) Meat substitution in burgers: nutritional scoring, sensorial testing, and life cycle assessment. Future Foods 4:100042. https://doi.org/10.1016/j.fufo.2021.100042

Sun C, Fu J, Chang Y, Li S, Fang Y (2022) Structure design for improving the characteristic attributes of extruded plant-based meat analogues. Food Biophys 17(2):137–149. https://doi.org/10.1007/s11483-021-09692-w

Tonsor GT, Lusk JL (2022) U.S. perspective: meat demand outdoes meat avoidance. Meat Sci 190:108843. https://doi.org/10.1016/j.meatsci.2022.108843

Trinci APJ (1994) Evolution of the Quorn® myco-protein fungus, Fusarium graminearum A3/5. Microbiology (Reading) 140:2181–2188

United Nations (1987) Sustainability. United Nations. https://www.un.org/en/academic-impact/sustainability

Urade R (2011) Chapter 38—fortification of bread with soy proteins to normalize serum cholesterol and triacylglycerol levels. In: Preedy VR, Watson RR, Patel VB (eds) Flour and breads and their fortification in health and disease prevention. Academic, pp 417–427. https://doi.org/10.1016/B978-0-12-380886-8.10038-8

Varghese KS, Pandey MC, Radhakrishna K, Bawa AS (2014) Technology, applications and modelling of ohmic heating: a review. J Food Sci Technol 51(10):2304–2317. https://doi.org/10.1007/s13197-012-0710-3

Wang Y, Tuccillo F, Lampi A-M, Knaapila A, Pulkkinen M, Kariluoto S, Coda R, Edelmann M, Jouppila K, Sandell M, Piironen V, Katina K (2022) Flavor challenges in extruded plant-based meat alternatives: a review. Compr Rev Food Sci Food Saf 21(3):2898–2929. https://doi.org/10.1111/1541-4337.12964

Wang T, Kaur L, Beniwal AS, Furuhata Y, Aoyama H, Singh J (2023) Physico-chemical and textural properties of 3D printed plant-based and hybrid soft meat analogs. Plant Foods Hum Nutr 78(2):375–382. https://doi.org/10.1007/s11130-023-01068-4

Wen Y, Chao C, Che QT, Kim HW, Park HJ (2023) Development of plant-based meat analogs using 3D printing: status and opportunities. Trends Food Sci Technol 132:76–92. https://doi.org/10.1016/j.tifs.2022.12.010

Wu YFG, Cadwallader KR (2002) Characterization of the aroma of a meatlike process flavoring from soybean-based enzyme-hydrolyzed vegetable protein. J Agric Food Chem 50(10):2900–2907. https://doi.org/10.1021/jf0114076

Wu YF, Baek HH, Gerard PD, Cadwallader KR (2000) Development of a meat-like process flavoring from soybean-based enzyme-hydrolyzed vegetable protein (E-HVP). J Food Sci 65(7):1220–1227. https://doi.org/10.1111/j.1365-2621.2000.tb10269.x

Xing Z, Li J, Zhang Y, Gao A, Xie H, Gao Z, Chu X, Cai Y, Gu C (2022) Peptidomics comparison of plant-based meat alternatives and processed meat after in vitro digestion. Food Res Int 158:111462. https://doi.org/10.1016/j.foodres.2022.111462

Younis K, Ashfaq A, Ahmad A, Anjum Z, Yousuf O (2023) A critical review focusing the effect of ingredients on the textural properties of plant-based meat products. J Texture Stud 54(3):365–382. https://doi.org/10.1111/jtxs.12704

Yuliarti O, Kiat Kovis TJ, Yi NJ (2021) Structuring the meat analogue by using plant-based derived composites. J Food Eng 288:110138. https://doi.org/10.1016/j.jfoodeng.2020.110138

Zhang W, Li S, Zhang B, Drago SR, Zhang J (2016) Relationships between the gelatinization of starches and the textural properties of extruded texturized soybean protein-starch systems. J Food Eng 174:29–36. https://doi.org/10.1016/j.jfoodeng.2015.11.011

Zhang J, Liu L, Liu H, Yoon A, Rizvi SSH, Wang Q (2019) Changes in conformation and quality of vegetable protein during texturization process by extrusion. Crit Rev Food Sci Nutr 59(20):3267–3280. https://doi.org/10.1080/10408398.2018.1487383

Zhang T, Dou W, Zhang X, Zhao Y, Zhang Y, Jiang L, Sui X (2021) The development history and recent updates on soy protein-based meat alternatives. Trends Food Sci Technol 109:702–710. https://doi.org/10.1016/j.tifs.2021.01.060

Zhang J, Chen Q, Kaplan DL, Wang Q (2022) High-moisture extruded protein fiber formation toward plant-based meat substitutes applications: science, technology, and prospect. Trends Food Sci Technol 128:202–216. https://doi.org/10.1016/j.tifs.2022.08.008

Zhu L-J, Shukri R, de Mesa-Stonestreet NJ, Alavi S, Dogan H, Shi Y-C (2010) Mechanical and microstructural properties of soy protein—high amylose corn starch extrudates in relation to physiochemical changes of starch during extrusion. J Food Eng 100(2):232–238. https://doi.org/10.1016/j.jfoodeng.2010.04.004

Zhu H-G, Tang H-Q, Cheng Y-Q, Li Z-G, Tong L-T (2021) Potential of preparing meat analogue by functional dry and wet pea (Pisum sativum) protein isolate. LWT 148:111702. https://doi.org/10.1016/j.lwt.2021.111702

# Chapter 5
# Proteins for Analogue Foods

**Elif Meltem Işçimen and Mehmet Hayta**

## 5.1 Introduction

By the year 2050, it is anticipated that there will be 9.8 billion people on the planet, which would result in a surge in the need for both macronutrients and micronutrients (Nations 2019). For humans, protein is a crucial macronutrient and it offers amino acids, which are necessary for growing and maintaining health (Galanakis 2019). Growing world population and the growing luxury of people everywhere, meaning that in order to meet demand, more protein-rich meals must be created (Searchinger et al. 2019). Given that many of the consumer-preferred and technologically useful proteins available today are currently derived from animals, the availability of adequate protein is of significant importance (Grossmann and Weiss 2021). Moreover, numerous studies have shown that regularly replacing animal-based protein containing cholesterol and saturated fat with a plant-based protein food rich in fiber, cholesterol-free, and low in saturated fat can lessen diabetes risk factors, cardiac disease, and certain cancer types (Manickavasagan et al. 2022). Currently, in developing countries where the cost or availability of animal proteins is high, an alternative protein source is derived from plants (Akhtar et al. 2022). In the last 20 years, plant proteins have drawn a lot of interest as a particularly nourishing and sustainable supply of vital amino acids (Lonnie et al. 2020). Also, as a replacement for plant proteins, it can be obtained from algae, fungi, and insects, which have proven to have little adverse environmental impact (Clune et al. 2017; Grossmann and Weiss 2021).

The demand for alternative proteins necessitates the application of innovative technology, especially in meat, fish, egg, and dairy analogues and vegetable drinks (Dekkers et al. 2018). The agricultural and food sectors have produced an abundant

E. M. Işçimen (✉) · M. Hayta
Erciyes University, Engineering Faculty, Food Engineering Department, Kayseri, Türkiye
e-mail: eliferen@erciyes.edu.tr; mhayta@erciyes.edu.tr



Ö. P. CAN et al. (eds.), *Food Analogues: Emerging Methods and Challenges*,
https://doi.org/10.1007/978-3-031-69872-9_5

supply of nutritious foods during the past century or more, which has contributed to a decrease in hunger and malnutrition worldwide (McClements and Grossmann 2021). There has been an increment in the investigation of and creation of dairy and meat analogue products intended for consumption in the last few years. There are many different food analogues produced with different protein sources including pulses, insects, tempeh, seeds, algae, and nuts (Lima et al. 2022). Numerous research projects were fulfilled to ensure the production of meat (De Angelis et al. 2020; Xia et al. 2023; Zahari et al. 2020), fish (Shi et al. 2023), egg (Lu et al. 2022, 2023), and dairy food (Ferawati et al. 2021; Levy et al. 2022) analogues from non-animal protein sources. In this chapter, production methods and current studies for food analogues produced using various plant protein sources will be discussed.

## 5.2 Protein Sources for Food Analogues

The most critical element for functionality in many plant-based diets is protein, because they have unique structural, texturizing, foaming, water-oil holding, emulsifying, and nutritive qualities. Choosing the right protein source is frequently one of the most crucial actions in creating a finished product of great quality (Loveday 2020). Numerous proteins that have been extracted from plant sources are substantial globular storage proteins that are present in food as part of supramolecular complexes (McClements and Grossmann 2021). However, it has been discovered that plant-based proteins are less digestible and lack several important amino acids than animal-based proteins (Sarwar 1997). For this reason, protein from insects with its high protein content (up to 40% and 70% on average dry weight), has drawn interest from the food sector together with vitamin and mineral contents, additionally fascinating polyunsaturated and saturated fatty acids (Gravel and Doyen 2020). Similarly, algae attract attention because they enable more and cheaper protein supply in less space (Van Krimpen et al. 2013). The food sources of protein mainly are plants, insects, and algae investigated for the production of food analogues (Fig. 5.1).

### *5.2.1 Cereals*

The most popular cereals, which are proteinaceous and fibrous, include wheat, rice, barley, and oats (Shewry 2009). Prolamins and/or glutelins, which are water-insoluble proteins make up the majority of the proteins in rice, maize, and wheat (Veraverbeke and Delcour 2002). Due to its viscoelastic properties and economic viability as a co-product in a number of food applications, the distinctive cereal protein is wheat gluten that is entirely used to prepare meat alternatives (Kyriakopoulou et al. 2019). Due to the fact that wheat gluten binds the fiber in the matrix together for meat analogues (Rizvi et al. 1980). Gliadin and glutenin are two of the protein groups that make up wheat gluten. Glutenins are significantly larger

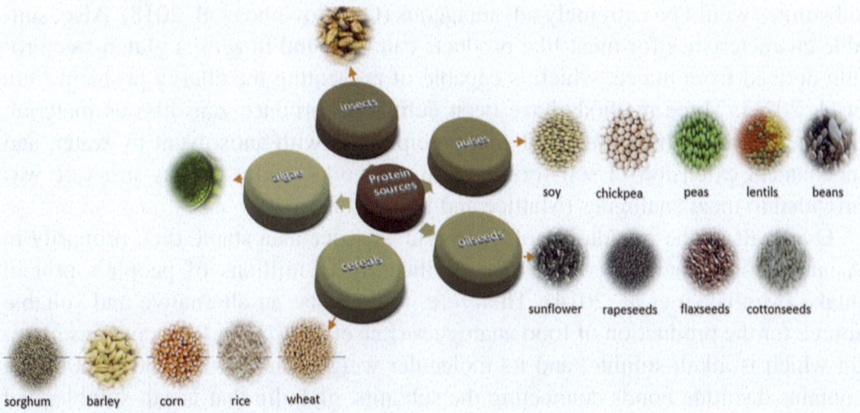


**Fig. 5.1** Protein sources for food analogues

molecules made up of several polypeptides joined by intermolecular disulfide bonds, whereas gliadins are monomeric proteins of low to medium molecular weight (Wieser 2007). Due to the viscoelastic qualities of gluten, which give meat substitutes their fibrous-like texture and firmness, gluten is a frequently used element in meat substitutes (Kyriakopoulou et al. 2021). A study reports the effects of fibrous structure during extrusion under varying humidity conditions on the protein digestibility of meat analogues produced from textured wheat protein (Lin et al. 2022). Other research examined the consequences of protein isolate from soy and its ratio to wheat gluten on the physicochemical characteristics of meat analogues produced by extrusion and found that meat analogues including 30% wheat gluten displayed the best texture, fibrous structure, firmness, and chewiness using sensorial and instrumental analysis (Chiang et al. 2019). In another similar study, the impact of the wheat-to-soy protein ratio on fibrous structure in meat analogues was evaluated and as a result, the ultrastructure was not significantly impacted by the relative concentration of wheat gluten, on the other hand, the rising SPI ratio improved the lysine content of meat analogues (Chiang et al. 2021). The usability of a combination of wheat and pea proteins in different ratios in the production of chicken-like meat analogues was also evaluated. The ratio of 4:13 pea protein: wheat protein has been supposed to be the most popular analog compared to other formulations. It has been observed that the inclusion of pea protein improves the viscoelastic features as well as the chewiness, hardness of the analogue. Therefore, it has been reported that the textural and viscoelastic properties of the analogue decreased when the proportion of the formulation of wheat protein was raised (Yuliarti et al. 2021). Considering the available studies in the literature, the potential of using cereal and plant protein composites in the formulation of effective meat analogues can be highlighted. Not only meat but fish analogues have been produced with the help of various celluloses using wheat protein (Sánchez-Alonso et al. 2007).

On the other hand, for regions where corn is extensively farmed and/or wheat import costs are high, the successful incorporation of zein into dough goods or meat substitutes would be extremely advantageous (Oguntoyinbo et al. 2018). Also, suitable characteristics for meat-like products can be found in zein, a gluten-free protein derived from maize, which is capable of preventing the allergy problem (Sun et al. 2021). Three methods have been defined to produce zein fibrous material: electrospinning, ethanol-induced zein precipitation with antisolvent by water, and mechanical extension of self-forming zein networks, and a fibrous structure was provided to meat analogues (Mattice and Marangoni 2020).

Over half of the population of the world eats rice as a staple diet, primarily in Asian nations where it contributes significantly to millions of people's protein intake (Muthayya et al. 2014). Therefore, rice can be an alternative and suitable source for the production of food analogues (Lee et al. 2022b). Rice contains glutelin which is alkali-soluble, and its molecular weight change from 60 to 600 kDa contains disulfide bonds connecting the subunits, globulin that is salt-soluble, and its molecular weight change from 12 to 20 kDa, water-soluble protein that is albumin, and prolamin that is fractions soluble in alcohol (Amagliani et al. 2017). The results of the inclusion of protein isolate from rice on the physicochemical characteristics of a meat alternative have been investigated via mixing rice protein isolate with soy protein isolate at different ratios (0:50, 5:45, 10:40 15:35). The integrity index, texturization degree, and chewiness of the meat analogue rise with rising up to 10% of rice protein isolate (Kim et al. 2022). In another study, rice protein isolates were mixed with varying amounts of soy protein isolate (25:75, 50:50, 75:25, and 100:0), and found that the addition of rice protein isolate reduced the particular mechanical energy, porosity, and propensity to absorb water of the meat analogues (Lee et al. 2022b).

Cereal proteins have been employed not only in the creation of meat analogues, but also directly or as an auxiliary in the production of dairy analogues. The cheddar cheese analogue was produced with whey protein isolate and corn milk (Sumarmono et al. 2021). A cheese analog made from corn milk with the best properties has been fabricated with the ideal ratio of maltodextrin, papain, and citric acid (Aini et al. 2019). The sensory properties of this cheese analog obtained from the extract of sweet corn were like those of cheese obtained from milk obtained from cow; that is, it has been reported to have a yellowish white color, a distinctive cheese flavor, soft texture and sour taste (Aini et al. 2019). The proteins in the wheat starch used during the cheese analog production were found as the main component that moistens the environment, and the free fat is emulsified by the soluble proteins, which also make the aqueous phase more viscous (Mounsey 2009).

### *5.2.2 Legumes/Pulses*

Plant-based materials (for example flour, protein, fiber, and starch) with advantages over animal-based proteins in terms of cost, sustainability, and nutrition, pulse crops can offer remedies for the gluten-free market (Khazaei et al. 2019). To produce meat alternatives, pulse crops like peas, lentils, lupines, chickpeas, and beans are frequently utilized (Osen and Schweiggert-Weisz 2016). The amount of protein in legume seeds is more preferable than grains (Shewry 2009). Globulins and albumins are the primary keeping proteins in pulses and beans, however, globulins are typically the primary storage proteins (often >50%) (Boye et al. 2010).

One of the biggest obstacles to promoting meat alternatives made from pulse proteins is the unwillingness of consumers to change from traditional meat to meat substitutes because they enjoy eating traditional meat and discover that meat substitutes don't have a satisfying sensory appeal (Kyriakopoulou et al. 2021). Pulse proteins play a role as emulsifying, water and oil binding, gelling, texturing, oil emulsifying agents, and a nutrient enhancers (Nadeeshani et al. 2022). Due to their superior gelling characteristics and leading in the development of an interwoven, fibrous matrix, nutritional qualities, bioavailability, and technological functionality, soybean components in particular have a considerable position purchasing plant-based proteins (Day and Swanson 2013; Samard and Ryu 2019). The major proteins of soybean are such as the 11S globulin glycinin and the 7S globulin β-conglycinin globulins (Phillips and Williams 2011). In many studies, soy protein has been used to manufacture meat analogues (Herz et al. 2021; Lin et al. 2000; Liu and Hsieh 2007; Ran et al. 2022; Zahari et al. 2020).

Another important protein source of pulses is pea. The pea protein consists of globulins especially 26–60 kDa vicilin-type 8S, albumins, and other globulins including basic-type 7S and legumin-type 11S (Lam et al. 2018). In a study investigating using pea proteins as a substitute for soy protein, pea proteins were mixed with wheat gluten. When wheat gluten was combined with pea protein isolate, fibrous morphology was obtained (Schreuders et al. 2019). It has been proven that meat analogues with a suitable texture are likely to produce from the mixture of different legume proteins using extrusion by blending oat protein ($O_P$), pea protein isolated ($P_{Is}$), dry-fractionated pea protein ($P_{DF}$), and soy protein isolated ($S_{Is}$) (Fig. 5.2). In general, the study reported that stable production of texturized proteins was achieved. On a dry matter basis, the $P_{DF}$-$O_P$ mixture had a protein content of 55 g to 100 $g^{-1}$ and was appropriate for use in the creation of meat-analogue goods, as starch, fat, and minerals with nutritional and practical qualities were present in large amounts (De Angelis et al. 2020).

Legume proteins are not only used to produce meat analogues but also have great importance to dairy products because they are inexpensive and healthiness of protein and fat (Moradi et al. 2021). Studies have been performed on producing cheese analogues using pulse milk because of their protein content and fat value (Ahmed 2014; Ferawati et al. 2021; Moradi et al. 2021). An animal-free cheese alternative with a clear label and good nutritional value utilizing pea protein obtained by dry fractioned and an emulsion-filled gel has been reported (Mefleh et al. 2022). There are studies to produce analogues for egg, which is an important animal protein source. The eggless omelets were formed using soy protein isolate, chickpea flour, baking powder, shortening, transglutaminase, mono, diglycerides, flour, potassium chloride, and hydrocolloids or gellan gum (Lu et al. 2022).

**Fig. 5.2** Meat analogues from different legume proteins (De Angelis et al. 2020)

### 5.2.3 *Oilseeds*

There are more than 200 different varieties of oilseed crops worldwide. Oilseeds are a useful natural plant protein source in addition to being used to make vegetable oils (Zhang et al. 2023). Storage proteins make up the majority of the proteins in oilseeds (Asgar et al. 2010). 11S globulins and albumins are the two main protein classes in oilseeds like rapeseed and sunflower, respectively, but a significant glutelin fraction may also be present in pumpkin seeds (Pham et al. 2017). Additionally, safflower seeds have 13–17% protein, cotton seeds 23%, and rapeseed 25% (Asgar et al. 2010). When added to specific foods, these proteins also enhance beneficial functional qualities like viscosity, emulsifying ability, and the capability to absorb oil and water. Protein isolates from rapeseed and soybean are better at whipping than those from sunflower, peanut, sesame, cotton, and safflower. The ability to whip of the oilseed proteins are also enhanced by the sugar addition (Kurek et al. 2022). Proteins from oilseeds and meals can be added to food to give it the required nutritional content and functional properties. Pasta or noodles, beverages, bakery goods, and meat substitutes are just a few of the food products that have already been incorporated as functional components (Zhang et al. 2023).

Sensory preference is the key to the market success of food analogues. A study was conducted on the attitudes of female consumers from five European countries, who also regularly consume meat, toward a rapeseed-protein-based meat analogue (Banovic and Sveinsdóttir 2021). Rapeseed protein isolates alone or mixed with wheat gluten to produce a meat analogue (Jia et al. 2021). Sunflower seeds, another protein-rich oilseed source, are perceived as a substitute plant protein supply for applications involving meat analogues. Degreased sunflower seeds and concentrate have been reported to form a fibrous item of 40% by weight by shearing and heating at 140 °C in a shear cell (Jia et al. 2022) (Table 5.1).

**Table 5.1** Main sources of plant-based proteins, properties, and processing methods to produce food analogues

| | Protein sources | Properties | Produced food analogue | Processing method | References |
|---|---|---|---|---|---|
| **Cereals** | **Wheat** | Wheat gluten<br>9.3% moisture and 77.5% protein<br>Commercial wheat gluten | **Meat** | Extrusion | Lin et al. (2022) |
| | **Wheat** | Wheat gluten (75% protein, 12.5% carbohydrate, 10% moisture, 1% fat, and 1.5% ash)/soy protein isolate (70.2% protein, 18.8% carbohydrate, 4.8% moisture, 1.8% fat, and 4.4% ash)<br>Commercial protein | **Meat** | High-moisture extrusion | Chiang et al. (2019) |
| | **Wheat** | Wheat gluten (70% protein, 20% carbohydrate, 3% fat, 3.5% moisture, and 3.5% ash)/soy protein isolate (85% protein, 4% fat, 5% ash, and 6% moisture)<br>Commercial protein | **Chicken** | Mechanical elongation | Chiang et al. (2021) |
| | **Wheat/Pea** | Wheat protein (75.0% protein, 1% fat, 8% ash, and 11% moisture)/pea protein (77.6% protein, 6%w/w fat, 7% ash, and 10% moisture)<br>Commercial protein | **Chicken (nugget)** | Mixed food processor | Yuliarti et al. (2021) |
| | **Wheat** | Wheat fiber (26% hemicellulose, 74% cellulose, and < 0.5 of lignin) | **Fish** | Meat mincer | Sánchez-Alonso et al. (2007) |
| | **Sorghum** | Zein (total zein comprising α-, β-, δ- and γ-zein) and Kafirin<br>Extracted from white sorghum | **Gluten** | Viscoelastic mass formation by coacervation | Oguntoyinbo et al. (2018) |
| | **Corn/Whey** | Corn milk and whey protein concentrate<br>Commercial protein | **Cheddar cheese** | Mixture and pasteurization | Sumarmono et al. (2021) |
| | **Soy/Rice** | Rice protein isolate/soy protein isolate<br>0:50, 5:45, 10:40; and 15:35 | **Meat** | Low-moisture extrusion | Kim et al. (2022) |
| | **Soy/Rice** | Rice protein isolate (92% purity)/soy protein isolate (100% purity)<br>Commercial protein | **Meat** | Low-moisture extrusion | Lee et al. (2022b) |

(continued)

**Table 5.1** (continued)

| | Protein sources | Properties | Produced food analogue | Processing method | References |
|---|---|---|---|---|---|
| **Legumes** | **Soy** | Soy protein<br>Commercial protein | **Fibrous Meat** | High-moisture extrusion | Liu and Hsieh (2007) |
| | **Soy** | Soy protein<br>Commercial protein | **Meat** | High-moisture extrusion | Herz et al. (2021) |
| | **Soy/Hemp** | Soy protein isolate (91% protein, 3% fat, 1% carbohydrate)/hemp protein isolate (71% protein, 7% fat, 15% carbohydrate)<br>Commercial protein | **Meat** | High-moisture extrusion | Zahari et al. (2020) |
| | **Soy** | Soy protein (90.0% protein, 1.5% fat, and 1.8% carbohydrate)/Konjac glucomannan<br>Commercial protein | **Fishball** | Food processor | Ran et al. (2022) |
| | | Soy protein isolate/ chickpea flour<br>Commercial protein | **Egg** | Electric blender | Lu et al. (2022) |
| | **Pea/Wheat** | Pea protein isolate (78.6 wt% protein)/wheat gluten (72.4 wt% protein)<br>Commercial protein | **Chicken** | High-temperature conical shear cell | Schreuders et al. (2019) |
| | **Soy/Pea/Oat** | Soy protein isolate, pea protein isolated, dry-fractionated pea protein, and oat protein<br>Commercial protein | **Meat** | Low-moisture extrusion | De Angelis et al. (2020) |
| | **Pea** | Dry-fractioned pea protein (560 g protein/kg dry matter)/inulin–olive oil emulsion-filled gel<br>Commercial protein | **Cheese** | Mixed and heated | Mefleh et al. (2022) |
| **Oilseeds** | **Rapeseed** | Rapeseed protein (55.9% protein, 2.2% oil, 33.6% carbohydrate)/wheat gluten (79.4% protein, 12.6% carbohydrate)<br>60/100 | **Meat** | High-temperature conical shear cell | Jia et al. (2021) |
| | **Sunflower seed** | De-oiled sunflower kernel (55% protein, 7.4% oil, 25.6% carbohydrate) | **Meat** | Shear cell | Jia et al. (2022) |

### 5.2.4 Algae

Algae are eukaryotic photosynthetic organisms that contain proteins but also provide vitamins and amino acids. They posses preferable physicochemical characteristics necessary to be used as a food component (Bhuva et al. 2021), such as the capability to gelate, bind oil and water, emulsify, and foam, microalgae can be used in a variety of food categories (Caporgno and Mathys 2018). Furthermore, compared to terrestrial plants including pulse/legumes, soybean, and wheat (0.6–1.2 tons/Ha/year, 1–2 tons/Ha/year, and 1.1 tons/Ha/year, respectively), greater protein output per unit area is found in seaweed and microalgae (2.5–7.5 tons/Ha/year and 4–15 tons/Ha/year, respectively) (Van Krimpen et al. 2013).

Algae are frequently on a scale with other sources of protein, such as soybean and egg, and their essential conform to the amino acid composition with FAO standards, making them usually recognized as a viable source of protein (Organization 1991). However, lysine and tryptophan are frequently the only amino acids present in the majority of algal species (Dawczynski et al. 2007; Kolb et al. 1999). Additionally, lysine, methionine, and cysteine are frequently in short supply in brown algae species, however, isoleucine and leucine are frequently found in low percentages in red species of algae (Dawczynski et al. 2007). Some algae and protein contents (dry matter %) are as follows: *Spirulina platensis* 46–63%, *Aphanizomenon flos-aquae* 62%, *Synechococcus sp.* 63%, *Euglena gracilis* 39–61%, *Arthrospira maxima* 60–71%, *Anabaena cylindrical* 43–56%, *Chlorella pyrenoidosa* 57%, *Dunaliella salina* 57%, *Chlamydomonas rheinhardii* 48%, *Chlorella vulgaris* 51–58%, *Scenedesmus obliquus* 50–56%, *Porphyridium cruentum* 28–39%, *Spirogyra sp.* 6–20%, (Becker 2007).

A variety of microalgae are combined with rice, grains, thickening agents, and edible oils to create a food product that is suitable for human ingestion and has a high digestibility rating (Bleakley and Hayes 2017). Recently, studies on the production of food analogues of algae have attracted attention. In a study, *Haematococcus pluvialis* residue and pea protein isolate were mixed to create meat-like substitutes through high moisture extrusion. Thanks to the crimson in hue of the residue used, the extrudates seemed to be dried red meat as a result (Xia et al. 2022). In another study, meat analogues were produced by mixing *Spirulina platensis* flour and lupine protein. In the study using high humidity extrusion cooking to create meat analogues, different *Spirulina* concentrations (15, 30, and 50%) were tried and it was stated that antioxidant properties increased with increasing algae concentration (Palanisamy et al. 2019). *Chlorella vulgaris* is a microalga regarded as a good source of fatty acids, fiber, ash, and protein. The cheese analogue was produced with added *Chlorella vulgaris* (1, 2, and 3%). Cheeses made from *Chlorella* biomass compared to control samples, contained higher concentrations of protein, carbs, and fiber. (Mohamed et al. 2013). Although there are existing studies in the literature, there is a need for more comprehensive studies on the use of protein from algae in the manufacturing of food analogues.

### 5.2.5 Insects

Sustainable meat production will pose a serious challenge for the future, as the world population grows, consumer demand for protein is rising, yet there is a shortage of farmland. Insects rich in protein can be thought of as a different protein source with fewer negative effects on the environment (Van Huis 2013). Furthermore, it has been discovered that proteins from plants are not as easily absorbed by the intestinal system and are deficient in several essential amino acids as animal-based proteins (Sarwar 1997). Because of the high protein content (up to 70% on a dry weight basis and an average of 40%), vitamin, and mineral content, and intriguing ratios of insect proteins'polyunsaturated to saturated fatty acids have recently received more attention (Zielińska et al. 2015). Insect proteins mainly contain myofibrillar proteins that are soluble in salt and sarcoplasmic proteins that are water soluble, and connective tissue which is either insoluble or soluble in acid or alkaline (Grossmann and Weiss 2021). The edible insects for human consumption are Coleoptera (beetles), odonatan (dragonflies), Hymenoptera (ants, wasps, and bees), Hemiptera (cicadas), Lepidoptera (butterflies), isopters (termites), Diptera (flies), and orthopterans (locust and crickets) (Kim et al. 2019b). Table 5.2 gives a summary of the research on the characteristics of insect proteins.

The edible insect industry is currently expanding at the quickest rates in Europe and the United States, where a trend is linked to high meat consumption (Jiri et al. 2014). Additionally, it is acknowledged that the insect industry has a growing global market, with uses extending beyond food to include material and pharmaceutical development (Van Thielen et al. 2019). Despite these developments, consumers who have never eaten edible insects believe that they are filthy, revolting, and harmful, which leads them to reject insects as a food source (Van Huis 2013). The consumption of edible insects has major obstacles because of these phenomena, particularly in Western nations (Van Thielen et al. 2019). Only a small percentage of customers (12.8% men and 6.3% women) consider edible insects as food (Verbeke 2015). Although the acceptance of insects is still quite low in western cultures, studies regarding the usage of insect proteins in the manufacture of food analogues are available in the literature (Kim et al. 2016; Smetana et al. 2018). Coagulation properties of insect protein in different temperature and $CaCl_2$ concentration has been reported for production of a meat analogue Azzollini, Wibisaphira, Lakemond, and Fogliano (Azzollini et al. 2019). The rise in $CaCl_2$ concentration, resulted in the a smoother network at 100 °C. Insect proteins can also be mixed with other sources to produce food analogues. The insect protein concentrate (both *Tenebrio molitor* and *Alphitobius diaperinus*) was mixed with soy protein to produce meat analogue. It has been reported that the mixture that contains the higher insect biomass (40% on dry matter basis) gives a meat-like texture (Smetana et al. 2019).

**Table 5.2** Investigations on the properties of proteins from edible insects

| Type of insects | Evaluated properties | Reference |
|---|---|---|
| *Tenebrio molitor*<br>*Zophobas morio*<br>*Alphitobius diaperinus*<br>*Acheta domesticus*<br>*Blaptica dubia* | Protein content<br>Protein digestability<br>Foam capacity and stability<br>Gel formation | Yi et al. (2013) |
| *Tenebrio molitor*<br>*Gryllus bimaculatus*<br>*Bombyx mori* | Amino acid profiles | Choi et al. (2017) |
| *Allomyrina dichotoma*<br>*Tenebrio molitor*<br>*Protaetia brevitarsis seulensis* | Protein solubility<br>Amino acid profiles<br>Foam capacity and stability<br>Emulsifying stability and emulsion capacity | Kim et al. (2019a) |
| *Hermetia illucens* | Optimization of protein extraction<br>Amino acid profiles<br>Water/oil absorption capacities<br>Nitrogen solubility<br>Foam capacity and stability | Mintah et al. (2020) |
| *Protaetia brevitarsis seulensis* | Protein solubility<br>Amino acid profiles<br>Foam stability and foam capacity<br>Emulsifying stability and emulsion capacity | Kim et al. (2021) |
| *Tussah pupa* | Protein solubility<br>Emulsifying stability and emulsion capacity<br>Foam stability and foam capacity | Ding et al. (2022) |
| *Hermetia illucens L.* | Protein solubility<br>Foam capacity and foam stability<br>Emulsion capacity and emulsion stability<br>Secondary and tertiary structural characteristics | Kim et al. (2023) |
| *Rhynchophorus ferrugineus* | Amino acid profiles<br>Foam capacity and foam stability<br>Emulsion capacity and emulsion stability<br>Water/oil absorption capacities<br>Thermal stability | Kavle et al. (2023) |
| *Hermetia illucens* | Amino acid profiles | Lee et al. (2023a) |
| *Protaetia brevitarsis* | ACE inhibitory activity<br>Cell culturing | Lee et al. (2023b) |

## 5.3 Techno-functional Properties of Proteins

### *5.3.1 Solubility*

The solubility of many plant-based compounds in water, oil, or other solvents is a crucial physicochemical property (McClements and Grossmann 2021). Likewise, the solubility of protein is crucial in food system applications since it is a prerequisite for other textural qualities including gelation, foaming, and emulsifying (Islam et al. 2023). The composition and sequence of amino acids, protein molecular weight, and the shape and amount of nonpolar and polar groups affect the solubility of protein (Zayas and Zayas 1997). While albumins can dissolve in water, globulins can dissolve in salt solutions. Gliadins are soluble in alcohol/water mixes, additionally glutelins are solubility in weak acid or weak alkali solutions (Fox et al. 2004).

Proteins such as gliadin and zein have relatively low water solubility due to their hydrophobic proteins (Davidov-Pardo et al. 2015). Zein is especially abundant in glutamic acid (21–26%), leucine (20%), alanine (10%), and proline (10%), despite being deficient in basic and acidic amino acids. Zein has a nitrogen balance that is adverse because of the noteworthy absence of tryptophane and lysine. The solubility characteristic of zein is caused by a large percentage of residues of nonpolar amino acids and a shortage of basic and acid amino acids (Shukla and Cheryan 2001). For this reason, zein and other water-insoluble proteins have been utilized to provide fibers or other desired structural elements to meat products produced from plants (Mattice and Marangoni 2020). It was indicated that the texture of soy protein isolate gels with 3% zein was comparable to chicken meat (Mattice and Marangoni 2020). On the other hand, in a study investigating the ability of globular proteins to be transported by zein and gelatin, electrospinning was unsuccessful because zein is poorly soluble in water and these globular proteins are poorly soluble in ethanol. This was probably due to the poor quality of the solvent used for the solving of proteins to enable spinning (Nieuwland et al. 2014). The low solubility of the protein provides smoother forms in the creation of meat analogues with the aid of the extruder, and the firmness, chewiness and chewiness of the product increase (Lin et al. 2000). In the soy protein isolate gel, which has been heated, Shimada and Cheftel (1988) discovered that the hardness of the gel rose as the solubility of the protein fell (Shimada and Cheftel 1988). Good indices of crosslink density and formation during processing are water-holding capacity and solubility, respectively. On the other hand, covalent crosslinks are unfavorable before or during thermomechanical processing because they impact other sensory qualities and decrease chain mobility while raising viscosity and avoiding material homogenization (Verbeek and van den Berg 2010). The majority of research on the impact of parameters for the procedure on the formation of chemical bonds took place while the moisture content was low to moderate, which may have distinct impacts on protein structure than when the moisture content was high (Lin et al. 2000). In a study to determine the correlation between extruder reactions, examination of the textural profile, and protein solubility of meat-like soy proteins in high-humidity extrusion, it has been reported that the overall texture of the food is more influenced by moisture content than by cooking temperature. In addition, it produced a stronger texture, higher die pressure, reduced total protein solubility, and lower moisture content (Lin et al. 2000).

The high molecular size and intermolecular aggregation of wheat gluten, which are mostly to blame for its insufficient ability to dissolve in water and are the result of powerful non-covalent interactions, like hydrogen bonds and hydrophobic interaction (Mejri et al. 2005). Because of these properties, gluten has been investigated for its use in the manufacture of meat alternatives because, for flesh analogues, it maintains the fiber in the matrix together (Chiang et al. 2019; Pietsch et al. 2017). It is possible to obtain a meat-like structure by using high-humidity extrusion with wheat gluten due to its functional properties (Kendler et al. 2021).

Environmental variables such as ionic strength, solvent type, pH, temperature, and processing conditions have an effect on the solubility of proteins (Zayas and Zayas 1997). The solubility of globulins particularly 11S globulins is significantly influenced by the ionic strength and pH (Lakemond et al. 2000). At low ionic strengths, pH 4.5–7, 7S globulin solubility has been reported as modest; whereas above and below this range, it was considerably high (>90%). In contrast, 11S globulins were more soluble over a pH of 4–7.5. Moreover, the majority of 11S globulins had negligible solubility, except for soy 11S, which had a strong solubility above pH 3–4 at high ionic strengths and no apparent minimum solubility down to pH 5–6. At pH 3 to 9, 7S globulins normally had a high solubility (Kimura et al. 2008). Because they have reduced molecular weights (10–18 kDa) and increased hydrophilicity than globulins, albumins are more soluble across a wider pH range. For instance, albumins from kidney beans, chia, pea, locust bean, pumpkin seeds, buckwheat seeds, hemp seeds, and sunflower seeds showed better solubility at various pH values from neutral to acidic pH than globulins (Djemaoune et al. 2019; Julio et al. 2019).

### *5.3.2 Emulsifying and Foaming*

Due to its beneficial effects on texture, water-holding capacity, and lower fat content, many studies have shown that an emulsion may be utilized as a fat replacement in meat goods (Wi et al. 2020). Many proteins are molecules with surface activity that can be used as emulsifiers. Because they can help with emulsion formation, stability, and the production of desirable physicochemical properties (Norde 2003). Soy proteins have been reported to have emulsifying and foaming characteristics, which are extensively studied in the literature, as well as additional plant proteins made from peas, rice, lupine, chickpeas, lentils, broad beans, sunflower, canola, and pumpkin (Grossmann and Weiss 2021; Ho et al. 2017). Due to their greater solubility, less weight, and greater flexibility at the molecular level, albumins derived from flaxseed, ginkgo biloba, hemp, and kidney beans adsorbed quickly at interfaces and quickly diffused to and rearranged at oil-water or air-water interfaces (Deng et al. 2011; Mundi and Aluko 2012; Nwachukwu and Aluko 2018; Osemwota et al. 2021). In contrast, globulins had generally higher emulsifying capabilities than albumins from peas, chickpeas, sunflowers, and canola possibly doe to their decreased hydrophobicity and/or molecular flexibility (Papalamprou et al. 2009; Patil and Benjakul 2017; Tan et al. 2014).

The two basic steps in the traditional manufacture of meat analogues are emulsion preparation and chunk creation. The emulsion is heated while being compressed in a particular way. In order to construct a three-dimensional network, the pressure organizes and orients protein chains (Malav et al. 2015). Food analogues, especially the main problem with meat analogue is the acquisition of meat texture, and structural fibers. This problem can be fixed by modifying the processing parameters or choosing the right plant-based protein sources (Palanisamy et al. 2018). In a study, the protein emulsion frozen in the freeze-structuring method was used to create a distinctive fibrous structure using different proportions of wheat and soy proteins. It was reported that the inclusion of pea protein raised the viscoelastic characteristics of the analogue, hardness, and chewiness, whereas when the levels of wheat protein in the production mixture were rised, it decreased the viscoelastic and textural properties (Yuliarti et al. 2021). In another study, a meat analogue has been prepared by using soy protein isolate and lecithin and it has been observed that the emulsion process affects tissue parameters positively (Wi et al. 2020).

Another important techno-functional property is the gelation. A study reported that potato protein and zein were covalently linked by tyrosinase to create a stable gel-like structure (Glusac et al. 2018). The use of zein in meat substitutes is improved by tyrosinase cross-linking since it gives the products a distinctive texture. This could eventually result in a diverse array of non-meat meals such as spreads, sauces, mousses, and cheeses (Glusac et al. 2018).

## 5.4 Processing Methods for Structure Formation of Food Analogues from Plant-Based Proteins

The basic requirement to produce a food analogue is to obtain products with an appropriate textural appearance that resembles the original (Singh et al. 2021). It is possible to replicate the original food using plant-based ingredients by using a variety of processing methods like extrusion, electrospinning, freeze structuring, shear cell technology, in-vitro animal cell culture, and to a lesser extent spinning (Bhuva et al. 2021; Dekkers et al. 2018).

### *5.4.1 Extrusion*

The process of wet, expansile, starchy, and/or proteinaceous materials being plasticized in a tube by a combination of heat, humidity, mechanical shear, and pressure is known as extrusion cooking of vegetable/plant proteins. Extrusion cooking raises the temperature of the food inside the tube, gelatinizes starchy components, denaturates proteins, stretches or restructures tactile components, and causes the extrudate to expand exothermically (Riaz 2011). A typical extruder has four primary sections: a mixing zone (kneading and paddle blocks), a feeding zone (conveying elements), a melting zone (with materials conveyed, kneaded, and reversed to up the mechanical energy input components), and a die (Osen et al. 2014). The characteristic texture, form, and functional features of the food particle can all be regulated by the feed formulation, cooling die design and extruder operation parameters. The procedures can be streamlined into three simple phases: The material must first be prepared outside the extruder, then it must be mixed and cooked inside the extruder barrel, and finally it must be cooled in the die (Singh et al. 2021). The extrusion techniques can be divided into two classes according to the amount of water used in the extrusion machine: high-moisture processes and low-moisture processes (Vatansever et al. 2020).

### 5.4.2 High-moisture Processes

The transport, mixing, shearing, heating, shaping, and cooling of raw materials under high moisture (above 40%), temperature, pressure, and shear conditions causes physicochemical transformations such as denaturing and crosslinking, aggregation and disaggregation of proteins, gelatinization and starch degradation, oxidation of lipids, and degradation of vitamins, antinutrients, and phytochemicals (Akdogan 1999).

The process steps in high-moisture extrusion are as follows (Liu et al. 2005; Maurya and Said 2014; Osen et al. 2014);

1. During the extrusion process, the texturized chunks are thoroughly hydrated, and then the moisture content is maintained by freezing.
2. Due to the waste of mechanical energy, the twin-screw rotator raises the processing temperature after preparing the raw material at 130–180 °C.
3. The protein eventually begins to melt in the extruder barrel, and the viscoelastic mass is then exposed to the protracted cooling die.
4. This results in the creation of structurally aligned texturized proteins, stops food material from expanding into newly created structures and makes it simple to pack food in moist conditions into pouches, cans, or frozen meals.

The most well-liked and commonly employed technique has been processing for high-moisture extrusion, due to its capacity to scale up and the consistently high caliber of the textured products it produces (Wild et al. 2014). The textured protein with a solid structure, chewy texture, and high moisture content can be ingested right away without having to go through a hydration procedure (Lee et al. 2022b).

The source of the protein, such as soy, pea, whey, and peanut proteins, or a protein blend, as well as the temperature, moisture content, and feed rate, are all extrusion characteristics that affect how fibrous is the extrudate (Pietsch et al. 2019; Ran et al. 2022; Zhang et al. 2020). For example, heat treatment during extrusion has been reported to play a dominant role in protein polymerization (Pietsch et al. 2019). The physical parameters of the meat analogue produced by high moisture extruder were considerably impacted by extruder water feed, screw rate, and barrel temperature (Palanisamy et al. 2019). On the other hand, there are studies conducted with

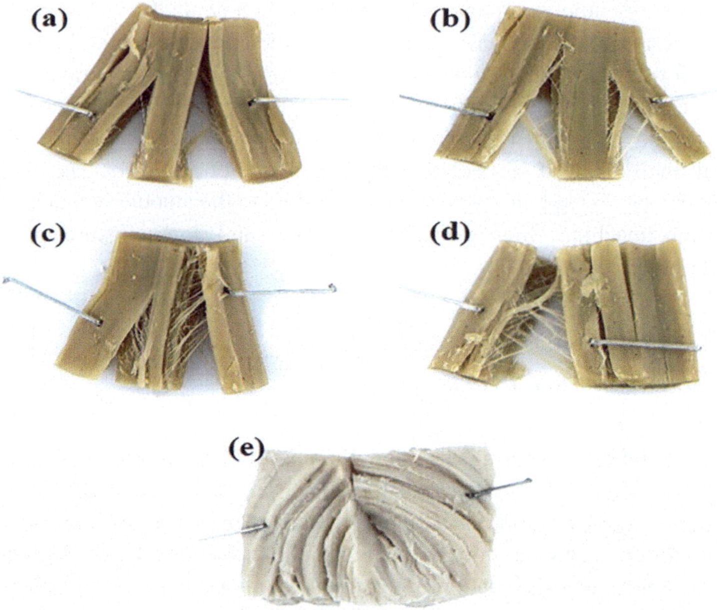


**Fig. 5.3** Visual representations of boiling chicken breast and meat analogues from extrusion with various ratios of soy protein isolate to wheat gluten (**a**) 0% wheat gluten, (**b**) 10% wheat gluten, (**c**) 20% wheat gluten, and (**d**) 30% wheat gluten, (**e**) visual image of boiled chicken breast, (**e**) visual image of boiled chicken breast (Chiang et al. 2019)

high moisture extrusion to examine the effect of mixing ratios of different proteins on the physicochemical, textural, and structural features of meat analogues (Chiang et al. 2019, 2021; Xia et al. 2022; Zhang et al. 2022a, b). In a study in which extrusion conditions and protein mixing ratios were examined, the effect of different moisture content (65%, 70%, 75%) and different protein ratios (20%, 40%, 60%) affect the final product's color and texture was investigated. It has been stated that the product produced with 60% hemp protein with 75% moisture content has a very soft and fragile texture (Zahari et al. 2020). The physical, chemical, and textural characteristics of created meat alternatives from combinations of soy protein and wheat gluten were examined by high moisture extruder. The image of manufactured products is given in Fig. 5.3 (Chiang et al. 2019).

### *5.4.3 Low-moisture Processes*

There are two forms of extrusion cooking, and the quantity of water pumped into the extruder throughout the operation influences the end product. High-moisture (wet; >50% of water) and low-moisture (dry; 35% of water) types are available (Osen et al. 2015). Within the sector of food, low moisture extrusion is a method that is frequently used (Dobraszczyk et al. 2005). Because of the extreme heat and

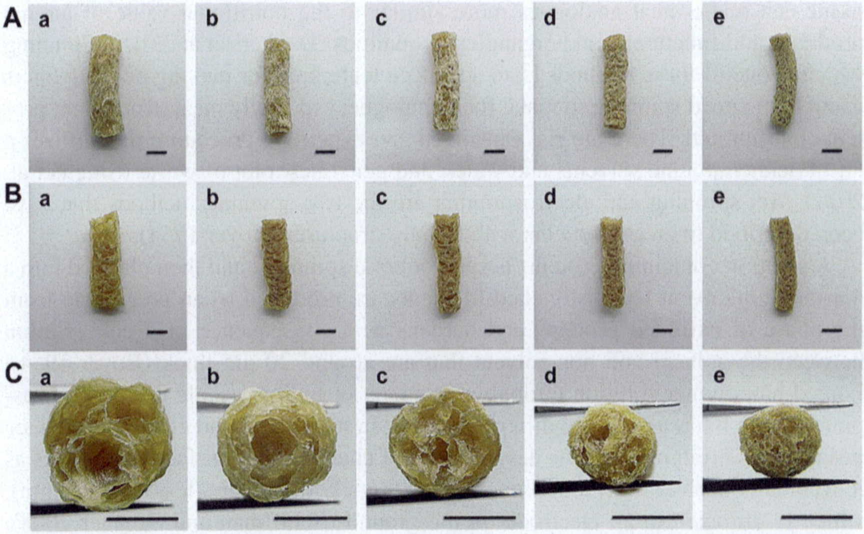


**Fig. 5.4** Illustration of the meat analogues (**A**) surface, (**B**) longitudinal section, and (**C**) transverse section [rice protein: soy protein, 0:100 (**a**) 25:75 (**b**) 50:50 (**c**) 75:25 (**d**) 100:0 and (**e**) (Scale bar: 0.5 cm)] (Lee et al. 2022a)

shear present during low moisture extrusion, the ingredients may experience structural changes that could alter the properties of the product like microstructure or expansion (Brennan et al. 2013). Due to its expansion and low moisture content, textured vegetable protein has a hard texture and a porous structure, making hydration necessary before eating (Lee et al. 2022b). For instance, the textured meat analogue is rehydrated during low-moisture extrusion before being fried or cooked (Beck et al. 2018). Currently, the most popular method is the low moisture extrusion process for producing meat substitutes commercially, although it can cause loss of nutrients due to poor rate of rehydration of the product, high energy consumption, and high temperatures (McClements and Grossmann 2021). Meat analogues were produced using various rice protein blend ratios change from 0% to 100% (w/w) for soy protein. Depending on the rice protein ratio, while properties such as a reduction in particular mechanical energy, expansion rate, porosity, water holding capacity, chewiness, springiness, stickiness, and flexibility, properties such as part density, true density and hardness increase (Fig. 5.4) (Lee et al. 2022a).

### 5.4.4 Spinning

It is difficult to replicate the feel of meat using plant elements since they are different from those found in meat. For instance, fibrillar orientation is not a characteristic of plant proteins (Taherian et al. 2011). Studies are carried out to make it possible for fibrous material obtained from plant-derived components to resemble muscle tissue can make meat analogues more similar to the nutritional value of animal products, and functional, and organoleptic qualities (Dekkers et al. 2018). Spinning which is one of these methods is an advanced technique for making delicate fibers using high-speed spinning to make food analogue, especially meat, from plant protein concentrated. There are risks involved and expensive procedure that utilizes a lot of acidic/alkaline solvents and water, and generates a lot of waste (Singh et al. 2021). Wet-spinning and electrospinning are the two spinning methods that have been described to create proteins with fibrous structures (Boyer 1954).

A solution containing proteins is spun out of a spinneret and then plunged into a non-protein-solvent bath. Stretched filaments are produced when precipitate from the phase of extruded proteins and hardens as a consequence of a conversation between the solvent and non-solvent that are around 20 μm thick (Boyer 1954). Forced biopolymer solution through a spinneret or hollow needle during electrospinning which has a potential difference between it and a ground electrode. Surface instability is brought on by the development of charge on the surface of particles as it eventually derives from the spinneret as extremely thin fibers (less than 100 nm), which re attracted to an electrode on the ground (Schiffman and Schauer 2008). Fibers for the use of meat analogues can be made via electrospinning, although most applications where nanofibers are employed as delivery mechanisms or carriers or bioactive components are covered for food-grade electrospinning, such as polyphenols and probiotics (Dekkers et al. 2018). Numerous research demonstrated the manufacture of food-grade fibers from plant-based sources such as soy, pea, and faba bean (Cui et al. 2022; Gallant et al. 1984). The electrospinning technology has been used to create nanofibers, but it is also a promising method for creating meat substitutes from entire muscle (Zhang et al. 2021). In particular, the wet spinning technique can be said to be receiving increasing attention as a potential way to replicate the fibers found in animal muscles (Cui et al. 2023). In the wet spinning technique, a fiber bundle is obtained by extruding the protein solution with a measuring pump at a certain temperature and flow rate. The fiber bundle is then cleaned in a wash bath before being picked up by a collector (Fig. 5.5) (Cui et al. 2022).

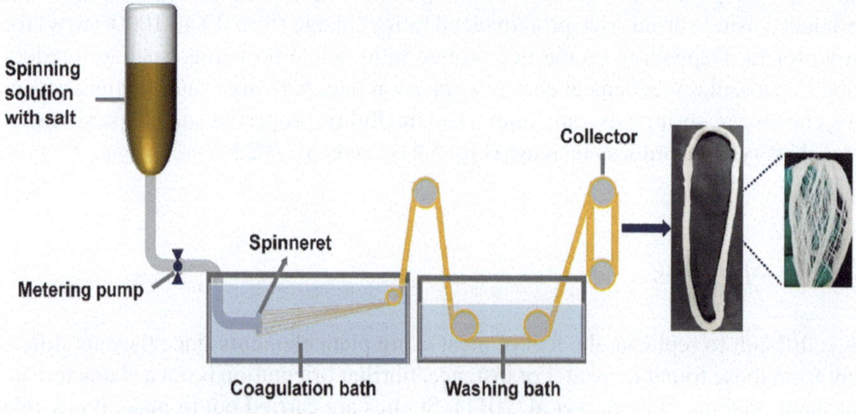


**Fig. 5.5** Process flow diagram wet-spinning (Cui et al. 2023)

### 5.4.5 Shear Cell and Couette Cell Technique

The first research on shear cell technology was published by (Manski et al. 2007). Alternatives to batch-process beef using shear cell technology are as effective as high-moisture extrusion (Grabowska et al. 2016). Conical shear cells or cylindrical Couette cells, which both rely on the concentric rheometer theory, are used for the technique. Shear cell device's upper cone is stationary while the lower cone rotates. To regulate the processing temperature, a heated bath or electric steamer is utilized. When the dough is heated and simply sheared, clearly defined fibrous structures are created. These structures are stable after cooling and do not subsequently get twisted (Manski et al. 2008). This method is a cost-effective technology that creates larger, softer, and more fibrous meat substitutes (Singh et al. 2021). Three fundamental phases are needed to create a meat analog in the shear cell process: cooling, thermo-mechanical treatment, and hydration. To prepare the dough, the protein isolate or concentrate is blended with water with a spatula, depending on the size of the process, either by hand or using a Z-blade mixer. At room temperature, the water-protein solution is allowed to hydrate. After hydration, a different protein or polysaccharide is injected, and the mixture is then continued to create dough. The dough is moved to the heated shredding machine, and the process of shredding is initiated (Fig. 5.6a). (Cornet et al. 2022). Production of meat analogues from pea protein isolate and wheat gluten mixtures using the shear cell technique was studied (Taghian Dinani et al. 2023). A sample provided at 120 °C, 20 rpm, and 30 min in a study in which a soy protein isolate-gluten mixture was handled by simple shear flow and heat under variable processing conditions are shown in Fig 5.6b (Krintiras

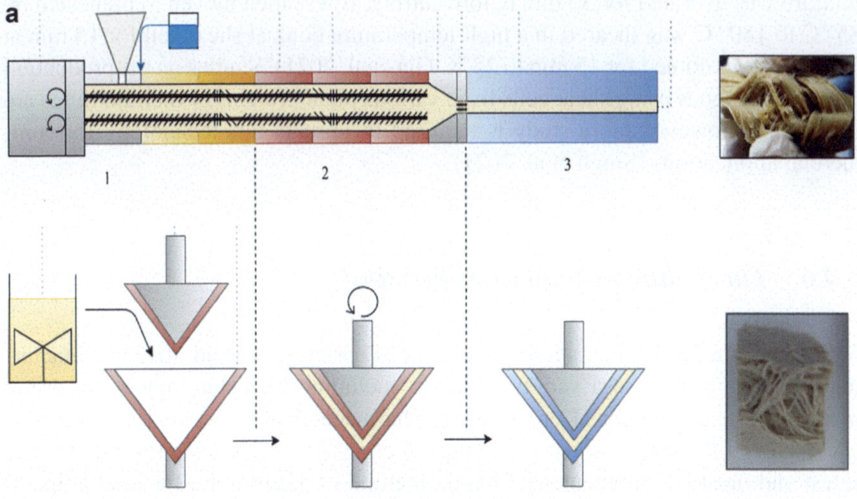


**Fig. 5.6** (**a**) A conical shear cell and an extruder (Cornet et al. 2022) (1) mixing and hydration; (2) thermomechanical processing; (3) cooling. (**b**) Meat analogue produced from wheat gluten: pea protein isolate using shear cell technique (Krintiras et al. 2016)

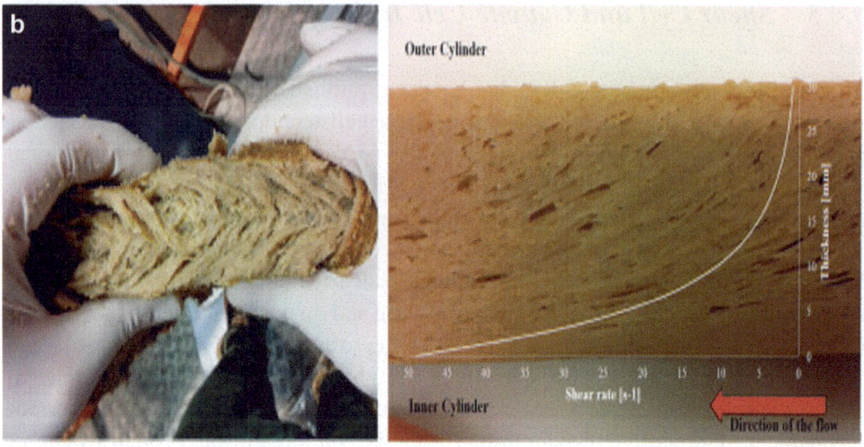


**Fig. 5.6** (continued)

et al. 2016). In another study, the mixture of pea protein isolate and wheat gluten was used to produce food analogue which have fibrous morphology by shear-cell. The fibers were visually examined and analyzed at different temperatures. When processed at 95 °C, the pea protein-wheat gluten mixture produced a weak and fragile dough. Higher temperatures produced different results: 110 °C produced a weak gel with small fibers, whereas 120 °C produced a fibrous product (Schreuders et al. 2019). Jia et al. (2021) reported that rapeseed protein concentrate-wheat gluten blends can be converted into fibrous products when processed with 40% by weight dry matter at 140 °C and 150 °C by high-temperature conical shear cell. The dough mixture was hydrated for 30 min before cutting. It was then moved to preheated by 85 °C to 150 °C was sheared in a high-temperature conical shear cell for 15 min at 30 rpm. It was cooled for 15 min to 25 °C (Jia et al. 2021). Studies on the production of food analogues using shear cells from various plant proteins or their mixtures are increasing. However, more study is required to upscale this technology for commercial applications (Singh et al. 2021).

### *5.4.6 Three-dimensional (3D) Printing*

Solid freeform fabrication, another name for 3D printing, and additive manufacturing, has several uses in a variety of fields, including biotechnology engineering, mechanical and biomedical engineering, pharmaceutical, food, and avionics science, as well as design science (Portanguen et al. 2019). Extrusion printing is the safest and most environmentally friendly method of 3D printing for food preparation (Zhang et al. 2022a). A detailed study of the used parts and process ideas to improve printing efficiency is required to develop the 3D printing method as an efficient tool to manufacture 3D food structures. In this way, the rheological characteristics of printed inks may have a direct impact on the printing effectiveness and form accuracy of printed objects. In order to let nozzle tip extrusion occur properly, typically, a printed biopolymeric solution must exhibit a well-defined shear-thinning tendency, and possess a viscoelastic nature and thixotropic properties. The ability of components to enhance the performance of the printed 3D objects in the end is determined by their compatibility with the printing process based on the flow behavior of the 3D printed inks, which is frequently regarded as the quality indicator of the end goods (Godoi et al. 2016; Shahbazi et al. 2021). The materials used in the four main 3D printing techniques are cultured muscle cells in vitro, animal byproducts and waste, insects, and plants. Despite the fact that several studies have concentrated on using 3D printing to make food products, including meat, there have been not many studies on the use of this technology (Ramachandraiah 2021).

The popularity of plant-based meat substitutes produced using 3D food printing technology has grown recently. In order to test whether cricket (*Gryllus bimaculatus*) gel might be used as a meat substitute, it was fractionated, combined with various amounts of soybean protein isolate, and its properties were assessed in a study. 5–20% soy protein isolate was added to the filtrate, supernatant, and pellet fractions of *Gryllus bimaculatus,* and the properties of the produced 3D meat analogues were investigated. As a result, a mixture of 5–15% has been reported to meet precise dimensional requirements in the pellet fraction with high resolution (Nam et al. 2023). In another study, for the creation of fish-like 3D models, Shi et al. (2023) using soy protein isolate, xanthan gum, and rice starch as food ink. By changing the soy protein isolate concentration from 10 to 30 wt% and the xanthan gum concentration from 1 to 5 wt% while keeping the rice starch concentration at 15 wt%, the printable zone of food ink was identified. The 3D visual meat analogue was produced from 20% soy protein isolate, 3% xanthan gum, and 15% rice starch mixture at nozzle inner diameter 200 μm, extrusion flow rate 32.56 mL/min, and XY moving speed 1200–1500 mm/min are shown in Fig. 5.7 (Shi et al. 2023). In another study, a new generation hybrid meat analogue produced by mixing both plant and animal protein components was produced using chicken mince and pea protein isolate by the 3D printer (Wang et al. 2022). Utilizing waste materials for 3D printing can boost the sustainability of food supply systems based on plants and animals. The consequences for the sustainability of extensive food analogue production must be evaluated while this unique technology is still in its infancy (Ramachandraiah 2021).

## 5.5 Conclusion and Future Aspects

Animal-derived protein does seem to be significant for humans from infancy through adulthood. However, it is important to recognize any potential health risks linked to a diet high in protein from animal sources (Hoffman and Falvo 2004). The typical foods should have more high-quality protein, be readily available, and be practical financially. Especially plant proteins are an important alternative source of animal proteins. Typically, plant proteins are less functional than those derived from

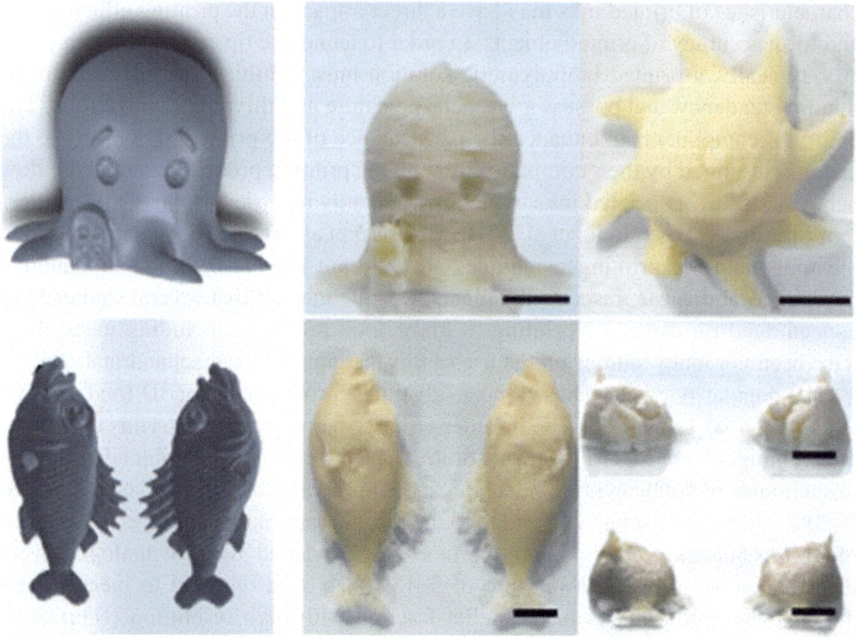

**Fig. 5.7** Meat analogues produced via high-precision 3D printing (Shi et al. 2023)

animals because they need more effort to process and are more vulnerable to external variables like temperature, pH, and ionic strength (Nasrabadi et al. 2021). Alternative protein sources mentioned in this study (plant-based, insects, and algae) may offer some advantages as protein from animal sources if the right sources are used in the right proportions and the right method. The protein-rich food analogues produced from the right source of protein and with true protein combination can successfully meet the daily protein needs of the consumer.

In general, the functional qualities of widely accessible protein-rich ingredients, the majority of which are highly refined from plant matter, are not ideal for applications involving food analogues especially. The need for additives to enhance the texture of the products serves as evidence of this (Kyriakopoulou et al. 2021). These problems can be solved by using appropriate additives as well as mixing alternative protein sources in the appropriate ratio. Recent food analogue research and development projects have demonstrated that using plant, insect, or algae-based proteins alone or in combination with other proteins and technologies like extrusion, electrospinning, shear cell, 3D printing, shredding, and mixing, it is feasible to create meat, cheese, and eggs analogues.

Appropriate improvements in marketing approaches, nutritional quality, shelf life, taste, and aroma need to be considered for food analogues to become widespread. In addition, it is an important element to ensure consumer acceptance by offering the product closest to the original.

## References


Ahmed NS (2014) Production of functional spreadable processed cheese analogue supplemented with chickpea. Int J Dairy Sci 9(1):1–14

Aini N, Prihananto V, Sustriawan B, Romadhon D, Ramadhan RN (2019) The formulation of cheese analogue from sweet corn extract. Int J Food Sci 2019:8624835

Akdogan H (1999) High moisture food extrusion. Int J Food Sci Technol 34(3):195–207

Akhtar S, Ismail T, Layla A, Hussain M, Qamar M (2022) An overview of plant-based protein rich products. In: Plant protein foods. Springer, Cham, pp 27–60

Amagliani L, O'Regan J, Kelly AL, O'Mahony JA (2017) The composition, extraction, functionality and applications of rice proteins: a review. Trends Food Sci Technol 64:1–12

Asgar M, Fazilah A, Huda N, Bhat R, Karim A (2010) Nonmeat protein alternatives as meat extenders and meat analogs. Compr Rev Food Sci Food Saf 9(5):513–529

Azzollini D, Wibisaphira T, Lakemond CMM, Fogliano V (2019) Toward the design of insect-based meat analogue: the role of calcium and temperature in coagulation behavior of Alphitobius diaperinus proteins. LWT 100:75–82. https://doi.org/10.1016/j.lwt.2018.10.037

Banovic M, Sveinsdóttir K (2021) Importance of being analogue: female attitudes towards meat analogue containing rapeseed protein. Food Control 123:107833. https://doi.org/10.1016/j.foodcont.2020.107833

Beck SM, Knoerzer K, Foerster M, Mayo S, Philipp C, Arcot J (2018) Low moisture extrusion of pea protein and pea fibre fortified rice starch blends. J Food Eng 231:61–71

Becker EW (2007) Micro-algae as a source of protein. Biotechnol Adv 25(2):207–210. https://doi.org/10.1016/j.biotechadv.2006.11.002

Bhuva V, Morya S, Borah A (2021) A review on meat analogue: is this time to see the algal proteins as a sustainable substitute for the meat proteins. Pharma Innov J 10(5):1160–1168

Bleakley S, Hayes M (2017) Algal proteins: extraction, application, and challenges concerning production. Food Secur 6(5):33

Boye J, Zare F, Pletch A (2010) Pulse proteins: processing, characterization, functional properties and applications in food and feed. Food Res Int 43(2):414–431

Boyer RA (1954) High protein food product and process for its preparation. Google Patents

Brennan MA, Derbyshire E, Tiwari BK, Brennan CS (2013) Ready-to-eat snack products: the role of extrusion technology in developing consumer acceptable and nutritious snacks. Int J Food Sci Technol 48(5):893–902

Caporgno MP, Mathys A (2018) Trends in microalgae incorporation into innovative food products with potential health benefits. Front Nutr 5:58

Chiang JH, Loveday SM, Hardacre AK, Parker ME (2019) Effects of soy protein to wheat gluten ratio on the physicochemical properties of extruded meat analogues. Food Struct 19:100102. https://doi.org/10.1016/j.foostr.2018.11.002

Chiang JH, Tay W, Ong DSM, Liebl D, Ng CP, Henry CJ (2021) Physicochemical, textural and structural characteristics of wheat gluten-soy protein composited meat analogues prepared with the mechanical elongation method. Food Struct 28:100183. https://doi.org/10.1016/j.foostr.2021.100183

Choi BD, Wong NA, Auh J-H (2017) Defatting and sonication enhances protein extraction from edible insects. Korean J Food Sci Anim Resour 37(6):955

Clune S, Crossin E, Verghese K (2017) Systematic review of greenhouse gas emissions for different fresh food categories. J Clean Prod 140:766–783

Cornet SH, Snel SJ, Schreuders FK, van der Sman RG, Beyrer M, van der Goot AJ (2022) Thermo-mechanical processing of plant proteins using shear cell and high-moisture extrusion cooking. Crit Rev Food Sci Nutr 62(12):3264–3280

Cui B, Liang H, Li J, Zhou B, Chen W, Liu J, Li B (2022) Development and characterization of edible plant-based fibers using a wet-spinning technique. Food Hydrocoll 133:107965. https://doi.org/10.1016/j.foodhyd.2022.107965

Cui B, Mao Y, Liu J, Liang X, Wu D, Chen X et al (2023) Effect of salt on solution behavior of spinning medium and properties of meat analogue fibers. Food Hydrocoll 139:108540. https://doi.org/10.1016/j.foodhyd.2023.108540

Davidov-Pardo G, Joye IJ, McClements DJ (2015) Encapsulation of resveratrol in biopolymer particles produced using liquid antisolvent precipitation. Part 1: preparation and characterization. Food Hydrocoll 45:309–316

Dawczynski C, Schubert R, Jahreis G (2007) Amino acids, fatty acids, and dietary fibre in edible seaweed products. Food Chem 103(3):891–899

Day L, Swanson BG (2013) Functionality of protein-fortified extrudates. Compr Rev Food Sci Food Saf 12(5):546–564

De Angelis D, Kaleda A, Pasqualone A, Vaikma H, Tamm M, Tammik M-L et al (2020) Physicochemical and sensorial evaluation of meat analogues produced from dry-fractionated pea and oat proteins. Food Secur 9(12):1754

Dekkers BL, Boom RM, van der Goot AJ (2018) Structuring processes for meat analogues. Trends Food Sci Technol 81:25–36

Deng Q, Wang L, Wei F, Xie B, Huang F, Huang W et al (2011) Functional properties of protein isolates, globulin and albumin extracted from Ginkgo biloba seeds. Food Chem 124(4):1458–1465

Ding Y, Gao Y, Sun J, Li C, Yue X, Shao J (2022) Influence of ultrasonic treatment on functional properties and structure of tussah pupa protein isolate. J Insects Food Feed 8(10):1133–1148

Djemaoune Y, Cases E, Saurel R (2019) The effect of high-pressure microfluidization treatment on the foaming properties of Pea Albumin aggregates. J Food Sci 84(8):2242–2249

Dobraszczyk BJ, Ainsworth P, Ibanoglu S, Bouchon P (2005) Baking, extrusion and frying. In: Food processing handbook. Wiley, Weinheim, pp 237–290

Ferawati F, Hefni M, Östbring K, Witthöft C (2021) The application of pulse flours in the development of plant-based cheese analogues: proximate composition, color, and texture properties. Food Secur 10(9):2208

Fox P, Kelly A, Yada R (2004) Proteins in food processing. CRC Press, Boca Raton

Galanakis CM (2019) Proteins: sustainable source, processing and applications. Academic

Gallant D, Bouchet B, Culioli J (1984) Ultrastructural aspects of spun pea and fababean proteins. Food Microstruct 3:175–183

Glusac J, Davidesko-Vardi I, Isaschar-Ovdat S, Kukavica B, Fishman A (2018) Gel-like emulsions stabilized by tyrosinase-crosslinked potato and zein proteins. Food Hydrocoll 82:53–63

Godoi FC, Prakash S, Bhandari BR (2016) 3d printing technologies applied for food design: status and prospects. J Food Eng 179:44–54

Grabowska KJ, Zhu S, Dekkers BL, de Ruijter NC, Gieteling J, van der Goot AJ (2016) Shear-induced structuring as a tool to make anisotropic materials using soy protein concentrate. J Food Eng 188:77–86

Gravel A, Doyen A (2020) The use of edible insect proteins in food: challenges and issues related to their functional properties. Innovative Food Sci Emerg Technol 59:102272. https://doi.org/10.1016/j.ifset.2019.102272

Grossmann L, Weiss J (2021) Alternative protein sources as technofunctional food ingredients. Annu Rev Food Sci Technol 12:93–117

Herz E, Herz L, Dreher J, Gibis M, Ray J, Pibarot P et al (2021) Influencing factors on the ability to assemble a complex meat analogue using a soy-protein-binder. Innovative Food Sci Emerg Technol 73:102806. https://doi.org/10.1016/j.ifset.2021.102806

Ho KK, Schroën K, San Martín-González MF, Berton-Carabin CC (2017) Physicochemical stability of lycopene-loaded emulsions stabilized by plant or dairy proteins. Food Struct 12:34–42

Hoffman JR, Falvo MJ (2004) Protein–which is best? J Sports Sci Med 3(3):118

Islam M, Huang Y, Jain P, Fan B, Tong L, Wang F (2023) Enzymatic hydrolysis of soy protein to high moisture textured meat analogue with emphasis on antioxidant effects: as a tool to improve techno-functional property. Biocatal Agric Biotechnol 50:102700

Jia W, Curubeto N, Rodríguez-Alonso E, Keppler JK, van der Goot AJ (2021) Rapeseed protein concentrate as a potential ingredient for meat analogues. Innovative Food Sci Emerg Technol 72:102758. https://doi.org/10.1016/j.ifset.2021.102758

Jia W, Sutanto IR, Ndiaye M, Keppler JK, van der Goot AJ (2022) Effect of aqueous ethanol washing on functional properties of sunflower materials for meat analogue application. Food Struct 33:100274. https://doi.org/10.1016/j.foostr.2022.100274

Jiri M, Otakar R, Marie B, Martina B (2014) A comprehensive look at the possibilities of edible insects as food in Europe–a review. Pol J Food Nutr Sci 64(3)

Julio LM, Ruiz-Ruiz JC, Tomás MC, Segura-Campos MR (2019) Chia (Salvia hispanica) protein fractions: characterization and emulsifying properties. J Food Meas Charact 13:3318–3328

Kavle RR, Pritchard ETM, Carne A, Bekhit AE-DA, Morton JD, Agyei D (2023) Nutritional composition and techno-functional properties of sago palm weevil (Rhynchophorus ferrugineus) larvae protein extract. J Asia Pac Entomol 26(2):102086. https://doi.org/10.1016/j.aspen.2023.102086

Kendler C, Duchardt A, Karbstein HP, Emin MA (2021) Effect of oil content and oil addition point on the extrusion processing of wheat gluten-based meat analogues. Food Secur 10(4):697

Khazaei H, Subedi M, Nickerson M, Martínez-Villaluenga C, Frias J, Vandenberg A (2019) Seed protein of lentils: current status, progress, and food applications. Food Secur 8(9):391

Kim H-W, Setyabrata D, Lee YJ, Jones OG, Kim YHB (2016) Pre-treated mealworm larvae and silkworm pupae as a novel protein ingredient in emulsion sausages. Innovative Food Sci Emerg Technol 38:116–123

Kim T-K, Yong HI, Jeong CH, Han SG, Kim Y-B, Paik H-D, Choi Y-S (2019a) Technical functional properties of water-and salt-soluble proteins extracted from edible insects. Food Sci Anim Res 39(4):643

Kim T-K, Yong HI, Kim Y-B, Kim H-W, Choi Y-S (2019b) Edible insects as a protein source: a review of public perception, processing technology, and research trends. Food Sci Anim Res 39(4):521

Kim T-K, Yong HI, Kang M-C, Jung S, Jang HW, Choi Y-S (2021) Effects of high hydrostatic pressure on technical functional properties of edible insect protein. Food Sci Anim Res 41(2):185

Kim J-W, Gu B-J, Ryu G-H (2022) Effects of isolated rice protein addition on the physicochemical characteristics of extruded low-moisture meat analog. J Korean Soc Food Sci Nutr 51(9):976–984

Kim Y, Kim T, Choi Y, Lee J, Jang H, Kang M, Choi Y (2023) Effects of hydrolysis and swelling on structural and functional properties of Hermetia illucens L.: insoluble protein in residue fraction. J Insects Food Feed 9:1–10

Kimura A, Fukuda T, Zhang M, Motoyama S, Maruyama N, Utsumi S (2008) Comparison of physicochemical properties of 7S and 11S globulins from pea, fava bean, cowpea, and french bean with those of soybean french bean 7S globulin exhibits excellent properties. J Agric Food Chem 56(21):10273–10279

Kolb N, Vallorani L, Stocchi V (1999) Chemical composition and evaluation of protein quality by amino acid score method of edible brown marine algae Arame (Eisenia bicyclis) and Hijiki (Hijikia fusiforme). Acta Aliment 28(3):213–222

Krintiras GA, Gadea Diaz J, van der Goot AJ, Stankiewicz AI, Stefanidis GD (2016) On the use of the Couette Cell technology for large scale production of textured soy-based meat replacers. J Food Eng 169:205–213. https://doi.org/10.1016/j.jfoodeng.2015.08.021

Kurek MA, Onopiuk A, Pogorzelska-Nowicka E, Szpicer A, Zalewska M, Półtorak A (2022) Novel protein sources for applications in meat-alternative products—insight and challenges. Food Secur 11(7):957

Kyriakopoulou K, Dekkers B, van der Goot AJ (2019) Plant-based meat analogues. In: Sustainable meat production and processing. Elsevier, pp 103–126

Kyriakopoulou K, Keppler JK, van der Goot AJ (2021) Functionality of ingredients and additives in plant-based meat analogues. Food Secur 10(3):600

Lakemond CM, de Jongh HH, Hessing M, Gruppen H, Voragen AG (2000) Soy glycinin: influence of pH and ionic strength on solubility and molecular structure at ambient temperatures. J Agric Food Chem 48(6):1985–1990

Lam A, Can Karaca A, Tyler R, Nickerson M (2018) Pea protein isolates: structure, extraction, and functionality. Food Rev Intl 34(2):126–147
Lee J-S, Choi I, Han J (2022a) Construction of rice protein-based meat analogues by extruding process: effect of substitution of soy protein with rice protein on dynamic energy, appearance, physicochemical, and textural properties of meat analogues. Food Res Int 161:111840. https://doi.org/10.1016/j.foodres.2022.111840
Lee J-S, Oh H, Choi I, Yoon CS, Han J (2022b) Physico-chemical characteristics of rice protein-based novel textured vegetable proteins as meat analogues produced by low-moisture extrusion cooking technology. LWT 157:113056. https://doi.org/10.1016/j.lwt.2021.113056
Lee JH, Kim T-K, Park S-Y, Kang M-C, Cha JY, Lim M-C, Choi Y-S (2023a) Effects of blanching methods on nutritional properties and physicochemical characteristics of hot-air dried edible insect larvae. Food Sci Anim Res 43(3):428
Lee JH, Kim T-K, Yong HI, Cha JY, Song K-M, Lee HG et al (2023b) Peptides inhibiting angiotensin-I-converting enzyme: isolation from flavourzyme hydrolysate of Protaetia brevitarsis larva protein and identification. Food Chem 399:133897. https://doi.org/10.1016/j.foodchem.2022.133897
Levy R, Okun Z, Shpigelman A (2022) Utilizing high-pressure homogenization for the production of fermented plant-protein yogurt alternatives with low and high oil content using potato protein isolate as a model. Innovative Food Sci Emerg Technol 75:102909
Lima M, Costa R, Rodrigues I, Lameiras J, Botelho G (2022) A narrative review of alternative protein sources: highlights on meat, fish, egg and dairy analogues. Foods 11(14):2053
Lin S, Huff H, Hsieh F (2000) Texture and chemical characteristics of soy protein meat analog extruded at high moisture. J Food Sci 65(2):264–269
Lin Q, Pan L, Deng N, Sang M, Cai K, Chen C et al (2022) Protein digestibility of textured-wheat-protein (TWP) -based meat analogues: (I) effects of fibrous structure. Food Hydrocoll 130:107694. https://doi.org/10.1016/j.foodhyd.2022.107694
Liu KS, Hsieh F-H (2007) Protein–protein interactions in high moisture-extruded meat analogs and heat-induced soy protein gels. J Am Oil Chem Soc 84(8):741–748. https://doi.org/10.1007/s11746-007-1095-8
Liu S, Peng M, Tu S, Li H, Cai L, Yu X (2005) Development of a new meat analog through twin-screw extrusion of defatted soy flour-lean pork blend. Food Sci Technol Int 11(6):463–470
Lonnie M, Laurie I, Myers M, Horgan G, Russell WR, Johnstone AM (2020) Exploring health-promoting attributes of plant proteins as a functional ingredient for the food sector: a systematic review of human interventional studies. Nutrients 12(8):2291
Loveday SM (2020) Plant protein ingredients with food functionality potential. Nutr Bull 45(3):321–327
Lu Z, Lee P-R, Yang H (2022) Chickpea flour and soy protein isolate interacted with κ-carrageenan via electrostatic interactions to form egg omelets analogue. Food Hydrocoll 130:107691. https://doi.org/10.1016/j.foodhyd.2022.107691
Lu Z, Liu Y, Lee YEJ, Chan A, Lee P-R, Yang H (2023) Effect of starch addition on the physicochemical properties, molecular interactions, structures, and in vitro digestibility of the plant-based egg analogues. Food Chem 403:134390
Malav O, Talukder S, Gokulakrishnan P, Chand S (2015) Meat analog: a review. Crit Rev Food Sci Nutr 55(9):1241–1245
Manickavasagan A, Lim L-T, Ali A (2022) Plant protein foods. Springer, Cham
Manski JM, van der Goot AJ, Boom RM (2007) Formation of fibrous materials from dense calcium caseinate dispersions. Biomacromolecules 8(4):1271–1279
Manski JM, van der Zalm EE, van der Goot AJ, Boom RM (2008) Influence of process parameters on formation of fibrous materials from dense calcium caseinate dispersions and fat. Food Hydrocoll 22(4):587–600
Mattice KD, Marangoni AG (2020) Comparing methods to produce fibrous material from zein. Food Res Int 128:108804
Maurya A, Said P (2014) Extrusion processing on physical and chemical properties of protein rich products-an overview. J Bioresource Eng Technol 2(4):61–67

McClements DJ, Grossmann L (2021) The science of plant-based foods: constructing next-generation meat, fish, milk, and egg analogs. Compr Rev Food Sci Food Saf 20(4):4049–4100
Mefleh M, Pasqualone A, Caponio F, De Angelis D, Natrella G, Summo C, Faccia M (2022) Spreadable plant-based cheese analogue with dry-fractioned pea protein and inulin–olive oil emulsion-filled gel. J Sci Food Agric 102(12):5478–5487
Mejri M, Rogé B, BenSouissi A, Michels F, Mathlouthi M (2005) Effects of some additives on wheat gluten solubility: a structural approach. Food Chem 92(1):7–15
Mintah BK, He R, Agyekum AA, Dabbour M, Golly MK, Ma H (2020) Edible insect protein for food applications: extraction, composition, and functional properties. J Food Process Eng 43(4):e13362
Mohamed A, Abo-El-Khair B, Shalaby SM (2013) Quality of novel healthy processed cheese analogue enhanced with marine microalgae Chlorella vulgaris biomass. World Appl Sci J 23(7):914–925
Moradi D, Ziarati P, Sawicka B (2021) Physicochemical and rheological properties of lentil milk/inulin blend—a feta cheese analogue. Legume Sci 3(4):e80
Mounsey JS (2009) Effect of wheat starch on imitation cheese texture. J Food Technol 7(2):30–33
Mundi S, Aluko R (2012) Physicochemical and functional properties of kidney bean albumin and globulin protein fractions. Food Res Int 48(1):299–306
Muthayya S, Sugimoto JD, Montgomery S, Maberly GF (2014) An overview of global rice production, supply, trade, and consumption. Ann N Y Acad Sci 1324(1):7–14
Nadeeshani H, Senevirathne N, Somaratne G, Bandara N (2022) Recent trends in the utilization of pulse protein in food and industrial applications. ACS Food Sci Technol 2(5):722–737. https://doi.org/10.1021/acsfoodscitech.1c00448
Nam HK, Kang TW, Kim I-W, Choi R-Y, Kim HW, Park HJ (2023) Physicochemical properties of cricket (Gryllus bimaculatus) gel fraction with soy protein isolate for 3D printing-based meat analogue. Food Biosci 102772. https://doi.org/10.1016/j.fbio.2023.102772
Nasrabadi MN, Doost AS, Mezzenga R (2021) Modification approaches of plant-based proteins to improve their techno-functionality and use in food products. Food Hydrocoll 118:106789
Nations U (2019) Growing at a slower pace, world population is expected to reach 9.7 billion in 2050 and could peak at nearly 11 billion around 2100: Department of Economic and Social Affairs. https://www.un.org/development/desa/en/news/population/world-population-prospects-2019.html
Nieuwland M, Geerdink P, Brier P, Van Den Eijnden P, Henket JT, Langelaan ML et al (2014) Reprint of "Food-grade electrospinning of proteins". Innovative Food Sci Emerg Technol 24:138–144
Norde W (2003) Colloids and interfaces in life sciences. CRC Press
Nwachukwu ID, Aluko RE (2018) Physicochemical and emulsification properties of flaxseed (Linum usitatissimum) albumin and globulin fractions. Food Chem 255:216–225
Oguntoyinbo SI, Taylor JR, Taylor J (2018) Comparative functional properties of kafirin and zein viscoelastic masses formed by simple coacervation at different acetic acid and protein concentrations. J Cereal Sci 83:16–24
Organization, W. H (1991) Protein quality evaluation: report of the joint FAO/WHO expert consultation, Bethesda 4–8 December 1989 (vol 51). Food & Agriculture Org
Osemwota EC, Alashi AM, Aluko RE (2021) Comparative study of the structural and functional properties of membrane-isolated and isoelectric pH precipitated green lentil seed protein isolates. Membranes 11(9):694
Osen R, Schweiggert-Weisz U (2016) High-moisture extrusion: meat analogues. https://doi.org/10.1016/B978-0-08-100596-5.03099-7
Osen R, Toelstede S, Wild F, Eisner P, Schweiggert-Weisz U (2014) High moisture extrusion cooking of pea protein isolates: raw material characteristics, extruder responses, and texture properties. J Food Eng 127:67–74

Osen R, Toelstede S, Eisner P, Schweiggert-Weisz U (2015) Effect of high moisture extrusion cooking on protein–protein interactions of pea (Pisum sativum L.) protein isolates. Int J Food Sci Technol 50(6):1390–1396

Palanisamy M, Töpfl S, Aganovic K, Berger RG (2018) Influence of iota carrageenan addition on the properties of soya protein meat analogues. LWT 87:546–552

Palanisamy M, Töpfl S, Berger RG, Hertel C (2019) Physico-chemical and nutritional properties of meat analogues based on Spirulina/lupin protein mixtures. Eur Food Res Technol 245:1889–1898

Papalamprou E, Doxastakis G, Biliaderis C, Kiosseoglou V (2009) Influence of preparation methods on physicochemical and gelation properties of chickpea protein isolates. Food Hydrocoll 23(2):337–343

Patil U, Benjakul S (2017) Characteristics of albumin and globulin from coconut meat and their role in emulsion stability without and with proteolysis. Food Hydrocoll 69:220–228

Pham TT, Tran TTT, Ton NMN, Le VVM (2017) Effects of pH and salt concentration on functional properties of pumpkin seed protein fractions. J Food Process Preserv 41(4):e13073

Phillips GO, Williams PA (2011) Handbook of food proteins. Elsevier

Pietsch VL, Emin MA, Schuchmann HP (2017) Process conditions influencing wheat gluten polymerization during high moisture extrusion of meat analog products. J Food Eng 198:28–35. https://doi.org/10.1016/j.jfoodeng.2016.10.027

Pietsch VL, Werner R, Karbstein HP, Emin MA (2019) High moisture extrusion of wheat gluten: relationship between process parameters, protein polymerization, and final product characteristics. J Food Eng 259:3–11. https://doi.org/10.1016/j.jfoodeng.2019.04.006

Portanguen S, Tournayre P, Sicard J, Astruc T, Mirade P-S (2019) Toward the design of functional foods and biobased products by 3D printing: a review. Trends Food Sci Technol 86:188–198

Ramachandraiah K (2021) Potential development of sustainable 3D-printed meat analogues: a review. Sustain For 13(2):938

Ran X, Lou X, Zheng H, Gu Q, Yang H (2022) Improving the texture and rheological qualities of a plant-based fishball analogue by using konjac glucomannan to enhance crosslinks with soy protein. Innov Food Sci Emerg Technol 75:102910. https://doi.org/10.1016/j.ifset.2021.102910

Riaz M (2011) Texturized vegetable proteins. In: Handbook of food proteins. Elsevier, pp 395–418

Rizvi S, Blaisdell J, Harper W (1980) Thermal diffusivity of model meat analog systems. J Food Sci 45(6):1727–1731

Samard S, Ryu GH (2019) Physicochemical and functional characteristics of plant protein-based meat analogs. J Food Process Preserv 43(10):e14123

Sánchez-Alonso I, Haji-Maleki R, Borderias AJ (2007) Wheat fiber as a functional ingredient in restructured fish products. Food Chem 100(3):1037–1043. https://doi.org/10.1016/j.foodchem.2005.09.090

Sarwar G (1997) The protein digestibility–corrected amino acid score method overestimates quality of proteins containing antinutritional factors and of poorly digestible proteins supplemented with limiting amino acids in rats. J Nutr 127(5):758–764

Schiffman JD, Schauer CL (2008) A review: electrospinning of biopolymer nanofibers and their applications. Polym Rev 48(2):317–352

Schreuders FKG, Dekkers BL, Bodnár I, Erni P, Boom RM, van der Goot AJ (2019) Comparing structuring potential of pea and soy protein with gluten for meat analogue preparation. J Food Eng 261:32–39. https://doi.org/10.1016/j.jfoodeng.2019.04.022

Searchinger T, Waite R, Hanson C, Ranganathan J, Dumas P, Matthews E, Klirs C (2019) Creating a sustainable food future: a menu of solutions to feed nearly 10 billion people by 2050. Final report: WRI

Shahbazi M, Jäger H, Ettelaie R, Chen J (2021) Construction of 3D printed reduced-fat meat analogue by emulsion gels. Part I: flow behavior, thixotropic feature, and network structure of soy protein-based inks. Food Hydrocoll 120:106967. https://doi.org/10.1016/j.foodhyd.2021.106967

Shewry PR (2009) Wheat. J Exp Bot 60(6):1537–1553

Shi H, Li J, Xu E, Yang H, Liu D, Yin J (2023) Microscale 3D printing of fish analogues using soy protein food ink. J Food Eng 347:111436. https://doi.org/10.1016/j.jfoodeng.2023.111436

Shimada K, Cheftel JC (1988) Texture characteristics, protein solubility, and sulfhydryl group/disulfide bond contents of heat-induced gels of whey protein isolate. J Agric Food Chem 36(5):1018–1025

Shukla R, Cheryan M (2001) Zein: the industrial protein from corn. Ind Crops Prod 13(3):171–192. https://doi.org/10.1016/S0926-6690(00)00064-9

Singh M, Trivedi N, Enamala MK, Kuppam C, Parikh P, Nikolova MP, Chavali M (2021) Plant-based meat analogue (PBMA) as a sustainable food: a concise review. Eur Food Res Technol 247:2499–2526

Smetana S, Larki NA, Pernutz C, Franke K, Bindrich U, Toepfl S, Heinz V (2018) Structure design of insect-based meat analogs with high-moisture extrusion. J Food Eng 229:83–85

Smetana S, Pernutz C, Toepfl S, Heinz V, Van Campenhout L (2019) High-moisture extrusion with insect and soy protein concentrates: cutting properties of meat analogues under insect content and barrel temperature variations. J Insects Food Feed 5(1):29–34

Sumarmono J, Sustriawan B, Aini N, Prihananto V, Widiastuti A (2021) The effect of addition of whey protein concentrate and emulsifier on characteristics of cheddar cheese analogue from corn milk. Paper presented at the IOP Conference Series: Earth and Environmental Science

Sun C, Ge J, He J, Gan R, Fang Y (2021) Processing, quality, safety, and acceptance of meat analogue products. Engineering 7(5):674–678. https://doi.org/10.1016/j.eng.2020.10.011

Taghian Dinani S, Charles Carrillo MF, Boom R, van der Goot AJ (2023) Quality improvement of plant-based meat alternatives by addition of iota carrageenan to pea protein–wheat gluten blend. Eur Food Res Technol. https://doi.org/10.1007/s00217-023-04244-7

Taherian AR, Mondor M, Labranche J, Drolet H, Ippersiel D, Lamarche F (2011) Comparative study of functional properties of commercial and membrane processed yellow pea protein isolates. Food Res Int 44(8):2505–2514

Tan SH, Mailer RJ, Blanchard CL, Agboola SO (2014) Emulsifying properties of proteins extracted from Australian canola meal. LWT-Food Sci Technol 57(1):376–382

Van Huis A (2013) Potential of insects as food and feed in assuring food security. Annu Rev Entomol 58:563–583

Van Krimpen M, Bikker P, Van der Meer I, Van der Peet-Schwering C, Vereijken J (2013) Cultivation, processing and nutritional aspects for pigs and poultry of European protein sources as alternatives for imported soybean products. Wageningen UR Livestock Research

Van Thielen L, Vermuyten S, Storms B, Rumpold B, Van Campenhout L (2019) Consumer acceptance of foods containing edible insects in Belgium two years after their introduction to the market. J Insects Food Feed 5(1):35–44

Vatansever S, Tulbek MC, Riaz MN (2020) Low-and high-moisture extrusion of pulse proteins as plant-based meat ingredients: a review. Cereal Foods World 65(4):12–14

Veraverbeke WS, Delcour JA (2002) Wheat protein composition and properties of wheat glutenin in relation to breadmaking functionality. Crit Rev Food Sci Nutr 42(3):179–208

Verbeek CJ, van den Berg LE (2010) Extrusion processing and properties of protein-based thermoplastics. Macromol Mater Eng 295(1):10–21

Verbeke W (2015) Profiling consumers who are ready to adopt insects as a meat substitute in a Western society. Food Qual Prefer 39:147–155

Wang T, Kaur L, Furuhata Y, Aoyama H, Singh J (2022) 3D printing of textured soft hybrid meat analogues. Food Secur 11(3):478

Wi G, Bae J, Kim H, Cho Y, Choi M-J (2020) Evaluation of the physicochemical and structural properties and the sensory characteristics of meat analogues prepared with various non-animal based liquid additives. Food Secur 9(4):461

Wieser H (2007) Chemistry of gluten proteins. Food Microbiol 24(2):115–119

Wild F, Czerny M, Janssen AM, Kole A, Zunabovic M, Domig KJ (2014) The evolution of a plant-based alternative to meat. Agro FOOD Industry Hi Tech 25(1):45–49

Xia S, Xue Y, Xue C, Jiang X, Li J (2022) Structural and rheological properties of meat analogues from Haematococcus pluvialis residue-pea protein by high moisture extrusion. LWT 154:112756. https://doi.org/10.1016/j.lwt.2021.112756

Xia S, Song J, Li K, Hao T, Ma C, Shen S et al (2023) Yeast protein-based meat analogues: Konjac glucomannan induces fibrous structures formation by modifying protein structure. Food Hydrocoll 142:108798

Yi L, Lakemond CMM, Sagis LMC, Eisner-Schadler V, van Huis A, van Boekel MAJS (2013) Extraction and characterisation of protein fractions from five insect species. Food Chem 141(4):3341–3348. https://doi.org/10.1016/j.foodchem.2013.05.115

Yuliarti O, Kiat Kovis TJ, Yi NJ (2021) Structuring the meat analogue by using plant-based derived composites. J Food Eng 288:110138. https://doi.org/10.1016/j.jfoodeng.2020.110138

Zahari I, Ferawati F, Helstad A, Ahlström C, Östbring K, Rayner M, Purhagen JK (2020) Development of high-moisture meat analogues with hemp and soy protein using extrusion cooking. Food Secur 9(6):772

Zayas JF, Zayas JF (1997) Solubility of proteins. In: Functionality of proteins in food. Springer, pp 6–75

Zhang J, Liu L, Jiang Y, Shah F, Xu Y, Wang Q (2020) High-moisture extrusion of peanut protein−/carrageenan/sodium alginate/wheat starch mixtures: effect of different exogenous polysaccharides on the process forming a fibrous structure. Food Hydrocoll 99:105311. https://doi.org/10.1016/j.foodhyd.2019.105311

Zhang L, Hu Y, Badar IH, Xia X, Kong B, Chen Q (2021) Prospects of artificial meat: opportunities and challenges around consumer acceptance. Trends Food Sci Technol 116:434–444

Zhang C, Wang C-S, Therriault D, Heuzey M-C (2022a) Development of aqueous protein/polysaccharide mixture-based inks for 3D printing towards food applications. Food Hydrocoll 131:107742

Zhang X, Zhao Y, Zhang T, Zhang Y, Jiang L, Sui X (2022b) High moisture extrusion of soy protein and wheat gluten blend: an underlying mechanism for the formation of fibrous structures. LWT 163:113561. https://doi.org/10.1016/j.lwt.2022.113561

Zhang M, Wang O, Cai S, Zhao L, Zhao L (2023) Composition, functional properties, health benefits and applications of oilseed proteins: a systematic review. Food Res Int 171:113061

Zielińska E, Baraniak B, Karaś M, Rybczyńska K, Jakubczyk A (2015) Selected species of edible insects as a source of nutrient composition. Food Res Int 77:460–466

# Chapter 6
# Vegetable Oils for Analogue Food Production

**Tugba Dursun Capar**

## 6.1 Vegetable Oils

Vegetable oils have been extracted from plants and seeds. Oils and fats are essential food ingredients and have applications in food preparation, frying, and cooking. They contribute a significant impact to the taste, flavor, and overall quality of the food. Vegetable oil production has increased gradually since the beginning of the century; it peaked in 2021–2022 at 208.81 million metric tons and was estimated to reach over 217.62 million metric tons in the 2022–2023 marketing year (Food and Agriculture Organization of the United Nations 2023).

Saturated fats (SFs) and oils are thought to be the less healthy, while health benefits have related to UFA in recent decades. On the other side, as consumer awareness of healthy foods has grown, researchers and the food industry have begun to concentrate on these kinds of products. Sunflower, soybean, palm, canola, peanut, safflower, and corn oil are the most common source of vegetable oils (Han et al. 2021). Soybean oil is the largest extensively produced plant-based oil. The second highly produced oil is palm oil which is predicted to suppress soybean oil during the text 10–15 years. Rapeseed/canola oil now occupies the third and in terms of the production of oils and fats, sunflower comes in fourth place. Soybean oil consist of neutral lipids, free fatty acids, and polar lipids. Also, it contains phytosterols, tocopherols and hydrocarbons, such as squalene. Soybean oil could be used in margarine, shortening and cooking oil in food applications. Minor components found in palm oil include squalene, aliphatic alcohols, tocotrienols, sterols, carotenoids, tocopherols, and tocotrienols. Palm olein frequently serves as an oil for cooking in

T. Dursun Capar (✉)
Erciyes University, Engineering Faculty, Food Engineering Department, Kayseri, Turkey

University of Massachusetts Amherst, Department of Food Science, Amherst, MA, USA
e-mail: tugbadursun@erciyes.edu.tr



Ö. P. CAN et al. (eds.), *Food Analogues: Emerging Methods and Challenges*,
https://doi.org/10.1007/978-3-031-69872-9_6

household and as well as industrial applications. Palm oil and its solid fraction stearin are typically used as frying oils for snacks, crackers, cookies, pastries, and as well as cacao butter replacer. Since frying is a heat operation, oil deteriorates quickly. One of the main advantages associated with palm oil, olein, and stearin is their oxidative stability. Sunflower oil, which is among the more widely used plant-based oils, may be preferred soybean, cottonseed, and palm oils. However, the sunflower is cultivated in a few geographical areas, giving it at a competitive disadvantage with other regional crops. Furthermore, its oxidative instability is one of the disadvantages in frying of food preparations. In comparison to sunflower, corn, and soybean oils, olive oil exhibits exceptional resistance to oil oxidation during frying and a reduced degree of modification as shown by evaluation of viscosity, total polar compounds, and tocopherol. Olive oil prevents rapid degradation at high temperatures due to its fatty acid content and the existence of native antioxidants including tocopherols, squalene, and $\Delta^5$-avenasterol (Ghouila et al. 2019; Rahim et al. 2023; Rani et al. 2021).

### 6.1.1 Fatty Acid Composition and Nutritional Benefits of Plant-Based Oils

Fatty acids (FAs) have a variety of function in human body and other organisms. FAs is one of the primary components of biological matter, along with proteins and carbohydrates. It is known that many FAs have anticancer potential (Kumar et al. 2016). The commonly found fatty acids in plant-based oils include palmitic ($C_{16:0}$), stearic ($C_{18:0}$), oleic ($C_{18:1}$), linoleic ($C_{18:2}$) and linolenic ($C_{18:3}$) acids (Table 6.1).

Since palm oil has an almost identical proportion of SFs to UFA, it has an equitable fatty acid constitution. Along with $C_{18:2}$ (10–11%), $C_{16:0}$ (44–45%) and $C_{18:1}$ (39–40%) are the main fatty acid components of palm oil that has a greater oxidative stability due to the lower concentration of linoleic and linolenic acids. Despite overwhelming evidence that the fatty acid erucic acid ($C_{22:1}$; n-9) in canola oil causes a greater risk to rat than to people, the stigma around this substance continues to exist. It has been enough to point out that the finding of $C_{22:1}$'s chain-reducing behavior to $C_{18:1}$ by peroxisomes during the last several decades constituted a significant advancement in our understanding of how fatty acids are synthesized. When the compound converts into oleic acid, $C_{22:1}$ acid residue is just as easy for mitochondria to break down as $C_{16:0}$ and other lipids. As the level of $C_{22:1}$ reduced, it resulted in marked increase in oleic acid. This oil's frying abilities were enhanced, and fried foods like potato chips and French fries had better storage quality. However, soybean and sunflower oil has low oxidative stability in frying applications on behalf of the highest amount of linoleic acid. $C_{18:1}$ and $C_{18:2}$ are present in almost equal amounts in sesame seed oil, along with trace amounts of SFs and linolenic acid. $C_{18:1}$ (33–54%) and $C_{18:2}$ (35–59%) constitute the largest portion the oil's

**Table 6.1** Fatty acid profile of commonly used edible plant-based oils

| Fatty acids (FA%) | Corn oil | Soybean oil | Sunflower oil | Peanut oil | Canola oil | Sesame oil | Palm oil | Olive oil | Safflower oil |
|---|---|---|---|---|---|---|---|---|---|
| C8:0 (Caprylic) | ND | ND | ND | ND | ND | ND | ND | ND | ND |
| C10:0 (Capric) | ND | ND | ND | ND | ND | ND | ND | ND | ND |
| C12:0 (Lauric) | ND | ND | ND | ND | ND | ND | 0.4 | ND | ND |
| C14:0 (Myristic) | ND | ND | ND | 0.1 | 0.1 | 0.1 | 1.4 | 0.0 | ND |
| C16:0 (Palmitic) | 12.94 | 11.67 | 6.43 | 10.4 | 5.1 | 9.6 | 47.5 | 12.1 | 6.1 |
| C18:0 (Stearic) | 2.12 | 3.87 | 3.69 | 3.0 | 1.7 | 5.0 | 4.8 | 2.6 | 2.3 |
| C20:0 (Arachidic) | 0.68 | 0.27 | 0.32 | 1.2 | 0.6 | 0.6 | 0.4 | 0.4 | 0.4 |
| C22:0 (Behenic) | 0.39 | ND | 0.66 | 2.30 | 0.3 | 0.2 | ND | 0.1 | 0.3 |
| C24:0 (Lignoceric) | 0.47 | ND | 0.29 | 1.4 | 0.2 | 0.1 | ND | 0.1 | 0.1 |
| C16:1(Palmitoleic) | 1.70 | ND | ND | 0.2 | 0.2 | 0.1 | 0.4 | 0.8 | 0.1 |
| C18:1n9c (Oleic) | 31.97 | 24.77 | 25.92 | 47.9 | 60.1 | 30.0 | 41.2 | 72.5 | 13.4 |
| C20:1n9 (Eicosenoic) | ND | ND | ND | 1.3 | 1.4 | ND | ND | 0.3 | 0.2 |
| C22:1n9 (Erucic) | ND | ND | ND | 0.1 | 0.4 | ND | ND | ND | ND |
| C18:2n6c (Linoleic) | 48.97 | 54.17 | 62.69 | 30.3 | 21.5 | 45.0 | 11.6 | 9.4 | 76.0 |
| C18:3n3c (Linolenic) | 0.76 | 5.16 | ND | 0.4 | 9.9 | 0.4 | 0.6 | 0.6 | 0.3 |
| TSFA | 16.60 | 15.90 | 11.39 | 18.3 | 8.0 | 15.7 | 15.3 | 15.3 | 9.1 |
| TMUFA | 33.67 | 24.77 | 25.92 | 49.6 | 62.4 | 40.1 | 41.6 | 73.8 | 13.9 |
| TPUFA | 49.74 | 59.33 | 62.69 | 30.8 | 31.5 | 45.7 | 12.2 | 10.0 | 77.3 |
| TPUFA/TSFA | 3.0 | 3.73 | 5.51 | 1.68 | 3.94 | 2.91 | 0.79 | 0.65 | 8.49 |

(continued)

**Table 6.1** (continued)

| Fatty acids (FA%) | Corn oil | Soybean oil | Sunflower oil | Peanut oil | Canola oil | Sesame oil | Palm oil | Olive oil | Safflower oil |
|---|---|---|---|---|---|---|---|---|---|
| Omega 6/Omega 3 | 64.15 | 10.50 | | 76 | 2.2 | 113 | | 16 | 253 |
| Reference | Dorni et al. (2017) | Dorni et al. (2017) | Dorni et al. (2017) | Christopoulou et al. (2004), Wang (2011) | Lee et al. (1998), Przybylski et al. (2005), Wang (2011) | Abou-Gharbia et al. (2000), Alpaslan et al. (2001), Christopoulou et al. (2004), Chung et al. (1995), Jahaniaval et al. (2000), Kochar (2011), Lee et al. (1998), Youdim et al. (2000) | Dubois et al. (2007) | Gunstone (2011), Lee et al. (1998), Mannina et al. (2003), Ollivier et al. (2003), Ranalli et al. (2002), Youdim et al. (2000) | Anwar et al. (2002), Christopoulou et al. (2004), Rallidis et al. (2003), Youdim et al. (2000) |

fatty acids, along with $C_{16:0}$ (8–17%) and $C_{18:0}$ (3–9%). Sesame oil is significantly more oxidation-resistant than other vegetable oils.

FAs can be divided based on their structural similarities as SFs, MUFA, and PUFA (Halvorsen and Blomhoff 2011). Regular use of SFs has been observed to raise cholesterol levels, which is associated with an increment in the mortality percentage from coronary heart disease (CHD) (Astrup et al. 2011; Weisburger 2000). It is well known that a healthy diet should only include a little quantity of SFs. In contrast, MUFA may protect against cardiovascular disease through numerous types of mechanism (López-Miranda et al. 2010). $C_{18:2}$ (n-6) and $C_{18:3}$ (n-3) are both PUFAs and cannot by synthesized by human body. Therefore, regular dietary intake of these important fatty acids is required. Clinical studies demonstrated that omega-3 fatty acids distribute substantial improvements to human well-being and useful in the treatment of inflammatory and autoimmune illnesses (Calder and Grimble 2002). In contrast, there is controversy concerning the influence of $C_{18:2}$ fatty acids on well-being (Jurić et al. 2020). Another important factor in preventing the conflicting effects of PUFAs on human health is maintaining an equivalent proportion of omega-3/omega-6. As found in eating habits of West, excess long chain fatty acids and increased proportion of C n-6/n-3 stimulate the formation of many illnesses, such as cardiovascular (CVD), cancer, and inflammations and immune disorders, while elevated amounts of n-3 suppress the pathogenesis of most diseases.

Fatty acid compositions of edible plant-based oils presented in Table 6.1. Soybean oil contains high amount of $C_{18:2}$ and less concentration of $C_{18:3}$ which are necessary fatty acids for individuals. Canola and olive oil are rich in oleic acid whereas, sunflower, safflower and soybean contain high amount of linoleic acid. The primary constituent fatty acids of peanut oils are oleic (18:1) and linoleic (18:2).

The location of product, climate, variety, and degree of fruit ripeness may have an impact on the fatty acid composition of olive oil, that can be differed against sample to sample. The primary fatty acid profile of olive oil is $C_{18:1}$, which constitutes to 72.5% of the overall.

The global stock has shown a trend of reduction due to excessive harvesting. As a result, the development of vegetable and plant-based sources of PUFAs has received growing interest. An increasing amount of research is pointing to the influence of fatty acids (FAs) in human nutrition, including their role in brain function, improvement and development of the human embryo, and prophylactic and therapeutic prevention of diseases like cardiovascular, inflammation, etc.

### 6.1.2 Phytosterol Proportion and Nutritional Benefits

Oils include crystalline alcohols called sterols, which can be esterified or unesterified. Processing of vegetable oil affects its sterol concentration, which can be reduced during deodorization. The chemical structure of the sterols is altered during refining as well.

Soybean oil contains very low level of phytosterols, fatty acid esters of phytosterols and sterol glycosides, which are further diminished during refining. Sterols constitute a significant portion of the unsaponifiable component of palm oil. Its products contain prevalent plant-based oil sterols, which are sitosterol, stigmasterol, and campesterol, with cholesterol having only a slight influence. Free and esterified sterols are both present in equal levels in canola oil. In the esterified and free sterol fractions of sitosterol and campesterol in canola oil are found similar amounts. Compared to esterified sterols, free sterols contained twice as much brassicasterol. The total sterol content in canola oil ranges from 0.7% to 1.0%. Brassicasterol is the main and special sterol for canola oil which could be used the authentication of different oils with canola oil.

Brassicasterol, ß-sitosterol, campesterol, stigmasterol, $\Delta^5$-avenasterol, $\Delta^7$-sitosterol, and $\Delta^7$-avenasterol are found in peanuts. These sterols which are second-generation of alcoholic compounds with 27–29 carbon molecules, become a shape of crystals at ambient temperature. ß-sitosterol which is the main constituent of peanut sterols, has been demonstrated to prevent the expansion of cancer cells and could provide preservation from colon, prostate, and breast cancer (Awad et al. 2000). Approximately 200 mg of sitosterol per 100 g exists in unrefined peanut oil, which is equivalent to the 220 mg of sitosterol per 100 g found in soybean oil (Awad et al. 2000). The four types of sterols that may be found in olive oil are: 4-desmethylsterols (4-methylsterols), 4-methylsterols (4,4-dimethylsterols), and triterpene dialcohols. Total sterol content in the sesame oils ranged from 0.51% to 0.76%, with desmethyl sterols accounting for 85–89%, monomethyl sterols for 9–11%, and dimethyl sterols for 2–4%. ß-sitosterol (62–67%), campesterol (15–20%), stigmasterol (5–8%), and $\Delta^5$- avenasterol (7–10%) are the primary sterols that exist in both free and esterified forms.

### 6.1.3 *Tocopherol Proportion and Nutritional Benefits*

Tocopherols are fat-soluble vitamins and plant-based oils include low amounts of tocopherols, which are all-natural antioxidants with varying levels of efficiency. Antioxidants that protect the oil from oxidation include carotenoids and tocopherols.

Soybean oil consist of at least four types of tocopherols which are γ, ß, δ, and α-tocopherol. Palm oil is high in vitamin A and contains tocopherols and tocotrienols, which are forms of vitamin E as well. Research on tocotrienols has increased recently as part of results demonstrating their significant efficacy in preventing heart illnesses and various types of cancer.

Tocopherol concentrations in canola oil are comparatively. The principal tocopherols present in canola oil are α-tocopherol and γ-tocopherol. The amount of tocopherol could be changed during to the processing such as extraction, refining and deodorization. It has been demonstrated that the tocopherol content of extracted peanut oil may reach 650 ppm, subject to the kind of peanut and its growing circumstances. α-tocopherol, which makes up to 95% of the total tocopherol content, is the main component of olive oil. The other tocopherols, γ and ß comprise up the remaining 5% of the composition. During the olive oil processing (refining, bleaching, and deodorizing) the tocopherol content may decrease. Crude sesame oils typically include tocopherols in the range of 400–700 mg/kg (ppm), with γ-tocopherol accounting for the majority (96–98%) and δ-tocopherol constituting the remainder (2–3%).

### 6.1.4 Caroteneoid Proportion and Nutritional Benefits

Over 700 natural carotenoids are known to exist in nature. The most well-known carotenoids are β-carotene, lycopene, lutein, zeaxanthin, flavaxanthin, canthaxanthin, capsanthin, capsorubin, and β-cryptoxanthin. Carotenoids are water-insoluble, lipophilic and the structure of these molecules is diverse. Due to carotenoids' singlet oxygen quencher ability, they prevent photooxidation of lipids.

Palm oil contains α-carotene, γ-carotene, ß-carotene, δ-carotene, ß-zeacarotene, α-zeacarotene and lycopene. In addition to serving as a natural food coloring for margarines, biscuits, and confectionery, the orange-red coloration of palm carotenoids is also a good source of pro-vitamin A. Carotenoids are the precursors of vitamins and are also known to have anti-carcinogenic characteristics. ß-carotene and lutein are the two primary carotenes present in olive oil. Olive oil has a total of 1–20 mg of carotenoids per kilogram. High level of carotenoids is reported in corn kernels however, the level of the carotenoid is relatively low in maize oil due to their low concentration in germ and partially removing during bleaching. Lutein and zeaxanthin are the two carotenoids that are most prevalent in maize oil.

## 6.2 Vegetable Oils for Analogue Dairy Foods

In today's food and recipes, a wide range of cheeses, including fresh cheese, matured cheese, spreadable cheese, and processed cheese, are utilized as ingredients. Depending on how they will be utilized, these cheeses' technological characteristics may vary. For instance, cheese with remarkable melt-ability properties and softness is preferred for pizza (Guinee 2016); nevertheless, high degree of melting ability might not be preferred for other meal dishes, such Palak paneer. Depending on the processes used during production, how the milk is handled, and the manufacturing method, different cheese varieties have different qualities (Lamichhane et al. 2018). The dairy product's composition, particularly its dry matter, fat, protein, and calcium content affects how it behaves (El-Bakry et al. 2018). Therefore, when creating an analogue cheese by substituting vegetable oil for milk fat, one must consider how the conventional product's altered composition may impair its technological capabilities and, consequently, its intended application. A cheese substitute is an item made using non-milk-based ingredients in place of either or both milk fat or milk protein (Guinee et al. 2004). The preparation involves mixing various edible fats/oils, proteins, emulsifying salts, and water until they are combined smoothly and uniformly with the use of heat and mechanical shear (Cunha et al. 2010). The market for cheese analogue has recently expanded globally due to its ease of production, the possibility of substituting dairy ingredients with more healthy and more economical ones made from plant origin, and decreased production expenses (Bachmann 2001). Studies of the physicochemical properties of dairy fat, its fractions, and mixtures with plant-based oils have received a lot of attention as they support the idea that products made with dairy fat and plant-based fats should have similar physicochemical and sensory characteristics. A material's structural properties (microstructure) influence its macroscopic behavior, such as rheological, sensorial, and textural (O'Riordan et al. 2011).

Fat is a key component of cheese. Rogers et al. (2010) suggested that cheese analogues produced with different fat contents might not be acceptable overall. Since the fatty acid composition, mostly UFA, fat replacement might affect the ripening biochemistry, microstructure, and sensory qualities (Moatsou et al. 2019). In addition, the textural issues in low-fat cheeses are the ones that are most frequently reported (Lobato-Calleros et al. 1997). Since suitable plant-based oils are the primary origin of bioactive lipids, they could be included in a range of cheese analogues at varying quantities to replace milk fat (Table 6.2). Several research have been published that vegetable-based oils such as soybean oil, sesame oil, palm oil, corn oil, or olive oil could be used in the production of cheese analogues (Fig. 6.1) (Kavak and Karabiyik 2019; Leong et al. 2020; Myhan et al. 2020). The ability of such analogues to succeed similarly to the original cheese is one of the concerns, as changes to the cheese's lipid composition may influence its stability and physical characteristics. In general, vegetable oils are less expensive than milk fat and are free of cholesterol. Cunha et al. (2010) reported that cheese became stickier and firmer as a result of a partial substitution of the dairy fat with plant-based fat. Plant oil-based cheeses have smoother textures than cheeses manufactured with dairy fat due to the creation of a distinct structure and consistency of dairy fat distribution.

According to studies, unsaturated fatty acid has been used to enhance the lipid profiles of cheese (Livingstone et al. 2012). The existence of short and intermediate fatty acids supplies activity to various essential units, including the muscles, heart, and liver (Miciński et al. 2012). Along with anti-inflammatory and antibacterial activities, it has also been given preventive capabilities against several diseases like ulcerative colitis and cancer (Tan et al. 2014). Despite these characteristics, research has revealed that consuming unsaturated fats has certain health benefits, such as preventing the oxidation of low-density lipoprotein (LDL) in the endothelium of the cardiovascular system. According to research by Briggs et al. (2017), a diet high in unsaturated fatty acid-rich foods may lower the possibility of CVD and lower LDL and total cholesterol levels. The analogues are also almost cholesterol-free since dairy fat is replaced with plant-based oils (Bachmann 2001). The use of plant-based oils in place of conventional oils and/or the enrichment of cheeses appear to be

**Table 6.2** Summary of the studies related to the replacement of dairy fat by plant-based oils

| Cheese variety | Dairy fat replacer | Fat substitution | Specifications | References |
|---|---|---|---|---|
| Cheddar cheese | Canola oil | 100% | Rheological and textural properties | Leong et al. (2020) |
| Kashar cheese | Palm oil | 0%, 50%, 100% | Physicochemical, textural, and sensory properties | Kavak and Karabiyik (2019) |
| Edam-type cheese | Palm oil | – | Rheological properties | Myhan et al. (2020) |
| Model processed cheese imitation | Canola oil | 26% | Microstructure, rheological properties and TPA | Ramel and Marangoni (2018) |
| White brined cheese | Sunflower oil | 20%, 30%, 40% | Physical and sensory properties | Shabani et al. (2016) |
| Processed cheese | Olive oil, corn oil, sesame oil | 20% | Chemical and sensory properties | Al-Ismail et al. (2015) |
| Tallaga cheese | Sunflower oil | 25%, 50%, 100% | Structure, texture, and sensory properties | Abd El-Salam (2015) |
| Model processed cheese analogue | Sunflower oil | 0%, 10%, 20%, 30%, 40% | Microstructure and rheological properties | Lee et al. (2015) |
| White Turkish cheese analogue | Corn oil | 0%, 50%, 60%, 100% | Texture and sensory analysis | Arslan et al. (2014) |
| White brined cheese | Olive oil, canola oil | 50%, 100% | Physicochemical and sensorial properties | Achachlouei et al. (2013) |
| Kasseri cheese | Soybean oil | 0%, 2.3%, 4.96%, 9.8% | Microbiological, physicochemical, rheological, and sensory properties | Kaminarides and Stachtiaris (2000) |
| Cheese analogues | Soybean oil, soybean fat | 17%, 33%, 50%, 66%, 100% | Texture and sensory | Lobato-Calleros et al. (1997) |

beneficial for the food industry and advantageous in avoiding the spread of several non-transmissible illnesses.

A cheese identical to Edam was created by Myhan et al. (2020) by substituting dairy fat with palm oil. The rheological characteristics of control Edam cheese and the alternative were examined throughout a 16-week maturation time. The substitution of palm oil for dairy fat had no effect on the rheological qualities of Edam-type cheese and the cheese-like product. There were no significant variations in the rheological qualities of the products at later stages of ripening.

In different research, Shabani et al. (2016) made a cheese analog based on ultrafiltered (UF) white-brined cheese by substituting sunflower oil for the milk fat and boiling it for varying amounts of time and temperatures. The final product's

**Fig. 6.1** Summary of the vegetable oils used in cheese analogues

suitability was determined using hardness, adhesiveness, meltability, and oiling-off. It was observed that partially replacing dairy milk with sunflower oil increased the cheese's adhesiveness, oiling-off, and spreadability whereas reducing its hardness.

Shaaban and Shaaban (2020) examined the oxidative stabilities of processed cheese analogues in which milk fat substituted with edible butter oil, rice bran oil, and palm oil. Rice bran oil exceeded other samples, such as butter oil and palm oil, in terms of its ability to scavenge free radicals. However, rice bran oil-containing samples had the greatest acid value and peroxide value after 3 months of storage at 5 °C and 25 °C.

The shape and position of fat globules have been studied in relation to converting from dairy fat to plant-based oils (Cunha et al. 2013; Lee et al. 2015; Leong et al. 2020). In the protein structure of cheese, fat globules are generally spherical and equally dispersed. When fat undergoes modification, either by replacement or fortification, it alters the matrix. For instance, certain cheese analogs had an appearance of increment in the dimension of their lipid globules, which might alter the long-term stability of the curd (Everett and Olson 2003). Regarding the meltability, some of the research reported the decrement in the meltability when dairy fat is substituted with plant-based oil. However, a study indicated the increment in the meltability after substitution (Totosaus et al. 2017).

As a result, the usage of vegetable oils may be beneficial in the substitution of milk fat. Their usage in cheese substitutes could be constrained, though, by possible "oiling-off" and off-flavor concerns. Regarding the findings of the studies, with the goal to build analogues that preserve conventional qualities, it is important to consider how full or limited substitution of dairy fat by plant-based oils influences the rheological characteristics of dairy products.

## 6.3 Vegetable Oils for Analogue Meat Products

Changes in consumer demands caused significant movements of the meat industry, like other food-related industries toward a healthy diet. Consumer interest in the potential benefits of diet-related improvements in health is one of the major themes driving changes in the use of meat analogues. The development of meat-based functional food is considered as a chance to update dietary targets for nutrients, improve their "image," and meet customer demands (Jiménez-Colmenero 2007).

Due to their significance, especially (in quantitative and qualitative terms) in the preparation of nutritious animal food, lipids are some of the bioactive constituents (essential components) that have attracted the most interest (Arihara 2006; Fernández-Ginés et al. 2005; Jiménez-Colmenero 2007). An increasing number of research suggests that dietary fat could be helpful in the management and avoidance of several chronic illnesses, including coronary heart disease. Numerous scientific institutions and dietary associations, including the World Health Organization (WHO and FAO 2003), have suggested recommendations for the ideal consumption of total and UFAs. Fifteen percent to 30% of total daily energy intake is the recommended range for dietary fat consumption. According to dietary recommendations, no more than 10% of total diet energy should be from SFAs, 6–10% from PUFAs (omega-6, 5–8%; omega-3, 1–2%), 10–15% from MUFAs, and less than 1% from trans fatty acids. Additionally, it is advised that daily cholesterol intake be controlled at 300 mg (WHO and FAO 2003). Certain recommendations are still based on these ratios considering the medical effects associated with eating oil vary by fatty acid levels. Thus, the n-6/n-3 PUFA proportion should not be more than 4 and the suggested range for PUFA to SFA is between 0.4 and 1.0 (Wood et al. 2004). Greater levels of n-3 PUFAs (and low n-6/n-3 PUFA ratios) have detrimental effects, whereas excessive levels of n-6 PUFAs (and high n-6/n-3 PUFA ratios) encourage the pathogenesis of a variety of conditions, such as cancer, inflammatory and autoimmune disorders, as well as CVD. Therefore, despite the common belief that a diet low in fat is preferable, there are other factors to consider, such as considering that some fatty acids are essential for human nutrition.

Many fatty acids found in a human diet are omega-6 polyunsaturated, monounsaturated, and SFAs. α-$C_{18:3}$, one of the n-3 PUFAs, is prevalent in significant amounts in vegetable products such oils (corn, soybean, cottonseed, canola, flaxseed, walnut, etc.), whereas other PUFAs (EPA, docosapentaenoic acid-DPA $C_{22:5}$, and DHA) are mostly exist in aquatic food. The optimal amount for n-3 intake recommended as 3–5.5 g per day. The proportion of n-6/n-3 in the West dietary is 15–20, which is quite high compared to the ideal level of 1–4 and indicates a deficiency in n-3 (Simopoulos 2002). Additionally, consumption patterns for n-3 PUFA-containing foods are now constant or falling (Lee et al. 2006). Therefore, in order to encourage a decrease in the n-6/n-3 PUFA ratio and enhance the health condition of the population, health authorities, and medical associations have advised increasing the intake of foods high in omega-3 PUFAs (particularly for specific sections of the population) (Garg et al. 2006).

Vegetable oils (olive, canola, sunflower, corn etc.) are good source of MUFAs and PUFAs and cholesterol free. The major concerns of the animal fat are high concentration of saturated fatty acid content. High intake of saturated fatty acid associated with adiposity, type 2 diabetes, elevated blood fats, elevated triglyceride level and cardiovascular illnesses (Islam et al. 2019; Kumar et al. 2016). Numerous research investigated the limited or total substitution of animal fat (saturated) with unsaturated fats (from diverse sources of vegetable oils), in an attempt to meet the changing customer expectations for healthful meat-based foods (Heck et al. 2017; Pires et al. 2019). In addition to providing a supply of several different health-promoting fatty acids, plant-based oils have been applied due to their various bioactive chemical content, some of which are antioxidants (Jiménez-Colmenero 2007).

Figure 6.2 demonstrates various types of vegetable oils that are high in MUFA, PUFA and can be utilized to replace the animal fat in meat products. According to Table 6.1, healthier meat products may be created using these vegetable oils, which have high MUFAs and PUFAs values whereas low in SFA. However, substitution of animal fat by vegetable oil may result in texture, flavor, and rancidity problems in the final product (Heck et al. 2017). Therefore, adjustments required in meat industry to maintain the desired sensory attributes of the final product.

The kind of the plant-based oil may affect the fatty acid profile of the substituted animal products. They have been reformulated to include oils from sources such as olive, sunflower, linseed, canola, corn, soy, peanut, sesame, etc. to offer healthier meat products. Some of these oils are rich in MUFAs while others promote the content of PUFAs. Among them olive oil gains much more attention due to being a high MUFA source (around 75%) and containing high ratio of vitamin E. Consuming of olive oil linked to lower potential of CVD, colon and breast cancer, lowered LDL value and cholesterol. Olive oil cab be utilized to substitute animal fat in a variety of cooked meat items, fermented sausages, and sausages to create reduced-fat formulations (Table 6.3). According to studies, olive oil increased the MUFA whereas decreased the SFA content of meat products (Ansorena and Astiasarán 2004; Koutsopoulos et al. 2008; Shin et al. 2020).

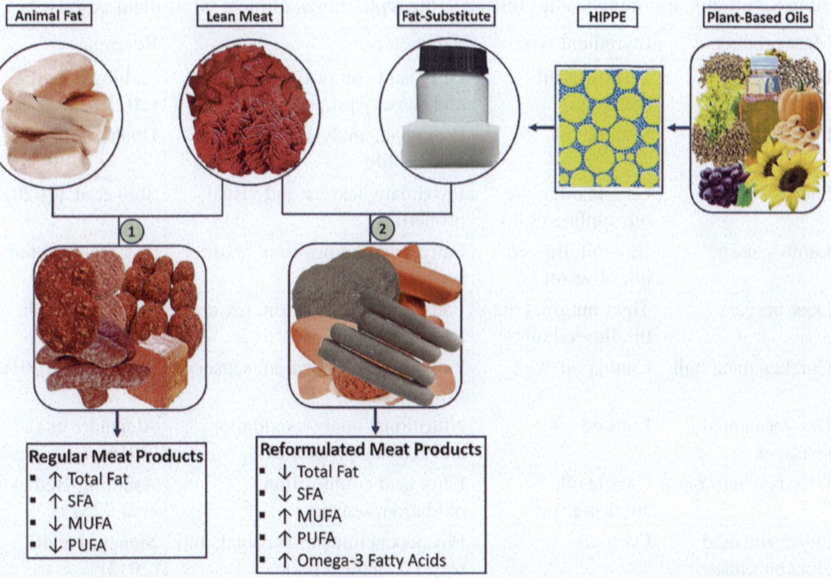


**Fig. 6.2** Schematic presentation of substitution of animal fat with vegetable oils (Badar et al. 2023). HIPPE high internal phase Pickering emulsions

Safflower oil contains significant concentrations of $C_{18:2}$ (n-6, 76%), thus reduce the PUFA/SFA, but it leads to unwanted raising of the ratio of n-6/n-3 in meat products. Fadiloglu et al. (2023) examined the use of safflower and walnut oil to reduce the SFA content of burger patties. Their results showed that safflower addition increased the MUFA+PUFA content at the end of the storage and safflower-added burger patties had the highest sensory evaluation scores, lower hardness, oxidation values and SFA content. Shin et al. (2020) reported that replacement of pork backfat with olive oil did not affect the texture and pointed to the possibility of olive oil. de Carvalho et al. (2020) described the preparation of lamb sausages with the replacement of animal fat with linseed, olive, and chia oils. They found that incorporation of vegetable oils improved fatty acid profile and nutritional properties of lamb sausages. Also, they mention the sensory properties of linseed oil substituted samples did not differ from control samples. In another study (Alejandre et al. 2016), investigated the linseed oil incorporation to dry fermented sausages found increment in the fatty acid composition and reduction in n-6/n-3 proportion. In order to prevent lipid oxidation during storage and increment in the monounsaturated fatty acids fraction (MUFA), Ansorena and Astiasarán (2004) found that adding olive oil to sausages was more successful than utilizing vacuum-storing techniques. As to increase the PUFA/SFA proportion and decrease the n-6/n-3 proportion of chicken meatballs without affecting the surface, variety boundaries, oxidation, and visible properties, Vieira et al. (2019) demonstrated the substitution of animal oil with canola oil.

**Table 6.3** Summary of the studies related to the replacement of meat fat by plant-based oils

| Meat product | Ingredient type | Parameters | References |
|---|---|---|---|
| Beef patty | Safflower oil, walnut oil | Fatty acid composition, texture, and sensory properties | Fadiloglu et al. (2023) |
| Low-fat burger | Sesame oil | Proximate analysis and fatty acid profile | Dhahir et al. (2021) |
| Sausage | Canola oil, olive oil, sunflower oil | Proximate, texture and visual properties | Shin et al. (2020) |
| Lamb sausage | Chia oil, linseed oil, olive oil | Fatty acid composition, texture, sensory | de Carvalho et al. (2020) |
| Deer burger | Tiger nut oil, chia oil, linseed oil | Fatty acid composition, texture, sensory properties | Vargas-Ramella et al. (2020) |
| Chicken meat ball | Canola oil | Fatty acid composition, sensory properties | Vieira et al. (2019) |
| Dry-fermented sausages | Linseed oil | Nutritional quality, oxidation, and sensory properties | Alejandre et al. (2016) |
| UK-style sausage | Canola oil, sunflower oil | Fatty acid composition, oxidation, sensory | Asuming-Bediako et al. (2014) |
| Dry-fermented chicken sausage | Corn oil | Physicochemical, microbial, and textural characteristics | Menegas et al. (2013) |
| Low-fat fermented sausages | Olive oil | Physicochemical, microbiological, and sensory properties | Koutsopoulos et al. (2008) |
| Dry-fermented sausages | Olive oil | Cholesterol level, fatty acid profile and oxidation properties | Ansorena and Astiasarán (2004) |
| Dry-fermented sausages | Soy oil | Fatty acid composition | Muguerza et al. (2003) |
| Frankfurter | Palm oil, cottonseed oil, olive oil | Fatty acid composition, sensory properties | Vural et al. (2004) |
| Cooked meat products | Olive oil | Sensory properties | Pappa et al. (2000) |

## 6.4 Conclusion

Vegetable oil is an important element of the individual nutrition and a significant origin of nourishing lipids. Also, it accounts for higher than 75% of the overall amount of fats eaten globally. Fats and lipids are an excellent source of fatty acids, especially those that are necessary. Vegetable oil is a crucial cooking medium that not only provides energy, but also protects human tissues, maintains normal body temperature, transports vitamins that are liposoluble, and operates other vital functions. Recently, there has been a rise in customer preference for healthful and nutritionally balanced food, which has resulted in the rise of numerous fat-free or low-fat products. Modern food technologists allow the replacement of highly saturated milk or beef fat with vegetable oils in analogue food production. In recent years, meat industry tries to design and develop new food formulations with a potential health benefit preventing the saturated fat and trans fats. Studies showed that high amount of saturated fat consumption correlated with the cardiovascular disease, high blood lipid content, high cholesterol level and other health implications. High consumption of meat, especially red meat, should be avoided due to the impression that animal fat is high in SFA. Therefore, replacement of animal fat with a plant-based oil which is rich in MUFA and PUFAs are studied and recommended to improve human health and consumer satisfaction.

## References


Abd El-Salam BA (2015) Effect of milk fat replacement with vegetable oil and/or whey protein concentrate on microstructure, texture and sensory characteristics of fresh soft cheese. Int J Dairy Sci 10(3):117–125. https://doi.org/10.3923/IJDS.2015.117.125

Abou-Gharbia HA, Shehata AAY, Shahidi F (2000) Effect of processing on oxidative stability and lipid classes of sesame oil. Food Res Int 33(5):331–340. https://doi.org/10.1016/S0963-9969(00)00052-1

Achachlouei BF, Hesari J, Damirchi SA, Peighambardoust SH, Esmaiili M, Alijani S (2013) Production and characterization of a functional Iranian white brined cheese by replacement of dairy fat with vegetable oils. Food Sci Technol Int 19(5):389–398. https://doi.org/10.1177/1082013212455341

Alejandre M, Poyato C, Ansorena D, Astiasarán I (2016) Linseed oil gelled emulsion: a successful fat replacer in dry fermented sausages. Meat Sci 121:107–113. https://doi.org/10.1016/J.MEATSCI.2016.05.010

Al-Ismail K, Al-Hiary B, Al-Dabbas M (2015) Evaluation of some chemical and sensory properties of processed cheese analogue with selected vegetable oils. Int J Chem Process Eng Res 2(6):75–85. https://ideas.repec.org/a/pkp/ijcper/v2y2015i6p75-85id1806.html

Alpaslan M, Boydak E, Hayta M, Gercek S, Simsek M (2001) Effect of row space and irrigation on seed composition of Turkish sesame (*Sesamum indicum* L.); Effect of row space and irrigation on seed composition of Turkish sesame (*Sesamum indicum* L.). J Am Oil Chem Soc 78:933–935. https://doi.org/10.1007/s11746-001-0366-0

Ansorena D, Astiasarán I (2004) Effect of storage and packaging on fatty acid composition and oxidation in dry fermented sausages made with added olive oil and antioxidants. Meat Sci 67(2):237–244. https://doi.org/10.1016/J.MEATSCI.2003.10.011

Anwar F, Bhanger MI, Nasir MKA, Ismail S (2002) Analytical characterization of Salicornia bigelovii seed oil cultivated in Pakistan. J Agric Food Chem 50(15):4210–4214. https://doi.org/10.1021/JF0114132

Arihara K (2006) Strategies for designing novel functional meat products. Meat Sci 74(1):219–229. https://doi.org/10.1016/J.MEATSCI.2006.04.028

Arslan S, Topcu A, Saldamli I, Koksal G (2014) Use of corn oil in the production of Turkish white cheese. J Food Sci Technol 51(10):2382. https://doi.org/10.1007/S13197-012-0752-6

Astrup A, Dyerberg J, Elwood P, Hermansen K, Hu FB, Jakobsen MU, Kok FJ, Krauss RM, Lecerf JM, Legrand P, Nestel P, Rise'rus U, Sanders T, Sinclair A, Stender S, Tholstrup T, Willett WC (2011) The role of reducing intakes of saturated fat in the prevention of cardiovascular disease: where does the evidence stand in 2010? 93:684–688. https://doi.org/10.3945/ajcn.110.004622. Role of Saturated Fat in Cardiovascular Disease

Asuming-Bediako N, Jaspal MH, Hallett K, Bayntun J, Baker A, Sheard PR (2014) Effects of replacing pork backfat with emulsified vegetable oil on fatty acid composition and quality of UK-style sausages. Meat Sci 96(1):187–194. https://doi.org/10.1016/J.MEATSCI.2013.06.031

Awad AB, Chan KC, Downie AC, Fink CS (2000) Peanuts as a source of β-sitosterol, a sterol with anticancer properties. Nutr Cancer 36(2):238–241. https://doi.org/10.1207/S15327914NC3602_14

Bachmann HP (2001) Cheese analogues: a review. Int Dairy J 11(4–7):505–515. https://doi.org/10.1016/S0958-6946(01)00073-5

Badar IH, Wang Z, Liu H, Chen Q, Xia X, Liu Q, Kong B (2023) Future prospects of high internal phase pickering emulsions stabilized by natural modified biopolymers as a potential fat substitute in meat products. Trends Food Sci Technol 140:104176. https://doi.org/10.1016/j.tifs.2023.104176

Briggs MA, Petersen KS, Kris-Etherton PM (2017) Saturated fatty acids and cardiovascular disease: replacements for saturated fat to reduce cardiovascular risk. Healthcare (Basel) 5(2):29. https://doi.org/10.3390/HEALTHCARE5020029

Calder PC, Grimble RF (2002) Polyunsaturated fatty acids, inflammation and immunity. Eur J Clin Nutr 56(3):14–19. https://doi.org/10.1038/sj.ejcn.1601478

Christopoulou E, Lazaraki M, Komaitis M, Kaselimis K (2004) Effectiveness of determinations of fatty acids and triglycerides for the detection of adulteration of olive oils with vegetable oils. Food Chem 84(3):463–474. https://doi.org/10.1016/S0308-8146(03)00273-5

Chung CH, Yeea YJ, Kimb DH, Kimb HK, Chungb DS (1995) Changes of lipid, protein, RNA and fatty acid composition in developing sesame (Sesamum indicum L.) seeds. Plant Sci 109:237–243

Cunha CR, Dias AI, Viotto WH (2010) Microstructure, texture, colour and sensory evaluation of a spreadable processed cheese analogue made with vegetable fat. Food Res Int 43:723–729. https://doi.org/10.1016/j.foodres.2009.11.009

Cunha CR, Grimaldi R, Alcântara MR, Viotto WH (2013) Effect of the type of fat on rheology, functional properties and sensory acceptance of spreadable cheese analogue. Int J Dairy Technol 66(1):54–62. https://doi.org/10.1111/j.1471-0307.2012.00876.x

de Carvalho FAL, Munekata PES, Pateiro M, Campagnol PCB, Domínguez R, Trindade MA, Lorenzo JM (2020) Effect of replacing backfat with vegetable oils during the shelf-life of cooked lamb sausages. LWT 122:109052. https://doi.org/10.1016/J.LWT.2020.109052

Dhahir O, Aali A, El-Bashar A, Al-Mossawi HJ (2021) Effect of replaced animal fat with sesame oil on the quality of frozen beef burgers. Nat Vol Essent Oil 8(6):1993–2009. https://www.nveo.org/index.php/journal/article/view/3859

Dorni C, Sharma P, Saikia G, Longvah T (2017) Fatty acid profile of edible oils and fats consumed in India. Food Chem. https://doi.org/10.1016/j.foodchem.2017.05.072

Dubois V, Breton S, Linder M, Fanni J, Parmentier M (2007) Fatty acid profiles of 80 vegetable oils with regard to their nutritional potential. Eur J Lipid Sci Technol 109(7):710–732

El-Bakry M, Sanchez A, Mehta BM (2018) Microstructure of dairy products. https://www.wiley.com/en-us/Microstructure+of+Dairy+Products-p-9781118964200

Everett DW, Olson NF (2003) Free oil and rheology of Cheddar cheese containing fat globules stabilized with different proteins. J Dairy Sci 86(3):755–763. https://doi.org/10.3168/JDS.S0022-0302(03)73656-X

Fadiloglu EE, Yildiz Turp G, Celebioglu C, Sel G (2023) Influence of different cooking methods on quality characteristics and nutritional value of gluten-free beef burger patties formulated with walnut oil, safflower oil and buckwheat. Meat Sci 204:109251. https://doi.org/10.1016/j.meatsci.2023.109251

Fernández-Ginés JM, Fernández-López J, Sayas-Barberá E, Pérez-Alvarez JA (2005) Meat products as functional foods: a review. J Food Sci 70(2). https://doi.org/10.1111/j.1365-2621.2005.tb07110.x. Institute of Food Technologists

Food and Agriculture Organization of the United Nations (2023) FAOSTAT

Garg ML, Wood LG, Singh H, Moughan PJ (2006) Means of delivering recommended levels of long chain n-3 polyunsaturated fatty acids in human diets. J Food Sci 71(5):R66–R71. https://doi.org/10.1111/J.1750-3841.2006.00033

Ghouila Z, Sehailia M, Chemat S (2019) Vegetable oils and fats: extraction, composition and applications. In: Plant based "Green Chemistry 2.0" moving from evolutionary to revolutionary, pp 339–375

Guinee TP (2016) Protein in cheese and cheese products: structure-function relationships. In: Advanced dairy chemistry: volume 1B: proteins: applied aspects: fourth edition, pp 347–415. https://doi.org/10.1007/978-1-4939-2800-2_14/COVER

Guinee TP, Carić M, Kaláb M (2004) Pasteurized processed cheese and substitute/imitation cheese products. Cheese: Chem Phys Microbiol 2(C):349–394. https://doi.org/10.1016/S1874-558X(04)80052-6

Gunstone F (2011) Chapter 1: Production and trade of vegetable oils. In: Gunstone F (ed) Vegetable oils in food technology: composition, properties and uses, Second edn. Wiley, Hoboken

Halvorsen BL, Blomhoff R (2011) Determination of lipid oxidation products in vegetable oils and marine omega-3 supplements. Food Nutr Res 55. https://doi.org/10.3402/FNR.V55I0.5792

Han K, Willams KJ, Goldberg AC (2021) Plant-based oils. In: Prevention and treatment of cardiovascular disease: nutritional and dietary approaches, pp 115–127

Heck RT, Vendruscolo RG, de Araújo Etchepare M, Cichoski AJ, de Menezes CR, Barin JS, Lorenzo JM, Wagner R, Campagnol PCB (2017) Is it possible to produce a low-fat burger with a healthy n − 6/n − 3 PUFA ratio without affecting the technological and sensory properties? Meat Sci 130:16–25. https://doi.org/10.1016/J.MEATSCI.2017.03.010

Islam MA, Amin MN, Siddiqui SA, Hossain MP, Sultana F, Kabir MR (2019) Trans fatty acids and lipid profile: a serious risk factor to cardiovascular disease, cancer and diabetes. Diabetes Metab Syndr Clin Res Rev 13(2):1643–1647. https://doi.org/10.1016/J.DSX.2019.03.033

Jahaniaval F, Kakuda Y, Marcone MF (2000) Fatty acid and triacylglycerol compositions of seed oils of five Amaranthus accessions and their comparison to other oils; Fatty acid and triacylglycerol compositions of seed oils of five Amaranthus accessions and their comparison to other oils. J Am Oil Chem Soc 77:847–852. https://doi.org/10.1007/s11746-000-0135-0

Jiménez-Colmenero F (2007) Healthier lipid formulation approaches in meat-based functional foods. Technological options for replacement of meat fats by non-meat fats. Trends Food Sci Technol 18(11):567–578. https://doi.org/10.1016/j.tifs.2007.05.006

Jurić S, Jurić M, Siddique AB, Fathi M (2020) Vegetable oils rich in polyunsaturated fatty acids: nanoencapsulation methods and stability enhancement. Food Rev Int. https://doi.org/10.1080/87559129.2020.1717524

Kaminarides S, Stachtiaris S (2000) Production of processed cheese using kasseri cheese and processed cheese analogues incorporating whey protein concentrate and soybean oil. Int J Dairy Technol 53(2):69–74. https://doi.org/10.1111/J.1471-0307.2000.TB02661.X

Kavak DD, Karabiyik H (2019) Quality evaluation of kashar cheese: influence of palm oil and ripening period. Food Sci Technol 40(2):354–360. https://doi.org/10.1590/FST.39618

Kochar SP (2011) Minor and speciality oils. In: Gunstone F (ed) Vegetable oils in food technology: composition, properties and uses. Wiley, Hoboken

Koutsopoulos DA, Koutsimanis GE, Bloukas JG (2008) Effect of carrageenan level and packaging during ripening on processing and quality characteristics of low-fat fermented sausages produced with olive oil. Meat Sci 79(1):188–197. https://doi.org/10.1016/J.MEATSCI.2007.08.016

Kumar A, Sharma A, Upadhyaya KC (2016) Vegetable oil: nutritional and industrial perspective. Curr Genomics 17(3). https://doi.org/10.2174/1389202917666160202220107

Lamichhane P, Kelly AL, Sheehan JJ (2018) Symposium review: structure-function relationships in cheese. J Dairy Sci 101(3):2692–2709. https://doi.org/10.3168/JDS.2017-13386

Lee D-S, Noh B-S, Bae S-Y, Kim K (1998) Characterization of fatty acids composition in vegetable oils by gas chromatography and chemometrics. Anal Chim Acta 358(20):163–175

Lee S, Faustman C, Djordjevic D, Faraji H, Decker EA (2006) Effect of antioxidants on stabilization of meat products fortified with n À 3 fatty acids. Meat Sci 72(1):18–24. https://doi.org/10.1016/j.meatsci.2005.05.022

Lee SK, Klostermeyer H, Anema SG (2015) Effect of fat and protein-in-water concentrations on the properties of model processed cheese. Int Dairy J 50:15–23. https://doi.org/10.1016/j.idairyj.2015.06.001

Leong TSH, Ong L, Gamlath CJ, Gras SL, Ashokkumar M, Martin GJO (2020) Formation of cheddar cheese analogues using canola oil and ultrasonication—a comparison between single and double emulsion systems. Int Dairy J 105:104683. https://doi.org/10.1016/J.IDAIRYJ.2020.104683

Livingstone KM, Lovegrove JA, Givens DI (2012) The impact of substituting SFA in dairy products with MUFA or PUFA on CVD risk: evidence from human intervention studies. Nutr Res Rev 25(2):193–206. https://doi.org/10.1017/S095442241200011X

Lobato-Calleros C, Vernon-Carter EJ, Guerrero-Legarreta I, Soriano-Santos J, Escalona-Beundia H (1997) Use of fat blends in cheese analogs: influence on sensory and instrumental textural characteristics. J Texture Stud 28(6):619–632. https://doi.org/10.1111/J.1745-4603.1997.TB00142.X

López-Miranda J, Pérez-Jiménez F, Ros E, De Caterina R, Badimón L, Covas MI, Escrich E, Ordovás JM, Soriguer F, Abiá R, Alarcón de la Lastra C, Battino M, Corella D, Chamorro-Quirós J, Delgado-Lista J, Giugliano D, Esposito K, Estruch R, Fernandez-Real JM et al (2010) Olive oil and health: summary of the II international conference on olive oil and health consensus report, Jaén and Córdoba (Spain) 2008. Nutr Metab Cardiovasc Dis 20(4):284–294. https://doi.org/10.1016/J.NUMECD.2009.12.007

Mannina L, Dugo G, Salvo F, Cicero L, Ansanelli G, Calcagni C, Segre A (2003) Study of the cultivar-composition relationship in sicilian olive oils by GC, NMR, and statistical methods. J Agric Food Chem 51(1):120–127. https://doi.org/10.1021/JF025656L/ASSET/IMAGES/LARGE/JF025656LF00006.JPEG

Menegas LZ, Pimentel TC, Garcia S, Prudencio SH (2013) Dry-fermented chicken sausage produced with inulin and corn oil: physicochemical, microbiological, and textural characteristics and acceptability during storage. Meat Sci 93(3):501–506. https://doi.org/10.1016/j.meatsci.2012.11.003

Miciński J, Zwierzchowski G, Kowalski IM, Szarek J, Pierozyński B, Raistenskis J (2012) The effects of bovine milk fat on human health. Pol Ann Med 19(2):170–175. https://doi.org/10.1016/J.POAMED.2012.07.004

Moatsou G, Zoidou E, Choundala E, Koutsaris K, Kopsia O, Thergiaki K, Sakkas L (2019) Development of reduced-fat, reduced-sodium semi-hard sheep milk cheese. Food Secur 8(6):204. https://doi.org/10.3390/FOODS8060204

Muguerza E, Ansorena D, Astiasarán I (2003) Improvement of nutritional properties of Chorizo de Pamplona by replacement of pork backfat with soy oil. Meat Sci 65(4):1361–1367. https://doi.org/10.1016/S0309-1740(03)00058-5

Myhan R, Jeliński T, Białobrzewski I, Sadowska J, Jachimczyk E (2020) The effect of milk fat substitution on the rheological properties of Edam-type cheese. Eur Food Res Technol 246(12):2443–2450. https://doi.org/10.1007/s00217-020-03587-9

O'Riordan ED, Duggan E, O'Sullivan M, Noronha N (2011) Production of analogue cheeses. In: Processed cheese and analogues. Wiley, Chichester, pp 219–244. https://doi.org/10.1002/9781444341850.CH9

Ollivier D, Artaud J, Pinatel C, Durbec JP, Guérère M (2003) Triacylglycerol and fatty acid compositions of French virgin olive oils. Characterization by chemometrics. J Agric Food Chem 51(19):5723–5731. https://doi.org/10.1021/JF034365P/SUPPL_FILE/JF034365PSI20030621_023551.JPG

Pappa IC, Bloukas JG, Arvanitoyannis IS (2000) Optimization of salt, olive oil and pectin level for low-fat frankfurters produced by replacing pork backfat with olive oil. Meat Sci 56(1):81–88. https://doi.org/10.1016/S0309-1740(00)00024-3

Pires MA, dos Santos IR, Barros JC, Trindade MA (2019) Effect of replacing pork backfat with Echium oil on technological and sensory characteristics of bologna sausages with reduced sodium content. LWT 109:47–54. https://doi.org/10.1016/J.LWT.2019.04.009

Przybylski R, Mag T, Eskin NAM, McDonald BE (2005) Canola oil. In: Bailey's industrial oil and fat products. https://doi.org/10.1002/047167849X.BIO004

Rahim MA, Ayub H, Sehrish A, Ambreen S, Khan FA, Itrat N, Nazir A, Shoukat A, Shoukat A, Ejaz A (2023) Essential components from plant source oils: a review on extraction, detection, identification, and quantification. Molecules 28(19):6881

Rallidis LS, Paschos G, Liakos GK, Velissaridou AH, Anastasiadis G, Zampelas A (2003) Dietary a-linolenic acid decreases C-reactive protein, serum amyloid A and interleukin-6 in dyslipidaemic patients. Atherosclerosis 167(2):237–242. https://doi.org/10.1016/S0021-9150(02)00427-6

Ramel PR, Marangoni AG (2018) Processed cheese as a polymer matrix composite: a particle toolkit for the replacement of milk fat with canola oil in processed cheese. https://doi.org/10.1016/j.foodres.2018.02.019

Ranalli A, Pollastri L, Contento S, Di Loreto G, Iannucci E, Lucera L, Russi F (2002) Acylglycerol and fatty acid components of pulp, seed, and whole olive fruit oils. Their use to characterize fruit variety by chemometrics. J Agric Food Chem 50(13):3775–3779. https://doi.org/10.1021/JF011506J/ASSET/IMAGES/LARGE/JF011506JF00005.JPEG

Rani H, Sharma S, Bala M (2021) Technologies for extraction of oil from oilseeds and other plant sources in retrospect and prospects: a review. J Food Process Eng 44(11):e13851

Rogers NR, Mcmahon DJ, Daubert CR, Berry TK, Foegeding EA (2010) Rheological properties and microstructure of Cheddar cheese made with different fat contents. J Dairy Sci 93(10):4565–4576. https://doi.org/10.3168/jds.2010-3494

Shaaban A, Shaaban H (2020) The effect of milk fat replacement by some edible oils on chemical composition, antioxidant activity and oxidative stability of spreadable processed cheese analogues. Int Res J Food Nutr 2:6–14

Shabani J, Sarfarazi M, Mirzaei H, Jafari SM (2016) Influence of the sunflower oil content, cooking temperature and cooking time on the physical and sensory properties of spreadable cheese analogues based on UF white-brined cheese. Int J Dairy Technol 69(4):576–584. https://doi.org/10.1111/1471-0307.12305

Shin D-J, Lee HJ, Lee D, Jo C, Choe J (2020) Fat replacement in chicken sausages manufactured with broiler and old laying hens by different vegetable oils. Poult Sci. https://doi.org/10.1016/j.psj.2020.01.008

Simopoulos AP (2002) The importance of the ratio of omega-6/omega-3 essential fatty acids. Biomed Pharmacother 56(8):365–379. https://doi.org/10.1016/S0753-3322(02)00253-6

Tan J, McKenzie C, Potamitis M, Thorburn AN, Mackay CR, Macia L (2014) The role of short-chain fatty acids in health and disease. Adv Immunol 121:91–119. https://doi.org/10.1016/B978-0-12-800100-4.00003-9

Totosaus A, Rojas-Nery E, Franco-Fernández MJ (2017) Soya bean oil/soya protein isolate and carrageenan emulsions as fat replacer in fat-reduced Oaxaca-type cheese. Int J Dairy Technol 70(4):499–505. https://doi.org/10.1111/1471-0307.12397

Vargas-Ramella M, Munekata PES, Pateiro M, Franco D, Campagnol PCB, Tomasevic I, Domínguez R, Lorenzo JM (2020) Physicochemical composition and nutritional properties of deer burger enhanced with healthier oils. Food Secur 9(5):571. https://doi.org/10.3390/FOODS9050571

Vieira ASP, De Souza XR, Rodrigues EC, Sousa DC (2019) Replacement of animal fat by canola oil in chicken meatball. Braz J Poultry Sci 21(3):eRBCA-2019-1013. https://doi.org/10.1590/1806-9061-2019-1013

Vural H, Javidipour I, Ozbas OO (2004) Effects of interesterified vegetable oils and sugar-beet fiber on the quality of frankfurters. Meat Sci 67(1):65–72. https://doi.org/10.1016/j.meatsci.2003.09.006

Wang T (2011) Soybean oil. In: Vegetable oils in food technology: composition, properties and uses, Second edn. Wiley, Hoboken, pp 59–105. https://doi.org/10.1002/9781444339925.CH3

Weisburger JH (2000) Eat to live, not live to eat. Nutrition 16(9):767–773. https://doi.org/10.1016/S0899-9007(00)00400-7

WHO and FAO (2003) Diet, nutrition, and the prevention of chronic diseases (Report of a joint WHO and FAO expert consulation). WHO Technical Report Series 916:1–160. https://apps.who.int/iris/bitstream/handle/10665/42665/WHO_TRS_916.pdf?sequence=1

Wood JD, Richardson RI, Nute GR, Fisher AV, Campo MM, Kasapidou E, Sheard PR, Enser M (2004) Effects of fatty acids on meat quality: a review. Meat Sci 66(1):21–32. https://doi.org/10.1016/S0309-1740(03)00022-6

Youdim KA, Martin A, Joseph JA (2000) Essential fatty acids and the brain: possible health implications. Int J Dev Neurosci 18:383–399. https://doi.org/10.1016/S0736-5748(00)00013-7

# Chapter 7
# Spirulina: Exploring Its Promising Role in Food Industry and Overcoming Challenges as a Food Analogue

Ayse Burcu Aktas

## 7.1 A Brief Overview of Spirulina

Microalgae are single-cell photosynthetic microorganisms. They may evolve and create biomass utilizing $CO_2$, light, and certain nutrients including nitrogen, phosphorus, and potassium. Microalgae are categorized into 16 taxonomic classes and green algae, diatoms, and cyanobacteria are being the most prominent species in terms of biotechnology (Borowitzka 2018).

The microalgae were regarded as a protein supply for consumption by people in the 1950s. *Chlorella* and *Arthrospira* species were commercially cultivated for protein source in the 1960s and 1970s. In the 1980s, *Dunaliella* and *Haematococcus* were grown in order to manufacture some pigments as food additives. In the early 1990s, the production of polyunsaturated fatty acids from microalgae was started, with a focus on omega-3 fatty acids (Fig. 7.1) (Hosseinkhani et al. 2022).

It is critical to decide on the appropriate growing methods for Spirulina in order to adjust the chemical components that are closely associated with biological activity. However, optimal Spirulina cultivation will have also an impact on Spirulina biomass. As Spirulina powder production rises, so will the income and commercial value of Spirulina biomass (Ahda et al. 2023).

The two species of cyanobacteria: *Arthrospira platensis* and *Arthrospira maxima* are the most well-known genus of Spirulina. *Spirulina* sp. and *Arthrospira* sp. contain cyanobacterial species that are remarkably similar to one another. Many species previously categorized as *Spirulina* sp. have lately been included in the genera of *Arthrospira* sp., which includes all commercialized Spirulina (Abreu et al. 2023). Spirulina is the most cultivated microalgae in the world, accounting for more than 30% of global microalgal biomass production (Costa et al. 2019). The American

A. B. Aktas (✉)
Sivas Cumhuriyet University, Faculty of Science, Biochemistry Department, Sivas, Turkey
e-mail: burcuaktas@cumhuriyet.edu.tr



Ö. P. CAN et al. (eds.), *Food Analogues: Emerging Methods and Challenges*,
https://doi.org/10.1007/978-3-031-69872-9_7

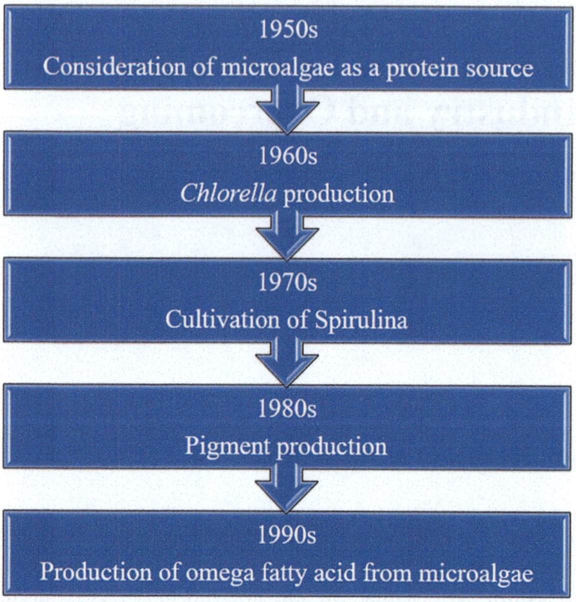


**Fig. 7.1** Chronology of commercializing microalgae as a food product

Food and Drug Administration (FDA) and the European Commission (EU, Novel Food Catalogue) have declared *A. platensis* "generally recognized as safe" (GRAS) for food applications (Montevecchi et al. 2022).

## 7.2 Nutritional Properties of Spirulina

Spirulina is a single-cell protein which contains higher amounts of crucial minerals and vitamins and could be employed in the production of functional foods (Fig. 7.2.) (Lafarga et al. 2020). It is well-known for its high protein content, which is approximately 60% dry weight and is capable of reaching as much as 70% dry weight when grown in nitrogen-free environments. This exceptional substance, on the other hand, is a rich source of other valuable components including chlorophylls, carotenoids, and phycobiliproteins. Since these are known as colored compounds, Spirulina could be used as pigments in the food industry (Da Rosa et al. 2015). The pigments derived from Spirulina may have certain health benefits, such as enhancing the immune system and preventing cancer, destructive chronic and heart diseases (Rodriguez-Concepcion et al. 2018).

The nutritional components of Spirulina were provided in Table 7.1 (AlFadhly et al. 2022; Dewi et al. 2016). Proteins, carbohydrates, and lipids are the primary components of Spirulina. Proteins provide a function in biosynthesis and cell division, whereas carbohydrates and lipids predominantly contribute to internal carbon

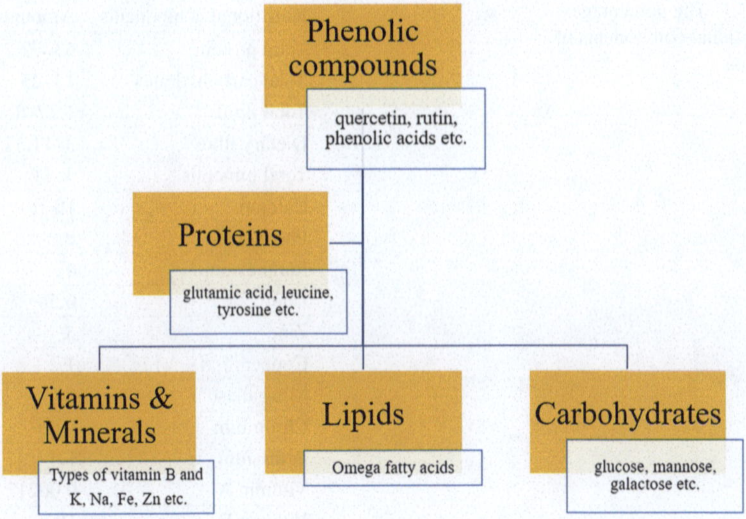


**Fig. 7.2** Major nutrients of Spirulina

and energy reserves (Table 7.1). The primary macromolecules (proteins, carbohydrates, and lipids) may differ significantly between microalgae phyla. According to a recent study, the average macromolecular composition of several phyla of *Cyanobacteria* was 2.4% protein, 4.2% fat, and 2.3% carbohydrate on a dry weight basis (Cuellar-Bermudez et al. 2015; Finkel et al. 2016).

Spirulina is high in protein and contains several amino acids comprising aspartic acid, glutamic acid, serine, glycine, histidine, arginin, threonin, alanine, proline, valine methionine, isoleucine, leucine, phenylalanine, and lysine. Spirulina's protein content fluctuates based on growing conditions and processing procedures. On average, dried spirulina comprises approximately 53–72% protein by weight. (Table 7.1) (Dewi et al. 2016).

Carbohydrates, along with proteins, lipids, are one of the macronutrients that compose Spirulina. The carbohydrate composition of Spirulina mainly consists of glucose fructose, xylose, galactose, ribose, sucrose, rhamnose and complex polysaccharides, with a significant quantity of a unique polysaccharide known as glycogen. Since cellulose is not present in Spirulina, its carbohydrates could be easily digested. Moreover, the low level of free sugars contributes to Spirulina being a valuable nutritional supplement for diabetes, obesity, and other disorders (Anvar and Nowruzi 2021; Braga da et al. 2018).

Spirulina contains a variety of fatty acids, including palmitic, oleic, and linoleic acid. These fatty acids have a number of activities in the body, including being components of cell membranes and serving as precursors for several signaling molecules. Spirulina consists of a significant amount of omega-3 (ω-3) fatty acid alpha-linolenic acid (ALA). The ω-3 fatty acids are widely recognized for their potential cardiovascular benefits as well as their role in brain health. ALA is

**Table 7.1** The percentages of nutritional components of Spirulina

| Nutritional components | Amounts (%) |
|---|---|
| Total protein | 53–72 |
| Total carbohydrates | 13–25 |
| Total lipid | 1.27–8 |
| Dietary fibre | 3–11.37 |
| Total minerals | 3–11 |
| Calcium | 10 |
| Phoshorus | 8 |
| Magnesium | 4 |
| Iron | 0.58 |
| Zinc | 3 |
| Copper | 1.2 |
| Manganese | 0.5 |
| Chromium | 0.05 |
| Potassium | 1.4 |
| Vitamin A | 0.00212 |
| Vitamin E | 10 |
| Vitamin B1 | 5.61 |
| Vitamin B2 | 0.4 |
| Vitamin B3 | 1.3 |
| Vitamin B5 | 0.2 |
| Vitamin B6 | 6 |
| Vitamin B8 | 0.005 |
| Vitamin B9 | 0.05 |
| Vitamin B12 | 0.35 |
| Chlorophyll | 0.6 |
| Gamma-linoleic acid | 8.87 |
| Linoleic acid | 17.43 |

classified as an essential fatty acid because the human body can not synthesize it and needs to get it from daily diet. Spirulina additionally consists of gamma-linolenic acid, an omega-6 fatty acid (ω-6). GLA is uncommon in many plant-based sources, hence Spirulina stands out in that regard (Li et al. 2018, 2019).

The majority of the important vitamins included in Spirulina, vitamin B12 is particularly remarkable. While Spirulina is commonly regarded as a vegetarian source of vitamin B12, its bioavailability and efficacy with regard to human nutritional requirements are still questioned. Spirulina is also an adequate supply of other B-vitamins, including B1 (thiamine), B2 (riboflavin), B3 (niacin), B6 (pyridoxine), and folate, all of which serve essential functions in the human body's metabolic processes. The vitamin A content of Spirulina is mostly derived from its high concentration of provitamin A-carotenoids, specifically beta-carotene. Beta-carotene is a precursor to vitamin A, and the body converts it into this crucial vitamin. Spirulina is regarded as a great plant-based source of provitamin A, making it especially beneficial for vegetarians and vegans who may have refuse to consume animal-derived vitamin A. Spirulina includes vitamin E, as well. Vitamin E is a fat-soluble antioxidant component that protects cells from oxidative damage. Spirulina has been discovered to be constituted of tocopherols, a group of substances which resemble Vitamin E. However, the actual vitamin concentrations of Spirulina can vary depending on factors including the individual strain, growth conditions, and processing procedures (Aouir et al. 2017; Grosshagauer et al. 2020; Soni et al. 2017).

Spirulina is an excellent source of several critical minerals. Spirulina contains iron, an important mineral required for oxygen delivery in the blood and overall cellular function. Calcium is essential for bone health and many physiological processes, and Spirulina includes significant levels of this mineral. Spirulina contains magnesium and phosphorus, which promotes the operation of muscles and nerves, generation of energy, and the preservation of bones. Phosphorus is essential for bone health, energy metabolism, and various cellular processes. Spirulina possesses significant quantities of potassium, which plays an important roles in regulating fluid equilibrium, neuronal signals, and skeletal muscle contractions. Spirulina is also a source of zinc, an essential trace element important for immune function. Spirulina contains manganese, a trace element involved in bone formation, blood clotting, and reducing oxidative stress. Spirulina also includes copper, which has a role in the antioxidant defense, the production of hemoglobin and collagen. Selenium is an essential trace element with antioxidant properties, and Spirulina may contribute to selenium intake (Almeida et al. 2021; Janda-Milczarek et al. 2023; Seghiri et al. 2019).

Spirulina also includes several bioactive chemicals, specifically phenolic compounds. Phenolic molecules are secondary metabolites with antioxidant characteristics and have some potential health advantages. Phenolic compounds found in Spirulina might consist of phenolic acids, flavonoids, and other polyphenols. These substances contribute to total antioxidant activity of Spirulina by neutralizing free radicals and reducing oxidative stress. Previous studies has revealed a presence of phenolic compounds in Spirulina, highlighting its potential as a natural antioxidant. However, the specific phenolic content differs among Spirulina strains and can be influenced by environmental conditions during cultivation (Bortolini et al. 2022).

## 7.3 Producing Food Analogues from Spirulina

The production of food analogues from spirulina, a nutrient-rich blue-green algae, has attracted widespread attention as a sustainable solution for satisfying the increasing desire for different plant-based foods. Spirulina has an unique nutritional profile, based on its high protein content, essential fatty acids, vitamins, and minerals. Food analogues from spirulina are generated through an advanced method that optimizes the dietary value of this organism while promising consumers preference and acceptance. Spirulina can be dried, milled, and blended with additional ingredients to achieve the desired taste and texture. The combination of spirulina with other plant-based ingredients like textured vegetable protein (TVP), grains, and flavorings are also possible. These analogues must be properly formulated in order to provide products that are both nutritional and attractive to consumers. Furthermore, strong procedures for quality control should be adopted to ensure the safety and quality of the final products (Hernandez et al. 2022; Lafarga et al. 2020).

The extraction of protein-rich biomass from spirulina is an essential part of its analogues. Spray drying is a common method for transforming spirulina into a smooth powder while maintaining the essential amino acids. The produced protein powder is utilized as a substrate for a wide range of products, including protein bars, burgers, and sausages. Proper control of drying conditions is required for keeping the stability of the protein structure and preventing the breakdown of delicate components, accordingly boosting the nutritional value of the end products (Bocker et al. 2021).

Extrusion, a prominent process in meat analogue production, is used to create the fibrous texture typical of meat. The objective is to optimize the ingredient balance to resemble the taste and aroma of usual meat while maintaining the nutritious nutritional value of spirulina (Fu et al. 2021). Furthermore, novel techniques, such as 3D printing for plant-based meat, have recently emerged as potential ways to improve the manufacture of spirulina-based meat analogues, bringing up new opportunities for customization and texture alteration (Letras et al. 2022).

## 7.4 Spirulina: A Food Analogue for Different Products

*Spirulina platensis* (also known as Spirulina) is available in tablets, capsules, and liquid form. These ingredients are used in many items such as candies, snacks, gum, wine, noodles, and beverages, including breakfast cereals. This section categorizes and summarizes the potential applications of Spirulina in different kinds of food products along with as food analogue.

## 7.5 Spirulina as a Food Analogue of Cereal and Cereal-Based Products

The nutritional value and possible health benefits of cereal and cereal-based products could be improved by incorporating Spirulina into their recipe. The supplementation of cereal products with Spirulina may boost their protein levels, and meet daily nutritional requirements in terms of vitamins and minerals, provide antioxidant properties with attractive appearance due to its chlorophyll content (Lafarga et al. 2020).

Spirulina has recently gained popularity for bakery products due to its higher protein and mineral contents. The addition of Spirulina into wheat cookies at the percentages of 2% and 6% by weight/weight enhanced protein content respectively

(Batista et al. 2019). Brownies enhanced with encapsulated and nonencapsulated Spirulina were evaluated in terms of micronutrients. The brownies with nonencapsulated or encapsulated Spirulina exhibited increased iron bioaccessibility. Moreover, Na (sodium), B (boron), Al (aluminum), P(phosphorus), Fe (iron), Zn (zinc), Ca (calcium), Cu (copper), Mg (magnesium), Mn (manganese), and S (sulfur) were the most common elements discovered in brownie samples processed with Spirulina (Penalver et al. 2023). Biscuits manufactured with microencapsulated Spirulina biomass had more 40% of protein than typical ones. Spirulina-enriched biscuits contain an adequate amount of iron (up to 35% of the RDA (Recommended Dietary Allowance)) (Paula da Silva et al. 2021). The hydrolyzed protein obtained from Spirulina used in the production of cake samples. The oil extracted from cakes had reduced the oxidative stability and microbial spoilage of the cakes was lower than regular samples during the storage period (Taghdiri et al. 2024).

Spirulina powder was blended with wheat flour at various concentrations (2, 4, 6, and 8%) to make bread more nutritious. The findings revealed that bread containing 6% Spirulina had greater total phenolic content (1.79 mg GAE/g) and antioxidant activity (16.51%) (Saharan and Jood 2021). The possibility of using Spirulina as a constituent in the manufacturing of sourdough crostini was investigated. While doughs reached a lesser volume than the regular types, products supplemented by Spirulina exhibited higher content of protein and polyphenols, as well as a stronger antioxidant capacity (Niccolai et al. 2019). The influence of Spirulina substitution on the physical and organoleptic attributes of bread were examined and the optimum Spirulina concentration of breads was found to be between 2% and 4%. While the bread samples had a decreased specific volume, crumb hardness, and degree of lightness at these concentrations, bread porosity and gumminess increased significantly. Moreover, in terms of organoleptic parameters, Spirulina substitution was acceptable at 2–4%, but the hedonic evaluation got worse since the Spirulina concentration was increased (Rahimah et al. 2024). The addition of Spirulina powder into the bread recipe increased the essential amino acid content (Montevecchi et al. 2022). New gluten-free (GF) breads with spirulina fermented by sourdough were produced, and digestion and colonic fermentation of GF bread were studied using an in vitro gut model. Sourdough GF breads with spirulina significantly improved beneficial colon microbiota markers (Nissen et al. 2024).

The nutritional value of snacks were maximized by addition of Spirulina (AlFadhly et al. 2022). A study discovered that supplementing snacks with Spirulina at varying concentrations (0%, 2.5%, 5%, 7.5%, 10%, 12.5%) enhanced the protein (9.43–18.11%) and ash (1.31–2.67%) percentages. The results revealed that snacks were microbiologically safe, and sensory analysis indicated that adding Spirulina to snacks at a concentration of 12.5% should be avoided (Haoujar et al. 2022). A vegan snack was made with Spirulina, and the results showed that the snack fortified with Spirulina had higher amounts of protein, ash, and a lower fat content compared to standard snack. It was determined that the more Spirulina powder used in the preparation, the lower the fat content of the snacks. The final product could be offered as a proteinaceous snack food (Pandey et al. 2023). Biscuits are a widely consumed as snack foods and biscuit manufacturers are gradually incorporating healthy components into their products. Spirulina provide proteins and antioxidants for wheat biscuits. A study aimed to create a unique gluten-free cereal snack which could be nutritionally enhanced by the addition of Spirulina by using 3D printing technology. Nutritional analysis revealed that snacks with Spirulina had greater protein and critical mineral content. The snacks containing 5% Spirulina exhibited the highest nutritional and sensory achievement with increased antioxidant activity, mineral, and protein content. These Spirulina-snacks gathered higher sensorial ratings from panelists (Letras et al. 2022). The snacks subjected to 2.6% Spirulina was both physically and microbiologically stabile after a 12-month storage period (Lucas et al. 2018).

Spirulina has been included into fresh pasta to strengthen its nutritional and functional properties as well as to keep the quality characteristics of pasta after cooking. The pasta with Spirulina was approved by consumers and the presence of Spirulina had no effect on the cooking quality characteristics of pasta. It was stated that Spirulina could be an innovative ingredient of fresh pasta (Zen et al. 2020). Pasta containing semolina with varying percentages (5, 10, and 15%) of Spirulina was manufactured, and some quality parameters including weight and volume gain, cooking loss, total mold and yeast count, and color values were identified. Spirulina-enriched pasta exhibited better technological properties and was microbiologically safe. The pasta samples enhanced with 10% Spirulina had best sensorial results according to the panelists (Ozyurt et al. 2015). The amino acid profile of the Spirulina-based pasta fullfilled the amino acid requirements of pregnant and breast-feeding women, the elderly children particularly from low-income regions (Ramírez-Rodrigues et al. 2021). The leaf extracts of basil and Spirulina were formulated for the production of noodles, and it was discovered that noodle samples became more chewy and dense than regular noodles (Al-Baarri et al. 2021). The quinoa flour was combined with Spirulina powder in order to develop a new gluten-free pasta for patients with celiac disease or obesity. The end product had higher nutritional value and rheological properties were also enhanced (Hussein et al. 2023). A study looked for the foods in question and discovered that the most likely spirulna-enriched products in the participants' diets were pasta and cereal bars (Lucas et al. 2023).

## 7.6 Spirulina as a Food Analogue of Dairy Products

The addition of Spirulina into dairy products is a novel approach for creating unique and high-value products enhanced with vegetable-derived components, which are regarded as particularly excellent sources of bioactive compounds (Hernandez et al. 2022). This section explored how the presence of Spirulina affected the physico-chemical, color and antioxidant textural and rheological characteristics, sensorial properties, and survival of starter cultures and probiotics in various dairy products (Fig. 7.3).

**Fig. 7.3** Spirulina in dairy products

**Table 7.2** Spirulina in milk and dairy products regulated by different authorities

| Food authority | Source of Spirulina | Dairy product category | Maximum daily intake |
|---|---|---|---|
| EFSA (EU) | *Arthrospira platensis* | Milk and dairy products | Unlimited |
| | *Chlorella luteoviridis* | | |
| | *Chlorella pyrenoidosa* | | |
| | *Chlorella vulgaris* | | |
| FDA (USA) | *Arthrospira platensis* | Milk and dairy products | 3 g per serving 0.208 g/100 g |
| | *Chlorella protothecoides* | Milk and dairy products | 1.35 g/100 g |
| | *Chlorella vulgaris* | Milk | 1.5 g/100 g |
| FSSAI (India) | *Arthrospira platensis* | Milk and dairy products | 4 g per day |
| | *Chlorella vulgaris* | Milk and dairy products | 5 g per day |
| | *Arthrospira platensis* dried powder | Milk and dairy products | 50 mg per day |

Food fortification with novel ingredients is regulated by various worldwide authorities including the European Food Safety Authority (EFSA), Food and Drug Administration (FDA), Food Safety and Standards Authority of India (FSSAI) in order to ensure food safety. Regarding the incorporation of Spirulina in milk and dairy products, the product categories and maximum daily intake values were explained in Table 7.2.

The use of Spirulina in yogurt formulations has been widely studied. Cultured buttermilk is reformulated by fortifying it with Spirulina, and different concentrations of Spirulina in buttermilk lead to a considerable increase in chlorophyll and carotenoid content, raising the antioxidant capacities (Rose et al. 2023). Spirulina was suggested as a natural ingredient for developing a new type of yogurt with higher nutritional values due to its greater protein, fat, and dietary fiber composition. The yogurt supplemented with 0.25 percentage of Spirulina exhibited greater water holding capacity and diminished whey syneresis throughout storage (Barkallah et al. 2017). The Spirulina was spray-dried and encapsulated with maltodextrin and utilized in yogurt fermentation. The yogurts fortified with encapsulated Spirulina had better homogeneous appearance and color, the nutritional value, improved antioxidant activity throughout storage (Da Silva et al. 2019). Fortifying yogurt with Spirulina raised its protein content. This finding was associated with higher protein content of Spirulina itself (around 70% in many cases) (Hernandez et al. 2022). Low-fat yogurt was supplemented with whey protein concentrates, calcium caseinates, and Spirulina. The results demonstrated that low-fat yogurt with Spirulina powder had the highest total phenolic content. However, low-fat yogurt with whey protein displayed higher viscosity and water holding capacity values. Furthermore, the gel formation of whey protein-fortified low-fat yogurts formed a fine network with very small pores in size when compared to other samples (Atallah et al. 2020). Spirulina-fortified probiotic yogurts were manufactured without any food additives. The probiotic yogurts enriched with 7% (w/w) Spirulina established to be quite acceptable product (Patel et al. 2019).

Spirulina has been investigated as a potential raw material for cheese fortification. Spirulina was utilized in the manufacturing of Traditional Greek Soft Cheese, and the presence of spiriluna was evaluated in terms of physicochemical, textural, and sensory properties of the cheese samples. In general, Spirulina improved the protein level of the cheeses. The cheese samples with 0.25 and 0.5% Spirulina were considered as an acceptable product by the panelists due to less prominent characteristic odor and taste (Bosnea et al. 2021). The effect of different concentrations of powdered *S. Platensis* on the protein and iron contents, color and hardness values of probiotic feta cheeses were examined. The study demonstrated that supplementing with Spirulina significantly increased the iron and protein concentrations as well as the hardness of feta cheeses. *S. platensis* has been shown to have a favorable impact on the survival of probiotic bacteria during cheese storage (Mazinani et al. 2016). The chemical, textural, and sensorial characteristics of feta cheeses supplemented by Spirulina (*A. platensis*) were analyzed. Spirulina containing cheesed had higher results for titratable acidity, dry matter, and protein content than the regular cheeses. The feta cheeses fortified with Spirulina had smoother textures which made the cheese samples much simpler to disintegrate and chew (Golmakani et al. 2019). The ricotta cheeses enriched Spirulina powder were produced, and the effects of spirulina on the physicochemical, sensory, and microbiological properties throughout a 21-day storage period at 5 °C were investigated. Incorporating 0.75 g dry spirulina powder into 100 g ricotta cheese yielded optimal color and sensory properties. Yeasts and mold were detected in control samples on the 14th day of storage and on the 21st day for Ricotta cheese enriched with spirulina powder (Ismail et al. 2023). Spirulina incorporation improved the sensory characteristics of mozzarella cheeses which led to a more appealing appearance, stronger butter odor and whey flavor, and increased sweetness, bitterness, juiciness, tenderness, oiliness, and buttermilk syneresis than the control samples (Lambiase et al. 2023). Cheeses enhanced with Spirulina platensis (3%), Chlorella vulgaris (1.467%), and curcumin (1%), exhibited the highest antioxidant activity and iron content. The cheese received the highest rating from panelists (Jalili et al. 2024). The edible film containing 2% Spirulina was used to package kefalotyri cheese. The efficiency of the film which was employed as package for kefalotyri cheese throughout 2 months under cold storage was evaluated. The GC-MS analysis of the edible film containing 2% spirulina revealed that the film does not maintain the unique odor. This film could be taken in with kefalotyri cheese and regarded a functional food, as proved by microbiological analysis (Kontogianni et al. 2023).

The ice cream production was reformulated by addition of *Spirulina platensis* in order to of reduce fat and sugar content. A combination of inulin (2%) and Spirulina (1%) to the ice cream recipe resulted in a 50% reduction in fat and 25% reduction in sugar content in the final product. The new ice cream formulation achieved sensory acceptance scores of up to 74% (Da Silva Faresin et al. 2022). Microalgae were dried using a spray drying technique to prepare the biomass for use as a natural colorant in ice cream production. It was concluded that microalgae might be employed as an alternative component to improve the color and functional features of ice cream (Durmaz et al. 2020). Spirulina was encapsulated in maltodextrin and Arabic gum and used to make homemade ice cream. Spirulina encapsulation with maltodextrin masks the taste and aroma of seaweed, but does not improve the acceptability of ice creams (Tiepo et al. 2021).

## 7.7 Production of Meat Analogues from Spirulina

Rising human population, urbanisation, and economic expansion in emerging nations have resulted in increased consumption of animal-derived protein during the last few decades. If the present worldwide consumption of meat trend persists, animal-derived protein requirement will be multiplied by 2050, which is a major concern for sustainability (Boland et al. 2013).

A meat-like product ought to resemble meat as a matter of texture, mouthfeel, flavor, and nutritional content. Although utilizing microalgal biomass is cost-effective, some species have hard cell walls and a substantial amounts of oil, which may restrict texture and interfere with nutritional absorption. Protein extracts may serve as a further component for extrusion. Emulsification allows for the assembly of fibrous products and fat mimetics after texturing (Fu et al. 2021). Flavoring is one of the key parameters in the development of new meat analogues During processing, two primary methods for creating meat-like flavors should be arranged: seasoning or spices, and chemical conversion component (amino acids and sugars-Maillard reaction). Seasonings and colorants can be examined at various stages of food production to identify potential beneficial chemical interactions. (Kyriakopoulou et al. 2019). The production of meat analogues from Spirulina can include an additional step to remove chlorophyll. Saponification is a typical method for removing chlorophyll, and ethanol can assist dissolve and remove chlorophyll from Spirulina, which is frequently removed via saponification (Li et al. 2016). The unfavorable odors and taste of microalgae, including earthy/musty, grassy, and fishy, may limit their use in food applications. Moreover, the smells which are mostly caused by volatile organic compounds (VOCs), which vary based on species, growth phase, and cultivation conditions. There are multiple methods to regulating odors in the final product. The addition of nitrogen to the growing medium can lower the formation of VOCs. Changing the harvesting time and cooking technique may also help to reduce unwanted odors (Fu et al. 2021). It was mentioned that more research is needed to make fat-mimetic products employing microalgae materials, as the quality of viscous meat emulsions and baked gels differ depending on protein functional properties (Kumar 2021).

The meat substitute industry intends to provide more than simply a choice for vegetarian and vegan customers. Therefore, researchers focused on microalgae as a protein alternative to soy protein concentrate, which is relevant for the growing industry. A meat analogue with fibrillary structure created from microalgae was established to be colorless (Caporgno et al. 2020). The previous researches revealed that high moisture extrusion (HME) processing of Spirulina showed that the end products had limited fibrous structure Therefore, Spirulina biomass needs to be combined with other plant proteins able to provide desired texture (Singh and Sit 2022). A preliminary study indicated that Spirulina can be combined with lupin protein up to 50% to produce HME-based meat analogues. Extrusion parameters like as temperature, water supply, and screw speed can be adjusted to provide the physical characteristics required for meat analogues. (Palanisamy et al. 2019).The organoleptic attributes of HME-based soya meat analogues including Spirulina have been examined in order to expand the application range of Spirulina biomass and increase the diversity of meat alternatives. It was concluded that Spirulina, partially substituted with, soy was suitable for extruded meat replacements as long as moisture was limited. (Grahl et al. 2018).

## 7.8 Spirulina for Formulations of Different Food Products

Spirulina is also utilized in the processing of several fermented foods. It was stated that probiotic fermentation has been shown to benefit Spirulina and can be used to create novel Spirulina-containing food products (Yu et al. 2020). The presence of spirulina (0.25% w/v) to craft beer fermentation increased the cytoprotective properties of the end product. The beer was discovered to exhibit antioxidant characteristics, with spirulina providing a small effect; however, spirulina-treated yeast cells had more riboflavin (Taiti et al. 2023). Spirulina has a synergistic effect on reducing power in Japanese quince and cranberry syrups. A blend of spirulina and Japanese quince syrup showed additive antiradical activity, nevertheless a mixture of spirulina and cranberry syrup was antagonistic (Stunda-Zujeva et al. 2023). C-phycocyanin (C-PC) extract from the cyanobacterium Arthrospira platensis is utilized as a natural coloring agent in several beverages, including different tonic drinks, and wine. It was demonstrated that C-PC extract was a natural colorant with desirable characteristics and could have the potential to be used as a natural coloring additive (Garcia et al. 2021). The possibility of Spirulina biomass in a vegetarain soybean drink as substrate for lactic acid fermentation by *Lactiplantibacillus plantarum* were analyzed, and spirulina biomass was proved to be an ideal substrate for this probiotic bacteria (Niccolai et al. 2020). The addition of 0.5% Spirulina substantially accelerated the fermentation of whey-based beverages. The whey-based sports drink containing Spirulina demonstrated improved textural, rheological, and sensory attributes (Elkot et al. 2024). The extraction conditions of phycocyanins from Spirulina were optimized for their integration into soft drinks. The use of phycocyanins extract in soft drinks formed a novel functional soft drink with blue natural colorants. As a result, phycocyanin extracts could be employed as natural colorants and antioxidants instead of synthetic ones (Benchikh et al. 2021).

## 7.9 Conclusions

Spirulina offers a diverse range of nutritious components, making it a promising candidate for different application in the food industry. Spirulina is considered as a long-term supplier of nutrients in terms of increasing numbers of people on conventional food supplies. Furthermore, it has some inherent advantages over traditional plants, such as the ability to be grown on non-agricultural while using less freshwater than typical crops for agriculture. Proteins, phenolic compounds, polyunsaturated fatty acids, vitamins, polysaccharides, minerals, and pigments of Spirulina improve the functional properties of food. Despite the fact that Spirulina has the potential to contribute in food formulation and used in food analogues, large-scale commercial development has been delayed by a number of challenges, including sensory difficulties, costly production system and downstream process costs. However, a lack of understanding and accessibility to algal food products is a key impediment to consumption and consumer demand, and enhancing disclosure of this information from Spirulina enterprises is one way for increasing customer confidence. As a result, additional studies and work are needed to properly comprehend and deal with practical and economic problems regarding the implementation of Spirulina in the food industry, as well as customer acceptance of food analogues.

## References

Abreu AP, Martins R, Nunes J (2023) Emerging applications of Chlorella sp. and Spirulina (Arthrospira) sp. Bioengineering 10(8):955. https://doi.org/10.3390/bioengineering10080955

Ahda M, Suhendra, Permadi A (2023) Spirulina platensis microalgae as high protein-based products for diabetes treatment. Food Rev Intl:1–9. https://doi.org/10.1080/87559129.2023.2238050

Al-Baarri AN, Aulia R, Prahasiwi EK, Mawarid AA, Pangestika W, Lestari FP (2021) The hardness analysis of noodles made from modified cassava flour, spirulina (*Spirulina platensis*) and basil leaves extract (*Ocimum sanctum L.*). IOP Conf Ser Earth Environ Sci 653:012051. https://doi.org/10.1088/1755-1315/653/1/012051

AlFadhly NK, Alhelfi N, Altemimi AB, Verma DK, Cacciola F, Narayanankutty A (2022) Trends and technological advancements in the possible food applications of Spirulina and their health benefits: a review. Molecules 27(17):5584. https://doi.org/10.3390/molecules27175584

Almeida LMR, da Silva Cruz LF, Machado BAS, Nunes IL, Costa JAV, de Souza Ferreira E, França LPV, Druzian JI, de Souza CO (2021) Effect of the addition of Spirulina sp. biomass on the development and characterization of functional food. Algal Res 58:102387. https://doi.org/10.1016/j.algal.2021.102387

Anvar AA, Nowruzi B (2021) Bioactive properties of spirulina: a review. Microb Bioact 4:134–142. https://doi.org/10.25163/microbbioacts.412117B0719110521

Aouir A, Amiali M, Bitam A, Benchabane A, Raghavan VG (2017) Comparison of the biochemical composition of different Arthrospira platensis strains from Algeria, Chad and the USA. J Food Meas Charact 11:913–923. https://doi.org/10.1007/s11694-016-9463-4

Atallah AA, Morsy OM, Gemiel DG (2020) Characterization of functional low-fat yogurt enriched with whey protein concentrate, Ca-caseinate and spirulina. Int J Food Prop 23(1):1678–1691. https://doi.org/10.1080/10942912.2020.1823409

Barkallah M, Dammak M, Louati I, Hentati F, Hadrich B, Mechichi T, Ayadi MA, Fendri I, Attia H, Abdelkafi S (2017) Effect of Spirulina platensis fortification on physicochemical, textural, antioxidant and sensory properties of yogurt during fermentation and storage. LWT 84:323–330. https://doi.org/10.1016/j.lwt.2017.05.071

Batista AP, Niccolai A, Bursic I, Sousa I, Raymundo A, Rodolfi L, Biondi N, Tredici MR (2019) Microalgae as functional ingredients in savory food products: application to wheat crackers. Food Secur 8(12):611. https://doi.org/10.3390/foods8120611

Benchikh Y, Filali A, Rebai S (2021) Modeling and optimizing the phycocyanins extraction from Arthrospira platensis (Spirulina) algae and preliminary supplementation assays in soft beverage as natural colorants and antioxidants. J Food Process Preserv 45(2):e15170. https://doi.org/10.1111/jfpp.15170

Bocker L, Bertsch P, Wenner D, Teixeira S, Bergfreund J, Eder S, Fischer P, Mathys A (2021) Effect of Arthrospira platensis microalgae protein purification on emulsification mechanism and efficiency. J Colloid Interface Sci 584:344–353. https://doi.org/10.1016/j.jcis.2020.09.067

Boland MJ, Rae AN, Vereijken JM, Meuwissen MP, Fischer AR, van Boekel MA, Rutherfurd SM, Gruppen H, Moughan PJ, Hendriks WH (2013) The future supply of animal-derived protein for human consumption. Trends Food Sci Technol 29(1):62–73. https://doi.org/10.1016/j.tifs.2012.07.002

Borowitzka MA (2018) Biology of microalgae. In: Microalgae in health and disease prevention. Academic, London, pp 23–72. https://doi.org/10.1016/B978-0-12-811405-6.00003-7

Bortolini DG, Maciel GM, Fernandes IDAA, Pedro AC, Rubio FTV, Branco IG, Haminiuk CWI (2022) Functional properties of bioactive compounds from Spirulina spp.: current status and future trends. Food Chem-Mol Sci 5:100134. https://doi.org/10.1016/j.fochms.2022.100134

Bosnea L, Terpou A, Pappa E, Kondyli E, Mataragas M, Markou G, Katsaros G (2021) Incorporation of *Spirulina platensis* on traditional Greek soft cheese with respect to its nutritional and sensory perspectives. PRO 70(1):99. https://doi.org/10.3390/foods_2020-07600

Braga da S, da Silveira Mastrantonio DJ, Costa JAV, de Morais MG (2018) Cultivation strategy to stimulate high carbohydrate content in Spirulina biomass. Bioresour Technol 269:221–226. https://doi.org/10.1016/j.biortech.2018.08.105

Caporgno MP, Böcker L, Müssner C, Stirnemann E, Haberkorn I, Adelmann H, Handschin S, Windhab EJ, Mathys A (2020) Extruded meat analogues based on yellow, heterotrophically cultivated Auxenochlorella protothecoides microalgae. Innovative Food Sci Emerg Technol 59:102275. https://doi.org/10.1016/j.ifset.2019.102275

Costa JAV, Freitas BCB, Rosa GM, Moraes L, Morais MG, Mitchell BG (2019) Operational and economic aspects of Spirulina-based biorefinery. Bioresour Technol 292:121946. https://doi.org/10.1016/j.biortech.2019.121946

Cuellar-Bermudez SP, Garcia-Perez JS, Rittmann BE, Parra-Saldivar R (2015) Photosynthetic bioenergy utilizing CO2: an approach on flue gases utilization for third generation biofuels. J Clean Prod 98:53–65. https://doi.org/10.1016/j.jclepro.2014.03.034

Da Rosa GM, Moraes L, Cardias BB, Costa JAV (2015) Chemical absorption and CO2 biofixation via the cultivation of Spirulina in semicontinuous mode with nutrient recycle. Bioresour Technol 192:321–327. https://doi.org/10.1016/j.biortech.2015.05.020

Da Silva Faresin L, Devos RJB, Reinehr CO, Colla LM (2022) Development of ice cream with reduction of sugar and fat by the addition of inulin, Spirulina platensis or phycocyanin. Int J Gastron Food Sci 27:100445. https://doi.org/10.1016/j.ijgfs.2021.100445

Da Silva SC, Fernandes IP, Barros L, Fernandes Â, Alves MJ, Calhelha RC, Pereira C, Barreira JCM, Manrique Y, Colla E, Ferreira ICFR, Barreiro MF (2019) Spray-dried Spirulina platensis as an effective ingredient to improve yogurt formulations: testing different encapsulating solutions. J Funct Foods 60:103427. https://doi.org/10.1016/j.jff.2019.103427

Dewi EN, Amalia U, Mel M (2016) The effect of different treatments to the amino acid contents of micro algae Spirulina sp. Aquat Procedia 7:59–65. https://doi.org/10.1016/j.aqpro.2016.07.008

Durmaz Y, Kilicli M, Toker OS, Konar N, Palabiyik I, Tamtürk F (2020) Using spray-dried microalgae in ice cream formulation as a natural colorant: effect on physicochemical and functional properties. Algal Res 47:101811. https://doi.org/10.1016/j.algal.2020.101811

Elkot WF, Elmahdy A, Talaat H, Alghamdia OA, Alhag SK, Al-Shahari EA, Al-Farga A, Ismail HA (2024) Development and characterization of a novel flavored functional fermented whey-based sports beverage fortified with Spirulina platensis. Int J Biol Macromol 258:128999. https://doi.org/10.1016/j.ijbiomac.2023.128999

Finkel ZV, Follows MJ, Liefer JD, Brown CM, Benner I, Irwin AJ (2016) Phylogenetic diversity in the macromolecular composition of microalgae. PLoS One 11(5):e0155977. https://doi.org/10.1371/journal.pone.0155977

Fu Y, Chen T, Chen SHY, Liu B, Sun P, Sun H, Chen F (2021) The potentials and challenges of using microalgae as an ingredient to produce meat analogues. Trends Food Sci Technol 112:188–200. https://doi.org/10.1016/j.tifs.2021.03.050

Garcia AB, Longo E, Bermejo R (2021) The application of a phycocyanin extract obtained from Arthrospira platensis as a blue natural colorant in beverages. J Appl Phycol 33(5):3059–3070. https://doi.org/10.1007/s10811-021-02522-z

Golmakani MT, Soleimanian-Zad S, Alavi N, Nazari E, Eskandari MH (2019) Effect of Spirulina (Arthrospira platensis) powder on probiotic bacteriologically acidified feta-type cheese. J Appl Phycol 31:1085–1094. https://doi.org/10.1007/s10811-018-1611-2

Grahl S, Palanisamy M, Strack M, Meier-Dinkel L, Toepfl S, Mörlein D (2018) Towards more sustainable meat alternatives: how technical parameters affect the sensory properties of extrusion products derived from soy and algae. J Clean Prod 198:962–971. https://doi.org/10.1016/j.jclepro.2018.07.041

Grosshagauer S, Kraemer K, Somoza V (2020) The true value of Spirulina. J Agric Food Chem 68(14):4109–4115. https://doi.org/10.1021/acs.jafc.9b08251

Haoujar I, Haoujar M, Altemimi AB, Essafi A, Cacciola F (2022) Nutritional, sustainable source of aqua feed and food from microalgae: a mini review. Int Aquat Res 14(3):1–9. https://doi.org/10.22034/IAR.2022.1958713.1278

Hernandez H, Nunes MC, Prista C, Raymundo A (2022) Innovative and healthier dairy products through the addition of microalgae: a review. Food Secur 11(5):755. https://doi.org/10.3390/foods11050755

Hosseinkhani N, McCauley JI, Ralph PJ (2022) Key challenges for the commercial expansion of ingredients from algae into human food products. Algal Res 64:102696. https://doi.org/10.1016/j.algal.2022.102696

Hussein AS, Mostafa S, Fouad S, Hegazy NA, Zaky AA (2023) Production and evaluation of gluten-free pasta and pan bread from *Spirulina* algae powder and quinoa flour. PRO 11(10):2899. https://doi.org/10.3390/pr11102899

Ismail HA, El-Sawah TH, Ayyash M, Adhikari B, Elkot WF (2023) Functionalization of Ricotta cheese with powder of spirulina platensis: physicochemical, sensory, and microbiological properties. Int J Food Prop 26(1):1968–1983. https://doi.org/10.1080/10942912.2023.2238916
Jalili S, Aryan S, Mousavinezhad SA, Moeini H, Dehnad D (2024) Optimizing Spirulina platensis, Chlorella vulgaris microalgae and curcumin application in functional cheese production and investigating its physicochemical properties and sensory evaluation by RSM. J Food Meas Charact 18(2):1144–1157. https://doi.org/10.1007/s11694-023-02231-w
Janda-Milczarek K, Szymczykowska K, Jakubczyk K, Kupnicka P, Skonieczna-Żydecka K, Pilarczyk B, Tomza-Marciniak A, Ligenza A, Stachowska E, Dalewski B (2023) Spirulina supplements as a source of mineral nutrients in the daily diet. Appl Sci 13(2):1011. https://doi.org/10.3390/app13021011
Kontogianni VG, Kosma I, Mataragas M, Pappa E, Badeka AV, Bosnea L (2023) Innovative intelligent cheese packaging with whey protein-based edible films containing spirulina. Sustain For 15(18):13909. https://doi.org/10.3390/su151813909
Kumar Y (2021) Development of low-fat/reduced-fat processed meat products using fat replacers and analogues. Food Rev Intl 37(3):296–312. https://doi.org/10.1080/87559129.2019.1704001
Kyriakopoulou K, Dekkers B, van der Goot AJ (2019) Plant-based meat analogues. In: Sustainable meat production and processing. Academic, London, pp 103–126. https://doi.org/10.1016/B978-0-12-814874-7.00006-7
Lafarga T, Fernández-Sevilla JM, González-López C, Acién-Fernández FG (2020) Spirulina for the food and functional food industries. Food Res Int 137:109356. https://doi.org/10.1016/j.foodres.2020.109356
Lambiase C, Braghieri A, Barone CMA, Di Francia A, Pacelli C, Serrapica F, Lorenzo JM, De Rosa G (2023) Use of cyanobacterium Spirulina (Arthrospira platensis) in buffalo feeding: effect on mozzarella cheese quality. Food Secur 12(22):4095. https://doi.org/10.3390/foods12224095
Letras P, Oliveira S, Varela J, Nunes MC, Raymundo A (2022) 3D printed gluten-free cereal snack with incorporation of Spirulina (Arthrospira platensis) and/or Chlorella vulgaris. Algal Res 68:102863. https://doi.org/10.1016/j.algal.2022.102863
Li T, Xu J, Wu H, Wang G, Dai S, Fan J, He H, Xiang W (2016) A saponification method for chlorophyll removal from microalgae biomass as oil feedstock. Mar Drugs 14(9):162. https://doi.org/10.3390/md14090162
Li TT, Liu YY, Wan XZ, Huang ZR, Liu B, Zhao C (2018) Regulatory efficacy of the polyunsaturated fatty acids from microalgae Spirulina platensis on lipid metabolism and gut microbiota in high-fat diet rats. Int J Mol Sci 19(10):3075. https://doi.org/10.3390/ijms19103075
Li TT, Tong AJ, Liu YY, Huang ZR, Wan XZ, Pan YY, Jia RB, Liu B, Chen XH, Zhao C (2019) Polyunsaturated fatty acids from microalgae Spirulina platensis modulates lipid metabolism disorders and gut microbiota in high-fat diet rats. Food Chem Toxicol 131:110558. https://doi.org/10.1016/j.fct.2019.06.005
Lucas BF, de Morais MG, Santos TD, Costa JAV (2018) Spirulina for snack enrichment: nutritional, physical and sensory evaluations. LWT 90:270–276. https://doi.org/10.1016/j.lwt.2017.12.032
Lucas BF, Costa JAV, Brunner TA (2023) Attitudes of consumers toward Spirulina and açaí and their use as a food ingredient. LWT 178:114600. https://doi.org/10.1016/j.lwt.2023.114600
Mazinani S, Fadaei V, Khosravi-Darani K (2016) Impact of Spirulina platensis on physicochemical properties and viability of Lactobacillus acidophilus of probiotic UF feta cheese. J Food Process Preserv 40(6):1318–1324. https://doi.org/10.1111/jfpp.12717
Montevecchi G, Santunione G, Licciardello F, Köker Ö, Masino F, Antonelli A (2022) Enrichment of wheat flour with Spirulina. Evaluation of thermal damage to essential amino acids during bread preparation. Food Res Int 157:111357. https://doi.org/10.1016/j.foodres.2022.111357
Niccolai A, Venturi M, Galli V, Pini N, Rodolfi L, Biondi N, D'Ottavio M, Batista AP, Raymundo A, Granchi L, Tredici MR (2019) Development of new microalgae-based sourdough "crostini": functional effects of Arthrospira platensis (spirulina) addition. Sci Rep 9(1):19433. https://doi.org/10.1038/s41598-019-55840-1

Niccolai A, Bažec K, Rodolfi L, Biondi N, Zlatić E, Jamnik P, Tredici MR (2020) Lactic acid fermentation of Arthrospira platensis (Spirulina) in a vegetal soybean drink for developing new functional lactose-free beverages. Front Microbiol 11:560684. https://doi.org/10.3389/fmicb.2020.560684

Nissen L, Casciano F, Chiarello E, Di Nunzio M, Bordoni A, Gianotti A (2024) Sourdough process and spirulina-enrichment can mitigate the limitations of colon fermentation performances of gluten-free breads in non-celiac gut model. Food Chem 436:137633. https://doi.org/10.1016/j.foodchem.2023.137633

Ozyurt G, Uslu L, Yuvka I, Gokdoğan S, Atci G, Ak B, Isik O (2015) Evaluation of the cooking quality characteristics of pasta enriched with Spirulina platensis. J Food Qual 38(4):268–272. https://doi.org/10.1111/jfq.12142

Palanisamy M, Töpfl S, Berger RG, Hertel C (2019) Physico-chemical and nutritional properties of meat analogues based on Spirulina/lupin protein mixtures. Eur Food Res Technol 245:1889–1898. https://doi.org/10.1007/s00217-019-03298-w

Pandey VK, Singh P, Srivastava S, Zanwar S, Dar AH, Singh R, Lal A (2023) Box–Behnken design based statistical modelling to study the effects of spirulina (Arthrospira platensis) incorporation on nutritional standards of vegan snack product. J Agric Food Res 14:100700. https://doi.org/10.1016/j.jafr.2023.100700

Patel P, Jethani H, Radha C, Vijayendra SVN, Mudliar SN, Sarada R, Chauhan VS (2019) Development of a carotenoid enriched probiotic yogurt from fresh biomass of Spirulina and its characterization. J Food Sci Technol 56:3721–3731. https://doi.org/10.1007/s13197-019-03844-0

Paula da Silva S, do Valle AF, Perrone D (2021) Microencapsulated Spirulina maxima biomass as an ingredient for the production of nutritionally enriched and sensorially well-accepted vegan biscuits. LWT 142:110997. https://doi.org/10.1016/j.lwt.2021.110997

Penalver R, Martínez-Zamora L, Lorenzo JM, Ros G, Nieto Martínez G (2023) Effect of hydroxytyrosol, Moringa, and spirulina on the physicochemical properties and nutritional characteristics of gluten-free brownies. Food Sci Nutr. https://doi.org/10.1002/fsn3.3778

Rahimah S, Sonjaya MF, Andoyo R, Satya A, Nurhasanah S, Chrismadha T (2024) Physical and organoleptic characteristic of bread substituted with *Spirulina platensis*. In: Biomass conversion and sustainable biorefinery: towards circular bioeconomy. Springer Nature Singapore, Singapore, pp 295–306. https://doi.org/10.1007/978-981-99-7769-7_14

Ramírez-Rodrigues MM, Estrada-Beristain C, Metri-Ojeda J, Pérez-Alva A, Baigts-Allende DK (2021) Spirulina platensis protein as sustainable ingredient for nutritional food products development. Sustain For 13(12):6849. https://doi.org/10.3390/su13126849

Rodriguez-Concepcion M, Avalos J, Bonet ML, Boronat A, Gomez-Gomez L, Hornero-Mendez D, Limon CM, Meléndez-Martínez AJ, Olmedilla-Alonso B, Palou A, Ribot J, Rodrigo MJ, Zacarias L, Zhu C (2018) A global perspective on carotenoids: metabolism, biotechnology, and benefits for nutrition and health. Prog Lipid Res 70:62–93. https://doi.org/10.1016/j.plipres.2018.04.004

Rose H, Bakshi S, Kanetkar P, Lukose SJ, Felix J, Yadav SP, Gupta PK, Paswan VK (2023) Development and characterization of cultured buttermilk fortified with Spirulina plantensis and its physico-chemical and functional characteristics. Dairy 4(2):271–284. https://doi.org/10.3390/dairy4020019

Saharan V, Jood S (2021) Effect of storage on Spirulina platensis powder supplemented breads. J Food Sci Technol 58(3):978–984. https://doi.org/10.1007/s13197-020-04612-1

Seghiri R, Kharbach M, Essamri A (2019) Functional composition, nutritional properties, and biological activities of Moroccan Spirulina microalga. J Food Qual 2019:1–11. https://doi.org/10.1155/2019/3707219

Singh A, Sit N (2022) Meat analogues: types, methods of production and their effect on attributes of developed meat analogues. Food Bioprocess Technol 15(12):2664–2682. https://doi.org/10.1007/s11947-022-02859-4

Soni RA, Sudhakar K, Rana RS (2017) Spirulina–From growth to nutritional product: a review. Trends Food Sci Technol 69:157–171. https://doi.org/10.1016/j.tifs.2017.09.010

Stunda-Zujeva A, Berele M, Lece A, Šķesters A (2023) Comparison of antioxidant activity in various spirulina containing products and factors affecting it. Sci Rep 13(1):4529. https://doi.org/10.1038/s41598-023-31732-3

Taghdiri S, Emtyazjoo M, Azizi MH, Ariaii P, Sedaghati M (2024) The effect of hydrolyzed protein obtained from *Chlorella vulgaris* on the shelf life and quality of oil cake during the storage period. J Food Meas Charact 18:582–592. https://doi.org/10.1007/s11694-023-02186-y

Taiti C, Stefano G, Percaccio E, Di Giacomo S, Iannone M, Marianelli A, Sotto AD, Garzoli S (2023) Addition of spirulina to craft beer: evaluation of the effects on volatile flavor profile and cytoprotective properties. Antioxidants 12(5):1021. https://doi.org/10.3390/antiox12051021

Tiepo CBV, Gottardo FM, Mortari LM, Bertol CD, Reinehr CO, Colla LM (2021) Addition of Spirulina platensis in handmade ice cream: Phisicochemical and sensory effects/Adição de Spirulina platensis em sorvete caseiro: Efeitos físico-químicos e sensoriais. Braz J Dev 7:88106–88123. https://doi.org/10.34117/bjdv7n9-121

Yu J, Ma D, Qu S, Liu Y, Xia H, Bian F, Zhang Y, Huang C, Wu R, Wu J, You S, Bi Y (2020) Effects of different probiotic combinations on the components and bioactivity of Spirulina. J Basic Microbiol 60(6):543–557. https://doi.org/10.1002/jobm.201900699

Zen CK, Tiepo CBV, da Silva RV, Reinehr CO, Gutkoski LC, Oro T, Colla LM (2020) Development of functional pasta with microencapsulated Spirulina: technological and sensorial effects. J Sci Food Agric 100(5):2018–2026. https://doi.org/10.1002/jsfa.10219

# Chapter 8
# Edible Marine Biological Materials in Food Analogues

**Xiaoming Jiang, Songgang Xia, Yukun Hou, Jian Song, Chengxin Ma, Tingting Hao, and Qingyi Wan**

## 8.1 Introduction

At present, food analogues mainly include meat analogues and milk analogues (Brooker et al. 2023; Sha and Xiong 2020). Especially, plant protein meat analogues have received widespread attention and been commercialized. The main business brands that are widely known are Beyond Meat, Impossible Food, and ATM (Sha and Xiong 2020). The increase in the consumption of plant protein meat analogues is due to several factors. First, the United Nations statistics suggest that by 2050, the world's population is predicted to exceed ten billion (UN 2019). As a result, the demand for meat consumption will continue to increase, which cannot be satisfied by current livestock farming (Xia et al. 2023; Chantanuson et al. 2022). Second, animal husbandry in its current form is not sustainable for the environment. Meat production not only leads to large amounts of greenhouse gases but also contributes to the heavy use of water and land resource as well as other pollution problems. Animal husbandry emits around 14.5% of human-induced GHG emissions and consumes 40% of cultivated land and 56% of agricultural water consumption (Weindl et al. 2017; Gerber et al. 2013). Next, there are several potential harmful impacts on human health associated with conventional meat production. The growth of animal husbandry would cause an increase in the prevalence of zoonotic diseases and the source of antibiotic resistance (Bryant 2022). In addition, overconsumption of red meat is closely associated with most acute health issues, such as heart disease, and cancer (Willett et al. 2019; Bryant 2022; Fu et al. 2021). As a result, plant-based

X. Jiang (✉)
State Key Laboratory of Marine Food Processing and Safety Control, College of Food Science and Engineering, Ocean University of China, Qingdao, China
e-mail: jxm@ouc.edu.cn

S. Xia · Y. Hou · J. Song · C. Ma · T. Hao · Q. Wan
College of Food Science and Engineering, Ocean University of China, Qingdao, China



Ö. P. CAN et al. (eds.), *Food Analogues: Emerging Methods and Challenges*,
https://doi.org/10.1007/978-3-031-69872-9_8

meat analogue is recognized as a potentially viable strategy to address the above problems. However, in addition to plant-based products, marine foods (also as blue foods) can offer new solutions to address the problems.

Studies show that blue foods, including finfish, crustaceans, cephalopods, other molluscs, aquatic plants, algae and other aquatic animals, are usually undervalued as a nutritional solution and have great potential to reduce food security and various forms of malnutrition (Golden et al. 2021). In addition, marine foods contain rich protein, healthy lipids (phospholipid omega-3 polyunsaturated fatty acids) and a large number of micronutrients which are healthier for humans than other animal-source foods; In particular, omega-3 fatty acids from marine foods can improve cardiovascular, brain and visual health in humans (Tigchelaar et al. 2022; Golden et al. 2021). Among the various fish species, cod is one of the most abundant fish species in terms of annual yield which is over ten million tons; it contains over 15% protein, a balanced amino acid composition, and antioxidant and anti-inflammatory bioactivities (Wang et al. 2021; Jensen et al. 2013). Besides, Antarctic krill (*Euphausia superba*), a crustacean distributed in the Antarctic, has attracted increasing attention in recent times. Antarctic krill is the most abundant oceanic food resource, about 500 million metric tons; in terms of nutrition, it has more than 10% protein, containing all the amino acids that the human being needs (Li et al. 2022b). Therefore, Antarctic krill is considered to be the largest and under-exploited reservoir of marine biological protein. Shellfish and cephalopods are mollusks. In 2014, the total production of mollusks (including 104 species) was 16.1 million metric tons (Venugopal and Gopakumar 2017). Moreover, Short et al. (2021) reported that small-scale fisheries and aquaculture can supply the livelihoods for more than 0.1 billion people and sustenance for approximately 1 billion people. According to Gephart et al. (2021), blue foods have considerable potential to reduce environmental stresses on the food system; for example, water and land use could be halved by the increase of fish yield, and emissions from capture fisheries could be reduced by more than half simply by improving fishing gear. Accordingly, marine foods are an important part of the food field.

However, the current attention on food analogues is mainly focused on plant-based products, while ignoring the substantial role played by marine biomaterials in food analogues. The main technologies used to process marine biological resources into food analogues are molding and cooking processes, extrusion technology, and three-dimensional (3D) printing technology (Li et al. 2023b; Hou et al. 2023; Harris and Johnson 1987). Surimi is the most widely produced ingredient used in processed food simulants, and it is primarily made from Alaskan pollock (Harris and Johnson 1987). Through the molding and cooking process, surimi can be used to simulate crab, shrimp, lobster, and scallop meat (Park et al. 1997). 3D printing technology does not require molds and is suitable for a wide range of materials; therefore, it has been used to prepare food analogues (Tibrewal et al. 2023). In addition, surimi has been used to produce various types of meat analogues through 3D printing technology (Kim et al. 2021). Moreover, extrusion is also a common technology for producing the meat analogues, divided into high-moisture extrusion (feeding material moisture > 40%) and low-moisture extrusion (feeding material moisture ≤ 40%) (Zhang et al. 2020). The extrusion process can produce extrudates with a rich fiber structure by processing high-protein raw materials. It combines the processes of material mixing, cooking and molding into one, with the merits of high efficiency and low energy consumption (Sha and Xiong 2020). Through high-moisture extrusion technology, a blend of surimi and vegetable proteins can be processed into meat analogues with a rich fiber structure (Hou et al. 2023). In addition, fish muscle and some other excipients can be used to produce meat analogues containing fibrous structure with expansion ratio of 16.6 through low-moisture extrusion technology (Choudhury and Gogoi 1996). Therefore, marine biomaterials have great potential for the production of food analogues, and this chapter focuses on reviewing a synthesis on the application of marine biomaterials (not including algae) in meat analogues.

## 8.2 Applications of the Edible Marine Biological Sources in Food Analogues

### *8.2.1 Fish as an Ingredient to Produce Meat Analogues*

The two primary fish species utilized as raw materials to process meat analogues are marine and freshwater fish. Fish has a rich nutritional value, a delicious flavor, and a high protein and phospholipid content. It is easily digestible and absorbable by the human body, having a major impact on the growth of both physical and intellectual strength. Compared to poultry and livestock, fish has low fat content and low heat energy, so fish meat is a superior high-protein and low-calorie food. In addition, fish are rich in various vitamins and minerals, including vitamin A, vitamin D, calcium, iodine, copper, zinc, and so on.

As a fish meat product, it can be divided into high-economic value fish and low-economic value fish. For seafood with a high economic value, like tuna and Atlantic salmon, etc., fresh food is the main consumption form, which maximizes the original taste and nutritional value of the fish meat. Low-economic-value fish and small miscellaneous fish undergo deep processing, such as traditional drying, pickling, smoking, canned processing, and processing into minced fish and its products. For fish rich in functional ingredients, it can be made into fish liver oil, fish glue, fish meal, etc. The formation of new meat analogues through tissue fibrosis of fish is a new research trend. Muscle protein in fish can be compounded with other plant proteins to prepare livestock and poultry meat analogues, and collagen can be used as an additive to improve its texture and nutrition. The lipid components (DHA and EPA) rich in fish can enrich the nutrition and flavor of food analogues.

## *8.2.2 High-Moisture Extrusion of Fish Meat to Produce Meat Analogues*

### 8.2.2.1 High-Moisture Extrusion Technology

High-moisture extrusion technology is a fast-acting, high-temperature processing technique where raw materials have a moisture content of over 40%. (Prabha et al. 2021; Zhang et al. 2022a). High-moisture extrusion technology is one of the best options for creating meat analogues with fiber structure because of its many benefits, which include no waste discharge, lower energy input and cost, higher efficiency and quality of tissue and organized products (Xia et al. 2022). During the high-moisture extrusion process, protein molecules unfold, associat, aggregat, and cross-link to form a fiber-oriented structure (Zhang et al. 2021). A twin-screw extruder, which has a feeding system and a cooling system and two co-rotating screws that make up the extruder's main body, is used for high moisture extrusion processing, as shown in Fig. 8.1.

### 8.2.2.2 Determination of Process Parameters in High-Moisture Extrusion of Fish

For determining the appropriate high-moisture extrusion parameters, single factor analysis or response surface analysis is an effective means. A two-level factorial experimental design was used by Thiébaud et al. (1996) to investigate the relationship between process and system parameters as well as the effects of process parameters on the color, tensile characteristics, and final moisture content (after immersion) of extruded raw fish products. This study confirmed that protein foods with fiber structure could be produced by extrusion cooking at high moisture content (>600 g/kg). The main factors that determine the texture of the extrudate are the temperature of the barrel, the screw speed, and the moisture content of the feed, in that order.

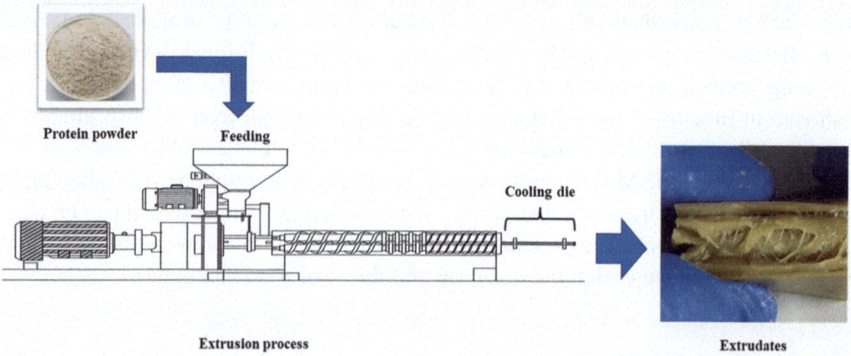


**Fig. 8.1** Schematic diagram of high moisture extrusion process

While the temperature of the barrel and screw speed have the greatest effects on the final moisture content of the extrudate (after immersion), the moisture content of the raw material and barrel temperature have the greatest effects on the color of the extrudate. The extrudate's properties were not significantly affected by the feed speed or average residence time within the experimental range. He et al. (2014) investigated the blends of soy protein isolate (SPI) and hairtail surimi (HS) through high-moisture extrusion. The water retention and the structure of the extruded material are related to the feed moisture and barrel temperature, respectively, through the investigation of the response surface method. The Meat analogues' in vitro protein digestibility may also be enhanced by high-moisture extrusion. Wang et al. (2012) explored how the organization of fish protein and its high-moisture meat is affected by the parameters of the operations. As the more significant process parameters in the high-moisture extrusion process, the results demonstrate that the moisture content, cavity temperature, and feed speed have distinct effects on the fiber degree, hardness, chewiness, and color of the products. Zhao et al. (2019) also investigated, using a single factor experiment, how process parameters affected the texture of freshwater fish mirror carp extruded by twin-screw extrusion. The ideal high-moisture extrusion process was ultimately identified. Numerous related studies have demonstrated that the moisture content of raw materials, barrel temperature, screw speed, and feeding speed are the main determinants of the tissue structure and texture of meat analogues made by high-moisture extrusion of fish.

#### 8.2.2.3 Application of Surimi in High-Moisture Extrusion

As an important product of fish processing, surimi not only promotes the development of the whole aquatic product processing industry but also enriches consumers' diets. As we all know, surimi products are mainly divided into fish cake, simulated crab meat, fish sausage, etc. Extruding and cooking surimi with high moisture to form a fibrous structure is undoubtedly a good idea for the deep processing industry of surimi. Aoki et al. (1989) conducted the initial investigation on surimi extrusion at high moisture content. Using surimi as raw material and twin-screw extruder for extrusion cooking, the fiber textured products were prepared. The direction of the fiber structure is directly related to the extrusion direction. According to these fibrous texture products, melting conditions should be applied to rearrange the proteins in surimi. It is imperative to fully melt the raw material when extruding and cooking surimi, as this is confirmed when the steady-state extrusion is abruptly stopped to examine the conditions inside the cylinder and confirm that the texture begins at the screw's exit before being transferred to the mold. In latest research, *Alaska pollock* surimi was used for high-moisture extrusion (Hou et al. 2023). Common plant proteins used in the high-moisture extrusion of surimi are soybean protein isolate and wheat protein. This process can enhance the nutritional composition of surimi extrudates and the stability of the final products. The findings demonstrated that varying amounts of plant protein had distinct effects on surimi's high-moisture extrusion, and that adding wheat protein enhanced the surimi extrudate's fiber degree, color value, thermodynamic characteristics, and other physical parameters. This work offers a theoretical foundation for the synthesis of fibrous animal imitation meat from marine aquatic protein as well as a novel application idea for marine biological simulation animal products. Kitagawa et al. (2007) also tried to prepare fiber-extruded products from raw fish by extrusion cooking. A twin-screw extruder was utilized for the high-moisture extrusion of defatted soy protein concentrate (SPC) and Walleye pollack surimi (WPS). When the moisture content is adjusted to about 60%, the final tube temperature is 170 °C, and the twin-screw extruder runs at a screw speed of 50 rpm. The WPS-SPC mixture (WPS/SPC = 70:30) deforms continuously, and the extrudate shows a strong fiber structure in the extrusion direction. These findings demonstrate the potential of Walleye pollack surimi as a high-moisture extrusion raw material. Zhang et al. (2022b) investigated the relationship between the texture properties of high-moisture extruded products (70%) and the gel properties of various ratios of mixed protein and soybean protein isolate. The extrudate has greater hardness, chewiness, gel strength, and fiber content during high-moisture extrusion when the ratio of SPI to surimi is 80:20. However, too much surimi in the blend will prevent the formation of fiber orientation structure. It's interesting to note that this study discovered a non-linear relationship between the specific mechanical energy (SME) and texture properties of the products and the gel strength of SPI-surimi blends. Yujie et al. (2022) investigated how hydrothermal parameters affected the regulation of extrusion quality as well as unit mechanical energy consumption and die pressure during high-moisture extrusion of a blend of soy protein isolate (SPI) and *Nemipterus virgatus* surimi. This study demonstrates that the quality of the SPI-surimi-mixed extrudate is significantly influenced by the combination of hydrothermal parameters. When the material moisture content is 70% and the extrusion temperature is 145 °C, the effect of the material moisture content is greater than the extrusion temperature.

#### 8.2.2.4 High-Value Utilization of Low-Value Fish in High-Moisture Extrusion

The use of fish with low economic value or by-products from fish processing for human consumption has gained popularity due to its potential benefits to the environment and the pursuit of economic maximization. However, the method of processing low-value fish into usable products is limited, and high-moisture extrusion may be a good way to solve this kind of problem. The high-moisture extrusion of fish was studied earlier in Japan, as reported by Isobe et al. (1990), using a twin-screw extruder to convert salmon into textured fish products, the simple extrusion of fish meat is difficult, while the addition of 10% wheat flour can achieve continuous extrusion cooking of salmon fillets and enhance the formation of fibrin matrix. At the same time, the addition of wheat flour promoted the decrease of salmon muscle protein solubility. Reduce the viscoelasticity of internal materials, spread polysaccharides between protein components, prevent adhesion between fibrin parts, and further stretch protein parts, which helps to improve fiber properties. Nisov et al.

(2020) investigated the viability of utilizing whole, eviscerated Baltic herring as a raw material for squeezing and cooking at a high moisture content to create plant-fish analogues. Research findings indicate that the extrusions made from high-moisture extrusions of whole fish, gutted fish, and pea protein isolates exhibit comparable microbiological quality. Furthermore, the entire fish and gutted fish had similar flavors, odors, and color values; however, the whole fish extrusion product has a comparatively weak structure. These findings clearly indicate that it can take the place of traditional products made with minced meat or fish meat. Kitagawa et al. (2008) pulverized the clam salmon and skipjack tuna meats containing skin and bone at 130 °C, then cooked by extrusion with high moisture, and examined how the composition of fatty and amino acids had changed. The results showed that taurine, methionine, and histidine decreased slightly during extrusion, but had no significant impact on the contents of other amino acids. Extrusion cooking also had no significant effect on fatty acid composition. This result means that fish with skin and bone can be used for high-moisture extrusion without significant nutritional loss. Kong et al. (2012) made further processing and utilization of the by-product of smoked salmon, that is, the dark meat on the side line of smoked salmon. The smoked ingredients were mixed with fresh salmon, while basic food ingredients such as cassava starch were added, the blend was pushed through a twin-screw extruder at a high moisture level, and then hot fumigated for 5 h to evaluate the changes of basic nutritional composition in the product. The sensory evaluation results show that 70 consumers' evaluation of all sensory characteristics is between "slightly like" and "moderate like", indicating that through some improvements in formula and processing, the by-products of smoked salmon can also increase economic value and avoid waste through high-moisture extrusion.

### *8.2.3 3D Printing of Fish to Produce Meat Analogues*

With the use of a digitally controlled extrusion process, 3D printing can be used to create food structures with a high surface-to-volume ratio, intricate designs, geometry, and customized nutrients (Prabha et al. 2021). Three categories of 3D printing technology exist. (i) selective laser sintering, (ii) melt deposition modeling, and (iii) stereolithography (Achinas et al. 2020). The most widely used 3D printing technology for biological printing applications, such as food printing, is extrusion molding, also known as melt deposition molding (Paxton et al. 2017). This technology, with its multiple print heads that can adjust custom shapes, bracket designs, fill densities, and use multiple biological inks at once, has been used for all major research on 3D printing for meat and its analogues. There are three categories into which extrusion technology falls: screw-based, air-driven, and syringe-based, screw-based 3D printing is suitable for materials with low viscosity and low tensile strength, air-driven 3D printing is suitable for printing low viscosity liquids with good fluidity, and syringe-based 3D printing is suitable for materials with high viscosity and high tensile strength (Tibrewal et al. 2023).

Fish can also use 3D printing technology to produce meat analogues, of which surimi gel is more studied in 3D printing. Wang et al. (2018) investigated the viability of surimi as a material for 3D-printed food in a study utilizing surimi. Overall, the results demonstrate that adding sodium chloride helps the slurry flow out of the nozzle on schedule, allowing for the creation of viscous post-deposition that keeps its shape. Stated differently, the surimi gel system-based 3D printing technique shows promise as a means of producing meat substitutes. Kim et al. (2021) sought to enhance the texture of 3D printed surimi products. Conducted additional research on coaxial nozzles using an outer layer of adhesive based on starch. To put it simply, the starch-based adhesive is put on the outside of the surimi to help it keep its shape during printing. This study validates the possibility of using coaxial nozzles to assist 3D printing in producing filamentous surimi-based products in starch-based adhesives. Dong et al. (2019) investigated how sweet potato starch affected the surimi gel's physical characteristics and how it behaved during 3D printing. The findings indicate that the viscosity of the starch–surimi mixture decreases as the starch content increases, which is advantageous for the preserved fish to flow from the printer nozzle. In other words, sweet potato starch can be used as a structural enhancer to effectively 3D print complex-shaped surimi. Gudjonsdottir et al. (2019) used low-field NMR and chemometrics to examine the effects of various processing techniques, salt addition, cold storage duration, and cooking on the printability and characteristics of surimi paste, a cod byproduct. The analysis shows that the two processing methods have their advantages and disadvantages. Increasing salt content has a stable effect on surimi and can be printed with fresh raw materials.

## 8.3 The Applications of Crustaceans in Food Analogues

In aquatic ecosystems, crustaceans are well known for their rich protein content and are widely used as high-quality food supplements. Marine crustacean resources include shrimps, crabs, gill-footed subclasses, and mesopods, etc. Several of these crustaceans, such as shrimps and crabs, are consumed as food and are further processed on a large scale. Shrimp surimi, a highly valued product of shrimp known for its rich flavor and nutritious content, is one of the most extensively farmed seafood species in the world (Nanda et al. 2021). The development of 3D printing technology has enabled the enrichment of the form of shrimp surimi products. At the same time, the beneficial nutrients that were present in crustaceans (astaxanthin, chitosan, etc.) are also widely used in food analogue products as favorable factors. For example, the creation of puffed snacks with a seafood flavor and high calcium content by using by-products from the extrusion of crabs (Murphy et al. 2003).

### 8.3.1 The Main Components in Crustaceans

Crustaceans are a class of food ingredients that contain macronutrients and a variety of beneficial nutrients. Shrimp/crab meat contains 22.78 g/19.35 g of protein per 100 g (Service 2018), and their protein density is higher than that of livestock and poultry meat, making it one of the good sources of dietary protein. Fatty acids that are polyunsaturated PUFA, particularly docosahexaenoic acid (DHA) and eicosapentaenoic acid (EPA), are the main constituents of total fatty acids in shrimp and have advantageous physiological roles (Gulzar and Benjakul 2018). Based on prior reports, the dry weight of crustaceans ranged from 14% to 35%, whereas the deproteinized dry shell waste of Antarctic krill had almost 40% crude chitin (Lordan et al. 2011). Chitin is commonly used as a raw material for the production of chitosan, oligosaccharides and glucosamine (Ferraro et al. 2010). Chitin, chitosan and their derivatives have a variety of food applications, e.g., as food additives, as nutritional supplements to increase dietary fiber, reduce lipid absorption, etc. (Shahidi et al. 1999). Crustacean exoskeleton and cephalothorax are rich in carotenoid pigments, particularly astaxanthin, which demonstrate antioxidant action in vitro by scavenging free radicals, chelating pro-oxidant metals, and preventing lipid peroxidation (Azelee et al. 2023). Messina et al. (2021) implemented a green and efficient method for extracting bioactive compounds from crustaceans, which has also made it possible to use astaxanthin in a wide range of applications such as food processing. Based on previous studies, Table 8.1 summarized the potential of some of the major components of crustaceans for use in analogues.

### 8.3.2 Crustaceans Protein as a Main Ingredient in Food Analogues

Surimi products were extensively dependent on the gelation properties of proteins. However, since shrimp or crab proteins differ significantly from fish proteins and because shrimp have a greater endogenous protease activity, it is difficult to make surimi-based goods that have the same gel strength and superior texture as fish surimi products (Jiang et al. 2023). Therefore, a number of supplementary means were often chosen to improve the gel properties of shrimp proteins and thus developed shrimp-based food analogues. For example, gel strength, texture quality, and water holding capability of salt-reduced white leg shrimp protein gels can be enhanced significantly by combining L-arginine and microbial transglutaminase (Jiang et al. 2023). Guo et al. (2017) induced secondary structure changes in shrimp surimi proteins by dense phase carbon dioxide (DPCD) technology, which altered protein interactions and cross-linking, and consequently enhanced gel strength. High pressure processing (HPP) technology used to improve blue crab (Callinectes sapidus) meat heating-induced gelling capacity (Martínez-Maldonado et al. 2020). Furthermore, 3D printing technology can create geometries that are challenging to

**Table 8.1** Potential application of the main components of crustaceans in analogues

| Main components | Sources of components or applications | Effects | References |
|---|---|---|---|
| Protein | White leg shrimp (*Litopenaeus vannamei*) | Combined treatment with L-Arg and MTGase improved the gel properties of salt-reduced shrimp surimi-based products. | Jiang et al. (2023) |
| | Shrimp surimi (*Litopenaeus vannamei*) | Protein interactions and chemical cross-linking induced by dense phase carbon dioxide (DPCD) technology to improve the strength of shrimp surimi gels. | Guo et al. (2017) |
| | Blue crab (*Callinectes sapidus*) | High-pressure technology and thermal induction for the production of high-quality crab gels. | Martínez-Maldonado et al. (2020) |
| | White shrimp (*Litopenaeus vannamei*) | Different starches were employed to improve 3D printing properties of shrimp surimi. | Pan et al. (2021) |
| Lipids | Extruded snacks | Shrimp head exudates improved protein content and flavor of extruded snacks. | Baskar et al. (2022) |
| Chitosan | Extrusion 3D printing for mayonnaise | Chitosan contributed to good extrudability and printability of extrudates | Zhao et al. (2023) |
| Chitosan | Extrusion processing | Thermoplastic chitosan promotes extrudate stability. | Mendes et al. (2016) |
| | Functional extruded snacks | Chitosan was supplemented in acetate-based extruded snacks to enrich their functional properties | Kumar et al. (2018) |

create with conventional technology, and it has the potential to significantly impact the creation of food analogs. Pan et al. (2021) effectively improved the precision and stability of 3D printing of minced shrimp by adding a certain amount of cross-linked starch (<10%), which advanced the development and production of shrimp-based food analogues.

### 8.3.3 Application of Crustacean Lipids in Food Analogues

The unusable parts of shrimps and crabs, such as the cephalothorax, were usually discarded in the food processing industry. However, its extrudate (shrimp oil) is gaining attention because it contains n-3 fatty acids such as EPA and DHA, as well as carotenoids. Shrimp oil from natural sources has beneficial physiological functions, and its functional properties are gradually recognized by consumers, with good prospects for the future market. Low-value shrimp head extrudates were added to extruded snack foods to enhance their flavor and nutritional value (Baskar et al. 2022). Meat analogues were a research hotspot in recent years. Lipids have multiple functions such as texture modifiers, juiciness enhancers, flavour providers satiety, etc., and hence are considered to be the main components of plant-based meat analogues (Ahmad et al. 2022). Shrimp head extrudates had the potential to be employed in fish, shrimp, and other seafood analogues. Furthermore, according to previous studies, shrimp oil stabilized by fish myofibrillar proteins possessed certain viscoelastic properties (Rajasekaran et al. 2022), and it could be used for further development of fat analogues.

### *8.3.4 Potential Application of Chitosan in Food Analogues*

Chitosan, as a dietary fiber, has a variety of functional, technological and physiological properties that contribute to food quality. Chitosan was widely used as a nutraceutical in the form of tablets and capsules due to its fat-absorbing ability (Ruiz Matute et al. 2013). Chitosan, which is produced by deacetylating chitin, is regarded as one of the crucial functional and technological elements (Chen et al. 2023b). By incorporating chitosan into snacks, it is possible not only to improve the functionality of the product, but also to create new functional foods with health benefits (Borderías et al. 2005). Chitosan exhibited great potential for use in food analogues through 3D printing, extrusion, or other technologies. Chitosan with anions could be combined with other substances to form polyelectrolyte composite gels, which had tremendous latent 3D printing properties and could be used in 3D printing-based food analogue products (Yang et al. 2021). Zhao et al. (2023) optimized the printability of 3D printing inks by applying chitosan to high internal phase pickering emulsions (HIPPEs). A high internal phase emulsion (HIPE) was developed for the fabrication of a chitosan (CS)-modified β-cyclodextrin (β-CD) via electrostatic interactions and hydrogen bonding for use in food 3D printing (Zhao et al. 2023). In addition, chitosan could be incorporated with other substances in the extrusion process to modify the physical properties of the product such as textural properties, thermal stability and so on (Mendes et al. 2016). Kumar et al. (2018) applied chitosan to acetate-based extruded snacks improved their fiber content and functional properties, while chitosan had a protective effect against lipid oxidation.

## 8.4 The Applications of Shellfish in Food Analogues

Shellfish are nutritious and high in protein, but they also contain high-quality lipids, vitamins, minerals, and natural colors. These rich and nutritious components provide ample possibilities for the use of shellfish in food analogues, as shown in Fig. 8.2. The proteins, lipids, polysaccharides, and other nutrients in shellfish can be used directly in high-moisture extrusion to produce meat analogues. In addition, proteins and polysaccharides, can be used as emulsifiers to prepare emulsions with shellfish oils for direct use in high-moisture extrusion or as inks for 3D printing of meat, and some nutrients can be encapsulated to prepare fat analogues and simulated liquid egg products.

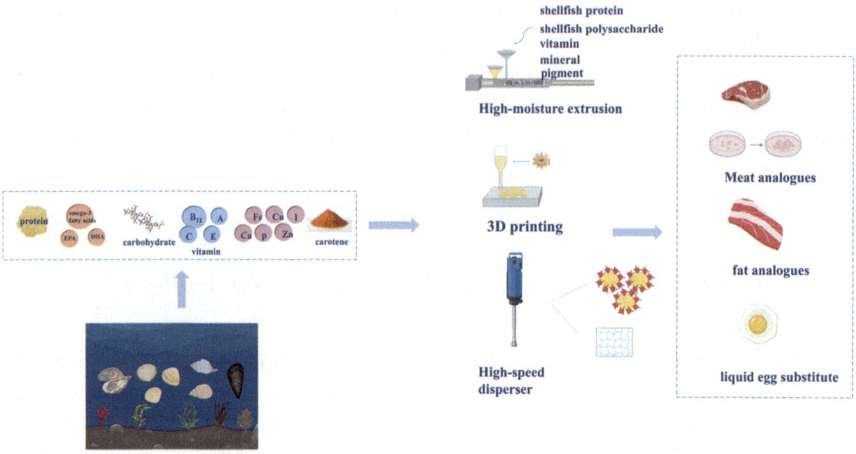


**Fig. 8.2** Schematic diagram of the and utilization of shellfish in food analogs

## 8.4.1 *The Nutrition of Shellfish*

Throughout history, shellfish have been associated with the daily lives of humans as a source of shells, food, dyes, medicines, and pigments (Benkendorff 2010). Shellfish are a rich biological resource with a wide variety and distribution, and are widely cultured around the world (Xiang et al. 2022). According to China Fisheries Statistical Yearbook 2023, marine shellfish aquaculture accounts for about 69% of seawater aquaculture production, which is a significant component.

Commonly eaten bivalves consist mainly of oysters, clams, mussels, and scallops, which are considered protein-rich, low in total and saturated fat. Moreover, omega-3 fatty acids can be found in abundance in mussels and oysters (Wright et al. 2018). Incorporating shellfish into dietary patterns helps provide omega-3 polyunsaturated fatty acids, bioactive protein, bioactive peptides, essential amino acids, vitamins, and minerals (Aru et al. 2018). Shellfish are rich in nutrients, with a high percentage of essential amino acids and a protein content of 10 g/100 g or more, far exceeding that of milk and comparable to eggs. In addition, shellfish contain more omega-3 fatty acids, which are essential unsaturated fatty acids. It can prevent cardiovascular disease and protect brain health.

In addition, shellfish are a great source of vitamins, choline, zinc, and iron (Wright et al. 2018). Therefore, shellfish are considered to be a low-fat, low-calorie food ingredients with high nutritional value. In recent decades, the demand and production of shellfish have increased, resulting in a large number of nutrient-rich by-products (Naik and Hayes 2019). Some shellfish with low meat yield, broken shellfish, adductor muscles, byssus threads, and shells can be classified as by-products, and these plentiful shellfish resources offer an excellent raw material basis for its application in food processing.

Among food analogues (including meat analogues, fat substitutes, egg substitutes, etc.), meat analogues, as a new product form, are in a developing trend. High-moisture extrusion meat, cell culture meat, and 3D-printed meat are some of the important innovations in the research of meat analogues. Currently, there is a lack of application of shellfish raw materials in the development of food analogues, while shellfish products should be more emphasized in the research of food analogues due to their large cultivation amount, rich by-products, nutrient-rich and low-fat content. Subsequent research should focus primarily on the bioactive potential of shellfish proteins, lipids, and carbohydrates and their potential applications in the processing of food analogues, which would be beneficial to both industry and academia.

### *8.4.2 Utilization of Proteins from Shellfish for Processing Food Analogues and Improving Texture and Nutrition*

Shellfish are a significant source of proteins, containing high levels of high-quality as well as low-cost protein (Cunha and Pintado 2022). Xia et al. (2022) investigated the compounding of microalgae protein and pea protein for high-moisture extrusion, and found that adding microalgae significantly improved the physical properties of meat analogues. This suggests that shellfish proteins may also show potential for application in compounding with plant proteins for high-moisture extrusion. The bioactive peptides and essential amino acids contained in shellfish proteins will also improve the nutritional value of the combined products. Chiang et al. (2019) found that wheat gluten promoted the production of fiber structures by rising disulfide bonds in meat analogues. Therefore, it is necessary to characterize the structural properties of shellfish proteins to guide future experiments. In addition, some amino acids, reducing sugars, and ribose are considered to be important precursors for the production of chicken flavor and beef flavor (Raza et al. 2020; Farmer 2005).

Marine shellfish are also a sustainable source of collagen, which is significant for the texture of meat (Naik and Hayes 2019). Pascual and Pla (2008)'s studies have shown that collagen affects the tenderness of meat, therefore, the addition of collagen to meat analogues has a potential positive effect on improving meat analogues. Furthermore, shellfish proteins can also be combined with polysaccharides to form hydrogels or prepared into emulsion gels for the production of fat analogues.

3D printing has shown certain application prospects in the production of customized meat (Wen et al. 2023). Inks that can be used for 3D printing have certain requirements, so emulsions that flow appropriately and encapsulate certain substances are potentially advantageous. For example, some nutrients (e.g., vitamins, minerals, antioxidants, probiotics, and flavor components, etc.) are embedded in emulsions to improve the nutrition of meat analogues and mimic the flavor of meat. Moreover, embedded colorants can give products a color similar to that of meat, and loaded fish oils, etc., give meat analogues the aroma and nutrients of fish, therefore, it can be applied to create personalized and customized meats. Cha et al. (2019) used the myofibrillar proteins found in mussels to create stable oil-in-water emulsions, which showed the potential of shellfish proteins to prepare emulsions. Moreover, some protein modification means, such as ultrasound, can enhance the emulsification and emulsion stability of shellfish proteins, which provides the research basis for shellfish proteins to prepare emulsions (Wu et al. 2019).

Cell-cultured meat is an innovative method of producing clean meat (Li et al. 2022a). Zernov et al. (2022) prepared chitosan-collagen hydrogel particles as palatable cellular microcarriers for cultured meat. The feasibility of using proteins for the preparation of cell scaffolds has also been reported in many works of literature, therefore, shellfish protein, as a high-quality protein, can be used as a protein supplement or alternative to preparing cell scaffolds (Chien and Shah 2012; Zheng et al. 2022a, b).

### *8.4.3 Improvement of Flavor Characteristics and Nutrition of Food Analogues by Lipids in Shellfish*

Lipids contribute to the tenderness and juiciness of meat analogues (Wang et al. 2023b). Shellfish oils, such as mussel oil, clam oil, and conch oil, can be combined with protein and directly applied to high-moisture extrusion. Kendler et al. (2021) studied that the addition of oil during wheat gluten extrusion could affect the polymerization of protein during extrusion. (Wang et al. 2022a) investigated the use of O/W emulsions for high-moisture extrusion to improve the fiber structure of meat analogues. Furthermore, different oil-to-water ratios also affect the color of high-moisture extrudates. Wang et al. (2023b) noted that the addition of oil made the meat analogues softer in texture and facilitated improved digestibility of the meat analogues. The addition of lipids to meat analogues could also provide components and acceptable sensory characteristics similar to animal meat.

There are ample studies showing that including omega-3 unsaturated fatty acids in the diet is beneficial to the human body. It is important to note that shellfish is one of the excellent sources of omega-3 unsaturated fatty acids, and the high level of nutrients contained in shellfish oils could also improve the nutritional properties of high-moisture extrudates, which lays the foundation for future applications of shellfish in high-moisture extrusion. Meanwhile, emulsions loaded with shellfish oils could be used for the development of 3D-printed meat products. Microencapsulation of oils is an effective tool for adding oils to meat products and preventing oil oxidation (Heck et al. 2021). Currently, there is limited research on adding microcapsules to high-moisture extrusion and 3D printing, which could be addressed in the future.

Shellfish oils such as emulsion gels or oleo gels could be used to prepare fat analogues. Wang et al. (2024) and Wang et al. (2022b) studied mung bean protein emulsion as a liquid egg substitute, with positive results. As a healthy oil, shellfish oil will have potential advantages in preparing liquid egg substitutes, which offers an innovative strategy for the development and application of stable egg substitutes.

Lipids play a significant role in determining the juiciness, softness, flavor, nutritional value, and storage stability of meat analogues (Chen et al. 2023a). Adding lipid components to the recipes of meat analogues contributes to a succulent eating experience. The smell and flavor of food result from aldehydes (Hanuš et al. 2009). Lipids play two main functions in the development of flavor in meat products: solvents and precursors of flavor substances. Lipolysis, lipid oxidation, and Meladic reactions produce a variety of metabolites, such as fatty acids and amino acids, which further react to generate a variety of flavor compounds that give meat its aroma (Fu et al. 2022). Phospholipids, in particular, have a significant impact on the properties of volatiles in cooked meat (Hanuš et al. 2009). Consequently, the addition of shellfish oils or shellfish lipids during processes such as high-moisture extrusion and 3D printing, etc. helps to enhance the flavor characteristics of the meat analogues obtained and to give them a desirable aroma.

### 8.4.4 *Utilization of Carbohydrates from Shellfish to Improve Texture and Nutrition of Food Analogues*

Carbohydrates play a key role in enhancing the flavor, texture, and appearance of meat analogues (Huang et al. 2022). The polysaccharide structure and composition of shellfish are diverse due to different species, habits, and environments. Studies have revealed that shellfish polysaccharides exhibit antibacterial, immunomodulatory, antioxidant, hypolipidemia, hypertension, and other biological activities (Xiang et al. 2022). Xiang et al. (2022) pointed out that shellfish are rich in chitosan. Chitosan is a polycationic compound that can interact with negatively charged materials, such as anionic polysaccharides, phospholipids, and proteins (Agulló et al. 2003). Porous materials, such as chitosan and collagen can be acted as cell scaffolding materials (Handral et al. 2020). Li et al. (2022a) successfully prepared a 3D edible scaffold using chitosan and sodium alginate. Many studies have indicated that chitosan-based Pickering emulsions prepared from composite particles formed by chitosan with proteins and polyphenols have better stability and suitable flowability (Song et al. 2023; Meng et al. 2023; Wang et al. 2023a). Therefore, chitosan-based Pickering emulsions may be an effective ink in the study of 3D-printed meat products. Embedding astaxanthin and Monascus pigment can give meat analogues a color similar to that of real meat, embedding carotenoids can give a yellow color similar to that of egg and chicken, and embedding some spice components can give meat analogues a flavor profile closer to that of animal meat. Li et al. (2023a) studied the preparation of emulsion gels from chitosan emulsions for the production of fat substitutes, and Brito et al. (2022) studied chitosan-based oleo gels for the production of fat substitutes. The fat substitutes obtained have potential health properties.

It was shown that the addition of polysaccharides in high-moisture extrusion could improve the fiber structure of extrudates by affecting protein aggregation. Zhang et al. (2020)‘s study concluded that the addition of exogenous polysaccharides to high-moisture extrusion can produce meat analogues with a rich fiber structure. Xia et al. (2023) produced meat analogues with a rich fiber structure by combining yeast protein with Konjac glucomannan. These studies all give us enlightenment: In future research on the preparation of meat analogues with rich fiber structure by high-moisture extrusion, research could favor the use of shellfish polysaccharides and explore the role shellfish polysaccharides play in high-moisture extrusion as well as the nutritional and digestive properties of high-moisture extrudates. Carbohydrates in meat are mainly used for flavoring and coloring through the Maillard reaction (Huang et al. 2022). Therefore, the addition of shellfish polysaccharides during the extrusion process facilitates the production of appetizing colors and flavors in high-moisture extrudates. In addition, the addition of polysaccharides during high-moisture extrusion could also help improve water-holding capacity.

### *8.4.5 Utilization of Pigments from Shellfish to Improve Color and Nutrition in Food Analogues*

With the popularity of meat analogue products in the market, the demand for various meat flavoring and coloring agents for meat analogue products is also increasing (Huang et al. 2022). The main pigments in shellfish are carotenoids, melanin, and tetrapyrrole (Williams 2016). Carotenoids have long served as colorants and antioxidants in food (Naik and Hayes 2019). This pigment gives the meat an orange color, also prevents the meat from oxidizing and spoiling, and its applications to food analogues can improve product preservation and value addition. Carotenoids play a vital role in human health with their antioxidant and anti-inflammatory properties, which can enhance the body's immunity and prevent various chronic diseases (Boonlao et al. 2022). The embedding of carotenoids to prepare Pickering emulsion could be directly used for high-moisture extrusion and the production of 3D printed meat, which can not only provide nutritional properties of meat analogues, but also give the product the required color, prevent product oxidation, and improve the shelf life of the product.

## 8.5 The Applications of Cephalopod in Food Analogues

### *8.5.1 Cephalopod Nutritional Composition*

Cephalopods, a class of marine mollusks, have existed for over 500 million years, making them one of the oldest and most sophisticated marine animal groups. They play vital roles in marine ecosystems and are significant in commercial fisheries; the first known cephalopods appeared in the late Cambrian period (Chen 2013). The subclasses *Nautiloidea* and *Coleoidea encompass* all extant cephalopods. These creatures can be found in every ocean on earth, from the Arctic and Antarctic to the tropics, spanning coastal waters to the open ocean, and ranging from the surface down to 500 m below (Chen 2013). The Food and Agriculture Organization of the United Nations reports that 3.7 million tons of cephalopods will be harvested in 2020, and these exports are expected to generate $10.2 billion in revenue, representing 6.8% of the total value of global aquatic exports. These cephalopods are thought to be in a sustainable supply to meet the demand for human consumption (FAO 2022).

Therefore, for consumers who hold a sustainable diet concept, but still require meat in their diet, cephalopod meat may offer an alluring alternative to livestock meat. In the future, with the advancement of food science and technology, cephalopod meat may be transformed into meat-substitute products through various food production and processing methods, ultimately leading to a highly competitive market (Mouritsen and Vinther Schmidt 2020).

Polyunsaturated fatty acid and high-quality protein are among the many components of seafood that are recognized to be good for human health. About 75% of the lipid composition of squid (*Loligo vulgaris*) is phospholipids, the most important component of phospholipids is phosphatidylcholine, which accounts for 56% of the total phospholipids. The main saturated fatty acids of phospholipids and non-phosphorylated lipids were C16:0, and the main unsaturated fatty acids of lipids were C22:6 Ω-3, followed by C20:5 Ω-3 (De Koning 1993). Yang analyzed and compared the nutrient content of Japanese sea squid and Peruvian squid, respectively, and the results showed that the protein content of these two kinds of squid reached 17.25%, consisting of 18 kinds of amino acids, of which the proportion of essential amino acids accounted for more than 30% of the total amino acids, which is the source of high-quality protein (Yang et al. 2013). Wen determined the amino acid and fatty acid composition of the kisslip cuttlefish (*Sepia Lycidas*). The crude protein and crude fat contents of the cuttlefish were 16.36 and 0.66%, respectively. The fatty acids were mainly polyunsaturated fats, accounting for 45.01% of the total fatty acids, and docosahexaenoic acid (DHA) was the dominant fatty acid, accounting for 24.54% of the total fatty acids (Wen et al. 2015). The polyunsaturated fatty acid level was greater than that of saturated and monounsaturated fatty acids. Polyunsaturated fatty acids make up 51.9–59.0% of all fatty acids (Torrinha et al. 2014).

By incorporating high-protein, low-fat cephalopods as a meat substitute, individuals may experience a range of benefits, including enhanced muscle gain, reduced fat accumulation, improved heart health, strengthened cognitive function, and reinforced immune support.

### 8.5.2 Extrusion Defects and Processing Techniques of Cephalopods

Cephalopods are often dried directly to produce air-dried food, giving squid products a firm texture. In addition, because the muscle of squid has 11% insoluble myosin, even after heating the squid is difficult to soften, seriously affecting the elderly and infants to eat this kind of food (Jun-hui et al. 2020).

#### 8.5.2.1 Extrusion

Research has indicated that the process of thermoplastic extrusion cooking has the potential to greatly enhance the nutritional qualities and digestibility of protein in inexpensive aquatic products. The squid muscle protein's tertiary and quaternary structures will break down during the extrusion process as the temperature rises and the molecular motion becomes more intense, resulting in the formation of largely linear macromolecules; improved *in vitro* digestibility results from these linear macromolecules' ability to disintegrate into several readily digested tiny proteins and the maximum temperature at which extrudates can be digested in vitro was 140 °C (Yang et al. 2010).

Using North Pacific squid as raw material, the SYSLG30-IV twin-screw extruder can produce low-value aquatic protein tissue products (Yang et al. 2011). As the material humidity increases, the structure of the extruded material fluctuates and decreases, resulting in a gradual decrease in chewability. Additionally, changes in screw speed have little influence on texture. It is the appropriate barrel temperature that is a prerequisite for achieving better texture and chewability (Yang et al. 2011). When applying squid to extrusion, it is necessary to control the extrusion temperature not only to ensure the taste and texture of the product but also to balance the nutritional content.

Squid–potato–corn flour mixture was analyzed using Fourier transform infrared spectroscopy and differential scanning calorimetry; in addition, the findings demonstrated that extrusion altered the protein and starch conformations in the squid-potato cornmeal mixture, which benefited the protein-starch interaction (Valenzuela-Lagarda et al. 2021). Along with altering the texture of the extrusion, hydrophobic interactions also play a role in the creation of new peptide bonds between the amino and carboxyl groups of the material, which are responsible for cross-linking. Additionally, this research opens the door to using extrusion technology to develop squid snacks that have the same physical and chemical features as conventional snacks but with a higher nutritional value (Valenzuela-Lagarda et al. 2021).

Ganesan et al. (2021) found that heating temperature, screw rotation speed, squid meal content, and other extrusion process parameters had significant effects on the expansion rate and finished weight of squid meal ($P \leq 0.05$). When the temperature of the No.1 heater is 64.10 °C, the temperature of the No. 2 heater is 129.98 °C, the screw speed is 317.87 rpm, and the content of squid meal is 6.56%, the physical and nutritional properties of the prepared squid and Mickey extrusion are the best. The essential fatty acids such as linolenic acid, eicosapentaenoic acid, docosahexaenoic acid, and arachidonic acid accounted for 0.98%, 0.42%, 0.98%, and 0.18% of the total fatty acids, respectively. It contains essential amino acids such as histidine, threonine, lysine, tryptophan, methionine, valine, isoleucine, leucine and phenylalanine (Ganesan et al. 2021).

High-quality protein, minerals, and unsaturated fatty acids are found in squid, while complex carbohydrates, dietary fiber, and a variety of vitamins are found in grains. These nutrients are combined to create a more balanced nutritional composition and to propose a novel combination for the creation of innovative, functional plant-based protein products.

#### 8.5.2.2 3D Printing

To provide comprehensive and balanced nutrition, squid processing powder can also be combined with plant powder in the new food processing technology and used as an edible ink in 3D printing production; additionally, 3D-printed food is easier to chew and digest than raw carrots and squid (Jeon et al. 2023). Hard, solid food is difficult for the elderly and those suffering from dysphagia to chew. However, cephalopod 3D-printed molded foods offers an excellent option when the human body wants to consume high-quality, low-calorie meals (Jeon et al. 2023).

### *8.5.3 Advantages and Possibilities of Cephalopod Extracts and Analogs*

Since only 50–60% of the giant squid is used in the production process, the by-products could be utilized to create useful foods. During the traditional processing of cephalopods, a large number of by-products are often produced, including skins, offal, and bones, which are rich in nutrients and functional components and have high utilisation value (Ezquerra-Brauer and Aubourg 2019).

These by-products can be converted into high value-added food analogue raw materials through scientifically sound extraction techniques and processing methods. For example, squid skin, as one of the main by-products, excels in high-value utilization, as it is rich in collagen, has excellent elasticity and viscosity, and can be used to improve and innovate foodstuffs such as extruded seafood products. Squid offal is also another important by-product resource. It is rich in protein, fat and minerals, making it an ideal raw material for the production of high-protein foods and fishmeal. In summary, squid by-products in the field of food extrusion processing can not only improve food quality and nutritional value, but also reduce production costs and promote the sustainable development of the seafood processing industry. As technology progresses and research deepens, it is believed that the high-value utilization of squid by-products will usher in a broader development prospect.

#### 8.5.3.1 Protein Hydrolyzed

The hydrolyzed protein of squid by-product can produce squid-extruded products with high protein content, low price, and good availability. The hydrolyzed protein of the squid by-product is compressed, mixed, cut, crumbled, and heated in a screw with Canihua (*Chenopodium pallidicaule* Aellen) to produce an extruded snack with a light beige color on the sense and without the residual smell of squid, and the nutrition of the product is also improved (Tapia et al. 2023). On the one hand, due to its high phenolic content and antioxidant activity, it boasts a shelf life of 283–630 days at 25 °C, achieving the effect of reducing food waste and economic loss. On the other hand, this hydrolyzed protein from squid by-product, when extruded with Chenopodium pallidicaule Aellen, also exhibits high in vitro protein digestibility and a low acid index (Tapia et al. 2023).

#### 8.5.3.2 Gelatin

Squid gelatin, which is naturally high in collagen, has great gel qualities that enhance the flavor, suppleness, and texture of meat products in the tissue while also adding more utility to the product. When compared to fish collagen, cephalopod collagen has a higher content of hydroxylysine, a higher content of hydroxyproline, and a higher degree of crosslinking. The addition of cuttlefish skin gelatin can obviously improve the physicochemical properties of food, increase and decrease the cooking loss rate, and increase the protein and water content, while also improving the texture parameters of food and product elasticity. The food industry receives advantages from the discovery that cephalopod gelatin enhances the texture of meat products. It not only helps deal with the by-product issue but helps solidify the texture of meat during the production of plant-tissue meat. Additionally, it resolves the issue of plant proteins' non-adhesion and creates opportunities for the creation of new products (Souissi et al. 2016).

#### 8.5.3.3 Fat

Oil is a necessary ingredient in meat products because it can improve food flavor while simultaneously preserving the moisture content of meat processing products. When creating a formula for plant tissue meat, oils should also be considered in the formulation. One type of fat that has a lot of nutritional value is squid lipid. The most significant components linked to the advantages of consuming marine foods are the high concentrations of EPA, DHA, and arachidonic acid (ARA) found in gigas PE. These components may also have some efficacy in treating Alzheimer's disease and promoting infant neurodevelopment (Saito et al. 2014). Tissue meat can taste more like real fish meat if the right amount of squid fat is added. This discovery may give these neglected cephalopods a nutritious marine diet.

#### 8.5.3.4 Polysaccharides

Abdelmalek studied the effects of polysaccharide sulfate derived from squid skin on angiotensin inhibition, antioxidant activity, and antimicrobial properties. The findings demonstrate the potential of sulfated polysaccharide in squid skin, as a natural source of ace inhibitory biomolecules, natural antibacterial agents, and antioxidants. These compounds can be utilized to improve the biological qualities of functional foods and avoid oxidation issues during food preparation (Abdelmalek et al. 2015). The function of thickening, moisturizing, and preserving will be strengthened by adding polysaccharides to the tissue meat with plant protein components, improving the palatability and edibility of the plant-based meat.

#### 8.5.3.5 Miso

Another crucial step in processing meat is seasoning. Rice koji is mixed with squid meat in a 1:1 (w/w) ratio to produce squid miso; besides, the activity of carbohydrate cracking/hydrolase during fermentation encourages the release of a number of phenolic chemicals, which contribute to the distinctive flavor of squid miso. Furthermore, there is a great chance that squid miso extract will stop lipid oxidation (Giri et al. 2011). Squid miso offers a plant-based alternative for vegetarians seeking to replicate the distinctive flavor of seafood. This allows them to continue enjoying the delectable taste of marine dishes without compromising their dietary choices.

## 8.6 Conclusion

In recent years, as people pay more and more attention to personal health issues, marine food has received more and more attention due to its advantages of high protein, low fat and rich in polyunsaturated fatty acids (EPA and DHA). Food analogues are an important part of marine food. The current marine raw materials used for the food analogues are becoming increasingly abundant, such as fish, shrimp, algae, shellfish, and mollusks. In addition, with the enhancement of food processing technology, the number of technologies applied to the production of food analogues are gradually increasing, mainly including molding and cooking process technology, extrusion technology, and 3D printing technology. The molding and cooking process technology is still the mainstream technology, but with the development of extrusion technology and 3D technology, these technologies will be more widely applied. In the future, more studies need to focus on the structural and nutritional properties of food analogues from marine raw materials and developing advanced processing technologies.

## References

Abdelmalek BE, Sila A, Krichen F, Karoud W, Martinez-Alvarez O, Ellouz-Chaabouni S, Ayadi MA, Bougatef A (2015) Sulfated polysaccharides from Loligo vulgaris skin: potential biological activities and partial purification. Int J Biol Macromol 72:1143–1151

Achinas S, Heins JI, Krooneman J, Euverink GJW (2020) Miniaturization and 3D printing of bioreactors: a technological mini review. Micromachines (Basel) 11(9):853

Agulló E, Rodríguez MS, Ramos V, Albertengo L (2003) Present and future role of chitin and chitosan in food. Macromol Biosci 3(10):521–530

Ahmad M, Qureshi S, Akbar MH, Siddiqui SA, Gani A, Mushtaq M, Hassan I, Dhull SB (2022) Plant-based meat alternatives: compositional analysis, current development and challenges. Appl Food Res 2(2):100154

Aoki K, Hara F, Ohmichi M, Nakatani N, Hosaka H (1989) Texturization of surimi using a twin-screw extruder. Nippon Shokuhin Kogyo Gakkaishi 36(9):748–753

Aru V, Khakimov B, Sørensen KM, Engelsen SB (2018) The foodome of bivalve molluscs: from hedonic eating to healthy diet. J Food Compos Anal 69:13–19

Azelee NIW, Dahiya D, Ayothiraman S, Noor NM, Rasid ZIA, Ramli ANM, Ravindran B, Iwuchukwu FU, Selvasembian R (2023) Sustainable valorization approaches on crustacean wastes for the extraction of chitin, bioactive compounds and their applications - a review. Int J Biol Macromol 253:126492

Baskar D, Dhanapal K, Madhavan N, Madhavi K, Kumar GP, Manikandan V, Sushma M (2022) Proximate composition and sensory evaluation of extruded snacks enriched with fish flour and shrimp head exudate during storage conditions. J Food Process Preserv 46(7):e16589

Benkendorff K (2010) Molluscan biological and chemical diversity: secondary metabolites and medicinal resources produced by marine molluscs. Biol Rev 85:757–775

Boonlao N, Ruktanonchai UR, Anal AK (2022) Enhancing bioaccessibility and bioavailability of carotenoids using emulsion-based delivery systems. Colloids Surf B: Biointerfaces 209:112211

Borderías AJ, Sánchez-Alonso I, Pérez-Mateos M (2005) New applications of fibres in foods: addition to fishery products. Trends Food Sci Technol 16(10):458–465

Brito GB, Peixoto VODS, Martins MT, Rosário DKA, Ract JN, Conte-Júnior CA, Torres AG, Castelo-Branco VN (2022) Development of chitosan-based oleogels via crosslinking with vanillin using an emulsion templated approach: structural characterization and their application as fat-replacer. Food Struct 32:100264

Brooker PG, Anastasiou K, Smith BPC, Tan R, Cleanthous X, Riley MD (2023) Nutrient composition of milk and plant-based milk alternatives: a cross-sectional study of products sold in Australia and Singapore. Food Res Int 173:113475

Bryant CJ (2022) Plant-based animal product alternatives are healthier and more environmentally sustainable than animal products. Future Foods 6:100174

Cha Y, Shi X, Wu F, Zou H, Chang C, Guo Y, Yuan M, Yu C (2019) Improving the stability of oil-in-water emulsions by using mussel myofibrillar proteins and lecithin as emulsifiers and high-pressure homogenization. J Food Eng 258:1–8

Chantanuson R, Nagamine S, Kobayashi T, Nakagawa K (2022) Preparation of soy protein-based food gels and control of fibrous structure and rheological property by freezing. Food Struct 32:100258

Chen X (2013) World cephalopod resources and their fisheries. China Ocean Press, China

Chen Q, Chen Z, Zhang J, Wang Q, Wang Y (2023a) Application of lipids and their potential replacers in plant-based meat analogs. Trends Food Sci Technol 138:645–654

Chen Y, Liu Y, Dong Q, Xu C, Deng S, Kang Y, Fan M, Li L (2023b) Application of functionalized chitosan in food: a review. Int J Biol Macromol 235:123716

Chiang JH, Loveday SM, Hardacre AK, Parker ME (2019) Effects of soy protein to wheat gluten ratio on the physicochemical properties of extruded meat analogues. Food Struct 19:100102

Chien KB, Shah RN (2012) Novel soy protein scaffolds for tissue regeneration: material characterization and interaction with human mesenchymal stem cells. Acta Biomater 8(2):694–703

Choudhury GS, Gogoi BK (1996) Extrusion processing of fish muscle. J Aquat Food Prod Technol 4(4):37–67

Cunha SA, Pintado ME (2022) Bioactive peptides derived from marine sources: biological and functional properties. Trends Food Sci Technol 119:348–370

De Koning AJ (1993) Phospholipids of marine origin—the squid (Loligo vulgaris). J Sci Food Agric 61(1):129–132

Dong X, Huang Y, Pan Y, Wang K, Prakash S, Zhu B (2019) Investigation of sweet potato starch as a structural enhancer for three-dimensional printing of Scomberomorus niphonius surimi. J Texture Stud 50(4):316–324

Ezquerra-Brauer JM, Aubourg SP (2019) Recent trends for the employment of jumbo squid (Dosidicus gigas) by-products as a source of bioactive compounds with nutritional, functional and preservative applications: a review. Int J Food Sci Technol 54(4):987–998

FAO (2022) The state of world fisheries and aquaculture 2022. In: Towards blue transformation. FAO, Rome

Farmer MALJ (2005) Precursors of chicken flavor. II. Identification of key flavor precursors using sensory methods. J Agric Food Chem 53:6455–6462

Ferraro V, Cruz IB, Jorge RF, Malcata FX, Pintado ME, Castro PML (2010) Valorisation of natural extracts from marine source focused on marine by-products: a review. Food Res Int 43(9):2221–2233

Fu Y, Chen T, Chen SHY, Liu B, Sun P, Sun H, Chen F (2021) The potentials and challenges of using microalgae as an ingredient to produce meat analogues. Trends Food Sci Technol 112:188–200

Fu Y, Cao S, Yang L, Li Z (2022) Flavor formation based on lipid in meat and meat products: a review. J Food Biochem 46(12):e14439

Ganesan V, Subbaiah B, Stephen NM, Dhanushkodi M, Kolandaivadivel R, Veeramani R (2021) Optimization of extrusion process using response surface methodology for producing squid and millet based extrudate. Turk J Fish Aquat Sci 22(2):TRJFAS20041

Gephart JA, Henriksson PJG, Parker RWR, Shepon A, Gorospe KD, Bergman K, Eshel G, Golden CD, Halpern BS, Hornborg S, Jonell M, Metian M, Mifflin K, Newton R, Tyedmers P, Zhang W, Ziegler F, Troell M (2021) Environmental performance of blue foods. Nature 597(7876):360–365

Gerber PJ, Steinfeld H, Henderson B, Mottet A, Opio C, Dijkman J, Falcucci A & Tempio G (2013) Tackling climate change through livestock – A global assessment of emissions and mitigation opportunities. Food and Agriculture Organization of the United Nations (FAO), Rome

Giri A, Osako K, Okamoto A, Okazaki E, Ohshima T (2011) Antioxidative properties of aqueous and aroma extracts of squid miso prepared with Aspergillus oryzae-inoculated koji. Food Res Int 44(1):317–325

Golden CD, Koehn JZ, Shepon A, Passarelli S, Free CM, Viana DF, Matthey H, Eurich JG, Gephart JA, Fluet-Chouinard E, Nyboer EA, Lynch AJ, Kjellevold M, Bromage S, Charlebois P, Barange M, Vannuccini S, Cao L, Kleisner KM, Rimm EB, Danaei G, DeSisto C, Kelahan H, Fiorella KJ, Little DC, Allison EH, Fanzo J, Thilsted SH (2021) Aquatic foods to nourish nations. Nature 598(7880):315–320

Gudjonsdottir M, Napitupulu RJ, Petty Kristinsson HT (2019) Low field NMR for quality monitoring of 3D printed surimi from cod by-products: effects of the pH-shift method compared with conventional washing. Magn Reson Chem 57(9):638–648

Gulzar S, Benjakul S (2018) Ultrasound waves increase the yield and carotenoid content of lipid extracted from cephalothorax of Pacific white shrimp (Litopenaeus vannamei). Eur J Lipid Sci Technol 120(5):1700495

Guo M, Liu S, Ismail M, Farid MM, Ji H, Mao W, Gao J, Li C (2017) Changes in the myosin secondary structure and shrimp surimi gel strength induced by dense phase carbon dioxide. Food Chem 227:219–226

Handral HK, Hua Tay S, Wan Chan W, Choudhury D (2020) 3D printing of cultured meat products. Crit Rev Food Sci Nutr 62(1):272–281

Hanuš LO, Levitsky DO, Shkrob I, Dembitsky VM (2009) Plasmalogens, fatty acids and alkyl glyceryl ethers of marine and freshwater clams and mussels. Food Chem 116(2):491–498

Harris CH, Johnson JM (1987) Select nutrient comparison of imitation shellfish with real shellfish. J Food Qual 10(5):335–342

He T, Mo B, Huang J, Fan D, Zhang W, Wang L, Zhao J, Chen W, Zhang H (2014) Twin-screw extrusion of hairtail surimi and soy protein isolate blends. Food Sci Technol Res 20(3):517–527

Heck RT, Lorenzo JM, Dos Santos BA, Cichoski AJ, de Menezes CR, Campagnol PCB (2021) Microencapsulation of healthier oils: an efficient strategy to improve the lipid profile of meat products. Curr Opin Food Sci 40:6–12

Hou Y, Xia S, Ma C, Xue C, Jiang X (2023) Effects of the soy protein to wheat gluten ratio on the physicochemical and structural properties of Alaska pollock surimi-based meat analogs by high moisture extrusion. Food Res Int 173:113469

Huang M, Mehany T, Xie W, Liu X, Guo S, Peng X (2022) Use of food carbohydrates towards the innovation of plant-based meat analogs. Trends Food Sci Technol 129:155–163

Isobe S, Uemura K, Noguchi A (1990) Texturization of chum salmon muscle harvested during spawning migration (Bunazake). Nippon Shokuhin Kogyo Gakkaishi 37(12):965–970

Jensen I-J, Larsen R, Rustad T, Eilertsen K-E (2013) Nutritional content and bioactive properties of wild and farmed cod (Gadus morhua L.) subjected to food preparation. J Food Compos Anal 31(2):212–216

Jeon EY, Chun YG, Kim B-K (2023) Investigation of carrot/squid blends as edible inks for extrusion 3D printing: effect of hydrocolloids incorporation. J Food Eng 364:111777

Jiang Q, Chen N, Gao P, Yu D, Yang F, Xu Y, Xia W (2023) Influence of L-arginine addition on the gel properties of reduced-salt white leg shrimp (Litopenaeus vannamei) surimi gel treated with microbial transglutaminase. LWT-Food Sci Technol 173:114310

Jun-hui X, Hui-juan C, Bin Z, Hui Y (2020) The mechanistic effect of bromelain and papain on tenderization in jumbo squid (Dosidicus gigas) muscle. Food Res Int 131:108991

Kendler C, Duchardt A, Karbstein HP, Emin MA (2021) Effect of oil content and oil addition point on the extrusion processing of wheat gluten-based meat analogues. Food Secur 10(4):697

Kim SM, Kim HW, Park HJ (2021) Preparation and characterization of surimi-based imitation crab meat using coaxial extrusion three-dimensional food printing. Innovative Food Sci Emerg Technol 71:102711

Kitagawa M, Iida T, Saeki H (2007) Fibrous texturization of walleye pollack surimi by extrusion cooking using a twin-screw extruder. Nippon Suisan Gakkaishi 73(5):905–915

Kitagawa M, Iid T, Nobuta S, Kishimura H, Saeki H (2008) Fibrous texturization and amino acid/fatty acid changes with the extrusion cooking of fish meat containing skin and bone. Nippon Suisan Gakkaishi 74(1):55–60

Kong J, Dougherty MP, Perkins LB, Camire ME (2012) Utilization of smoked salmon trim in extruded smoked salmon jerky. J Food Sci 77(6):S211–S215

Kumar R, Xavier KAM, Lekshmi M, Balange A, Gudipati V (2018) Fortification of extruded snacks with chitosan: effects on techno functional and sensory quality. Carbohydr Polym 194:267–273

Li L, Chen L, Chen X, Chen Y, Ding S, Fan X, Liu Y, Xu X, Zhou G, Zhu B, Ullah N, Feng X (2022a) Chitosan-sodium alginate-collagen/gelatin three-dimensional edible scaffolds for building a structured model for cell cultured meat. Int J Biol Macromol 209:668–679

Li S, Lin S, Jiang P, Bao Z, Li S, Sun N (2022b) Insight into the gel properties of Antarctic krill and Pacific white shrimp surimi gels and the feasibility of polysaccharides as texture enhancers of Antarctic krill surimi gels. Food Secur 11(16):2517

Li C, Xie W, Zhang X, Liu J, Zhang M, Shao J-h (2023a) Pickering emulsion stabilized by modified pea protein-chitosan composite particles as a new fat substitute improves the quality of pork sausages. Meat Sci 197:109086

Li G, Zhan J, Hu Z, Huang J, Yao Q, Yuan C, Chen J, Hu Y (2023b) Effects of nano starch-lutein on 3D printing properties of functional surimi: enhancement mechanism of printing effects and anti-oxidation. J Food Eng 346:111431

Lordan S, Ross RP, Stanton C (2011) Marine bioactives as functional food ingredients: potential to reduce the incidence of chronic diseases. Mar Drugs 9(6):1056–1100

Martínez-Maldonado MA, Velazquez G, Ramírez de León JA, Borderías AJ, Moreno HM (2020) Effect of high pressure processing on heat-induced gelling capacity of blue crab (Callinectes sapidus) meat. Innovative Food Sci Emerg Technol 59:102253

Mendes JF, Paschoalin RT, Carmona VB, Sena Neto AR, Marques ACP, Marconcini JM, Mattoso LHC, Medeiros ES, Oliveira JE (2016) Biodegradable polymer blends based on corn starch and thermoplastic chitosan processed by extrusion. Carbohydr Polym 137:452–458
Meng W, Sun H, Mu T, Garcia-Vaquero M (2023) Pickering emulsions with chitosan and macroalgal polyphenols stabilized by layer-by-layer electrostatic deposition. Carbohydr Polym 300:120256
Messina CM, Manuguerra S, Arena R, Renda G, Ficano G, Randazzo M, Fricano S, Sadok S, Santulli A (2021) In vitro bioactivity of astaxanthin and peptides from hydrolisates of shrimp (Parapenaeus longirostris) by-products: from the extraction process to biological effect evaluation, as pilot actions for the strategy "From waste to profit". Mar Drugs 19(4):216
Mouritsen OG, Vinther Schmidt C (2020) A role for macroalgae and cephalopods in sustainable eating. Front Psychol 11:1402
Murphy MG, Skonberg DI, Camire ME, Dougherty MP, Bayer RC, Briggs JL (2003) Chemical composition and physical properties of extruded snacks containing crab-processing by-product. J Sci Food Agric 83(11):1163–1167
Naik AS, Hayes M (2019) Bioprocessing of mussel by-products for value added ingredients. Trends Food Sci Technol 92:111–121
Nanda PK, Das AK, Dandapat P, Dhar P, Bandyopadhyay S, Dib AL, Lorenzo JM, Gagaoua M (2021) Nutritional aspects, flavour profile and health benefits of crab meat based novel food products and valorisation of processing waste to wealth: a review. Trends Food Sci Technol 112:252–267
Nisov A, Aisala H, Holopainen-Mantila U, Alakomi HL, Nordlund E, Honkapaa K (2020) Comparison of whole and gutted Baltic herring as a raw material for restructured fish product produced by high-moisture extrusion cooking. Food Secur 9(11):1541
Pan Y, Sun Q, Liu Y, Wei S, Xia Q, Zheng O, Liu S, Ji H, Deng C, Hao J (2021) The relationship between rheological and textural properties of shrimp surimi adding starch and 3D printability based on principal component analysis. Food Sci Nutr 9(6):2985–2999
Park JW, Lin TM, Yongsawatdigul J (1997) New developments in manufacturing of surimi and surimi seafood. Food Rev Intl 13(4):577–610
Pascual M, Pla M (2008) Changes in collagen, texture and sensory properties of meat when selecting rabbits for growth rate. Meat Sci 78(4):375–380
Paxton N, Smolan W, Bock T, Melchels F, Groll J, Jungst T (2017) Proposal to assess printability of bioinks for extrusion-based bioprinting and evaluation of rheological properties governing bioprintability. Biofabrication 9(4):044107
Prabha K, Ghosh P, S A, Joseph RM, Krishnan R, Rana SS, Pradhan RC (2021) Recent development, challenges, and prospects of extrusion technology. Future Foods 3:100019
Rajasekaran B, Singh A, Zhang B, Hong H, Benjakul S (2022) Changes in emulsifying and physical properties of shrimp oil/soybean oil-in-water emulsion stabilized by fish myofibrillar protein during the storage. Eur J Lipid Sci Technol 124(10):2200068
Raza A, Song H, Raza J, Li P, Li K, Yao J (2020) Formation of beef-like odorants from glutathione-enriched yeast extract via Maillard reaction. Food Funct 11(10):8583–8601
Ruiz Matute AI, Cardelle-Cobas A, García-Bermejo AB, Montilla A, Olano A, Corzo N (2013) Synthesis, characterization and functional properties of galactosylated derivatives of chitosan through amide formation. Food Hydrocoll 33(2):245–255
Saito H, Sakai M, Wakabayashi T (2014) Characteristics of the lipid and fatty acid compositions of the Humboldt squid, Dosidicus gigas: the trophic relationship between the squid and its prey. Eur J Lipid Sci Technol 116(3):360–366
Service AR. (2018) Crustaceans, shrimp, mixed species, cooked, moist heat (may contain additives to retain moisture). https://fdc.nal.usda.gov/fdc-app.html#/food-details/171971/nutrients
Sha L, Xiong YL (2020) Plant protein-based alternatives of reconstructed meat: science, technology, and challenges. Trends Food Sci Technol 102:51–61
Shahidi F, Arachchi JKV, Jeon Y-J (1999) Food applications of chitin and chitosans. Trends Food Sci Technol 10(2):37–51

Short RE, Gelcich S, Little DC, Micheli F, Allison EH, Basurto X, Belton B, Brugere C, Bush SR, Cao L, Crona B, Cohen PJ, Defeo O, Edwards P, Ferguson CE, Franz N, Golden CD, Halpern BS, Hazen L, Hicks C, Johnson D, Kaminski AM, Mangubhai S, Naylor RL, Reantaso M, Sumaila UR, Thilsted SH, Tigchelaar M, Wabnitz CCC, Zhang W (2021) Harnessing the diversity of small-scale actors is key to the future of aquatic food systems. Nat Food 2(9):733–741

Song S, Zhong L, Wei Y, Li Y, Tao L, Yu L (2023) Highly stable solid-like Pickering emulsions stabilized by kafirin-chitosan complex particles. LWT Food Sci Technol 177:114591

Souissi N, Jridi M, Nasri R, Ben Slama R, Njeh M, Nasri M (2016) Effects of the edible cuttlefish gelatin on textural, sensorial and physicochemical quality of octopus sausage. LWT Food Sci Technol 65:18–24

Tapia M, Marimón SJ, Salazar N (2023) Development of extruded snacks with protein hydrolysed from jumbo squid (Dosidicus gigas) by-product and Cañihua (Chenopodium pallidicaule Aellen). Biol Life Sci Forum 2023(25):4

Thiébaud M, Dumay E, Cheftel JC (1996) Influence of process variables on the characteristics of a high moisture fish soy protein mix texturized by extrusion cooking. LWT-Food Sci Technol 29:526–535

Tibrewal K, Dandekar P, Jain R (2023) Extrusion-based sustainable 3D bioprinting of meat & its analogues: a review. Bioprinting 29:e00256

Tigchelaar M, Leape J, Micheli F, Allison EH, Basurto X, Bennett A, Bush SR, Cao L, Cheung WWL, Crona B, DeClerck F, Fanzo J, Gelcich S, Gephart JA, Golden CD, Halpern BS, Hicks CC, Jonell M, Kishore A, Koehn JZ, Little DC, Naylor RL, Phillips MJ, Selig ER, Short RE, Sumaila UR, Thilsted SH, Troell M, Wabnitz CCC (2022) The vital roles of blue foods in the global food system. Glob Food Sec 33:100637

Torrinha Á, Cruz R, Gomes F, Mendes E, Casal S, Morais S (2014) Octopus lipid and vitamin E composition: interspecies, interorigin, and nutritional variability. J Agric Food Chem 62(33):8508–8517

UN (2019) World population prospects 2019. https://population.un.org/wpp/Download/Standard/Population/. Accessed 1 Sept 2020

Valenzuela-Lagarda JL, Pacheco-Aguilar R, Gutiérrez-Dorado R, Mendoza JL, López-Valenzuela JÁ, Mazorra-Manzano MÁ, Muy-Rangel MD (2021) Interaction of squid (Dosidicus giga) mantle protein with a mixtures of potato and corn starch in an extruded snack, as characterized by FTIR and DSC. Molecules 26(7):2103

Venugopal V, Gopakumar K (2017) Shellfish: nutritive value, health benefits, and consumer safety. Compr Rev Food Sci Food Saf 16(6):1219–1242

Wang S, Liu J-r, Fu R-z, Xue C-h (2012) Effects of process parameters on fish protein texturization under high moisture. J Fish China 36(11):1776–1784

Wang L, Zhang M, Bhandari B, Yang C (2018) Investigation on fish surimi gel as promising food material for 3D printing. J Food Eng 220:101–108

Wang R, Wang T, Feng W, Wang Q, Wang T (2021) Rice proteins and cod proteins forming shared microstructures with enhanced functional and nutritional properties. Food Chem 354:129520

Wang H, Zhang L, Czaja TP, Bakalis S, Zhang W, Lametsch R (2022a) Structural characteristics of high-moisture extrudates with oil-in-water emulsions. Food Res Int 158:111554

Wang Y, Zhao J, Zhang S, Zhao X, Liu Y, Jiang J, Xiong YL (2022b) Structural and rheological properties of mung bean protein emulsion as a liquid egg substitute: the effect of pH shifting and calcium. Food Hydrocoll 126:107485

Wang C, Jiang H, Li Y (2023a) Water-in-oil Pickering emulsions stabilized by phytosterol/chitosan complex particles. Colloids Surf A Physicochem Eng Asp 657:130489

Wang H, Wang R, Zhang L, Zhang W, Bakalis S, Li Y, Lametsch R (2023b) Physicochemical properties, texture, and in vitro protein digestibility in high-moisture extrudate with different oil/water ratio. Food Res Int 163:112286

Wang Y, Wang L, Zhang S, Li P, Jiang J (2024) Influence of pH-shifting and transglutaminase on the freeze-thaw stability and thermal gel properties of mung bean protein-based liquid egg substitute prepared with two different oil phases. Food Hydrocoll 146:109182

Weindl I, Bodirsky BL, Rolinski S, Biewald A, Lotze-Campen H, Müller C, Dietrich JP, Humpenöder F, Stevanović M, Schaphoff S, Popp A (2017) Livestock production and the water challenge of future food supply: implications of agricultural management and dietary choices. Glob Environ Chang 47:121–132

Wen J, Chen D, Zeng L, Wang Y, Tian L, Lai J (2015) Nutritional quality and safety of the important commercial cuttlefish Sepia lycidas. J Aquat Food Prod Technol 24(2):101–111

Wen Y, Chao C, Che QT, Kim HW, Park HJ (2023) Development of plant-based meat analogs using 3D printing: status and opportunities. Trends Food Sci Technol 132:76–92

Willett W, Rockström J, Loken B, Springmann M, Lang T, Vermeulen S, Garnett T, Tilman D, DeClerck F, Wood A, Jonell M, Clark M, Gordon LJ, Fanzo J, Hawkes C, Zurayk R, Rivera JA, De Vries W, Majele Sibanda L, Afshin A, Chaudhary A, Herrero M, Agustina R, Branca F, Lartey A, Fan S, Crona B, Fox E, Bignet V, Troell M, Lindahl T, Singh S, Cornell SE, Srinath Reddy K, Narain S, Nishtar S, Murray CJL (2019) Food in the Anthropocene: the EAT–Lancet Commission on healthy diets from sustainable food systems. Lancet 393(10170):447–492

Williams ST (2016) Molluscan shell colour. Biol Rev 92(2):1039–1058

Wright AC, Fan Y, Baker GL (2018) Nutritional value and food safety of bivalve molluscan shellfish. J Shellfish Res 37(4):695–708

Wu D, Wu C, Ma W, Wang Z, Yu C, Du M (2019) Effects of ultrasound treatment on the physicochemical and emulsifying properties of proteins from scallops (Chlamys farreri). Food Hydrocoll 89:707–714

Xia S, Xue Y, Xue C, Jiang X, Li J (2022) Structural and rheological properties of meat analogues from Haematococcus pluvialis residue-pea protein by high moisture extrusion. LWT-Food Sci Technol 154:112756

Xia S, Song J, Li K, Hao T, Ma C, Shen S, Jiang X, Xue C, Xue Y (2023) Yeast protein-based meat analogues: konjac glucomannan induces the fibrous structure formation by modifying protein structure. Food Hydrocoll 142:108798

Xiang X, Jiang Q, Yang H, Zhou X, Chen Y, Chen H, Liu S, Chen L (2022) A review on shellfish polysaccharides: extraction, characterization and amelioration of metabolic syndrome. Front Nutr 9:974860

Yang T, Xin J, Xu Q, Luo H (2010) Preliminary study on processing of texturized by extrusion in low value North Pacific squid protein. Fishery Modernization 37(01):52–55

Yang T, Xin J, Xu Q, Luo H (2011) Effect of extrusion cooking process parameter on functional properties of texturized products in low value aquatic protein. J China Inst Food Sci Technol 11(01):183–189

Yang X, Wang L, Li X, Huang H, Chi M (2013) Analysis and evaluation of nutritional compositions of Dosidicus gigas and Onychoteuthis borealijaponicus okada. Modern Food Sci Technol 29(09):2247–2251+2293

Yang W, Tu A, Ma Y, Li Z, Xu J, Lin M, Zhang K, Jing L, Fu C, Jiao Y, Huang L (2021) Chitosan and whey protein bio-inks for 3D and 4D printing applications with particular focus on food industry. Molecules 27:173

Yujie Z, Jinchuang Z, Qiongling C, Haodong L, Ning H, Qiang W (2022) Effect of hydrothermal parameter combination on the high-moisture extrusion process and the product quality of soy protein-surimi blends. Food Sci Human Wellness 43(19):71–79

Zernov A, Baruch L, Machluf M (2022) Chitosan-collagen hydrogel microparticles as edible cell microcarriers for cultured meat. Food Hydrocoll 129:107632

Zhang J, Liu L, Jiang Y, Shah F, Xu Y, Wang Q (2020) High-moisture extrusion of peanut protein-/carrageenan/sodium alginate/wheat starch mixtures: effect of different exogenous polysaccharides on the process forming a fibrous structure. Food Hydrocoll 99:105311

Zhang J, Chen Q, Liu L, Zhang Y, He N, Wang Q (2021) High-moisture extrusion process of transglutaminase-modified peanut protein: effect of transglutaminase on the mechanics of the process forming a fibrous structure. Food Hydrocoll 112:106346

Zhang B, Kang X, Cheng Y, Cui B, Abd El-Aty AM (2022a) Impact of high moisture contents on the structure and functional properties of pea protein isolate during extrusion. Food Hydrocoll 127:107508

Zhang Y, Zhang J, Chen Q, He N, Wang Q (2022b) High-moisture extrusion of mixed proteins from soy and surimi: effect of protein gelling properties on the product quality. Food Secur 11(10):1397

Zhao G, Bi W, Zhang G, Yang C, Shi Y, Liu H, Li J, Zhao C, Sun X (2019) Effect of extrusion variables on the extrudate characteristics of fish muscle-soy protein blend and optimization of these variables. J Environ Biol 40(3(SI)):409–417

Zhao Q, Fan L, Liu Y, Li J (2023) Mayonnaise-like high internal phase Pickering emulsions stabilized by co-assembled phosphorylated perilla protein isolate and chitosan for extrusion 3D printing application. Food Hydrocoll 135:108133

Zheng Y-Y, Chen Y, Zhu H-Z, Li C-B, Song W-J, Ding S-J, Zhou G-H (2022a) Production of cultured meat by culturing porcine smooth muscle cells in vitro with food grade peanut wire-drawing protein scaffold. Food Res Int 159:111561

Zheng Y-Y, Shi Y-F, Zhu H-Z, Ding S-J, Zhou G-H (2022b) Quality evaluation of cultured meat with plant protein scaffold. Food Res Int 161:111818

# Chapter 9
# Edible Insects in Food Analogues

## Edible Insects as a Potential Ingredient: Challenges and Opportunities

**Haydee Eliza Romero-Luna, Denis Cantú-Lozano, Maribel Jiménez-Fernández, and Guadalupe Luna-Solano**

## 9.1 Introduction

Population growth has triggered a demand for healthy, nutritious, and tasty foods with valuable protein content. However, the proteins considered par excellence are those from animal sources, which have been questioned owing to the lack of sustainability of these sources. Therefore, the search for analogous sources of proteins has been a topic of interest for researchers and consumers. In this sense, edible insects offer nutrients such as proteins (>60%) and essential amino acids (>5%), energy (>700 kcal), fats (>40%), fiber, minerals (>500 mg), and vitamins (Castro-López et al. 2020). Furthermore, edible insects require fewer water and land resources, have better feed conversion efficiency, and generate fewer greenhouse gases, unlike conventional animal proteins, so they can sustainably offset protein demand (Poma et al. 2017).

Entomophagy refers to the ingestion of insects and has been a long-standing practice in about 100 countries belonging to the African, Asian, and American continents. Some ethnic groups consume insects regularly as part of their diet, this translates into the consumption of 2100 species of insects by a total of 2 billion people (Tang et al. 2019; Van Huis et al. 2022). However, it is not widely accepted by culture, in addition to the development of possible allergies, lack of microbiological safety, and bad taste, flavor, and texture. In this regard, the Food and Agriculture Organization (FAO) establishes that edible insects are a potential solution for the rising need for nutritious foods and calls for further research into the

H. E. Romero-Luna · D. Cantú-Lozano · G. Luna-Solano (✉)
Instituto Tecnológico de Orizaba, Orizaba, Veracruz, Mexico

M. Jiménez-Fernández
Centro de Investigación y Desarrollo en Alimentos, Universidad Veracruzana, Xalapa, Veracruz, Mexico
e-mail: maribjimenez@uv.mx



Ö. P. CAN et al. (eds.), *Food Analogues: Emerging Methods and Challenges*,
https://doi.org/10.1007/978-3-031-69872-9_9

nutritional properties, processing, and preservation of food products with insects, as well as their environmental impact, and the formulation of a legal structure for the comprehensive development of edible insect product production (from artisanal to industrial level), which is expected to drive growth in trade, consumption, and consumer acceptance (Van Huis et al. 2013).

Acceptance is an important challenge in terms of insect consumption; therefore, researchers have tried to find processing options that in turn allow extended shelf life, thus increasing availability and acceptance by consumers. These solutions can range from removing legs and wings before consumption, processing them into a powder or paste, or isolating insect proteins and lipids for use as ingredients in other food formulations (Sosa and Fogliano 2017; Lange and Nakamura 2021). In view of the above, the objective of this chapter is to give an overview of the main edible insects as well as their nutritional properties, and to discuss the challenges represented by their processing and potential application in the production of similar foods, ending with problems and opportunities represented by their incorporation into the current diet.

## 9.2 Generalities of Edible Insects

### *9.2.1 Edible Insect Species*

In general, insects belong to the class Insecta and phylum Arthropoda, which are invertebrate animals that represent 95% of the animal kingdom, which is why they contribute significantly to the biodiversity of the planet (Devi et al. 2023; Sales et al. 2023). In different places of the world, including European (11), Oceanian (14), American (23), Asian (29), and African countries (35), they are consumed. Three Asian countries (China, Thailand, and India), and one from Latin American (Mexico) are the consumers and have a greater diversity of edible species (Jongema 2017).

Edible insects can be eaten as eggs, larvae, pupae, or adults; larvae or pupae are the most consumed stages of growth. Edible insects such as beetles, caterpillars, bees, wasps, and ants, grasshoppers, locusts, and crickets, cicadas, aphids, leafhoppers, planthoppers, true bugs, termites, dragonflies, flies, roaches, and spiders are the most widely consumed (Jongema 2017; Baiano 2020; da Silva Lucas et al. 2020).

### *9.2.2 Edible Insect Production*

In the following years, entomophagy could be considered fundamental to human nutrition because of the nutritional characteristics of insects. It is expected that between 2000 and 2050, the need for conventional protein products such as meat will double, adding to the growing controversy over their consumption (Baiano 2020). This would mean a greater production of greenhouse gases because meat accounts for 14.5% of the emissions, as well as larger extensions of land and water required. In contrast, research has shown that producing 1 kg of protein from insects such as mealworms requires 10% of the land required to produce beef (Oonincx and de Boer 2012; Baiano 2020). Nonetheless, only approximately 8% of edible insects come from farms, which could become a severe problem because if more than 90% are taken from the wild, the species could be in danger of extinction, in addition to not ensuring the quality and safety of the insects (Yen 2015a, b). Therefore, large-scale production of edible insects is already taking place in the western world, either for food or feed (Van Huis 2022). Although not all species can be fully cultivated, there are two approaches to breeding edible insects: first, the insects are completely tamed and bred in captivity, and second, the wild habitat of insects is modified so that the insects breed in captivity but do not separate from their wild populations, which is also called semi-cultivation. This could prevent the extinction of insect species, conserve their ecosystems, and contribute to food security (Feng et al. 2018).

Insect breeding can be performed on a small scale to an industrial level because it does not require expensive materials; therefore, it can be implemented in a backyard. However, to increase the industrial-scale production of edible insects, greater knowledge, better rearing conditions, and diet formulations are necessary. Small-scale insect-producing companies are generally located in developing countries, where most of their production is destined for consumption in a local market. In contrast, large-scale facilities are a growing market in Western and Asian countries, and most are used to obtain food for pets, zoo animals, pest control, and medical purposes (Baiano 2020).

Apiculture and sericulture have been practiced for a long time and are the most significant examples of large-scale insect farming. However, the selection of insects for mass production and consumption by humans or animals requires a slightly more intricate process. In this regard, the ideal insect should demonstrate plentiful egg production, rapid egg hatching, brief larval time, optimal pupation timing, high larval and pupal weights, high efficiency in production, affordable feed costs, immunity to diseases, capacity to exist in large populations, and, above all, high-quality protein content (Dobermann et al. 2017). For his part, van Huis (2022) mentions that edible insects can be selected based on four criteria: (1) Automated: the use of devices that automate tasks such as feeding, cleaning, and handling is critical for scaling the insect's production. (2) Use cheap substrates: Employing inexpensive substrates for insect breeding is necessary to not raise the production cost; and (3) avoid diseases. Insects are susceptible to disease, and on a farm with a high production of insects, this could be a problem. Therefore, various strategies have been used to prevent diseases, depending on the insect. *H. illucens*, which is reared worldwide, has not shown any signs of pathogens, and (4) potential markets.

## 9.3 Nutritional and Functional Value of Edible Insects

### *9.3.1 Nutritional Contribution*

As mentioned above, the rise in edible insects has been mainly due to their rapid reproduction rate, sustainability, and nutrients. In general, insects possess important macronutrients (proteins and fats) and micronutrients (vitamins and minerals) similarly to chicken eggs (Table 9.1) (Zielińska et al. 2015; da Silva Lucas et al. 2020).

Edible insects have high protein content, remarkable quality, and minimal environmental costs (Imathiu 2020), so they have gained attention as potential future food. Insect-derived proteins have a significant percentage of essential amino acids, are highly digestible, and are classified as high-value proteins, even more compared to proteins derived from plants, including cereals, and legumes (Schlüter et al. 2017), and proteins from pork, chicken, eggs, and meat. Nonetheless, insect proteins have potential applications as food industry ingredients if they meet functionality requirements. Studies have shown that proteins present in insects have functional properties, such as water and lipid holding, emulsifying, and thickening capacities, as well as foaming, gelling, and structuring abilities (Sosa and Fogliano 2017). Nevertheless, to achieve the successful use of protein as an ingredient, proteins can be extracted and fractionated through enzymatic modifications, since it has been observed that this process improves its functionality (Hall et al. 2018).

Fat is another major component in insects. Compared to fatty fish, insects are considered a significant source of healthy fats (polyunsaturated fatty acids). In addition, insects accumulate less cholesterol than foods of animal origin, making them a healthier food option (Ritter 1990).

Regarding the mineral content, insects can provide the necessary amount to meet daily phosphorus (P) requirements. Copper (Cu), potassium (K), magnesium (Mg), manganese (Mn), selenium (Se), and zinc (Zn) can also be found in significant quantities. Compared to beef, pork, and chicken, they can provide more calcium (Ca) and iron (Fe), providing one more reason to practice entomophagy. If it is not enough, with its low sodium (Na) content, it can be incorporated into diets with minimal sodium consumption without problems (Rumpold and Schlüter 2013a; Köhler et al. 2019).

Different vitamins are found in edible insects, including vitamin A (β-carotene), and B complex vitamins (thiamin, riboflavin, pantothenic acid, pyridoxine, biotin, folic acid, and cobalamin), as well as vitamin C and E (ascorbic acid and alpha-tocopherol, respectively) (Devi et al. 2023; Imathiu 2020).

These nutritional characteristics make edible insects an alternative source of food to combat malnutrition, not only because of their protein composition but also because of their fatty acids, vitamins, and minerals. However, nutrient content varies according to the type of insect, species, climate, feeding, metamorphosis phase, gender, and environment in which the insects are found (Ojha et al. 2021). Kouřimská and Adámková (2016) mention that the fat and energy content increase in the larval and pupal stages. Likewise, it has been observed that the presence of certain

**Table 9.1** Nutrimental content of edible insects

| Edible insect | Moisture (%) | Protein (g/kg dw) | Fat (g/kg dw) | Carbohydrates (g/kg dw) | Ash (g/kg dw) | Minerals | References |
|---|---|---|---|---|---|---|---|
| *Tenebrio molitor* larvae | 2.8–63.18 | 383–584.74 | 253.39–371 | 62.6–262.5 | 35–44.1 | B, Ca, Cu, Fe, Mg, Mn, Mo, P, Zn | Kim et al. (2017, 2021), Ghosh et al. (2017), Kuntadi et al. (2018), |
| *Oxya chinensis sinuosa* | 8.7 | 742.8 | 30.3 | 182.9 | 44 | B, Ca, Cu, Fe, Mg, Mn, Mo, P, Zn | Kim et al. (2017) |
| *Bombyx mori* | 7.9 | 225.78–600.3 | 190.81–294.7 | 9.2–234.2 | 48–68.85 | B, Ca, Cu, Fe, Mg, Mn, Mo, P, Zn | Kim et al. (2017), Kuntadi et al. (2018), Omotoso (2015) |
| *Protaetia brevitarsis* larvae | 7.8–79.73 | 442.3–574.4 | 104.1–211.64 | 110.6–237.1 | 57.22–84.5 | B, Ca, Cu, Fe, Mg, Mn, Mo, P, Zn | Kim et al. (2017, 2021), Ghosh et al. (2017) |
| *Verlarifictorus aspersus* | 1.4 | 463.1 | 285.2 | 221.1 | 30.6 | B, Ca, Cu, Fe, Mg, Mn, Mo, P, Zn | Kim et al. (2017) |
| *Gonimbrasia belina* | 5.68–9.1 | 467–626.51 | 110.01–140.4 | 85.8–221 | 77–113.8 | Ca, Cu, Fe, K, Mg, Mn, Na, Pb | Siulapwa et al. (2014), Vanqa et al. (2022) |
| *Ruspolia differens* | 4.5 | 490.53 | 539.05 | 92.41 | 24.2 | Ca, Cu, Fe, K, Mg, Mn, Na, Pb | Siulapwa et al. (2014) |

(continued)

**Table 9.1** (continued)

| Edible insect | Moisture (%) | Protein (g/kg dw) | Fat (g/kg dw) | Carbohydrates (g/kg dw) | Ash (g/kg dw) | Minerals | References |
|---|---|---|---|---|---|---|---|
| *Macrotermes falciger* | 4.1 | 475.91 | 473.05 | 360.84 | 80.3 | Ca, Cu, Fe, K, Mg, Mn, Na, Pb | Siulapwa et al. (2014) |
| *Valanga nigricornis* | 7.05 | 766.9–779.99 | 69–101.66 | 85. 64–96.2 | 28–32.70 | Ca, Fe | Kuntadi et al. (2018), Paulin and Purwanto (2020) |
| *Nomadacris succincta L.* | – | 654.2 | 109.1 | 154.8 | 32.6 | Ca, Fe | Kuntadi et al. (2018) |
| *Zophobas morio* | – | 468.0–499.6 | 289.8–436.4 | 13.9–142.1 | 34.1–81.7 | Ca, Mg, Fe, Zn | Kuntadi et al. (2018), Soares Araújo et al. (2019) |
| *Allomyrina dichotoma* larvae | 88.06 | 245.39–541.8 | 202.4–245.39 | 40.3–475.71 | 33.5–38.8 | Ca, Cu, Fe, K, Mg, Mn, Na, P, Zn | Ghosh et al. (2017), Kim et al. (2021) |
| *Teleogryllus emma* adult | | 556.5 | 251.4 | 103.7 | 81.7 | Ca, Cu, Fe, K, Mg, Mn, Na, P, Zn | Ghosh et al. (2017) |
| *Gryllus bimaculatus* adult | 3.0 | 583.2 | 118.8 | 95.3 | 96.9 | Ca, Cu, Fe, K, Mg, Mn, Na, P, Zn | Ghosh et al. (2017), Udomsil et al. (2019) |
| *Hermetia illucens* larvae | 5.76–6.41 | 349 | 279.3 | 236.6 | 75 | Ca, Cu, Fe, K, Mg, Mn, Na, P, Zn | Kamau et al. (2018), Vanqa et al. (2022), Romano et al. (2023) |

This table was created by the authors of the chapter

components, such as humidity and light, can be influenced by the management and conditions in which insects are found. It has been observed that the content of vitamin D increases in yellow mealworms and house crickets when exposed to low ultraviolet radiation (Oonincx et al. 2018). Similarly, a diet could imply an increase in the levels of certain nutritional components. For example, vitamin A and omega-3 fatty acid levels could increase if carotenoids and fatty acids were provided in the diet, respectively (Oonincx and Finke 2021).

### *9.3.2 Health Benefits of Edible Insect*

Insect consumption can produce health benefits in addition to the nutritional content. For example, bioactive compounds have shown antioxidant capacity and ACE (angiotensin-converting enzyme) inhibitory activity as well as antidiabetic properties and human fat reduction (Table 9.2). These benefits may be mostly attributed to bioactive peptides, which have also shown antimicrobial activity (Van Huis et al. 2021; Hadj Saadoun et al. 2022). These peptides can be obtained after the digestive process of the insect or through hydrolysis with enzymes; that is, hydrolysates of *Amphiacusta annulipes* obtained from the *in vitro* digestion process showed high antioxidant activity by various methods; they showed an $IC_{50}$ of 19.1 g/mL by the DPPH method, as well as a 58.82% $Fe^{2+}$ chelation ability (Zielińska et al. 2017). Peptides with a DPP-IV (dipeptidyl peptidase-4) inhibitory activity after *in vitro* digestion of crickets (*Gryllodes sigillatus*) have been identify (Hall et al. 2018). Other authors have used the enzyme alcalase to produce peptides with antioxidant, antihypertensive, and antidiabetic properties from insects such as *Gryllodes sigillatus* and *Bombyx mori*. From these studies, the authors concluded that there is a correlation between the exposed bioactivity and the degree of hydrolysis of proteins from edible insects (Yang et al. 2013; Hall et al. 2018; Nongonierma et al. 2018). Nevertheless, information on insect-derived peptides and health benefits claims is limited. Therefore, its study is an unexplored and promising topic that must be investigated in depth.

Another functional component of edible insects is chitin, which is equivalent to 10% (d.b.) of insects as it is part of the exoskeleton of insects. Benefits such as the capacity to modulate the microbiota, that is, to act as a prebiotic, allowing greater development of good bacteria and avoiding the presence of pathogenic bacteria, as well as immune response stimulation, anti-inflammatory effects, and diabetes and obesity prevention, have been reported (Lopez-Santamarina et al. 2020; Stull 2021). Therefore, components such as bioactive peptides or chitin obtained after hydrolysis from insects can be used as functional food ingredients and incorporated to food systems and.

**Table 9.2** Health benefits and distribution of edible insects

| Scientific name | Common name | Health benefit | Distribution | References |
|---|---|---|---|---|
| Coleoptera | | | | |
| *Allomyrina dichotoma* | Japanese rhinoceros beetle | Antioxidant properties | Thailand, Japan | Pyo et al. (2020) |
| *Protaetia brevitarsis* | | Prevent obesity and obesity-related metabolic diseases, antihypertensive properties | Korea, Thailand | Ahn et al. (2019), Lee et al. (2023) |
| *Protaetia brevitarsis* larvae | | Antioxidant and immune-enhancing properties; *EPI* treatment of macrophages induced the activation of macrophages and consequently promoted the production of pro-inflammatory mediators (NO, TNF-α, and IL-1β) | Korea, Thailand | Lee et al. (2023), Pyo et al. (2020) |
| *Tenebrio molitor* larvae | Mealworm | Antioxidant, antithrombotic properties, and DPP-IV inhibiting peptides | China, Laos, Mexico | Dávalos Terán et al. (2020) |
| Blattaria | | | | |
| *Periplaneta americana* | Cockroaches | Hyperglycemia-preventive effect | Mexico Brazil China | Lu et al. (2022) |
| Diptera | | | | |
| *Hermetia illucens* larvae | Black soldier fly | Antioxidant properties | Malaysia | Vanqa et al. (2022) |
| *Hemiptera* | | | | |
| *Aspongopus chinensis* adult | | Antiinflammatory effects | India | Shi et al. (2014) |
| *Lethocerus indicus* | Giant water bug | Antioxidant properties | China, Thailand Myanmar, MalaysiaVietnam, India | Yeerong et al. (2021) |
| *Tessaratoma papillosa* adult | Lichi stink bug | Anticancerigen activity | Thailand, China | Li et al. (2023) |

(continued)

**Table 9.2** (continued)

| Scientific name | Common name | Health benefit | Distribution | References |
|---|---|---|---|---|
| Hymenoptera | | | | |
| *Apis mellifera* larva | Honey bee | Antioxidant, antothrombotic properties | Japan, Congo, Tanzania, Zambia, Senegal, Nigeria, Sierra Leone, Thailand, Malaysia, Vietnam, India, Nepal, Sri Lanka | Pyo et al. (2020) |
| *Polyrhachis vicina* | | Antioxidant and manage obesity properties | China | Zhang et al. (2022) |
| *Vespa affinis* larva | | Antioxidant properties | Malaysia, Thailand | Dutta et al. (2016) |
| Isoptera | | | | |
| *Bombyx mori* pupa | Silkworm moth | Neuroprotective effects | China, Japan, Korea, Mexico, Thailand, Myanmar, Vietnam, India, Laos | Baek et al. (2020) |
| *Macrotermes subhylanus* adult | | Antioxidant properties | Angola, Zambia, Kenya, Togo | Vanqa et al. (2022) |
| Lepidoptera | | | | |
| *Antheraea assama* | Muga silk-moth | ACE-inhibitory and antioxidant properties | India | Sarkar et al. (2023) |
| *Bombyx batryticatus* | | Antioxidant properties | | Anuduang et al. (2020) |
| *Clanis bilineata* larvae | | Hypolipidemic activity | China | Xia et al. (2013) |
| *Gonimbrasia belina* larvae | Mopane moth | Antioxidant properties | Malawi, South Africa, Sambia y Zimbabwe | Vanqa et al. (2022) |
| Orthoptera | | | | |
| *Acheta domesticus* adult | Cricket | Antioxidant and anti-ageing properties | Thailand, Mexico | Lucas-González et al. (2019), Yeerong et al. (2021) |
| *Brachytrupes orientalis* | | Antioxidant and antidiabetic properties | India | Dutta et al. (2017) |
| *Gryllus bimaculatus* adult | Two spotted cricket | Antioxidant properties and increase muscle synthesis by activating the Akt/mTOR/S6K pathway and prevent the activation of muscle degradation factors (MuRF1, atrogin1) by dexamethasone. | Thailand, India, Laos, China, Zambia, Sierra Leone, Guinea, Liberia, Togo, Nigeria, Congo | Kim et al. (2023), Pyo et al. (2020) |

(continued)

**Table 9.2** (continued)

| Scientific name | Common name | Health benefit | Distribution | References |
|---|---|---|---|---|
| *Locusta migratoria* adult | Locust migratoria | Therapeutic effects on dyslipidemia | Madagascar, Cameroon, Congo, Zimbabwe, South Sudan, Thailand, China | Ochiai et al. (2022) |
| *Oxya chinensis sinuosa* adult | | Anti-platelet aggregation activity, anti-inflammatory and anti-cancer activity | Indonesia, China, Korea | Lee et al. (2017), Bahuguna et al. (2022) |
| *Sphenarium purpurascens* adult | Grasshopper | Inhibit proinflammatory enzymes and antioxidant properties | Mexico | Villaseñor et al. (2022) |
| *Teleogryllus emma* | | Antioxidant, anti-thrombotic properties | Thailand | Pyo et al. (2020) |

This table was created by the authors of the chapter

## 9.4 Processing Methods

Other factors that affect nutritional value include the manner in which insects are killed, and the preparation process. Consequently, it has become a challenge to find an adequate transformation method to avoid the loss of nutrients, extend shelf life, and be accepted and consumed.

During the first part of processing, the insect should be collected, washed, inactivated, and wings or legs removed, if required (Rumpold and Schlüter 2013b). Larvae can be collected by sieving, whereas crickets and grasshoppers can be picked up (Fig. 9.1). In some cases, a starvation period may result in the development of a better taste and a decrease in the count of microorganisms present (Fernandez-Cassi et al. 2019). However, this process is not effective in reducing microorganisms, but rather results in a reduction in lipid content in *Acheta domesticus* (Inacio et al. 2021). This was verified by Wynants et al. (2017), who also rinsed *T. molitor* larvae to decrease their microbial count. Neither method, either alone or in combination, was useful in reducing microbial load. Therefore, researchers recommend the use of hot water or diluted chemicals to rinse insects and reduce the microbial load. Antimicrobial agents, such as ethanol, hydrogen peroxide, and sodium hypochlorite, have been successfully used (Crippen and Sheffield 2006).

Subsequently, the insect must be killed; some of the most commonly used methods are freezing, drowning in hot water, or steaming (Fig. 9.1). Studies have shown that killing is an important step during processing because parameters such as color, nutritional composition, taste, and microbial load can vary. Leni et al. (2019) compared two methods to kill black soldier fly larvae: freezing at −20 °C and blanching in boiling water for 40 s. They reported that blanching inhibits phenol oxidase, browning, and other stress-activated metabolic pathways that can cause slow-freezing kill. Another advantage of blanching is that proteins are more extractable

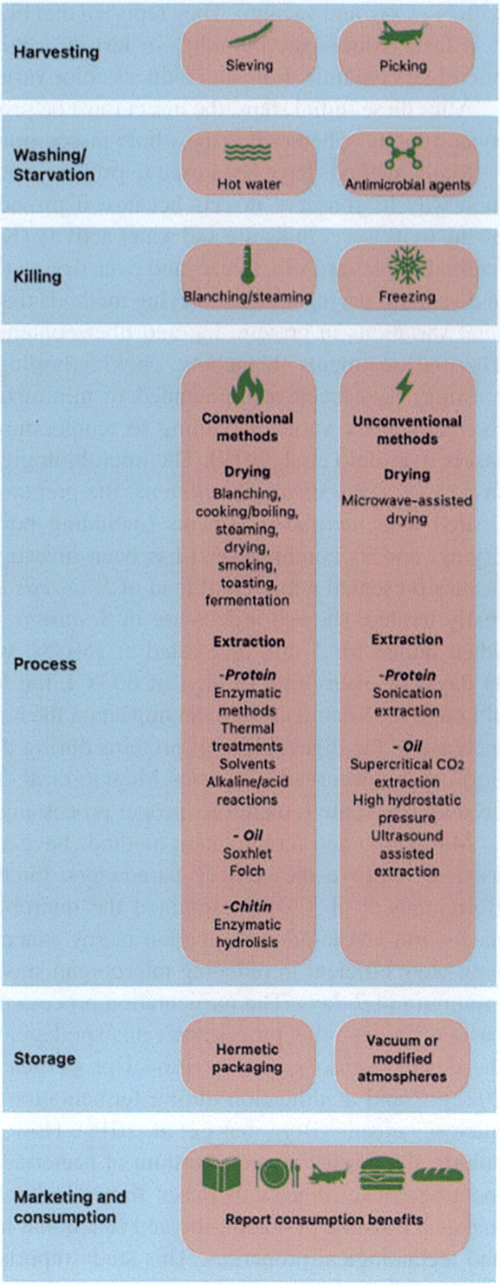


**Fig. 9.1** Methods used in each of the stages of processing edible insects. (This figure was created by the authors of the chapter)

under milder conditions, more susceptible to enzymatic hydrolysis, and less nutritionally deteriorated compared to the freezing method. Similarly, Larouche et al. (2019) evaluated ten different techniques of killing *H. illucens* larvae, including blanching, desiccation, grinding, freezing, high hydrostatic pressures, and asphyxia with $N_2$, $CO_2$, and vacuum. They reported that blanching is the ideal method because it is fast, reduces the humidity in larvae, minimizes the oxidation of lipids and microbial contamination, and reduces color variation.

After these initial steps, the insect must be processed carefully to guarantee quality and safety, whether it is the whole insect, powder, or single-component extract.

Many methods have been used to process whole insects; drying is the most common way to conserve insects because it provides stable products during storage owing to its lower moisture and water activity (Kamau et al. 2018). The main drying methodologies are sun, freeze, and oven drying (Yisa et al. 2023). Microwave, bed-, and vacuum-drying are other drying methods that help to preserve insects with minimal variations in protein, fat, and fiber content (Hernández-Álvarez et al. 2021). Thermal treatments (blanching, cooking/boiling, steaming, drying, smoking, and toasting) have been recommended to minimize microbiological risks. However, their efficiency varies according to temperature, time of thermal treatment, and insect (Garofalo et al. 2019). The microbiological quality of four species of edible insects (*A. domesticus, R. differens*, the prepupae of *H. illucens,* and *S. littoralis*) treated with thermal treatments (including boiling, toasting, solar-drying, oven-drying, and its combinations) has been investigated (Nyangena et al. 2020). The insects presented a microbial load of *S. aureus* and *Salmonella*, which, when thermally treated, showed a decrease in *S. aureus* and the elimination of *Salmonella* when boiled for 5 min or roasted at 150 °C, whereas when they were sun-dried (3 days) or oven-dried (3 days at 60 °C), the load only decreased. Nevertheless, although the treatments have an impact on the nutritional composition of insects, an increase in the digestibility of proteins during thermal processes (boiling or cooking) can be observed (Caparros Megido et al. 2018). In addition, an overall safe product is essential; therefore, proper processing selection is necessary.

Marination and fermentation methods have also been reported. They have been used to improve the sensory parameters, microbiological safety, and shelf life. Borremans et al. (2018) examined the microbiological characteristics of yellow mealworm larvae after marination in soy sauce and red wine. They reported that both were efficient in reducing microorganisms and increased their shelf life by a minimum of 7 days. The fermentation process must be carried out with powder or paste of insects; this process prevents spoilage, improves digestibility, and obtains the desired sensory and nutritional properties using microorganisms (Van Huis 2022). Rapid acidification during fermentation is essential for the elimination of microorganisms (Borremans et al. 2018). However, this requires the addition of a culture starter, that is, a consortium of bacteria, yeasts, or fungi. For example, fermenting cricket flour with yeasts *Yarrowia lipolytica* and *Debaryomyces hansenii* makes it possible to obtain safe and functional hydrolysates with desirable sensory and technological properties. This study reports a reduction in chitin content and compounds with antimicrobial activity, including organic acids, gamma aminobutyric acid, and chitosan, in addition to health-promoting compounds, such as arachidonic and linolenic acids, and alpha- and beta- aminobutyric acids (Ojha et al. 2021; Patrignani et al. 2020).

As mentioned above, insects can also be processed to obtain powder, paste, or even extract nutrients such as proteins, fat, or chitin to be incorporated as ingredients in food preparation (insects as food ingredients or as food analogs will be addressed in the next section). In this regard, powder is obtained through a drying process, followed by grinding or milling (dry milling), and sieving to obtain flour that has a long shelf life, is stable at room temperature, and is ready to be marketed. When grinding is performed when the insect is fresh or after blanching (wet milling), the material obtained, called paste, is fluid and is preferred at an industrial level but unstable because of its high humidity; therefore, it must be refrigerated (De Smet et al. 2019; Ojha et al. 2021). However, chitin can interact with proteins, limiting its use in foods (Liceaga 2021). Therefore, protein extraction from edible insect flour was performed. The first step in protein extraction is defatting, which depends on the protein extraction yield (Sales et al. 2023). During this step, solvents such as hexane, methanol, or ethanol can be used, although more advanced techniques, such as extraction with supercritical fluids or high hydrostatic pressure, have gained popularity (Bolat et al. 2021). Once defatted, enzymatic methods, thermal treatments, solvents, or alkaline/acid reactions can be used to extract the proteins. Subsequently, recovery, purification, and drying were performed to obtain a hydrolyzed edible insect protein powder (Liceaga 2021). Nonetheless, insect proteins have structural limitations; therefore, they sometimes have poor processing properties. Therefore, electron-beam irradiation and $\gamma$-rays have been used to modify them, resulting in an improvement in the techno functional properties of proteins, such as foaming capacity, solubility, emulsion stability, and hydrophobicity, which are important for its application in the food industry (Lee et al. 2023).

Conventional techniques have been used to extract oils, such as Soxhlet and Folch, in addition to more innovative methods, such as ultrasound or supercritical $CO_2$ extractions, which are expensive but allow the extraction of solutes, avoiding their oxidation, as well as thermosensitive compounds (Laroche et al. 2019; Purschke et al. 2018). Chitin extraction mainly consists of three steps: demineralization, deproteinization, and deacetylation. The first step involved the dissolution of calcium carbonates and proteins with strong acids and bases. The latter can also be eliminated by enzymatic hydrolysis with alcalase, and after separating the precipitate, it undergoes deacetylation, which results in thermoalkaline hydrolysis. Once chitin is obtained, it is dried to obtain a dry powder as the final product ready for use (da Silva Lucas et al. 2021).

Once the insect is processed, storage conditions are crucial for prolonging its useful life, which will vary according to the insect and the presentation of the food product, that is, whether it is a powder, or the insect is whole (Grabowski and Klein 2017). Among the main methods used are hermetic packaging (Adámek et al. 2018) as well as vacuum and modified atmosphere packaging, which have shown to be promising options for storing edible insects subjected to drying for a long time, either whole or in powder form. In some cases, vacuum packaging can increase shelf life if combined with refrigeration or if it is kept in the dark at room temperature (Ssepuuya et al. 2016). Application of modified atmospheres (60% $CO_2$ and 40% $N_2$) diminished microbial growth in a meat analog based on *T. molitor* and *Alphitobuous diaperinus* larvae, compared to standard atmospheric conditions.

## 9.5 Edible Insects as Food Ingredient

The edible insects' market is valued at $3.2 billion and is estimated to reach $7.8 billion by 2028 (Mordor Intelligence 2023). According to Moorthy Eswaran et al. (2023), there are currently over 250 commercially available products, which include flours, dehydrated larvae and pupae, canned and extruded products, protein isolates, snacks, sweets, chocolates, bakery products, pasta, burgers, nuggets, milk, and ice cream, among others (Bekhit et al. 2022).

### *9.5.1 Bakery Products*

In this regard, processed products begin with insects in powder form, mostly derived from mealworms, crickets, and locusts. The addition of insect powder, in addition to increasing the nutritional properties, also modifies the sensory and technological properties of food (Biró et al. 2020). A general improvement in the nutritional properties has been observed when replacing between 5% and 25% of the raw material with insect flour. For instance, there was a rise in the concentration of fat, proteins, minerals, and vitamins (A and C), and minerals (Yazici and Ozer 2021). In addition, some bakery products contain saturated fats (caproic and arachidic acids) and some essential amino acids (lysine and leucine) (Kowalski et al. 2022).

In contrast, the incorporation of insects in powder or extract form into foods improves the consumer acceptance (Mishyna et al. 2020); for instance, the addition of cricket flour (*A. domesticus* and *G. bimiculatus*) to biscuits, fritters, samosa, and pancakes has been accepted by children (Ayieko et al. 2016). However, its addition modifies the sensory properties. The physical properties of the extruded brown rice were modified by the integration of cricket flour because a darker color was observed and compared to commercial brown rice, it had a softer and more adhesive texture, its aroma and taste were fishier, and its stability during shelf life was major (Tao 2016). Nevertheless, species, stage of development, and processing have an impact on the flavor and texture of the food product; therefore, the correct selection of these factors implies the acceptance of the novel foods based on insects and allows the acceptability of insects as new foods with a wide nutritional and functional value. For instance, Pérez-Rodríguez et al. (2023) evaluated the addition of *S. purpurascens* and *T. molitor* powders fermented with *Aspergillus oryzae* to dough to make bread. They initially described that the amount of protein increased with fermentation, which, when introduced to bread, could improve the nutritional composition of the product as well as some properties, such as texture and sensory characteristics, compared to the addition of non-fermented insects.

### 9.5.2 Pasta

Unlike in bakery products, the use of insect flour to wheat pasta has not been thoroughly studied. Duda et al. (2019) replaced semolina with cricket powder (5, 10, and 15%) and observed that this replacement influenced the color and reduced the water loss and absorption of pasta. Likewise, the textural properties were modified as the firmness of the paste was strengthened. Regarding sensory properties, the authors mentioned that the change in flavor was pronounced but did not have a negative impact on acceptance.

On the other hand, to take advantage of the protein content of insects, the pastes have been added with protein hydrolysates from insects (Pasini et al. 2022). In this study, a color change is observed, resulting in a darker paste with the addition of insects and firmer. In addition, a higher water absorption index was obtained for the paste containing *A. domesticus*, for which the authors identified insect proteins as functional ingredients with multiple potential uses in food. However, consumer acceptance, an important parameter, was not evaluated in this study because it is one of the main limitations of insect consumption. Thus, sensory studies are recommended.

### 9.5.3 Meat and Milk Analogs

Meat analogs have been developed based on edible insects, although research has focused on formulations in which a certain percentage of meat is changed with whole insect powder, protein, or defatted form. The development of meat analogs is influenced by the techno functional properties of proteins, such as water and lipid holding capacity, emulsifying capacity and stability, solubility, and gel formation. The incorporation of insect powder to meat products has improved the nutritional and technological characteristics, resulting in higher stability of emulsions, protein digestibility, increased water absorption, and reduced losses during cooking (Martins Borges et al. 2022). Analogous meat products based on edible insects are more thermally stable, have greater protein digestibility, and have a desirable texture depending on the formulation. However, these products have technological limitations, such as a decrease in textural properties and integrity index, in addition to the fact that their cutting properties are not similar to those of meat (Smetana et al. 2018; Cho and Ryu 2021; Kim et al. 2022). Another food of animal origin with a high demand is milk, which has been studied to formulate its analog based on insects. Tello et al. (2021) developed a milk prototype based on *Tenebrio molitor* larvae, which had a comparable nutritional value and minimal environmental impact.

### 9.5.4 Ice Cream

Ice cream is widely accepted worldwide by both adults and children. Therefore, this product was also fortified with silkworm pupae flour. In this study, the results showed an effect on color that reduced its visual appeal at higher flour content, and the panelists mentioned a strong aftertaste. The hardness and adhesiveness were altered when more than 1% flour was added. Therefore, the authors recommend including no more than 1% flour in ice cream to maintain functional properties and palatability (David-Birman et al. 2022).

Mariod and Fadul (2015) previously added gelatin from *Coridius viduatus* and *Agonoscelis versicoloratus* to ice cream. However, the texture and taste showed important differences compared with the addition of commercial gelatin.

These and other studies have demonstrated that the addition of powder or extracts from insects into food improves its nutritional quality, texture, and sensory properties (Azzollini et al. 2019; Kang et al. 2023; Kim et al. 2022; Nam et al. 2023). Furthermore, the incorporation of insects into processed foods could normalize their consumption and, therefore, increase acceptance among consumers.

## 9.6 Challenges and Opportunities

The topic of edible insects has become popular among scientists and at an industrial level because of the opportunities that their use and exploitation represent; that is, one of their main advantages is their nutritional value, followed by their sustainability, factors that would allow global food security. However, the development of food based on insects has been hindered mainly by consumer acceptance and the risks that insect consumption could represent, such as microbiological, allergenic, antinutritive, and xenobiotic presence. In addition, processing and scaling at the industrial level also represent one of the main obstacles, as does the absence of regulations (Fig. 9.2).

### 9.6.1 Challenges

#### 9.6.1.1 Consumer Acceptance

Insects, with their nutritional benefits, seem to be a favorable answer to the food security that we seek to achieve worldwide. However, the acceptance of consumer is the main obstacle in the consumption of insects. Because insects are not viewed favorably, they are considered unpleasant, dirty, dangerous, and undesirable, especially in Western countries (Looy et al. 2014). Furthermore, meat consumption is strongly rooted (Eckl et al. 2021). Therefore, the social background is decisive in

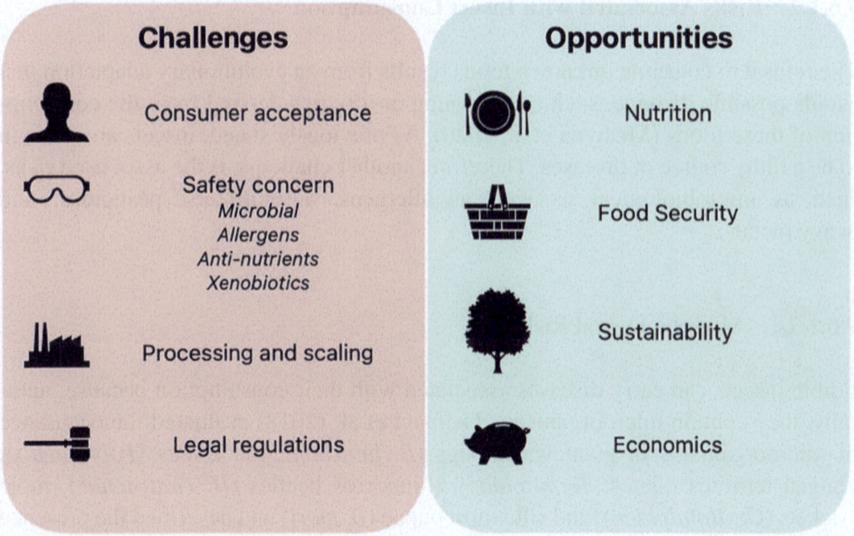


**Fig. 9.2** Challenges and opportunities facing the edible insect's market. (This Figure was created by the authors of the chapter)

the acceptance of new foods derived from edible insects. Despite its low acceptance, Protifarm (the Netherlands), Kreca Ento-Feed BV (the Netherlands), Ynsect (France), AgriProtein (South Africa), Flying Spark (Israel), Thailand Unique (Thailand), Entomo Farms, Ltd. (Canada), Exo Inc. (USA), and Enviro Flight LLC (USA) are industries that show that insects are a growing market worldwide, with startups and industries dedicated to the farming and marketing of edible insects; the latter considered large leading industries in the market. Some of these companies sell dehydrated insects. Others sell products that include insects in their list of ingredients, which could promote entomophagy and increase acceptance, normalizing insect consumption in Western countries. This has been observed in studies conducted in Belgium and Germany (Orsi et al. 2019; Van Thielen et al. 2019).

There are some factors that positively and negatively influence acceptance and willingness to eat insects. Age is an important factor because young people (18–30 years old) show greater acceptance of consuming foods with incorporated insects. Recent studies have reported that unfamiliarity, revulsion, and food neophobia are the principal factors that negatively affect insect acceptance, whereas curiosity, environmental sustainability, nutritional content, and health benefits are the main motivators that lead consumers to eat insects (Kasza et al. 2023; Ordoñez-López et al. 2023; Orsi et al. 2019; Ribeiro et al. 2022). Puteri et al. (2023) mention that a strategy to increase the acceptance of these products is by emphasizing the presence of insects, informing them, and highlighting the significant benefits of their consumption from commercialization. In addition, this can be strengthened through publicity and degustation campaigns.

#### 9.6.1.2 Risks Associated with Insect Consumption

The refusal to consume unknown foods results from an evolutionary adaptation that avoids possible damage, such as poisoning or diseases derived from the consumption of these foods (Mishyna et al. 2020). As previously stated, insects are through to be a filthy source of diseases. Therefore, another challenge is the associated risks, such as microbiological as well as allergens, antinutrients, pesticides, and heavy metals.

##### 9.6.1.2.1 Microbiological Risks

Edible insects can carry diseases associated with their consumption because, naturally, they contain microorganisms. Osimani et al. (2018) evaluated the occurance of microorganisms in giant water bugs (*B. lutarium*), black ants (*Polyrhachis*), winged termites (alates, *Termitoidae*), rhinoceros beetles (*H. contractum*), mole crickets (*Gryllotalpidae*), and silkworm pupae (*B. mori*) and described the presence of the various genera including *Pseudomonas, Lactobacillus, Staphylococcus, Streptococcus, Vibrio,* among others. Although the load of pathogenic microorganisms in humans is low in insects, enteric pathogens (*Campylobacter* spp., *Salmonella* spp., *S. aureus,* and *E. coli*) have been reported (Grabowski and Klein 2017; Marshall et al. 2016). The presence of these microorganisms may be due to their intrinsic microbiota, although they can also contaminate depending on the collection method, hygiene practices, processing, and handling used during preparation (Rumpold and Schlüter 2013a). Another study reported that *Salmonella* spp. or *E. coli* were not present in any insect-based products sold in Canada, indicating adequate sanitary conditions in the food production chain (Kolakowski et al. 2021). However, spore-forming bacteria can occur in processed foods with simple interventions, such as blanching, where live bacteria can be eliminated but spores remain, which, after germination, can cause contamination of the food and subsequent illness (Marshall et al. 2016). The presence of microorganisms with the ability to form spores (*Bacillus* and *Clostridia*) has been reported in edible insects (Osimani and Aquilanti 2021). Therefore, by performing hygienic and adequate handling and processing, products will be suitable and safe for human consumption.

Parasitic diseases have rarely been considered; however, when consuming edible insects, they must be considered because insects are natural vectors of parasites. A study on insect farms detected the presence of parasites in 244 of 300 farms, with only 91 of the cases being pathogenic in humans. Parasites include *Cryptosporidium* spp., which are associated with chronic diarrhea and can be found in cockroaches and crickets; *Isospora* spp., which cause isosporiasis, an intestinal infection found in cockroaches, flies, and beetles; *Balantidium* spp., which causes balantidiasis, an intestinal infection, is found on farms with closed habitats; *Entameba* spp. can cause dysentery; cysticercoids and their eggs have also been described in insects. Certain nematodes can also be found in insects, such as crickets and cockroaches, the latter being one of the main vectors of parasites and transmitters of diseases in humans. In this case, the authors recommend keeping farms at low temperatures to avoid the proliferation of certain parasites, and mention that simple preservation methods such as drying, cooking, and freezing may be sufficient to avoid the transmission of parasites (Gałęcki and Sokół 2019).

The intake of edible insects can increase the presence of mycotoxins, which are secondary metabolites generated by *Fusarium, Aspergillus,* and *Penicillium* food spoilage molds. Mycotoxins are a major concern because they can cause long-term negative health effects (Omotayo et al. 2019). Insects can become contaminated because of the substrate they are fed, improper handling, or poor storage conditions (Baiano 2020; Lange and Nakamura 2021). Mycotoxins Aflatoxin B1, deoxynivalenol, fumosin B1 and T-2, ochratoxin A, zearalenone, have been identify in certain Coleoptera and Diptera species (Bisconsin-Junior et al. 2023). Although all of these values were below what was recommended by the WHO (World Health Organization), the fasting process in insects could reduce the number of mycotoxins.

#### 9.6.1.2.2 Allergens

Another problem associated with eating insects is the presence of allergens in insects. In sensitive people, the consumption of certain foods or even contact and inhalation, such as insects, could trigger reactions as a response from the immune system, which in the extreme could end in death (Imathiu 2020). Arginine kinase, α-amylase, glyceraldehyde 3-phosphate dehydrogenase, hemocyanin, and tropomyosin are proteins present in insects that are considered allergens (Murefu et al. 2019; Srinroch et al. 2015). Studies have shown that insects such as mealworms, sago worms, caterpillars, bees, cicadas, *Cruchus lentis*, *Clanis bilineata*, silkworms, grasshoppers, and locusts are responsible for the development of food allergies, the latter being the cause of anaphylactic shock (Baiano 2020; de Gier and Verhoeckx 2018). Likewise, reports of allergies have been described for the additive carmine derived from carminic acid obtained from cochineal (*Dactylopius coccus*), which generally includes proteins that could be the cause of allergic reactions, some of which are severe (Takeo et al. 2018). Consequently, there is a need to generate additional information about insects and the allergies they could cause, in addition to strategies that include correct labeling to safeguard consumer health.

#### 9.6.1.2.3 Antinutrients

Anti-nutritional compounds can reduce intake, digestion, absorption, and the use of macro- and micronutrients, which could lead to adverse effects in consumers. Many of these compounds are generated by insects as a defense mechanism and can cause irritation (carbonic acids, alcohols, aldehydes, and phenols) or exert systemic effects (alkaloids, steroids, alkenes, and quinones) (Belluco et al. 2018; Lange and Nakamura 2021). However, these compounds can generate other adverse effects depending on their type and concentration. These two factors depend on the insects. As demonstrated by a study that showed the analysis of the antinutritional composition of *Oryctes rhinoceros* larvae and *Zonocerous variegatus,* the first showed a lower concentration of hemagglutinin (0.002 mg/100 g), while the second presented hemagglutinin 1000 times higher (2.18 mg/100 g), besides to elevated concentrations of hydrogen cyanide (5.21 mg/100 g) (Anaduaka et al. 2021). Nevertheless, these can be reduced to safe levels by boiling or cooking (Kunatsa et al. 2020).

As mentioned in Sect. 9.3, chitin is a part of the insect exoskeleton, and although it has been considered a functional component, it has also been considered an antinutrient. Regarding this, Moruzzo et al. (2021b) described that it affects the bioavailability of minerals, although other studies have suggested that it may be a potential allergen (da Silva Lucas et al. 2020). Therefore, additional research on the anti-nutritional properties of chitin is required.

#### 9.6.1.2.4 Xenobiotics

Another risk that must be considered is that xenobiotics, which can cause concern from rearing due to environmental contaminants that can accumulate and remain in the insect, can be processed where there may be cross-contamination by other raw materials or migration from packaging materials (Poma et al. 2022). Pesticides and heavy metals are the contaminants accumulated in edible insects.

Insects, especially those collected from the wild, may accumulate pesticide residues because they could be in direct contact or because they are feeding on crops irrigated with insecticides, herbicides, or fungicides (Lange and Nakamura 2021). This could cause food poisoning in those who consume insects with an accumulation of pesticide residues, as is the case in Thailand, where insects contaminated with pesticide residues cause poisoning in consumers (DeFoliart 1999). Likewise, other studies have reported that a single contact with a contaminated crop could be sufficient for the insect to contain substantial pesticide residues (Houbraken et al. 2016). However, the amount of pesticide residue is insufficient to cause such problems (De Paepe et al. 2019). Nonetheless, it is necessary to take action to reduce the amount of pesticides further. In this case, insect farms could be an answer, as the control exercised over their feeding could help generate insects free of pesticide residues (Imathiu 2020). Furthermore, it has been reported that fasting also has an impact on reducing the concentration of various pesticides (Houbraken et al. 2016), so it could be a strategy to reduce them because these could cause long-term effects such as diabetes, cancer, respiratory diseases, and oxidative stress (Aguilar-Toalá et al. 2022).

Heavy metals are metallic elements that can be toxic and have a negative impact on animals' and humans' health (Jan et al. 2015). Plants can absorb these heavy metals from contaminated soil or air; therefore, insects accumulate heavy metals when they feed on plants in this environment (Aguilar-Toalá et al. 2022). Mwelwa et al. (2023) assessed the levels of heavy metals in plants and insects in a mining area in Africa. This study revealed that there was a greater accumulation of cadmium, copper, nickel, iron, lead, and zinc in plants, while insects show a species-dependent accumulation; that is, *C. forda* has a greater accumulation of nickel, *I. obscura* of cadmium and lead, and *M. falciger* and *I. rubra* of zinc and iron. Eating wild insects could increase the daily intake of heavy metals above the limit allowed by the WHO, depending on insects that are harvested from a wild environment. Therefore, this risk could be reduced through the controlled production, processing, and storage of edible insects (Imathiu 2020).

#### 9.6.1.3 Processing and Industrial Scaling

In a previous section, research on the processing that insects have undergone was discussed; however, the existing information shows artisanal or laboratory-level processes, so processing and its subsequent scaling could be one of the main challenges.

As mentioned in Sect. 9.4, one of the problems of processing selection is that it allows the generation of a safe product while maintaining its nutritional properties; that is, the selected process allows inhibiting the presence of microorganisms, which could compromise the health of the consumer, and likewise does not substantially affect the nutrients contained in the edible insect (Garofalo et al. 2019).

One of the factors involved in scaling is the selection of the ideal insect for domestication, which has been developed in different insects with success; however, it changes from insect to insect, and as mentioned in Sect. 9.2.2, the insect must meet high production requirements and performance, with low costs and low vulnerability to diseases, as well as having high-quality proteins, and it must also allow the automation of breeding (Dobermann et al. 2017).

One of the main interests at the industrial level is the extraction of proteins from insects, whether for use as additives or to create new products with high protein content; however, the protein content and functionality vary with the species of insect, rearing methods, and processing technique (Sosa and Fogliano 2017). Therefore, in this sense, it is necessary to focus more research on this regard in order to produce protein isolates for use in food formulations with adequate functional properties that are profitable in a sustainable environment (Gravel and Doyen 2020).

For all the above, coupled with the fact that the current market trend indicates the need for large quantities of edible insects with standardized quality and large-scale production, scaling these processes to an industrial level will only be possible under the necessary regulations.

#### 9.6.1.4 Regulations

Regulations are an extremely important requirement for farm development and the marketing of insect food (Żuk-Gołaszewska et al. 2022), which aim to guarantee food safety. Safety is evaluated by competent authorities; for example, the EFSA (European Food Safety Authority) oversees this task in Europe, although some countries on this continent have their own agencies, such as Belgium with the Federal Agency for Safety of the Food Chain and the United Kingdom with the Food Standards Agency. In North America, Canada has the Canadian Food Inspection Agency and Health Canada to approve the safety of new foods, and in the United States, the Food and Drug Administration does the same. The evaluation must be carried out periodically from the insect harvest since some factors, such as the contaminants, are critical for food safety and quality, so guidelines must be followed, for example, the Codex Alimentarius (Grabowski et al. 2020). However, this does not consider insects as food but only as impurities that can be found in some cereals. So, an update is suggested.

Regulations are essential in countries with regular consumption of insects, as well as in developed countries where it is not common to observe entomophagy, to stay away from the risks that come with eating insects. The nations where insects are frequently consumed do not have legal frameworks that allow for regulation of their production, marketing, and consumption. On the contrary, European countries have strong guidelines that imply an obstacle to the application of insects to food. In this sense, the EFSA establishes that insects and insect-based products must be called "novel food". However, the approval of novel foods has a transition period of 2 years, which implies a barrier to the development of insect-based foods (Dobermann et al. 2017). The objective of Regulation (EU) No. 2015/2283 is to safeguard the health of consumers by offering guidelines that guarantee the safety of innovative foods that come onto the market. These regulations allow the approval of foods that are safe for consumption and have clear labeling. To date, EFSA has recognized whole insects, their parts, powders, and their extracts of lesser mealworms (*A. diaperinus*), crickets (*A. domesticus*), and yellow mealworms (*T. molitor*) as novel foods (IPIFF 2023). This reduced number of accepted insects may be due to the demanding requirements necessary to accept novel foods, which must encourage the food safety through rigorous scientific research, which is economically expensive and requires time. However, more scientific investigation is being generated on the safety and processing of insects for food, so more data is being obtained, which could lead to more edible insects being accepted as novel foods (Imathiu 2020).

The lack of regulations and the strict regulations are an imposing barrier that prevents the utilization of edible insects. On the one hand, a lack of regulation could cause concern among consumers in western countries (Mancini et al. 2019). On the other hand, overly stringent regulations do not allow adequate growth of the edible insect market, especially in Europe (Shelomi 2015). Therefore, it is necessary to find a middle point in the use of regulations, which must consider the entire production chain, including insect farming, as well as the use and marketing of products with insects. This will allow the market for edible insects to grow, as well as the acceptance of consumers, but above all, it will allow the consumer's health to be safeguarded.

### 9.6.2 Opportunities

Its main opportunities lie in the high nutritional value that insects have, although factors such as sustainability and low cost imply greater attractiveness for those who wish to market insect-based products. Furthermore, these factors cement the idea of achieving food security in the near future.

#### 9.6.2.1 Nutritional Value

As mentioned in Sect. 9.3, insects provide energy, fat, minerals, vitamins, and especially proteins, which can be on par with other sources of animal origin, such as meat or eggs. Nutritional profiles vary by species, diet, stage of metamorphosis, habitat, and environmental conditions (Lange and Nakamura 2021). It is essential to consider that insects are a very significant globally distributed source of nutrients that everyone can access and benefit from.

#### 9.6.2.2 Food Security

Food security is achieved when every person on the planet has adequate physical and financial access to safe, nutritious, and healthy food. However, rapid population growth and declining crop productivity and food availability make this a problem, especially in developing nations (Abril et al. 2022). In fact, it has been established that in the coming decades, food security will be a major problem due to the fact that, by 2050, there will be 9.8 billion people on the planet, requiring 100% more food production (Lange and Nakamura 2021). For this reason, the FAO/WHO has proposed an initiative in order to feed people with alternative protein sources. In this sense, insects have been considered since their nutritional content is high, consisting of protein, vitamins, and minerals essential for meeting physiological needs and leading a healthy life (Orkusz 2021). In addition, insects are profitable, easy to grow, and environmentally sustainable, so all of this may help ensure the world's food supply (Dickie et al. 2019).

#### 9.6.2.3 Sustainability and Economics

Demographic growth has brought with it a high demand for food production and, therefore, a severe environmental impact, mainly an excess of $CO_2$ emissions as well as in water and land. So, currently, sustainability has become an important and increasingly common term, even in the world of food, where efforts have been made to replace traditional food sources, such as cattle, with sources with less environmental impact, such as insects (da Silva Lucas et al. 2020).

Livestock farming has many environmental disadvantages, including that it uses more than 70% of agricultural land, and the rise in need for food owing to population growth would cause an increase in the area dedicated to it, which translates into more contamination with heavy metals, pathogens, and ammonia emissions. A greater area of cropland would also be required for feed, which would lead to deforestation (Lange and Nakamura 2021).

In contrast, insects present many sustainable and economic advantages, for example: (1) they have high feed conversion efficiency; (2) they can be fed with agro-industrial waste; (3) they produce fewer greenhouse gases and emissions of ammonia; (4) they require a smaller area of land and water; and (5) they transmit zoonotic diseases less frequently (Lange and Nakamura 2021). Insects can generate the same amount of proteins as animals with less food due to their high feed conversion efficiency. For example, for every kilogram of weight gained, a beef needs 10 kg of feed, a pig needs 5 kg, and a chicken needs 2.5 kg. In contrast, crickets only require 1.7 kg of feed, and their edible mass percentage is major (80%), compared to cattle (40%), chickens (55%), or pigs (70%) (Moruzzo et al. 2021a). To achieve this high efficiency, it is necessary to consider the substrate on which the insects feed, as its nutritional composition must be appropriate for each species and allow for maximum performance. In addition to being commendable, organic waste or by-products could be used for this purpose, especially those with high protein concentrations, as it has been determined in house crickets that a high-protein substrate shows greater feed conversion efficiency (Sorjonen et al. 2019). Nevertheless, feed conversion efficiency also varies on the capacity of the insect (Oonincx et al. 2019). Livestock farming has contributed approximately 18% to climate change owing to its high greenhouse gas emissions, while insects generate low quantities of these gases (Moruzzo et al. 2021a). In this regard, research has indicated that farming *A. domesticus*, *L. migratoria* and *T. molitor* produces less greenhouse gases (Oonincx et al. 2010). Something similar happens with water and land since meat production requires approximately 4300 L of water for each kg of chicken, 6000 L for pork, and 15,400 L for beef (Ibidhi and Salem 2020). While insects show a smaller water footprint than any of these (Miglietta et al. 2015), in addition, insects can contribute to water sanitation since some insects could be raised in animal and human feces. Although the safety of food derived from insects could be compromised but could be improved as part of the process, this fact would help reduce pollution (Moruzzo et al. 2021a). Regarding land use, insects have greater land use efficiency and efficiency in converting byproducts and organic waste into food, which in turn guarantees high profitability (Alexander et al. 2017). There is an absence of information regarding land use and water footprints concerning insects, so in-depth research on the subject is required.

Livestock is affected by important diseases that can cause the death of numerous animals. Animals can act as reservoirs of pathogens from bacteria, parasites, and viruses, among others, that can cause health problems for consumers, including resistance to antibiotics, or even generate pandemics such as that generated by the SARS-CoV-2 (Rahman et al. 2020). On the contrary, bovine spongiform encephalopathy and zoonotic infections such as H1N1 could be minimal due to the taxonomic distance of insects from humans (Belluco et al. 2023). However, as mentioned previously in this section, insects have also generated concern due to the possible presence of pathogens that could cause diseases in consumers. More information is required regarding the potential risks of insects transmitting zoonotic infections.

So, it is a priority to establish legal guidelines in all countries that allow the nutritional benefits of insects to be obtained safely for the consumer and to achieve food security that worsens as overpopulation occurs without compromising wild insect populations sustainably and economically. Therefore, it is essential to continue with research studies regarding insects' dietary requirements, farming, processing, and safety to not compromise consumers health.

## 9.7 Conclusion

Insects represent the majority of the planet's biodiversity and show important nutritional qualities, which have been associated with benefits to consumers. This has led to them being identified as the food of the future. Various processing methods have been tested in order to maintain these nutritional properties and extend the shelf life of insect-based products, which has allowed for an increase in the variety of products such as pasta, baking products, milk and meat analogues, and ice cream, among others. However, despite the advantages they represent, their use has been limited by various challenges, such as consumer acceptance as well as the microbiological, allergy, and antinutrient risks associated with their consumption. Furthermore, its industrial use has had difficulties scaling its production at an industrial level, mainly due to the lack or excess of present regulations.

Therefore, insects show great opportunities but also challenges that can be resolved with continuous research that allows their incorporation as ingredients in the preparation of various foods and, in this way, improves acceptance by consumers, generates appropriate laws, and protects above all the health of the consumer while allowing the growth and use of edible insects.

**Acknowledgement** This work was funded by the Veracruzano Council for Scientific Research and Technological Development with number 1111 1744/2023 and by the National Council of Humanities, Sciences and Technologies from Mexico through a postdoctoral scholarship provided to Dr. Haydee Eliza Romero Luna with CVU number 421967.

## References

Abril S, Pinzón M, Hernández-Carrión M, Sánchez-Camargo ADP (2022) Edible insects in Latin America: a sustainable alternative for our food security. Front Nutr 9:904812

Adámek M, Mlček J, Adámková A, Suchánková J, Janalíková M, Borkovcová M, Bednářová M (2018) Effect of different storage conditions on the microbiological characteristics of insect. Potravinarstvo 12:1

Aguilar-Toalá JE, Cruz-Monterrosa RG, Liceaga AM (2022) Beyond human nutrition of edible insects: health benefits and safety aspects. Insects 13(11):1007

Ahn EM, Myung NY, Jung HA, Kim SJ (2019) The ameliorative effect of *Protaetia brevitarsis* Larvae in HFD-induced obese mice. Food Sci Biotechnol 28:1177–1186

Alexander P, Brown C, Arneth A, Dias C, Finnigan J, Moran D, Rounsevell MD (2017) Could consumption of insects, cultured meat or imitation meat reduce global agricultural land use? Glob Food Sec 15:22–32

Anaduaka EG, Uchendu NO, Osuji DO, Ene LN, Amoke OP (2021) Nutritional compositions of two edible insects: *Oryctes rhinoceros* larva and *Zonocerus variegatus*. Heliyon 7(3):e06531

Anuduang A, Loo YY, Jomduang S, Lim SJ, Wan Mustapha WA (2020) Effect of thermal processing on physico-chemical and antioxidant properties in mulberry silkworm (*Bombyx mori* L.) powder. Food Secur 9(7):871

Ayieko MA, Ogola HJ, Ayieko IA (2016) Introducing rearing crickets (gryllids) at household levels: adoption, processing, and nutritional values. J Insects Food Feed 2(3):203–211

Azzollini D, Wibisaphira T, Lakemond CMM, Fogliano V (2019) Toward the design of insect-based meat analogue: the role of calcium and temperature in coagulation behavior of *Alphitobius diaperinus* proteins. LWT 100:75–82
Baek SY, Li FY, Kim JH, Ahn C, Kim HJ, Kim MR (2020) Protein hydrolysate of silkworm pupa prevents memory impairment induced by oxidative stress in scopolamine-induced mice via modulating the cholinergic nervous system and antioxidant defense system. Prev Nutr Food Sci 25(4):389
Bahuguna A, Khaket TP, Bajpai VK, Shukla S, Park I, Na M et al (2022) N-Acetyldopamine dimers from *Oxya chinensis sinuosa* attenuates lipopolysaccharides induced inflammation and inhibits cathepsin C activity. Comput Struct Biotechnol J 20:1177–1188
Baiano A (2020) Edible insects: an overview on nutritional characteristics, safety, farming, production technologies, regulatory framework, and socio-economic and ethical implications. Trends Food Sci Technol 100:35–50
Bekhit AEDA, Ahmed IAM, Al-Juhaimi FY (2022) Non-bovine milk: sources and future prospects. Food Secur 11(13):1967
Belluco S, Mantovani A, Ricci A (2018) Edible insects in a food safety perspective. In A. Halloran, R. Flore, P. Vantomme, & N. Roos (Eds.). Edible insects in sustainable food systems (pp. 109–128). Cham, Switzerland: Springer. https://doi.org/10.1007/978-3-319-74011-9_7
Belluco S, Bertola M, Montarsi F, Di Martino G, Granato A, Stella R et al (2023) Insects and public health: an overview. Insects 14(3):240
Biró B, Sipos MA, Kovács A, Badak-Kerti K, Pásztor-Huszár K, Gere A (2020) Cricket-enriched oat biscuit: technological analysis and sensory evaluation. Food Secur 9:1561
Bisconsin-Junior A, Feitosa BF, Silva FL, Mariutti LRB (2023) Mycotoxins on edible insects: should we be worried? Food Chem Toxicol 177:113845
Bolat B, Ugur AE, Oztop MH, Alpas H (2021) Effects of high hydrostatic pressure assisted degreasing on the technological properties of insect powders obtained from *Acheta domesticus* & *Tenebrio molitor*. J Food Eng 292:110359
Borremans A, Lenaerts S, Crauwels S, Lievens B, Van Campenhout L (2018) Marination and fermentation of yellow mealworm larvae (*Tenebrio molitor*). Food Control 92:47–52
Caparros Megido R, Poeleart C, Ernens M, Liotta M, Blecker C, Danthine S et al (2018) Effect of household cooking techniques on the microbiological load and the nutritional quality of mealworms (*Tenebrio molitor* L. 1758). Food Res Int 106:503–508
Castro-López C, Santiago-López L, Vallejo-Cordoba B, González-Córdova AF, Liceaga AM, García HS, Hernández-Mendoza A (2020) An insight to fermented edible insects: a global perspective and prospective. Food Res Int 137:109750
Cho SY, Ryu GH (2021) Effects of mealworm larva composition and selected process parameters on the physicochemical properties of extruded meat analog. Food Sci Nutr 9(8):4408–4419
Crippen TL, Sheffield C (2006) External surface disinfection of the lesser mealworm (Coleoptera: Tenebrionidae). J Med Entomol 43(5):916–923
da Silva Lucas AJ, de Oliveira LM, Da Rocha M, Prentice C (2020) Edible insects: an alternative of nutritional, functional, and bioactive compounds. Food Chem 311:126022
da Silva Lucas AJ, Oreste EQ, Costa HLG, López HM, Saad CDM, Prentice C (2021) Extraction, physicochemical characterization, and morphological properties of chitin and chitosan from cuticles of edible insects. Food Chem 343:128550
Dávalos Terán I, Imai K, Lacroix IM, Fogliano V, Udenigwe CC (2020) Bioinformatics of edible yellow mealworm (*Tenebrio molitor*) proteome reveal the cuticular proteins as promising precursors of dipeptidyl peptidase-IV inhibitors. J Food Biochem 44(2):e13121
David-Birman T, Romano A, Aga A, Pascoviche D, Davidovich-Pinhas M, Lesmes U (2022) Impact of silkworm pupae (Bombyx mori) powder on cream foaming, ice cream properties and palatability. Innovative Food Sci Emerg Technol 75:102874
de Gier S, Verhoeckx K (2018) Insect (food) allergy and allergens. Mol Immunol 100:82–106
De Paepe E, Wauters J, Van Der Borght M, Claes J, Huysman S, Croubels S, Vanhaecke L (2019) Ultra-high-performance liquid chromatography coupled to quadrupole orbitrap high-resolution

mass spectrometry for multi-residue screening of pesticides, (veterinary) drugs and mycotoxins in edible insects. Food Chem 293:187–196

De Smet J, Lenaerts S, Borremans A, Scholliers J, Van der Borght M, Van Campenhout L (2019) Stability assessment and laboratory scale fermentation of pastes produced on a pilot scale from mealworms (*Tenebrio molitor*). LWT Food Sci Technol 102:113–121

DeFoliart GR (1999) Insects as food: why the western attitude is important. Annu Rev Entomol 44(1):21–50

Devi WD, Bonysana R, Kapesa K, Mukherjee PK, Rajashekar Y (2023) Edible insects: as traditional medicine for human wellness. Future Foods 7:100219

Dickie F, Miyamoto M, Collins CMT (2019) The potential of insect farming to increase food security. In: Mikkola H (ed) Edible insects. IntechOpen, London, pp 1–10

Dobermann D, Swift JA, Field LM (2017) Opportunities and hurdles of edible insects for food and feed. Nutr Bull 42(4):293–308

Duda A, Adamczak J, Chełmińska P, Juszkiewicz J, Kowalczewski P (2019) Quality and nutritional/textural properties of durum wheat pasta enriched with cricket powder. Food Secur 8(2):46

Dutta P, Dey T, Manna P, Kalita J (2016) Antioxidant potential of Vespa affinis L., a traditional edible insect species of North East India. PLoS One 11(5):e0156107. https://doi.org/10.1371/journal.pone.0156107

Dutta P, Dey T, Dihingia A, Manna P, Kalita J (2017) Antioxidant and glucose metabolizing potential of edible insect, *Brachytrupes orientalis* via modulating Nrf2/AMPK/GLUT4 signaling pathway. Biomed Pharmacother 95:556–563

Eckl MR, Biesbroek S, Van't Veer P, Geleijnse JM (2021) Replacement of meat with non-meat protein sources: a review of the drivers and inhibitors in developed countries. Nutrients 13(10):3602

Feng Y, Chen XM, Zhao M, He Z, Sun L, Wang CY et al (2018) Edible insects in China: utilization and prospects. Insect Sci 25:184–198

Fernandez-Cassi X, Supeanu A, Vaga M, Jansson A, Bogyist S, Vagsholm I (2019) The house cricket (*Acheta domesticus*) as a novel food: a risk profile. J Insects Food Feed 5:137–157. https://doi.org/10.3920/Jiff2018.0021

Gałęcki R, Sokół R (2019) A parasitological evaluation of edible insects and their role in the transmission of parasitic diseases to humans and animals. PLoS One 14(7):e0219303. https://doi.org/10.1371/journal.pone.0219303

Garofalo C, Milanović V, Cardinali F, Aquilanti L, Clementi F, Osimani A (2019) Current knowledge on the microbiota of edible insects intended for human consumption: a state-of-the-art review. Food Res Int 125:108527

Ghosh S, Lee SM, Jung C, Meyer-Rochow VB (2017) Nutritional composition of five commercial edible insects in South Korea. J Asia Pac Entomol 20(2):686–694

Grabowski NT, Klein G (2017) Microbiological analysis of raw edible insects. J Insects Food Feed 3(1):7–14

Grabowski NT, Tchibozo S, Abdulmawjood A, Acheuk F, M'Saad Guerfali M, Sayed WA, Plötz M (2020) Edible insects in Africa in terms of food, wildlife resource, and pest management legislation. Food Secur 9(4):502

Gravel A, Doyen A (2020) The use of edible insect proteins in food: challenges and issues related to their functional properties. Innovative Food Sci Emerg Technol 59:102272

Hadj Saadoun J, Sogari G, Bernini Vet al. (2022) A critical review of intrinsic and extrinsic antimicrobial properties of insects. Trends Food Sci Technol 122:40–48

Hall F, Johnson PE, Liceaga A (2018) Effect of enzymatic hydrolysis on bioactive properties and allergenicity of cricket (*Gryllodes sigillatus*) protein. Food Chem 262:39–47

Hernández-Álvarez AJ, Mondor M, Piña-Domínguez IA, Sánchez-Velázquez OA, Melgar Lalanne G (2021) Drying technologies for edible insects and their derived ingredients. Dry Technol 39(13):1991–2009

Houbraken M, Spranghers T, De Clercq P, Cooreman-Algoed M, Couchement T, De Clercq G et al (2016) Pesticide contamination of *Tenebrio molitor* (Coleoptera: Tenebrionidae) for human consumption. Food Chem 201:264–269

Ibidhi R, Salem HB (2020) Water footprint of livestock products and production systems: a review. Anim Prod Sci 60(11):1369–1380

Imathiu S (2020) Benefits and food safety concerns associated with consumption of edible insects. NFS J 18:1–11

Inacio AC, Vågsholm I, Jansson A, Vaga M, Boqvist S, Fraqueza MJ (2021) Impact of starvation on fat content and microbial load in edible crickets (*Acheta domesticus*). J Insects Food Feed 7(7):1143–1147

IPIFF, EU Novel Food Legislation (2023) The international platform of insects for food and feed. http://ipiff.org/insects-novel-food-eu-legislation/. Accessed 12 Dec 2023

Jan AT, Azam M, Siddiqui K, Ali A, Choi I, Haq QMR (2015) Heavy metals and human health: mechanistic insight into toxicity and counter defense system of antioxidants. Int J Mol Sci 16(12):29592–29630

Jongema Y (2017) List of edible insects of the world (April 1, 2017)—WUR. https://www.wur.nl/en/research-results/chair-groups/plant-sciences/laboratory-of-entomology/edible-insects/worldwide-species-list.htm. Accessed 14 July 2023

Kamau E, Mutungi C, Kinyuru J, Imathiu S, Tanga C, Affognon H, Ekesi S, Nakimbugwe D, Fiaboe KKM (2018) Moisture adsorption properties and shelf-life estimation of dried and pulverised edible house cricket *Acheta domesticus* (L.) and black soldier fly larvae *Hermetia illucens* (L.). Food Res Int 106:420–427. https://doi.org/10.1016/j.foodres.2018.01.012

Kang TW, Choi RY, Kim IW, Lee EJ, Kim HW, Park HJ (2023) Evaluation of feasibility of *Tenebrio molitor* larval fractions as a meat analog using 3D printing. Innovative Food Sci Emerg Technol 89:103446

Kasza G, Izsó T, Szakos D, Nugraha WS, Tamimi MH, Süth M (2023) Insects as food-Changes in consumers' acceptance of entomophagy in Hungary between 2016 and 2021. Appetite 188:106770

Kim SK, Weaver CM, Choi MK (2017) Proximate composition and mineral content of five edible insects consumed in Korea. CyTA-J Food 15(1):143–146

Kim TK, Yong HI, Jung S, Sung JM, Jang HW, Choi YS (2021) Physicochemical and textural properties of emulsions prepared from the larvae of the edible insects *Tenebrio molitor, Allomyrina dichotoma*, and *Protaetia brevitarsis seulensis*. J Anim Sci Technol 63(2):417

Kim TK, Yong HI, Cha JY, Park SY, Jung S, Choi YS (2022) Drying-induced restructured jerky analog developed using a combination of edible insect protein and textured vegetable protein. Food Chem 373:131519

Kim MH, Kim SJ, Kim SH, Park WJ, Han JS (2023) *Gryllus bimaculatus*-containing diets protect against dexamethasone-induced muscle atrophy, but not high-fat diet-induced obesity. Food Sci Nutr 4(11):2787–2797

Köhler R, Kariuki L, Lambert C, Biesalski HK (2019) Protein, amino acid and mineral composition of some edible insects from Thailand. J Asia Pac Entomol 22(1):372–378

Kolakowski BM, Johaniuk K, Zhang H, Yamamoto E (2021) Analysis of microbiological and chemical hazards in edible insects available to Canadian consumers. J Food Prot 84(9):1575–1581

Kouřimská L, Adámková A (2016) Nutritional and sensory quality of edible insects. NFS journal 4:22–26. https://doi.org/10.1016/j.nfs.2016.07.001

Kowalski S, Oracz J, Skotnicka M, Mikulec A, Gumul D, Mickowska B et al (2022) Chemical composition, nutritional value, and acceptance of nut bars with the addition of edible insect powder. Molecules 27(23):8472. https://doi.org/10.3390/molecules27238472

Kunatsa Y, Chidewe C, Zvidzai CJ (2020) Phytochemical and anti-nutrient composite from selected marginalized Zimbabwean edible insects and vegetables. J Agric Food Res 2:100027

Kuntadi K, Adalina Y, Maharani KE (2018) Nutritional compositions of six edible insects in Java. Indones J For Res 5(1):57–68

Lange KW, Nakamura Y (2021) Edible insects as future food: chances and challenges. J Future Foods 1(1):38–46

Laroche M, Perreault V, Marciniak A, Gravel A, Chamberland J, Doyen A (2019) Comparison of conventional and sustainable lipid extraction methods for the production of oil and protein isolate from edible insect meal. Food Secur 8(11):572

Larouche J, Deschamps MH, Saucier L, Lebeuf Y, Doyen A, Vandenberg GW (2019) Effects of killing methods on lipid oxidation, colour and microbial load of black soldier fly (*Hermetia illucens*) larvae. Animals 9(4):182

Lee W, Lee H, Kim MA, Choi J, Kim KM, Hwang JS et al (2017) Evaluation of novel factor Xa inhibitors from *Oxya chinensis sinuosa* with anti-platelet aggregation activity. Sci Rep 7(1):7934

Lee JH, Kim YJ, Choi YJ, Kim TK, Cha JY, Park MK, Jung S, Choi YS (2023) Effect of gamma-ray and electron-beam irradiation on the structural changes and functional properties of edible insect proteins from *Protaetia brevitarsis* larvae. Food Chem 434:137463

Leni G, Caligiani A, Sforza S (2019) Killing method affects the browning and the quality of the protein fraction of Black Soldier Fly (*Hermetia illucens*) prepupae: a metabolomics and proteomic insight. Food Res Int 115:116–125

Li H, Chumroenphat T, Boonarsa P, Yahuafai J, Wrigley C, Siriamornpun S (2023) Evaluation of roasting and grilling effects on chemical composition, volatile profiles, and toxicity of stink bugs (*Tessaratoma papillosa*): implications for utilization as functional food ingredients. Food Secur 12(16):3053

Liceaga AM (2021) Processing insects for use in the food and feed industry. Curr Opin Insect Sci 48:32–36

Looy H, Dunkel FV, Wood JR (2014) How then shall we eat? Insect-eating attitudes and sustainable foodways. Agriculture and human values 31:131–141. https://doi.org/10.1007/s10460-013-9450-x

Lopez-Santamarina A, Mondragon ADC, Lamas A et al (2020) Animal-origin prebiotics based on chitin: an alternative for the future? A critical review. Foods 9(6):782

Lu K, He Y, Wu C, Bao J (2022) Moderate hyperglycemia-preventive effect and mechanism of action of *Periplaneta americana* oligosaccharides in streptozotocin-induced diabetic mice. Nutrients 14(21):4620

Lucas-González R, Fernández-López J, Pérez-Álvarez JA, Viuda-Martos M (2019) Effect of drying processes in the chemical, physico-chemical, techno-functional and antioxidant properties of flours obtained from house cricket (*Acheta domesticus*). Eur Food Res Technol 245(7):1451–1458

Mancini S, Moruzzo R, Riccioli F, Paci G (2019) European consumers' readiness to adopt insects as food. A review. Food Res Int 122:661–678

Mariod AA, Fadul H (2015) Extraction and characterization of gelatin from two edible Sudanese insects and its applications in ice cream making. Food Sci Technol Int 21(5):380–391

Marshall DL, Dickson JS, Nguyen NH (2016) Ensuring food safety in insect-based foods: mitigating microbiological and other foodborne hazards. In: Insects as sustainable food ingredients. Academic Press, United Kingdom, pp 223–253

Martins Borges M, Vicente da Costa D, Machado Trombete F, Ferreira Ignácio Câmara AK (2022) Edible insects as a sustainable alternative to food products: an insight into quality aspects of reformulated bakery and meat products. Curr Opin Food Sci 46:100864

Miglietta PP, De Leo F, Ruberti M, Massari S (2015) Mealworms for food: a water footprint perspective. Water 7(11):6190–6203

Mishyna M, Chen J, Benjamin O (2020) Sensory attributes of edible insects and insect-based foods–Future outlooks for enhancing consumer appeal. Trends Food Sci Technol 95:141–148. https://doi.org/10.1016/j.tifs.2019.11.016

Moorthy Eswaran UG, Karunanithi S, Gupta RK et al (2023) Edible insects as emerging food products–processing and product development perspective. J Food Sci Technol 60:2105–2120

Mordor Intelligence (2023). https://www.mordorintelligence.com/industry-reports/edible-insects-market. Accessed 28 Sept 2023

Moruzzo R, Mancini S, Guidi A (2021a) Edible insects and sustainable development goals. Insects 12(6):557

Moruzzo R, Riccioli F, Espinosa Diaz S, Secci C, Poli G, Mancini S (2021b) Mealworm (*Tenebrio molitor*): potential and challenges to promote circular economy. Animals 11(9):2568

Murefu TR, Macheka L, Musundire R, Manditsera FA (2019) Safety of wild harvested and reared edible insects: a review. Food Control 101:209–224
Mwelwa S, Chungu D, Tailoka F, Beesigamukama D, Tanga C (2023) Biotransfer of heavy metals along the soil-plant-edible insect-human food chain in Africa. Sci Total Environ 881:163150
Nam HK, Kang TW, Kim IW, Choi RY, Kim HW, Park HJ (2023) Physicochemical properties of cricket (*Gryllus bimaculatus*) gel fraction with soy protein isolate for 3D printing-based meat analogue. Food Biosci 53:102772
Nongonierma AB, Lamoureux C, FitzGerald RJ (2018) Generation of dipeptidyl peptidase IV (DPP-IV) inhibitory peptides during the enzymatic hydrolysis of tropical banded cricket (*Gryllodes sigillatus*) proteins. Food Funct 9:407–416
Nyangena DN, Mutungi C, Imathiu S, Kinyuru J, Affognon H, Ekesi S et al (2020) Effects of traditional processing techniques on the nutritional and microbiological quality of four edible insect species used for food and feed in East Africa. Food Secur 9(5):574
Ochiai M, Tezuka K, Yoshida H, Akazawa T, Komiya Y, Ogasawara H et al (2022) Edible insect *Locusta migratoria* shows intestinal protein digestibility and improves plasma and hepatic lipid metabolism in male rats. Food Chem 396:133701
Ojha S, Bekhit AED, Grune T, Schlüter OK (2021) Bioavailability of nutrients from edible insects. Curr Opin Food Sci 41:240–248
Omotayo OP, Omotayo AO, Mwanza M, Babalola OO (2019) Prevalence of mycotoxins and their consequences on human health. Toxicol Res 35:1–7
Omotoso OT (2015) An evaluation of the nutrients and some anti-nutrients in silkworm, *Bombyx mori* L. (Bombycidae: lepidoptera). Jordan J Biol Sci 8(1):45–50
Oonincx DG, de Boer IJM (2012) Environmental impact of the production of mealworms as a protein source for humans - a life cycle assessment. PLoS One 7(12):e51145
Oonincx DGAB, Finke MD (2021) Nutritional value of insects and ways to manipulate their composition. J Insects Food Feed 7(5):639–659
Oonincx DG, Van Itterbeeck J, Heetkamp MJ, Van Den Brand H, Van Loon JJ, Van Huis A (2010) An exploration on greenhouse gas and ammonia production by insect species suitable for animal or human consumption. PLoS One 5(12):e14445
Oonincx DGAB, Van Keulen P, Finke MD, Baines FM, Vermeulen M, Bosch G (2018) Evidence of vitamin D synthesis in insects exposed to UVb light. Sci Rep 8(1):10807
Oonincx DG, van Broekhoven S, van Huis A, van Loon JJ (2019) Correction: feed conversion, survival and development, and composition of four insect species on diets composed of food by-products. PLoS One 14(10):e0222043
Ordoñez-López MF, Ghnimi S, Liu C (2023) Willingness to consume insect-based food in France: determinants and consumer perspectives. LWT - Food Sci Technol 185:115179
Orkusz A (2021) Edible insects versus meat—Nutritional comparison: knowledge of their composition is the key to good health. Nutrients 13(4):1207
Orsi L, Voege LL, Stranieri S (2019) Eating edible insects as sustainable food? Exploring the determinants of consumer acceptance in Germany. Food Res Int 125:108573
Osimani A, Aquilanti L (2021) Spore-forming bacteria in insect-based foods. Curr Opin Food Sci 37:112–117
Osimani A, Milanović V, Garofalo C, Cardinali F, Roncolini A, Sabbatini R et al (2018) Revealing the microbiota of marketed edible insects through PCR-DGGE, metagenomic sequencing and real-time PCR. Int J Food Microbiol 276:54–62
Pasini G, Cullere M, Vegro M, Simonato B, Dalle Zotte A (2022) Potentiality of protein fractions from the house cricket (*Acheta domesticus*) and yellow mealworm (*Tenebrio molitor*) for pasta formulation. LWT - Food Sci Technol 164:113638
Patrignani F, Del Duca S, Vannini L, Rosa M, Schlüter O, Lanciotti R (2020) Potential of *Yarrowia lipolytica* and *Debaryomyces hansenii* strains to produce high quality food ingredients based on cricket powder. LWT - Food Sci Technol 119:108866
Paulin IG, Purwanto MGM (2020) Nutritional characteristics of teak grasshopper (*Valanga nigricornis*Burmeister), cricket (*Brachytrupes portentosus* L.), and mealworm (*Tenebrio molitor*) as alternative food sources in Indonesia. Indones J Biotechnol Biodiv 4(1):52–61

Pérez-Rodríguez E, Ibarra-Herrera CC, Pérez-Carrillo E (2023) Effect of incorporation of solid-state fermented edible insects *Tenebrio molitor* and *Sphenarium purpurascens* with *Aspergillus oryzae* in the elaboration of bread. LWT - Food Sci Technol 184:115003

Poma G, Cuykx M, Amato E, Calaprice C, Focant JF, Covaci A (2017) Evaluation of hazardous chemicals in edible insects and insect-based food intended for human consumption. Food Chem Toxicol 100:70–79. https://doi.org/10.1016/j.fct.2016.12.006

Poma G, Cuykx M, Da Silva KM, Iturrospe E, van Nuijs AL, van Huis A, Covaci A (2022) Edible insects in the metabolomics era. First steps towards the implementation of entometabolomics in food systems. Trends Food Sci Technol 119:371–377

Purschke B, Brüggen H, Scheibelberger R, Jäger H (2018) Effect of pre-treatment and drying method on physico-chemical properties and dry fractionation behaviour of mealworm larvae (Tenebrio molitor L.). Eur Food Res Technol 244:269–280. https://doi.org/10.1007/s00217-017-2953-8

Puteri B, Jahnke B, Zander K (2023) Booming the bugs: how can marketing help increase consumer acceptance of insect-based food in western countries? Appetite 187:106594

Pyo SJ, Kang DG, Jung C, Sohn HY (2020) Anti-thrombotic, anti-oxidant and haemolysis activities of six edible insect species. Food Secur 9(4):401

Rahman MT, Sobur MA, Islam MS, Ievy S, Hossain MJ, El Zowalaty ME et al (2020) Zoonotic diseases: etiology, impact, and control. Microorganisms 8(9):1405

Ribeiro JC, Gonçalves ATS, Moura AP, Varela P, Cunha LM (2022) Insects as food and feed in Portugal and Norway–cross-cultural comparison of determinants of acceptance. Food Qual Prefer 102:104650

Ritter KS (1990) Cholesterol and insects. Food Insects Newsletter 3(1)

Romano N, Sinha A, Powell A, Fischer H (2023) Mineral composition in black soldier fly (*Hermetia illucens*) larvae and resulting frass from fruit and their peels. J Insects Food Feed 9(1):43–53

Rumpold BA, Schlüter OK (2013a) Nutritional composition and safety aspects of edible insects. Mol Nutr Food Res 57(5):802–823

Rumpold BA, Schlüter OK (2013b) Potential and challenges of insects as an innovative source for food and feed production. Innovative Food Sci Emerg Technol 17:1–11

Sales L, Nogueira Silva NF, Jessen F, Mohammadifar MA, Stephani R, Fernandes de Carvalho A, Tuler Í, Casanova F (2023) Edible insect as an alternative protein source: a review on the chemistry and functionalities of proteins under different processing methods. Heliyon 9(4):e14831

Sarkar P, Pecorelli A, Woodby B, Pambianchi E, Ferrara F, Duary RK, Valacchi G (2023) Evaluation of anti-oxinflammatory and ACE-inhibitory properties of protein hydrolysates obtained from edible non-mulberry silkworm pupae (*Antheraea assama* and *Philosomia ricinii*). Nutrients 15(4):1035

Schlüter O, Rumpold B, Holzhauser T, Roth A, Vogel RF, Quasigroch W et al (2017) Safety aspects of the production of foods and food ingredients from insects. Mol Nutr Food Res 61:1600520. https://doi.org/10.1002/mnfr.201600520

Shelomi M (2015) Why we still don't eat insects: assessing entomophagy promotion through a diffusion of innovations framework. Trends Food Sci Technol 45(2):311–318

Shi YN, Tu ZC, Wang XL, Yan YM, Fang P, Zuo ZL et al (2014) Bioactive compounds from the insect *Aspongopus chinensis*. Bioorg Med Chem Lett 24(22):5164–5169

Siulapwa N, Mwambungu A, Lungu E, Sichilima W (2014) Nutritional value of four common edible insects in Zambia. Int J Sci Res 3:876–884

Smetana S, Larki NA, Pernutz C, Franke K, Bindrich U, Toepfl S, Heinz V (2018) Structure design of insect-based meat analogs with high-moisture extrusion. J Food Eng 229:83–85

Soares Araújo RR, dos Santos Benfica TAR, Ferraz VP, Santos EM (2019) Nutritional composition of insects *Gryllus assimilis* and *Zophobas morio*: potential foods harvested in Brazil. J Food Compos Anal 76:22–26

Sorjonen JM, Valtonen A, Hirvisalo E, Karhapää M, Lehtovaara VJ, Lindgren J et al (2019) The plant-based by-product diets for the mass-rearing of *Acheta domesticus* and *Gryllus bimaculatus*. PLoS One 14(6):e0218830

Sosa DAT, Fogliano V (2017) Potencial of insect-derived ingredients for food applications. In: Schields VDC (ed) Insect physiology and ecology. InTech, London, pp 215–231

Srinroch C, Srisomsap C, Chokchaichamnankit D, Punyarit P, Phiriyangkul P (2015) Identification of novel allergen in edible insect, *Gryllus bimaculatus* and its cross-reactivity with Macrobrachium spp. allergens. Food Chem 184:160–166

Ssepuuya G, Aringo RO, Mukisa IM, Nakimbugwe D (2016) Effect of processing, packaging and storage-temperature based hurdles on the shelf stability of sautéed ready-to-eat *Ruspolia nitidula*. J Insects Food Feed 2(4):245–253

Stull VJ (2021) Impacts of insect consumption on human health. J Insects Food Feed 7:695–713

Takeo N, Nakamura M, Nakayama S, Okamoto O, Sugimoto N, Sugiura S, Matsunaga K (2018) Cochineal dye-induced immediate allergy: review of Japanese cases and proposed new diagnostic chart. Allergol Int 67(4):496–505

Tang C, Yang D, Liao H, Sun H, Liu C, Wei L, Li F (2019) Edible insects as a food source: a review. Food Prod Process Nutr 1:8

Tao J (2016) Potential utilization of edible insects in extruded rice products to address malnutrition issues in developing countries. Master's thesis, California State Polytechnic University

Tello A, Aganovic K, Parniakov O, Carter A, Heinz V, Smetana S (2021) Product development and environmental impact of an insect-based milk alternative. Future Foods 4:100080

Udomsil N, Imsoonthornruksa S, Gosalawit C, Ketudat-Cairns M (2019) Nutritional values and functional properties of house cricket (*Acheta domesticus*) and field cricket (*Gryllus bimaculatus*). Food Sci Technol Res 25(4):597–605

Van Huis A (2022) Edible insects: challenges and prospects. Entomol Res 52(4):161–177

Van Huis A, Van Itterbeeck J, Klunder H, Mertens E, Halloran A, Muir G, Vantomme P (2013) Edible insects: future prospects for food and feed security (No. 171). Food and agriculture organization of the United Nations

Van Huis A, Rumpold B, Maya C et al (2021) Nutritional qualities and enhancement of edible insects. Annu Rev Nutr 41:551–576

Van Huis A, Halloran A, Van Itterbeeck J, Klunder H, Vantomme P (2022) How many people on our planet eat insects: 2 billion? J Insects Food Feed 8(1):1–4

Van Thielen L, Vermuyten S, Storms B, Rumpold B, Van Campenhout L (2019) Consumer acceptance of foods containing edible insects in Belgium two years after their introduction to the market. J Insects Food Feed 5(1):35–44

Vanqa N, Mshayisa VV, Basitere M (2022) Proximate, physicochemical, techno-functional and antioxidant properties of three edible insect (*Gonimbrasia belina, Hermetia illucens* and *Macrotermes subhylanus*) flours. Food Secur 11(7):976

Villaseñor VM, Enriquez-Vara JN, Urías-Silva JE, del Carmen Lugo-Cervantes E, Luna-Vital DA, Mojica L (2022) Mexican grasshopper (*Sphenarium purpurascens*) as source of high protein flour: techno-functional characterization, and *in silico* and *in vitro* biological potential. Food Res Int 162:112048

Wynants E, Crauwels S, Lievens B, Luca S, Claes J, Borremans A et al (2017) Effect of post-harvest starvation and rinsing on the microbial numbers and the bacterial community composition of mealworm larvae (Tenebrio molitor). Innov Food Sci Emerg Technol 42:8–15. https://doi.org/10.1016/j.ifset.2017.06.004

Xia Z, Chen J, Wu S (2013) Hypolipidemic activity of the chitooligosaccharides from Clanis bilineata (Lepidoptera), an edible insect. Int J Biol Macromol 59:96–98

Yang R, Zhao X, Kuang Z, Ye M, Luo G, Xiao G, Xiong Z (2013) Optimization of antioxidant peptide production in the hydrolysis of silkworm (*Bombyx mori* L.) pupa protein using response surface methodology. J Food Agric Environ 11:952–956

Yazici GN, Ozer MS (2021) Using edible insects in the production of cookies, biscuits, and crackers: a review. Biol Life Sci Forum 6(1):80

Yeerong K, Sriyab S, Somwongin S, Punyoyai C, Chantawannakul P, Anuchapreeda S et al (2021) Skin irritation and potential antioxidant, anti-collagenase, and anti-elastase activities of edible insect extracts. Sci Rep 11(1):22954

Yen AL (2015a) Insects as food and feed in the Asia Pacific region: current perspectives and future directions. J Insects Food Feed 1:33–55

Yen AL (2015b) Conservation of Lepidoptera used as human food and medicine. Curr Opin Insect Sci 12:102–108

Yisa NK, Osuga IM, Subramanian S, Ekesi S, Emmambux MN, Duodu KG (2023) Effect of drying methods on the nutrient content, protein, and lipid quality of edible insects from East Africa. J Insects Food Feed 9(5):647–659

Zhang Z, Chen S, Wei X, Xiao J, Huang D (2022) Characterization, antioxidant activities, and pancreatic lipase inhibitory effect of extract from the edible insect *Polyrhachis vicina*. Front Nutr 9:860174

Zielińska E, Baraniak B, Karaś M, Rybczyńska K, Jakubczyk A (2015) Selected species of edible insects as a source of nutrient composition. Food Res Int 77:460–466

Zielińska E, Baraniak B, Karaś M (2017) Antioxidant and anti-inflammatory activities of hydrolysates and peptide fractions obtained by enzymatic hydrolysis of selected heat-treated edible insects. Nutrients 9(9):970. https://doi.org/10.3390/nu9090970

Żuk-Gołaszewska K, Gałęcki R, Obremski K, Smetana S, Figiel S, Gołaszewski J (2022) Edible Insect Farming in the Context of the EU Regulations and Marketing—An Overview Insects 13(5):446. https://doi.org/10.3390/insects13050446

# Chapter 10
# Hybrid Meat Products: Using Plant, Fungi, and Insect Sources for Flexitarian Diets—Technological, Nutritional and Sensory Insights

**Ana Karoline Ferreira Ignácio Câmara, Maristela Midori Ozaki, Camila de Souza Paglarini, Mirian dos Santos, and Marise Aparecida Rodrigues Pollonio**

## 10.1 Introduction

Meat is widely consumed globally due to its flavor, aroma, texture, and nutritional value. It is a high-quality protein source, with significant content of vitamin B12, and highly bioavailable iron, zinc, and phosphorus (Bohrer 2017; Olmedilla-Alonso et al. 2013). Furthermore, meat proteins have notable functional properties, such as emulsification, gelation, and water holding capacity (WHC) providing structure and cohesion to meat products (Yang et al. 2016), making their replacement a notable challenge for the industry and researchers.

On the other hand, meat products contain specific components in their composition (biogenic amines, polycyclic aromatic hydrocarbons, and N-nitroso compounds) that have been frequently correlated with a higher risk for some types of cancer (Huang et al. 2021). Additionally, livestock contributes significantly to climate change (Grossi et al. 2019). At the same time, in the current scenario, an exceptional rise in the worldwide demand for high-quality proteins is expected in the next 20 years, primarily due to population expansion, increasing earnings, urbanization of developing countries, and the growing awareness of society about the binomial diet and health, with a more significant requirement for more nutritious and protein-rich diets (Anzani et al. 2020; Milford et al. 2019).

---

A. K. F. I. Câmara (✉)
Department of Food Engineering, Federal University of São João Del Rei, Sete Lagoas, Minas Gerais, Brazil
e-mail: ana.karoline@ufsj.edu.br

M. M. Ozaki · M. dos Santos · M. A. R. Pollonio
School of Food Engineering, State University of Campinas, Campinas, Brazil

C. de Souza Paglarini
Luiz de Queiroz College of Agriculture, University of São Paulo, Piracicaba, Brazil



Ö. P. CAN et al. (eds.), *Food Analogues: Emerging Methods and Challenges*,
https://doi.org/10.1007/978-3-031-69872-9_10

Given that individuals are more prone to decrease rather than eliminate their meat consumption, flexitarian (reduced meat) diets caught the attention of several people. Many consumers view the flexitarian diet as a feasible and desirable option to diminish their carbon footprint and prioritize their health (Corrin and Papadopoulos 2017; Kemper and White 2021). Introducing alternative protein sources and enhancing the efficiency of meat production stand as feasible approaches to fulfilling this demand in an environmentally sustainable manner (Henchion et al. 2017). Meat consumption involves many social relationships and positive beliefs, so a transition to veganism is very challenging (Grasso and Jaworska 2020; Spencer et al. 2018). However, adopting a flexitarian or semi-vegetarian diet, characterized primarily by other sources of proteins such as mushrooms or vegetable proteins with reduced meat consumption, is less restrictive and could be more easily implemented.

With the increase of this flexitarian public the development of meat substitutes can be categorized as hybrid meat products, blending animal protein with others protein sources such as vegetable, edible insects, and fungi. These products could satisfactorily serve consumers who wish to reduce their meat consumption without giving up the taste, familiarity, and convenience attributes of traditional meat products (Grasso et al. 2022).

Hybrid meat products processes can be adapted from existing industrial ones and bring a series of advantages from a nutritional and environmental point of view, along with sensory and stability challenges. Figure 10.1 outlines and summarizes the main characteristics of alternative protein sources that can be introduced into the formulations of these hybrids.

A relevant aspect pointed out by Baune et al. (2021) is that in producing hybrids, the health aspects must be achieved to select the raw materials must be clearly defined. Furthermore, a higher degree of raw material processing must be considered as it can provide greater sensory acceptance but also greater environmental impact. Therefore, this approach requires a more careful evaluation of a wide range of raw materials and additional analysis to find a balance. In addition to a careful assessment of the raw materials to be used, introducing other protein sources to the meat matrix requires an in-depth investigation into the stability of these products. The interactions between meat proteins and alternative protein sources require further elucidation, and studies are still incipient. Choices of process conditions (ionic strength, temperature, time, pH, heating) and an assertive experimental design regarding protein concentrations/proportions are crucial to optimize the resulting hybrid meat products and ensure adequate processing and quality characteristics (Grasso and Goksen 2023).

Another point of discussion concerns the nutritional quality of hybrid meat products. An important question that must be asked is whether, with the reduction of meat protein, the products will have an adequate nutritional profile since the composition of essential amino acids (AA) can vary according to the type of protein that will partially replace the meat protein, considered complete. The essential AA composition of the main protein sources, traditional and alternative are shown in Table 10.1.

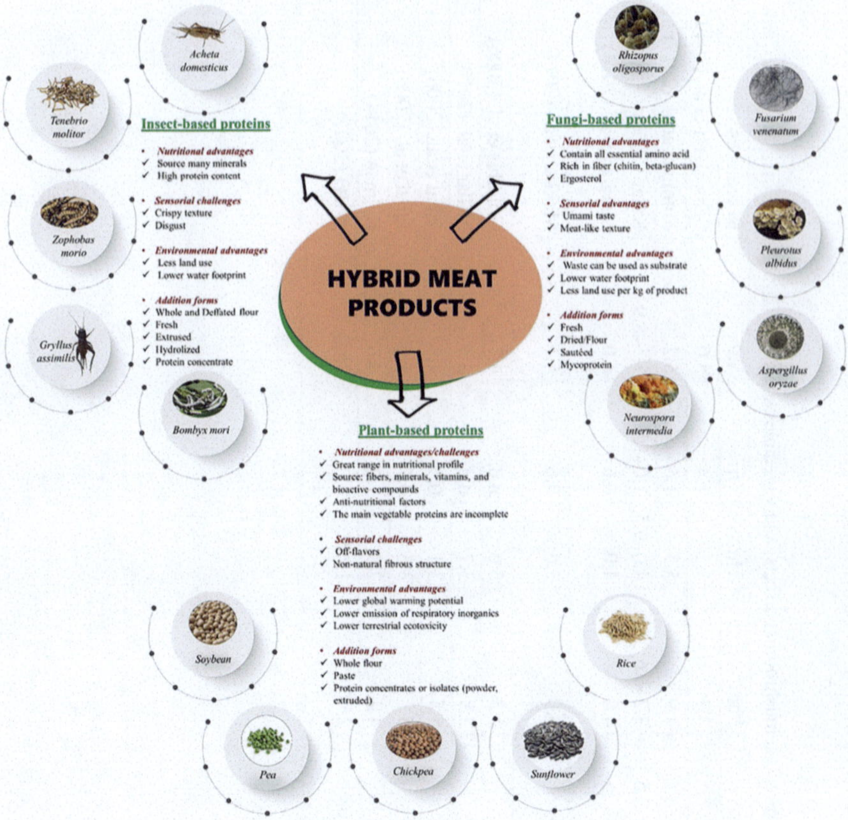


**Fig. 10.1** Main characteristics of alternative protein sources for hybrid meat products

According to Table 10.1, leucine is the most abundant AA in mushrooms compared to other proteins, even compared to meat protein. On the other hand, tryptophan stood out as a limiting AA when compared to all the proteins described in the Table 10.1, with its minimal presence in vegetable proteins. Insect proteins, mycoproteins, and mushrooms presented lower amounts of methionine than meat protein. It is crucial to highlight that hybrid meat products can have an attractive nutritional advantage considering, for example, a product with 50% meat and 50% another protein, the AA profiles can complement each other and result in a product with quantities significantly adjusted to the daily requirements recommended by health institutions.

Given the context presented, this review provides a broad and critical view of alternative protein sources with the main strategies for reformulating hybrid meat products and their effects on techno-functional, nutritional, and sensory properties. The focus was also to discuss how vegetable, fungi, and insect proteins interact with meat proteins, addressing the most recent studies and how reformulations with these proteins can impact the quality of hybrid meat products.

**Table 10.1** Essential AA content (g/100 g protein) of pork, mycoproteins, mushrooms, insects, and vegetables sources

| Source | Ile | Leu | Lys | Met | Phe | Thr | Trp | Val | References |
|---|---|---|---|---|---|---|---|---|---|
| Pork loin[a] | 4.8 | 7.8 | 8.1 | 2.5 | 3.7 | 5.2 | 1.4 | 5.0 | Schweigert et al. (1949) |
| Mycoproteins | 5.2 | 8.6 | 8.3 | 2.1 | 4.9 | 5.5 | 1.6 | 6.2 | Finnigan et al. (2017) |
| Mushrooms | 3.2–5.3 | 7.0–16.3 | 4.2–8.6 | 1.0–3.5 | 3.2–6.6 | 4.0–9.5 | 1.0–1.7 | 3.5–8.9 | Kalač (2009) |
| *Tenebrio molitor* larvae | 4.6 | 8.6 | 5.4 | 1.5 | 4.0 | 4.0 | 0.6 | 6.0 | Makkar et al. (2014) |
| *Acheta domesticus* adult | 4.4 | 9.8 | 5.4 | 1.4 | 3.0 | 3.6 | 0.6 | 5.1 | Makkar et al. (2014) |
| Soy | 4.4 | 7.8 | 6.4 | 1.1 | 5.3 | 3.6 | n.d. | 4.7 | Goldflus et al. (2006) |
| Yellow pea | 3.5 | 7 | 6.3 | 1.1 | 3.8 | 3.7 | n.d. | 3.9 | Millar et al. (2019) |
| Chickpea | 4.8 | 8.7 | 7.2 | 1.1 | 5.5 | 3.1 | 0.9 | 4.6 | Iqbal et al. (2006) |
| Lentil | 4.1 | 7.8 | 7 | 0.8 | 5 | 3.5 | 0.7 | 5.0 | Iqbal et al. (2006) |
| Rice | 3.5 | 6.4 | 2.4 | 2.3 | 4.4 | 2.9 | 1.2 | 4.6 | Kalman (2014) |
| Daily requirements (mg/kg) | 20.0 | 39.0 | 30.0 | 10.0 | 25.0[b] | 15.0 | 4.0 | 26.0 | WHO (2007) |

*n.d.* non determined
[a]Longissimus dorsi
[b]Total daily recommendation for phenylalanine + tyrosine

## 10.2 Hybrid Meat Products with Vegetable Protein Sources: Technological, Sensory, and Nutritional Properties

The consumption of processed products based on vegetable protein (e.g., tofu and Tempeh) is practiced since ancient civilizations in countries such as China and India. During the 1960s, the concept of plant-based meat became evident with the use of textured vegetable protein, mainly soy-based, to produce plant-based meat products, such as bacon and patties. With the increase in the number of vegetarians, particularly in developed countries, others plant proteins have been used to replace animal-based proteins producing analogues or hybrid meat products with different vegetable proteins (peas, wheat, rice, corn, lentils, soy, etc.) (He et al. 2020).

As well as meat products analogues, some vegetable proteins can also be used in hybrid meat products. Considering that the appeals to develop a hybrid meat product is different from meat products analogues the idea of hybrid meat products is to incorporate favorable associations such as well-being, reduce environmental impact, and the generally positive outcomes linked to reducing meat consumption.

Plant-based hybrid meat products can be developed using a diverse array of more sustainable protein sources. These include, but are not limited to, flours or concentrated proteins derived from legumes, oilseeds, and cereals. These ingredients can be utilized either singularly or in various combinations, offering a range of reformulation possibilities, which can affect the functional properties of the reformulated products.

The aim of this section is to present findings from studies involving different types of vegetable proteins as potential meat replacers, even though hybrid products are little explored yet, and to provide information into the technological, sensory, and nutritional attributes of these emerging products.

## 10.3 Proteins Sources for Plant-Based Hybrid Meat Products

Legumes are one of the most promising for meat substitution (due to the relatively high protein content in most legume seeds); they are second only to cereals in providing food crops for world agriculture (Foyer et al. 2016; Petrat-Melin and Dam 2023). Soy protein, a well-established non-meat ingredient within the meat industry, is widely recognized for its functional attributes, including water-binding and fat-binding capabilities. It has been utilized to enhance emulsion stability and yield, offering a cost-effective alternative to lean meat (Chin et al. 2000).

Pulses possess functional attributes, including solubility, WHC, fat absorption, solubility, emulsifying activity, and gel forming, enable them act as binders, fillers, and functional enhancers in hybrid meat products, contributing for transformation into innovative products such as meat analogues and hybrid meat products (Bessada et al. 2019; Havemeier et al. 2017; Kyriakopoulou et al. 2021). Some examples of pulses are peas, several types of beans, chickpea, and lentil.

Cereals are another source of macro components, mainly carbohydrates, and they have been considered as the backbone of agriculture and the most prolific food source (Serna-Saldivar 2010). Rice is one of the main cereals produced in the world; it is mainly composed of starch and minor amounts of protein (6–7%), lipid (0.3–0.5%), fiber (0.2–0.5%), minerals, vitamins, and phenolic compounds, considering the rice endosperm. Rice protein can be obtained from broken grain and rice bran, byproducts obtained from the processing of milled rice (De Souza et al. 2016). Oat is another example of cereal with high nutritional value, mainly due to protein (15–20%) and dietary fiber contents; the fractionation of β-glucan from oat grains provide concentrated protein fractions as a side stream, offering a new good-quality protein source (Mäkinen et al. 2016).

Oilseed proteins, in turn, are particularly attractive due to their lower cost as compared with other protein sources. The protein content of defatted meals depends on the seed and ranges between 35% and 60% (dry base) (Moure et al. 2006). Canola, flaxseed, and sunflower meals obtained from oil extraction can be used to produce protein concentrates with suitable nutritional and functional properties.

## 10.4 Techno-Functional Effects of Meat Replacement in Plant-Based Hybrid Meat Products

Several techno-functional parameters, such as emulsion stability, WHC, and gelling ability can be affected by partial substitution of meat, consequently, the cooking yield and texture related parameters of the product could be impacted. Several studies have evaluated the use of legumes as flour, pastes and protein concentrates in hybrid meat products.

Parameters such as emulsion stability, cooking yield, and WHC tend to increase with the inclusion of legumes due the high WHC and emulsification properties of leguminous proteins, mainly globulins. Additionally, the carbohydrate fraction present in larger or smaller quantities in flours and legume proteins can also contribute to the increased WHC. On the other hand, some textural parameters like hardness and chewiness tend to decrease in these products due to the poor gelation ability of these proteins compared to the myofibrillar ones. Several researches have evaluated the use of plant proteins as partial substitute of meat in the production of hybrid products. The main impact on the technological, sensory, and nutritional properties is highlighted in Table 10.2.

The evaluation of leguminous flours in the hybrid meat products development has been accessed by several researchers. In general, leguminous flours are rich in protein content, and also carry other macronutrients, such as vitamins, dietary fibers, and minerals, which resulted in hybrid meat products with dense nutrients. On the other hand, the texture and flavor may be pronounced affected. Leonard

**Table 10.2** Main techno-functional, sensory, and nutritional effects of hybrid meat products with vegetable sources

| Hybrid meat products | Reformulation strategy | Techno-functional effects | Sensory effects | Nutritional effects | References |
|---|---|---|---|---|---|
| Sausages | Substitution of 20% meat by pea protein (isolate and extruded at low and high moisture). | Emulsion stability was not improved; hardness and shear force reduced with pea proteins inclusion. Cohesiveness, springiness, and chewiness did not change. Color was slightly affected. | The use of isolated pea protein did not affect sensory profile of hybrid products, the other treatments were not evaluated. | Hybrid products with an adequate amount of essential AA were obtained. | Broucke et al. (2022) |
| Sausages | Replacement of beef with 0% to 36% (w/w) hydrated lupin flour. | Hybrid sausages were softer and more adhesive than control. | Appearance, aroma, flavor, and overall liking was not affected in hybrid products with 12% of meat replacement by lupin. | The content of fiber was increased, while fat was reduced in hybrid products. | Leonard et al. (2019) |
| Sausages | Substitution of 25% pork by hydrated oats or tofu. | WHC and cooking yield increased in treatments with hydrated oats. The same was not observed to tofu products. | Tofu and oat hybrid sausages presented a better tenderness, juiciness, and overall acceptability. | Hybrid sausages contained less fat and the same content of protein when compared to control. | Yang et al. (2007) |
| Sausages | Substitution of 50% of beef by rice or soy proteins with 2 levels of transglutaminase | Significantly increasing in the textural parameters, with denser microstructure, significant increase in water loss during thermal treatment for rice-based sausages. | Soy-based sausages with TG did not differ from meat sausage regarding consumer preference and received desirable sensory attributes, the opposite results were observed in rice-based sausages. | *Not evaluated in this study.* | dos Santos et al. (2023) |

(continued)

**Table 10.2** (continued)

| Hybrid meat products | Reformulation strategy | Techno-functional effects | Sensory effects | Nutritional effects | References |
|---|---|---|---|---|---|
| Dry-cured sausages | Replacement of meat by 12.5, 25, 37.5, and 50% of texturized pumpkin seed proteins. | The acidification behavior was compared to control; all treatments achieved suitable decreasing in pH and Aw. Texture decreased as pumpkin protein increased. Color changed significantly. | *Not evaluated in this study.* | Protein and fat content did not differ among treatments before and after ripening. | Ebert et al. (2022) |
| Low-fat burgers | Replacement of 10–44% of meat with pulse flours (lentil, chickpea, pea, and bean) with different levels of hydration. | Hybrid products presented lower shrinkage and cooking loss. | Hybrid burgers with 44% of pulse flours and a medium level of hydration were acceptable sensorially. | *Not evaluated in this study.* | Argel et al. (2020) |
| Chicken patties | Chicken patties with rice bran (0–8%) as chicken meat replacer. | Decrease in moisture, pH, cooking loss, expressible moisture, and color attributes as rice brain increased. Texture parameters increased. | Treatments with 2% rice bran were considering juice and did not differ from control regarding acceptability. | Only the nutritional effect of stabilized rice bran was evaluated and showed a great nutritional value (15.83% dietary fiber, 11.43% protein, 17.78% fat). | Aviles et al. (2023) |
| Beef patties | Beef patties were processed with 40% meat and 4% pea, fava bean protein, and extruded milled pea starch. | Hybrid patties presented a firmer texture and a higher cooking loss than control. | Hybrid patties were described as juicy. Overall liking was similar between hybrid and control patties. | *Not evaluated in this study.* | Petrat-Melin and Dam (2023) |
| Meatballs | Substitution of beef by 15% or 30% texturized soy protein (TSP) using yeast. | The inclusion of 15% texturized soy protein was not different from control, considering hardness, and products with 30% soy decreased hardness. | Considering overall acceptability and flavor, hybrid meatball with 15% TSP and yeast were the most acceptable products. | Hybrid patties presented a balanced protein content. The fat content decreased in hybrid patties with 30% TSP. | Grasso et al. (2019) |

et al. (2019), for instance, evaluated the impact of adding hydrated roasted lupin flour (0–36%) as meat replacer in beef sausages on various characteristics. The stability of the meat emulsion improved with the addition of lupin flour. The cooking loss in the sausage formulations was reduced significantly when lupin flour was included. Otherwise, the texture parameters decrease with the addition of lupin flour. The sausages became softer and exhibited greater adhesiveness compared to the traditional formulation. According to the authors, both the protein fraction and the fibers present in the lupin flour contributed to this effect. The textural parameters significantly decreased with the partial replacement of meat by lupin flour, relating to the poor gelling capacity of lupin proteins compared to meat proteins. According to the authors lupin flour could be added in the sausages up to 12% without compromise the sensory evaluation. Similar results were observed in sausages with partial substitution of beef by okara flour (0–40%) (Noriham et al. 2016).

In hybrid burgers, a study of Argel et al. (2020) exploring the substitution of 10–44% of pork meat with pulse flours (chickpea, lentil, bean, and pea) in low-fat burgers. Hybrid burgers demonstrated higher cooking yield, lower shrinkage, and expressible liquids compared to control with 100% meat. Similarly, Chandler and McSweeney (2022) demonstrated that hybrid chicken burgers with 25%, 50%, or 75% of chickpea, yellow pea, and lentil flours presented better cooking yield than control chicken burger, interesting, the level of meat substitution by the pulses did not affect the cooking yield. Otherwise, diameter and thickness reductions were less in hybrid burgers compared to the control burger, with the reductions decreasing as the pulse content increased. Both studies demonstrated a significant reduction in texture parameters, mainly in hardness, the springiness and cohesiveness, otherwise, varied regarding the meat substitute used and its level.

Another study by Argel et al. (2022) assessed the effects of substituting lean pork meat with different levels of white bean flour ranging from 8% to 15% and water/bean flour ratios ranged from 1.2 to 1.8 g/g in hybrid burgers. The incorporation of bean flours led to cooking yield higher than 88%. There was a decrease in hardness and cohesiveness as the bean flour level increased, with a more pronounced effect at higher water/bean flour ratios. The interactions between meat proteins and bean flour resulted in more thermal resistance of starch and protein fractions. The desirability optimization predicted cooking yield comparable of pork burgers an optimum formulation of 15% bean flour and a water/bean flour ratio of 1.36 g/g.

The partial substitution of meat by concentrated proteins has also been evaluated, for instance, dos Santos et al. (2022) showed 50% red meat substitution by soy, pea and sunflower concentrate proteins resulting in hybrid meat emulsions with higher emulsion stability than control treatment with meat. Otherwise, the rice and fava bean protein resulted in a lower emulsion stability. The authors noted a challenge with the color of sunflower-based hybrid meat emulsions due to the brownish color of the sunflower protein. Despite this, they suggested that the color issue might be less significant in other hybrid beef products such as burgers, meat patties, and uncured sausages, where the darker color could be more acceptable or less noticeable.

Broucke et al. (2022) demonstrated the replacement of 20% pork meat by pea protein (isolate and extruded at low and high moisture) in emulsified cooked sausages did not improve emulsion stability. Besides, hardness and shear force reduced with pea proteins, although no change in springiness, cohesiveness, and chewiness were found. Color parameters were slightly affected by the strategies employed. Similarly, Das et al. (2008) reported that texturized soy and soy paste used to replace 15% of goat meat in nuggets did not affect the product yield, pH, moisture, and fat percentage.

The reduction in texture parameters can be overcome by the incorporation of transglutaminase, which covalently binds proteins, increasing gel formation capacity, water retention, and textural parameters, as demonstrated by dos Santos et al. (2023). According to their results, the use of 0.25% transglutaminase significantly increased the textural parameters, without negatively affecting the stability of the emulsion, color, and pH of soy-based hybrid emulsions made with a 50% reduction of meat.

Cereals, seeds, and oilseeds, although less frequently assessed in hybrid meat products compared to legumes, offer substantial potential due to their nutritional profiles and functional properties. Yang et al. (2007) replaced up to 25% of pork with hydrated oats or tofu in sausage formulations. A notable enhancement in WHC and cooking yield was observed in treatments with hydrated oats, tofu otherwise did not demonstrate similar performance. Conversely, when considering texture, an increased oat content significantly reduced the sausage hardness and other texture parameters more than tofu. Despite the reduction in textural parameters, there was an observed increase in the sensory scores for juiciness, tenderness, and overall acceptability of the sausages. This demonstrates that both hydrated oat and tofu can serve as promising strategies for reducing meat content in hybrid products.

Aviles et al. (2023) investigated the potential of using stabilized rice bran (2–8%) as a substitute for meat in the production of hybrid chicken patties. Rice bran impact water retention of the patties, as its concentration increases, moisture, weight loss, and expressible water lowered in the formulations. The higher fiber fraction might be associated with these results. Color and pH also decreased with rice bran. Texture parameters such as cohesiveness, hardness, and gumminess, in general, increased with rice bran addition. The addition of up to 2% of rice bran may not affect the sensory quality of the product.

Tahmasebi et al. (2016) evaluated the effects of pigeon pea flour, corn flour, walnut, and sesame pastes, and sunflower oil as meat and fat substitutes in sausages using response surface methodology. Both pigeon pea and corn flour were associated with high emulsion stability, and consequently, higher cooking yield, these results were associated to fiber and starch content. Pigeon pea flour was the main responsible for increasing the texture parameters, and it was correlated to its protein and fiber fraction. The optimized formulation showed a total desirability value of 0.732, demonstrating that optimization methodologies can be an excellent choice for developing hybrid meat products that closely mimic the characteristics of traditional meat.

Ebert et al. (2022) examined the replacement of pork meat with texturized pumpkin seed proteins (from 12.5% to 50%) in dry-cured hybrid meat sausages. The authors demonstrated quite promising results regarding the development of acidification and drying, which was very similar to the control treatment with 100% meat. However, the textural parameters springiness, cohesiveness, and chewiness were lower than control and decreased as the content of pumpkin texturized protein raised in the formulations. Furthermore, the color of the products was quite different from the control, which could be a challenge in sensory acceptability. However, the use of natural colorants such as beetroot, hibiscus extracts, among others, may mitigate this undesired effect.

## 10.5 Sensory and Consumer Willingness of Plant-Based Hybrid Meat Products

When it comes to meat consumption, there is a general agreement that consumer choices are significantly influenced by the sensory attributes of products (Hoek et al. 2011). And successfully substituting part of the meat with vegetable proteins while preserving both sensory characteristics and technological properties is a significant challenge. Considering the sensory aspects of hybrid meat products, this section will focus on results of sensory experiments on these products and consumers' perceptions of it, mainly because they are new products and a potential market to be explored.

In a recent study, a survey was applied in three different countries of Europe—Spain, Denmark, and England—to evaluate the consumers perceptions of hybrid meat products (Ryder et al. 2023). The first question refers to consumers' perception of hybrid meat products, regarding ethics, sustainability, quality, sensory, etc. One of the results indicates that one emotion, alone, does not represent what the consumers feel, for example, the sensory aspect was also related to production, ethics, and sustainability.

In another survey, applied with German consumers, the authors investigated the preferences and attitudes related to hybrid meat (Profeta et al. 2021). There was no sensory evaluation on this study, but the consumers were asked to compare meat products and hybrid meat products imagining what they think about taste, health, and environment. Even though meat hybrids replace only a portion of the meat with plant-based proteins, the results showed that hybrid meat is perceived as more health, better for environment and animal welfare. However, when it comes to taste, there is a preference for traditional meat, and this parameter could be a barrier to the consumers to accept these new products.

The acceptance of meat substitutes has been linked especially to the attributes of taste and texture (Hoek et al. 2013). Neville et al. (2017) used a check-all-that-apply (CATA) questionnaire to evaluate the sensory attributes of hybrid sausages, hybrid burgers, and commercial products—meat-free and meat sausage or beef burgers.

The results indicate that hybrid products were well accepted by the consumers and meat analogues (meat-free products) were the least liked. With this, the authors observed that hybrid meat products are important to build this transition from meat consumers to meat-reduced diet, since they represent the middle way.

Using the same sensory evaluation (CATA questionnaire) and hedonic scale, Petrat-Melin and Dam (2023) developed patties (with hybrid meat and beef burgers). The hybrid patties were processed using about 40% minced meat and 4% pea, fava bean protein, and extruded milled pea starch. The authors tried to mimic the texture properties of the beef burgers, and textured vegetable protein was blended with gluten to serve as a binder and, along with extruded pea starch functioning as a binder for water and fat. However, the texture profile analysis results indicated that comparable texture properties to a beef patty were not similar. The CATA results showed that hybrid burgers were different from beef burgers, but the new products were characterized as 'soft' and 'juicy', however, there is a challenge to develop hybrid products without 'off-flavor', which is one of the characteristics cited by the assessors and it is not desirable.

Regarding the sensory profile of the hybrid meat products, in the study of Baune et al. (2023), the authors evaluated the sensory aspects of hybrid meatballs with 30% meat substitution by soybean, pumpkin, or pea proteins as dried textured, or high moisture extrudate vegetable proteins. In this study, two types of sensory analysis were performed, with and without a trained panel. The results of the sensory analysis—with a trained panel – showed that the highest score was attributed to all-meat products (regarding the meat taste). And an interesting result was obtained to products with 30% textured vegetable protein of pea with untrained consumers, were overall acceptance and odor presented good results in a hedonic scale.

In another study, Broucke et al. (2022) used a trained panel to compare a traditional pork sausage and hybrid sausage with pea protein isolate and no significant differences were observed, considering the taste of the products. These findings show the positive results to continue to develop hybrid products that ally sensory approval and nutritional benefits.

Grasso et al. (2019) studied the effects of adding texturized soy protein (15% and 30%) with or without nutritional yeast as a flavor enhancer in beef meatballs. All hybrid meatballs received significantly higher acceptability scores for texture compared to the control. Specifically, the 15% texturized soy protein inclusion with yeast achieved the highest scores for flavor and overall acceptability. Also, according to the Penalty-lift analysis of CATA terms, attributes such as moist looking, juicy, soft, and crumbly and easy to cut were identified as the main drivers for liking the TSP-containing meatballs. The control samples were significantly more often associated with the term hard a key driver for dislike.

Finally, dos Santos et al. (2023) demonstrated the significant effect of transglutaminase in enhancing the textural parameters of soy-based hybrid sausages, which resulted in a sensory preference indistinguishable from that of traditional meat sausages. The authors also demonstrated that projective mapping was efficient in separating treatments with similarities and differences. Furthermore, in the multiple frequency analysis, the soy-based hybrid sausage treated with transglutaminase was correlated with positive attributes such as juicy, pleasant aroma, and tasty.

## 10.6 Nutritional Appeal of Plant-Based Hybrid Meat Products

Legume is rich in proteins with good AA profile; however, they are limiting in the sulfur-containing AA, such as methionine, cystine, and cysteine, and also in tryptophan (Maphosa and Jideani 2017). Carbohydrates present in legumes have a low glycemic index, and they are good sources of dietary fibers, both soluble and insoluble ones (Rebello et al. 2014). Furthermore, they have a good lipid profile, with a high content of unsaturated fatty acids (UFA), such as oleic, linoleic, and linolenic acids; and low content of saturated fatty acids (SFA), although they are generally low in fat, except for peanuts (± 45%), and soybeans (± 18%) (Sánchez-Chino et al. 2015). Legumes also have several bioactive compounds such as phytochemicals and antioxidants, including isoflavones, carotenoids, and alkaloids.

Pulses, belonging to the *Leguminosae* family, represent a significant source of dietary protein, since present substantial levels of essential AA, such as lysine, and leucine, with significant amounts of several vitamins, minerals, and they are sources of polyunsaturated fatty acids (PUFA) (Asgar et al. 2010; Havemeier et al. 2017).

The nutritional composition of soybeans varies depending on the specific variety and growing conditions. Typically, they contain about 40% protein, 20% fat, 30% carbohydrates, and from 10% to 30% moisture (Chen et al. 2012; Nishinari et al. 2014). In addition to being a rich source of fat and carbohydrates, they present the highest protein content among grains and legumes (Asgar et al. 2010). Broucke et al. (2022) demonstrated that substituting up to 20% pork meat in sausages with extruded pea products did not compromise the AA profile. Also, the essential AA were present in adequate quantities to all treatments.

Hybrid plant-meat products offer a balance between nutritional and sustainability aspects (Baune et al. 2021). The hybrid meatballs, formulated with textured proteins from peas, sunflowers, or pumpkins, contained a higher concentration of essential fatty acids and displayed an improved ω-6: ω-3 ratio. Additionally, these hybrid products showed an increased dietary fiber content. Although the protein digestibility-corrected amino acid score for children was slightly lower in the hybrid meatballs, it was still high compared to the plant-based raw materials used.

Londero et al. (2015) focusing on develop a beef meatball formulation with 30% meat replacement and 100% textured soy protein replacement by landrace bean flour, they showed a reduction in fat content (from 18% to 4.2%) with a good n6/n3 ratio, and high protein and fiber content. Besides, the hybrid meatball contained no limiting essential AA compared to the FAO/WHO standard. According to the authors the nutritional quality of the reformulated meatball was better than the commercial meatball treatment.

Given the scenario described, hybrid meat products elaborate with vegetables offer significant nutritional benefits. These include the incorporation of dietary fibers, vitamins, minerals, and a variety of bioactive compounds typically found in vegetables but absent in meat. Also, the high-quality protein content of meat ensures the product maintains an excellent AA profile. Hybrid meat products with vegetables not only diversify their nutritional value but also enhances their overall health benefits.

## 10.7 Hybrid Meat Products with Insect Proteins: Technological, Sensory, and Nutritional Properties

In a current context, faced with numerous challenges related to food security, population growth, and environmental impacts, edible insects have stood out as favorable option sources of nutrients and other functional compounds (FAO 2021; van Huis and Rumpold 2023). Among the nutritional advantages of edible insects, we can mention the high protein content with well-adjusted amounts of AA, source of several minerals for instance as magnesium, iron, manganese, zinc, phosphorus, and selenium, besides to being rich in PUFA (Rumpold and Schlüter 2013), depending on rearing conditions, feeding, developmental stage and environmental factors (Payne et al. 2016). Furthermore, various environmental advantages are correlated with insect rearing, such as lower land, water, and energy requirements, besides lower greenhouse gas emissions than conventional livestock farming (Oonincx and de Boer 2012).

In Western societies, despite the numerous advantages highlighted in insect production and their nutritional significance, there persists a widespread reluctance towards consuming insects due to food neophobia and a sense of disgust (Megido et al. 2016). One strategy to promote greater acceptance and integration of edible insects into the diets, involves incorporating insect powders or protein isolates into familiar and widely accepted products, for example, sausages, burgers, nuggets, and others (Choi et al. 2017). The concept of hybrid meat products, integrating insects as a partial replacement for meat, represents a recent and significant approach aimed at aligning with the sustainable development goals outlined by the United Nations, facilitating the reduction of meat protein consumption (United Nations 2015) and achieving the flexitarian consumer public, currently growing (Charlebois et al. 2020; Sheen et al. 2023).

Meat products are complex matrices containing a combination of myofibrillar protein, fat, salt, water, and additives, with a greater or lesser degree of comminuted raw materials depending on the product type (Gordon et al. 1992). The functionality of myofibrillar protein influences the texture, moisture retention, and tenderness of processed meats (Xiong 1994). The characteristics of hybrid meat products are influenced by the interplay between proteins from both meat and insects. The topics below discuss the main properties and challenges insect proteins can provide regarding techno-functional, nutritional, and sensorial effects, as well as the influence of processing on product quality.

## 10.8 Edible Insects and Its Way of Addition in Meat Products

Out of more than one million species of insects, around two thousand are edible, making them food and feed options. The most common edible insect species are caterpillars and beetles (49%), furthermore wasps, ants, and bees (14%); crickets and grasshoppers (13%); hemiptera (10%); termites (3%); dragonflies (3%); and flies (2%) (Ordoñez-Araque and Egas-Montenegro 2021).

According to van Huis and Rumpold (2023) the species that are reared for human consumption are many mealworm species, primarily the *Tenebrio molitor*, and the *Alphitobius diaperinus*; silkworm species, especially the domesticated silkworm (*Bombyx mori*), several cricket species, such the house cricket (*Acheta domesticus*); locust species, such as the migratory locusts (*Locusta migratoria*). In their comprehensive review, Borges et al. (2022) specifically examined the inclusion of insects into meat product formulations, highlighting *Tenebrio molitor, Acheta domesticus, Bombyx mori*, and *Zophobas morio* as the most researched species within meat matrices.

Several technologies have been developed focusing on using insects as ingredients, which could increase consumer acceptance, especially in the West, where entomophagy is not traditionally practiced in most countries. Emerging technologies such as cold atmospheric processing, pressure plasma treatment, and ultrasound-assisted extraction are primarily tailored for the extraction and drying of protein, fat, and chitin. Various drying methods like freeze-drying, oven-drying, fluidized bed drying, and microwave-drying are utilized in this context (Melgar-Lalanne et al. 2019). Studies on hybrid meat products with edible insects are still limited. Despite this, it is possible to observe growth of scientific publications, and the main forms of incorporation in these products are whole flour (Choi et al. 2017), defatted flour (Zhang et al. 2022b), fresh, extruded in blends with vegetable protein concentrates (Kiiru et al. 2020), protein concentrates (Yong et al. 2023), prior acid hydrolysis for chitin degradation (Kim et al. 2016).

Kim et al. (2016) evaluated the nutritional and technological properties of *Bombyx mori* and *Tenebrio molitor* flours defatted and subjected to acid hydrolysis as a new ingredient for application in meat emulsions. Defatting and acid hydrolysis of silkworm and mealworm flours decreased redness, pH, yellowness, and protein solubility, but considerably increased the protein content. Nevertheless, irrespective of the pretreatment methods employed, the impact on sausage color and cooking yield remained resembling across all pretreated insect flours. Conversely, Zhang et al. (2022b) compared the effects of partially replacing meat with *Tenebrio molitor* flour obtained through freeze-drying and microwaving on the quality attributes of sausages. They found lower cooking losses, greater stability of the meat emulsion, and better textural and sensorial properties in treatments with flour pre-dried in microwaves.

Most studies on hybrid meat products used flours (Cavalheiro et al. 2023; Park et al. 2017; Walkowiak et al. 2019), where insects go through different dehydration processes that affect both the functional properties of their proteins, the composition of reformulated meat products, and their technological properties. For this reason, Scholliers et al. (2020a) studied a different strategy: the incorporation of fresh *Zophobas morio* larvae in cooked sausage, considering the crude protein content (16.6–19.7%) similar to meat (16–22%) and the adequate properties of gelling of this insect. The partial replacement of meat by insects positively affected the stabilization of fat and water in the meat matrix during heating and refrigerated storage. Nevertheless, a rise in cooking losses was observed for all hybrid products at elevated temperatures.

On the other hand, Smetana et al. (2018) reaffirm the importance of the similarity of new ingredients in physical properties and flavor with meat and how extrusion technology can help mimic its fibrous texture in their research with analogs. However, as far as we know, no studies have assessed the potential of extruded insect-based ingredients in hybrid meat products.

In general, different forms of insect pre-processing (defatting, drying, and extraction process) for incorporation as ingredients in meat products have been shown to affect the functionality of proteins and the interaction between the various components that are part of the constitution of a meat product and therefore need to be better understood (Queiroz et al. 2023).

## 10.9 Techno-Functional Effect of Edible Insect Proteins and Flour in Hybrid Meat Products

The utilization of edible insects within the food industry is poised for significant growth in the forthcoming years. FAO has been pivotal in fostering and endorsing this trend, notably since the release of the document "Edible Insects: Future Perspectives for Food Security" in 2013 (Tavares et al. 2022; van Huis and Rumpold 2023). Since then, it is possible to observe an increase in scientific publications and patents for developing insect-based ingredients and products (Borges et al. 2022; Han et al. 2023; Tavares et al. 2022). According to Grasso et al. (2022), many consumers are familiar with the concept of hybrid meat products, which can be very beneficial in introducing these products into the diet and reducing phobia and disgust towards edible insects. Added to this is that industrial processes already used in the food industry can be adapted to create hybrid products. Table 10.3 summarizes insect-based ingredients main techno-functional, sensory, and nutritional effects in hybrid meat products.

**Table 10.3** Main techno-functional, sensory and nutritional effects of hybrid meat products with edible insects

| Hybrid meat products | Reformulation strategy | Techno-functional effects | Sensory effects | Nutritional effects | References |
|---|---|---|---|---|---|
| Tteokgalbi, a cooked meat product | Edible insect larvae extracts (*T. molitor*, *P. brevitarsis*, and *A. dichotoma*) at 1% or 2% concentration. | Extracts of *A. dichotoma* (2%) enhanced the product's quality attributes without negatively impacting microbial counts, while also exhibiting inhibitory effects on lipid oxidation. | *Not evaluated in this study* | The antioxidant capacity and the high phenolic content can benefit the human body. | Yong et al. (2023) |
| Beef burgers | Lean meat was replaced by 10, 15, and 20% of cricket (*Gryllus assimilis*) flour (CF). | Adding CF resulted in softer meat products with lower shrinkage, cooking loss, and higher moisture and fat retention. | FC and F10% had greater sensory acceptance. CATA test consumers linked FC and F10% with the descriptors burger aroma, meat aroma and juicy. | Higher CF levels increase the fat content of products. | Martins et al. (2023) |
| Frankfurters | Lean meat was replaced by 5, 10, and 15% of yellow mealworm larvae (*Tenebrio molitor* L.) flour (YMLF) obtained by two pre-drying methods (freeze-dried – FD and microwave drying – MD). | Hybrids frankfurters produced with MD flours presented a more stable emulsion, with higher cooking yield and better texture properties than products with FD flours. | The higher the YMLF content, the lower the acceptability of the products. MD processing was more adequate than FD one. | The reformulation did not alter the protein and fat content. However, as the replacement ratio of FD-YMLF increased, the moisture content of the samples decreased. | Zhang et al. (2022b) |
| Hybrid model systems | *Zophobas morio* larvae (ZM) in different insect:meat ratios (75:25, 50:50 and 25:75) | The hybrids (80 °C) were similar to the meat model systems (70 °C) for viscoelastic properties and water retention but presented worse texture attributes. | *Not evaluated in this study* | *Zophobas morio* and pork shoulder had similar nitrogen contents, but the meat showed a higher moisture content and lower fat than insect. | Scholliers et al. (2020b) |

(continued)

**Table 10.3** (continued)

| Hybrid meat products | Reformulation strategy | Techno-functional effects | Sensory effects | Nutritional effects | References |
|---|---|---|---|---|---|
| Frankfurters | Replacement of lean meat with cricket (*Acheta domesticus*) flour (CF) at 2.5%, 5% and 7.5% levels. | The samples with CF were darker, less cohesive, elastic, and firmer than the control. CF reduced the processing loss of products. | Hybrid products with 7.5% of CF were described with a bitter taste and had the lower acceptability. | The content of proline, glycine, and alanine were higher in CF5.0 and CF7.5, and CF7.5 also presented the higher level of SFA and PUFA. | Cavalheiro et al. (2023) |
| Cooked meat product | Replacement of lean meat with yellow melworms (*Tenebrio molitor)* and black soldier flies (*Hermetia illucens*) whole and autoclaved at 10% and 20% levels (T10, T20, H10, H20 respectively) | Adding insects resulted in higher pH and yellowness, darker and softer in the H10 and H20, higher cooking losses and pH (products with 20% insects), lower hardness values, and higher yellowness in the T10 and T20. | The products with Hermetia had lower sensory values than the Tenebrio products. After opening the packages, the samples with insects showed lower sensory values than the control (day 7 of storage). | Adding the insects reduced the protein content in the products, and the fat contents were comparable between the treatments. | Lemke et al. (2023) |
| Meat emulsion | Replacement a portion of lean pork and/or back fat with cricket flour CF (*Acheta domesticus)* at 5% and 10%. | Levels of up to 10% replacement by CF did not interfere with the technological properties of texture and cooking yield. | *Not evaluated in this study* | The meat emulsion with 10% CF increased micronutrients (potassium, magnesium, and phosphorus) and proteins. | Kim et al. (2017) |
| Sausages | The impact of blending broccoli (BR) upcycled brewer's spent grain (BSG), and *Tenebrio molitor* flour (IF). | Every formulation containing IF or BSG exhibited a notable rise in hardness and chewiness. The reddest products were those where BSG and IF replaced 15% of the broccoli. | The commercial Bratwurst and the hybrid meat sausage (22% BR, 3% BSG, and 10% IF) did not differ in odor and juiciness in the sensory analysis. | The samples with IF showed higher fat and protein content. The addition of IF and BSG powders reduced moisture content and enhanced iron and zinc levels in the samples. | Talens et al. (2022) |

| | | | | | |
|---|---|---|---|---|---|
| Frankfurters | Replacement of lean pork with yellow mealworm powder YMP (*Tenebrio molitor*) at 10% up to 60% levels. | pH and yellowness of products with insects were higher than those of the control. The products with YMP showed lower values than the control for lightness, hardness, and chewiness. | The control treatment attained the highest scores in color, flavor, off-flavor, and juiciness. There was no difference in overall acceptability between the control and the treatments with up to 20% replacement. | Samples containing YMP exhibited higher protein and ash content than the control while showcasing lower moisture content. | Choi et al. (2017) |
| Beef burgers | Partial/full beef substitution with Black soldier fly larvae BSFL (*Hermetia illucens*) at 25, 50, 75, and 100%. | The BSFL burgers were darker, with a softer texture and lower cooking losses than the control. | *Not evaluated in this study* | BSFL treatments were lower in fat, protein, and moisture than control. | Bessa et al. (2023) |
| Meat emulsion | Locust (*locusta migratoria*) protein hydrolysates (LoProHs) pre-processed with microwave (Mic-LoProHs) and ultrasonication (Ult-LoProHs) replaced 1.5% of mutton. | For 14 days, the meat emulsions with LoProHs were better in all quality parameters (microbial quality, antioxidant potential, lipid stability) than control. | Meat emulsions with LoProHs showed higher scores for texture, flavor, overall acceptability, appearance, and color (day 7). | In the gastrointestinal simulation, the digested samples with LoProHs showed higher DPPH ABTS, and FRAP means than the control. | Singh et al. (2023) |
| Sausages | Cricket flour CF (*Acheta domesticus*) was added at 1%, 2.5%, and 5% levels. | During storage, the CF expedited the oxidation of both lipids and proteins. The textural properties of the samples with CF were lower than the control. | *Not evaluated in this study* | The moisture content decreased while the protein content increased with the increase of CF. There was no effect on the fat content and other components among all the sausage groups. | Han et al. (2023) |

Reducing meat proteins in products is a considerable challenge, whether due to the essential role of myofibrillar proteins in texture and stability properties or because of the perception of flavor, aroma, and juiciness that these proteins help to provide (Gordon et al. 1992; Xiong 1994). In all the studies shown in Table 10.3, the impact of replacing meat proteins with insect-based ingredients on textural properties is evient.

Several authors observed inferior textural properties in products reformulated with edible insects, with reduced hardness, chewiness, elasticity, and cohesiveness, regardless of the substitution levels evaluated, the type of insect used, and the meat product (Bessa et al. 2023; Choi et al. 2017; Lemke et al. 2023). The enhanced structural integrity observed in the control meat products in these studies may be attributed to the intermolecular interactions among myosin molecules. Myosin can form heat-induced gels, with hydrophobic interactions, disulfide bonds that contribute to the characteristics of elasticity, hardness, and gumminess, and with electrostatic interactions due to polar groups, depending on pH conditions, the concentration of salt and temperature (Zhang et al. 2022c). Furthermore, another factor that may have contributed to the inferior textural properties is the interference of insect cuticular components, such as epicuticles and scleroproteins, which are highly hydrophobic and can decrease the protein gel structure formed by myofibrillar proteins (Han et al. 2023).

Conversely, Kim et al. (2017) and Cavalheiro et al. (2023), found divergent results for emulsified hybrid meat products with lean meat reduction and replacement with cricket flour (*Acheta domesticus*). According to Cavalheiro et al. (2023), adding cricket proteins to the meat emulsion increased the hardness of the products, probably because of the formation of a denser protein matrix which exhibited higher resistance to compression. The authors ascribe these findings to variances in the protein-moisture interaction between the control and reformulated samples.

Regarding the physicochemical property of pH, in different studies that investigated hybrid meat products, both emulsified, such as sausages, and restructured, such as hamburgers, higher pH values were found when using protein sources from the species *Tenebrio molitor*, *Gryllus assimilis*, and *Hermetia illucens* (Bessa et al. 2023; Choi et al. 2017; Martins et al. 2023; Yong et al. 2023). Insects such as T. *molitor, H. illucens*, and *G. assimilis* have higher pH values, from 6.4 to 7.4 (Bessa et al. 2023; Kim et al. 2016; Martins et al. 2023) than pork, with a final pH of 5.7 (Purchas et al. 2009), and beef, with a pH of around 5.6 (Patinho et al. 2024). According to Bednářová et al. (2013), the quantity and type of AA present in the protein structure of insects directly influence the pH values found.

However, considering other species, Singh et al. (2023), Kim et al. (2017), and Cavalheiro et al. (2023) did not find a significant increase in pH values with the addition of *Locusta migratoria* and *Acheta domesticus*, respectively, in meat emulsions. The cited authors used concentrations below 10%; in addition, according to Kim et al. (2017), the pH of house cricket flour (5.82) was similar to the pH of lean meat (5.63) and fat (6.03).

Cooking losses, a critical parameter often assessed in meat products, are a crucial indicator closely linked with meat matrices' water and fat retention capabilities.

Lower cooking losses were presented in burgers with *Gryllus assimilis* at levels of up to 20% replacement (Martins et al. 2023), in frankfurters with microwave-dried defatted *Tenebrio molitor* flour at levels of up to 15% (Zhang et al. 2022b), in hamburgers with *Hermetia illucens* at levels of up to 75% replacement for lean meat (Bessa et al. 2023) and in cooked meat sausages (Tteokgalbi) with extracts of *Tenebrio molitor*, *Allomyrina dichotoma*, and *Protaetia brevitarsis* at levels of 2%. pH values are positively related to water binding in meat products. As the pH deviates further from the isoelectric point, meat proteins' repulsive forces augment, enhancing water retention capacity (Puolanne et al. 2001). Scholliers et al. (2020b) also demonstrated that higher heating temperatures (80 °C) in hybrid model systems with *Zophobas morio* contribute to obtaining viscoelastic properties and water retention capacity more similar to traditional meat products.

In addition to all the techno-functional parameters mentioned, one of the most significant impacts on hybrid meat products with edible insects is their high capacity to change the color of the product due to the inherent characteristic of the insect itself color, which can considerably influence consumer acceptance. The black, brown, and yellow colors typical of insects are attributed to the cuticular protein melanin (Wittkopp and Beldade 2009). In all the studies mentioned in Table 10.3, it is possible to observe a decrease in luminosity, regardless of the insect species used, from 2% concentration in the formulations. Concerning the b* parameter, an increase in the yellowness is also noticeable in all the research evaluated. Regarding parameter a*, divergent values are found depending on the species of insects, concentration, and type of product (Cavalheiro et al. 2023; Han et al. 2023; Lemke et al. 2023).

## 10.10 Sensory Aspects of Edible Insect Proteins in Hybrid Meat Products

Entomophagy, or insect consumption, although practiced in many regions of the world, tends to provoke negative reactions among Western consumers, and consumer acceptance continues to be a crucial barrier to the adoption of diets based on edible insects (van Huis and Rumpold 2023). Neophobia is closely related to the emotion of repulsion or disgust, and both emotions contribute to individuals' reluctance to include specific foods in their daily diet (La Barbera et al. 2018; Melgar-Lalanne et al. 2019).

However, although achieving consumer acceptance remains a considerable challenge for industries, researchers have observed a shift in consumer preferences and social attitudes towards entomophagy in Western societies as more people have sought healthier diets with more sustainable protein sources (Gómez-Corona and Valentin 2023; Megido et al. 2016). Moreover, sensory attributes can undergo enhancement across various technological phases, from insect production to the elaboration of novel products like hybrid meat alternatives, catering to a growing demand for sustainable dietary options (Mishyna et al. 2020). Adding insects in the form of powders or extracts so that they become imperceptible in the product's appearance and presentation helps increase their acceptance and inclusion in the diet.

Table 10.3 describes the main sensory effects of reformulations of hybrid meat products with edible insects and shows different results, even when the same types of meat products are considered. Zhang et al. (2022b) and Choi et al. (2017) evaluated various levels of replacement of lean meat with *T. molitor* flour in Frankfurters. In the study by Zhang et al. (2022b), treatments with up to 5% replacing lean meat with microwave-dried mealworm flour had good sensory acceptance. In contrast, Choi et al. (2017) found global acceptance with no differences from the control in treatments with up to 20% replacement. Lemke et al. (2023) compared the sensory effects of replacing lean meat with *T. molitor* and *H. illucens* flours in cooked meat products at 10 and 20% levels, with all insect treatments having lower scores than the control. However, the acceptance of treatments with *H. illucens*, regardless of the levels used, was worse when compared to those products with *T. molitor.*

Variations in outcomes are typically anticipated due to the influence of cooking methods and the incorporation of additional ingredients like salt, sugar, additives, and flavorings. Moreover, the black soldier fly (*H. illucens*) is renowned for its efficient conversion of food waste into insect biomass, with its lipid composition often imparting an earthy flavor (Wang and Shelomi 2017). *Tenebrio molitor* larvae, commonly fed with cereal bran or flour, exhibit a sweet flavor reminiscent of nutty and cocoa notes, contributing to the observed sensory differentiation. The substrates provided to insects during their breeding directly influence the sensory profile found (Ruby et al. 2015).

There are few studies, still incipient, regarding meat products reformulated with cricket flour, and the results differ considerably. Martins et al. (2023) found greater sensory acceptance in hamburger treatments with up to 10% replacement of lean meat with *Gryllus assimilis* flour. Frankfurters incorporating *Acheta domesticus* flour, with up to 5% replacement, did not exhibit significant differences in the overall acceptance parameter compared to the control. Recent research has shown that using proteases on *Acheta domesticus* proteins yields peptides with heightened sensory appeal, presenting a promising avenue for their incorporation as ingredients in many formulations (Luna et al. 2021).

## 10.11 Nutritional Aspects of Edible Insect Proteins in Hybrid Meat Products

As alternative protein sources, edible insects boast rich nutritional profiles comprising high-quality nutrients and an array of bioactive compounds. They have outstanding globally in food security thematic debates owing to their notable protein content ranging from 35% to 61% and lipid content from 13% to 33%. Furthermore, they serve as valuable sources of vitamins, minerals, and fiber, partly attributed to their chitin content (Ojha et al. 2021). The nutritional profile of edible insects depends on the species, diet, metamorphic stage, habitat, environmental conditions, and processing methods; therefore, it can be highly variable (Lange and Nakamura 2021).

Due to the high protein values in their composition, the number of research using insects as alternative protein sources in hybrid meat products has increased, with the main form of incorporation in these products being flour. For this reason, when evaluating the nutritional composition of reformulated samples, a reduction in moisture content and an increase in protein content is often found (Cavalheiro et al. 2023; Choi et al. 2017; Kim et al. 2017; Talens et al. 2022), since the lean meat that has a high moisture content (around 60%) is removed, and insect meal is added, with a considerably lower moisture content (3–5%) and a higher concentration of the value protein.

From a nutritional point of view, a relevant characteristic to highlight is that in addition to the increase in protein values, some authors also found higher levels of amino acids. Cavalheiro et al. (2023) demonstrated higher levels of the AA proline, glycine, and alanine than the control in Frankfurters with 5 and 7.5% cricket flour (*Acheta dosmesticus*). Oliveira et al. (2024) examined various nutritional aspects of *Tenebrio molitor* and *Gryllus assimilis* powder across different developmental stages. Their findings indicated that both species serve as significant sources of essential AA, including histidine, threonine, sulfur AA, and aromatic AA, that meet the requirements proposed by FAO/WHO for adults. Lysine was the limiting AA in both species; however, as shown in Table 10.1, beef and pork are good sources of this AA, so in hybrid meat product formulations, the nutritional profiles can complement each other.

Fat is the second largest component of edible insects and can present a wide variation, according to the factors above, highlighting that the metamorphic larval stage has the highest fat content compared to the adult stage. Among edible insects, such as worms, crickets and flies, it has been observed that they are rich in UFA, such as α-linolenic and linoleic acids, which are crucial for maintaining the body's normal functions (Kolobe et al. 2023). According to Table 10.3, significantly higher lipid values were found in hybrid products with whole meal flours at levels starting at 10% (Martins et al. 2023; Talens et al. 2022). At lower substitution levels, such as that used by Han et al. (2023) (5%), there were no effects on the lipid content in sausages with cricket flour (*Acheta domesticus*).

Very few studies still evaluate the modification of the lipid profile of hybrid meat products with edible insects. Cavalheiro et al. (2023) found a significant increase in PUFA content, especially in linoleic acid content in Frankfurters with 7.5% cricket flour (*Acheta domesticus*). Lucas-González et al. (2019) also found a high percentage of UFA in freeze-dried cricket flour, around 45%, demonstrating the favorable lipid profile that whole insect flours can present.

Insects serve as an excellent source of various micronutrients, including copper (Cu), iron (Fe), magnesium (Mg), manganese (Mn), phosphorus (P), selenium (Se), and zinc (Zn) (Lu et al. 2024), and can be a very relevant nutritional enrichment alternative for hybrid products. Kim et al. (2017) found considerable changes in the mineral composition of meat emulsions by replacing 10% of lean meat with cricket flour (*Acheta domesticus*), with higher levels of phosphorus, potassium, magnesium, zinc, and manganese. Hybrid meat sausage with 10% *Tenebrio molitor* flour showed higher iron and zinc concentrations than other formulations evaluated (with broccoli) (Talens et al. 2022). Similar findings were reported by Oliveira et al. (2024). Their evaluation of *T. molitor* and *G. assimilis* flours categorized the examined powders as exceptional sources of phosphorus, copper, iron, zinc, and manganese and as favorable sources of magnesium, potassium, and niacin.

However, although many insects have a very favorable mineral profile, there is still a large gap, and few studies address their absorption and transformation together with other nutrients during processing. The bioavailability of minerals in edible insects varies with the food matrix, breeding methods, digestive process, absorption, and numerous metabolic processes that regulate the transport of nutrients in the body (Lu et al. 2024).

Another very positive effect of incorporating edible insects into meat product formulations is the antioxidant and antimicrobial properties provided by bioactive peptides, which can become bio-preservative for foods of animal origin that are vulnerable to lipid and protein oxidation and microbial deterioration (Rivas-Navia et al. 2023; Singh et al. 2023). *Locusta migratoria* protein hydrolysates (1.5%) evaluated by Singh et al. (2023) in meat emulsions successfully improved the storage quality of products. Furthermore, insect extracts can be sources of phenolic compounds with antioxidant properties, as demonstrated by Yong et al. (2023) in cooked meat sausage (Tteokgalbi). Adding 2% *A. dichotoma* to Tteokgalbi delayed lipid oxidation during 14 days of storage without adversely affecting microbial counts.

## 10.12 Proteins from Kingdom Fungi: Nutritional, Techno-Functional, and Sensorial Properties

Kingdom fungi includes yeasts (unicellular organisms) and filamentous fungi (mushrooms, and moulds) (Barzee et al. 2021). Fungi as truffles and mushrooms have been safely consumed for thousand years by humans. Filamentous fungi and yeast are essential for the processing of gorgonzola cheese, soy sauce, sake, and the obtention of organic acids, and enzymes for food and feed production (Bartholomai et al. 2022). Recently, filamentous fungi, mushroom, mycelia or mycoprotein, has been studied as a potential meat replacer in the formulation of meat analogues or hybrid meat products due to its high-quality nutritional profile, and productive, economical, and sustainable obtention.

Mycoprotein can be described as the ingredient/food obtained by fermentation of filamentous fungi (*e.g., Fusarium venenatum*) on a carbohydrate substrate, to produce a high-protein, low-fat product that can be an interesting meat substitute

(Souza Filho et al. 2019). In 2002, it was designated as Generally Recognized As Safe (GRAS) by the FDA (FDA 2002).

Mycoproteins are more sustainable than plant and animal proteins when land use efficiency, energy use, and global warming potential are evaluated (Awasthi et al. 2020, 2023). In addition, fungi can be easily cultivated on a large scale, its growth does not depend on the seasons and can be cultivated all year round (Awasthi et al. 2023), and are less dependent on fertilizers and pesticides (Hellwig et al. 2020). Another factor that makes fungi cultivation environmentally friendly is that waste such as agricultural and food processing byproducts can be used as substrate (Awasthi et al. 2023). Furthermore, several health benefits have been linked with the replacement of meat by mycoprotein, including improvements in glycemic response and blood cholesterol concentration (Souza Filho et al. 2019).

## 10.13 Nutritional Characteristics of Filamentous Fungi

The good nutritional profile of mushrooms and mycelia are associated with its content of protein (up to 90% in dry matter), essential AA (see Table 10.1), lipids and UFA, vitamins, fiber, and umami flavor (Strong et al. 2022). In a dry basis, mushrooms present a high content of protein (16.5–59.4%), and carbohydrates (16.4–75.0%), depending on species. A disadvantage for food industry is that the dry matter content of mushroom is very low, normally about 100 g/kg.

Dry mycoproteins typically contains 45% of protein, 13% of fat, 25% of fiber, 10% of carbohydrates, and 3.4% of ash, besides it contains all of essential AA (Table 10.1) (Finnigan et al. 2017) specially lysine, which is a limiting AA in plant proteins. Furthermore, it is an interesting meat substitute because fungi protein can naturally simulate meat fiber not requiring processes such as extrusion and shearing as in plant-based ones (K. Zhang et al. 2023).

Fungi also contain ergosterol, a biological precursor of vitamin D2 (Ahlborn et al. 2019), are low in fat and free of cholesterol (Awasthi et al. 2023). Mycoprotein consumption is also associated with a high satiety due to its high fiber content (chitin and beta-glucan) (Denny et al. 2008). Besides, they are safe from a toxicological perspective and the allergenic risk is lower than animal and plant-based products (e.g., legumes, soy, milk, eggs, and crustacean shellfish) (Molfetta et al. 2022).

A disadvantage of mycoprotein is its low digestibility (45–78%), and it may be related to the presence of fiber in the cell wall of filamentous fungi that are hard to digest. The application of enzymatic treatment, ultrasonication, and high-pressure homogenization may be strategies to improve the digestibility of mycoproteins (Zeng et al. 2023).

## 10.14 Mycoproteins Production

Several fungi species have been studied to produce mycoproteins as potential meat replacer, obtaining a product with a protein content of up to 60% (Table 10.4). The product from *Fusarium venenatum* presents the higher content of protein, as well as market support. Traditional substrates used to fungi growth are food-grade carbohydrates, salts (K, P, Mg), and vitamins (*e.g.*, biotin) (Hashempour-Baltork et al. 2020). However, wastes from animal husbandry, agriculture and forestry can be used for mycoprotein production, encouraging the use of waste, reducing costs, and making the process more sustainable (Zhang et al. 2022a).

Processing can be conducted by submerged or solid-state fermentation, and the time and temperature of the process in dependent on the fungi needs (Stoffel et al. 2019). After fermentation, biomass is heated (72 °C for 30 min) to reduce RNA content (to avoid uric acid accumulation in the human body) (Moura et al. 2022), and centrifugated or filtrated to produce pastes with a meat-like texture or bread dough. The success of mycoprotein production is affected by the accessibility of the substrates during all year and its distance to manufacturing plants (Hashempour-Baltork et al. 2020).

Despite the significant content, most fungi are not a good source of protein as meat, and a strategy to increase their protein content is the solid-state fermentation

**Table 10.4** Potential fungi, substrates, fermentation conditions, and protein content of mycoproteins as potential meat substitute in hybrid meat products

| Fungi | Substrate | Fermentation conditions | Protein (%) | References |
|---|---|---|---|---|
| *Rhizopus oligosporus* | Residual water from tempeh factories | Submerged fermentation 30 °C and 110 rpm for 2 days | 19.44 (wm) | Wikandari et al. (2023) |
| *Penicillium maximae* | Glucose, peptone, $KH_2PO_4$, $MgSO_4$ | Submerged fermentation 27 °C and 120 rpm for 15 days | 34.8 (dm) | Moura et al. (2022) |
| *Aspergillus oryzae* | Oat flour | Submerged fermentation 35 °C, 0.45 vvm for 2 days | 37.0 (dm) | Rousta et al. (2021) |
| *Neurospora intermedia* | Stale sourdough bread | Solid-state fermentation 35 °C, 90–95% RH under continuous light (3000 lux) for 6 days | 37.5 (dm) | Gmoser et al. (2020) |
| *Pleurotus albidus* | Brewer-spent grain | Solid-state fermentation 28 °C for 15 days | 22.6 (dm) | Stoffel et al. (2019) |
| *Pleurotus sapidus* | Apple pomace | Submerged fermentation 24 °C and 150 rpm in the dark for 4 days | 25.4 (dm) | Ahlborn et al. (2019) |
| *Fusarium venenatum* | Pea-processing byproduct | Submerged fermentation 35 °C and 150 rpm for 3 days with α-amylase | 59.75 (dm) | Souza Filho et al. (2018) |

dm, dry matter; wm, wet matter

using agroindustrial waste from breweries or vitiviniculture (Stoffel et al. 2019). It is also a good way to increase the content of essential AA (Xiao et al. 2015).

The high content of acid glutamic in mushrooms and mycoprotein confer them a meat-like flavor. In the same way, their structure similar to muscle fibers of meat is a positive characteristic to mimetic meat products. The addition of fungi protein also can improve functional properties of meat products, making it an attractive meat replacer (Kurek et al. 2022).

## 10.15 Techno-Functional Properties of Fungi Proteins

Techno-functional properties (water absorption index – WAI, water solubility index – WSI, water holding capacity – WHC, foaming capacity and stability, and fat absorption capacity – FAC) of fungi proteins depend mainly on the type of substrate, the type of fungi, and fermentation conditions.

The WHC of mycelium of *Plearotus sapidus* grown on apple pomace was dependent of the addition of chlorides and heating, ranging from 5.0 to 7.0 mL $g^{-1}$, presenting similar WHC to soy protein isolate (Ahlborn et al. 2019). The type of substrate and macrofungi also affected WHC, obtaining values from 1.4 to 6.3 g $g^{-1}$ (Stoffel et al. 2019).

The FAC of mushrooms and mycoproteins is dependent on specie and substrate and can range from 45% to 154% (Cha et al. 2014; Stoffel et al. 2019). WAI and WSI can be affected by protein denaturation and starch damage (Xiao et al. 2015).

The functional properties also can differ among biomass and mycoprotein of the same fungi. The protein isolate of *Penicillium maximae* presented similar WHC but stronger FAC than biomass powder. The high WHC and FAC of biomass powder may be due to its higher fiber content (37.3% dm). The powdered biomass solution was not able to form foam, while 120% foam with 91% stability was formed in the protein isolate solution (Moura et al. 2022). Foaming capacity of mycoproteins can be improved by sonication.

## 10.16 Fungi Proteins as Potential Meat Replacer in Hybrid Meat Products

The use of edible mushrooms is not new in meat area. They have been added in meat products as antimicrobial, antioxidant, flavor and texture enhancer, and sodium, phosphates, and fat replacers (Pérez-Montes et al. 2021; Qing et al. 2021; Wang et al. 2019b). They have also been used to produce meat analogues, such as sausage (Yuan et al. 2021), nuggets (Husain and Huda-Faujan 2020), and patties (Hellwig et al. 2020; Rousta et al. 2021). Although fungi-based meat analogue presenting a sensorial acceptance closest to meat products than plant-based meat analogues

(Hashempour-Baltork et al. 2020; Stephan et al. 2018) they can be salty, bitter and sour with an undesirable texture (Hellwig et al. 2020; Rousta et al. 2021), and for this reason, the production of hybrid products is more interesting.

Few studies have reported the use of mycoproteins as meat replacer in hybrid meat products, as presented in Table 10.5. Most of them have used fresh mushroom as partial meat replacer, which may cause undesirable nutritional effects such as an increase in moisture content and consequently a decreasing in protein content which may affect shelf life of the products (Wan Rosli et al. 2011; Wan Rosli and Solihah 2014; Wang et al. 2019a).

On the other hand, hybrid products can have a higher fiber content (Banerjee et al. 2020; Wan Rosli and Solihah 2012; Wang et al. 2019a) and phenolic and antioxidants compounds (Banerjee et al. 2020; Wang et al. 2019a; Wan-Mohtar et al. 2020).

Table 10.5 shown that hardness, color, and cooking yield are also affected by partial replacement of meat by mushrooms or mycoproteins. Fresh mushrooms in general decrease hardness in the products (Wang et al. 2019a; Wong et al. 2017), while dried fungi can increase (Wan-Mohtar et al. 2020) or decrease (Banerjee et al. 2020) it. In general, hybrid products with fungi are darker and less red than traditional meat products (Wan-Mohtar et al. 2020; Wong et al. 2017). Related to cooking yield, hybrid products with fresh mushrooms presented a lower cooking yield than control (Wan Rosli and Solihah 2012; Wang et al. 2019a; Wong et al. 2017) and the use of dried fungi cause an improvement in emulsion stability, cooking water holding capacity and oxidative stability (Banerjee et al. 2020).

## 10.17 Sensory Properties of Fungi-Hybrid Meat Products

The sensorial properties of fungi-hybrid meat products are dependent on the type of fungi and meat product, the composition of the substrate, the percentage of meat replacement, and the processing steps to product formulation (freezing, heating, extrusion, comminution, and others).

Few studies have evaluated sensorial properties of fungi-hybrid meat products (Table 10.5), and most of them only performed affective tests. Hybrid beef taco may be acceptable to consumers in a meat replacement amount of up to 80% (Guinard et al. 2016; Myrdal Miller et al. 2014), 50% for hybrid beef and chicken patties, while for pork sausage and chicken nugget a meat replacement higher than 25% and 10%, respectively, affected negatively sensorial properties.

The use of mycoprotein as meat replacer is a good strategy in the formulation of reduced-salt products due to its high umami content. A reduction of 45% of salt did not alter liking scores of hybrid beef taco (Wong et al. 2017).

When consumers see the word "mushroom" they associate it with an innovative and healthiness burger (Patinho et al. 2021). Consumer acceptance of fungi-hybrid meat products has a big influence of their rating of product taste, sustainability, cost, health, and novelty. Other factors such as food values and lifestyle (healthy eating

**Table 10.5** Main techno-functional, sensory, and nutritional effects of hybrid meat products with fungi proteins

| Hybrid meat products | Reformulation strategy | Techno-functional effects | Sensory effects | Nutritional effects | References |
|---|---|---|---|---|---|
| Chicken nuggets | Meat was replaced by 10, 20 and 30% of fruiting-body-base flour (FBBF) from *Pleurotus sapidus* | The hardness and chewiness increased in hybrid nuggets. Hybrid products were darker and redder than control. | The addition of FBBF affected negatively taste, texture, and aftertaste perception. Products with 10% of meat replacement were acceptable by consumers. | FBBF is a good source of antioxidants compounds. | Wan-Mohtar et al. (2020) |
| Goat meat nuggets | Hybrid were produced substituting levels of 2, 4, and 6% by dried *Flammulina velutipes,* enoki mushroom stem waste. | Hybrid products presented better emulsion stability than control. The addition of mushroom waste improved the WHC, oxidative stability and slightly decreased the hardness. Color was not negatively affected. | Sensorial properties (appearance, texture, flavor, juiciness, overall liking) were not affected in hybrid nuggets. | Hybrid products presented a better dietary fiber, ash and phenolics content compared to control. | Banerjee et al. (2020) |
| Pork sausage | Replacement of meat (25, 50, 75, and 100%) by dried shiitake (*Lentinula edodes*) | Hybrid sausages were softer, darker, with lower values of pH and higher moisture and cooking loss compared to control. | Consumers scored higher control e hybrid product with 25% of meat replacement in texture, flavor, and overall liking attributes. Related to appearance, samples were judged equal. Meat analogue (100% of meat replacement) had lower scores for aroma, flavor, texture and overall liking. | Total dietary fiber, cysteine, glutamic, methionine, total phenolic content; and antioxidant activity were improved in hybrid products, while a reduction of protein, ash, pH, energy level and hardness were observed. | Wang et al. (2019a) |

(continued)

**Table 10.5** (continued)

| Hybrid meat products | Reformulation strategy | Techno-functional effects | Sensory effects | Nutritional effects | References |
|---|---|---|---|---|---|
| Beef taco | Substitution of 25, 50, 75, and 100% of beef by fresh (90% moisture) white button mushrooms (immature *Agaricus bisporus)* | Cooking yield and hardness was lower in hybrid taco.<br>Hybrid taco was darker and less red than control.<br>An increase in moisture content with increasing mushroom level was observed. | *Not evaluated in this study* | *Not evaluated in this study* | Wong et al. (2017) |
| Beef taco | Replacement of 15, 30, and 45% of beef by white button (immature *Agaricus bisporus)* or portobello mushrooms (mature *Agaricus bisporus)* | Mushroom type did not affect moisture, cooking yield and texture of hybrid taco, but affected color parameters, where hybrid taco with white mushroom presenting a similar color to control.<br>Blanching treatment and mushroom particle size slightly affected physical parameters of hybrid taco.<br>Color, moisture, and texture were affected by mushroom concentration. | Overall liking was not affected and color acceptability was reduced in hybrid taco. Salt reduction (up to 45%) did not alter liking scores of hybrid products. | *Not evaluated in this study* | Wong et al. (2017) |

| | | | | | |
|---|---|---|---|---|---|
| *Carne asada* | 50% of meat replacement by sautéed crimini mushrooms (*Agaricus bisporus*) | *Not evaluated in this study* | Overall liking, appearance, flavor and texture/mouth decreased significantly in hybrid *carne asada.*<br>The descriptors mushroom, onion, veggie, earthy, salty, and rancid/stale oil flavors, oily/greasy mouthfeel and moisture were more intense in hybrid products. | *Not evaluated in this study* | Guinard et al. (2016), Myrdal Miller et al. (2014) |
| Beef taco | Replacement of beef by 50 and 80% of sautéed white mushrooms (*Agaricus bisporus*) | *Not evaluated in this study* | Consumer acceptance of hybrid beef taco was equal to control, and for some segments of the consumer population was even higher than control.<br>Descriptive analysis shown that an enhancement in overall flavor as well as mushroom, veggie, onion, garlic and earthy flavors, and umami and sweet tastes was observed in hybrid products. | *Not evaluated in this study* | Guinard et al. (2016), Myrdal Miller et al. (2014) |

(continued)

**Table 10.5** (continued)

| Hybrid meat products | Reformulation strategy | Techno-functional effects | Sensory effects | Nutritional effects | References |
|---|---|---|---|---|---|
| Chicken patty | Chicken was replaced by 25 and 50% blanched grey oystermushroom(*Pleurotussajor-caju* – 90.2% moisture) | Moisture content was higher in hybrid patties comparing to control. | Sensory attributes (aroma, color, springiness, juiciness, flavor, and overall liking) of hybrid products were scored equal to control. | Ash, and carbohydrate content was higher, while protein, and fat values were lower in hybrid patties comparing to control. The same behavior was observed for moisture, fat, and protein after 6 month of chilled storage (−18 °C). | Wan Rosli and Solihah (2014) |
| Pork patty | Pork was replaced by 10, 20 and 30% fresh white jelly mushroom (*Tremella fuciformis* – 85.87% moisture) | Hybrid products shown a higher moisture, cooking yield and oil biding capacity than control. Hybrid patties were darker e less red than control. | Appearance, color, flavor, and texture were rated equal in control and hybrid patties. Hybrid patties were more accepted related to oilness, mushroom flavor, and overall liking. | Raw and cooked hybrid products presented higher ash content, and lower protein content than control. Fat content was not affected by reformulation. | Cha et al. (2014) |
| Beef patty | Beef was replaced by 25 and 50% fresh grey oyster mushroom *(Pleurotus sajor-caju)* (PSC) | Hybrid products presented lower cooking yield, moisture retention, fat retention, and shrinkage than control.<br>Moisture values increased in hybrid patties. | Hybrid products had the same sensorial profile (color, juiciness, flavor elasticity, and overall liking) that control. | Hybrid patties had a higher fiber content than control. Protein, and fat content decreased in hybrid patties. | Wan Rosli and Solihah (2012) |
| Chicken patty | Chicken was replaced by 25 and 50% fresh grey oyster mushroom *(Pleurotus sajor-caju)* | Moisture values increased in hybrid patties. | Sensorial attributes (aroma, color, flavor springiness, juiciness, and overall liking) of hybrid patties do not differ for control. | Protein, fat, ash, and carbohydrates content decreased in hybrid patties. | Wan Rosli et al. (2011) |

and food innovativeness) also have influence on the acceptability of these products (Lang 2020). For young consumers (18–35 years) the purchase intention of fungi-hybrid meat products is positively affected by food innovation, the desire of a sustainable diet, and information about sustainability and nutrition (Sogari et al. 2021).

The behavior of consumers of fungi-hybrid meat products was also evaluated by (Sogari et al. 2022). Their findings suggest that previous knowledge about hybrid meat and plant-based products is a positive factor to purchase these products in the near future. Strategies such as nudge and education interventions can increase the desire to order hybrid patties nevertheless of right-wing ideology or meat-eating attitudes or behaviors (Prusaczyk et al. 2021).

The use of sustainability appeal can be a strategy to improve the acceptance of fungi-based hybrid meat products (Hellwig et al. 2020). A previous study showed that providing information about mushroom sustainability has more of an impact on acceptability than information linked to satisfaction and nutrition (Sogari et al. 2022). Consequently, disseminating this information can be a good marketing strategy for the hybrid meat products industry. The optimization of fungi inclusion to produce hybrid meat products is also a good strategy to improve acceptance by consumers (Wong et al. 2017).

Finally, it's important to highlight that further study and enhanced technology are necessary to process edible fungi into meat substitutes, such as improving sensorial properties, production efficacy, separation, and recovery processing technologies (Zhang et al. 2022a). Also, the high number of edible mushrooms and filamentous fungi for mycoprotein production available all over the word increase the range of opportunities to use these natural products as meat alternatives to produce hybrid meat products (Pérez-Montes et al. 2021).

## 10.18 Challenges and Perspectives

The challenges associated with diminishing meat consumption, while simultaneously promoting animal welfare and mitigating environmental impacts, are significant, as elucidated in this chapter. The development of hybrid products involves nutritional, sensory, and technological modifications. One of the perspectives is the development of products that includes a new or altered molecular composition, using different technologies, such as produced through animal cell or tissue cultures (cultured meat), microalgae proteins, or high-moisture extrusion, for example.

Cultured meat, produced by culturing animal cells in scaffolds within bioreactors that contain culture media, represents a significant advancement in meat production. The cells then multiply and differentiate into muscle and fat cells, which can be processed into various forms of meat. Cultured meat has the potential to address several issues associated with traditional meat production, mainly regarding the environmental impact, however, there are a lot of concerns regarding the potential health benefits, safety, environmental benefits, ethical aspects, these products (Chriki and Hocquette 2020).

Besides, they are still facing a challenge regarding legislative issues and up-scaling production to reduce the price. Considering this last aspect, the combination of cultured meat with traditional meat or plant-based ingredients is a way to reduce the price and improve the nutritional quality of the products (Schnettler 2020).

The hybrid cultured meat approach is seen as a potential bridge for advancement in the alternative protein sector, particularly among flexitarians. The Food companies Meatable and Love Handle Singapore plan to open the world's first hybrid meat kitchen and innovation center in Singapore by 2024, aiming to supply restaurants and supermarkets with hybrid meat products from cultured meat and plant-based ingredients (Meatable 2022). Also, according to a recent study focusing on understanding consumer attitudes towards different types of meat alternatives, over one-third of UK consumers would likely purchase or consume hybrid plant-based/cultivated meats if available on the market (Coucke et al. 2023).

Considering the development of hybrid meat products that use high-moisture extrusion technology, the interesting perspective is the formation of a texture that is as close as possible to the experience of eating products with a meat texture. The high-moisture extrusion process constitutes a procedure characterized by elevated temperatures and shear forces. During this process, protein unfolding, aggregation, and cross-linking occur, culminating in the development of fibrous structures resembling meat. This transformation is facilitated by a substantial decrease in temperature at the cooling die (Knoch 2016; Nisov et al. 2022). The fibers form within the cooling die as the hot protein melt is processed and solidifies during flow in the extruder. While meat structure and proteins exhibit a predominantly conserved nature, plant proteins are characterized by their heterogeneity and diverse structures and properties, applying this extrusion forms a different structure and texture to the product, comparable to a meat product (Knoch 2016; Xiong 2023).

Pöri et al. (2023) evaluated this new approach in the processing of hybrid meat products. The authors assessed the high-moisture extrusion of a 1:1 mixture of minced beef with varied fat contents (7% and 17%) combined with either isolated pea protein or textured pea protein. The extrudate-containing meat and textured protein exhibited a significant increase in cross-sectional shear force and compression compared to those with isolated pea protein. Interestingly, the use of high- or low-fat meat did not lead to any differences in these parameters. Moreover, extrudates with textured protein demonstrated more meat-like properties than those with isolated pea protein. The authors suggest that prior texturization might have contributed to fiber formation and reduced the volatile flavor compounds in the pea protein, which could explain the observed results.

Finally, evaluating new protein sources and their production viability is a crucial area of research, in this way, microalgae proteins stand out significantly. Microalgae offer several advantages over conventional protein sources, including a high protein content that can reach up to 50% of their dry mass. They also have a diverse nutritional profile, containing minerals, vitamins, fibers, and UFA, and exhibit high-efficiency growth. However, to be competitive, microalgae protein production must overcome technological limitations and high production costs (Amorim et al. 2021). Despite these challenges, microalgae are an important and favorable protein to be considered in the next years and indicating this great opportunity to develop hybrid meat products.

## 10.19 Conclusion

This chapter aimed to critically review the preeminent sources of alternative proteins for developing hybrid meat products, addressing their techno-functional, nutritional, and sensory effects, as well as the main reformulation strategies. Plant-based proteins, insects, and fungi may play a significant role in the food industry, as their incorporation into hybrid meat products can offer nutritional advantages and a strong sustainability appeal. Hybrids could serve as an essential strategy for gradually transitioning society's diet to reduce meat consumption while still leveraging meat's composition and functionality benefits, along with those demonstrated in the alternative protein chapter.

The wide array of alternative protein sources was evident, with numerous studies, many of them highly advanced and informative, particularly regarding technological properties such as emulsion stability, water retention capacity, texture, and sensory studies. However, the adjustment of process and pre-processing conditions will be vital tools for overcoming the challenges still encountered and for achieving proteins with improved digestibility and nutritional quality.

## References

Ahlborn J, Stephan A, Meckel T, Maheshwari G, Rühl M, Zorn H (2019) Upcycling of food industry side streams by basidiomycetes for production of a vegan protein source. Int J Recycl Org Waste Agric 8(S1):447–455. https://doi.org/10.1007/s40093-019-00317-4

Amorim ML, Soares J, Coimbra JS dos R, Leite MdeO, Albino LFT, Martins MA (2021) Microalgae proteins: production, separation, isolation, quantification, and application in food and feed. Crit Rev Food Sci Nutr 61(12):1976–2002. https://doi.org/10.1080/10408398.2020.1768046

Anzani C, Boukid F, Drummond L, Mullen AM, Álvarez C (2020) Optimising the use of proteins from rich meat co-products and non-meat alternatives: nutritional, technological and allergenicity challenges. Food Res Int 137(May):109575. https://doi.org/10.1016/j.foodres.2020.109575

Argel NS, Ranalli N, Califano AN, Andrés SC (2020) Influence of partial pork meat replacement by pulse flour on physicochemical and sensory characteristics of low-fat burgers. J Sci Food Agric 100(10):3932–3941. https://doi.org/10.1002/jsfa.10436

Argel NS, Lorenzo G, Domínguez R, Fraqueza MJ, Fernández-López J, Sosa ME, Cesar P, Campagnol B, Lorenzo JM, Andrés SC (2022) Hybrid meat products: incorporation of white bean flour in lean pork burgers. Appl Sci 12(15):7571. https://doi.org/10.3390/APP12157571

Asgar MA, Fazilah A, Huda N, Bhat R, Karim AA (2010) Nonmeat protein alternatives as meat extenders and meat analogs. Compr Rev Food Sci Food Saf 9(5):513–529. https://doi.org/10.1111/j.1541-4337.2010.00124.x

Aviles MV, Naef EF, Abalos RA, Lound LH, Gómez MB, Olivera DF (2023) Use of a rice industry by-product as a meat replacer in a hybrid chicken patty: technological and sensory impact. Int J Gastron Food Sci 31(January):100674. https://doi.org/10.1016/j.ijgfs.2023.100674

Awasthi MK, Sarsaiya S, Patel A, Juneja A, Singh RP, Yan B, Awasthi SK, Jain A, Liu T, Duan Y, Pandey A, Zhang Z, Taherzadeh MJ (2020) Refining biomass residues for sustainable energy and bio-products: an assessment of technology, its importance, and strategic applications in circular bio-economy. Renew Sust Energ Rev 127:109876. https://doi.org/10.1016/j.rser.2020.109876

Awasthi MK, Kumar V, Hellwig C, Wikandari R, Harirchi S, Sar T, Wainaina S, Sindhu R, Binod P, Zhang Z, Taherzadeh MJ (2023) Filamentous fungi for sustainable vegan food production systems within a circular economy: present status and future prospects. Food Res Int 164:112318. https://doi.org/10.1016/j.foodres.2022.112318

Banerjee DK, Das AK, Banerjee R, Pateiro M, Nanda PK, Gadekar YP, Biswas S, McClements DJ, Lorenzo JM (2020) Application of enoki mushroom (Flammulina Velutipes) stem wastes as functional ingredients in goat meat nuggets. Food Secur 9(4):432. https://doi.org/10.3390/foods9040432

Bartholomai BM, Ruwe KM, Thurston J, Jha P, Scaife K, Simon R, Abdelmoteleb M, Goodman RE, Farhi M (2022) Safety evaluation of Neurospora crassa mycoprotein for use as a novel meat alternative and enhancer. Food Chem Toxicol 168:113342. https://doi.org/10.1016/j.fct.2022.113342

Barzee TJ, Cao L, Pan Z, Zhang R (2021) Fungi for future foods. J Future Foods 1(1):25–37. https://doi.org/10.1016/j.jfutfo.2021.09.002

Baune MC, Jeske AL, Profeta A, Smetana S, Broucke K, Van Royen G, Gibis M, Weiss J, Terjung N (2021) Effect of plant protein extrudates on hybrid meatballs – Changes in nutritional composition and sustainability. Future Foods 4:100081. https://doi.org/10.1016/J.FUFO.2021.100081

Baune MC, Broucke K, Ebert S, Gibis M, Weiss J, Enneking U, Profeta A, Terjung N, Heinz V (2023) Meat hybrids–An assessment of sensorial aspects, consumer acceptance, and nutritional properties. Front Nutr 10. https://doi.org/10.3389/fnut.2023.1101479

Bednářová M, Borkovcová M, Mlček J, Rop O, Zeman L (2013) Edible insects – Species suitable for entomophagy under condition of Czech Republic. Acta Univ Agric Silvic Mendel Brun 64(3):587–594

Bessa LW, Pieterse E, Marais J, Hoffman LC (2023) Black soldier fly larvae (Hermetia illucens) as a meat replacer in a burger patty. J Insects Food Feed 9(9):1211–1222. https://doi.org/10.3920/JIFF2021.0208

Bessada SMF, Barreira JCM, Oliveira MBPP (2019) Pulses and food security: dietary protein, digestibility, bioactive and functional properties. Trends Food Sci Technol 93:53–68. https://doi.org/10.1016/j.tifs.2019.08.022

Bohrer BM (2017) Review: Nutrient density and nutritional value of meat products and non-meat foods high in protein. Trends Food Sci Technol 65:103–112. https://doi.org/10.1016/j.tifs.2017.04.016

Borges MM, da Costa DV, Trombete FM, Câmara AKFI (2022) Edible insects as a sustainable alternative to food products: an insight into quality aspects of reformulated bakery and meat products. Curr Opin Food Sci 46:100864. https://doi.org/10.1016/j.cofs.2022.100864

Broucke K, Van Poucke C, Duquenne B, De Witte B, Baune M-C, Lammers V, Terjung N, Ebert S, Gibis M, Weiss J, Van Royen G (2022) Ability of (extruded) pea protein products to partially replace pork meat in emulsified cooked sausages. Innov Food Sci Emerg Technol 78:102992. https://doi.org/10.1016/j.ifset.2022.102992

Cavalheiro CP, Ruiz-Capillas C, Herrero AM, Pintado T, Cruz TdaMP, da Silva MCA (2023) Cricket (Acheta domesticus) flour as meat replacer in frankfurters: nutritional, technological, structural, and sensory characteristics. Innov Food Sci Emerg Technol 83(November 2022):103245. https://doi.org/10.1016/j.ifset.2022.103245

Cha M-H, Heo J-Y, Lee C, Lo YM, Moon B (2014) Quality and sensory characterization of white jelly mushroom (*Tremella fuciformis*) as a meat substitute in pork patty formulation. J Food Process Preserv 38(4):2014–2019. https://doi.org/10.1111/jfpp.12178

Chandler SL, McSweeney MB (2022) Characterizing the properties of hybrid meat burgers made with pulses and chicken. Int J Gastron Food Sci 27:100492. https://doi.org/10.1016/j.ijgfs.2022.100492

Charlebois S, Somogyi S, Music J, Caron I (2020) Planet, ethics, health and the new world order in proteins. J Agric Stud 8(3):171. https://doi.org/10.5296/jas.v8i3.16535

Chen K-I, Erh M-H, Su N-W, Liu W-H, Chou C-C, Cheng K-C (2012) Soyfoods and soybean products: from traditional use to modern applications. Appl Microbiol Biotechnol 96(1):9–22. https://doi.org/10.1007/s00253-012-4330-7

Chin KB, Keeton JT, Miller RK, Longnecker MT, Lamkey JW (2000) Evaluation of konjac blends and soy protein isolate as fat replacements in low-fat bologna. J Food Sci 65(5):756–763. https://doi.org/10.1111/j.1365-2621.2000.tb13582.x

Choi YS, Kim TK, Choi HD, Park JD, Sung JM, Jeon KH, Paik HD, Kim YB (2017) Optimization of replacing pork meat with yellow worm (Tenebrio molitor L.) for frankfurters. Korean J Food Sci Anim Resour 37(5):617–625. https://doi.org/10.5851/kosfa.2017.37.5.617

Chriki S, Hocquette J-F (2020) The myth of cultured meat: a review. Front Nutr 7. https://doi.org/10.3389/fnut.2020.00007

Corrin T, Papadopoulos A (2017) Understanding the attitudes and perceptions of vegetarian and plant-based diets to shape future health promotion programs. Appetite 109:40–47. https://doi.org/10.1016/j.appet.2016.11.018

Coucke N, Slabbinck H, Vermeir I (2023) Consumer preferences towards plant-based, hybrid and cultivated meat analogues offered in different meal contexts and at various consumption moments: a choice-based conjoint experimental design and an online survey. Food Qual Prefer 112:105006. https://doi.org/10.1016/J.FOODQUAL.2023.105006

Das AK, Anjaneyulu ASR, Gadekar YP, Singh RP, Pragati H (2008) Effect of full-fat soy paste and textured soy granules on quality and shelf-life of goat meat nuggets in frozen storage. Meat Sci 80(3):607–614. https://doi.org/10.1016/j.meatsci.2008.02.011

De Souza D, Sbardelotto AF, Ziegler DR, Marczak LDF, Tessaro IC (2016) Characterization of rice starch and protein obtained by a fast alkaline extraction method. Food Chem 191:36–44. https://doi.org/10.1016/j.foodchem.2015.03.032

Denny A, Aisbitt B, Lunn J (2008) Mycoprotein and health. Nutr Bull 33(4):298–310. https://doi.org/10.1111/j.1467-3010.2008.00730.x

dos Santos M, da Rocha DAVF, Bernardinelli OD, Oliveira Júnior FD, de Sousa DG, Sabadini E, da Cunha RL, Trindade MA, Pollonio MAR (2022) Understanding the performance of plant protein concentrates as partial meat substitutes in hybrid meat emulsions. Food Secur 11(21). https://doi.org/10.3390/foods11213311

dos Santos M, Ribeiro WO, Monteiro JdeS, dos Santos BA, Campagnol PCB, Pollonio MAR (2023) Effect of transglutaminase treatment on the structure and sensory properties of rice- or soy-based hybrid sausages. Food Secur 12(23). https://doi.org/10.3390/foods12234226

Ebert S, Jungblut F, Herrmann K, Maier B, Terjung N, Gibis M, Weiss J (2022) Influence of wet extrudates from pumpkin seed proteins on drying, texture, and appearance of dry-cured hybrid sausages. Eur Food Res Technol 248(6):1469–1484. https://doi.org/10.1007/S00217-022-03974-4/FIGURES/4

FAO (2021) Looking at edible insects from a food safety perspective. In: Challenges and opportunities for the sector. Rome. https://doi.org/10.4060/cb4094en

FDA (2002) Marlow foods. GRAS notification for mycoprotein (submitted as: US FDA. Marlow Foods Ltd. Stokesley (UK). GRN 091)

Finnigan T, Needham L, Abbott C (2017) Mycoprotein. In: Sustainable protein sources. United States: Academic Press, Elsevier, pp 305–325. https://doi.org/10.1016/B978-0-12-802778-3.00019-6

Foyer CH, Lam HM, Nguyen HT, Siddique KHM, Varshney RK, Colmer TD, Cowling W, Bramley H, Mori TA, Hodgson JM, Cooper JW, Miller AJ, Kunert K, Vorster J, Cullis C, Ozga JA, Wahlqvist ML, Liang Y, Shou H et al (2016) Neglecting legumes has compromised human health and sustainable food production. Nature Plants 2(8):1–10. https://doi.org/10.1038/NPLANTS.2016.112

Gmoser R, Fristedt R, Larsson K, Undeland I, Taherzadeh MJ, Lennartsson PR (2020) From stale bread and brewers spent grain to a new food source using edible filamentous fungi. Bioengineered 11(1):582–598. https://doi.org/10.1080/21655979.2020.1768694

Goldflus F, Ceccantini ML, Santos W (2006) Amino acid content of soybean samples collected in different Brazilian States - Harvest 2003/2004. Braz J Poult Sci 8(2):105–111. https://doi.org/10.1590/S1516-635X2006000200006

Gómez-Corona C, Valentin D (2023) The crispy cricket – attitudes, habits, and tradition in insect consumption. Food Qual Prefer 110(March). https://doi.org/10.1016/j.foodqual.2023.104960

Gordon A, Barbut S, Schmidt G (1992) Mechanisms of meat batter stabilization: a review. Crit Rev Food Sci Nutr 32(4):299–332. https://doi.org/10.1080/10408399209527602

Grasso S, Goksen G (2023) The best of both worlds? Challenges and opportunities in the development of hybrid meat products from the last 3 years. LWT 173(December 2022):114235. https://doi.org/10.1016/j.lwt.2022.114235

Grasso S, Jaworska S (2020) Part meat and part plant: are hybrid meat products fad or future? Food Secur 9(12). https://doi.org/10.3390/foods9121888

Grasso S, Smith G, Bowers S, Ajayi OM, Swainson M (2019) Effect of texturised soy protein and yeast on the instrumental and sensory quality of hybrid beef meatballs. J Food Sci Technol 56(6):3126–3135. https://doi.org/10.1007/s13197-018-3552-9

Grasso S, Asioli D, Smith R (2022) Consumer co-creation of hybrid meat products: a cross-country European survey. Food Qual Prefer 100(October 2021):104586. https://doi.org/10.1016/j.foodqual.2022.104586

Grossi G, Goglio P, Vitali A, Williams AG (2019) Livestock and climate change: impact of livestock on climate and mitigation strategies. Anim Front 9(1):69–76. https://doi.org/10.1093/af/vfy034

Guinard J-X, Myrdal Miller A, Mills K, Wong T, Lee SM, Sirimuangmoon C, Schaefer SE, Drescher G (2016) Consumer acceptance of dishes in which beef has been partially substituted with mushrooms and sodium has been reduced. Appetite 105:449–459. https://doi.org/10.1016/j.appet.2016.06.018

Han X, Li B, Puolanne E, Heinonen M (2023) Hybrid sausages using pork and cricket flour: texture and oxidative storage stability. Food Secur 12(6). https://doi.org/10.3390/foods12061262

Hashempour-Baltork F, Khosravi-Darani K, Hosseini H, Farshi P, Reihani SFS (2020) Mycoproteins as safe meat substitutes. J Clean Prod 253:119958. https://doi.org/10.1016/j.jclepro.2020.119958

Havemeier S, Erickson J, Slavin J (2017) Dietary guidance for pulses: the challenge and opportunity to be part of both the vegetable and protein food groups. Ann N Y Acad Sci 1392(1):58–66. https://doi.org/10.1111/nyas.13308

He J, Evans NM, Liu H, Shao S (2020) A review of research on plant-based meat alternatives: driving forces, history, manufacturing, and consumer attitudes. Compr Rev Food Sci Food Saf 19(5):2639–2656. https://doi.org/10.1111/1541-4337.12610

Hellwig C, Gmoser R, Lundin M, Taherzadeh MJ, Rousta K (2020) Fungi burger from stale bread? A case study on perceptions of a novel protein-rich food product made from an edible fungus. Food Secur 9(8):1112. https://doi.org/10.3390/foods9081112

Henchion M, Hayes M, Mullen AM, Fenelon M, Tiwari B (2017) Future protein supply and demand: strategies and factors influencing a sustainable equilibrium. Food Secur 6(7):1–21. https://doi.org/10.3390/foods6070053

Hoek AC, Luning PA, Weijzen P, Engels W, Kok FJ, de Graaf C (2011) Replacement of meat by meat substitutes. A survey on person- and product-related factors in consumer acceptance. Appetite 56(3):662–673. https://doi.org/10.1016/j.appet.2011.02.001

Hoek AC, Elzerman JE, Hageman R, Kok FJ, Luning PA, de Graaf C (2013) Are meat substitutes liked better over time? A repeated in-home use test with meat substitutes or meat in meals. Food Qual Prefer 28(1):253–263. https://doi.org/10.1016/j.foodqual.2012.07.002

Huang Y, Cao D, Chen Z, Chen B, Li J, Guo J, Dong Q, Liu L, Wei Q (2021) Red and processed meat consumption and cancer outcomes: umbrella review. Food Chem 356(March):129697. https://doi.org/10.1016/j.foodchem.2021.129697

Husain H, Huda-Faujan N (2020) Potential application of grey oyster mushroom stems as Halal meat replacer in imitation chicken nuggets. Food Res 4(S1):179–186. https://doi.org/10.26656/fr.2017.4(S1).S18

Iqbal A, Khalil IA, Ateeq N, Sayyar Khan M (2006) Nutritional quality of important food legumes. Food Chem 97(2):331–335. https://doi.org/10.1016/J.FOODCHEM.2005.05.011

Kalač P (2009) Chemical composition and nutritional value of European species of wild growing mushrooms: A review. Food Chem 113(1):9–16. https://doi.org/10.1016/j.foodchem.2008.07.077

Kalman DS (2014) Amino acid composition of an organic brown rice protein concentrate and isolate compared to soy and whey concentrates and isolates. Food Secur 3(3):394–402. https://doi.org/10.3390/FOODS3030394

Kemper JA, White SK (2021) Young adults' experiences with flexitarianism: the 4Cs. Appetite 160:105073. https://doi.org/10.1016/j.appet.2020.105073

Kiiru SM, Kinyuru JN, Kiage BN, Martin A, Marel AK, Osen R (2020) Extrusion texturization of cricket flour and soy protein isolate: influence of insect content, extrusion temperature, and moisture-level variation on textural properties. J Food Sci Nutr 8(8):4112–4120. https://doi.org/10.1002/fsn3.1700

Kim HW, Setyabrata D, Lee YJ, Jones OG, Kim YHB (2016) Pre-treated mealworm larvae and silkworm pupae as a novel protein ingredient in emulsion sausages. Innov Food Sci Emerg Technol 38:116–123. https://doi.org/10.1016/j.ifset.2016.09.023

Kim HW, Setyabrata D, Lee YJ, Jones OG, Kim YHB (2017) Effect of house cricket (Acheta domesticus) flour addition on physicochemical and textural properties of meat emulsion under various formulations. J Food Sci 82(12):2787–2793. https://doi.org/10.1111/1750-3841.13960

Knoch A (2016) Production of restructured meatlike products by high moisture extrusion technology. In: Reference module in food science. United States: Elsevier. https://doi.org/10.1016/B978-0-08-100596-5.03280-7

Kolobe SD, Manyelo TG, Malematja E, Sebola NA, Mabelebele M (2023) Fats and major fatty acids present in edible insects utilised as food and livestock feed. Veter Anim Sci 22(September):100312. https://doi.org/10.1016/j.vas.2023.100312

Kurek MA, Onopiuk A, Pogorzelska-Nowicka E, Szpicer A, Zalewska M, Półtorak A (2022) Novel protein sources for applications in meat-alternative products—insight and challenges. Food Secur 11(7):957. https://doi.org/10.3390/foods11070957

Kyriakopoulou K, Keppler JK, van der Goot AJ (2021) Functionality of ingredients and additives in plant-based meat analogues. In: Foods, vol 10, Issue 3. https://doi.org/10.3390/foods10030600

La Barbera F, Verneau F, Amato M, Grunert K (2018) Understanding Westerners' disgust for the eating of insects: the role of food neophobia and implicit associations. Food Qual Prefer 64(September 2017):120–125. https://doi.org/10.1016/j.foodqual.2017.10.002

Lang M (2020) Consumer acceptance of blending plant-based ingredients into traditional meat-based foods: evidence from the meat-mushroom blend. Food Qual Prefer 79:103758. https://doi.org/10.1016/j.foodqual.2019.103758

Lange KW, Nakamura Y (2021) Edible insects as future food: chances and challenges. J Future Foods 1(1):38–46. https://doi.org/10.1016/j.jfutfo.2021.10.001

Lemke B, Siekmann L, Grabowski NT, Plötz M, Krischek C (2023) Impact of the addition of Tenebrio molitor and Hermetia illucens on the physicochemical and sensory quality of cooked meat products. Insects 14(5). https://doi.org/10.3390/insects14050487

Leonard W, Hutchings SC, Warner RD, Fang Z (2019) Effects of incorporating roasted lupin (Lupinus angustifolius) flour on the physicochemical and sensory attributes of beef sausage. Int J Food Sci Technol 54(5):1849–1857. https://doi.org/10.1111/ijfs.14088

Londero PMG, Prestes RC, Roman SS, Kubota EH, Hübscher GH, Silva CG, Fischer L, Alves JS (2015) Development of functional beef meatball using landrace common bean flour as a substitute for meat and textured soy protein. Int Food Res J 22(5):1615–1623

Lu M, Zhu C, Smetana S, Zhao M, Zhang H, Zhang F, Du Y (2024) Minerals in edible insects: a review of content and potential for sustainable sourcing. Food Sci Human Wellness 13(1):65–74. https://doi.org/10.26599/FSHW.2022.9250005

Lucas-González R, Fernández-López J, Pérez-Álvarez JA, Viuda-Martos M (2019) Effect of drying processes in the chemical, physico-chemical, techno-functional and antioxidant properties of flours obtained from house cricket (Acheta domesticus). Eur Food Res Technol 245(7):1451–1458. https://doi.org/10.1007/s00217-019-03301-4

Luna GC, Martin-Gonzalez FS, Mauer LJ, Liceaga AM (2021) Cricket (Acheta domesticus) protein hydrolysates' impact on the physicochemical, structural and sensory properties of tortillas and tortilla chips. J Insects Food Feed 7(1):109–120. https://doi.org/10.3920/JIFF2020.0010

Mäkinen OE, Sozer N, Ercili-Cura D, Poutanen K (2016) Protein from oat: structure, processes, functionality, and nutrition. In: Sustainable protein sources, pp 105–119. United States: Academic Press. https://doi.org/10.1016/B978-0-12-802778-3.00006-8

Makkar HPS, Tran G, Heuzé V, Ankers P (2014) State-of-the-art on use of insects as animal feed. Anim Feed Sci Technol 197:1–33. https://doi.org/10.1016/j.anifeedsci.2014.07.008

Maphosa Y, Jideani VA (2017) The role of legumes in human nutrition. In: IntechOpen, pp 103–121. https://doi.org/10.5772/intechopen.69127

Martins VMG, Milano P, Pollonio MAR, dos Santos M, de Oliveira AP, Savay-da-Silva LK, Câmara AKFI, Paglarini CdeS (2023) Adding cricket (Gryllus assimilis) flour in hybrid beef burger: physicochemical, technological and sensory challenges. Int J Food Sci Technol 1–10. https://doi.org/10.1111/ijfs.16826

Meatable and plant-based butcher Love Handle partner to build world's first hybrid meat innovation center in Singapore (2022) Press release: Meatable and plant-based butcher Love Handle partner to build world's first hybrid meat innovation center in Singapore. https://meatable.com/meatable-and-love-handle/

Megido RC, Gierts C, Blecker C, Brostaux Y, Haubruge É, Alabi T, Francis F (2016) Consumer acceptance of insect-based alternative meat products in Western countries. Food Qual Prefer 52:237–243. https://doi.org/10.1016/j.foodqual.2016.05.004

Melgar-Lalanne G, Hernández-Álvarez AJ, Salinas-Castro A (2019) Edible insects processing: traditional and innovative technologies. Compr Rev Food Sci Food Saf 18:1166–1191. https://doi.org/10.1111/1541-4337.12463

Milford AB, Le Mouël C, Bodirsky BL, Rolinski S (2019) Drivers of meat consumption. Appetite 141(January):104313. https://doi.org/10.1016/j.appet.2019.06.005

Millar KA, Gallagher E, Burke R, McCarthy S, Barry-Ryan C (2019) Proximate composition and anti-nutritional factors of fava-bean (Vicia faba), green-pea and yellow-pea (Pisum sativum) flour. J Food Compos Anal 82:103233. https://doi.org/10.1016/J.JFCA.2019.103233

Mishyna M, Chen J, Benjamin O (2020) Sensory attributes of edible insects and insect-based foods – Future outlooks for enhancing consumer appeal. Trends Food Sci Technol 95(November 2019):141–148. https://doi.org/10.1016/j.tifs.2019.11.016

Molfetta M, Morais EG, Barreira L, Bruno GL, Porcelli F, Dugat-Bony E, Bonnarme P, Minervini F (2022) Protein sources alternative to meat: state of the art and involvement of fermentation. Food Secur 11(14):2065. https://doi.org/10.3390/foods11142065

Moura MAF, Alves VS, Takahashi JA (2022) Nutritional quality, techno-functional characteristics, and safety of biomass powder and protein isolate produced from Penicillium maximae. Food Secur 11(22):3621. https://doi.org/10.3390/foods11223621

Moure A, Sineiro J, Domínguez H, Parajó JC (2006) Functionality of oilseed protein products: a review. Food Res Int 39(9):945–963. https://doi.org/10.1016/j.foodres.2006.07.002

Myrdal Miller A, Mills K, Wong T, Drescher G, Lee SM, Sirimuangmoon C, Schaefer S, Langstaff S, Minor B, Guinard J-X (2014) Flavor-enhancing properties of mushrooms in meat-based dishes in which sodium has been reduced and meat has been partially substituted with mushrooms. J Food Sci 79(9). https://doi.org/10.1111/1750-3841.12549

Neville M, Tarrega A, Hewson L, Foster T (2017) Consumer-orientated development of hybrid beef burger and sausage analogues. Food Sci Nutr 5(4):852–864. https://doi.org/10.1002/fsn3.466

Nishinari K, Fang Y, Guo S, Phillips GO (2014) Soy proteins: a review on composition, aggregation and emulsification. Food Hydrocoll 39:301–318. https://doi.org/10.1016/j.foodhyd.2014.01.013

Nisov A, Nikinmaa M, Nordlund E, Sozer N (2022) Effect of pH and temperature on fibrous structure formation of plant proteins during high-moisture extrusion processing. Food Res Int 156:111089. https://doi.org/10.1016/j.foodres.2022.111089

Noriham A, Muhammad Ariffaizuddin R, Noorlaila A, Faris Zakry AN (2016) Potential use of okara as meat replacer in beef sausage. J Teknol 78(6–6):13–18. https://doi.org/10.11113/jt.v78.9015

Ojha S, Bekhit AE-D, Grune T, Schlüter OK (2021) Bioavailability of nutrients from edible insects. Curr Opin Food Sci 41:240–248. https://doi.org/10.1016/j.cofs.2021.08.003

Oliveira LA, Pereira SMS, Dias KA, Paes SdaS, Grancieri M, Jimenez LGS, de Carvalho CWP, de Oliveira EE, Martino HSD, Della Lucia CM (2024) Nutritional content, amino acid profile, and protein properties of edible insects (Tenebrio molitor and Gryllus assimilis) powders at different stages of development. J Food Compos Anal 125(November 2023):105804. https://doi.org/10.1016/j.jfca.2023.105804

Olmedilla-Alonso B, Jiménez-Colmenero F, Sánchez-Muniz FJ (2013) Development and assessment of healthy properties of meat and meat products designed as functional foods. Meat Sci 95(4):919–930. https://doi.org/10.1016/j.meatsci.2013.03.030

Oonincx DGAB, de Boer IJM (2012) Environmental impact of the production of mealworms as a protein source for humans - a life cycle assessment. PLoS One 7(12):1–5. https://doi.org/10.1371/journal.pone.0051145

Ordoñez-Araque R, Egas-Montenegro E (2021) Edible insects: a food alternative for the sustainable development of the planet. Int J Gastron Food Sci 23(September 2020). https://doi.org/10.1016/j.ijgfs.2021.100304

Park YS, Choi YS, Hwang KE, Kim TK, Lee CW, Shin DM, Han SG (2017) Physicochemical properties of meat batter added with edible silkworm pupae (Bombyx mori) and transglutaminase. Korean J Food Sci Anim Resour 37(3):351–359. https://doi.org/10.5851/kosfa.2017.37.3.351

Patinho I, Saldaña E, Selani MM, Teixeira ACB, Menegali BS, Merlo TC, Rios-Mera JD, Dargelio MDB, Rodrigues H, Contreras-Castillo CJ (2021) Original burger (traditional) or burger with mushroom addition? A social representation approach to novel foods. Food Res Int 147:110551. https://doi.org/10.1016/j.foodres.2021.110551

Patinho I, Cavalcante CL, Saldaña E, Gagaoua M, Behrens JH, Contreras-castillo CJ (2024) Assessment of beef sensory attributes and physicochemical characteristics: a comparative study of intermediate versus normal ultimate pH striploin cuts. Food Res Int 175(November 2023). https://doi.org/10.1016/j.foodres.2023.113778

Payne CLR, Scarborough P, Rayner M, Nonaka K (2016) Are edible insects more or less "healthy" than commonly consumed meats? A comparison using two nutrient profiling models developed to combat over- and undernutrition. Eur J Clin Nutr 70(3):285–291. https://doi.org/10.1038/ejcn.2015.149

Pérez-Montes A, Rangel-Vargas E, Lorenzo JM, Romero L, Santos EM (2021) Edible mushrooms as a novel trend in the development of healthier meat products. Curr Opin Food Sci 37:118–124. https://doi.org/10.1016/j.cofs.2020.10.004

Petrat-Melin B, Dam S (2023) Textural and consumer-aided characterisation and acceptability of a hybrid meat and plant-based burger patty. Food Secur 12(11). https://doi.org/10.3390/foods12112246

Pöri P, Aisala H, Liu J, Lille M, Sozer N (2023) Structure, texture, and sensory properties of plant-meat hybrids produced by high-moisture extrusion. LWT 173:114345. https://doi.org/10.1016/j.lwt.2022.114345

Profeta A, Baune M-C, Smetana S, Bornkessel S, Broucke K, Van Royen G, Enneking U, Weiss J, Heinz V, Hieke S, Terjung N (2021) Preferences of German consumers for meat products blended with plant-based proteins. Sustain For 13(2). https://doi.org/10.3390/su13020650

Prusaczyk E, Earle M, Hodson G (2021) A brief nudge or education intervention delivered online can increase willingness to order a beef-mushroom burger. Food Qual Prefer 87:104045. https://doi.org/10.1016/j.foodqual.2020.104045

Puolanne EJ, Ruusunen MH, Vainionpää JI (2001) Combined effects of NaCl and raw meat pH on water-holding in cooked sausage with and without added phosphate. Meat Sci 58(1):1–7. https://doi.org/10.1016/S0309-1740(00)00123-6

Purchas RW, Morel PCH, Janz JAM, Wilkinson BHP (2009) Chemical composition characteristics of the longissimus and semimembranosus muscles for pigs from New Zealand and Singapore. Meat Sci 81(3):540–548. https://doi.org/10.1016/j.meatsci.2008.10.008

Qing Z, Cheng J, Wang X, Tang D, Liu X, Zhu M (2021) The effects of four edible mushrooms (Volvariella volvacea, Hypsizygus marmoreus, Pleurotus ostreatus and Agaricus bisporus) on physicochemical properties of beef paste. LWT 135:110063. https://doi.org/10.1016/j.lwt.2020.110063

Queiroz LS, Nogueira Silva NF, Jessen F, Mohammadifar MA, Stephani R, Fernandes de Carvalho A, Perrone ÍT, Casanova F (2023) Edible insect as an alternative protein source: a review on the chemistry and functionalities of proteins under different processing methods. Heliyon 9(4). https://doi.org/10.1016/j.heliyon.2023.e14831

Rebello CJ, Greenway FL, Finley JW (2014) A review of the nutritional value of legumes and their effects on obesity and its related co-morbidities. 5. https://doi.org/10.1111/obr.12144

Rivas-Navia DM, Dueñas-Rivadeneira AA, Dueñas-Rivadeneira JP, Aransiola SA, Maddela NR, Prasad R (2023) Bioactive compounds of insects for food use: potentialities and risks. J Agric Food Res 14(June). https://doi.org/10.1016/j.jafr.2023.100807

Rousta N, Hellwig C, Wainaina S, Lukitawesa L, Agnihotri S, Rousta K, Taherzadeh MJ (2021) Filamentous fungus Aspergillus oryzae for food: from submerged cultivation to fungal burgers and their sensory evaluation—a pilot study. Food Secur 10(11):2774. https://doi.org/10.3390/foods10112774

Ruby MB, Rozin P, Chan C (2015) Determinants of willingness to eat insects in the USA and India. J Insects Food Feed 1(3):215–225. https://doi.org/10.3920/JIFF2015.0029

Rumpold BA, Schlüter OK (2013) Nutritional composition and safety aspects of edible insects. Molec Nutr Food Res 57(5):802–823. https://doi.org/10.1002/mnfr.201200735

Ryder C, Jaworska S, Grasso S (2023) Hybrid meat products and co-creation: what do consumers say, feel and think? Front Nutr 10. https://doi.org/10.3389/fnut.2023.1106079

Sánchez-Chino X, Jiménez-Martínez C, Dávila-Ortiz G, Álvarez-González I, Madrigal-Bujaidar E (2015) Nutrient and nonnutrient components of legumes, and its chemopreventive activity: a review. Nutr Cancer 67(3):401–410. https://doi.org/10.1080/01635581.2015.1004729

Schnettler D (2020) A hybrid approach to cultivated meat. Ind Biotechnol 16(6):346–348. https://doi.org/10.1089/ind.2020.29231.dsc

Scholliers J, Steen L, Fraeye I (2020a) Partial replacement of meat by superworm (Zophobas morio larvae) in cooked sausages: effect of heating temperature and insect:Meat ratio on structure and physical stability. Innov Food Sci Emerg Technol 66(June):102535. https://doi.org/10.1016/j.ifset.2020.102535

Scholliers J, Steen L, Fraeye I (2020b) Structure and physical stability of hybrid model systems containing pork meat and superworm (Zophobas morio larvae): the influence of heating regime and insect: meat ratio. Innov Food Sci Emerg Technol 65(July):102452. https://doi.org/10.1016/j.ifset.2020.102452

Schweigert BS, Guthneck BT, Kraybill HR, Greenwood DA (1949) The amino acid composition of pork and lamb cuts. J Biol Chem 180(3):1077–1083. https://doi.org/10.1016/S0021-9258(19)51222-4

Serna-Saldivar SO (2010) Cereal grains: properties, processing and nutritional attributes. In: Food preservation technology series. United States: CRC PRESS, p 1699.

Sheen F, Lim AJY, Forde CG (2023) Diversity among flexitarian consumers; stratifying meat reducers by their underlying motivations to move to a plant-based diet. Food Qual Prefer 112(September):105022. https://doi.org/10.1016/j.foodqual.2023.105022

Singh S, Bhat HF, Kumar S, Lone AB, Aadil RM, Aït-Kaddour A, Hassoun A, Proestos C, Bhat ZF (2023) Ultrasonication and microwave pre-treated locust protein hydrolysates enhanced the storage stability of meat emulsion. Ultrason Sonochem 98(June):106482. https://doi.org/10.1016/j.ultsonch.2023.106482

Smetana S, Pernutz C, Toepfl S, Heinz V, Van Campenhout L (2018) High-moisture extrusion with insect and soy protein concentrates: cutting properties of meat analogues under insect content and barrel temperature variations. J Insects Food Feed 5(1):29–34. https://doi.org/10.3920/JIFF2017.0066

Sogari G, Li J, Wang Q, Lefebvre M, Gómez MI, Mora C (2021) Factors influencing the intention to purchase meat-mushroom blended burgers among college students. Food Qual Prefer 90:104169. https://doi.org/10.1016/j.foodqual.2020.104169

Sogari G, Li J, Wang Q, Lefebvre M, Huang S, Mora C, Gómez MI (2022) Toward a reduced meat diet: University North American students' acceptance of a blended meat-mushroom burger. Meat Sci 187:108745. https://doi.org/10.1016/j.meatsci.2022.108745

Souza Filho PF, Nair RB, Andersson D, Lennartsson PR, Taherzadeh MJ (2018) Vegan-mycoprotein concentrate from pea-processing industry byproduct using edible filamentous fungi. Fungal Biol Biotechnol 5(1):5. https://doi.org/10.1186/s40694-018-0050-9

Souza Filho PF, Andersson D, Ferreira JA, Taherzadeh MJ (2019) Mycoprotein: environmental impact and health aspects. World J Microbiol Biotechnol 35(10):147. https://doi.org/10.1007/s11274-019-2723-9

Spencer M, Cienfuegos C, Guinard J-X (2018) The Flexitarian Flip™ in university dining venues: student and adult consumer acceptance of mixed dishes in which animal protein has been partially replaced with plant protein. Food Qual Prefer 68:50–63. https://doi.org/10.1016/j.foodqual.2018.02.003

Stephan A, Ahlborn J, Zajul M, Zorn H (2018) Edible mushroom mycelia of Pleurotus sapidus as novel protein sources in a vegan boiled sausage analog system: functionality and sensory tests in comparison to commercial proteins and meat sausages. Eur Food Res Technol 244(5):913–924. https://doi.org/10.1007/s00217-017-3012-1

Stoffel F, Santana WdeO, Gregolon JGN, Kist TBL, Fontana RC, Camassola M (2019) Production of edible mycoprotein using agroindustrial wastes: influence on nutritional, chemical and biological properties. Innovative Food Sci Emerg Technol 58:102227. https://doi.org/10.1016/j.ifset.2019.102227

Strong PJ, Self R, Allikian K, Szewczyk E, Speight R, O'Hara I, Harrison MD (2022) Filamentous fungi for future functional food and feed. Curr Opin Biotechnol 76:102729. https://doi.org/10.1016/j.copbio.2022.102729

Tahmasebi M, Labbafi M, Emam-Djomeh Z, Yarmand MS (2016) Manufacturing the novel sausages with reduced quantity of meat and fat: the product development, formulation optimization, emulsion stability and textural characterization. LWT Food Sci Technol 68:76–84. https://doi.org/10.1016/j.lwt.2015.12.011

Talens C, Llorente R, Simó-Boyle L, Odriozola-Serrano I, Tueros I, Ibargüen M (2022) Hybrid sausages: modelling the effect of partial meat replacement with broccoli, upcycled brewer's spent grain and insect flours. Food Secur 11(21). https://doi.org/10.3390/foods11213396

Tavares PPLG, Lima MdosS, Pessôa LC, Bulos RBdeA, de Oliveira TTB, da Silva Cruz LF, Assis DdeJ, Morte ESdaB, Ribeiro CVDM, Souza COde (2022) Innovation in alternative food sources: a review of a technological state-of-the-art of insects in food products. Food Secur 11(3792):1–29. https://doi.org/10.3390/foods11233792

United Nations (2015) Resolution adopted by the General Assembly on 25 September 2015 70/1. Transforming our world: the 2030 Agenda for Sustainable Development

van Huis A, Rumpold B (2023) Strategies to convince consumers to eat insects? A review. Food Qual Prefer 110(June):104927. https://doi.org/10.1016/j.foodqual.2023.104927

Walkowiak K, Kowalczewski PL, Kubiak P, Baranowska HM (2019) Effect of cricket powder addition on 1h Nmr mobility and texture of pork pate. J Microbiol Biotechnol Food Sci 9(2):191–194. https://doi.org/10.15414/jmbfs.2019.9.2.191-194

Wan Rosli WI, Solihah MA (2012) Effect on the addition of Pleurotus sajor-caju (PSC) on physical and sensorial properties of beef patty. Int Food Res J 19(3):993–999

Wan Rosli WI, Solihah MA (2014) Nutritional composition and sensory properties of oyster mushroom-based patties packed with biodegradable packaging. Sains Malays 43(1):65–71

Wan Rosli WI, Solihah MA, Mohsin SSJ (2011) On the ability of oyster mushroom (Pleurotus sajor-caju) confering changes in proximate composition and sensory evaluation of chicken patty. Int Food Res J 18(4):1463–1469

Wang YS, Shelomi M (2017) Review of black soldier fly (Hermetia illucens) as animal feed and human food. Food Secur 6(10). https://doi.org/10.3390/foods6100091

Wang L, Guo H, Liu X, Jiang G, Li C, Li X, Li Y (2019a) Roles of Lentinula edodes as the pork lean meat replacer in production of the sausage. Meat Sci 156:44–51. https://doi.org/10.1016/j.meatsci.2019.05.016

Wang L, Li C, Ren L, Guo H, Li Y (2019b) Production of pork sausages using *Pleaurotus eryngii* with different treatments as replacements for pork back fat. J Food Sci 84(11):3091–3098. https://doi.org/10.1111/1750-3841.14839

Wan-Mohtar WAAQI, Halim-Lim SA, Kamarudin NZ, Rukayadi Y, Abd Rahim MH, Jamaludin AA, Ilham Z (2020) Fruiting-body-base flour from an Oyster mushroom waste in the development of antioxidative chicken patty. J Food Sci 85(10):3124–3133. https://doi.org/10.1111/1750-3841.15402

WHO (2007) Joint WHO/FAO/UNU Expert Consultation. Protein and amino acid requirements in human nutrition. World Health Organ Tech Rep Ser. PMID: 18330140.:1–265

Wikandari R, Tanugraha DR, Yastanto AJ, Manikharda Gmoser R, Teixeira JA (2023) Development of meat substitutes from filamentous fungi cultivated on residual water of tempeh factories. Molecules 28(3):997 https://doi.org/10.3390/molecules28030997

Wittkopp PJ, Beldade P (2009) Development and evolution of insect pigmentation: genetic mechanisms and the potential consequences of pleiotropy. Semin Cell Dev Biol 20(1):65–71. https://doi.org/10.1016/j.semcdb.2008.10.002

Wong KM, Decker EA, Autio WR, Toong K, DiStefano G, Kinchla AJ (2017) Utilizing mushrooms to reduce overall sodium in taco filling using physical and sensory evaluation. J Food Sci 82(10):2379–2386. https://doi.org/10.1111/1750-3841.13838

Xiao Y, Xing G, Rui X, Li W, Chen X, Jiang M, Dong M (2015) Effect of solid-state fermentation with Cordyceps militaris SN-18 on physicochemical and functional properties of chickpea (Cicer arietinum L.) flour. LWT Food Sci Technol 63(2):1317–1324. https://doi.org/10.1016/j.lwt.2015.04.046

Xiong YL (1994) Myofibrillar protein from different muscle fiber types: implications of biochemical and functional properties in meat processing. Crit Rev Food Sci Nutr 34(3):293–320. https://doi.org/10.1080/10408399409527665

Xiong YL (2023) Meat and meat alternatives: where is the gap in scientific knowledge and technology? Ital J Anim Sci 22(1):482–496. https://doi.org/10.1080/1828051X.2023.2211988

Yang HS, Choi SG, Jeon JT, Park GB, Joo ST (2007) Textural and sensory properties of low fat pork sausages with added hydrated oatmeal and tofu as texture-modifying agents. Meat Sci 75(2):283–289. https://doi.org/10.1016/j.meatsci.2006.07.013

Yang H, Zhang W, Li T, Zheng H, Khan MA, Xu X, Sun J, Zhou G (2016) Effect of protein structure on water and fat distribution during meat gelling. Food Chem 204:239–245. https://doi.org/10.1016/j.foodchem.2016.01.053

Yong HI, Kim TK, Cha JY, Lee JH, Kang MC, Jung S, Choi YS (2023) Effects of edible insect extracts on the antioxidant, physiochemical, and microbial properties of Tteokgalbi during refrigerated storage. Food Bioscience 52(November 2022):102377. https://doi.org/10.1016/j.fbio.2023.102377

Yuan X, Jiang W, Zhang D, Liu H, Sun B (2021) Textural, sensory and volatile compounds analyses in formulations of sausages analogue elaborated with edible mushrooms and soy protein isolate as meat substitute. Food Secur 11(1):52. https://doi.org/10.3390/foods11010052

Zeng B, Nilsson K, Teixeira PG, Bergenståhl B (2023) Study of mycoprotein extraction methods and its functional properties. Colloids Surf A Physicochem Eng Asp 659:130800. https://doi.org/10.1016/j.colsurfa.2022.130800

Zhang C, Guan X, Yu S, Zhou J, Chen J (2022a) Production of meat alternatives using live cells, cultures and plant proteins. Curr Opin Food Sci 43:43–52. https://doi.org/10.1016/j.cofs.2021.11.002

Zhang F, Cao C, Kong B, Sun F, Shen X, Yao X, Liu Q (2022b) Pre-dried mealworm larvae flour could partially replace lean meat in frankfurters: effect of pre-drying methods and replacement ratios. Meat Sci 188(November 2021):108802. https://doi.org/10.1016/j.meatsci.2022.108802

Zhang Y, Bai G, Jin G, Wang Y, Wang J, Puolanne E, Cao J (2022c) Role of low molecular additives in the myofibrillar protein gelation: underlying mechanisms and recent applications. Crit Rev Food Sci Nutr 0(0):1–19. https://doi.org/10.1080/10408398.2022.2133078

Zhang K, Zang M, Wang S, Zhang Z, Li D, Li X (2023) Development of meat analogs: focus on the current status and challenges of regulatory legislation. Compr Rev Food Sci Food Saf 22(2):1006–1029. https://doi.org/10.1111/1541-4337.13098

# Chapter 11
# Rheological and Textural Properties of Food Analogues

**Ilkem Demirkesen, Behic Mert, Rui Costa, and Tomasz Szablewski**

## 11.1 Introduction

The global food situation is changing dramatically, driven by a growing desire for environmentally friendly and health-conscious eating options. As a result, there has been a rise in the creation and introduction of new plant-based product alternatives to traditional animal-based products in the food industry. This seismic shift is demonstrated by the rise of meat, dairy, egg, and fish analogues, for example, which are accurately engineered to replicate the textural, rheological attributes as well as nutritional profile of traditional animal-based products while also addressing pressing concerns about environmental sustainability, health, and animal welfare (McClements et al. 2021).

The environmental imperative of shifting towards plant-based diets is a central driving force behind the rise of analogue foods. A substantial portion of plant-based

---

I. Demirkesen
Department of Animal Health, Food and Feed Research, General Directorate of Agricultural Research and Policies, Ministry of Agriculture and Forestry, Ankara, Turkey

B. Mert (✉)
Department of Food Engineering, Middle East Technical University, Ankara, Turkey
e-mail: bmert@tarimorman.gov.tr

R. Costa
Polytechnic Institute of Coimbra, Coimbra Agriculture School, Coimbra, Portugal

Research Centre for Natural Resources, Environment and Society (CERNAS), Coimbra Agriculture School, Coimbra, Portugal
e-mail: ruicosta@esac.pt

T. Szablewski
Department of Food Quality and Safety Management, Faculty of Food Science and Nutrition, Poznań University of Life Sciences, Poznan, Poland
e-mail: tomasz.szablewski@up.poznan.pl



Ö. P. CAN et al. (eds.), *Food Analogues: Emerging Methods and Challenges*,
https://doi.org/10.1007/978-3-031-69872-9_11

resources is diverted to feed livestock, resulting in significant energy losses in the conversion of plant nutrients into animal products. Redirecting these resources towards direct human consumption can yield up to 70% more calories, potentially feeding billions more people. Unsustainable livestock production further aggravates the global challenges of climate change, resource depletion, biodiversity loss, and greenhouse gas emissions. In a world where animal products contribute to only 18% of global caloric intake, but demand 83% of cropland and are responsible for 56% of food-related greenhouse gas emissions, transitioning to sustainable diets could have a profound impact on mitigating these crises (Cassidy et al. 2013). Due to the increasing need for analogue products, considerable research and development efforts have resulted in new processing methods and technologies, altering the food production environment.

Recently, the research has focused on novel formulating ingredients, structuring technologies, and product formulations. However, there is still a significant deficit in standardized procedures for analyzing, evaluating, and comparing food analogues (McClements et al. 2021). Rheological and textural properties are essential tools to compare the quality of food products, and in the case of analogues, as these compare with original animal origin products, as they directly influence consumer acceptance and overall product quality. This chapter provides a comprehensive overview of the rheological and textural properties of meat, dairy and egg analogues, focusing on key parameters related to ingredients and measurement techniques used in their characterization. Furthermore, the chapter explores the relationship between rheological and textural properties, highlighting how changes in rheology can influence the final texture of food analogues. Emphasis is placed on the importance of rheological and textural profiling, which combines rheological and textural data to develop a comprehensive understanding of overall the quality attributes of food analogues. By elucidating the rheological and textural properties of these alternative food options, this chapter also aims to facilitate the creation of high-quality products that meet consumer expectations in terms of sensory experience and nutritional value.

## 11.2 Rheological and Textural Properties of Meat Analogues

### *11.2.1 The Effect of Formulation on Rheological and Textural Properties of Meat Analogues*

The unique texture of meat arises from the complicated structure of its muscle tissue, composed of repetitive units called sarcomeres that unite to form muscle fibers. This framework is enveloped by connective tissue and cell membranes and may include intramuscular fat in certain meat varieties (McClements et al. 2021). The sensory qualities of meat, which are perceived by our senses, are closely tied to its protein matrix architecture, protein aggregates, fats or emulsions, and the retention of water. Consequently, terms such as firmness, tenderness, juiciness, and overall mouthfeel describe these textural attributes (Sha and Xiong 2020).

The creation of texture in processed meat hinges on the emulsifying and gelling and abilities of myofibrillar proteins, such as myosin (pre-rigor) or actomyosin (post-rigor). A crucial role of myofibrillar proteins is their capacity to emulsify fats. Myofibrillar protein gels, due to their adhesive properties, bind meat particles together and retain water, thereby contributing to the overall structure and integrity of meat products like patties and frankfurters. Myosin, characterized by its distinctive fibrous structure, excels at forming viscoelastic and adhesive gels during cooking. In meat processing, mechanical actions break down large fat particles into smaller granules. Myosin and actomyosin then adhere to the interface between fat and water, with nonpolar regions in the fat and polar regions in the watery phase, further enhancing the structure and integrity of products like patties and frankfurters (Sha and Xiong 2020). Consequently, a meat protein gel is a viscoelastic structure formed by cross-linked proteins that have unfolded due to heat, low pH and bivalent ions. As the gel cools, it gains elasticity due to strengthened hydrogen bonding. This phenomenon is responsible for the characteristic juiciness and ease of slicing found in cooked meat products (Sha and Xiong 2020).

The meat analogues can be formulated by vegetable protein sources such as soy, pea, etc., as well as microalgae, by using texturing agents and a shearing process to give fibrous texture. The resulting texture, however, has been described as highly elastic, rubbery, and inflexible, resulting in a bad mouthfeel. Furthermore, it has been suggested that animal-based meat analogues are more digestible during the digestive phase than their plant-based counterparts. These variations were directly linked to protein secondary structure, disulfide bond creation, and the perceived viscosity of the digesting solution (Xie et al. 2022). Therefore, imitating meat texture poses a formidable challenge due to the inherent differences in plant-based materials compared to meat. The primary obstacle lies in forming a fibrous, meat-like structure using globular proteins as well as arranging the fat component with plant-based oils and fats. In the case of plant proteins, they must be denatured, aligned, unfolded, and cross-linked since they are usually globular in their original form (Sha and Xiong 2020). On the other hand, the fine-scale structure of meat muscle comprises myofibrillar protein and myoglobin arranged in a hierarchical fibrillar pattern that is challenging to replicate in plant-based meat analogues. Therefore, the development of suitable rheological and textural properties of meat analogues is a frequently difficult task for food manufacturers, and these quality attributes are contingent on the ingredients used in formulation (Kyriakopoulou et al. 2019).

Meat analogues, typically consist of various ingredients, with around 50–80% water content, 10–25% textured vegetable proteins, 4–20% non-textured proteins, up to 15% fat, 3–10% flavor enhancers, coloring agents in proportions of up to 0.5% and 1–5% binding agents. (Kyriakopoulou et al. 2019). The combination of these ingredients is carefully balanced to produce meat analogues with appealing rheological, textural as well as sensorial qualities. For example, the elevated water content serves multiple purposes, contributing to enhancing juiciness, acting as a plasticizer throughout processing, and facilitating emulsification. Since simply the utilization of protein for nutrition purposes doesn't provide the desired rheology and texture, textured proteins are preferred choice in meat analogues. Legume/pulse proteins like beans, chickpeas, lupines, lentils, peanut, peas, and cereal proteins like wheat, rice, barley, corn, sorghum, oats, and amaranth, oilseed proteins like soybeans, sesame, rapeseed (canola), cottonseed, safflower, linseed, flaxseed, and sunflower seed, or their combinations can be used as protein sources (Table 11.1). On the other hand, they are primarily composed of globular proteins, which are not favorable to the formation of the desired structure. As a result, plant proteins frequently necessitate extensive processing and the addition of ingredients in order to create a suitable product structure. It has been reported that soy protein is one of the most preferred protein ingredients in meat analogues due to their gelling, and emulsifying capacities. Extruded soy proteins can be texturally improved by adding sulfur-containing adjuncts or elemental sulfur throughout extrusion, because disulfide bonding plays a crucial role in both maintaining stiff structure and generating the fibrous texture of extruded products (Chiang et al. 2019). In the study of Zahari et al. (2020), it has been reported that the meat analogues prepared with soya protein and extrusion process provided desired fibrous texture due to disulfide bonding of soya.

Pea, chickpea, lupine and lentil proteins have also been preferred because of their foam stabilization and emulsification characteristics. Yuliarti et al. (2021) found that the textural characteristics of chicken analogues were discovered to be related to their viscoelastic capabilities and to be substantially influenced by the degree of cross-linking between protein molecules. While the utilization of pea protein boosted the textural characteristics such as hardness and chewiness, and viscoelastic properties (viscous ($G''$) and elastic ($G'$) moduli values) of meat analogue, the textural and viscoelastic properties of the analogue were reduced with the elevated levels of wheat protein used in the formulation. The variation in elastic moduli values of samples containing different amounts of wheat and pea protein could be attributed to differences in dough pH values. It was probable that protein in the analogues with a higher pH (the analogues containing the highest pea protein, 17%w/w, had a pH of 8.4) had more total negative charge groups. This increase in negative charge groups could have resulted in more intramolecular repulsion, ultimately leading to an enhancement of the dough gel. Thus, the composite of wheat protein and pea protein in a 3:1 ratio displayed the best texture profile, resulting in a fibrous structure and texture profiles similar to commercial products reflecting the critical balance between of wheat and pea protein to create meat analogues.

Because pea proteins have a softer texture than soy proteins, salts have been added to boost gel strength by protein hydrogen bonding modification (Kyriakopoulou et al. 2019). Except for chickpeas, pea, lupine and lentil proteins have also been shown to have lesser gelling capacities than soy. To improve the gelling behavior of legumes, fractionation procedures and extra pretreatments have been applied (Kyriakopoulou et al. 2019). In a study on plant proteins, five types, including mung bean, peanut, pea, soy, and wheat gluten, were extruded at 50% moisture. Among them, peanut protein isolate stood out with excellent textural properties and good oil

**Table 11.1** The impact of formulation and production method on rheological and textural properties of meat and fish analogues

| The type of food analogue | Production Method | The type of main ingredient | The applied rheological method | The applied textural method | The findings | References |
|---|---|---|---|---|---|---|
| Meat analogue | Blending and steaming | Soybean, konjac glucomannan (0, 2, 4 and 6%), curdlan (6%) and methylcellulose (6%) | SAOS (Frequency sweep, Temperature sweep) | TPA | The viscoelastic characteristics, which were significantly impacted by the addition of konjac gum, were closely linked to its textural characteristics.<br>The meat analogs with konjac showed mostly elastic behavior when subjected to temperature changes, both heating and cooling.<br>The addition of konjac gum enhanced hardness, cohesiveness, and decreased stickiness, suggesting that molecular interactions mediated by konjac reinforced the patty's overall structure. | Yuliarti et al. (2023) |
| Meat analogue | Freeze structuring technique | Pea protein and wheat protein at different ratios (17:0; 13:4; 8.5:8.5; 4:13 and 0:17) | SAOS (Frequency sweep, Temperature sweep) LAOS (Strain sweep) | TPA | The textural traits and configuration of plant-based substitutes are intricately tied to their viscoelastic characteristics, where the extent of intermolecular protein bonding assumes a pivotal role.<br>The introduction of pea protein elevated textural (chewiness, hardness) and rheological (and viscoelastic properties) properties. Conversely, heightened wheat protein levels diminished the textural and viscoelastic aspects of the analog.<br>The most favorable textural profile was achieved through a 3:1 blend of wheat and pea protein, closely mirroring commercial products.<br>The scale of the protein network formed among molecules is a critical determinant of meat analogue texture. | Yuliarti et al. (2021) |
| Meat analogue | Freeze structuring technique | Pea protein isolate-wheat gluten | SAOS (Frequency sweep) LAOS (Strain sweep) | TPA | Wheat gluten-pea protein isolate, with a higher protein content, exhibited higher complex viscosity, while Dry-pea protein concentrate lacked proteins, resulting in lower batter homogeneity.<br>The increased protein content and cross-linking in wheat gluten-pea protein isolate enhanced hardness and chewing performance.<br>Wheat gluten-pea protein isolate nuggets had more compact, structure, while Dry-pea protein concentrate nuggets exhibit a fibrous structure with interconnected layers held together by fibers. | Zhu et al. (2021) |

(continued)

**Table 11.1** (continued)

| The type of food analogue | Production Method | The type of main ingredient | The applied rheological method | The applied textural method | The findings | References |
|---|---|---|---|---|---|---|
| Meat analogue | High moisture extrusion | Pea protein isolate-wheat gluten | – | Tensile strength | Shearing during cooling in the High-Temperature Short-Time Cooling method harmed the production of fibrous products, weakening them and causing structural issues.<br>Importantly, the energy required for creating these fibers in High-Temperature Short-Time Cooling is much lower than that in High-Moisture Extrusion. | Köllmann et al. (2023) |
| Meat analogue | – | Fusarium venenatum biomass (mycoprotein), egg white protein | SAOS (Frequency sweep) | – | The storage modulus (G′) of mycoprotein showed notable increases with higher mycoprotein solids, along with the presence of NaCl and $CaCl_2$.<br>Specifically, $CaCl_2$ had a more pronounced impact at pH 5 and 7. Conversely, at pH 3, a different pattern emerged, where G′ decreased as ion concentration rose.<br>The highest G′ values for mycoprotein were observed at pH 3 without added salts, highlighting the significant influence of electrostatic interactions among mycoprotein components on the overall texture. | Okeudo-Cogan et al. (2023) |
| Meat analogue | High-temperature conical shear cell | Rapeseed protein concentrate (RPC), wheat gluten, soy protein concentrate (SPC) | – | Tensile strength | RPC can be used to produce fibrous materials even without wheat gluten.<br>RPC outperformed SPC in generating more prominent fibers.<br>The combination of RPCs with wheat gluten demonstrated a greater potential for creating diverse fibrous materials at 140 °C. This is attributed to the complete melting of rapeseed protein concentrate particles.<br>The incorporation of wheat gluten into RPCs led to an enhanced anisotropic structure, resulting in increased tensile stress across all ratios. | Jia et al. (2021) |
| Meat analogue | Mechanical elongation method | The different ratios of wheat gluten:soy protein isolate (100:0, 80:20, 60:40 and 40:60) | – | TPA | The least hardness and chewiness were obtained from meat analogues including only wheat gluten compared to the mixtures of soy protein isolate and wheat gluten.<br>The mechanical elongation technique, employing blends of wheat gluten and soy protein isolate, provided suitable meat analogues. | Chiang et al. (2021) |

| | | | | | | |
|---|---|---|---|---|---|---|
| Meat analogue | Extrusion | Soy protein isolate, vital wheat gluten | – | TPA and Cutting strength | Textured vegetable protein exhibited comparable chewiness and cutting strength to the chicken sample in contrast to other meats.<br>Its structure featured non-uniform air cells, contributing to its fibrous quality. | Samard and Ryu (2019b) |
| Meat analogue | High moisture extrusion cooking | Soy protein isolate, iota carrageenan (0.75%, 1.5%, 2.25% and 3%) | – | TPA | The incorporation of iota carrageenan led to a notable enhancement in cutting force and elasticity.<br>It resulted in a denser network within the meat analogues, contributing to the altered texture.<br>Formulating soy-based meat analogs with 1.5% iota carrageenan can yield improved physical, textural, and sensory characteristics. | Palanisamy et al. (2018) |
| Meat analogue | High-moisture extrusion | Faba bean protein concentrate | – | Cutting strength and Compression | Moisture content had a more substantial impact compared to temperature and feed rate.<br>A correlation was identified between water binding capacity and characteristics like firmness (both instrumental and sensory), juiciness, fibrousness, and gumminess. | do Carmo et al. (2021) |
| Meat analogue | Shear cell technology | Pea protein isolate, wheat gluten, xanthan, guar gum, sodium alginate, iota-carrageenan, low acyl gellan gum, carboxymethyl cellulose (CMC), low methylated locust bean gum, and pectin | – | Tensile Test | Xanthan also enhanced sample anisotropy but did not enhance tensile stress. On the other hand, gellan gum substantially increased tensile stress, especially at higher concentrations.<br>Xanthan and gellan gum have significant potential to enhance the textural characteristics of meat analogs. | Dinani et al. (2023) |

(continued)

**Table 11.1** (continued)

| The type of food analogue | Production Method | The type of main ingredient | The applied rheological method | The applied textural method | The findings | References |
|---|---|---|---|---|---|---|
| Meat analogue | Extrusion | Soy protein isolate, transglutaminase | – | TPA and Tensile Test | Gels formed through covalent, isopeptide bonds were essential for serving as effective binders in a bacon analogue, while gels created through physical bonds lacked the necessary adhesive properties. | Herz et al. (2021) |
| Meat analogue | High-moisture extrusion | Oat fiber concentrate and pea protein | – | TPA | Higher levels of oat fiber concentrate had an adverse influence on the structural strength of the meat analogues, particularly in terms of chewiness and hardness, while long cooling die temperature had a strengthening effect on the structure.<br>Meat analogues with increased oat fiber concentrate had smaller void thickness, which in turn reduced their water holding capacity.<br>Substantial amounts of oat fiber concentrate (ranging from 30% to 50%) can be incorporated into analogues while preserving meat-mimicking fibrous structures and the viscous properties of oat fiber. | Diaz et al. (2022) |
| Meat analogue | High-temperature and shear-induced structure formation | Soy protein isolate (SPI), vital wheat gluten, pea protein isolate (PPI) | – | Tensile strength | Blending PPI with wheat gluten at 120 °C created distinctive fibrous structures, with lower temperatures resulting in weaker products and higher temperatures yielding robust and layered ones.<br>However, the blends of wheat gluten and SPI produced consistent fibrous materials over a wider shearing temperature range (110–140 °C). At 120 °C, soy blends were three times stronger than pea protein blends, while at 140 °C, both soy and pea protein blends showed similar strength.<br>Both soy and pea-based blends saw a notable strength decrease at 130 °C. Compared to chicken meat, soy blends exhibited greater tensile strength at 110 and 120 °C, whereas PPI based blends fell within the range of chicken meat.<br>Wheat gluten and pea protein isolate offer the potential to produce fibrous meat analogs with matrix strength akin to cooked chicken meat. | Schreuders et al. (2019) |

| | | | | | | |
|---|---|---|---|---|---|---|
| Meat analogue | High moisture extrusion cooking<br>Dry extrusion | Microalgae powder and a soy protein concentrate | – | Cutting strength and TPA | Extrudates containing both soy and microalgae can enhance their fibrous texture by decreasing moisture content.<br>Favorable textural attributes can be achieved with 30% microalgae incorporation at a 60% moisture level. | Caporgno et al. (2020) |
| Meat analogue | 3D-dimensional printing technology | Pea protein isolate | SAOS (Frequency sweep, Temperature sweep)<br>Flow curve | Compression | Both pea protein isolate paste and pea protein isolate chicken paste demonstrated weak gel-like behavior in rheological tests.<br>The addition of maize starch increased paste viscosity and minimized changes in moduli during temperature variations.<br>Combining chicken with pea protein isolate paste decreased the viscosity alterations with increasing shear rate.<br>Taking into account both printing performance and fiber formation, the optimal formulation selected was pea protein isolate paste with 20% chicken. | Wang et al. (2022) |
| Fish Analogue | Extrusion cooking | Soy protein, isolate, konjac glucomannan | SAOS (Frequency sweep, Strain sweep)<br>Linear Creep-recovery test | TPA | Konjac glucomannan progressively increased the gel strength of plant-based fishball analogues.<br>Higher konjac glucomannan levels promoted hydrogen bond formation between konjac glucomannan and plant proteins.<br>Increased konjac glucomannan levels contributed to a denser microstructure in the plant-based fishball analogues. | Ran et al. (2022a) |
| Fish Analogue | Extrusion cooking | Soy protein, isolate, konjac glucomannan | SAOS (Strain sweep, Frequency sweep)<br>Flow curve | – | Konjac glucomannan decreased the release of amino acids, glucose, and fatty acids during in vitro digestion.<br>The presence of Konjac glucomannan reduced the enzymatic action sites, delaying the digestion process.<br>Higher levels of konjac glucomannan significantly increased the apparent viscosity, the viscoelasticity and structural strength of plant-based fishball analogues digesta. | Ran et al. (2022b) |

absorption and emulsifying abilities, making it a top choice for high-quality meat analogues. However, in case of structure, mung bean and peanut protein isolates had a dense texture with no pores, while wheat gluten displayed fewer air cells, and soy protein isolate and pea protein isolate had more air pores and a sponge-like structure (Samard and Ryu 2019a). Rapeseed protein is particularly well-suited for use in meat analogues due to its ability to induce gel formation. This may promote meat-like textures under high pressure or heat (Kyriakopoulou et al. 2019). At higher ionic strength, canola protein fractions, primarily albumin and globulin formed cohesive gels. However, at lower salt concentrations, globulins exhibited the least frequency-dependent activity and may hold more promise for building cohesive gels (Kyriakopoulou et al. 2019). The gelling behavior of pulse proteins are linked to their molecular structure, protein particle size, and other factors, like the presence of protein and phenolic compounds (Shrestha et al. 2023).

Microalgae have also become recognized as a viable-raw material for high-moisture extrusion cooking when combined with soy protein concentrates. Microalgae offer various functional properties, including the ability to form gels, absorb water and fat, emulsify, and create foam in food products. When combined with soy protein, this fusion significantly influenced the formation of a fibrous structure. This effect is likely due to the resistant cell walls of microalgae and their relatively high fat content. The findings of the study by Caporgno et al. (2020), which included fiber-level assessment and mechanical texture analysis, demonstrated that extruded products with promising characteristics could be achieved by incorporating 30% microalgae at a moisture level of 60%.

By aligning protein filaments with fibrous or flaky texture, fungal protein produced plausible meat (Vandelook et al. 2021). Quorn®, was the first meat alternative made from fungus (*Fusarium venenatum*), started in 1985. When cooked, the fungal proteins had a smooth and soft texture, but they could also be cooked to a firmer and juicier meat-like texture (Rousta et al. 2021; Okeudo-Cogan et al. 2023).

In the creation of a high-performance bacon analogue, researchers have found that extruded plant-based proteins and fat mimetics can be effectively combined using a binder system primarily based on soy protein isolate. Notably, the most successful adhesion was achieved through the inclusion of transglutaminase in the binding suspension. The resulting gels, which were cross-linked by transglutaminase and possessed the highest hardness, demonstrated acceptable cohesion (Herz et al. 2021). Combining protein ingredients with starches and cellulose has also proven to be an efficient approach for replicating a meat-like texture. Mixing soy protein isolate with starch and wheat gluten, especially when there was a greater increase in enthalpy change, specific mechanical energy, and pressure during extrusion, resulted in extruded products that are harder and have higher tensile strength. However, this combination led to extrudates with a lower fibrous texture (Zhang et al. 2016). On the other hand, Chiang et al. (2019) stated that mixing soy protein isolate with starch and wheat gluten at the appropriate amounts (20% and 30% wheat gluten) provided the closest textural characteristics of meat analogues to the conventional meat since wheat gluten played a noticeable role by assisting in the formation of more disulphide bonds in meat analogues to produce fibrous structure. Similarly, Smetana et al. (2018) stated that introducing soy fiber at levels ranging from 5% to 10% resulted in meat analogues with a more distinct and refined texture.

Even though meat analogues have a low fat content, the inclusion of oil or fat during processing influence the creation of fibrous structures, hence the rheological as well as textural properties of the products (Table 11.1). The use of high oil content (higher than 15%) negatively affects the alignment of the macromolecules. Since the use of high oil content creates very slippery material, it has been reported to negatively impact the shear forces experienced during the extrusion process. Conversely, the addition of appropriate amount of oil or fat in a meat-analogue formulation provides juiciness, tenderness, and flavor release; hence, the desired textural and rheological properties can be obtained from the products (Kyriakopoulou et al. 2019). Thus, in order to develop fibrous meat analogues, high-fat legumes such as peanut or soybean can be included in the meat analogue formulation (Geerts et al. 2018). These compounds can serve as substitutes or supplements for the fat content in the product. They can be integrated into the final product formulation or infused into the meat-like structures. Sunflower oil, rapeseed oil, corn oil, palm oil, canola oil, soya oil, etc. are also acceptable in the formulations (Kyriakopoulou et al. 2019).

Hydrocolloids such as xanthan gum, carrageenan, cellulose, guar gum and pectin have been proposed as extenders for meat analogues (Table 11.1) (Kyriakopoulou et al. 2019). They offer gelling and thickening properties and have been proven to enhance oil-binding and water-binding capacity and hence the rheological characteristics of these products. The improvement of the textural characteristics of meat analogues with the use of carrageenan has been shown (Arora et al. 2017; Palanisamy et al. 2018). The study conducted by Arora et al. (2017) demonstrated the positive impact of ingredients like carrageenan and xanthan gum in sausage analogues made from mushrooms with the addition of a 5% saturated fat content. Dinani et al. (2023) has highlighted the potential of xanthan gum and low acyl gellan gum hydrocolloids to enhance the textural qualities of meat analogues. In their research, various hydrocolloids, including xanthan gum, sodium alginate, carboxymethyl cellulose, iota-carrageenan, low acyl gellan gum, locust bean gum guar gum and low methylated pectin were evaluated at different concentrations (1–3%) in blends including wheat gluten and pea protein isolate. The processing method employed was shear cell technology. The findings of study indicated that incorporating xanthan gum, guar gum, and carboxymethyl cellulose resulted in the creation of a structure that exhibited better flow orientation and fiber development in comparison to samples lacking these hydrocolloids. However, the increase in water holding capacity was minimal, except for xanthan gum. Additionally, while xanthan gum enhanced the anisotropy index of the samples, it did not significantly enhance the tensile stress of the products. Interestingly, the inclusion of low acyl gellan gum, especially at higher concentrations, resulted in a substantial increase in tensile stress. This underscores the promising role of hydrocolloids, particularly xanthan and low acyl gellan gum, in enhancing the textural characteristics of meat analogues.

### *11.2.2 The Effects of Technological Methods on Rheological and Textural Properties of Meat Analogues*

The processing parameters also influence the formation of appropriate texture and rheological qualities in meat analogue (Table 11.1). Recent advancements in texture replication have been driven by two primary approaches: top-down and bottom-up.

The bottom-up approach, seen in cultured meat, mycoprotein, electrospinning, *in vitro* cell culturing and wet spinning, aims to mimic the complete hierarchical meat structure but it is resource-intensive. It involves building these structures from small, anisotropic elements. This approach mimics the structural organization of muscle, which spans various scales, starting from the nanoscale with muscle cells and myofibrils, then extending to the macroscale with sarcomeres containing proteins like myosin and actin, all interconnected by connective tissue (Dekkers et al. 2018).

The top-down approach, conversely, employed in extrusion, shear cell technology, protein-hydrocolloid mixing, and freeze structuring, replicates the fibrousness of meat analogues primarily on larger length scales. However, they do not fully reflect the hierarchical organization of meat. These processes involve forces like shearing or unidirectional freezing to align and structure biopolymers, ultimately affecting the mechanical characteristics of the material (Dekkers et al. 2018). For example, Yuliarti et al. (2021) achieved a structure with fibrous layers in nuggets using freeze structuring, but the success of this method depended on the specific type of protein used in creating the meat analogues.

In bottom-up methods, such as wetspinning, thicker fibers are created, but they generate more waste due to the use of chemical reagents. On the other hand, electrospinning, also a bottom-up approach, produces thinner fibers and is considered more environmentally friendly. In contrast, the top-down approach involves the use of the shear cell technique, where plant-based materials are structured through the application of intense shearing forces, often using devices like Couette cells. This method holds promise for scaling up meat analogue production, but further research and development are needed to enhance its environmental friendliness and efficiency (Kołodziejczak et al. 2021).

To create meat analogues, mostly the extrusion technique is used. The extrusion process for plant-based proteins induces several crucial transformations. Initially, the proteins shift from their natural globular shape to a textured, linear structure because of a combination of heating, shearing, compression, and cooling. This transformation is essential for attaining the appropriate texture in meat analogues. Samard and Ryu (2019b) found that high-moisture extrusion cooking, along with the inclusion of wheat gluten, resulted in fibrous and compact extrudates resembling real muscle meat, with enhanced structural integrity and texture stability. The screw speed had only a slight effect on the springiness of the product.

The quality of meat analogues produced through extrusion is heavily dependent on precise control of various factors, including screw speed, barrel temperature, cooling die design, flow rate, matrix geometry, and feed moisture content. Adjusting these parameters significantly influences the resulting texture and structure of the products (Sun et al. 2022). For instance, elevating both temperature (ranging from 130 to 180 °C) and screw speed (ranging from 400 to 1800 rpm) while reducing water flow (40–68%) led to denser textures with more layers of fibers (Palanisamy et al. 2019). Higher cooking temperatures, like 160 °C, produced more organized structures, while lower temperatures did not support texturization. Toasting the soy protein fraction at 150 °C led to the formation of fibrous structures. Pea protein-gluten mixtures exhibited similar strength to soy protein blends at 140 °C but weaker strength at 110 and 120 °C (Schreuders et al. 2019). Elevated barrel temperatures promote protein polymerization reactions by intensifying the extent of protein denaturation, while shear forces induce the ongoing aggregation and fragmentation of these denatured proteins. Reducing the cooling die temperature supports the development of a layered structure, which in turn promotes molecular rearrangement within the denatured protein molecules. It is crucial to combine alternative protein sources with flavors, coloring agents and other biopolymers to achieve the appropriate product properties (Sengar et al. 2023).

Moisture content plays a crucial role in texture development. Research by do Carmo et al. (2021) highlighted that the impact of moisture content was generally more critical than temperature and feed rate in producing meat analogues containing faba bean protein concentrate through high-moisture extrusion. Higher moisture content reduced the viscosity, elasticity, and dough temperature, resulting in softer textures. It also affected protein interactions, thus influencing texturization (Sun et al. 2022). In extrusion, low-moisture conditions (up to 30% water content) result in textured vegetable protein known for its porous structure and high water-binding capacity. However, it requires additional hydration to attain a meat-like texture. On the other hand, high-moisture extrusion, with a moisture content between 40% and 80%, results in meat analogues that exhibit a fibrous structure similar to that of whole muscle or restructured meat products. This method allows for more complex formulations and is cost-effective. For instance, meat analogues with 52% moisture content exhibited the greatest texturization, while those with 49% moisture content resembled the texture of boiled chicken breast (Chiang et al. 2020).

### *11.2.3 Methods for Testing the Rheological and Textural Properties of Meat Analogues*

Understanding the rheological and textural behavior of meat analogues is crucial since their fibrous structure develops during processing, which is often conducted at high temperatures and pressures, posing analytical challenges. Therefore, employing a combination of various textural and rheological analysis methods is a valuable approach (Schreuders et al. 2019). One key method for understanding how meat analogues respond to shear stress or shear rate during processing is known as flow curve measurements. These measurements provide insights into food component behavior, especially in processes like extrusion. The obtained shear stress *vs.* shear rate profile helps determine apparent shear viscosity and yield stress (Yazar and Demirkesen 2023).

Viscoelastic materials can be classified into two types based on their behavior. Linear viscoelastic materials remain consistent in their properties regardless of strain and strain rate, while non-linear viscoelastic materials exhibit properties that change with variations in strain and strain rate, eventually leading to material breakdown. Most food products display linear viscoelasticity under low shear strain. Linear viscoelastic characteristics are predominantly determined by Small Amplitude Oscillatory Shear (SAOS) tests. These tests assess material responses to small strains and frequencies without significantly altering the three-dimensional structure of the food material (Yazar and Demirkesen 2023). SAOS tests include frequency sweeps, stress/strain sweeps, and temperature sweeps. In many studies, SAOS tests have been performed to assess the rheological behavior of meat analogues (Table 11.1). Large Amplitude Oscillatory Shear (LAOS) tests are also instrumental in exploring the non-linear rheological properties of materials like meat analogues. These tests provide insights into how materials behave under substantial deformations, complementing the information obtained from SAOS tests. In the production of meat analogues, especially during processes like extrusion, the material is subjected to shear outside the linear viscoelastic range (Yuliarti et al. 2023). Thus, LAOS studies are employed to determine how the material behaves under high strains, similar to what real products might experience during shearing, extrusion, chewing, or cutting. These studies involve shearing the food sample in oscillatory mode at a fixed frequency while applying elevated levels of strain.

Amplitude sweeps are valuable in understanding the rheological behavior of materials like meat analogues. These tests help define the linear viscoelastic range (LVR), a critical parameter indicating the range in which material properties like storage (G′), loss (G″), and complex (G*). The complex modulus combines both the storage modulus (G′) and the loss modulus (G″) and is expressed as a complex number: $G^* = G' + iG''$; where "i" represents the imaginary unit. The complex modulus remain constant without material breakdown. When the material experiences strains beyond this range, it starts to degrade and flow, leading to irreversible structural breakdown at a constant frequency. Comparing the LVR amplitudes between different materials provides insights into the conditions that result in structural breakdown. There are two modes for setting deflection: the strain sweep (controlled-strain mode) and stress sweep (controlled-stress mode). The amplitude sweep test is vital for determining LVR, which establishes the % strain (or stress) value used in subsequent frequency sweep experiments. It's crucial to conduct a strain sweep or stress sweep test before the main measurement to determine the maximum allowable amplitude of oscillation. In the context of meat analogues, amplitude testing has found applications, particularly in the development of fat replacement systems. This method helps evaluate the behavior of meat analogues under different processing conditions and understand the limits of their structural stability.

Frequency sweeps, conducted under non-destructive conditions, involve oscillatory testing in the linear viscoelastic range. This method reveals the time-dependent (relaxation) behavior of food materials, providing insights into their properties. By subjecting the sample to constant strain or stress and varying the angular frequency, these sweeps can uncover information about relaxation periods, material behavior under processing and storage conditions, and the nature of crosslinks present in meat analogues. In essence, frequency sweeps help examine how structuring approaches affect the protein- and fat-phases in meat analogues (Yazar and Demirkesen 2023).

Temperature sweeps involve subjecting the sample to both heating and cooling cycles while simultaneously applying shear in oscillatory conditions at a constant strain or frequency, typically at 1 Hz. This method reflects thermal changes in a food material, such as gelation and aggregation. The sample can be in either a liquid or solid form before analysis to observe any thermal changes or phenomena. The process starts with heating a liquid protein dispersion, leading to denaturation and a subsequent sol-gel transition at a specific temperature. This temperature must be achieved through extrusion or other structuring procedures to promote protein crosslinking. Then, upon re-heating the gelled substance, the sort of bonds responsible for structure development (e.g., hydrogen bonds, hydrophobic bonds, or covalent bonds) becomes evident. In the study of Yuliarti et al. (2023), a temperature sweep test was used to examine how the properties of the meat analogue changed concerning time and temperature during thermal processing. To conduct this test, the formulations underwent a three-phase temperature sweep: The first phase involved heating the samples from 25 to 90 °C. Subsequently, the samples were held at a constant temperature of 90 °C for 3.75 min. Finally, the last phase consisted of cooling the samples from 90 °C back to 25 °C. Throughout these phases, a heating and cooling rate of 10 °C per minute was applied. Rheological analysis revealed that the patties exhibited primarily elastic behavior ($G' > G''$), and this behavior continued to develop during both heating and cooling temperature changes. During heating, an initial reduction in $G'$ and $G''$ was observed, likely because of the weakening of hydrogen bonds. Continuing the heating process led to a gradual increase in both modulus values, indicating the progressive formation of protein gels (Fig. 11.1).

Subsequent cooling cycles further enhanced gel strength, indicating the reorganization of gel networks and the development of hydrogen bonds. Notably, the addition of konjac gum reduced the overall gel strength of the patties, possibly due to the high-water affinity of konjac gum, which could hinder protein gel formation. However, with increasing konjac gum concentration, the gel strength improved again, potentially because of the formation of konjac gel (Fig. 11.1). The tan $\delta$ values during heating indicated the melting of ingredients in the patties. Konjac-containing patties exhibited the lowest tan $\delta$ values, indicating enhanced solid characteristics and stabilizing effects on the structure of the patties. In contrast, the commercial beef patty had the lowest gel strength and exhibited no gel network formation, likely due to fat melting in the formulation (Fig. 11.1).

Additionally, textural measurement showed that the incorporation of curdlan and methylcellulose in the patties significantly increased hardness, while the konjac patty was the most cohesive. These hydrocolloids improved patty hardness and cohesiveness, potentially by forming more compact protein gel networks. In terms

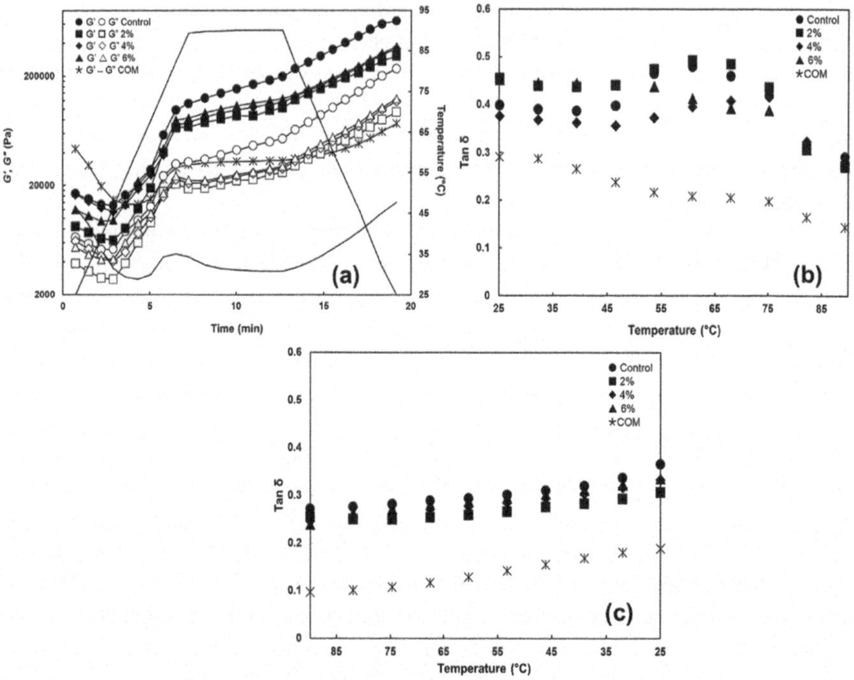


**Fig. 11.1** (**a**) Temperature-dependent behavior of meat analogue patties with varying concentrations of konjac glucomannan over time, (**b**) tan δ values from 25 to 90 °C, and (**c**) tan δ values from 90 to 25 °C (at 0.5% strain and 1 Hz). (Reproduced from Yuliarti et al. (2023) with permission from the publisher)

of dough rheology (figure is not presented), konjac patty dough was the strongest and relatively harder to deform, which is important for processing steps such as molding and steaming. In contrast, curdlan-containing dough was less elastic and had a weaker network, while methylcellulose had a slight improvement in dough gel strength (figure is not presented).

During LAOS investigations, stress is plotted against strain or strain rate, creating Lissajous plots. These plots provide insights into a material's viscous and elastic behavior at different strain levels, which is particularly useful for understanding properties relevant to extrusion and shearing processes. The dissipation ratio, calculated from the area in the Lissajous plots representing viscoelastic and perfect plastic responses, offers a way to represent the balance between a material's elastic and viscous components as strain increases. Lower dissipation ratios show predominantly elastic behavior, while higher ratios (>0.8) suggest the presence of more viscous components. This information is valuable for gaining a deeper understanding of how meat analogues behave under actual processing conditions, facilitating the optimization of their quality and performance (McClements et al. 2021).

Texture measurements are crucial for understanding and quantifying the physical properties of meat analogues (Table 11.1). These measurements serve two primary purposes in meat analogue research. Firstly, they enable a direct comparison of the textural attributes of meat analogue materials with those of traditional meat products helping in modifying composition and processing to achieve meat-like textures, including aspects like bite strength and mouthfeel. Secondly, texture measurements provide insights into anisotropic structures within meat analogues, where the response of material varies with the direction of applied force (McClements et al. 2021).

To conduct texture analysis for meat analogues, various mechanical methods are employed. Tensile and compression tests are two such methods that measure a material's resistance to compression or elongation, involving large deformations. Hence, they quantify the force required to compress or extend a material over a specified length. These tests secure a sample between two grips and subject it to controlled extension or compression, providing parameters like maximum rupture force, breaking strength, tensile strength, fracture energy, etc. Tensile and compression tests are versatile and applicable to a wide range of products, including meat analogues, and can reveal anisotropic characteristics, shedding light on how textural properties vary with the direction of applied force.

The importance of texture measurements is evident in processes like cooling during high-temperature, short-time processing to create fibrous food products. As the material cools, it becomes more viscous and may even solidify. The degree of structure formation in high-temperature shear cell (HTSC) processing is influenced by the processing parameter of rotational speed. During cooling, applying shear to the material can enhance structure formation by inducing mechanical deformation. Shear forces are effectively transmitted to the two phases of the product as it solidifies. Nevertheless, excessive shear stress during cooling can lead to structural breakdown. Therefore, the balance between shear-induced deformation and solidification is crucial for optimal fibrous structure formation. Köllmann et al. (2023) examined the effect of cooling and shearing on the formation of fibrous structures in plant-based materials during HTSC processing, particularly focusing on the role of shear rate and duration. For this purpose, tensile tests were performed in both perpendicular and parallel directions to the shear flow, enabling the quantification of anisotropy in the textural characteristics of the HTSC products. In general, HTSC products exhibited greater strength in the parallel direction compared to the perpendicular direction (Fig. 11.2). Products subjected to shearing during the cooling process displayed lower tensile strain and stress, particularly in the perpendicular direction when processed at rotational speeds under 30 rpm (Fig. 11.2). However, at rotational speeds exceeding 30 rpm, the parallel tensile strain of products sheared during cooling increased (Fig. 11.2c). Mechanical anisotropy was more pronounced at higher rotational speeds, although no visual fibrous structures were observed (Fig. 11.2a, b). Products sheared during heating showed a higher variation in strains, particularly notable when sheared at high rotational speeds for 15 min. This suggests that the structure of these products exhibited increased heterogeneity on the measurement's length scale.

In this study, it was found that the Young's modulus, which reflects elasticity, was lower in products subjected to shearing during cooling compared to those sheared

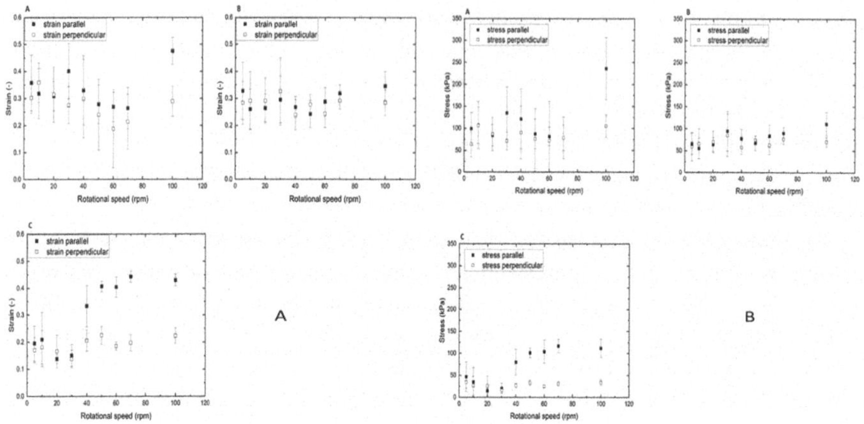


**Fig. 11.2** In panel **a**, tensile strain at fracture was plotted against shear rate, while in panel **b**, it was plotted against rotational speed. The data is provided for products sheared for 15 min during heating, 8 min during heating, and 8 minutes during cooling, in both parallel (■) and perpendicular (□) direction. (Reproduced from Köllmann et al. (2023) with permission from the publisher)

during heating. The Young's modulus was affected by the rotational speed, with products sheared at higher speeds during heating showing increased rigidity. Shearing during cooling had a weakening effect on the products, as evidenced by reduced tensile stress and strain and a lower Young's modulus (Fig. 11.3). This study revealed that shearing during the cooling phase had a negative impact on achieving the desired visual fibrousness in the products. However, it also emphasized the potential to create products with diverse mechanical and visual properties by modifying the processing parameters, particularly during the cooling stage. Importantly, predicting product properties solely based on a single processing parameter, like rotational speed, remains a complex challenge. The study concluded that both improved structure formation and structural breakdown are integral to shaping the properties of plant-based materials during HTSC processing.

Texture Profile Analysis (TPA) is a comprehensive compression method that integrates various textural parameters, such as hardness, springiness, adhesiveness, gumminess, cohesiveness, and chewiness, into a single measurement. In a TPA test, a sample often obtained using a coring tool, is subjected to two cycles of compression and relaxation between plates larger in diameter than the sample, at a specified speed. The sample is compressed (downward) and then relaxed (upward) to achieve a relative deformation typically within the range of 50–75% of its initial height. This "double compression test" is commonly used in meat analogue studies and mimics the chewing process, making it invaluable for assessing the sensory attributes of both meat analogues and traditional meat products, offering a comprehensive view of their textural characteristics. Another mechanical method for assessing food texture is the single compression test, typically conducted as an axial compression method between two flat plates. Single compression tests are less commonly utilized (Schreuders et al. 2021).

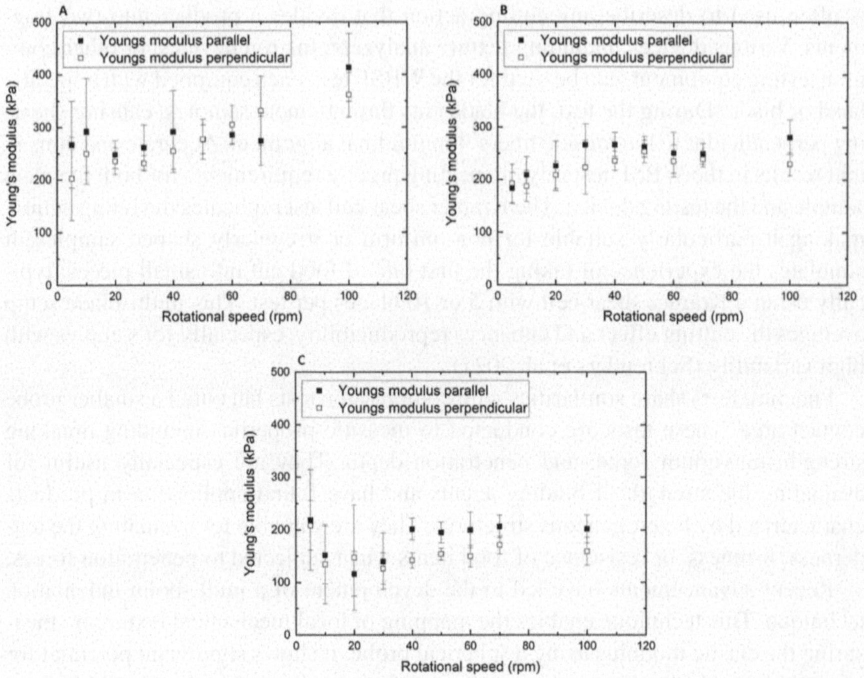


**Fig. 11.3** Young's modulus was plotted against rotational speed in both parallel (■) and perpendicular (□) directions for products sheared for (**a**) 15 min during heating, (**b**) 8 min during heating, and (**c**) 8 min during cooling. (Reproduced from Köllmann et al. (2023) with permission from the publisher)

In meat analogue evaluation, shear tests and cutting force are commonly employed to measure hardness and assess structural and textural characteristics. These tests entail cutting samples into specific shapes and placing them under a knife-shaped apparatus. The knife is then lowered at a consistent speed, typically between 1 and 10 mm per second, while recording both the force applied and the path taken to cut the specimen. Results from these tests are typically reported as either the maximum cutting force experienced or the force measured at a predefined sample depth. Various tools and geometries, such as knife configurations, Kramer shear cells, and Warner-Bratzler shear cells, have been utilized in meat analogue research to measure cutting force, providing valuable insights into product texture and hence structural properties. One commonly used method, the Warner-Bratzler Shear Force (WBSF) test, originally developed to assess meat tenderness, has been adapted for meat analogue texture evaluation. It quantifies the force required to cut or tear through a sample. Standardization in the execution of this test is emphasized to ensure consistent and reliable results. The WBSF test measures hardness or toughness by determining the maximum force required to cut a meat sample as it moves a specified distance. This test involves shearing, where the meat slides parallel to the contact plane with the applied force tangential to the segment, although it is often used to describe any cutting action that divides a product into two fragments. Various devices, including texture analyzers, Instron devices, or other common testing equipment, can be used for the WBSF test, each equipped with a specific head or blade. During the test, the blade cuts through meat samples, causing shearing perpendicular to the muscle fibers' longitudinal alignment. Accurate and consistent results in the WBSF test rely on meeting precise requirements for both the meat sample and the testing device. The Kramer shear cell test replicates the biting action, making it particularly suitable for non-uniform or irregularly shaped samples. It simulates the experience of taking the first bite of food cut into small pieces, typically using a Kramer shear cell with 5 or 10 blades per test. This multi-blade setup averages the cutting effect and enhances reproducibility, especially for samples with high variability (Schreuders et al. 2021).

Puncture tests share similarities with compression tests but entail a smaller probe contact area. These tests are conducted to measure properties including breaking strength, maximum force, and penetration depth. They are especially useful for evaluating the strength of binding agents and have found application in products characterized by heterogeneous structures. They are valuable for evaluating the tenderness, firmness, or resistance of food items when subjected to penetration forces.

Recent advancements have led to the development of a multi-point indentation technique. This technique enables the mapping of local mechanical texture by measuring the elastic modulus using a spherical probe. It shows significant potential for characterizing the heterogeneous structures of meat and meat analogues. It has been applied to meat patties, meat analogue sausages, and chicken breast highlighting differences in breaking forces (Schreuders et al. 2021).

## 11.3 Rheological and Textural Properties of Fish Analogues

### *11.3.1 The Effect of Formulation on Rheological and Textural Properties of Fish Analogues*

Fish flesh is predominantly composed of water, proteins, lipids, and smaller amounts of carbohydrates, minerals, and vitamins. The key structural element in fish muscle is myotomes, sheets of muscle held together by myocommata, a type of connective tissue. These myotomes are mainly composed of actin and myosin, forming muscle fibers that further aggregate into myofibrils with distinct structures (Kazir and Livney 2021).

Fish texture is influenced by myofibrillar protein and collagen (3–10% of total proteins) concentrations, with higher levels resulting in a firmer texture. Microbiological and autolysis processing can weaken myofibril structures. Various factors, including those related to the fish itself and the cooking method, also impact fish texture (Kazir and Livney 2021). In a manner similar to meat analogues, plant-based proteins from various sources are used to create fish analogues. These proteins can be used individually or in combinations to improve nutritional and functional properties. Additional ingredients like salts, lipids, and dietary fibers play a vital role in shaping the texture of these analogues, mirroring their animal-based counterparts. Salts affect protein solubility and denaturation, lipids contribute to the gel matrix, and dietary fibers enhance texture and gel strength (Kazir and Livney 2021).

There is limited research on fish analogues, but a study by Ran et al. (2022a) demonstrated that konjac glucomannan significantly enhanced the texture and rheological properties of fishball analogues. Higher glucomannan content enhanced the hardness, chewiness, gel strength, and textural properties. It also led to more ordered protein structures and a firmer texture. The study suggested that increased glucomannan content contributed to enhanced viscoelasticity, structural strength, and potential health benefits, such as controlled blood glucose levels and appetite (Ran et al. 2022b).

### *11.3.2 The Effects of Technological Methods on Rheological and Textural Properties of Fish Analogues*

The structuring methods employed for fish analogues share common ground with those used in the development of meat analogues. These methods aim to create a fibrous structure that mimics the texture and mouthfeel of real fish. Structuring techniques involve various forces and interactions, including hydrogen bonding, van der Waals forces, hydrophobic forces, disulfide bonds, enzymatic reactions, electrostatic forces, and more. The goal is to create a network of unfolded proteins that can provide the desired texture. Common structuring techniques used for plant-based proteins, which are predominantly applied in the production of meat analogues, include electrospinning, wet spinning, extrusion and 3D printing.

Traditional approaches to creating plant-based meat analogues primarily involve high-temperature extrusion shearing and spinning. However, these techniques face challenges in replicating the realistic texture of fish meat with microscale precision. Lately, 3D printing has gained attention as a method for producing fish analogues, offering benefits such as design flexibility, efficiency, and minimal waste generation. Among the various 3D printing technologies, extrusion printing is a food-friendly approach. In their study, Shi et al. (2023) produced a high-precision food ink for 3D printing fish analogues using soy protein isolate. Soy protein isolate, which is challenging to work with due to its susceptibility to breaking under high shearing forces and poor shape fidelity, was modified by adding xanthan gum to adjust its rheology and enhance printability. Rice starch was also incorporated to improve the texture, making it more comparable to real fish meat. The resulting 3D-printed fish analogues achieved a high level of precision, with a resolution of approximately 200 μm. These analogues displayed similar texture properties to real fish meat, including springiness, hardness, gumminess, chewiness, and cohesiveness.

### *11.3.3 Methods for Testing the Rheological and Textural Properties of Fish Analogues*

Similar methods are employed for the characterization of rheological and textural properties in fish analogues as those used for meat analogues, as described previously. In their study, Ran et al. (2022a) used a creep-recovery test in addition to a frequency sweep in order to characterize the rheological properties of fishball analogues. A creep test evaluates the deformation of food over time under a constant load, showing its long-term stability. A recovery test gauges the capacity of food to revert to its previous shape following deformation, offering information on its elastic properties. The creep compliance $J_{(t)}$ of the samples was examined through a creep recovery test, which involved slight deformation during rheological measurements and provided insights into the textural attributes of food over a longer time scale. Higher creep compliance ($J_{(t)}$) typically indicates a sequential breakage of crosslinks during loading, particularly for molecular fragments with weaker crosslinks. The creep curves for the samples increased during the creep phase and then reduced gradually during the recovery phase, with the trend showing that the creep compliance decreased with the elevated levels of konjac glucomannan, suggesting that konjac glucomannan improved the strength of crosslinks in the gels. The Burgers model was employed to fit the creep curves, and it provided a good fit to the data. Parameters, the instantaneous compliance ($J_0$) and viscoelastic compliance ($J_1$), derived from the model, were found to be negatively correlated with gel firmness. In this case, $J_0$ and $J_1$ reduced significantly with higher konjac glucomannan concentrations, indicating a firmer and more compact gel network. This could be related to the higher deacetylation degree of konjac glucomannan with higher concentrations, leading to stiffened molecular aggregations and denser crosslinks within the gel network. These results align with the texture profile analysis results, which indicated similar hardness between plant-based fishballs containing 6.5% konjac glucomannan and conventional ones (fishballs made from Itoyori surimi). However, fishballs containing 8% konjac glucomannan exhibited significantly lower $J_0$ and a very hard and chewy texture compared to fishballs made from Itoyori surimi. Furthermore, the resistance to flow, denoted as $\eta 0$ (the dynamic viscosity of a fluid), increased with higher konjac glucomannan levels, indicating a more solid-like behavior. Retardation time $\lambda$, representing the time required for delayed strain to achieve 63.2% of the final value, decreased with increased konjac glucomannan. A fishball containing 3.5% konjac glucomannan showed lower elasticity, likely due to its lower viscosity, which facilitated rapid gelation, resulting in a less homogeneous and more sparse gel. Incorporating konjac glucomannan into plant fish balls increased elasticity significantly, with fishball 6.5% showing elasticity similar to fishballs made from Itoyori surimi.

## 11.4 Rheological and Textural Properties of Plant-Based Milk and Dairy Analogues

### *11.4.1 The Effect of Formulation on Rheological and Textural Properties of Plant-Based Milk and Cheese Analogues*

Plant-based milk alternatives aim to replicate the composition, functional, physicochemical, microstructural, and sensorial attributes of cow's milk. Thus, understanding cow's milk properties is crucial in designing these alternatives. Cow's milk has a consistent composition despite variations caused by factors like maturity, species, habitat, diet, and health. Cow's milk is a naturally occurring colloidal suspension that comprises fat globules and casein micelles dispersed in an aqueous solution comprising a variety of nutrients and bioactive compounds. It is designed to provide essential nutrients to support the growth and immune system of the animal. Thus, the unique composition and microstructure of cow's milk make it challenging to replicate using plant-based milk analogues (McClements and Grossmann 2022).

Plant-based dairy drinks are transforming the landscape of non-dairy milk and dairy products, utilizing sources like soy, almond, coconut, oat, and rice (Table 11.2). While they imitate the appearance and consistency of traditional dairy milk, their nutritional content differs significantly. Plant-based dairy drink analogues often rely on fermentation for taste and flavor enhancement and may not require additional thickening agents. Solubility is a key criterion in selecting plant materials to ensure a desirable mouthfeel, making an understanding of their rheological properties essential for process control (McClements and Grossmann 2022).

The development of plant-based milk analogues aims to replicate the physicochemical characteristics of cow's milk, including appearance, rheology, texture, mouthfeel, and flavor. In the plant tissue disruption approach, plant materials like oats, almonds, soybeans, cashews, or coconut flesh are mechanically or biochemically broken down into small fragments. These plant materials often contain natural oil bodies, similar in composition and structure to milk fat globules. The emulsification approach involves creating oil-in-water emulsions with fat droplets dispersed in an aqueous medium. The fat droplets simulate the characteristics of milk fat globules. To mimic the characteristics of casein micelles as well as other components present in cow's milk, plant-based milk alternatives often incorporate additional ingredients such as oligosaccharides, salts, and soluble proteins. The preparation of this emulsion involves homogenizing water, emulsifiers, plant-based oils, and various other components. Additional ingredients, such as vitamins, minerals, colors, flavors, or preservatives, may also be added to enhance the sensory attributes and nutritional profile (McClements and Grossmann 2022).

Plant-based yogurt-like products are also typically produced by fermenting aqueous extracts from sources like hazelnut, oats, peas, cashews, almonds, coconuts, soy, etc. (Demirkesen et al. 2018). However, replicating the thick texture of traditional dairy yogurt with non-fermented plant-based alternatives is challenging.

**Table 11.2** The impact of formulation and production method on rheological and textural properties of milk and cheese analogues

| The type of food analogue | The type of main ingredient | The applied rheological method | The applied textural method | The findings | References |
|---|---|---|---|---|---|
| Milk Analogue | Peanut, soybean, buckwheat, adlay, oat, and adzuki bean | Flow curve | | All the various types of samples displayed stable systems, with the exception of buckwheat milk. Buckwheat milk may undergo gelatinization, which can result in alterations to its microstructure and potentially impact its rheological and sensory characteristics.<br>The rheological properties of these plant-based milks exhibited non-Newtonian behavior.<br>Consumer acceptance of plant-based milk appeared to be closely related to aroma and flavor characteristics, as revealed by sensory assessments. | Yao et al. (2022) |
| Yogurt Analogue | Hazelnut | SAOS (Time sweep, Frequency sweep) | TPA | Gel formation was indicated by a rapid increase in complex modulus ($G^*$) around 120 min into fermentation.<br>Reduced moisture content led to higher $G^*$ values. Hazelnut products required lower moisture content than conventional yogurt for similar $G^*$ values.<br>Shearing affected elastic and viscous modulus values, reducing them with higher product concentration.<br>The hardest texture was obtained from microfluized sample containing the highest hazelnut concentration and microfluized sample containing the lowest hazelnut concentration was the softest | Demirkesen et al. (2018) |
| Mozzarella Cheese Analogue | Palm kernel oil, acid casein, pre-gelatinized starch | – | TPA | Significant differences observed in texture, functionality, and sensory properties between natural Mozzarella cheese and its analogue.<br>Natural cheese exhibited higher hardness, cohesiveness, springiness, gumminess, and chewiness but lower adhesiveness.<br>The analogue had better shredability but took longer to melt at 230¬∞C.<br>Natural cheese demonstrated superior meltability, stretchability, and less fat leakage. | Dharaiya et al. (2021) |

| | | | | | |
|---|---|---|---|---|---|
| Yogurt Analogue | Lupine seed | Flow curve | – | All samples exhibited shear-thinning behavior.<br>Viscosity was determined at a shear rate of 50 $s^{-1}$, selected to reflect differences in mouthfeel.<br>The addition of starch increased the viscosity of unfermented samples.<br>Unfermented prototypes had low viscosities, while thickened prototypes exhibited higher viscosities.<br>The thickened prototypes approached the viscosity of a commercial soy yogurt product.<br>The choice of LAB strains during fermentation either increased or decreased viscosity compared to the respective unfermented samples. | Laaksonen et al. (2021) |
| Feta cheese analogue | Lentil milk, inulin | – | TPA | A high amount of lentil milk (20%) resulted in a sandy texture. The formulation containing 10% lentil milk and 3.5% inulin achieved the best texture and sensory scores. | Moradi et al. (2021) |
| Cheese analogue | Boiled and roasted yellow peas and faba beans, κ-carrageenan and iotacarrageenan, κ-carrageenan, and xanthan gum | – | TPA | A firm and sliceable analogue was successfully created utilizing different types of pulse flour at a ratio of 1:4 flour:water, along with the incorporation of 1% κ-carrageenan.<br>The hardness of the resulting legume-based cheese analogues was greater than that of standard Gouda cheese but less than that of commercially available vegan cheese analogues. | Ferawati et al. (2021) |

Thickening agents are used in many commercial non-fermented plant-based yogurts, which may have an impact on consumer approval (Demirkesen et al. 2018).

Cheese analogues are products created by partially or completely substituting natural cheese components with vegetable proteins and fats derived from various sources (such as rapeseed, sunflower, palm, and soybean oil), along with the addition of emulsifying salts. Plant-based cheese analogues can be classified into two main types: non-fermented and fermented options. Non-fermented alternatives use ingredients like vegetable oils and polysaccharides to mimic the texture of dairy cheese, but they may lack certain characteristics like stretchability and flow. On the other hand, fermented plant-based cheese alternatives, while less common, are gaining popularity. They are created through the fermentation of plant-based extracts with lactic bacteria, which enhances their rheology, texture and sensory properties, making them more similar to traditional dairy cheese. The protein content in these alternatives varies, with fats and polysaccharides often playing a larger role in replicating the desired textural properties. Some options have higher protein and lower fat content and are desired for their color and flavor. Thus, the success of these plant-based cheese alternatives depends on achieving a balance between nutritional, textural, and rheological as well as sensorial properties by utilizing diverse plant-based ingredients and fermentation techniques (Kamath et al. 2022).

The formulation of cheese analogues can be adjusted by varying the quantities of caseinates, acid, stabilizers, and vegetable oil, as well as by altering processing parameters, to achieve different functional properties and meet specific requirements. For example, to reduce salt content, various approaches have been explored, including salt substitution, flavor enhancers, and bitter blockers (Kamath et al. 2022).

### 11.4.2 The Effect of Technological Methods on Rheological and Textural Properties of Plant-Based Milk and Cheese Analogues

In the formation of plant-based milk analogues, the emulsification approach is crucial. Various homogenization methods, including colloid mills, high-shear mixers, microfluidizers, high-pressure valve homogenizers, and ultrasonic homogenizers are used to create fine, uniform fat droplets. Then, milk analogues typically undergo thermal processing, like pasteurization or sterilization, using specific emulsifiers to coat the fat droplets and considering factors like salt content and pH values to ensure heat stability (McClements and Grossmann 2022).

Cheese analogue production has several routes. Either are prepared from entire seeds, nuts or grains, or from concentrates, which can be polysaccharide (starch) based or protein based, which confer quite different nutritional properties. The processing of cheese analogues involves careful ingredient selection and manufacturing methods. The manufacturing procedure often involves melting the fat, raising its temperature, adding a stabilizer system, blending in water to create an emulsion, slowly adding protein to initiate texture development, or using starch, and incorporating salt, flavor compounds, and acid. The drop in pH significantly influences texture development. An innovative approach involves using ultrafiltration and high hydrostatic pressure to slow down acid development and enhance texture (Kamath et al. 2022). Cheese analogues are essentially oil-in-water emulsions, similar to natural cheese, with fat droplets embedded in a protein gel matrix acting as an emulsifier (Sołowiej and Nastaj 2016). Thus, creating cheese analogues to replicate the properties of conventional cheese often involves adjustments to processing methods like heating times, temperatures, and physical factors like blending speed and ingredients. Extrusion-cooking methods, with or without emulsifying salts, have widely been employed to create cheese analogues. The extrusion process results in products with a range of textures, from hard blocks to soft spreads, depending on composition and process parameters. It is also possible to produce fibrous cheese analogues resembling mozzarella-based string cheese through extrusion (Bachmann 2001). In starch-based cheeses, the gelatinization properties (extent of retrogradation, gel strength and gelatinization temperature), dependent on the amylose and amylopectin ratio, and the starch treatment (for example, native or oxidized), affect mostly the texture. The texture of the protein-based cheeses is affected by the properties of the protein and the strategy used in the sol-gel transition, where temperate, acid, crosslinking with enzymes or ionic force can be used individually or simultaneously (McClements and Grossmann 2022).

### 11.4.3 Methods for Testing the Rheological and Textural Properties of Plant-Based Milk and Cheese Analogues

The determination of rheological properties of plant-based milk analogues is frequently performed by flow curve measurements through either shear stress or shear rate sweep, which provide the determination of apparent shear viscosity and provide insights into the rheological behavior of these products. For ideal fluids (Newtonian) like cow's milk, the apparent shear viscosity is independent of shear rate. On the other hand, plant-based drinks exhibit non-Newtonian pseudoplastic shear-thinning behavior ($n < 1$), with viscosity decreasing as shear rate increases. This behavior can be described using a power-law equation, where the consistency index (K) and power index (n) provide insights into the flow characteristics of fluid (Demirkesen et al. 2010).

The rheological properties of milk analogues show differences depending on their composition, use of hydrocolloid, and other ingredients (Table 11.2). The shear viscosity of milk analogues is strongly influenced by their fat content, with the relationship being characterized by a relatively slow increase in viscosity as the fat content rises from 0% to 20%. However, the viscosity increases much more rapidly when the fat content exceeds 20% and goes up to around 50%. This is in line with cow's milk products, which exhibit relatively lower viscosities due to their low-fat contents, ranging from 0.1% to 3.6%, while cream products, which have significantly higher fat contents ranging from 18% to over 36%, demonstrate notably higher viscosities. The viscosity of fluid dairy products, particularly cream, is also influenced by the aggregation state of colloidal particles. When a dairy product becomes acidic, colloidal particles like fat globules and casein micelles may flocculate, leading to a reduction in the electrostatic repulsion between them. Partially crystalline milk fat globules can also experience partial coalescence, resulting in clumping and increased viscosity. Additionally, milk analogues tend to exhibit higher viscosity due to the incorporation of hydrocolloids.

Effective volume fraction (the fraction of volume not occupied by solids) is crucial in determining rheological behavior, varying markedly across milk analogues due to ingredient and thickening agent disparities. Polysaccharides, like hydrocolloids with high molecular weights and extensive structures, are more effective in this regard. Since they encompass a substantial amount of water, this results in an increased effective volume fraction (McClements and Grossmann 2022).

While less common, oscillation tests can also provide insight into the viscoelastic properties of these drinks. Additionally, texture perception is relevant for understanding the overall sensory experience of plant-based drinks. Demirkesen et al. (2018) used a small deformation time sweep (oscillatory time sweeps) experiment to observe the rheological changes of hazelnut yogurt samples during fermentation. It has been observed that there is a consistent decrease in pH during fermentation for all hazelnut yogurt samples. The pH values had a reproducible plateau for the initial 100 min post-inoculation, likely due to lactic acid neutralization by bacterial-produced ammonia. After approximately 120 min of fermentation, there was a sudden rise in the complex modulus ($G^*$), signifying the formation of a gel structure. Samples with smaller hazelnut-to-water ratios had a longer onset of fermentation. The observed trend suggests that as moisture content decreased, there was a consistent increase in the complex modulus values. It's worth noting that the fermented hazelnut product reached similar complex modulus values as conventional yogurt but required a lower moisture content to do so. This indicates that the hazelnut product achieved the desired rheological properties with less moisture compared to yogurt.

To assess the influence of shear on the viscoelastic properties of the products, mechanical spectra were generated from two types of fermented hazelnut samples: set and stirred. The stirred products were prepared by first disrupting the gel structure using a plunger and then subjecting them to shearing. This allowed for a comparative evaluation of how shear affected the viscoelastic characteristics of the hazelnut products. The comparison of the results revealed that higher product concentrations resulted in a greater decrease in elastic ($G'$) as well as in viscous ($G''$) moduli values. Throughout the fermentation process, a 3-D gel matrix was created by the random aggregation of hazelnut's storage proteins and oleosins. This matrix was further reinforced by the presence of a network of both soluble and insoluble fibers. The firmness values of the fermented products were as expected, with homogenized samples having the highest concentration of hazelnut, being the hardest and homogenized samples having the lowest concentration, being the softest.

Microfluidized sample had a significantly higher hardness value despite having the same composition as a not microfluidized sample.

The rheological properties of cheese depend on its structure as well as on other factors, such as temperature and pH. Cheese analogues have viscoelastic properties that can be improved based on their composition, particularly fat content. The size, distribution, and type of fat droplets play a significant role in characterizing the texture and elasticity of these products, contributing to their unique characteristics. The viscoelastic properties of cheese analogues have been evaluated through tests like a frequency sweep test. Higher fat concentrations lead to a higher viscous modulus (G″) and a lower elastic modulus (G′) at low frequencies, while the reverse occurs at high frequencies. This phenomenon can be ascribed to alterations in protein interactions within the cheese matrix, which can be adjusted through the heating process (Lee et al. 2015). Similar findings were obtained in another study (Cunha et al. 2013), where cheese spread analogues underwent modifications due to the significant impact of the type of oil used for replacing milk fat, specifically hydrogenated soybean oil. At elevated frequencies, the cheese analogue exhibited solid-like behavior because there was inadequate disruption of intermolecular protein connections. In contrast, at lower frequencies, molecular chains had the chance to reconfigure and flow, leading to a more fluid-like behavior. This behavior can be associated with a high degree of emulsification, marked by the decrease in the size of fat globules. This reduction results in an increased surface area and a greater number of protein-protein bonds within the matrix.

The studies also indicated that analogue samples solidify at higher temperatures due to the removal of fat from the casein matrix, leading to a more substantial para-casein network (Urgu et al. 2019). The size of the fat globules in the cheese matrix is another critical factor affecting elasticity, with smaller fat globules contributing to greater elasticity. Elasticity in analogue cheese can also be influenced by the even distribution of fat droplets, which can improve the overall elasticity of the product.

In some cases, the presence of fatty acids with lower melting points, such as oleic acid, may also affect the viscoelastic properties of the cheese analogue when milk fat is replaced. Hanáková et al. (2013) examined processed cheese models with a substantial 99% substitution of milk fat with coconut oil or palm oil. Interestingly, they found that as the temperature decreased during cooling, there was a rise in the G″ modulus across all samples. The products containing palm oil exhibited the lowest G′ values. This observation might be connected to the existence of fatty acids with lower melting points, particularly oleic acid. These studies have collectively shown the complex relationship between fat replacement, emulsification, and the viscoelastic characteristics of cheese analogues.

Texture analysis of cheese analogues involves various tests, including tensile tests, TPA, penetrometer, compression tests, shear tests, etc. Texture analysis of cheese analogues are commonly conducted using techniques like TPA. Various factors, including the contents of fat, water, and emulsifying agents, the variety of cheese, and the manufacturing process, can influence the textural properties of cheese. Generally, higher fat content makes cheese softer, more cohesive, less elastic, and stickier. Thus, when milk fat is replaced with vegetable oils in cheese analogues, it typically leads to an increase in hardness. This phenomenon is often associated to the fact that the lack of fat results in the concentration of other components, leading to the formation of rigid protein bonds that enhance the physical strength of the cheese analogues (Leong et al. 2020). However, exceptions to this trend exist, where milk fat substitution can actually lead to a decrease in hardness (Kavak and Karabiyik 2019). For example, in a study involving cheddar cheese analogues, it was observed that replacing full-fat with canola oil resulted in a reduction in firmness compared to the control cheese. Interestingly, when a double emulsion was used, the cheddar cheese analogue displayed increased hardness (Leong et al. 2020). The softness observed in samples with a single emulsion can be related to its liquid form at room temperature (Leong et al. 2020). Additionally, the ratio of saturated fatty acids within the fat content significantly influences hardness. The presence of fat acids in breaking down protein bonds, ultimately lowering hardness and yielding a softer texture. In some cases, an increase in hardness is associated with a higher protein mass per unit area of fat, as observed when hydrogenated vegetable fat is used for milk fat replacement. This is because a higher protein density contributes to greater hardness due to the pivotal role of the protein matrix in the structural integrity of the product (Leong et al. 2020).

The textural properties of cheese analogues can also be influenced by a range of factors, including water content, pH range, and emulsifying agents. Springiness in cheese analogues is primarily determined by the protein matrix, and it typically does not show significant differences among samples with varying fat content. Differences in springiness in the analogues were attributed to variations in the protein's adsorption onto the fat surface, which depended on the level of unsaturation in the fat used. The high saturated fat concentration of butter encourages the protein's lipophilic groups to bind to nearby fat droplets in the emulsion, creating a reticulated structure that gives the finished product its springiness. Thus, some studies have reported that the springiness of cheese analogues can decrease when milk fat is replaced with emulsified soybean oil or palm oil. Additionally, both adhesiveness and gumminess tend to increase in cheese analogues when partial milk fat replacement is employed, often attributed to the presence of smaller fat globules and a higher protein density (Kavak and Karabiyik 2019). Chewiness, on the other hand, often exhibits higher values in cheese analogues with lower fat content due to inadequate breakdown of casein, resulting in a firmer texture (Abd El-Wahed and Hassanien 2019). The attribute of cohesiveness in cheese analogues can vary, with some studies indicating an increase and others a decrease after milk fat replacement, a variation influenced by factors like the type of oil used, cheese variety, and the method of fat substitution (Totosaus et al. 2017; Lamichhane et al. 2018; Abd El-Wahed and Hassanien 2019; Kavak and Karabiyik 2019).

Dynamic shear rheology, such as the amplitude sweep test, is employed to gain insights into the viscoelastic properties of cheese. In the linear viscoelastic region, the cheese structure is not permanently damaged, but beyond this region, it breaks down (yield point). Cheese analogues exhibit different moduli ($G'$ and $G''$) and yield points compared to their dairy counterparts. These reflect the differences in their responses to cutting, chewing, and processing. Plant-based cheddar cheese, for example, is more resistant to deformation, giving higher moduli (G′ and G″) values and requiring more force for deformation (Fox et al. 2017). Creep and stress relaxation tests, under normal force or shear conditions, are also used to assess cheese viscoelastic behavior. Various models, such as the Maxwell, Voigt-Kelvin, and Peleg models, have also been employed to define the rheological properties of different kinds of cheeses (Muthukumarappan and Swamy 2023).

## 11.5 Rheological and Textural Properties of Egg Analogues

### *11.5.1 The Effect of Formulation on Rheological and Textural Properties of Egg Analogues*

Creating egg analogues is a precise process that involves selecting plant proteins like peas, chickpeas, beans, soybeans, and sunflowers with thermal properties similar to those found in eggs, typically denaturing between 63 and 93 °C. As mentioned above, these plant protein sources have the ability to form gels by crosslinking at low pH, heat etc. However, it is crucial to control various factors such as protein structure, pH, concentration, salt levels, and temperature during the process to achieve the desired results. Various additional ingredients, such as emulsified plant-based oils, natural pigments like curcumin or carotenoids, and flavorings like sugars, salts, herbs, or spices, have been used. Amphiphilic plant-based proteins and phospholipids serve as emulsifiers or foaming agents to stabilize emulsions or foams, while plant-based globular proteins impact gelling, binding, and thickening characteristics during cooking. Egg analogues can be fortified with vitamins and minerals to match the micronutrient profile of real eggs. Additionally, bases, acidulants, or buffers may be incorporated to maintain the desired pH level. Additionally, plant-based preservatives, such as antimicrobials, chelating agents or antioxidants might be necessary to prevent microbial growth or unwanted chemical reactions. This comprehensive approach allows for the creation of plant-based egg analogues that closely resemble the textural, appearance as well as the flavor characteristics of real eggs (McClements and Grossmann 2022).

### *11.5.2 The Effects of Technological Methods on Rheological and Textural Properties of Egg Analogues*

The choice of processing steps depends on the ingredients as well as the used processes. A common approach involves creating liquid egg analogues comprising fat droplets coated with proteins dispersed in a thick aqueous solution including thickening polysaccharides and gelling globular proteins. Hydrophobic additives such as flavors, vitamins, or pigments, are typically dissolved in the oil phase. In order to prepare the aqueous phase, a hydrophilic emulsifier, usually plant-based proteins and/or phospholipids, is dissolved in water. Other hydrophilic additives, including gelling agents, thickeners, colors, pH adjusters, preservatives, and flavors may also be utilized into the aqueous phase. pH is carefully regulated to ensure that the ingredients perform as intended. High-shear mixing is employed during homogenization to blend the oil and water phases, resulting in a coarse emulsion featuring fat droplets coated with emulsifiers. The size of these fat droplets is reduced further using a homogenizer, which prevents aggregation or creaming. In some cases, pre-emulsified oil-soluble ingredients can be used, eliminating the need for oil phase preparation and homogenization. These powdered ingredients can be directly mixed with the aqueous phase ingredients. Additional water-soluble additives can be introduced to the aqueous phase following the homogenization process. Enzymatic crosslinking agents like transglutaminase may be introduced to increase viscosity. The product may undergo thermal processing to promote protein unfolding and aggregation and to deactivate enzymes or microorganisms. Careful temperature-time control is required, particularly for fluid egg analogues, to prevent premature gelation (McClements and Grossmann 2022).

### *11.5.3 Methods for Testing the Rheological and Textural Properties of Egg Analogues*

The formulation of plant-based egg analogues should be carefully designed to replicate the rheological and textural properties of real eggs in both uncooked and cooked states, as well as the changes that occur during cooking. This involves optimizing the selection and concentration of ingredients to achieve the desired properties. According to rheological properties, uncooked egg analogues should exhibit shear-thinning behavior at ambient temperatures. This can be achieved by incorporating thickening agents such as gums or protein aggregates derived from cross-linked plant proteins. It is necessary to carefully adjust these additives in order to precisely replicate the shear viscosity versus shear rate properties of raw eggs (McClements and Grossmann 2022).

To replicate the process of real eggs transitioning into gels upon heating, it is essential for egg analogues to include an adequate amount of globular proteins that unfold and aggregate when exposed to temperatures exceeding their thermal denaturation threshold. These proteins should have denaturation temperatures similar to those of egg proteins (e.g., ovalbumin, around 63–85 °C). The rheological properties of the formed gel, including hardness, gelling temperature, and fracture properties, should closely resemble those of hen's eggs. Various factors, including protein type and concentration, ionic strength, pH, ingredient interactions, and heating conditions must be considered when optimizing the formulation (McClements and Grossmann 2022).

Lu et al. (2023a) evaluated the influence of corn starch, glutinous rice starch, and potato starch on the physicochemical attributes of plant-based eggs. It was aimed at understanding the mechanisms responsible for the varying effects of these different starch types on the modification of plant-based eggs. The large deformation behaviors of the plant-based eggs were evaluated through strain sweep analysis. In the linear viscoelastic region, where predominantly elastic behavior was evident, the storage modulus (G′) surpassed the loss modulus (G″), signifying the strength of the gel. Potato and corn starch-modified samples exhibited significantly greater gel strength than glutinous rice starch-modified samples, corresponding to cohesiveness results that depended on tensile and compression strength. The increased amylose content in potato and corn starch, in comparison to glutinous rice starch, enabled greater proximity of hydroxyl groups, thus promoting the gelling process. Among different ratios, soy protein isolate/potato starch (8:6) had the most substantial impact on gel strength. The higher proportion of potato starch served as both a filling reinforcement and water-absorbing agent, resulting in the densification of the protein matrix and an increase in G′. However, further increasing the potato starch ratio weakened the gel strength, as the higher ratio may not sufficiently wrap the excess starch granules. In the same study, thermal behaviors were characterized by a temperature sweep over a heating-cooling cycle. G′ increased gradually with rising temperature, reaching peak values ($G'_{peak}$) around 85 °C, attributed to starch granule hydration and swelling. Further heating to 95 °C resulted in $G'_{decline}$ due to starch granule rupture. Soy protein isolate/potato starch 8:6 showed the lowest $G'_{breakdown}$, indicating well-embedded potato starch granules that retarded rupture and were consistent with their high gel strength. The associative protein-starch interactions during continuous heating at 95 °C contributed to the subsequent increase in G′ upon cooling from 95 to 50 °C. Soy protein isolate/corn starch 8:6 and soy protein isolate/glutinous rice starch 8:6 exhibited higher $G'_{breakdown}$, suggesting more ruptured granules, resulting in lower $G'f_{inal}$ and $G'_{setback}$ values compared to soy protein isolate/potato starch 8:6. This difference was likely because of the gel strengthening impacts of corn and glutinous rice starch fragments.

Cooked egg analogues should also exhibit textural properties similar to those of real cooked eggs. For example, the hardness values of cooked hen's eggs are around 1300 N, and plant-based egg analogues can aim to achieve similar properties. The presence of hydrocolloids and high fat content, which create a three-dimentional network of aggregated droplets, results in non-ideal plastic behavior in egg-based products like mayonnaise and salad dressings. The rheological properties of these kinds of products can be described by the Herschel-Bulkley model, including parameters like yield stress ($\tau_0$), consistency index (K), and power law index (n). Conventional egg-based mayonnaise and salad dressings typically display substantial yield stress and demonstrate pronounced shear-thinning characteristics ($n \approx 0.4$). The yield stress must be surpassed before it flows. By adjusting factors such as fat droplet concentration, size, molecular and physical interactions, and the types and quantities of thickening agents used, plant-based emulsified products can be formulated to have rheological properties similar to their egg-based counterparts (McClements and Grossmann 2022) (Table 11.3).

**Table 11.3** The impact of formulation and production method on rheological and textural properties of egg analogues

| The type of food analogue | The type of main ingredient | The applied rheological method | The applied textural method | The findings | References |
|---|---|---|---|---|---|
| Egg analogue | Chickpea flour, soy protein isolate, κ-carrageenan, gellan gum. | SAOS (Temperature sweep, Time sweep, Frequency sweep) | TPA | Among the formulations, one with 0.3% κ-carrageenan closely resembled the physiochemical properties of eggs in terms of hardness, specific volume, and gel strength.<br>In contrast, 0.1% κ-carrageenan led to some κ-carrageenan-protein interactions, but not as dominantly as 0.3% κ-C.<br>Nonetheless, when the κ-carrageenan concentration reached 0.5%, the degree of rheological synergism reduced, while electrostatic interactions rose, suggesting an increase in κ-carrageenan-κ-carrageenan interactions.<br>On the other hand, gellan gum systems showed segregated interactions with proteins in all cases, affecting the macroscale attributes of plant-based egg products thereby accounting for the differences in their physiochemical characteristics | Lu et al. (2022) |
| Egg analogue | Corn starch, potato starch, glutinous rice starch, different ratios of soy protein isolate/ starch (10:4, 8:6, 6:8) | SAOS (Temperature sweep, Time sweep, Amplitude oscillatory sweep)<br>LAOS (Strain sweep) | TPA | Modification with soy protein isolate/potato starch at a ratio of 8:6 resulted in improved springiness, hardness, chewiness, cohesiveness, and specific volume of the plant-based omelet, bringing it to a level comparable to traditional egg omelet.<br>On the other hand, corn and glutinous rice starch proved to be less effective in providing the necessary support for the gel structures and were unable to replicate the texture of the native egg omelet. | Lu et al. (2023a) |
| Egg analogue | Soy protein isolate, potato starch, chickpea flour, HPMC | Flow curve | TPA | The addition of 1.5% HPMC to egg analogue matched the foaming capacity and stability of liquid egg. This was attributed to thermodynamic incompatibility between biomacromolecules at the air-liquid interface, which accelerated the adsorption of proteins and lipids through a depletion mechanism.<br>As a result, this produced the highest levels of consistency, volumetric synergy, and elasticity in the foams. | Lu et al. (2023b) |

## 11.6 Conclusion

In conclusion, this chapter delves into the complex world of rheological and textural properties in the context of food. Food analogues, whether plant-based or designed to replicate specific properties of traditional foods, hold great promise for satisfying diverse consumer needs driven by health consciousness, environmental sustainability, and ethical considerations. The search for analogues that accurately mimic the complex rheological and texture characteristics of traditional foods goes far beyond taste and nutrition. It explores the basic knowledge of molecular interactions between ingredients and how these interactions might be used to make products that provide a sensory experience similar to that of their animal origin counterparts.

The rheological and textural profiles of food analogues are critical factors determining their acceptance by consumers. The fundamental science behind rheology and texture analysis has been explored, with a focus on key parameters, ingredients, and measurement techniques that are critical in characterizing the properties of food analogues. The basic concepts that underlie the texture and rheology of these products have been revealed, ranging from the comprehension of the viscoelastic characteristics of these materials to the complex interaction of ingredients and processing techniques. Furthermore, a wide range of food analogues was encompassed, from meat substitutes to dairy alternatives and egg analogues. Each category presents unique challenges and opportunities, and it has been demonstrated how the careful selection of ingredients, formulation parameters, and processing techniques can yield products that closely mimic their traditional counterparts. Through rigorous testing methods such as TPA, tensile, strain sweep, frequency sweep, temperature sweep methods, etc., a comprehensive understanding of the behavior of food analogues under various conditions has been established. These analytical tools provide insight into the complex nature of food analogue design, allowing for the optimization of their texture and rheology to closely mimic the foods they are modeled after.

This review, by consolidating current knowledge on the rheological and textural optimization of food analogues, underscores the scientific and technological expertise required to innovate in the field of food analogues. The knowledge compiled here will play a critical role in directing future research and innovation in the creation of food ingredients that satisfy consumer demand for a culinary experience that honors the traditional sensory qualities of food while also being ethically and environmentally conscious.

## References

Arora B, Kamal S, Sharma VP (2017) Effect of binding agents on quality characteristics of mushroom based sausage analogue. J Food Process Preserv 41(5):e13134

Bachmann HP (2001) Cheese analogues: a review. Int Dairy J 11(4–7):505–515

Caporgno MP, Böcker L, Müssner C, Stirnemann E, Haberkorn I, Adelmann H et al (2020) Extruded meat analogues based on yellow, heterotrophically cultivated Auxenochlorella protothecoides microalgae. Innovative Food Sci Emerg Technol 59:102275

Cassidy ES, West PC, Gerber JS, Foley JA (2013) Redefining agricultural yields: from tonnes to people nourished per hectare. Environ Res Lett 8(3):034015

Chiang JH, Loveday SM, Hardacre AK, Parker ME (2019) Effects of soy protein to wheat gluten ratio on the physicochemical properties of extruded meat analogues. Food Struct 19:100102

Chiang JH, Hardacre AK, Parker ME (2020) Extruded meat alternatives made from Maillard-reacted beef bone hydrolysate and plant proteins: part I–Effect of moisture content. Int J Food Sci Technol 55(2):649–659

Chiang JH, Tay W, Ong DSM, Liebl D, Ng CP, Henry CJ (2021) Physicochemical, textural and structural characteristics of wheat gluten-soy protein composited meat analogues prepared with the mechanical elongation method. Food Struct 28:100183

Cunha CR, Grimaldi R, Alcântara MR, Viotto WH (2013) Effect of the type of fat on rheology, functional properties and sensory acceptance of spreadable cheese analogue. Int J Dairy Technol 66(1):54–62

Dekkers BL, Boom RM, van der Goot AJ (2018) Structuring processes for meat analogues. Trends Food Sci Technol 81:25–36

Demirkesen I, Mert B, Sumnu G, Sahin S (2010) Rheological properties of gluten-free bread formulations. J Food Eng 96(2):295–303

Demirkesen I, Vilgis TA, Mert B (2018) Effect of microfluidization on the microstructure and physical properties of a novel yoghurt formulation. J Food Eng 237:69–77

Dharaiya C, Jana A, Patel A, Patel D (2021) Comparison of natural Mozzarella cheese with acid casein based Mozzarella cheese analogue. Indian J Dairy Sci 74(4):301–308

Diaz JR, Kantanen K, Edelmann JM, Suhonen H, Sontag-Strohm T, Jouppila K, Piironen V (2022) Fibrous meat analogues containing oat fiber concentrate and pea protein isolate: mechanical and physicochemical characterization. Innovative Food Sci Emerg Technol 77:102954

Dinani ST, Broekema NL, Boom R, van der Goot AJ (2023) Investigation potential of hydrocolloids in meat analogue preparation. Food Hydrocoll 135:108199

do Carmo CS, Knutsen SH, Malizia G, Dessev T, Geny A, Zobel H et al (2021) Meat analogues from a faba bean concentrate can be generated by high moisture extrusion. Future Foods 3:100014

El-Wahed EMA, Hassanien MF (2019) Chemical, rheological and sensory characteristics of processed cheese spread analogues. Carpathian J Food Sci Technol 11(1)

Ferawati F, Hefni M, Östbring K, Witthöft C (2021) The application of pulse flours in the development of plant-based cheese analogues: proximate composition, color, and texture properties. Food Secur 10(9):2208

Fox PF, Guinee TP, Cogan TM, McSweeney PL, Fox PF, Guinee TP et al (2017) Cheese: structure, rheology and texture. In: Fundamentals of cheese science. Springer, Boston, pp 475–532

Geerts ME, Dekkers BL, van der Padt A, van der Goot AJ (2018) Aqueous fractionation processes of soy protein for fibrous structure formation. Innovative Food Sci Emerg Technol 45:313–319

Hanáková Z, Buňka F, Pavlínek V, Hudečková L, Janiš R (2013) The effect of selected hydrocolloids on the rheological properties of processed cheese analogues made with vegetable fats during the cooling phase. Int J Dairy Technol 66(4):484–489

Herz E, Herz L, Dreher J, Gibis M, Ray J, Pibarot P et al (2021) Influencing factors on the ability to assemble a complex meat analogue using a soy-protein-binder. Innovative Food Sci Emerg Technol 73:102806

Jia W, Curubeto N, Rodríguez-Alonso E, Keppler JK, van der Goot AJ (2021) Rapeseed protein concentrate as a potential ingredient for meat analogues. Innovative Food Sci Emerg Technol 72:102758

Kamath R, Basak S, Gokhale J (2022) Recent trends in the development of healthy and functional cheese analogues-a review. LWT 155:112991

Kavak DD, Karabiyik H (2019) Quality evaluation of kashar cheese: influence of palm oil and ripening period. Food Sci Technol 40:354–360

Kazir M, Livney YD (2021) Plant-based seafood analogs. Molecules 26(6):1559

Köllmann N, Schreuders FK, Zhang L, van der Goot AJ (2023) On the importance of cooling in structuring processes for meat analogues. J Food Eng 350:111490
Kołodziejczak K, Onopiuk A, Szpicer A, Poltorak A (2021) Meat analogues in the perspective of recent scientific research: A review. Food Secur 11(1):105
Kyriakopoulou K, Dekkers B, Jan A, van der Goot. (2019) Plant-based meat analogues. In: Sustainable meat production and processing. Academic, pp 103–126
Laaksonen O, Kahala M, Marsol-Vall A, Blasco L, Järvenpää E, Rosenvald S et al (2021) Impact of lactic acid fermentation on sensory and chemical quality of dairy analogues prepared from lupine (Lupinus angustifolius L.) seeds. Food Chem 346:128852
Lamichhane P, Kelly AL, Sheehan JJ (2018) Effect of milk centrifugation and incorporation of high-heat-treated centrifugate on the composition, texture, and ripening characteristics of Maasdam cheese. J Dairy Sci 101(7):5724–5737
Lee SK, Klostermeyer H, Anema SG (2015) Effect of fat and protein-in-water concentrations on the properties of model processed cheese. Int Dairy J 50:15–23
Leong TS, Ong L, Gamlath CJ, Gras SL, Ashokkumar M, Martin GJ (2020) Formation of cheddar cheese analogues using canola oil and ultrasonication–A comparison between single and double emulsion systems. Int Dairy J 105:104683
Lu Z, Lee PR, Yang H (2022) Chickpea flour and soy protein isolate interacted with κ-carrageenan via electrostatic interactions to form egg omelets analogue. Food Hydrocoll 130:107691
Lu Z, Liu Y, Lee YEJ, Chan A, Lee PR, Yang H (2023a) Effect of starch addition on the physicochemical properties, molecular interactions, structures, and in vitro digestibility of the plant-based egg analogues. Food Chem 403:134390
Lu Z, Lee PR, Yang H (2023b) Synergistic adsorption of surface-active components at the air-liquid interface improves foaming properties of plant-based egg analogues. Food Hydrocoll 137:108414
McClements DJ, Grossmann L (2022) Eggs and egg products. In: Next-generation plant-based foods. Springer, Cham, pp 341–388. https://doi.org/10.1007/978-3-030-96764-2_7
McClements DJ, Weiss J, Kinchla AJ, Nolden AA, Grossmann L (2021) Methods for testing the quality attributes of plant-based foods: meat-and processed-meat analogs. Food Secur 10(2):260
Moradi D, Ziarati P, Sawicka B (2021) Physicochemical and rheological properties of lentil milk/inulin blend—A feta cheese analogue. Legume Sci 3(4):e80
Muthukumarappan K, Swamy GJ (2023) Rheology, microstructure, and functionality of cheese. In: Advances in food rheology and its applications. Woodhead Publishing, pp 297–334
Okeudo-Cogan MC, Murray BS, Ettelaie R, Connell SD, Radford SJ, Micklethwaite S, Sarkar A (2023) Understanding the microstructure of a functional meat analogue: demystifying interactions between fungal hyphae and egg white protein. Food Hydrocoll 140:108606
Palanisamy M, Töpfl S, Aganovic K, Berger RG (2018) Influence of iota carrageenan addition on the properties of soya protein meat analogues. LWT 87:546–552
Palanisamy M, Töpfl S, Berger RG, Hertel C (2019) Physico-chemical and nutritional properties of meat analogues based on spirulina/lupin protein mixtures. Eur Food Res Technol 245:1889–1898
Ran X, Lou X, Zheng H, Gu Q, Yang H (2022a) Improving the texture and rheological qualities of a plant-based fishball analogue by using konjac glucomannan to enhance crosslinks with soy protein. Innovative Food Sci Emerg Technol 75:102910
Ran X, Yang Z, Chen Y, Yang H (2022b) Konjac glucomannan decreases metabolite release of a plant-based fishball analogue during in vitro digestion by affecting amino acid and carbohydrate metabolic pathways. Food Hydrocoll 129:107623
Rousta N, Hellwig C, Wainaina S, Lukitawesa L, Agnihotri S, Rousta K, Taherzadeh MJ (2021) Filamentous fungus aspergillus oryzae for food: from submerged cultivation to fungal burgers and their sensory evaluation—a pilot study. Food Secur 10(11):2774
Samard S, Ryu GH (2019a) Physicochemical and functional characteristics of plant protein-based meat analogs. J Food Process Preserv 43(10):e14123

Samard S, Ryu GH (2019b) A comparison of physicochemical characteristics, texture, and structure of meat analogue and meats. J Sci Food Agric 99(6):2708–2715
Schreuders FK, Dekkers BL, Bodnár I, Erni P, Boom RM, van der Goot AJ (2019) Comparing structuring potential of pea and soy protein with gluten for meat analogue preparation. J Food Eng 261:32–39
Schreuders FK, Schlangen M, Kyriakopoulou K, Boom RM, van der Goot AJ (2021) Texture methods for evaluating meat and meat analogue structures: a review. Food Control 127:108103
Sengar AS, Beyrer M, McDonagh C, Tiwari U, Pathania S (2023) Effect of process variables and ingredients on controlled protein network creation in high-moisture plant-based meat alternatives. Food Secur 12:3830. https://doi.org/10.3390/foods12203830
Sha L, Xiong YL (2020) Plant protein-based alternatives of reconstructed meat: science, technology, and challenges. Trends Food Sci Technol 102:51–61
Shi H, Li J, Xu E, Yang H, Liu D, Yin J (2023) Microscale 3D printing of fish analogues using soy protein food ink. J Food Eng 347:111436
Shrestha S, van't Hag L, Haritos V, Dhital S (2023) Rheological and textural properties of heat-induced gels from pulse protein isolates: lentil, mungbean and yellow pea. Food Hydrocoll 143:108904
Smetana S, Larki NA, Pernutz C, Franke K, Bindrich U, Toepfl S, Heinz V (2018) Structure design of insect-based meat analogs with high-moisture extrusion. J Food Eng 229:83–85
Sołowiej B, Nastaj M (2016) Relevance and production of dairy analogues and restructured dairy products. In: Reference module in food science, pp 2010–2015
Sun C, Fu J, Chang Y, Li S, Fang Y (2022) Structure design for improving the characteristic attributes of extruded plant-based meat analogues. Food Biophys 17:137–149
Totosaus A, Rojas-Nery EF-F, M. J. (2017) Soya bean oil/soya protein isolate and carrageenan emulsions as fat replacer in fat-reduced Oaxaca-type cheese. Int J Dairy Technol 70(4):499–505
Urgu M, Türk A, Ünlütürk S, Kaymak-Ertekin F, Koca N (2019) Milk fat substitution by microparticulated protein in reduced-fat cheese emulsion: the effects on stability, microstructure, rheological and sensory properties. Food Sci Anim Resour 39(1):23
Vandelook S, Elsacker E, Van Wylick A, De Laet L, Peeters E (2021) Current state and future prospects of pure mycelium materials. Fungal Biol Biotechnol 8(1):1–10
Wang T, Kaur L, Furuhata Y, Aoyama H, Singh J (2022) 3D printing of textured soft hybrid meat analogues. Food Secur 11(3):478
Xie Y, Cai L, Zhao D, Liu H, Xu X, Zhou G, Li C (2022) Real meat and plant-based meat analogues have different in vitro protein digestibility properties. Food Chem 387:132917
Yao Y, He W, Cai X, Bekhit AEDA, Xu B (2022) Sensory, physicochemical and rheological properties of plant-based milk alternatives made from soybean, peanut, adlay, adzuki bean, oat and buckwheat. Int J Food Sci Technol 57(8):4868–4878
Yazar G, Demirkesen I (2023) Linear and non-linear rheological properties of gluten-free dough systems probed by fundamental methods. Food Eng Rev 15(1):56–85
Yuliarti O, Kovis TJK, Yi NJ (2021) Structuring the meat analogue by using plant-based derived composites. J Food Eng 288:110138
Yuliarti O, Ng L, Koh WM, Abdullah MFBMF, Sentana AD (2023) Structural properties of meat analogue with added konjac gels. Food Hydrocoll 142:108716
Zahari I, Ferawati F, Helstad A, Ahlström C, Östbring K, Rayner M, Purhagen JK (2020) Development of high-moisture meat analogues with hemp and soy protein using extrusion cooking. Food Secur 9(6):772
Zhang W, Li S, Zhang B, Drago SR, Zhang J (2016) Relationships between the gelatinization of starches and the textural properties of extruded texturized soybean protein-starch systems. J Food Eng 174:29–36
Zhu HG, Tang HQ, Cheng YQ, Li ZG, Tong LT (2021) Potential of preparing meat analogue by functional dry and wet pea (Pisum sativum) protein isolate. LWT 148:111702

# Chapter 12
# Microbiological Approaches for Food Analogues Production: Nutritional and Safety Aspects

**Audry Peredo-Lovillo, Haydee Eliza Romero-Luna, Francisco Erik González-Jiménez, and Maribel Jiménez-Fernández**

## 12.1 Introduction

Throughout human history, agriculture and industry have provided safe, tasty, convenient, and affordable food products to reduce hunger and malnutrition worldwide. However, the growing world population, predicted for 2050 with a total of 10 billion people, makes the supply of quality food products more difficult, leading people to consume industrialized foods products with not recommended concentrations of fat, sugar, and salt, which promote the incidence of chronic degenerative diseases such as obesity, cardiovascular disease, type 2 diabetes, cancer, hypertension, and stroke, mainly in developed countries (Hashempour-Baltork et al. 2020; McClements and Grossmann 2021; Molfetta et al. 2022). Example of these type of processed foods are the foods of animal origin, including egg, milk, fish, and their derivatives, whose consumption and production not only harms the consumer health but also has negative impacts on the environment, leading a pollution enhancement in land and water, greenhouse gas generation, and biodiversity loss (McClements and Grossmann 2021).

To counteract the adverse approaches of animal-based food products, traditional sources such as plants are used to design and develop healthier and more sustainable animal-based food analogues compared to animal-derived products (Fig. 12.1).

---

A. Peredo-Lovillo · F. E. González-Jiménez
Laboratorio de Ingeniería y Biotecnología de Alimentos, Facultad de Ciencias Químicas, Universidad Veracruzana, Orizaba, Veracruz, Mexico
e-mail: auperedo@uv.mx; franciscogonzales02@uv.mx

H. E. Romero-Luna
Instituto Tecnológico de Orizaba, Orizaba, Veracruz, Mexico

M. Jiménez-Fernández (✉)
Centro de Investigación y Desarrollo en Alimentos (CIDEA), Universidad Veracruzana, Xalapa-Enríquez, Veracruz, Mexico
e-mail: maribjimenez@uv.mx



Ö. P. CAN et al. (eds.), *Food Analogues: Emerging Methods and Challenges*,
https://doi.org/10.1007/978-3-031-69872-9_12

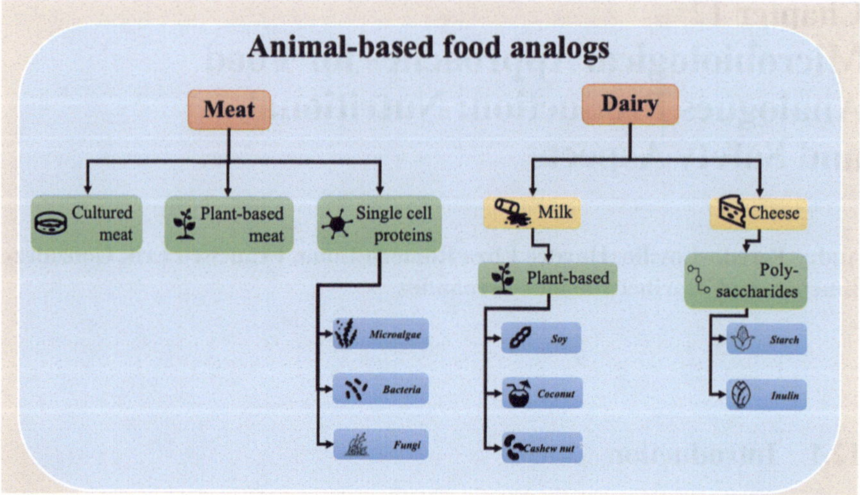


**Fig. 12.1** Raw materials used for animal-based food analogues. (Figure of own authorship to be published in this chapter)

These healthier alternatives seem to be efficient enough to satisfy the requirements of the growing world population, becoming analogues of animal-based foods with similar or even greater physicochemical, sensory, and nutritional characteristics (Razzaq et al. 2020). The production of plant-based analogues is transforming the modern food industry, which is now focused on generating more food products from cereals, legumes, nuts, and seeds, without depleting natural resources (Molfetta et al. 2022). In turn, the plant-based diet has been adopted by many consumers since it incorporates elements with a greater nutritional contribution and the possibility of having functional properties, including reduction in the content of cholesterol, lactose, and dairy allergens such as casein, hormones, antibiotic residues, and vegan-friendly labels (Pua et al. 2022).

Conversely, these desirable properties are dependent on the type of plant and the degree of processing of final product. For example, fruits, vegetables, nuts, and whole grains are considered fresh or minimally processed plant-based foods that improve nutrition, while highly processed plant products tend to show a lower nutritional value, because their high concentrations in of saturated fat, salt, and digestible carbohydrates (McClemens and Grossmann 2021).

In addition to the limitations of plant-based analogues, another disadvantage is their incompatibility for the production of complex compositions and structures such as those desirable in meats, cheeses, and other dairy analogues. Therefore, for the consumers to find this type of animal-based food analogues desirable and acceptable, auxiliary strategies to imitate the composition, as well as physicochemical, sensory, and nutritional features must be implemented. One of these strategies consists of the incorporation and use of fermentative microorganisms to produce specific metabolites that contribute to the food analogue design (Grossmann and

McClements 2021). In this context, fermentation is considered a biotechnological process where a food substrate is metabolized by a specific microorganism or group of them to obtain a final product with desirable properties. Among these properties can be find improvements in both flavor, texture, nutritional composition, and food safety of the final product (Pua et al. 2022). Hence, the use of the fermentation process can be considered a key point for the design and creation of novel food analogues with improved functional and organoleptic properties, along with adequate consumer acceptability and microbiological safety (Tamang et al. 2016; Pua et al. 2022).

The entire or partial replacement of food products of animal origin (including dairy and meat products) by plant or novel protein sources is an issue that cannot be ignored; and future research studies must expand the range of options for consuming food analogues based mainly on the substitution of proteins of animal origin, by those extracted both from plants and microbial species, which at the same time meet both the quality, sensory (mainly texture and mouthfeel), microbiological and nutritional benefits demanded by consumers. Thus, production of this new generation of food analogues needs a complete comprehension of the molecular attributes, molecular structural assembly, and their impact on physicochemical and sensory characteristics, as well as the microbiological aspects of the end product are considered key factors to design, formulate, and develop new food analogue alternatives (McClements and Grossmann 2021; Kamath et al. 2022). Based on the above, the present chapter addresses the microbiological and nutritional aspects of food analogues based on the use of microbial strains for the preservation and control, nutritional, as well as safety aspects, maintaining sensory properties and quality of final product.

## 12.2 Meat Analogues

Meat is considered an essential part of the daily human diet. Beef, pork, and chicken are the most demanded meat products worldwide, being the United States and Australia as the main consumers of meat (He et al. 2020). Recently, meat production and consumption have become topics of concern, due to the adverse effects both on the environment (releasing potential global warming gases such as methane and nitrous oxide), as well as human health (Marinova and Bogueva 2019; He et al. 2020). In this context, different alternatives have been tested to minimize the environmental footprints of meat production, improving at the same time the nutritional value of the meat-derived products, making the use of unconventional protein sources the most promising alternative to substitute traditional meat, either from cultured meat or plant-based meat. In addition to these meat analogues, other protein analogues include single-cell proteins, obtained from species of algae, fungi/yeast, and bacteria (He et al. 2020). Different studies have addressed the technological and industrial applications for meat analogues production, but the safety and microbiological aspects are not fully described yet (Hadi and Brightwell 2021).

It is known that the consumption of processed meat products increases the probability of diet-related disease incidence, including obesity, type 2 diabetes, cardiovascular disorders, strokes, or cancer (Kołodziejczak et al. 2022). Additionally, diseases including avian influenza, mad cow disease (human Creutzfeldt-Jakob disease associated with Bovine Spongiform Encephalopathy), Q fever, SARS (severe acute respiratory syndrome) and MERS (Middle East respiratory syndrome), are considered microbiological risks associated with meat products consumption, mainly in those with a prolonged storage in which the indiscriminate use of antibiotics promotes antibiotic resistance of pathogenic bacteria (i.e., *Staphylococcus aureus* resistant to methicillin) (Hadi and Brightwell 2021; Kołodziejczak et al. 2022). In this sense, in the following lines, the potential safety and microbiological aspects of the different meat analogues will be described.

### *12.2.1 Cultured Meat*

Exclusively produced at the laboratory level, cultured meat is considered as the less susceptible meat analogue to zoonotic diseases. Known as in vitro, lab-grown, or cell-based meat, cultured meat is obtained from animal stems cultivated under controlled conditions (Post et al. 2020; Hadi and Brightwell 2021). In vitro cultured meat is obtained through harvesting cells from living tissues and proliferated using cell engineering techniques, making possible its production with not large-scale livestock (Lima et al. 2022).

For the production of cultured meat, a small portion of animal tissue is removed and then, stem cells are proliferated separately in a liquid medium containing specific nutrients such as amino acids, lipids, vitamins, and salts (Lima et al. 2022). Then, stem cells are cultured in a suitable medium, mainly composed by fetal bovine serum (FBS) (Post et al. 2020; Hadi and Brightwell 2021). During proliferation process, cultivated cells doubles their population within 7–8 weeks and continues until millions of cells are produced in controlled environmental conditions, commonly using bioreactors. These cells fuse forming myotubes, and these myotubes are then submerged in a collagen gel to form muscle fiber. Finally, muscle maturation and protein production are achieved by stimulating the innate ability of muscle cells to contract (Lima et al. 2022).

To be used in meat cultivation, the FBS must be free of bovine viruses (including reovirus 3, adenovirus, parvovirus, respiratory syncytial virus, rabies virus, bluetongue virus) (Datar and Betti 2012; Hawkes 2015). Pathogenic and infectious prions are other of potential microbiological considerations in animal serum used for meat cultivation. Prions can be present in animal serum possibly due to cross-species and blood-related transmissions. In this case, prions are associated with transmissible encephalopathies, such as scrapie in sheep, chronic wasting disease in deer, BSE in cattle, and variant Creutzfeldt-Jakob disease in humans (Lee et al. 2013). The potential microbiological risks of BSE, used in cultured meat production, are based on its capacity to transmit pathogenic prions from infected cattle to humans, demonstrating that prions can use blood components as transmission vectors (Hadi and Brightwell 2021).

### 12.2.2 Plant-Based Meat

One of fastest-growing sectors both in agriculture and in the food industry is the plant-based food market. Plant-based products include extensive categories designed to emulate animal-origin food products such as meat, fish, eggs, milk, and their derivatives, which must achieve the physicochemical, sensory, and functional attributes claimed by the consumers (McClements and Grossmann 2021). The consumption of proteins from vegetable sources through their incorporation in food products has increased over the years, since these protein-rich products tend to be safer and more innocuos (Joshi and Kumar 2015). Traditionally, soybean protein is the most recurrent ingredient in plant-based food analogues, such as tempeh and tofu (both considered as a soybean fermented paste). These products are obtained from simple or artisanal processing, including fermentation techniques, have been consumed for centuries as traditional products mainly in the Asian southwest (Ismail et al. 2020). However, the adoption of soybean-derived food analogues in diet could be difficult for some consumers since these products are not capable of emulating the sensory properties of meat-derived food products. To counteract this disadvantage, texturized vegetable proteins (from soybean proteins, wheat glutens, or legume proteins) are used to replace the conventional meat, highlighting the strong off-flavor, or beany flavor of soybean-derived meat analogues (Hadi and Brightwell 2021). In this regard, seitan, a Chinese traditional plant-based meat analogue made from wheat gluten, is used to remove the starch granules from wheat flour to form an elastic mass from sticky insoluble gluten. Thus, the elastic mass obtained from wheat gluten can be manufactured and molded to fit with specific meat products (He et al. 2020).

Conversely, another important concern is the potentially allergic responses to soybeans, lentils, common beans, and peas present in the food analogue. Also, the proteins from cereals (wheat, rice, and barley) can be a risk factor for individuals who are gluten intolerant (Ismail et al. 2020; Hadi and Brightwell 2021). Based on the above, new trends has been interested in the study and application of non-traditional proteins obtained from unconventional sources in the development of food analogues with no limitations related to safety aspects. In this context, novel plant-based analogues, obtained through innovative applications of traditional biotechnological techniques such as fermentation, have emerged. Nevertheless, meat analogues obtained through fermentation are considered as single-cell proteins, since their natural source is the microbial metabolism of fermentative microorganisms.

As cultured meat, plant-based meat can also carry pathogenic bacteria of plant origin and found in the raw material. Most of these pathogenic bacteria can be inactivated or inhibited by the thermal processes, however, sporulated bacteria such as

*Clostridium* spp. or *Bacillus* spp. may resist these therma processes and then trigger infectious diseases (Wild et al. 2014). Although it has been observed that extrusion is capable of inactivating the spores from *Clostridium sporogenes* ATCC 19404, the presence of *Bacillus amyloliquefaciens* AB255669 has been reported in samples of extruded vegetable proteins, possibly due to re-infection after the extrusion process. Other re-infection cases of plant-based meats are also reported, identifying *Enterococcus durans*, *Exigobacterium acetylicum*, *Acinetobacter* spp., and *Staphylococcus* spp. in non-inoculated samples. Similarly, the presence of lactic acid bacteria species such as *Lactobacillus sakei*, *Enterococcus faecium* and *Carnobacterium divergens*, have been also reported in samples of plant-based meat (Geearerts et al. 2020; Hadi and Brightwell 2021).

In addition to the microbial spoilage or the presence of pathogenic bacteria, the presence of anti-nutrients poses a health risk of plant-based meat. These anti-nutrients including phytochemicals species such as protease inhibitors, $\alpha$-amylase inhibitors, lectins (phytohemagglutinin), polyphenols (tannins and phytic acid), allergens (such as prolamins, cupins), pathogenic proteins (Gly m 4 in soybean), prolifins (Gly m 3 in soybean), are related to negative physiological effects including altered gut function and endocrine disruption (Hadi and Brightwell 2021). Despite the fact that thermal processes inactivate the metabolism of undesirable microorganisms, as well as the bioactivities of different anti-nutrients and allergens, this same procedure may also induce the formation of carcinogens in samples of plant-based meat. Carcinogenic species of polycyclic aromatic hydrocarbons, nitrosamines, or heterocyclic aromatic amines have been identified in fermented plant-based meat products; however, the analysis of their potential safety risk is still limited and more in depth studies are needed in the future (He et al. 2020; Hadi and Brightwell 2021).

### 12.2.3 Single-Cell Proteins

The use of microorganisms in fermentations is an ancient biotechnological tool to obtain food with enhanced nutritional, sensory, and functional properties, however, in recent years such microorganisms are considered an emerging food source that must be more exploited. For example, different microbial proteins have been featured as potential replacements for animal proteins in food products. Microorganisms can be used as a source of ingredients for food analogs with a key role in the transition towards more sustainable production systems of foods with potential benefits related to a lower environmental impact and production costs, introducing the term “microbial urban farming”, that does not necessarily uses conventional fuels (such as fossil fuels), and offers an opportunity for the use of traditional technologies used in the food industry, at least for the generation of animal-based proteins (Ciani et al. 2021; Molfetta et al. 2022).

Single-cell proteins are obtained from the edible microbial biomass produced by fermentative unicellular microorganisms, such as algae, fungi, yeast, and bacteria.

Compared to animal proteins, the microbial biomass has many advantages for food analogues production. For example, its production is independent of agriculture or climate conditions, thus, seasonal variations do not affect them and arable land and potable water are not needed, making it possible to install anywhere and without interruptions. Also, the microorganisms used are able to grow in a wide variety of substrates even in nutrient poor water, $CO_2$, or methane. Hence, the microbial generation of single-cell proteins is considered an eco-friendly process that ensures food safety while minimizing the impact on global sustainability (Molfetta et al. 2022).

Commonly, the microorganisms used to produce single-cell proteins included microalgae such as *Arthrospira platensis* and *Chorella* vulgaris, as well as yeasts and fungal species such as *Saccharomyces cerevisiae*, *Cyberlindera jadinii*, (also identified as *Pichia jadinii* or *Candida utilis*), and *Fusarium venenatum* strains. Otherwise, fermentative bacteria, specifically those that can assimilate $CO_2$ as a carbon source, produce single-cell proteins mainly for feed applications, although in the future can be incorporated in food products (Ciani et al. 2021; Leger et al. 2021). So, the production of single-cell proteins depends mainly on the type and metabolism of microorganism used, however, other factors including the substrate, the culture conditions, and nutritional value should be considered. Based on these characteristics, single-cell protein products have been classified by order of preference as follows: microalgae > fungi > bacteria (Molfetta et al. 2022).

#### 12.2.3.1 Single-Cell Proteins from Microalgae

For many years, algae have been part of the human diet, being classified according to their chemical components, morphology, structure, as well as their size (either unicellular microalgae or macroalgae) (Lima et al. 2022). As a photosynthetic microorganism, prokaryotic microalgae convert inorganic cabron, nitrogen, and phosphorous nutrients in biomass (Molfetta et al. 2022). In addition to this advantage, producing marine algae for industrial applications is also used to produce food components such as agar-agar, carrageenan, and alginates, which can be considered natural additives with gelling, tickening, and stabilizing properties (Lima et al. 2022).

Regarding their nutritional value, algae are considered sources of unsaturated fatty acids such as eicosapentaenoic and docosahexaenoic acids and the bioavailability of algae proteins can be enhanced through pre-treatments that help to break down the cell wall. In this sense, the most consumed algae are spirulina and chlorella, which once subjected to dehydration provide *ca.* 70% of proteins along with the essential amino acids (Lima et al. 2022).

Specifically, microalgae strains are used in industrial processes highlighting those that belong to *Chlorella* sp., *Nannochloropsis* sp., *Tetraselmis* sp., *Tisochrisis lutea*, *Dunalliela salina*, *Phaeodactylum tricornutum*, *Porphyridium* sp., and *Scenedsmus* sp., species, or genus. So, the biomass obtained from heterotrophically cultured microalgae can present higher protein concentrations and an amino acid composition comparable to protein-rich vegetable substrates such as soybean (Molfetta et al. 2022).

#### 12.2.3.2 Single-Cell Proteins from Fungi and Yeast

Nowadays, different minced-type food analogues, produced from filamentous fungus mycoproteins are currently commercially available in different presentation such as such as chunks, sausages, and burgers (Wiebe 2002, Dekkers et al. 2018, Ismail et al. 2020). For example, Quorn™ product, a meat analogue elaborated from single-cell mycoproteins generated by the soil fungus *Fusarium*. This fibrous meat analogue has a texture similar to meat, due to the filamentous mycoproteins produced by fungus such as *Rhizopus oryzae*, *Rhizopus delemar*, *Aspergillus oryzae*, *Mucor indicus*, and *Aspergillus niger* (Yusuf and Setiarto 2022). This naturally produced mycoprotein is low in fat, however it has a high content of fiber and a significant contribution of high-quality protein composed by an excellent pattern of amino acids (Molfetta et al. 2022).

To obtain fibrous meat-analogue products, the filamentous fungus must be disintegrated by centrifugation, and consequently subjected to molding, steaming, cooling, and texturing processes (Lima et al. 2022). To mycoprotein production, filamentous fungi such as *Fusarium venenatum, Saccharomyces cerevisiae*, *Candida tropicalis*, *Kluyveromyces marxianus*, *Candida kruzei*, *Hanseniaspora uvarum*, *Zygosaccharomyces rouxii*, *Kluyveromyces lactis*, *Rhodotorula lactis*, and *Rhodorturula graminis*, are firstly subjected to continuous and controlled fermentation processes (with carefully controlled pH, and temperature). Once the fermentation is finished, centrifugation and thermal processes are applied to degraded fungal RNA into its constituent monomers. Fibrous product (obtained as residual biomass), is then molded, steamed, and texturized according to the requirements. Additionally, mycoproteins are commonly mixed with small amounts of egg albumin, as well as roasted barley, malt extract, and water in order to give it a flavorful character. Regarding its nutritional contribution in diet, the chemical composition of mycoproteins include proteins comparable in quality with those of conventional meats, also having an excellent protein bioavailability, low fat, and high content of fiber (Wiebe 2004; Lima et al. 2022).

Another microbial ingredient used in meat analogues production is soy leghemoglobin, a hemoprotein produced by *Pichia pastoris* expressing the leghemoglobin c2 gene firstly identified in soybean (*Glycine max*). According to bioinformatics analysis and *in vitro* pepsin digestibility, no genotoxic effects of recombinant soybean leghemoglobin were found on *Salmonella enterica* subsp. *enterica* ser. Typhimurium and *Escherichia*, likewise no adverse effects have been reported in Sprague Dawley rats to which this recombinant protein was administered (750 mg/kg/day for 28 days) (Fraser et al. 2018; Jin et al. 2018).

Moreover, the main safety hazard related to mycoproteins is allergens in individuals with mold allergy susceptibility. Specifically, it has been observed strong reactions to *Fusarium vasinfectum* and *Fusarium venenatum*, although IgE cross-reactivity was also detected with *Alternaria alternate*, *Aspergillus fumigatus,* and *Cladosporium herbarm* (Katona and Kaminski 2002). Additionally, new trends in mycoprotein research include the inclusion of new carbon sources to replace refined glucose syrup, as well as inhibit the potential production of mycotoxins. Hence, based on the mold's capacity to produce allergic reactions, the search for new carbon sources for mycoprotein generation must consider the possible production of mycotoxins and other potential allergenic reactions (Hadi and Brightwell 2021).

#### 12.2.3.3 Single-Cell Proteins from Bacteria

Bacterial strains can form biomass autotrophy or heterotrophy. The bacterial biomass can be obtained through bacterial inoculation into bioreactors with different grown substrates including waste products such as brewery wastewater, stick water, sewage sludge, cheese whey, potato starch processing water, and fructose (Lee et al. 2015). Among the bacterial species capable of producing single-cell proteins are *Lactobacillus acidophilus*, *Methylomonas*, *Methylophilus*, *Kefir* sp., *Bacilllus pumilis*, *Corynobacterium amminoagenes*, and *Methylococcus capsulatus*. *M. capsulatus* has the ability to promote health benefits in monogastric animals, alleviating mainly colitis and improving gut tract function in murine models (Kleiveland et al. 2013; Yusuf and Setiarto 2022). Bacteria are commonly cultured to produce single-cell proteins destined for animal feed. Bacterial yields 50–85% single-cell proteins by dry weight, with a relative content of essential amino acids compared with single-cell proteins from algae or fungi. Despite its amino acids content, bacterial single-cell protein contains up to 3% of methionine, a concentration greater than that allowed for human consumption. Also, the nucleic acid content is considered a risk factor for human health, since a diet high in nucleic acids leads to the production and accumulation of uric acid. So, to include the bacterial single-cell proteins in food products for human consumption, additional processing including enzymatic and chemical protocol are needed (Yusuf and Setiarto 2022).

Nowadays, there are few reports about bacterial single-cell proteins for human consumption, however, Solein® product commercialized by Solar Foods, is a bacterial single-cell protein that has shown interesting properties based on its water and oil binding capacities, as wel as its foaming, and emulsifying capacity, which potentialize its use as an ingredient in bacterial-based meat products (Molfetta et al. 2022).

In general, obtaining single-cell proteins from bacteria strains is based mainly on the fermentation process, more specifically on the substrate used and the possible products to be obtained, as well as other external (Yusuf and Setiarto 2022). In this sense, it is important to mention that the choice of the fermented microorganism has an effect on the yield of the protein generated and on its potential functional and nutritional characteristics.

#### 12.2.3.4 Functional and Nutritional Properties of Single-Cell Proteins

Single-cell proteins can be directly used either food products or as ingredients, additives, or nutraceuticals in products for human consumption. As mentioned above, microalgae are considered sources of dietary proteins, which have the essential amino acids needed for the correct infant nutrition. Moreover, microalgae are also rich in unsaturated fatty acids (omega-3 and omega-6) with potential functionalities as anticancer, immunomodulatory, and cardiovascular disease-preventing properties (Torres-Tiji et al. 2020). On the contrary, the use of microalgae biomass is considered an important factor in maintaining human health, since this biomass has a limited digestibility, and its structure can restrict the access of the digestive enzymes to the cell components. In addition to its limited digestibility, microalgae could have a lower vitamin (mainly B12 and D3 vitamins) and minerals (mainly iron) contribution than meat and dairy products. However, vitamins D3, B2, B3, B9, and B12 can be synthetized by the microalgae species such as *Dunaleilla salina*, *Nannochloropisis oceanica*, *Nannochloropisis limnetica*, *Chlorella*s sp., and *Nannochloropisis oceanica*, after exposition to external stimulus such as UV-B radiation. Regarding mineral content, it has been reported that *Bracteacoccus minor* and *Chlorococcum humicula* are considered good sources of iron as well as zinc and cobalt respectively (Santhakumaran et al. 2020).

For example, the single-cell proteins extracted from inactivated baker's yeast *S. cerevisiae* showed desirable and technological properties such as low bulk density, high water and oil retention capacity, excellent foaming, and emulsifying activity. These properties potentialize the use of *S. cerevisiae* single-cell proteins as ingredients for bakery products, desserts, and sauces. Similarly, gels made with single-cell proteins from kefir grains (a consortium composed of undefined yeast and bacteria strains), showed a harder texture profile than gels made with soy flour, which demonstrated that some sensory attributes such as texture, are closely related to the concentration and nature of the proteins (Razzaq et al. 2020). Based on data obtained from these fermentative microorganisms, it can be considered the fermentation as an important tool to produce desired protein analogues that contribute to imparting sensory characteristics to the final product, as well as to improve taste, aspect, and digestibility; thus, the use or design of microbial consortia with desirable metabolic properties could be an efficient route to obtain more acceptable single-cell proteins (Molfetta et al. 2022).

One of the main microorganisms used for single-cell protein production is fungi. When fungi are employed as microbial starters, they generate dietary proteins in the form of mycoproteins, edible mushrooms, and traditional foods. Fungal single-cell proteins exhibit distinct nutritional profiles, reduced production costs, and greater environmental advantages when juxtaposed with plant- and animal-based proteins. It is true that certain types of fungi are grown using the leftovers, byproducts, and wastes from farming, forestry, and food processing (Hashempour-Baltork et al. 2020).

As the same other ingredients, microbial ingredients used in the formulation of food analogues must be sensory evaluated before being incorporated into formulation of food analogues. So, the sensory evaluation of single-cell proteins should be performed just before application as food or food ingredient to guarantee consumer acceptance. This is why in Table 12.1 there are enlisted the texture and sensory characteristics of microbial-added food analogues, including single-cell properties.

**Table 12.1** Texture and sensory characteristics of microbial-added food analogues[a]

| Food analogue | Microbial specie | Texture and sensory characteristics | Reference |
|---|---|---|---|
| **Meat** | | | |
| | *Fusarium venenatum* | Generation of protein fibrils or fibers in meat analogue similar in size to the meat myofibrils | Dekkers et al. (2018) |
| | *Auxenochlorella protothecoides*, *Rhizopus* spp. | Increase of tenderness in extruded algae-based meat analogue | Caporgo et al. (2020) |
| **Single-cell proteins** | | | |
| | *Arthrospira platensis* (spirulina) | Improvement the mouthfeel, enhancement of syneresis and apparent viscosity in yogurt | Barkallah et al. (2017) |
| | | Generation of a softer feta cheese acidified by *Lacticaseibacillus casei* with an easier disintegration and chewing | Golmakani et al. (2019) |
| | *Spirulina*, *Chlorella* | Increase of hardness, adhesiveness, and gumminess of Spanish "chorizo" sausages | Thirumdas et al. (2018) |
| **Milk** | | | |
| | *Lactobacillus bulgaricus*, *Streptococcus thermophilus* | Reduction of beany and soapy flavor in fermented soybean milk | Obadina et al. (2013) |
| **Cheese** | | | |
| | *L. rhamnosus*, *Bifidobacterium animalis* susbp. *lactis* | Good body, creamy flavor, better springiness, and lower gumminess of sufu | Abd-Rabou et al. (2016) |
| | *G. candidum* | Flavor development, presence of rid and beany taste, with a significant reduction in hardness and chewiness in soybean cheese Induction of white and velvety appearance | Li et al. (2020) and Vong and Liu (2017) |
| | *C. vulgaris* | Firm and strong cohesive network of algae-based cheese with lower oiling off and meltability | Tohamy et al. (2018) |

(continued)

**Table 12.1** (continued)

| Food analogue | Microbial specie | Texture and sensory characteristics | Reference |
|---|---|---|---|
| | *M. purpureus, L. casei* | Provide aroma and color to the soybean milk-based cheese blend | Lorrungrua et al. (2014) |
| | *Aspergillus oryzae* | Decrease of hardness in sufu | Li et al. (2010) |
| | *Actinomucor elegans* | Production of few aroma components and mild taste in sufu | Ma et al. (2013) |
| | *Mucor recumosus* | Production of taste substances (Glu, Val, tartaric and acetic acids) after sufu fermentation and ripening. | Xi et al. (2022) |
| | *Mucor wutungkiao* 40477 | Decrease of hardness and elasticity in sufu, and generation of umami flavor | Chen et al. (2023) |
| | *Enterococcus italicus, Leuconostoc citreum, Leuconostoc mesenteroides, Leuconostoc pseudomesenteroides, Tetragenococcus halophilus, Cladisoporum colombiae, Hannaella oryzae* | Production of 26 high-quality umami peptides in spontaneously fermented sufu | Mao et al. (2023) |

[a]Table of own authorship to be published in this chapter

## 12.3 Dairy Analogues

Milk is considered a reservoir of energy, macro- and micronutrients including proteins, carbohydrates, lipids, minerals, and vitamins. Nevertheless, milk and its derivatives are indispensable products in the human diet around the world (Kamath et al. 2022). The formulation of dairy products-analogues encompasses alternative fat and protein sources that stimulate the flavor and structure characteristics of cow's milk, but at the same time serve as healthier ingredients with lower levels of calories, fat, and cholesterol (Lima et al. 2022). In general, common ingredients for dairy-analogue products also include organic acids, flavoring agents, salts, and preservatives, some of them such as lactic acid and nisin produced by fermentative species of lactic acid bacteria.

### *12.3.1 Milk Analogue*

Cow's milk is considered a natural and complete food providing major nutrients including macronutrients such as fat, proteins, and carbohydrates. Also, cow's milk micronutrients are based on the presence of minerals such as calcium, selenium; and complex B vitamins including riboflavin, vitamin $B_{12,}$ and pantothenic acid ($B_5$), which are involved in the overall growth and body health homeostasis (Paul et al. 2019). Among the few limitations of milk, the low or near absence of other essential micronutrients such as iron and vitamins like folate and amino acids are the most notable. Furthermore, lactose intolerance disease is still growing in developing countries reducing the consumption of whole milk and their derivative dairy products. These facts restrict milk consumption mainly in infants older than 12 months and the elderly population. Additionally, in arid regions, the availability of milk generates high prices, and the potential presence of undesirable bacteria (*Salmonella* spp. and *Escherichia coli* O157:H7), as well as other health concerns derived from fat content (cholesterol), cow's milk allergy (mainly suffered by immunocompromised infants and children), and antibiotic residues limit the consumption of milk and their respectively dairy derivatives (Vanga et al. 2015; Vanga and Raghavan 2018). In brief, all the previous limitations and concerns have served as a basis for the search, design, and development of alternatives for milk, including milk analogues.

#### 12.3.1.1 Soybean Milk

Recently, herbal milk analogues have gained interest from consumers due to their functional and health benefits. An example of this is soybean milk, considered a rich source of high-quality proteins (7 g of protein per 8 ounces), only comparable to those of cow's milk. Soybean milk also contains mono- and polyunsaturated fatty acids, isoflavones (genistein, daidzein, and glycation), phytic acid, sterols, saponins, fiber, minerals (iron, calcium, and zinc), and B complex vitamins (Rizzo and Baroni 2018; Paul et al. 2019).

Conversely, some phytochemicals from soybean milk (i.e., phytic acid, urease, proteinase-inhibitors, and oxalic acid) can induce anti-nutritive characteristics of the final product which can be minimized by employing bio-transformation processes such as fermentation. In this sense, fermentation reduces the anti-nutritional phytochemicals in raw material while releasing bioactive components through the breakdown of complex organic molecules. This last phenomenon solely depends on the fermentative culture used in fermentation. For example, soybean fermented by *Bacillus subtilis* MTCC5480 resulted in a liberation of bioactive amino acids promoted by an exhaustive protein hydrolysis. Meanwhile, using *B. subtilis* and *Rhizopus oligosporus* increased the liberation of glutamine and aspartic acid, and also the presence of smaller peptides content. Based on these findings, it has been reported that soymilk fermented by the same *B. subtilis* was able to demonstrate free radical scavenging properties (antioxidant activity) related to a potential hypotensive effect triggered by the inhibition of angiotensin I-converting enzyme (Sanjukta and Rai 2016; Rai et al. 2017; Paul et al. 2019).

#### 12.3.1.2 Coconut Milk

Coconut milk is used all over the world for ice cream, biscuits, bakeries, and confections. Medium-chain triglycerides make it easier to digest coconut milk, and soluble and insoluble fiber boosts its nutritional value and may even have useful properties like antioxidants (Paul et al. 2019).

#### 12.3.1.3 Oat Milk

As soybean milk, oat milk is considered an alternative source of nutrients and bioactive compounds such as phenolic compounds, avenanthramides, saponins, phytic acid, sterols, and antioxidants. Besides, oat milk contains dietary fibers defined as "the edible part of a plant or analogues carbohydrates that are resistant to the digestion and absorption in the human small intestine with complete or partial in the large intestine", by the Association of Cereal Chemist (AACC) (AACC 2001). Dietary fiber consumption is related to health benefits, promoting laxation, as well as blood cholesterol and glucose attenuation. Mainly composed of β-glucans, the dietary fiber of oats is fermented to produce functional food analogues with enhanced antioxidant activities derived from the liberation of phenolic compounds and glucosamine. Nevertheless, fermentation of oat milk is not always a beneficial process since some of these antioxidant compounds (such as synaptic acid) can be degraded by fermentative strains such as *Monascus anka*, although there may be a concurrent production of other phenolic compounds i.e., chlorogenic acid and quercetin. Thus, a controlled fermentation is required to obtain oat milk with well-defined functional properties (Bei et al. 2017; Paul et al. 2019).

#### 12.3.1.4 Almond Milk

Commonly considered one of the "brain foods", almonds are a source of vitamins such as E and B complex, protein, monounsaturated fatty acids, dietary fiber, phenolic compounds, and minerals. Additionally, almonds have bioactive compounds with anti-inflammatory, anti-hyperlipidemic, anti-tumoral, and antioxidant activities. However, regarding the negative properties of almonds, there is the recurrent presence of *E. coli* O157:H7 and *Listeria monocytogenes*, as well as the concentration of allergens (including amidine, legumin, prunin, and prunin), which can range from a slight oral allergy to fatal anaphylaxis. To counteract these health risks, mechanical treatments and fermentation are applied to remove allergens from almond milk, enhancing its position among other milk analogues (Paul et al. 2019).

#### 12.3.1.5 Other Milk Analogues

Other plant-based analogues for cow's milk include those obtained from rice, cocoa, kidney beans, peanuts, and hemp. Rice is a basic food for about half the world's population. Its consumption with bran provides beneficial polyphenols (Shao et al. 2018). Due to its high nutrient profile and underutilization, the food industry is focusing on producing a milk analogueue from rice bran, which has lower soluble solids and viscosity than soybean milk. Likewise, it has been observed that soaking increases bioactive components as well as minerals and vitamins ($B_6$, and $B_{12}$) and insoluble fiber. These beneficial approaches can be maintained through natural fermentation, in which lactic acid bacteria decrease the presence of anti-nutritional factors, enhancing mineral content, digestibility, and organ immune response (Sharma et al. 2018a).

Similarly, kidney beans are a source of macronutrients (complex carbohydrates and proteins) and micronutrients (vitamins and minerals), as well as bioactive phytochemicals, including phenolic compounds, polyphenols, and phytosterols. In the same way, kidney beans are usually recommended for diabetics because they have a low glycemic index. Some of the carbohydrates identified in kidney beans are raffinose, stachyose, and verbascose, which can act as prebiotics for *Bifidobacterium* and *Lactobacillus* strains in the colon (Los et al. 2018). For the preparation of milk analogues, kidney beans are first rinsed and then soaked until they swell completely, followed by grinding and boiling. Finally, the supernatant obtained after centrifugation is considered the milk (Chen et al. 2019). As with the previous plant-based milk analogues, the presence of anti-nutrients and allergens is one of the health concerns related to bean milk. To reduce these anti-nutritional factors, solid and liquid fermentations have been adopted using *B. subtilis* as a starter strain to produce fermented milk analogues with antioxidant activity through the production of isoflavone aglycones. In addition to the previous beneficial effects, it has been observed that the spontaneous fermentation of this kidney bean milk by endophytic lactic acid bacteria induces the presence of $\gamma$-aminobutyric acid (GABA), which is hypotensive, hypocholesterolemic, and decreases the proliferation of cancer cells (Limón et al. 2015; Nikmaram et al. 2017).

Plant-based food substitutes also emulate other dairy products such as butter and milk. This is the case of peanut butter, whose nutritional value is based on the presence of almost all 20 amino acids, proteins, essential unsaturated fatty acids, and minerals. However, this peanut butter presents stability problems due to its high fat content. Furthermore, peanut allergens, phytic acid, and insoluble fibers are issues faced by the peanut milk and butter industry. To minimize these problems, the peanut milk industry has used extraction methods, filtration, and lactic acid fermentation with specific starters. In this way, there is a reduction or elimination of anti-nutritional factors and food allergens (Arya et al. 2016).

### 12.3.2 Cheese Analogues

Considered one of the most popular and consumed dairy products, cheese is categorized based on its type of coagulation, ripening conditions, quantity of moisture and fat, and heat treatments (Kamath et al. 2022). Beside of its nutritional contribution, cheese has bioactive compounds produced during the fermentation process and ripened by lactic acid bacteria (including probiotics). These compounds are associated with positive health effects, mainly hypotensive effects. Despite its functionality, the conventional process to obtain cheese is considered expensive. So, a cheaper alternative is cheese analogues, in which the replacement of milk, fat, and protein with oils and starches from vegetable sources helps reduce production costs (Hao et al. 2021). Moreover, cheese analogues involve no ripening stage, and their nutritional quality can be improved by the addition of micronutrients such as vitamins and minerals, among other food additives (Kamath et al. 2022). Water, fats, and protein sources are key ingredients or components for cheese analogue formulation, although other additives (emulsifying salts, hydrocolloids, preservatives, flavors, and acidifying substances) can also be incorporated (Guinee et al. 2007; Ørskov et al. 2021).

As mentioned above, various polysaccharides such as starch and inulin have been used as ingredients in cheese analogues, mainly reducing the fat content. The use of prebiotics such as inulin and resistant starches and starter cultures allow to develop functional cheese analogues, acting as symbiotics. For example, Abd-Rabou et al. (2016), used 2% inulin for the elaboration of a low-fat and salt cheese analogue inoculated with *Lactobacillus rhamnosus* and *Bifidobacterium animalis* subsp. *lactis* as starter cultures. The sensory attributes of this symbiotic cheese analogue were enhanced by the inulin addition, promoting bacterial proteolysis and thus, the generation of better springiness and lower gumminess.

Additionally, cheese analogues can also be elaborated with plant-based milks such as soybean milk and cashew nut milk, as well as coconut oil and milk, which are good sources of medium-chain triglycerides effective in treating and controlling of Alzheimer's disease (Chatterjee et al. 2020). One of the most popular cheese analogues is tofu, a soft paste made from soybean milk fermented by lactic acid bacteria, fungi, and yeasts (Li et al. 2020). However, other cheese analogues made from fermented whole soybeans are douchi, natto, and tempeh, while sufu is prepared through the fermentation of soybean curd (Xie and Gänzle 2023).

Fermented soybean products are produced using different microbes. For example, *Bacillus subtilis* var. Natto is used to ferment natto, while *Rhizopus* spp. is used to ferment tempeh. Douchi and sufu are made using spontaneous fermentation followed by ripening (Xie and Gänzle 2023). In comparison to cow's milk, where lactose is the only fermentable substrate, in soybean milk and curd there are other fermentable carbohydrates such as fructose, sucrose, glucose, verbascose, maltose, stachyose, and raffinose. So, the difference in substrates also changes the fermentative microbes. While traditional cultures used to acidify cheese include *Lactobacillus delbrueckii*, *Lactobacillus helveticus*, *Lactococcus lactis*, *Lactococcus cremoris*,

*Leuconostoc* spp., and *Streptococcus thermophilus*, acidification of plant-based cheese is carried out spontaneously by certain species of lactic acid bacteria. These differences in fermentative cultures are also observed in the ripening step. Acidifying microorganisms such as non-starter lactic acid bacteria, propionibacteria, *Penicillium roqueforti*, and aerobic *Penicillium camemberti*, associated with the yeasts *Debaromyces hansenii* and *Geotrichum condidum,* are used during cheese ripening. While in fermented soybean-based cheese, the process is different; that is, hydrolytic cultures (mycelial fungi and bacilli) grow first, then proteolytic microorganisms (yeasts and lactic acid bacteria) determine the texture, taste, and odor of cheese-analogues (Xie et al. 2018; Xie and Gänzle 2023).

In this sense, probiotic strains are used to produce health benefits derived from their proteolytic enzymes that may produce bioactive peptides, including those with hypotensive and immunomodulatory properties (Jeewanthi and Paik 2018). For example, Sharma et al. (2018a, b) prepared a functional spread cheese analogue using peanut seeds and water for milk extraction and subsequent protein coagulation when *L. rhamnosus* NCDC18 strains were added, and the fermentation process was conducted at 37 °C for 24 h. Likewise, Łopusiewicz et al. (2020) obtained bioactive peptides from Camembert cheese analogue elaborated by fermentation with lactic acid bacteria, *Penicillium camemberti*, and *Geotrichum candidum* from flaxseed oil cake. This fermented cheese analogue presented improved antioxidant activity and greater oil-oxidative stability. Similarly, Moa-tofu (fermented by *Mucor micheli ex fries*) showed antioxidant and hypotensive properties, while soy-based red cheese (fermented by *Monascus purpureus*) exerted hypocholesterolemic properties derived from the generation of monacolin, a secondary fungal metabolite that reduces plasma cholesterol (Chen and Hu 2005; Li et al. 2020). Thus, cheese analogues, elaborated from the fermentation of plant-based ingredients, can be an interesting topic for developing new food analogues with desirable and programmed bioactivities.

Additionally, microorganisms not only increase the functional properties of analogues food but also contribute to improving its sensory properties, thus guaranteeing its acceptance by consumers (Lorrungrua et al. 2014; Zielinska et al. 2015). In this context, a challenge for cheese analogues is to achieve comparable functionality alongside the needed rheological, textural, and microstructural characteristics to expand the versatility and applications of the food. Hence, the formulation of cheese analogues must avoid modification in the original technological characteristics of conventional cheese after substituting milk constituents (i.e., milk fat replaced by vegetable oil) in order to maintain the rheological properties and sensory acceptability (Giha et al. 2021). Otherwise, the development of cheese analogues also contemplates the addition of microalgal biomass to cheese blends. In this sense, the incorporation of *Chlorella vulgaris* in powder with concentrations of 2–6% in cheese blends increases the mineral content (Se, Zn, Fe, Mg, and K) as well as its antioxidant activity (Tohamy et al. 2018).

On the other hand, the incorporation of non-pathogenic or probiotic strains into the formulation of a cheese analogue prevents the development of pathogens and microbial spoilage in storage (Kamath et al. 2022). Regarding this, the incorporation of probiotic *L. rhamnosus* NRC AM6, *Lactobacillus casei* NRC AM2, *Pediococcus pentosaceus* NRC AM4, and *Pediococcus acidilactici* NRC AM8 strains inhibits the development of coliforms, yeast/molds, *Salmonella, Enterococci*, and thermoduric bacteria during storage (Effat et al. 2018). Likewise, Matias et al. (2014) produced a soy-based petit-suisse cheese, added with probiotic microorganisms (*Lactobacillus acidophilus* La-5, *Bifidobacterium animalis* Bb-12), and a starter culture (*Streptococcus thermophilus*). The authors observed that the soymilk-based petit-suisse cheese reached the highest sensory score in comparison to other soy-based and milk-based chesses; however, hardness and gumminess were higher in the soy-based cheese, which indicates that probiotic soy-based cheese analogues are an option to obtain new functional foods without dairy ingredients. Nevertheless, there are a few negative aspects of lactic acid bacteria inoculation in cheese analogue production, including excessive acidity production (>pH 5), that might result in an unstable structure of the soybean milk gel network (Li et al. 2020).

Although milk and milk-derived products are considered fundamental foods in the human diet, there are medical complications such as lactose intolerance, allergens, and cardiovascular diseases that have motivated the population to look for new alternatives based on vegetarian products or plant-based analogues, mainly those with a similar protein contribution and that in turn allow the manufacture of probiotic and functionalized beverages and food stuffs (Paul et al. 2019). Nevertheless, plant-based milk analogues are highly varied around the world, and thus, their applications and uses still have technological and regulatory limitations.

## 12.4 Limitations of Plant-Based Food Analogues

Plant-based analogues, specifically plant-derived proteins, present some limiting microbial aspects. In this sense, biogenic amines such as histamine, tyramine, tryptamine, putrescine, cadaverine, and phenylethylamine are produced due to the decarboxylation of amino acids performed by microorganisms naturally present in protein-rich raw materials used in food analogue development. Ingestion of biogenic amines can cause nausea, headaches, palpitations, and oral burning, as well as other health complications for consumers. These and other antinutritional factors, such as lectins, phytic acid, protease inhibitors, alkaloids, phenolic compounds, and $\alpha$ -galactosides, can be reduced by soaking, fermentation, cooking, roasting, or sprouting techniques before including them in the formulation and subsequent consumption of analogueous foods (Fig. 12.2) (Licandro et al. 2020; Molfetta et al. 2022).

Regarding single-cell proteins, one of their main defects relies on their high nucleic acid content (*ca.* 6.12% dry basis). An elevated consumption of these single-cell proteins can result in the precipitation of uric acid and trigger health problems such as gout symptoms or kidney stone deposition. Therefore, the application of these microorganisms in foods requires treatments that allow the nucleic acid content to be reduced to below 2% dry basis, such as thermal (heat treatment at

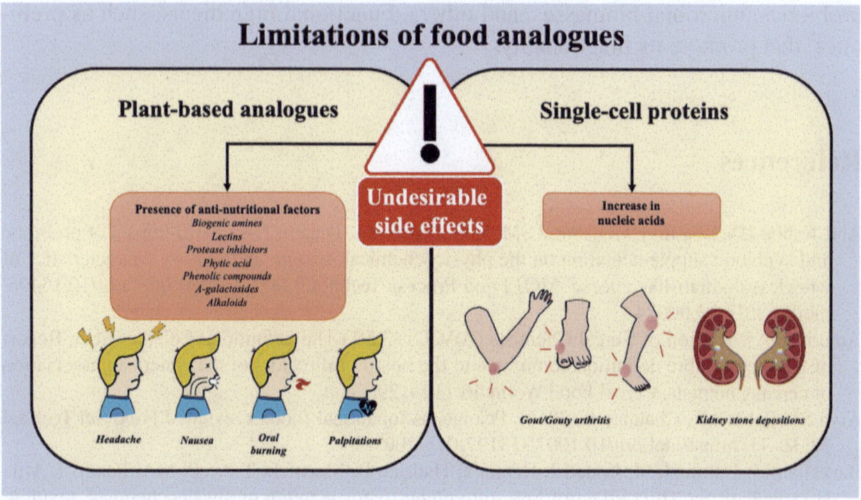


**Fig. 12.2** Limitations of food analogues based on their anti-nutritional phytochemicals (in plant-based analogues), and genetic material from microbial species producers of proteins analogues. (Figure of own authorship to be published in this chapter)

60–70 °C), chemical (NaCl, HCl, or NaOH), or enzymatic (ribonuclease or endonuclease) treatments (Khan et al. 2022; Molfetta et al. 2022).

## 12.5 Conclusions and Future Prospects

The growing interest in food analogues is related both to the constant search to reduce costs as well as to the demand for new, healthier, and more functional foods that have sensory and texture characteristics that are most similar to conventional products, which meet consumer preferences. Different strategies have been reported to obtain analogue foods that meet consumer demands; one of them is the incorporation of beneficial microorganisms such as probiotics, whose presence favorably impacts the functionality of food analogues, thus driving an increase in their consumption and trade. The other alternative is the addition of microbial biomass, which likewise favors the functionality of the food analogues while also modifying the sensory properties, mainly texture, of the final product. Both aspects—microbiological and sensory—are decisive for consumers' approval of the product and its inclusion in different markets worldwide. In this case, despite having a great demand from consumers and interest from the food industry, food analogues do not have clear legislation, mainly on topics related to their microbiological safety. Therefore, future work related to the development of new analogue foods should focus first on the formulation and development of these products and then address their legislation based on their stability and biological safety, considering the addition of probiotics, microbial biomasses, and others. Functional ingredients, such as prebiotics, that promote its functionality.

## References


Abd-Rabou HS, El-Ziney MG, Awad SM, El Sohaimy SA, Dabour NA (2016) Impact of probiotic and synbiotic supplementation on the physicochemical, texture and sensory characteristics of wheyless domiati-like cheese. MOJ Food Process Technol 3:00074. https://doi.org/10.15406/mojfpt.2016.03.00074

American Association of Cereal Chemists (AACC) (2001) The definition of dietary fiber. Report of the dietary fibre definition committee to the board of directors of the American association of cereal Chemists. Cereal Food World 46:112–129.

Arya SS, Salve AR, Chauhan S (2016). Peanuts as functional food: a review. J Food Sci Technol 53:31–41. https://doi.org/10.1007/s13197-015-2007-9

Barkallah M, Dammak M, Louati I, Hentai F, Hadrich B, Mechichi T, Ayadi MA, Fendri I, Attia H, Adbelkafi S (2017) Effect of *Spirulina platensis* fortification pf physicochemical, textural, antioxidant and sensory properties of yogurt during fermentation and storage. LWT-Food Sci Technol 84:323–330. https://doi.org/10.1016/j.lwt.2017.05.071

Bei Q, Liu Y, Wang L, Chen G, Wu., Z. (2017) Improving free, conjugated, and bound phenolic fractions in fermented oats (*Avena sativa* L.) with *Monascus anka* and their antioxidant activity. J Funct Foods 32:185–194. https://doi.org/10.1016/j.jff.2017.02.028

Caporgo MP, Böcker L, Müssner C, Stirnemann E, Haberkorn I, Adelmann H, Handschin S, Windhab EJ, Mathys A (2020) Extruded meat analogues based on yellow, heterotrophically cultivated *Auxenochlorella protothecoides* microalgae. Innov Food Sci Emerg Technol 59:102275. https://doi.org/10.1016/j.ifset.2019.102275

Chatterjee P, Fernando M, Fernando B, Dias CB, Shah T, Silva R, Williams SS, Pedrini S, Hillebrandt H, Goozee K, Barin E, Sohrabi HR, Garg ML, Cunnane SC, Martins RN (2020) Potential of coconut oil and medium chain triglycerides in the prevention and treatment of Alzheimer's disease. Mech Ageing Dev 186:111209. https://doi.org/10.1016/j.mad.2020.111209

Chen F, Hu, X (2005) Study on red fermented rice with high concentration of monacolin K and low concentration of citrinin. Int J Food Microbiol 103:331–337. https://doi.org/10.1016/j.ijfoodmicro.2005.03.002

Chen Y, Zhang H, Liu R, Mats L, Zhu H, Pauls KP, Deng Z, Tsao R (2019) Antioxidant and anti-inflammatory polyphenols and peptides of common bean (*Phaseolus vulgaris* L.) milk and yogurt in Caco-2 and HT-29 cell models. J Funct Foods 53:125–135. https://doi.org/10.1016/j.jff.2018.12.013

Chen Z, Song J, Ren L, Wang H, Zhang Y, Suo H (2023) Effect of the succession of the microbial community on physicochemical properties and flavor compounds of *Mucor wutungkiao* fermented sufu. Food Biosci 51:102345. https://doi.org/10.1016/j.fbio.2022.102345

Ciani M, Lippolis A, Fava F, Rodolfi L, Niccolai A, Tredici MR (2021) Microbes: foods for the future. Food Secur 10:971. https://doi.org/10.3390/foods10050971

Datar I, Betti M (2012) Possibilities for an in vitro meat production system. Innovative Food Sci Emerg Technol 11:13–22. https://doi.org/10.1016/j.ifset.2009.10.007

Dekkers BL, Boom RM, van der Goot AJ (2018) Structuring process for meat analogues. Trends Food Sci Technol 81:25–36. https://doi.org/10.1016/j.itifs.2018.08.011

Effat BAM, Hassan ZMR, Mabrouk AMM, Sadek ZIM, Magdoub MNI, Tawfik NF (2018) Properties of low salt soft cheese supplemented with probiotic cultures. Int J Adv Res Biol Sci 5:1–10. https://doi.org/10.22192/ijarbs.2018.05.02.001

Fraser RZ, Shitut M, Agrawal P, Mendes O, Klapholz S (2018) Evaluation of soy leghemoglobin protein preparation derived from *Pichia pastoris*, intended for use as a flavor catalyst in plant-based meat. Int J Toxicol 37:241–262. https://doi.org/10.1177/1091581818766318

Geearerts W, De Vuyst L, Leroy F (2020) Ready-to-eat meat alternatives, a study of their associated bacterial communities. Food Biosci 37:100681. https://doi.org/10.1016/j.fbio.2020.100681

Giha V, Ordoñez MJ, Villamil RA (2021) How does milk fat replacement influence cheese analogue microstructure, rheology, and texture profile? Concise Rev Hypotheses Food Sci 86:2802–2815. https://doi.org/10.1111/1750-3841.15799

Golmakani MT, Soleimanian-Zad S, Alavi N, Nazari E, Eskandari H (2019) Effect of Spirulina (*Arthrospira platensis*) powder on probiotic bacteriologically acidified feta-type cheese. J Appl Phycol 31:1085–1094. https://doi.org/10.1007/s10811-018-1611-2

Grossmann L, McClements DJ (2021) The science of plant-based foods: approaches to create nutritious and sustainable plant-based cheese analogs. Trends Food Sci Technol 118(Part A):207–229. https://doi.org/10.1016/j.tifs.2021.10.004

Guinee TP, Mulholland EO, Kelly J, Callaghan DJO (2007) Effect of proteinto-fat ratio of milk on the composition, manufacturing efficiency, and yield of Cheddar cheese. J Dairy Sci 90:110–123. https://doi.org/10.3168/jds.S0022-0302(07)72613-9

Hadi J, Brightwell G (2021) Safety of alternative proteins: technological, environmental and regulatory aspects of cultured meat, plant-based meat, insect protein and single-cell protein. Food Secur 10:1226. https://doi.org/10.3390/foods10061226

Hao X, Yang W, Zhu Q, Zhang G, Zhang X, Liu L, Li X, Hussain MA, Ni C, Jiang X (2021) Proteolysis and ACE-inhibitory peptide profile of Cheddar cheese: effect of digestion treatment and different probiotics. LWT-Food Sci Technol 145:111295. https://doi.org/10.1016/j.lwt.2021.111295

Hashempour-Baltork F, Khosravi-Darani K, Hosseini H, Farshi P, Reihani SFS (2020) Mycoproteins as safe meat substitutes. J Clean Prod 253:119958. https://doi.org/10.1016/j.clepro.2020.119958

Hawkes PW (2015) Fetal bovine serum: geographical origin and regulatory relevance of viral contamination. Bioresour Bioprocess 2:34. https://doi.org/10.1186/s40643-015-0063-7

He J, Evans NM, Liu H, Shao S (2020) A review of research on plant-based meat alternatives: driving forces, history, manufacturing, and consumer attitudes. Compr Rev Food Sci Food Saf 19:2639–2656. https://doi.org/10.1111/1541-4337.12610

Ismail I, Hwang Y-H, Joo S-T (2020) Meat analog as future food: a review. J Anim Sci Technol 62:111–120. https://doi.org/10.5187/jast.2020.62.2.111

Jeewanthi R, Paik HD (2018) Modifications of nutritional, structural, and sensory characteristics of non-dairy soy cheese analogs to improve their quality attributes. J Food Sci Technol 55:4384–4394. https://doi.org/10.1007/s13197-018-3408-3

Jin Y, He X, Andoh-Kumi K, Fraser RZ, Lu M, Goodman RE (2018) Evaluating potential risks of food allergy and toxicity of soy leghemoglobin expressed in *Pichia pastoris*. Mol Nutr Food Res 62:1700297. https://doi.org/10.1002/mnfr.201700297

Joshi VK, Kumar S (2015) Meat analogues: plant based alternatives to meat products: a review. Int J Food Ferment Technol 5:107–119. https://doi.org/10.5958/2277-9396.2016.00001.5

Kamath R, Basak S, Gokhale J (2022) Recent trends in the development of healthy and functional cheese analogues-a review. LWT 155:112991. https://doi.org/10.1016/j.lwt.2021.112991

Katona SJ, Kaminski ER (2002) Sensitivity to Quorn mycoprotein (*Fusarium venenetum*) in a mould allergic patient. J Clin Pathol 55:876–877. https://doi.org/10.1136/jcp.55.11.876-a

Khan MKI, Asif M, Razzaq ZU, Abulnaja KO (2022) Sustainable food industrial waste management through single cell protein production and characterization of protein enriched bread. Food Biosci 46:101406. https://doi.org/10.1016/j.fbio.2021.101406

Kleiveland CR, Hult LTO, Spetalen S, Kaldhusdal M, Christofferesen TE, Bengtsson O, Romarheim OH, Jacobsen M, Leaa T (2013) The noncommensal bacterium *Methylococcus capsulatus* (Bath) ameliorates dextran sulfate (sodium salt)-induced ulcerative colitis by influencing mechanisms essential for maintenance of the colonic barrier function. Appl Environ Microbiol 79:48–57. https://doi.org/10.1128/AEM.02464-12

Kołodziejczak K, Onopiuk A, Szpicer A, Poltorak A (2022) Meat analogues in the perspectives of recent scientific research: a review. Food Secur 11:105. https://doi.org/10.3390/foods11010105

Lee J, Kim SY, Hwang KJ, Ju YR, Woo H-J (2013) Prion diseases as transmissible zoonotic diseases. Osong Public Health Res Perspect 4:57–66. https://doi.org/10.1016/j.phrp.2012.12.008
Lee JZ, Logan A, Terry S, Spear JR (2015) Microbial response to single-cell protein production and brewery wastewater treatment. Microb Biotechnol 8:65–76. https://doi.org/10.1111/1751-7915.12128
Leger D, Matassa S, Noor E, Shepon A, Milo R, Bar-Even A (2021) Photovoltaic-driven microbial protein production can use land and sunlight more efficiently than conventional corps. Proc Natl Acad Sci USA 118:e2015025118. https://doi.org/10.1073/pnas2015025118
Li YY, Yu RC, Chou CC (2010) Some biochemical and physical changes during the preparation of the enzyme-ripening sufu, a fermented product of soybean curd. J Agric Food Chem 58:4888–4893. https://doi.org/10.1021/jf904600a
Li Y, Zhang X, Yang JJ, Ma XY, Jia XD, Du P, Li AL (2020) Influence of the addition of *Geotrichum candidum* on the microbial, chemical, textural, and sensory features of soft soy cheese. J Food Process Preserv 44:e14823. https://doi.org/10.1111/jfpp.14823
Licandro H, Ho PH, Nguyen TKC, Petchkongkaew A, Nguyen HV, Chu-Ky S, Nguyen TVA, Lorn D, WachéY (2020) How fermentation by lactic acid bacteria can address safety issues in legumes food products? Food Control 110:106957. https://doi.org/10.1016/j.foodcont.2019.106957
Lima M, Costa R, Rodrigues I, Lomeiras J, Botelho G (2022) A narrative review of alternative proteins sources: highlights on meat, fish, egg and dairy analogues. Food Secur 11:2053. https://doi.org/10.3390/foods11142053
Limón RI, Peñas E, Torino MI, Martínez-Villaluenga C, Dueñas M, Frias J (2015) Fermentation enhances the content of bioactive compounds in kidney bean extracts. Food Chem 172:343–352. https://doi.org/10.1016/j.foodchem.2014.09.084
Łopusiewicz Ł, Drozłowska E, Tarnowiecka-Kuca A, Bartkowiak A, Mazurkiewicz-Zapałowicz K, Salachna P (2020) Biotransformation of flaxseed oil cake into bioactive camembert-analogue using lactic acid bacteria, *Penicillium camemberti* and *Geotrichum candidum*. Microorganisms 8:1266. https://doi.org/10.3390/microorganisms8091266
Lorrungrua C, Sinma K, Pantagrud P, Wannasiris S, Mahabandha K, Khucharoen K (2014) Red cheese production from soymilk by *Monascus purpureus* and *Lactobacillus casei*. J Appl Sci 14:2819–2824. https://doi.org/10.3923/jas.2014.2819.2824
Los FGB, Zielinski AAF, Wojeicchowski JP, Nogueira A, Demiate IM (2018) Beans (*Phaseolus vulgaris* L.): whole seeds with complex chemical composition. Curr Opin Food Sci 19:63–71. https://doi.org/10.1016/j.cofs.2018.01.010
Ma YL, Wang JH, Cheng YQ, Yin LJ, Li LT (2013) Some biochemical and physical changes during manufacturing of Grey sufu, a traditional Chinese fermented soybean curd. Int J Food Eng 9:45–54. https://doi.org/10.1515/ijfe-2012-0204
Mao J, Zhou Z, Yang H (2023) Microbial succession and its effect on the formation of umami peptides during sufu fermentation. Front Microbiol 14:1181588. https://doi.org/10.3389/fmicb.2023.1181588
Marinova D, Bogueva DJSE (2019) Planetary health and reduction in meat consumption. Sustainable Earth 2:3. https://doi.org/10.1186/s42055-019-0010-0
Matias NS, Bedani R, Castro IA, Saad SMI (2014) A probiotic soy-based innovative product as an alternative to *petit-suisse* cheese. LWT-Food Sci Technol 59:411–417. https://doi.org/10.1016/j.lwt.2014.05.054
McClements DJ, Grossmann L (2021) The science of plant-based foods: constructing next-generation meat, fish, milk, and egg analogs. Compr Rev Food Sci Food Saf 20:4049–4100. https://doi.org/10.1111/1541-4337.12771
Molfetta M, Morais EG, Barreira L, Bruno GL, Porcelli F, Dugat-Bony E, Bonnarme P, Minervini F (2022) Protein sources alternative to meat: state of the art and involvement of fermentation. Food Secur 11:2065. https://doi.org/10.3390/foods11142065
Nikmaram N, Dar BN, Roohinejad S, Koubaa M, Barba FJ, Greiner R, Johnson SK (2017) Recent advances in c-aminobutyric acid (GABA) properties in pulses: an overview. J Sci Food Agric 97:2681–2689. https://doi.org/10.1002/jsfa.8283

Obadina AO, Akinola OJ, Shittu TA, Bakare HA (2013) Effect of natural fermentation on the chemical and nutritional composition of fermented soymilk nono. Nigerian Food J 31:91–97. https://doi.org/10.1016/S0189-7241(15)30081-3

Ørskov KE, Christensen LB, Wiking L, Hannibal T, Hammershøj M (2021) Imitation cheese-new insights to relation between microstructure and functionality. Food Struct 29:100206. https://doi.org/10.1016/j.foostr.2021.100206

Paul AA, Kumar S, Kumar S, Sharma R (2019) Milk analog: plant based alternatives to conventional milk, production, potential and health concerns. Crit Rev Food Sci Nutr 60:3005–3023. https://doi.org/10.1080/10408398.2019.1674243

Post MJ, Levenberg S, Kaplan DL, Genovese N, Fu J, Bryant CJ, Negowetti N, Verzijden K, Moutsatsou P (2020) Scientific, sustainability and regulatory challenges of cultured meat. Nat Food 1:403–415. https://doi.org/10.1038/s43016-020-0112-z

Pua A, Tang VCY, Goh RMV, Sun J, Lassabliere B, Liu SQ (2022) Ingredients, processing, and fermentation: addressing the organoleptic boundaries of plant-based dairy analogues. Food Secur 11:875. https://doi.org/10.3390/foods11060875

Rai AK, Sanjukta S, Chourasia R, Bhat I, Bhardwaj PK, Sahoo D (2017) Production of bioactive hydrolysate using protease, β-glucosidase and α-amylase of *Bacillus* spp. isolated from kinema. Bioresour Technol 235:358–365. https://doi.org/10.1016/j.biortech.2017.03.139

Razzaq ZU, Khan MKI, Maan AA, Rahman S, u. (2020) Characterization of single cell protein from *Saccharomyces cerevisiae* for nutritional, functional and antioxidant properties. J Food Meas Charact 14:2520–2528. https://doi.org/10.1007/s11694-020-00498-

Rizzo G, Baroni L (2018) Soy, soy foods and their role in vegetarian diets. Nutrients 10:43. https://doi.org/10.3390/nu10010043

Sanjukta S, Rai KA (2016) Production of bioactive peptides during soybean fermentation and their potential health benefits. Trends Food Sci Technol 50:1–10. https://doi.org/10.1016/j.tifs.2016.01.010

Santhakumaran P, Ayyappan SM, Ray JG (2020) Nutraceutical applications of twenty-five species of rapid-growing green-microalgae as indicated by their antibacterial, antioxidant and mineral content. Algal Res 47:101878. https://doi.org/10.1016/j.algal.2020.101878

Shao Y, Hu Z, Yu Y, Mou R, Zhu Z, Beta T (2018) Phenolic acids, anthocyanins, proanthocyanidins, antioxidant activity, minerals and their correlations in non-pigmented, red, and black rice. Food Chem 239:733–741. https://doi.org/10.1016/j.foodchem.2017.07.009

Sharma N, Gayathri R, Priya VV (2018a) Assessment of nutritional value of overnight soaked cooked rice over un-soaked cooked rice. Int J Res Pharm Sci 9:616–619

Sharma P, Sharma D, Amin A (2018b) Development of a functional fermented peanut-based cheese analogue using probiotic bacteria. *BioTechnologia*. J Biotechnol Comput Biol Bionanotechnol 99:435–441. https://doi.org/10.5114/bta.2018.79973

Tamang JP, Shin D-H, Jung S-J, Chae S-W (2016) Functional properties of microorganisms in fermented foods. Front Microbiol 7:578. https://doi.org/10.3389/fmicb.2016.00578

Thirumdas R, Brnčić M, Brnčić SR, Barba FJ, Gálvez F, Zamuz S, Lacomba R, Lorenzo JM (2018) Evaluating the impact of vegetal and microalgae protein source on proximate composition, amino acid profile, and physicochemical properties of fermented Spanish "chorizo" sausages. J Food Process Preserv 42:e13817. https://doi.org/10.1111/jfpp.13817

Tohamy MM, Ali MA, Shaaban HAG, Mohamad AG, Hasanain AM (2018) Production of functional spreadable processed cheese using *Chlorella vulgaris*. Acta Sci Pol Technol Aliment 17:347–358. https://doi.org/10.17306/J.AFS.2018.0589

Torres-Tiji Y, Fields FJ, Mayfield SP (2020) Microalgae as a future food source. Biotechnol Adv 41:107536. https://doi.org/10.1016/j.biotechadv.2020.107536

Vanga SK, Raghavan V (2018) How well do plant based alternatives fare nutritionally compared to cow's milk? J Food Sci Technol 55:10–20. https://doi.org/10.1007/s13197017-2915-y

Vanga SK, Singh A, Raghavan V (2015) Review of conventional and novel food processing methods on food allergens. Crit Rev Food Sci Nutr 57:2077–2094. https://doi.org/10.1080/10408398.2015.1045965

Vong WC, Liu SQ (2017) Changes in volatile profile of soybean residue (okara) upon solid state fermentation by yeasts. J Sci Food Agric 97:135–143. https://doi.org/10.1002/jsfa.7700

Wiebe M (2002) Myco-protein from *Fusarium venenatum*: a well-established product for human consumption. Appl Microbiol Biotechnol 58:421–427. https://doi.org/10.1007/s00253-002-0931-x

Wiebe MG (2004) Quorn™ myco-protein-overview of a successful fungal product. Mycologist 18:17–20. https://doi.org/10.1017/S0269915X04001089

Wild F, Czerny M, Janssen AM, Kole APW, Zunabovic M, Doming KJ (2014) The evolution of a plant-based alternative to meat: from niche markets to widely accepted meat alternatives. Agro Food Industry Hi Tech 25:45–49

Xi X, Ke J, Liu X, Gu X, Wang Y (2022) Physicochemical and taste characteristics of traditional Chinese fermented food sufu. J Food Process Preserv 46:e16845. https://doi.org/10.1111/jfpp.16845

Xie C, Gänzle M (2023) Microbiology of fermented soy food in Asia: can we learn lessons for production of plant cheese analogues? Int J Food Microbiol 407:110399. https://doi.org/10.1016/j.ijfoodmicro.2023.110399

Xie C, Zeng H, Li J, Qin L (2018) Comprehensive explorations of nutritional, functional and potential tasty components of various types of sufu, a Chinese fermented soybean appetizer. Food Sci Technol 39:105–114. https://doi.org/10.1590/fst.37917

Yusuf D, Setiarto RHB (2022) Quality aspects related to meat analogue based on microbiology, plants and insects protein. Rev Agric Sci 10:206–219. https://doi.org/10.7831/ras.10.0_206

Zielinska D, Kaminska A, Kolozyn-Krajewska D (2015) Development of tofu production method with probiotic bacteria addition. J Microbiol Biotechnol Food Sci 04:485–490. https://doi.org/10.15414/jmbfs.2015.4.6.485-490

# Chapter 13
# Application of Multi-criteria Decision Making Techniques in Sensory Evaluation of Food Analogues

**Ebuzer Arslan, Tuğba Dedebaş, and Emre Hastaoğlu**

## 13.1 Introduction

Sensory analysis or sensory evaluation is the discipline that investigates the acceptability of food products through sensory responses occurring during food consumption. In sensory analysis methods, food attributes are evaluated by panelists by using their senses of sight, smell, taste, touch, and hearing (Barallat Pérez et al. 2023; Ruiz-Capillas et al. 2021). Sensory studies in the development of new products are necessary for enhancing product features, revealing consumer expectations, and improving consumer acceptance (Sogari et al. 2023).

Various test methods are used in sensory analysis, including discrimination tests (to determine the discernible differences in specific sensory attributes between two or more products), descriptive tests (to rank perceived intensities of specific attributes on a predetermined scale), and hedonic tests (to assess liking of products).

**Discrimination tests**, are employed to ascertain the presence of a perceivable difference in identified sensory characteristics between two or more products.

E. Arslan
Management and Organization Department, Yıldızeli Vocational College, Sivas Cumhuriyet University, Sivas, Turkey
e-mail: ebuzerarslan@cumhuriyet.edu.tr

T. Dedebaş (✉)
Food Technology Department, Bolvadin Vocational College, Afyon Kocatepe University, Afyon, Turkey
e-mail: tdebas@aku.edu.tr

E. Hastaoğlu
Department of Gastronomy and Culinary Arts, Faculty of Tourism, Sivas Cumhuriyet University, Sivas, Turkey
e-mail: ehastaoglu@cumhuriyet.edu.tr



Ö. P. CAN et al. (eds.), *Food Analogues: Emerging Methods and Challenges*,
https://doi.org/10.1007/978-3-031-69872-9_13

**Descriptive tests**, specific attributes of foods are rated by panelists by using a predefined scale. In some cases, panelists might be requested to determine the most suitable characteristics to describe a product. The main difficulty in descriptive tests lies in identifying the most appropriate attributes and establishing the most suitable intensity scale for each attribute. As a result, ratings may vary significantly among individuals tasting the same food. To overcome this problem, descriptive sensory evaluations typically involve approximately 8–12 panelists who undergo training to calibrate the panel on the identified attributes before conducting sensory analysis. Intensity scales typically range from 1 to 9; each number is associated with a specific descriptor indicating increasing intensity, such as 1 = soft, 5 = hard, and 9 = very hard.

**Sensory tests,** are used to evaluate consumer preference for specific product attributes and overall product acceptance. In this method, untrained consumers are usually asked to rate how much they like a food. Sensory tests can provide information about the desirability that may influence the potential success of a product in the market. Typically, a 9-point hedonic scale ranging from 'dislike extremely' to 'like extremely' is used to evaluate a product's overall liking or specific attributes (McClements and Grossman 2022; Piggott et al. 1998; Short et al. 2021).

Subjective differences arise in the evaluation of foods due to characteristics among panelists, such as cultural background, education, environment, habits, and weaknesses. Therefore, careful consideration of factors such as personnel selection, training, preparation, informing the panel, location of sensory analysis, sample preparation and serving, random labeling of samples with three-digit codes, presentation of products in random sampling order, uniformity in plates, dishes, glasses, or bowls with similar size, shape, and distribution, conducting experiments under controlled environmental conditions (including lighting and temperature), and sample collection is required to eliminate these differences and limit subjectivity in the test. To help reduce bias, a three-digit code should be provided to blindfolded participants (McClements and Grossman 2022; Piggott et al. 1998). Additionally, appropriate data analysis, statistical analyses, and interpretation are essential components of sensory techniques due to potential variability in the results obtained from sensory analysis. This section will provide information on new techniques used in the sensory evaluation of foods and sensory evaluation of food analogues.

## 13.2 Use of Multicriteria Decision-Making Methods in Sensory Analysis of Foods

Multicriteria decision-making involves assigning values to alternatives by evaluating multiple criteria together, and it is divided into two categories: multi-attribute decision-making and multi-objective decision-making. If the problem involves evaluating alternatives by assigning scores to certain attributes and selecting the best based on those scores, it is referred to as a multi-attribute decision-making problem. On the other hand, the multi-objective decision-making problem is concerned with selecting the best alternative based on conflicting objectives. In both types of problems, there may be one or more decision-makers involved (Karaatlı et al. 2015; Phua and Minowa 2005). Multicriteria decision-making is a method that enables the selection of the best option from among multiple criteria applied simultaneously. In a rational decision-making environment, the preferred choice is often constrained by limitations and the objectives of management. Multicriteria decision-making showed rapid advancement both theoretically and in practical applications in the field of decision analysis. With its strong logical structure and success in decision-making, it found a wide range of applications (Güneş and Umarusman 2003; Karaatlı et al. 2015). The primary objective of multicriteria decision-making techniques is to identify the best alternative that provides the highest level of satisfaction across all relevant criteria (Chatterjee and Chakraborty 2012).

Multicriteria decision-making applications range from personal decision-making on a micro-scale, family budget planning on a small scale, organizational decision-making in businesses and the public sector on a medium scale, to government budget allocation decisions and investment decisions in holdings on a macro scale. These techniques enable the formulation of specific, agreed-upon, realistic, time-bound, and measurable decisions. Thus, the analysis of complex and holistic data and the systematic conduct of decision-making processes become feasible. One of the main challenges of multicriteria decision-making techniques is that personal evaluations may yield different results at different times.

Studies were carried out by using the Simple Additive Weighting (SAW) method of multicriteria decision-making with various research topics such as renewable and nuclear energy (Shakouri et al. 2014), personnel selection (Afshari et al. 2010), diligent student research (Manokaran et al. 2011), and gas sensor selection (Tsai et al. 2009). Additionally, studies were conducted by using the Technique for Order Preference by Similarity to an Ideal Solution (TOPSIS) for the financial rating of automotive companies (Yurdakul and Iç 2003), efficiency of participation banks (Yayar and Baykara 2012), and stakeholder relationship analysis (Öztürkoğlu and Türker 2013). The ranking of livable cities was achieved using various multicriteria decision-making techniques. Using TOPSIS, SAW, and Grey Relational Analysis techniques, data containing multiple criteria from economics, education, health, and security were evaluated simultaneously, resulting in the same ranking in all three techniques (Karaatlı et al. 2015).

In their study comparing the sensory properties of hot chocolate beverages containing different amounts of fat, Dogan et al. (2016) determined the optimal amount of fat in hot chocolate by using Analytic Hierarchy Process (AHP), Weighted Sum Model (SAW), Technique for Order Preference by Similarity to an Ideal Solution (TOPSIS), and Elimination and Choice Expressing Reality (ELECTRE) methods. Nine samples of hot chocolate containing varying amounts of fat were subjected to sensory evaluation by a panel. In the sensory analysis of hot chocolate samples, consistency, taste, color, and overall liking criteria were evaluated by 28 panelists using a 9-point hedonic scale. Given the results reported in their study, the hot chocolate sample with the highest fat content was the most preferred among the samples in all multicriteria decision-making techniques in all multicriteria decision-making techniques.

To develop a functional pudding product, the sensory properties of puddings produced by adding fiber to three different pudding varieties (vanilla, strawberry, chocolate) were determined using multicriteria decision-making techniques such as AHP, SAW, TOPSIS, and ELECTRE (Gurmeric et al. 2013). Sensory evaluation of six different pudding samples containing and not containing dietary fiber, including vanilla, strawberry, and chocolate flavors, was conducted using a 9-point hedonic type scale to assess appearance, taste and aroma, mouthfeel, appearance retention, stickiness, and overall acceptability criteria. Given the results obtained by using multicriteria decision-making techniques, the strawberry-flavored pudding was the most preferred in terms of sensory properties.

In a study aiming to facilitate the processing of a mixture of quantitative and qualitative information and generating an overall value for each alternative in the food supply chain using multicriteria decision-making techniques, it was noted that there are various criteria and decision steps due to the many stages from food production to consumer sales (Linnemann et al. 2015). Various criteria at key stages such as where the food is produced, raw material sourcing, processing techniques, distributors, wholesalers, transportation, cold chain, etc., affect the quality of the food supply chain. In a study conducted using Multiple Attribute Value Theory (MAVT) and Analytic Hierarchy Process (AHP) techniques, the resulting preference order differed primarily due to the way criterion weights emerged, the scoring of alternatives, and the use of scales in MAVT as opposed to pairwise comparison in AHP. On the other hand, in both methods, the preference order of the top-level criteria was the same, and the weights yielded similar results.

In a study evaluating the physicochemical, textural, bioactivity, and sensory properties of samples containing different proportions of persimmon fruit puree in ice cream, the optimal amount of persimmon puree in ice cream was determined (Karaman et al. 2014). The bioactivity and sensory properties of ice cream samples were evaluated by using the TOPSIS technique. In the study, the bioactivity criteria were determined as antioxidant activity and total phenolic content, while the sensory criteria were identified as color, aroma, taste, texture, and resistance to melting. Given the study results, an ice cream sample containing 24% persimmon puree was identified as the most optimal ratio based on the specified criteria.

In a study aiming to improve the functional properties of a dairy product known as mellorine, various proportions of beetroot, red cabbage, and broccoli juice were added to enhance its sensory and bioactivity properties, and the evaluations were conducted by using the TOPSIS technique (Ozturk et al. 2014). DPPH activities and total phenolic content were selected as bioactivity properties, while appearance, consistency, and taste were chosen as sensory criteria. Among the mellorine samples evaluated using the TOPSIS technique, the sample containing 10% red cabbage juice was determined to be the most preferred sensory-wise.

In a study employing multicriteria decision-making approaches in food engineering, Analytic Hierarchy Process (AHP) and Tabular Method (TM) techniques were indicated for potential use in the thermal processing of packaged foods. In food processing processes, the use of T-Choice and Mprioryty software was recommended for the comparison of AHP and TM techniques (Abakarov 2011).

In a study on the identification of potential risks for countries exporting grain products to European Union countries, TOPSIS, ELECTRE, and Cross Efficiency (CE) decision-making techniques were utilized. While the rankings obtained from the three methods used in the study were not identical, similar results were achieved in distinguishing countries based on their food risk levels (Puertas et al. 2020).

### 13.2.1 AHP

Analytic Hierarchy Process (AHP), used in solving complex decision-making problems, was first introduced to the literature by Myers and Alpert in 1968. In 1977, Thomas Saaty introduced the AHP model for solving decision-making problems by integrating components such as the hierarchical structure of complex units, comparison of criteria, determination of the importance levels of multiple criteria, eigenvectors, and consistency coefficients (Önder and Önder 2018). AHP is based on comparisons aimed at determining the factors influencing decisions and the degrees of importance of these factors using a pre-determined pairwise comparison survey related to a subject. These comparisons result in a proportional distribution of differences in importance and decision points.

There are 5 basic stages in the AHP method, and below, the stages and formulas related to the solution of the method are explained together (Saaty 1990; Aktaş et al. 2015; Uzun and Kazan 2016; Yüksel and Akın 2006).

1. **Stage: Defining the Problem and Objective**

   The first step in the AHP method involves defining the problem and determining the decision points. Then, factors influencing the decision-making process are ranked. It is important to correctly identify and explain these factors in detail to ensure the consistency and accuracy of the pairwise comparisons to be made.

2. **Stage: Establishing the Comparison Matrix for Factors**

   After defining the decision problem, assuming the number of decision points related to the decision m and the number of factors affecting any decision given is n, the comparison matrix for the factors is an $n$ x $n$ square matrix. The diagonal values of this matrix are equal to 1.

   The comparison matrix is shown below.

$$A = \begin{bmatrix} a_{11} & a_{1j} & \cdots & a_{1n} \\ a_{i1} & a_{ij} & \cdots & a_{in} \\ \vdots & \vdots & \vdots & \vdots \\ a_{n1} & a_{nj} & \cdots & a_{nn} \end{bmatrix}$$

The reason for the diagonal value in the comparison matrix being equal to 1 is the comparison of the relevant factor with itself. Each factor is compared with each of the other factors according to their levels of importance. In making these comparisons, Saaty's developed table of importance levels (1–9 scale) is used (Table 13.1).

The importance of criterion $i$ over criterion $j$ is determined through pairwise comparisons. These comparisons are made considering the table of importance levels. For instance, if the obtained value from the comparison is 7, it is understood that criterion $i$ is considerably more important than criterion $j$. In this case, criterion $j$ is important relative to criterion $i$ at a level of 1/7. Comparisons are made for values above the diagonal of the comparison matrix. To calculate values below the diagonal, the formula $a_{ji} = \frac{1}{a_{ij}}$ is utilized.

3. **Stage: Determining the Importance Distribution of Factors**

The importance levels of factors are presented in the comparison matrix. However, to ascertain the percentage distributions of each factor within the whole, the column vectors of the comparison matrix are utilized.

$$B_i = \begin{bmatrix} b_{11} \\ b_{21} \\ \cdot \\ \cdot \\ \cdot \\ b_{n1} \end{bmatrix}$$

**Table 13.1** Significance Degree table (Saaty 1990)

| Significance Degrees | Value explanations |
|---|---|
| 1 | "Both factors have **equal importance**." |
| 3 | "Given experiences and analyses, one factor is **slightly more important** than the other." |
| 5 | "One factor is **probably more important** than the other factor." |
| 7 | "One factor is **significantly more important** than the other factor." |
| 9 | "One factor is **absolutely more important** than the other factor" |
| 2,4,6,8 | "The intermediate values to be explained between the factors." |

Column vectors are created by dividing each value of the decision matrix by the column total. The following formula is employed in the calculation of column vectors.

$$b_{ij} = \frac{a_{ij}}{\sum_{i=1}^{n} a_{ij}}$$

When the generated column vectors are combined, the C matrix is formed.

$$C = \begin{bmatrix} c_{11} & c_{12} & \cdots & c_{1n} \\ c_{21} & c_{22} & \cdots & c_{2n} \\ \vdots & \vdots & \vdots & \vdots \\ c_{n1} & c_{n2} & \cdots & c_{nn} \end{bmatrix}$$

After obtaining the C matrix, the arithmetic mean of each row is calculated to obtain the "priority vector," also referred to as the W "column vector" values related to criteria. The following formula is used in calculating the W vector.

$$w_{ij} = \frac{\sum_{i=1}^{n} c_{ij}}{n}$$

Following the calculations, the W vector is obtained as follows.

$$W = \begin{bmatrix} w_1 \\ w_2 \\ \vdots \\ w_n \end{bmatrix}$$

4. **Stage: Measuring the Consistency in Factor Comparisons**
To calculate the consistency ratio (CR), the priority vector needs to be calculated. The priority vector is obtained by multiplying the C matrix with the W column vector.

$$D = \begin{bmatrix} c_{11} & c_{12} & \cdots & c_{1n} \\ c_{21} & c_{22} & \cdots & c_{2n} \\ \vdots & \vdots & \vdots & \vdots \\ c_{n1} & c_{n2} & \cdots & c_{nn} \end{bmatrix} x \begin{bmatrix} w_1 \\ w_2 \\ \vdots \\ w_n \end{bmatrix} = \begin{bmatrix} d_1 \\ d_2 \\ \vdots \\ d_n \end{bmatrix}$$

The sum of the values of the priority vector (D), calculated with the above formula, is divided by the number of factors, resulting in a value called the Fundamental Value ($\lambda$).

$$\lambda = \frac{\sum_{i=1}^{n} d}{n}$$

Subsequently, the Consistency Index (CI) should be computed. The consistency indicator is computed by using the formula:

$$CI = \frac{\lambda - n}{n - 1}$$

Once the consistency indicator is calculated using the formula above, the Random Index (RI) by the number of criteria is presented in Table 13.2.

The consistency value, CR, is calculated by dividing the determined RI value based on the number of criteria. The CR value is calculated using the following formula:

$$CR = \frac{CI}{RI}$$

If the calculated CR value is lower than 0.10, it suggests that the comparisons are consistent; if it is greater than 0.10, it indicates inconsistency.

5. **Stage: Determining the General Priority Order**

In the final step, the final priority values, which are determined based on the percentage weights of the criteria calculated according to the Analytic Hierarchy Process, are ranked, and the alternative with the highest importance degree in this ranking is selected.

### 13.2.2 TOPSIS

Developed in 1981 by Hwang and Yoon, TOPSIS is a multicriteria decision-making method designed to select the best alternative based on the concept of a compromise solution. This method was developed as an alternative to the ELECTRE method (Tzeng and Huang 2011).

The main objective aim of the method is to make a selection among alternative options by comparing specific criteria and the highest and lowest values these criteria can take. This selection is based on the assumption that it will be "closest to the positive ideal solution point; farthest from the negative ideal solution point." (Oprıovıc and Tzeng 2004).

**Table 13.2** Random Consistency Index (Saaty 2008)

| *n* | *1* | *2* | *3* | *4* | *5* | *6* | *7* | *8* | *9* | *10* | *11* | *12* | *13* | *14* | *15* |
|---|---|---|---|---|---|---|---|---|---|---|---|---|---|---|---|
| ***RI*** | ***0.0*** | ***0.0*** | ***0.52*** | ***0.89*** | ***1.11*** | ***1.25*** | ***1.35*** | ***1.40*** | ***1.45*** | ***1.49*** | ***1.52*** | ***1.54*** | ***1.56*** | ***1.58*** | ***1.59*** |

Due to its lack of complex solution paths and imposing mathematical models, the TOPSIS method has found applicability in various fields. It is an easy, understandable, and interpretable method.

There are six stages in the solution of the TOPSIS method, and these stages are as follows gerçekleştirilir (Hwang and Yoon 1981; Oprıovıc and Tzeng 2004; Özdemir 2018; Özbek 2017).

1. **Stage: Establishment of Decision Matrix**

The decision matrix to be formed by decision-makers is a matrix with $m \times n$ dimensions, where the decision points of the decision-maker are shown in rows and factors are shown in columns.

$$A_{ij} = \begin{bmatrix} a_{11} & a_{12} & \cdots & a_{1n} \\ a_{21} & a_{22} & \cdots & a_{2n} \\ \vdots & \vdots & \vdots & \vdots \\ a_{m1} & a_{m2} & \cdots & a_{mn} \end{bmatrix}$$

2. **Stage: Obtaining the Normalized Matrix**

The squares of each $a_{ij}$ value of the decision matrix created by the decision-maker are calculated to obtain column sums. Then, each $a_{ij}$ value is normalized by dividing it by the square root of its column sum. This process is performed using the following formula;

$$n_{ij} = \frac{a_{ij}}{\sqrt{\sum_{i=1}^{m} a_{ij}^2}} (i = 1,2,\ldots,m \; ve \; j = 1,2,\ldots,n)$$

As a result of the process, the normalization matrix is obtained as follows:

$$N = \begin{bmatrix} n_{11} & n_{12} & \cdots & n_{1n} \\ n_{21} & n_{22} & \cdots & n_{2n} \\ \vdots & \vdots & \ddots & \vdots \\ n_{m1} & n_{m2} & \cdots & n_{mn} \end{bmatrix}$$

3. **Stage: Obtaining the Weighted Normalized Matrix**

After obtaining the normalization matrix, each value such as $w_i$ is weighted by a value. The weighting process, done according to the importance level of the factors, reveals the subjective aspect of the method. One important point to note here is that the total of the weights, the sole subjective parameter of the TOPSIS method, must equal 1. The formula is as follows:

$$n_i = \sum_{i=1}^{n} w_i = 1$$

The weighted normalized matrix is obtained by multiplying the $n_{ij}$ values obtained with the normalization matrix by the weights $w_i$.

$$V = \begin{bmatrix} w_1 n_{11} & w_2 n_{12} & \cdots & w_n n_{1n} \\ w_1 n_{21} & w_2 n_{22} & \cdots & w_n n_{2n} \\ \vdots & \vdots & \ddots & \vdots \\ w_1 n_{m1} & w_2 n_{m2} & \cdots & w_n n_{mn} \end{bmatrix} = \begin{bmatrix} v_{11} & v_{12} & \cdots & v_{1n} \\ v_{21} & v_{22} & \cdots & v_{2n} \\ \vdots & \vdots & \ddots & \vdots \\ v_{m1} & v_{m2} & \cdots & v_{mn} \end{bmatrix}$$

4. **Stage: Determining the Ideal and Negative Solution Values**
At this stage, the maximum ($A^*$) and minimum ($A^-$) values in each column of the weighted normalized matrix to obtain an ideal solution.

$$A^* = \left\{ \max_j v_{ij} | j = 1,2,\ldots,n; i = 1,2,\ldots,m \right\}$$

$A^* = \left\{ v_1^*, v_2^*, \ldots, v_n^* \right\}$ maximum value for each column

$$A^- = \left\{ \min_i v_{ij} | j = 1,2,\ldots,n; i = 1,2,\ldots,m \right\}$$

$A^- = \left\{ v_1^-, v_2^-, \ldots, v_n^- \right\}$ minimum value for each column.

5. **Stage: Calculation of Distances from Ideal Maximum and Minimum Points**
After the calculation of maximum and minimum values, then the distances to the ideal maximum and ideal minimum points are calculated by using the formulas below;

$$S_i^* = \sqrt{\sum_{j=1}^{n} \left( v_{ij} - v_j^* \right)^2} \textit{ ideal maksimum uzakl}\mathbb{k}$$

$$S_i^- = \sqrt{\sum_{j=1}^{n} \left( v_{ij} - v_j^- \right)^2} \textit{ ideal minimum uzakl}\mathbb{k}$$

At this point, the number of ideal maximum ($S_i^*$) and ideal minimum $\left( S_i^- \right)$ values should be equal to the number of decision points.

6. **Stage: Calculation of Relative Closeness to Ideal Solution Point**
Closeness coefficient is symbolized with $C_i^*$. Here, $C_i^*$ takes values in the range $0 \le C_i^* \le 1$, with $C_i^*$ values closer to 1 indicating closeness to the maximum ideal solution point, whereas those closer to 0 indicating closeness to minimum ideal solution point. Closeness coefficient is calculated using the formula given below.

$$C_i^* = \frac{S_i^-}{S_i^- + S_i^*}$$

### 13.2.3 ELECTRE

First introduced in 1965 by Bernard Roy at a conference and initially published in 1968, ELECTRE is one of the methods of Multicriteria Decision Making (MCDM). There are multiple versions of the ELECTRE method (I, II, III, IV, V, IS, A) (Huang and Chen 2005).

Among these versions, ELECTRE I and ELECTRE IS are utilized to find solutions to selection problems; ELECTRE II, ELECTRE III, and ELECTRE IV are employed to rank criteria from good to bad, while ELECTRE V and ELECTRE A versions are used for classification purposes. The ELECTRE method relies on pairwise comparisons of alternatives based on criteria. Through this method, researchers and decision-makers can weight numerous quantitative and qualitative criteria and select the most suitable alternative by considering the total weights of each criterion (Şahin 2018).

The solution steps of the ELECTRE I method are summarized below: (Aktaş et al. 2015; Şahin 2018; Çelik and Ustasülayman 2015; Tzeng and Huang 2011; Figueira et al. 2005).

1. **Stage: Establishment and Normalization of Decision Matrix**

The rows of the binary comparison matrix created by the decision-maker from potential alternatives represent the superiority desired among decision points, while the columns represent the evaluation criteria to be used in decision-making. In matrix $A_{ij}$, m represents the number of decision points, and n represents the number of evaluation criteria. The decision matrix can be represented as follows:

$$A_{ij} = \begin{bmatrix} a_{11} & a_{12} & \cdots & a_{1n} \\ a_{21} & a_{22} & \cdots & a_{2n} \\ \vdots & \vdots & \vdots & \vdots \\ a_{m1} & a_{m2} & \cdots & a_{mn} \end{bmatrix}$$

The normalization of the formed decision matrix is carried out using separate formulas for cost and benefit criteria.

For cost criteria:

$$X_{ij} = \frac{\frac{1}{a_{ij}}}{\sqrt{\sum_{i=1}^{m} \frac{1}{a_{ij}}^{2}}}, i = 1,2,\ldots,m; j = 1,2,\ldots,n$$

For benefit criteria:

$$X_{ij} = \frac{a_{ij}}{\sqrt{\sum_{i=1}^{m} a_{ij}^2}}, i = 1,2,\ldots,m; j = 1,2,\ldots,n$$

The $X_{ij}$ matrix is formed as follows after calculations:

$$X_{ij} = \begin{bmatrix} x_{11} & x_{12} & \cdots & x_{1n} \\ x_{21} & x_{22} & \cdots & x_{2n} \\ \vdots & \vdots & \vdots & \vdots \\ x_{m1} & x_{m2} & \cdots & x_{mn} \end{bmatrix}$$

2. **Stage: Establishment of Weighted Standard Decision Matrix**

The factor weights ($w_i$) determined by the decision-maker are multiplied with the weights of each criterion in the normalization matrix to obtain the weighted decision matrix.

$$V = \begin{bmatrix} x_{11}.w_1 & x_{12}.w_2 & \cdots & x_{1n}.w_n \\ x_{21}.w_1 & x_{22}.w_2 & \cdots & x_{2n}.w_n \\ \vdots & \vdots & \vdots & \vdots \\ x_{m1}.w_1 & x_{m2}.w_2 & \cdots & x_{mn}.w_n \end{bmatrix}$$

3. **Stage: Determining the Concordance and Discordance Sets**

The determination of concordance and discordance sets relies on the $V$ matrix. Utilizing $A = \{a, b, c \ldots\}$ to represent a series of alternative sets, the establishment of the value set according to the concordance ($C_{ab}$) and discordance ($D_{ab}$) sets employs the following formulas. Each element of the concordance set ($C_{ab}$):

$$C_{ab} = \left\{ j \middle| x_{aj} \geq x_{bj} \right\}$$

is determined using the formula. After calculating the concordance set ($C_{ab}$), each element of the discordance set ($D_{ab}$) is determined using the formula:

$$D_{ab} = \left\{ j \middle| x_{aj} < x_{bj} \right\} = j - C_{ab}$$

4. **Stage: Establishment of Concordance and Discordance Matrices**

After the creation of compatibility and incompatibility matrices, initially, the elements of the compatibility matrix ($C_{ab}$) are measured as "how much a is at least as good as b," and these elements of the matrix are calculated to take values in the [0,1] interval as follows:

$$\left(C_{ab}\right) = \sum_{j \in C_{ab}} w_i$$

The compatibility matrix is as follows:

$$C = \begin{bmatrix} - & c_{12} & \cdots & c_{1m} \\ c_{21} & - & \cdots & c_{2m} \\ \vdots & \vdots & \ddots & \vdots \\ c_{m1} & c_{m2} & \cdots & - \end{bmatrix}$$

The elements of the incompatibility matrix ($D_{ab}$) measure "the degree to which b is preferable to a directly," and the elements of this matrix also take values in the [0,1] interval. The elements of the incompatibility matrix are calculated using the following formula:

$$D_{ab} = \frac{\max_{j \in D_{ab}} \left| v_j(a) - v_j(b) \right|}{\max_{j \in J} \left| v_j(a) - v_j(b) \right|}$$

And, the incompatibility matrix is as follows:

$$D = \begin{bmatrix} - & d_{12} & \cdots & d_{1m} \\ d_{21} & - & \cdots & d_{2m} \\ \vdots & \vdots & \ddots & \vdots \\ d_{m1} & d_{m2} & \cdots & - \end{bmatrix}$$

5. **Stage: Establishment of Concordance and Discordance Index Matrices**
To generate the concordance and discordance index matrices, firstly, the Indifference Threshold value should be calculated. For the concordance index matrix, the Indifference Threshold value is calculated using the following formula:

$$\overline{c} = \frac{\sum_{a=1}^{m} \sum_{b=1}^{m} c(a,b)}{m(m-1)}$$

The superiority of one alternative over another, expressed relative to the threshold value, is obtained as the elements of the E matrix as follows:

$$E = \begin{cases} e(a,b) = 1 \text{ ĕger } c(a,b) \geq \overline{c} \\ e(a,b) = 0 \text{ ĕger } c(a,b) < \overline{c} \end{cases}$$

For the discordance index matrix, the Indifference Threshold value is calculated using the following formula:

$$\bar{d} = \frac{\sum_{a=1}^{m}\sum_{b=1}^{m} d(a,b)}{m(m-1)}$$

Using the discordance threshold value, the elements of the incompatibility index matrix $F$ are obtained in a similar manner to the $E$ matrix as follows:

$$F = \begin{cases} f(a,b) = 1 \, \breve{e}ger \, d(a,b) \geq \bar{d} \\ f(a,b) = 0 \, \breve{e}ger \, d(a,b) < \bar{d} \end{cases}$$

6. **Stage: Establishment of Total Dominance Matrix and Determining the Importance Order**

The elements of dominance matrix G are equal to the mutual multiplication of elements $g_{kl}$, $e_{kl}$, and $f_{kl}$. The resulting G matrix, depending on matrices C and D, has $m$ x $m$ dimensions and is formed as follows, with values of 1 and 0:

$$G = \begin{bmatrix} - & 0 & 0 \\ 1 & - & 0 \\ 1 & 1 & - \end{bmatrix}$$

The rows and columns of the G matrix represent decision points. For instance, if the G matrix is calculated as above, it indicates the absolute superiority of the second alternative to the first alternative, the third alternative to the first alternative, and the second alternative.

### *13.2.4 Saw*

SAW, first introduced to the literature in 1954 by Churchman and Ackoff, was applied to portfolio selection problems. Also known as the Weighted Sum Model in the literature, the SAW method is one of the preferred and widely used MCDA methods due to its ease of calculation and simplicity of mathematical operations (Fishburn 1967; Yeh 2003;Tzeng and Huang 2011).

The solution steps of the SAW method are summarized below: (Çakır and Perçin 2013; Ömürbek et al. 2016; Hadi et al. 2019).

1. **Stage: Normalization of Decision Matrix**

The SAW method aims to create an index for each criterion by summing their contributions. Since addition cannot be performed on different units, as is the case with other methods, normalization must be performed in this method as well. The type of criterion should be determined before normalization. If the criterion type is maximization, the following formula is used for the benefit criterion:

$$r_{ij} = \frac{X_{ij}}{\max_{ij}} i = 1,2,\ldots,m; j = 1,2,\ldots,n$$

As seen in the formula, normalization is carried out by scaling each criterion value to the maximum value in that row. If the criterion type is minimization, the following formula is used for the cost criterion:

$$r_{ij} = \frac{\min_{ij}}{X_{ij}} i = 1,2,\ldots,m; j = 1,2,\ldots,n$$

In the minimization formula, normalization is performed by scaling each criterion value to the smallest value in that row. All calculated criteria should be positive. If there are negative values, these values should be converted to positive values using the formula $r_{ij} = r_{ij} + |minr_{ij}| + 1$.

2. **Stage: Calculation of Alternative Preference Values**
   The total preference values for each alternative are calculated by multiplying the values previously calculated with the criterion weights using the formula below.

$$V_i = \sum_{j=1}^{n} w_j . r_{ij} \; i = 1,2,\ldots,m$$

The notation $w_j$ here indicates the importance weight given to criterion j. A higher $V_i$ value implies that alternative $A_i$ is more preferred. In the SAW method, it should be noted that all criteria should have numerical data in the same unit of measurement.

### 13.2.5 BWM

The Best-Worst Method (BWM) was developed by Jafar Rezaei in 2015 (Rezaei 2015). This method allows decision-makers to make a more consistent evaluation by making fewer pairwise comparisons based on binary comparison logic. Rezaei stated in his study that the decision-maker must first determine the best (most important, most desired) and worst (least important, least desired) criteria, and then propose pairwise comparisons of these criteria with other criteria. He stated that, in this method, instead of pairwise comparisons of all criteria, comparing the best criterion with others and the worst criterion with others will facilitate the decision-maker's work and that this situation can solve the problem with fewer pairwise comparisons compared to other CPM methods based on the pairwise comparison (Rezaei 2015). In addition, the BWM method, by using two vectors instead of a single vector used by binary comparison methods, allows for the control of its consistency and, by not requiring a full comparison matrix, provides ease in terms of time and data (Rezaei 2020). The BWM method is analyzed by following the steps below: (Rezaei 2015; Çakır and Can 2019; Yücesan and Gül 2021).

1. **Stage: Determining the Assessment Criteria**

At this point, $n$ number of criteria $(k_1, k_2, k_3, \ldots, k_n)$ are determined to make a decision.

2. **Stage: Determining the Best and Worst Criteria:**

Decision-maker determines the best (the most desired) criterion "$k_B$" and the worst (the least desired) criterion "$k_W$" among the criteria.

3. **Stage: Determining the Preference Ratio of the Selected Best Criterion over All Other Criteria**

At this stage, the selected best criterion is compared to other criteria by using a scale between 1 and 9 and the best criterion preference is determined.

| Value | Significance Level |
|---|---|
| 1 | Equally important |
| 3 | Moderately more important |
| 5 | Highly important |
| 7 | Much more highly important |
| 9 | Significantly important |

At the end of this stage, the $A_B = (a_{B1}, a_{B2}, a_{B3}, \ldots, a_{Bn})$ vector, which is called *Best* − *Others* $(A_B)$, from the best to others is achieved. In $A_B$ vector, each $a_{Bj}$ indicates the preference of the best criterion B over criterion j. Moreover, $a_{BB} = 1$. It means that the most important criterion will be compared to itself.

4. **Stage: Determining the Preference Ratio of the Selected Best Criterion over All Other Criteria**

At this stage, comparing the selected worst criterion to all other criteria and determining the relative importance values between 1 and 9, the worst vector $A_W = (a_{W1}, a_{W2}, a_{W3}, \ldots, a_{Wn})$ is established. In $A_W$ vector, each $a_{jW}$ indicates the preference of criterion j over the worst criterion W. Moreover, $a_{WW} = 1$. It means that the worst criterion will be compared to itself.

5. **Stage: Determining the Suitable Weights for Each Criterion**

At this stage, the optimal criterion weights $\left(w_1^*, w_2^*, w_3^*, \ldots, w_n^*\right)$ are determined. The objective here is to determine the most suitable weight to achieve the maximum absolute difference. To have the most suitable weight for the criteria, each pair of $\frac{w_B}{w_j}$ and $\frac{w_j}{w_w}$ should have $\frac{w_B}{w_j} = a_{jw}$. To compare them for all $j$s, a solution minimizing the maximum absolute differences $\left|\frac{w_B}{w_j} - a_{Bj}\right|$ and $\left|\frac{w_j}{w_w} - a_{jw}\right|$ for all $j$s should be achieved. Considering the need for not being negative and to rule ot sum, the mathematical problem is as follows.

$$\min_j \text{maks} \left\{ \left| \frac{w_B}{w_j} - a_{Bj} \right|, \left| \frac{w_j}{w_w} - a_{jw} \right| \right\}$$

Under the following conditions,

$$\sum w_j = 1$$

$$w_j \geq 0 \, tüm \; j'ler \; i\breve{c}in$$

The problem should be converted to the form given below.

$$\min \xi$$

$$\left| \frac{w_B}{w_j} - a_{Bj} \right| \leq \xi \; tüm \; j'ler \; i\breve{c}in$$

$$\left| \frac{w_j}{w_w} - a_{jw} \right| \leq \xi \; tüm \; j'ler \; i\breve{c}in$$

$$\sum w_j = 1$$

$$w_j \geq 0 \, tüm \; j'ler \; i\breve{c}in$$

After completing and resolving the model, the most suitable weights $(w_1^*, w_2^*, w_3^*, \ldots, w_n^*)$ and the $\xi$ value are obtained. $\xi$ value indicates the consistency ratio of the analyses conducted. The higher levels of this value indicate poorer consistency levels and lower reliability, whereas lower levels indicate a higher consistency.

### 13.2.6 COPRAS

The Complex Proportional Assessment (COPRAS) method was initially developed in 1996 by Zavadskas and Kaklauskas. This method, meaning "Complex Proportional Evaluation," is a multicriteria technique capable of assessing qualitative and quantitative criteria considering both benefit and cost attributes. It is employed for ranking and evaluating alternatives, distinguishing itself from other methods by enabling the comparison of alternatives and demonstrating their superiority over each other in a percentage-based manner (Kaklauskas et al., 2005). Additionally, two different versions of the COPRAS method were developed: the fuzzy version COPRAS-F, developed by Zavadskas and Antucheviciene (2007), and the gray version COPRAS-G, developed by Zavadskas and colleagues (2008).

The procedural steps of the COPRAS method are outlined below: (Zavadskas et al. 2008; Kaklauskas et al. 2005; Özbek 2017).

1. **Stage: Establishing the Decision Matrix**

The decision matrix (X) is formed by decision-makers in the initial stage of the decision-making process. The rows of the decision matrix consist of the decision options to be ranked, while the columns contain evaluation factors referred to as criteria. The decision matrix is constructed as follows:

$$X_{ij} = \begin{bmatrix} x_{11} & x_{12} & \cdots & x_{1n} \\ x_{21} & x_{22} & \cdots & x_{2n} \\ x_{31} & x_{32} & \cdots & x_{3n} \\ \vdots & \vdots & \vdots & \vdots \\ x_{m1} & x_{m2} & \cdots & x_{mn} \end{bmatrix}$$

In the decision matrix, $i$ refers to the decision options ($i = 1, 2, 3, \ldots, m$), whereas $j$ refers to the criteria ($j = 1, 2, 3, \ldots, n$).

2. **Stage: Standardization of Decision Matrix**

Elements of the decision matrix (X);

$$x^*_{ij} = \frac{x_{ij}}{\sum_{i=1}^{m} x_{ij}} (i = 1,2,3,\ldots,m \, ve \, j = 1,2,3,\ldots,n)$$

After the calculations are made using the formula, the standardized decision matrix is obtained as follows.

$$X^* = \begin{bmatrix} x^*_{11} & x^*_{12} & \cdots & x^*_{1n} \\ x^*_{21} & x^*_{22} & \cdots & x^*_{2n} \\ x^*_{31} & x^*_{32} & \cdots & x^*_{3n} \\ \vdots & \vdots & \vdots & \vdots \\ x^*_{m1} & x^*_{m2} & \cdots & x^*_{mn} \end{bmatrix}$$

3. **Stage: Weighting the Standardized Decision Matrix**

The weighted standardized decision matrix is obtained by multiplying the $x^*_{ij}$ values by $w_i$ weights. The new weighted elements of the matrix are obtained using the formula given below.

$$d_{ij} = x^*_{ij} w_j$$

4. **Stage: Summing the Weighted Standardized Indices**

Given the minimization-oriented criteria for the costs by considering the utility and cost criteria in summing the indices, the lower the $S_{-i}$ value is, the easier it is to achieve the objective. Similarly, the higher the $S_{+i}$ value calculated by considering the maximization-oriented criteria is, the more possible it is to achieve the objective.

$$S_{+i} = \textit{Sum of the utility criteria in the weighted standardized decision matrix}$$

$$S_{-i} = \textit{Sum of the cost criteria in the weighted standardized decision matrix}$$

5. **Stage: Determining the Relative Significance Levels Values of Alternatives**
The relative significance values of the alternatives being compared are determined using the formula given below.

$$Q_i = \frac{S_{-\min}\sum_{i=1}^{m} S_{-i}}{S_{-i}\sum_{i=1}^{m}\frac{S_{-\min}}{S_{-i}}}, i = 1,2,3,\ldots,m; j = 1,2,3,\ldots,n$$

$Q_i$ values are ranked from the highest one to the lowest; higher $Q_i$ values indicate a higher level of relative significance.

6. **Stage: Determining the Utility Degrees of Alternatives**
The alternative with a utility degree of 100 is considered the best option. Other alternatives are ranked relative to the best option. The benefit degrees of alternatives are determined using the following equation.

$$N_i = \left(\frac{Q_i}{Q_{\max}}\right).100\%$$

## 13.3 Sensorial Analysis of Food Analogues

The demand for plant-based meat and milk alternatives, known as food analogues, started increasing with the increasing population and changes in consumer lifestyles in developing countries. As a reason for this increase in the demand for plant-based meats, dairy products, and their derivatives, the perception that the consumption of animal-derived foods has negative effects have negative effects mainly originates from the proliferation of veganism and vegetarianism (Godschalk-Broers et al. 2022; Han et al. 2023; Barallat Pérez et al. 2023).

In recent years, the preference for plant-based meat and milk analogues devoid of animal-derived nutrients by consumers paved the way for research and development in alternative foods (Zahari et al. 2023). However, despite the increasing availability of food analogues, the transition to plant-based diets is still hindered due to the insufficient resemblance of existing products to the sensory properties of animal-based counterparts (Godschalk-Broers et al. 2022). The sensory attributes of plant-based foods, such as their appearance, texture, mouthfeel, and taste, must be evaluated by individuals, who genuinely consume and rate them. Plant-based foods mainly aim to accurately offer the sensory properties that animal-derived foods do. Therefore, the objective of sensory evaluation is to reveal similarities and differences between these two types of products. While all these testing methods can be employed to assess plant-based foods, descriptive and sensory tests are most commonly used (McClements and Grossman 2022; Aydar 2023).

Literature review revealed that, in a study conducted by employing focus group techniques with three consumer groups representing perceptions of plant-based food consumption (food science professionals/students, vegans and vegetarians, and general consumers), three main themes representing participants' views on plant-based foods were proposed. These themes include sustainability and ethics, health, and sensory aspects. Similarities and differences in perception among groups were discussed. Overall, most participants in their study were found to be open and willing to try innovative plant-based products as long as they meet desired expectations (Batista et al. 2023).

In another study, sensory analysis was conducted on 159 individuals to examine consumers' sensory perception, motivation, and reasons for purchasing alternative products such as plant-based meat, milk, and cheese available in the market. Panelists participating in the tests provided information about whether they were vegetarian, vegan, flexitarian, or omnivorous. The study suggests that animal welfare and environmental concerns are key reasons for purchasing plant-based products across all dietary patterns, as they are considered substitutes for meat and cheese (Pointke et al. 2022).

### *13.3.1 Sensory analysis of Plant-Based Meat Analogues*

Due to factors such as health concerns, animal welfare, and sustainability, plant-based meat alternatives became more commonly available in developed countries. The global market for plant-based meat alternatives is estimated to reach $15.6 billion by 2026, with an annual compound growth rate of 7.2% (Usman et al. 2023). There is potential in formulations as long as resources can produce more desirable tastes through compounds that mimic meat-like flavor and contain plant-based proteins. Despite advancements in the knowledge of compounds that can mimic meat-like flavors and contain plant-based proteins, modulation and calibration of sensory attributes for specific purposes and target consumers remain necessary during the production of plant-based meat analogues. Sensory evaluation results in plant-based meat analogues will help improve the overall quality of the product (Fu et al. 2022; Miller et al. 2024).

There are numerous studies in the literature on the sensory evaluation of meat analogues. In one study aiming to make plant-based meat analogues more appealing to consumers, the importance of focusing on achieving a texture similar to meat was highlighted. The study investigated the appearance, texture, hardness, chewiness, taste, color, and overall acceptability of five different meat analogues prepared with hemp seed protein concentrate, wheat gluten, and chickpea protein concentrate (Zahari et al. 2023). In a ranking test involving 42 panelists, fried meat analogues prepared by mixing various pea/wheat protein ratios were evaluated on a scale from 1 to 5, with 1 representing the highest preference and 5 representing the lowest. Considering the results they reported, no significant difference was found in the taste of each sample, and it was noted that the composite meat analog did not contain any unpleasant taste or odor as a result of changes in the plant-based protein ratios (Yuliarti et al. 2021). In another study, De Angelis et al. (2020) emphasized that dried pea protein yielded a softer product in comparison to both pea and soy protein isolates when they prepared extruded meat analogues using dried pea protein, pea protein isolate, soy protein isolate, and soy proteins in different ratios, and conducted a descriptive sensory analysis with seven different panelists who evaluated the appearance, aroma, taste, and textural properties of the meat analogues on a structured scale ranging from 0 to 10 (De Angelis et al. 2020). In a separate study, Gomez et al. (2019) investigated the effects of 24 different sous vide cooking processes on the sensory parameters of ready-to-eat products made from marinated beef and meat analogues. A consumer hedonic evaluation test was conducted by 73 consumers, who evaluated the odor and color parameters on a 7-point hedonic category scale ranging from "disliked extremely" to "liked extremely" in a blind condition. The study concluded by indicating the potential of meat-like products, stating that they possessed sensory properties similar to those derived from beef. In a study quantitatively evaluating the differences in texture between chicken and meat analogues, participants were asked to rate the fibrous structure, hardness, and chewiness of the products using a 9-point scale. The study reported that meat analogues had a higher fibrous structure and hardness score compared to grilled chicken breast (Chiang et al., 2019).

### 13.3.2 Sensory analysis of Plant-Based milk Analogues

Due to the increasing negative effects of animal-derived foods on human health, animal welfare, and the environment, consumer interest and research on plant-based milk analogues have been increasing in recent years. However, the presence of various bioactive phytochemicals, the emergence of plant-based proteins, the absence of cholesterol, the high energy input required for production, limited availability of milk in some regions, vegan diets, increasing numbers of people with lactose intolerance, and limited resources such as soil and feed are accelerating the development of alternative milk products. Nevertheless, plant-based milk analogues face many challenges in mimicking the organoleptic properties of dairy products due to their unwanted taste and textures (Kamath et al. 2022; Pua et al. 2022; Aydar 2023).

Plant-based milk analogues are in strict competition with traditional dairy products from large ruminants from the aspects of taste, functionality, nutrition, and cost, which can be challenging considering the compositional limitations of the plant matrix. In addition, most of these products are formulated and produced with significant proportions of fats, starches, and additives (flavorings, colorings, and stabilizers) that are considered less nutritious than equivalent dairy products, and the yield of milk analogues (cheese) produced is low. The functions of these plant-based analogues, such as their behavior under high temperatures, gel stability, or probiotic activity, may not meet consumer expectations, and they may need to be further restructured to meet the textural demands of analogues (Pua et al. 2022; Moss et al. 2023). Aini et al. (2020) added papain and pineapple extracts to cheese analogues produced using corn extract to improve their physicochemical and sensory properties. Despite the positive sensory characteristics of the cheese analog, the study indicated that its yield was low.

In a literature review, it was aimed to develop a probiotic soy-based product similar to Petit-suisse cheese and evaluate perspectives on potential benefits for consumer health, sensory acceptability, and instrumental texture during storage. Within the scope of that study, three different soy-based milk analogues were examined: milk-based petit-suisse (control), mixed with milk cream and soy, and soy milk-based. In the sensory evaluation of milk and soy-based products, there was a consensus among consumers as no significant difference was observed among those who tasted the same product. During the analyzed period, more than 64% of consumers gave scores above 6 for the cheese analog made with soy milk, which corresponds to an expression between "I liked it a bit" and "I liked it very much," suggesting that consumers did not remain indifferent to the product but liked it (Matias et al. 2014). In another study carried out by Kailaje et al. (2022), cornstarch was used at the concentration of 2% for lord and 1% for kulfi, produced with coconut and oat milk instead of cow's milk, to achieve the desired texture and stability as sensory evaluation in the production of milk analogues. As a result of their study, it was reported that coconut has the potential to be an alternative milk source in the production of improved nutritional benefits and acceptable milk product analogues. Mattice and Marangoni (2020) developed a plant-based cheese analog using corn at a ratio of 0–30% and compared the analogues with gluten and pea protein-based substitutes. Corn was preferred by consumers more because of its mouthfeel similar to cheddar due to its high stretchability and viscosity.

## 13.4 Conclusion

In recent years, the growing population and the changes in consumer lifestyles in developing countries led to increasing demand for plant-based food analogues as a result of the adverse effects of consuming animal-derived foods. However, evaluating the acceptance of food analogues derived from products such as peas, wheat, peanuts, chickpeas, soy protein, and corn among consumers involves the use of various sensory tests in many studies. This complicates the comparison of results from different studies. Therefore, standardized procedures should be developed to characterize the sensory properties of plant-based foods. It is thought that standardizing sensory evaluation results and utilizing new sensory techniques in the future will help improve the quality of plant-based food analogues.

## Refrences


Abakarov A (2011) A multi-criteria decision making approach for food engineering. 11th International Congress on Engineering and Food (ICEF 11)

Afshari A, Mojahed M, Yusuff R (2010) Simple additive weighting approach to personnel selection problem. Int J Innov Manag Technol 1:511–515

Aini N, Sustriawan B, Prihananto V, Sumarmono J, Ramadan RN, Romadhon D (2020) Formulation of low-fat cheese analogue from sweet corn extract using papain and lime extract as coagulant. Food Res 4(4):1071–1081. https://doi.org/10.26656/fr.2017.4(4).395

Aktaş R, Doğanay MM, Gökmen Y, Türen U (2015) Sayısal Karar Verme Yöntemleri. Beta Yayıncılık, İstanbul

Aydar YA (2023) An overview of plant-based food alternatives (PBFAs): classification, textural and sensory characteristics. In: Plant-based foods: ingredients, technology and health aspects. Springer, Cham, pp 9–14

Barallat Pérez C, Oliviero T, Fogliano V, Janssen H-G, Martins SIFS (2023) Flavour them up! Exploring the challenges of flavoured plant-based foods. Flavour Fragrance J 38(3):125–134. https://doi.org/10.1002/ffj.3734

Batista LF, Rocha F, Dos Santos Dias MM, Santos Pires AC, Vidigal MCTR (2023) Comfort plant-based food: what do consumers want? – a focus group study with different consumers group. Int J Gastron Food Sci 34(100):810. https://doi.org/10.1016/j.ijgfs.2023.100810

Çakır E, Can M (2019) Best-Worst Yöntemine Dayalı ARAS Yöntemi ile Dış Kaynak Kullanım Tercihinin Belirlenmesi: Turizm Sektöründe Bir Uygulama. Turizm Sektöründe Bir Uygulama. Atatürk Üniversitesi Sosyal Bilimler Enstitüsü Dergisi 23(3):1273–1300

Çakır S, Perçin S (2013) Çok Kriterli Karar Verme Teknikleriyle Lojistik Firmalarında Performans Ölçümü. Ege Akademik Bakış Dergisi 13(4):449–460

Çelik P, Ustasülayman T (2015) ELECTRE I ve PROMETHEE Yöntemleri ile GSM Operatörlerinin Hizmet Kalitesinin Değerlendirilmesi, Uluslararası İktisadi ve İdari İncelemeler Dergisi (April); 137–160.

Chiang JH, Loveday SM, Hardacre AK, Parker ME (2019) Effects of soy protein to wheat gluten ratio on the physicochemical properties of extruded mest properties of extruded mest analogues. Food Struct 19:100102

Chatterjee P, Chakraborty S (2012) Material selection using preferential ranking methods. Mater Des 35:384–393

De Angelis D, Kaleda A, Pasqualone A, Vaikma H, Tamm M, Tammik M-L, Squeno G, Summo C (2020) Physicochemical and sensorial evaluation of meat analogues produced from dry-fractionated pea and oat proteins. Foods 9(12):1754. https://doi.org/10.3390/foods9121754

Dogan M, Aslan D, Aktar T, Goksel Sarac M (2016) A methodology to evaluate the sensory properties of instant hot chocolate beverage with different fat contents: multi-criteria decision-making techniques approach. Eur Food Res Technol 242:953–966

Figueira J, Mousseau V, Roy B (2005) Electre methods multiple criteria decision analysis: state of the art surveys (s. 133–153). Springer New York, New York

Fishburn PC (1967) Additive utilities with incomplete product set: applications to priorities and assignments. MD ORSA Publication, Baltimore

Fu J, Sun C, Chang Y, Li S, Zhang Y, Fang Y (2022) Structure analysis and quality evaluation of plant-based meat analogues. J Texture Stud 54(3):383–393

Godschalk-Broers L, Sala G, Scholten E (2022) Meat analogues: relating structure to texture and sensory perception. Foods 11(15):2227. https://doi.org/10.3390/foods11152227

Gomez I, Ibanez FC, Beriani MJ (2019) Physicochemical and sensory properties of sous vide meat and meat analog products marinated and cooked at different temperature-time combinations. Int J Food Prop 22(1):1693–1708. https://doi.org/10.1080/10942912.2019.1666869

Güneş M, Umarusman N (2003) Bir karar destek aracı bulanık hedef programlama ve yerel yönetimlerde vergi optimizasyonu uygulaması. Rev Soc Econ 2:242–255

Gurmeric VE, Dogan M, Toker OS, Senyigit E, Ersoz NB (2013) Application of different multi-criteria decision techniques to determine optimum flavour of prebiotic pudding based on sensory analyses. Food Bioprocess Technol 6:2844–2859

Hadi F, Permana R, Syafwan H (2019) Decision support system in determining structural position mutations using simple additive weighting (SAW) method. Int Conf Comput Sci Eng 1339:012015

Han C, Wang G, Guo J, Wnag J, Yang X (2023) Oral oil release improves lubrication and sensory properties of meat Analogues with protein-stabilized oleogel. Food Hydrocolloids 142(108):788. https://doi.org/10.1016/j.foodhyd.2023.108788

Huang W-C, Chen C-H (2005) Using the ELECTRE II method to apply and analyze the differentation theory. Proc Eastern Asia Soc Transp Stud 5:2237–2249

Hwang CL, Yoon K (1981) Multiple attribute decision making: methods and applications. Springer, New York

Kailaje J, Sonawane SK, George PAA, Dabade A, Bhushette P (2022) Development of a dairy analog with combinations of non-dairy milk. Ind Biotechnol 18(5). https://doi.org/10.1089/ind.2022.0018

Kaklauskas A, Zavadskas EK, Saulius R (2005) Multivariant design and multiple criteria analysis of building refurbishments. Energy Build 37(4):361–372

Kamath R, Basak S, Gokhale J (2022) Recent trends in the development of healthy and functional cheese analogues-a review. LWT-Food Sci Technol 155(112):991

Karaatlı M, Ömürbek N, Budak I, Dağ O (2015) Çok Kriterli Karar Verme Yöntemleri ile yasanabilir illerin Sıralanması. Selçuk Üniversitesi Sosyal Bilimler Enstitüsü Dergisi, pp 215–228

Karaman S, Toker ÖS, Yüksel F, Çam M, Kayacier A, Dogan M (2014) Physicochemical, bioactive, ve sensory properties of persimmon-based ice cream: technique for order preference by similarity to ideal solution to determine optimum concentration. J. Dairy Sci 97:97–110

Linnemann AR, Hendrix EMT, Apaiah R, Van Boekel TAJS (2015) Food chain design using multi criteria decision making, an approach to complex design issues. NJAS Wageningen J Life Sci 72:13–21

Manokaran E, Subhashini S, Senthilvel S, Muruganandham R, Ravichandran K (2011) Application of multi criteria decision making tools ve validation with optimization technique-case study using TOPSIS, ANN & SAW. undefined

Matias NS, Bedani R, Castro IA, Saad SMI (2014) A probiotic soy-based innovative product as an alternative to petit-suisse cheese. LWT- Food Sci Technol 59(1):411–417. https://doi.org/10.1016/j.lwt.2014.05.054

Mattice KD, Marangoni AG (2020) Physical properties of plant-based cheese products produced with zein. Food Hydrocolloids 105(105):746. https://doi.org/10.1016/j.foodhyd.2020.105746

McClements DJ, Grossman L (2022) Physicochemical and sensory properties of plant-based foods. In: Next-generation plant-based foods design, production, and properties. Springer, Cham, pp 218–222

Miller O, Scarlett CJ, Adhikari B, Akanbi TO (2024) Are plant-based meat analogues fulfilling their potentials? An Australian perspective. Future Foods 9:100305. https://doi.org/10.1016/j.fufo.2024.100305

Moss R, LeBlanc J, Gorman M, Ritchie C, Duizer L, McSweeney MB (2023) A prospective review of the sensory properties of plant-based dairy and meat alternatives with a focus on texture. Foods 12(8):1709. https://doi.org/10.3390/foods12081709

Ömürbek N, Karaatlı M, Balcı HF (2016) Entropi Temelli MAUT ve SAW Yöntemleri İle Otomotiv Firmalarının Performans Değerlemesi. Dokuz Eylül Üniversitesi, İktisadi ve İdari Bilimler Fakültesi Dergisi 31(1):ss. 227–255

Önder G, Önder E (2018) Operasyonel, Yönetsel ve Stratejik Problemlerin Çözümünde Çok Kriterli Karar Verme Yöntemleri. Dora Yayıncılık, Bursa

Oprıovıc S, Tzeng GH (2004) Compromise solutions by MCDM methods: a comparative analysis of VIKOR and TOPSIS. Eur J Oper Res 156:445–455

Özbek A (2017) Çok Kriterli Karar Verme Yöntemleri ve Excel İle Problem Çözümü. Seçkin Yayıncılık, Ankara

Özdemir M (2018) Operasyonel, Yönetsel ve Stratejik Problemlerin Çözümünde Çok Kriterli Karar Verme Yöntemleri. Dora Yayıncılık, Bursa
Ozturk G, Dogan M, Said Toker O (2014) Physicochemical, functional ve sensory properties of mellorine enriched with different vegetable juices ve TOPSIS approach to determine optimum juice concentration. Food Biosci 7:45–55
Öztürkoğlu Y, Türker D (2013) Application of TOPSIS to Analyze Stakeholder Relations. Int J Bus Soc Res 3:245
Phua MH, Minowa M (2005) A GIS-based multi-criteria decision making approach to forest conservation planning at a landscape scale: a case study in the Kinabalu Area, Sabah, Malaysia. Landsc Urban Plan 71:207–222
Piggott JR, Simpson SJ, Williams SAR (1998) Sensory analysis. Int J Food Sci Technol 33:7–18
Pointke M, Ohlau M, Risius A, Pawelzik E (2022) Plant-based only: investigating consumers' sensory perception, motivation, and knowledge of different plant-based alternative products on the market. Foods. 11. https://doi.org/10.3390/foods11152339
Pua A, Tang VCY, Goh RMV, Sun J, Lassabliere B, Liu SQ (2022) Ingredients, processing, and fermentation: addressing the organoleptic boundaries of plant-based dairy analogues. Foods 11(6):875. https://doi.org/10.3390/foods11060875
Puertas R, Marti L, Garcia-Alvarez-coque JM (2020) Food supply without risk: Multicriteria analysis of institutional conditions of exporters. Int J Environ Res Public Health 17
Rezaei J (2015) Best-worst multi-criteria decision making method. Omega 53:49–57
Rezaei J (2020) A concentration raito for nonlinear best-worst method. Int J Inf Technol Decis Mak 19(3):891–907
Ruiz-Capillas C, Herero AM, Pintado T, Delgado-Pando G (2021) Sensory analysis and consumer research in new meat products development. Foods 10(2):429. https://doi.org/10.3390/foods10020429
Saaty TL (1990) How to make a decision: the analytic hierarchy process. Eur J Oper Res 48:s. 9–26
Saaty TL (2008) Relative measurement and its generalization in decision making why pairwise comparisons are central in mathematics for the measurement of intangible factors the analytic hierarchy/network process. Rev R Spanish Acad Sci Ser A Math (RACSAM) 102(2):264
Şahin S (2018) Operasyonel, Yönetsel ve Stratejik Problemlerin Çözümünde Çok Kriterli Karar Verme Yöntemleri. Dora Yayıncılık, Bursa
Shakouri HG, Nabaee M, Aliakbarisani S (2014) A quantitative discussion on the assessment of power supply technologies: DEA (data envelopment analysis) ve SAW (simple additive weighting) as complementary methods for the "Grammar". Energy
Short EC, Kinchla AJ, Nolden AA (2021) Plant-based cheeses: a systematic review of sensory evaluation studies and strategies to increase consumer acceptance. Foods 10(4):725. https://doi.org/10.3390/foods10040725
Sogari G, Caputo V, Petterson AJ, Mora C, Boukid F (2023) A sensory study on consumer valuation for plant-based meat alternatives: What is liked and disliked the most? Food Res Int 169(112):813. https://doi.org/10.1016/j.foodres.2023.112813
Tsai HH, Wu DH, Chiang TL, Chen HH (2009) Robust design of SAW gas sensors by Taguchi dynamic method. Sensors
Tzeng GH, Huang JJ (2011) Multiple attribute decision making: methods and applications. CRC Press,Taylor & Francis Group, A Chapman & Hall
Usman M, Swanson G, Chen B, Xu M (2023) Sensory profile of pulse-based high moisture meat analogues: a study on the complex effect of germination and extrusion processing. Food Chem 426(136):585. https://doi.org/10.1016/j.foodchem.2023.136585
Uzun S, Kazan H (2016) Çok Kriterli Karar Verme Yöntemlerinden AHP TOPSIS ve PROMETHEE Karşılaştırılması: Gemi İnşada Ana Makine Seçimi Uygulaması. J Transp Logist 1(1):99–103
Yayar R, Baykara HV (2012) TOPSIS Yöntemi ile Katılım Bankalarının Etkinliği ve Verimliliği Üzerine Bir Uygulama An Implementation upon Efficiency ve Productivity of Participation Banks with TOPSIS Method. Bus Econ Res J 3:21–42

Yeh CH (2003) The Selection of Multi-attribute Decision Making Methods for Scholarship Student Selection. Int J Select Assess 11(4):289–296

Yücesan M, Gül M (2021) Nötrosofik BWM ile Hata Türü ve Etkileri Anlizi", Bulanık Çok Kriterli Karar Verme Yöntemleri-MS Excel® ve Software Çözümlü Uygulamalar/Editörler Mehmet Kabak ve Bebek Erdebili. Nobel Yayınevi, 1.Basım Ankara

Yüksel İ, Akın A (2006) Analitik Hiyerarşi Proses Yöntemiyle İşletmelerde Strateji Belirleme. Doğuş Üniversitesi Dergisi 7(2):254–268

Yuliarti O, Kiat Kovis TJ, Yi NJ (2021) Structuring the meat analogue by using plant-based derived composites. J Food Eng 288(110):138. https://doi.org/10.1016/j.jfoodeng.2020.110138

Yurdakul M, Iç YT (2003) Türk otomotiv firmalarinin performans ölçümü ve analizine yönelik TOPSIS yöntemini kullanan bir örnek çalişma. J Fac Eng Archit Gazi Univ

Zahari I, Rinaldi S, Ahlstrom C, Östbring K, Rayner M, Purhagen J (2023) High moisture meat analogues from hemp—the effect of co-extrusion with wheat gluten and chickpea proteins on the textural properties and sensorial attributes. LWT 189(115):494

Zavadskas EK, Antuchevičienė J (2007) Multiple criteria evaluation of rural building's regeneration alternatives. Build Environ 42(1):436–451

Zavadskas EK, Arturas K, Turkish Z, Tamosaitiene J (2008) Contractor selection multi-attribute model applying COPRAS method with grey interval numbers. J Bus Econ Manage 9(3):181–187

# Chapter 14
# 3D Printing for the Production of Food Analogues

**Deniz Günal-Köroğlu, Sevil Cikrikci Erunsal, Gulay Ozkan, Busra Gultekin Subasi, and Merve Yavuz-Düzgün**

## 14.1 Introduction

The rising global human population is leading to an upward trend in food consumption, posing a potential adverse impact on animal populations and environmental sustainability (Chao et al. 2023). To overcome these problems, the replacement of animal sources with other alternatives comes to the fore. At this point, 3DP takes place among the innovative technologies in the food field, especially for producing analogs showing similar texture, composition, and sensory properties to the original ones. On the other side, 3DP is a very promising manufacturing type for the production of high-quality foods for children with an attractive appearance and nutritious content (Yu et al. 2023b).

3DP has the ability to manufacture shapes with complex geometry in a short time. While traditional food processing techniques require several unit operations like weighing, mixing, cutting, slicing, etc. to obtain material in a desired size and shape, these operations could be eliminated using printing techniques as a single step that comprises unit operations (Lipton et al. 2015). Apart from those

D. Günal-Köroğlu · G. Ozkan (✉)
Department of Food Engineering, Faculty of Chemical and Metallurgical Engineering, Istanbul Technical University, Istanbul, Turkey
e-mail: ozkangula@itu.edu.tr

S. C. Erunsal
Department of Food Engineering, Konya Food and Agriculture University, Konya, Turkey

B. Gultekin Subasi
Department of Food Science, Center for Innovative Food (CiFOOD), Aarhus University, Aarhus, Denmark

M. Yavuz-Düzgün
Department of Gastronomy and Culinary Arts, Faculty of Applied Sciences, Altinbas University, Istanbul, Turkey



Ö. P. CAN et al. (eds.), *Food Analogues: Emerging Methods and Challenges*,
https://doi.org/10.1007/978-3-031-69872-9_14

advantages of emerging printed foods, there are also some drawbacks or limitations that should be taken into consideration and improved during printing processing. Some of them could be listed as required skilled personnel, additive manufacturing and digital gastronomy techniques, and high cost (Sun et al. 2018).

The purpose of the current chapter is to present a critical assessment in terms of 3DP in food science and technology. In this regard, firstly, techniques to perform 3DP processing as well as printable materials and controlling parameters for printed foods will be described deeply. Furthermore, applications of food printing have been covered by emphasizing the composition and optimum formulation of food products, and the consequences of 3DP on the formulations.

## 14.2 3D Printing Techniques

The overall methods to conduct 3DP processing technology are described within seven main categories, which are vat photopolymerization, direct energy deposition and material extrusion (Dassaults Systeme), sheet lamination, material and binder jetting, and powder bed fusion (ISO/ASTM 52900:2021 2021) (Fig. 14.1). Among them, the first three applications are not suitable to employ for the food-based

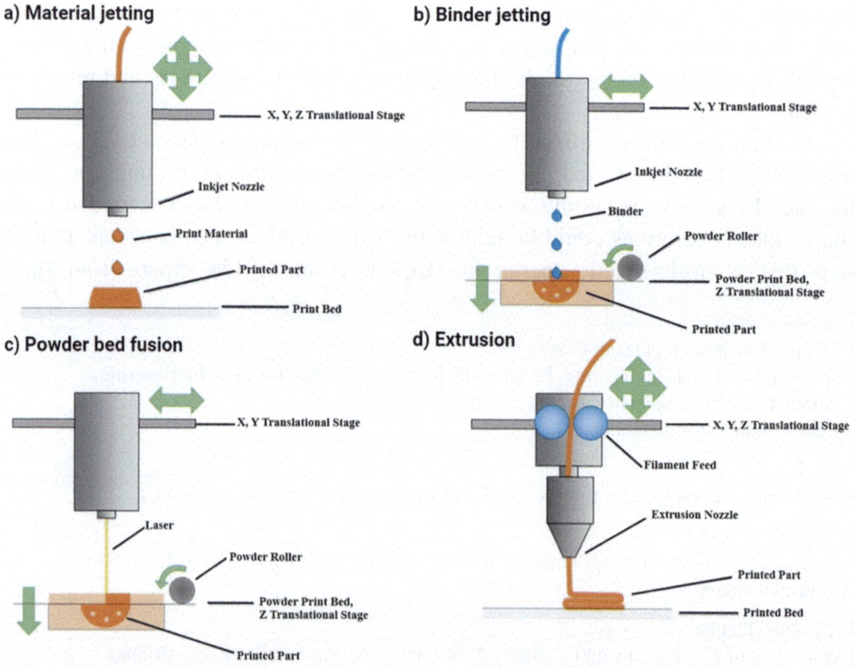


**Fig. 14.1** Methods for 3DP compatible for food ingredients; (**a**) Material jet (inkjet) printing, (**b**) Binder jetting printing, (**c**) Powder bed fusion printing, (**d**) Extrusion-based printing. (Reproduced from Kjar and Huang 2019)

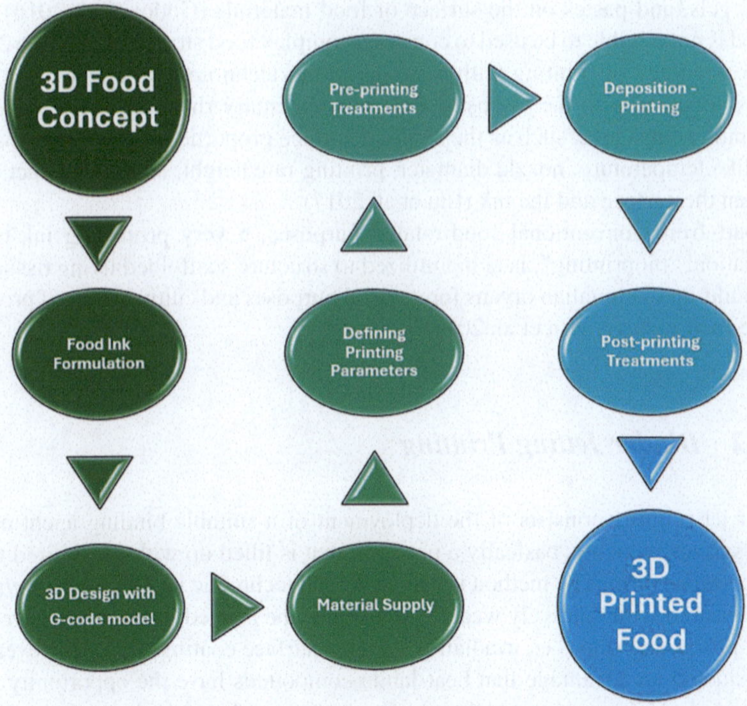


**Fig. 14.2** Schematic flow diagram for 3DP process of food materials

biomatrices, however, the last four applications allow the use of food-grade materials to be processed by the 3DP approach (Enfield et al. 2023). A common flow chart, which is proposed for the mentioned food 3DP approaches, is provided in Fig. 14.2. The discrimination among the methods that are eligible/non-eligible for food applications is mainly due to the incompatibility of the processing material with the proposed technological design and operational cost. In this subsection, the principles of commonly used 3DP food processing methods are discussed briefly.

### *14.2.1 Material Jet (Inkjet) Printing*

Photopolymer-containing droplets are dispersed throughout a thermal or piezoelectric head towards the target surface, like 2D inkjet printing, due to the common application feature of either continuous or drop-on-demand printing (Kruth et al. 2007). The method enables high accuracy and resolution with a printing capability of less than 20 μm and allows the printing of multiple compounds at a time. However, the method is suitable only for low-viscosity materials which is the main limiting factor collecting a 3D final structure. Due to this reason, this method is more suitable for printing 2D images, coatings, and edible films such as using chocolate, icings, gels, and pastes on the surface of food materials (Godoi et al. 2016). This method is not eligible to be used to construct complex food structures (Baiano 2022).

The accuracy of printing with material jetting technology is affected by many factors, together with the viscosity and complementary rheological parameters of the printing substance, such as the exposed surface properties and technical parameters like temperature, nozzle diameter, printing rate/height, and the contact angle between the surface and the ink (Liu et al. 2017).

Apart from conventional food-related purposes, a very promising ink-jetting application, "bioprinting", is being utilized to structure scaffolded living tissues for both build-up mammalian organs for medical purposes and cultivated meat products for the food industry (Jin et al. 2021).

### *14.2.2 Binder Jetting Printing*

Binder jet printing consists of the deployment of a suitable binding agent onto a target surface, which is basically a pool/bed that is filled up with the desired material in powder form. The method is a non-thermal technique in which binding/printing is ensured with relatively weaker bonds, thus the printed material might require a post-processing step (i.e., irradiation, baking, surface coating, etc.). However, this is considered an advantage that heat-labile compounds have the opportunity to be 3D printed while protecting their native states under non-thermal conditions (Miyanaji et al. 2018). For food systems, powder formed materials such as sugar, starch, and flour are eligible for binder jet printing using edible liquid binding agents like sugar and/or gum-based solutions (Holland et al. 2018).

The properties of the powdered material, the binding agent, and operational parameters are of extreme importance in order to gain a 3D structure following the printing procedure. Particle size, flowability, bulk density, and wettability without clumping and caking are the critical factors for powdered material while viscosity, surface tension, and density of binding agent are encountered as the main parameters for an accomplished and successfully structured 3DP final product. On the other hand, the type of head, printing velocity, droplet path, nozzle diameter, layer thickness, and resonance frequency are the critical independent variables for processing that need to be optimized for the customized end product (Enfield et al. 2023).

Advantageously, the printing of complex structures with a high printing speed is possible with binder jet printing; however, dryness and the rough surface appearance of the product might be the drawbacks of this method (Baiano 2022).

### *14.2.3 Powder Bed Fusion Printing*

The generated energy through a laser, an electron beam, or hot air is employed to melt the target powdered material following the fusion with powder bed fusion printing. Constructing complex matrices with high accuracy is possible with top-level customizability. However, it is a high-cost operation requiring significant time and labor due to the necessity for some pre- and post-operational conditions such as preheating, vacuum generation, and cooling periods. Food materials that are eligible to be powdered, like sugar, fat, and starch granules, might be used for powder bed fusion printing; however, the commercial usage for the food industry is still limited (Enfield et al. 2023; Liu et al. 2017). The printing can be applied with different methods such as electron beam melting, selective laser melting, and selective heat sintering, where the last one is more suitable for printing food-based biomasses (Baiano 2022).

The main factors that affect the printing process in terms of accuracy consist of material characteristics such as size of particles, melting point, glass transition temperature, bulk density, wettability, and flowability properties of the powdered ink food; operational conditions such as laser type, power, energy, spot diameter, thickness, and scanning speed; and post-processing treatments such as removing the excessive material from the printed structure (Godoi et al. 2016; Liu et al. 2017).

### 14.2.4 Extrusion-Based Printing

Material extrusion printing is the most preffered printing technology for food applications due to its allowance to print a variety of soft-structured edible materials as ink for different purposes. A wide selection of materials—to the contrary of other printing techniques—has been recently investigated for their potential to create 3D structures such as hydrogels, emulsions, cheeses, chocolate, various animal and plant-based proteins, fruit and vegetable purees, sugar frostings, starches, and gelatin/pectin (Ainis et al. 2023; Calton et al. 2023; Tay et al. 2023; Yu et al. 2023a). Despite the abundance and accessibility of this technique for many different food materials, it is possible only to print simple structures, performance yield is not feasible to create complex geometries (Keerthana et al. 2020).

A material extrusion 3D printer typically uses thermal energy to fuse the nozzle-extruded melted semi-solid thermoplastic material toward a suitable platform to construct the targeted geometry layer-by-layer, one to the top of another (Enfield et al. 2023). The fused layer is supposed to solidify following the extrusion and allow the next layer to attach. It is possible to conduct the extrusion-based printing with three extrusion mechanisms, which are screw-based, air pressure-based, and syringe-based extrusions (Liu et al. 2017). Obtaining an accomplished geometry with high precision, the properties of the printing material, and operational parameters are of great importance. Rheological, thermal, and mechanical properties of the material should be optimized and determined prior to the printing process. Operational conditions like extrusion temperature, printing height, nozzle diameter, moving rates of the nozzle, and printer, as well as optimized layer and printing heights, are critical parameters to create a high-quality printed 3D food structure (Pérez et al. 2019).

## 14.3 Printable Materials and Recipe Optimization

In this part of the chapter, commonly used ingredients for food printing, their functions, and factors affecting final product quality will be explained to provide insight for further studies. The literature review will help summarize past printing experiences, giving some detailed examples of remark points.

Extrusion-based printing is one of the most used types in food printing thanks to its safety, user-friendly mechanism, and portable design. Since the success of printability becomes the first criterion, it should be well-defined. "Printability (Pr)" represents the easy extrusion ability of the ink and the required stability (sustaining predesigned shape) of the sample to support its own weight after printing (Portanguen et al. 2019). It could be quantitatively expressed by dividing the actual area by the theoretical area (Emir and Erunsal 2022). If the Pr value is very close to 1, this will be a sign of success. Printability could also be assessed by means of shape fidelity, and it was similarly expressed according to the projection areas of the control frame and printed sample for lattice structures (Godoi et al. 2019).

The relation is given below as:

$$F(\%) = 100 \times (Ac - As) / Ac$$

$$Circularity = C = \frac{4\pi A}{L^2}$$

$$\mathrm{Pr} = \frac{\pi}{4}\frac{1}{C} = \frac{L^2}{16A}$$

F: The deviation from the expected shape
$A_C$: the projection areas (in pixels) of the control frame
$A_S$: the projection areas (in pixels) of the 3D-printed sample
L: perimeter
Pr: Printability
A: area of lattice

During test measurements for printability, the geometry given in Fig. 14.3 can be used as a predesigned object to print. *D* indicates nozzle diameter, so printed filament is expected to have the same width theoretically (it might be approximately 0.8xnozzle diameter in actual experiments) and the distance between two filaments can be planned to be the same value (*D*). Then, printability success can be examined after the deposition stage compared to theoretical expectations.

Lattice structures and their representation were given as an example due to their significant advantages for soft materials like hydrogels. They facilitate the assessment of printability by examining factors such as overlapping, circularity levels within square areas, and adherence to the intended geometry during the printing process.

Achieving a favorable result requires the careful optimization of both the ingredients used and the process conditions employed. Figure 14.4 displays some examples of the factors influencing printability, and each variable will be emphasized in

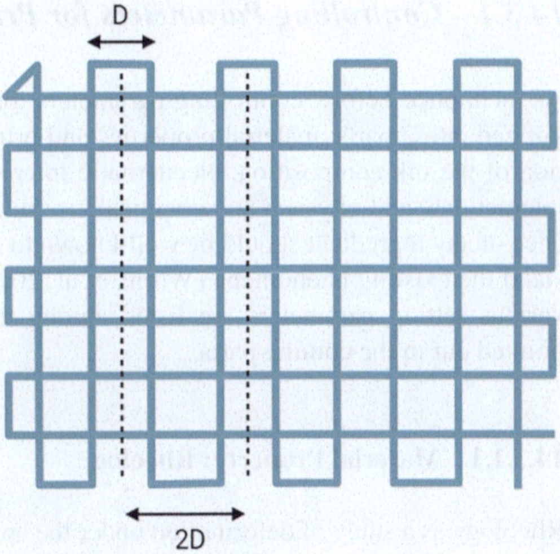


**Fig. 14.3** Representative lattice structure. (This figure has been created using the Biorender program "https://www.biorender.com/")

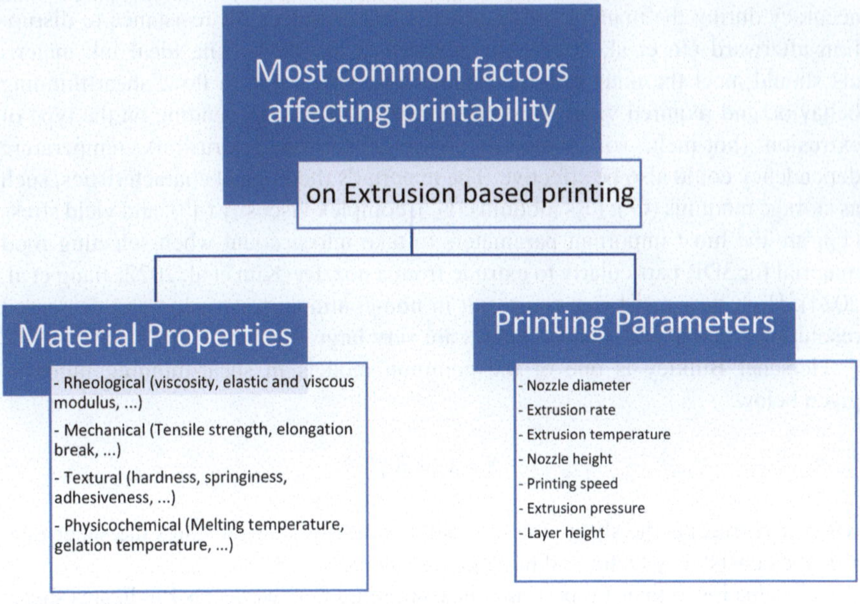


**Fig. 14.4** Some of the important factors for printability

the following parts. First, food ink should be flowable under pressure or stress, while maintaining stability to preserve the intended geometric shape after deposition (Pulatsu and Lin 2021). Hence, the rheological properties of the ink become one of the dominant factors for successful food printing. At this point, it will be beneficial to start an explanation from the rheology and the required flow behavior type for 3DP. Then, other factors will then be explored to understand food printing criteria.

### *14.3.1 Controlling Parameters for Printed Foods*

As mentioned before, controlling parameters for successful printing could be categorized into 2 parts: material properties and printing parameters. During the selection of the ink composition, electrostatic interactions, van der Walls forces, other interactions, and physical and chemical crosslinks that will be formed in the presence of any ingredient should be well-known to estimate printing results and understand the existing phenomena (Wilms et al. 2021). Depending on the printer type, various setting parameters are listed among the affecting factors. They will be pointed out in the coming parts.

#### 14.3.1.1 Material Property: Rheology

Rheology as a study of deformation under the applied stress/force is one of the critical aspects for printing performance, especially for extrusion-based ones. Eventually, a smooth material flow through the nozzle ensures more uniformity and printing accuracy during the manufacturing process and enhances the resistance to disruption afterward (Jo et al. 2021; Ahmadzadeh et al. 2023). The ideal ink materials should meet the main criteria: enough viscosity value to flow, shear-thinning behavior, and required yield stress (Boehm et al. 2023). Depending on the type of extrusion (hot-melt, soft materials, hydrogel-forming extrusion), temperature dependency could also be effective. The material's rheological characteristics, such as storage modulus (G′), loss modulus (G″), complex viscosity ($\eta^*$), and yield stress ($\tau_0$), are the most important parameters to take into account when selecting food material for 3DP, particularly to extrude from a nozzle. (Kim et al. 2022; Jiang et al. 2024). High flow resistance can result in non-continuous extrusion and decreased resolution when $\eta^*$, G', and G" values are very large (Kim et al. 2022).

Herschel Bulkley is one of the common models in shear-thinning fluids as given below:

$$\tau = \tau_y + K\gamma^n$$

where $\tau$ represents the shear stress, $\tau_y$ is the yield stress. $\gamma$ indicates the shear rate, K is the consistency factor and n is the flow index.

"n" value lower than 1 represents shear-thinning flow while $n > 1$ indicates shear-thickening fluid. Yield stress should also be checked for food deposition accuracy and shape retention. Other than viscosity and yield stress, storage (G′) and loss modulus (G″) values are crucial for printability classification, basically for buildability. The elastic modulus, or storage modulus (G′), is a measure of a material's capacity to resist structural deformation. Elastic modulus is essential because it shows mechanical strength and shape retention capacity. The viscous modulus, which reflects extrudability via a nozzle, is represented by the loss modulus (G″) (Kim et al. 2022; Letras et al. 2022; Wang et al. 2023; Jiang et al. 2024). Low G′ food inks were found to have poor printing stability and a propensity to spread into underlayers (Lim et al. 2023). However, G″, therefore, is the stress-to-strain ratio in vibratory circumstances (Wilson et al. 2020).

The angular frequency or dynamic mechanical loss tangents depends on the loss factor ($\tan \delta = G''/G'$) and a higher or lower value of $\tan(\delta)$ than 1 may also help to understand the viscoelastic property of the food ink (Pérez et al. 2019). Nevertheless, a $\tan \delta$ value higher than 1 indicates viscous behavior. Whereas lower $\tan \delta$ ($\tan \delta < 1$) indicated solid behavior (elastic), greater $\tan \delta$ indicated liquid behavior (viscous, more flowable) (Wilson et al. 2020; Bulut and Candoğan 2022; Huang et al. 2022; Niu et al. 2023a, b; Jiang et al. 2024). So, if G′ is greater than G″ ($G' \geq G''$), and both G′ and G″ increase with angular frequency, it suggests that the material behaved more like a solid than a liquid (Wilson et al. 2020; Jo et al. 2021; Liu et al. 2021b; Wang et al. 2023; Jiang et al. 2024). The molecular interactions between the components of food ink are mostly responsible for the gel structure formation (complex network) between macromolecules (for example between protein and starch) (Wilson et al. 2020; Cai et al. 2022). Otherwise, the network deformation of the food ink is the cause of the noticeable and monotonic decline in G′ values. When the G′ value drops below the G″ value ($G' \leq G''$) at high enough stresses, the natural structure of the food ink is harmed. Thereby, the flow is encouraged by the disturbance of the network structure, and the food ink behaves like a liquid, predominantly viscous property. When $G' \approx G''$, the food ink behaves like a semi-solid, like gel (Cai et al. 2022; Wang et al. 2023; Qiu et al. 2023a).

Extrusion-based printing is based on the continuous flow of the ink in a layer-by-layer manner. Basically, the ink flows through the nozzle upon the applied pressure gradient $\Delta P$ along with the length (l). Thus, shear stress ($\tau_r$) becomes the function of radial position, pressure, and length as given below (Godoi et al. 2019):

$$Tr = \frac{r\Delta P}{2l}$$

Assuming laminar flow, negligible end effects, and incompressible fluid, no slip condition (zero velocity) exists at the nozzle wall center ($r = R$) and velocity in the center ($r = 0$) becomes maximum. Then, a macroscopic force balance can be evaluated for the flowing ink. This expression might also be helpful to describe shear rate and shear stress to estimate characteristic flow properties.

Extrusion-based printing can work under three different printing mechanisms: pneumatic (air pressure driven), piston (mechanical force driven), and screw (thermal extrusion rotation). While screw-based extrusion is not recommended for high-viscous materials, a syringe-based extrusion unit is seen as suitable for pastes with high mechanical strength and viscosity. Pneumatic extrusion, alternatively, seems suitable for low viscous materials (Godoi et al. 2019).

Wilms et al. (2021) stated that product rheology is the function of so many variables such as sample particle size, process temperature, time, and shear. It was demonstrated in the aforementioned study that increasing particle size or decreasing particle volume fraction diminished the consistency factor and yield stress of ink (Wilms et al. 2021). Similar behavior was evaluated for temperature, as well. Secondly, extrusion pressure was described as a function of food rheology and processing parameters. In another study performed by Zhu et al. (2019), a linear correlation was observed between the ingredient's zero shear viscosity, low stress (positive relation with extrusion pressure), and corresponding printing stability in the tomato-based product. Importantly, they indicated altered printing behaviors on fat-based products while reasonable correlation of tomato paste-based ink with aqueous-based products (Zhu et al. 2019).

That's why, it could be recommended not to think of any parameter independent from the other parameters. Therefore, the recipe and process could be manipulated by the alteration of so many variables. Shear thinning behavior with time dependency (thixotropic) may be sometimes considered. For detailed information about other cases, empirical, and semi-empirical equations, the review of Wilms et al. (2021) could be checked.

#### 14.3.1.2 Other Material Properties

Thermal properties of the materials play a significant role, especially in melting-extrusion-based systems that apply temperature within the cartridge. Glass transition temperature and melting point are crucial to keeping a stable structure after deposition (Godoi et al. 2019). This concept was demonstrated under the section of "printer parameters" in detail.

Texture is one of the important quality features of printed foods. It does not only influence printability but also has an impact on final product quality and consumer acceptance. Since texture differentiates between conventional and printed products, oral sensory perception, surface characteristics, and physical attributes influenced by texture also give distinct results on differently processed foods (Waseem et al. 2023). Texture Profile Analysis (TPA) is mostly performed by the texture analyzer to obtain hardness (maximum force of the first compression), springiness (distance ratio of material going back to its first condition), adhesiveness (obtained from the second area in the graph), cohesiveness (the work required for the deformation of the gel in the down movement of the probe), gumminess (hardness multiplied by cohesiveness), and chewiness (gumminess multiplied by springiness) values (Godoi et al. 2019).

Mechanical properties of printed constructs are another key parameter for ideal printing conditions. The tensile strength (the maximum force that a sample can bear upon pulling) and elongation at break (change of sample length) are the main mechanical parameters tested for printed samples (Wen et al. 2022). Moreover, some of the rheological parameters are related to this concept, such as elastic modulus and yield stress. As illustrated in Fig. 14.3, yield stress is the stress that should be exceeded for the flow initiation of the ink. Tensile strength is defined as the maximum stress that a material can endure. Lastly, elastic modulus can be found from the linear portion of the stress–strain curve as displayed in Fig. 14.5. Depending on the material (semi-solid, soft, or hard material), the correct measurement type and experiment parameters can be determined.

Fourier transform infrared (FTIR) spectroscopy and scanning electron microscopy (SEM) are also mostly performed to comprehend textural changes in printed

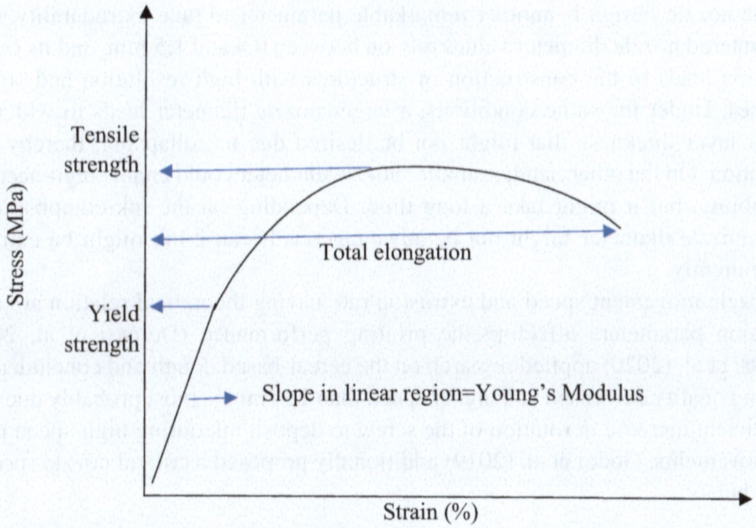


**Fig. 14.5** Representative mechanical properties

products even after post-processing (Wen et al. 2022). As an example, Wen et al. (2022) analyzed the effect of xylose (carbohydrate-based additive) on the texture and microstructure of 3D-printable mung bean protein by texture analyzer, SEM, and FTIR, respectively. Moisture content and water activity are among other material properties that might be set and controlled before and after deposition (Waseem et al. 2023).

#### 14.3.1.3 Printer Parameters

In addition to material characteristics, parameters in the process design play a crucial role in identifying opportunities for printing optimization. Therefore, both ink material and the printing process should be concurrently taken into account during the optimization stage (Wilms et al. 2021).

The thermal properties of the ink were mentioned in the previous section. They have remarkable material properties but are highly dependent on printing conditions. Chocolate is a perfect example at this point, having unique melting characteristics due to the presence of cocoa butter in the composition. To achieve the desired printing performance, process temperature becomes the dominant factor and is altered with respect to chocolate composition. The melting temperature of white chocolate is higher than that of dark chocolate, as milk chocolate has the lowest melting point due to sugar and cocoa butter concentrations (Godoi et al. 2019). As the cartridge temperature is set before the printing, this information should be considered for the perfect tempering and extruding process. For the fixed product, an extrusion temperature (inside the cartridge) increase might give an opportunity to reduce apparent viscosity and extrusion pressure.

The nozzle design is another remarkable parameter to tune extrudability. Most encountered nozzle diameter values rely on between 0.4 and 1.5 mm, and its correct selection leads to the construction of structures with high resolution and smooth surfaces. Under the same conditions, a larger nozzle diameter leads to wider and higher layer thickness that might not be desired due to collapsing, thereby poor resolution. On the other hand, a smaller nozzle diameter could enable high-accuracy printability, but it might take a long time. Depending on the ink composition, a small nozzle diameter might not be advantageous because ink might be extruded intermittently.

Nozzle movement speed and extrusion rate having theoretical relation are other extrusion parameters affecting the printing performance (Derossi et al. 2020). Derossi et al. (2020) applied research on the cereal-based dough and concluded that printing quality decreased at a higher speed than 100 mm/s, most probably due to an insufficient increase in rotation of the screw to deposit ink during high-speed printing movements. Godoi et al. (2019) additionally proposed a critical nozzle speed as given below:

$$V_N = \frac{4Q}{\pi D_N^2}$$

where $V_N$ represents the optimal nozzle speed, Q is the ink flow rate and $D_N$ is the nozzle diameter.

#### 14.3.1.4 Post-processing

As some of the food products (chocolate, etc.) do not require post-processing applications, most of them might need a treatment after the solidification stage. Demei et al. (2022) listed drying, frying, baking, cooling, and sterilization as post-process treatments to improve product characteristics. Stable printed products after post-processing can be perfectly obtained by controlling the chemical, physical, structural, rheological, and mechanical properties of the materials (Godoi et al. 2019).

Food materials with high fiber and protein content are among the products exposed to the combination of printing and drying (Dankar et al. 2018). Different drying types (oven drying, freeze-drying, microwave, etc.) also become very effective in modifying solid matter content and final product properties. As oven drying might lead to some spreadability by diminishing the material viscosity upon heating, freeze drying has the potential to enhance dry matter content and hardness, making it a preferable option for certain products (Lille et al. 2018). Frying is another method applied prior to the consumer presentation of the printed products. Potato, rice starch, and chicken-based samples are among such kinds of foods to adjust their internal structure and sensory perception (Demei et al. 2022). Baking is an inevitable process after printing bakery products such as cookies. It provides shape retention, improvement of rheology, texture, and microstructure as obtained in conventionally produced baked goods (Pulatsu and Lin 2021).

Sterilization is another applied post-treatment for food safety, providing microbial contamination control. There is a direct relationship between water activity and microbial growth (water activity higher than 0.85 tends to microbial growth) affecting food shelf life (Demei et al. 2022). It was reported in a study that water activity showed a slight increase after the printing of fiber-enriched snacks, most probably owing to the entanglements of water in the fiber network (Keerthana et al. 2020). For much more details about the effect of post-processing on final product quality, the review prepared by Demei et al. (2022) is suggested to read.

### *14.3.2 Printable Materials/Recipe Optimization*

Recipe optimization is necessary for 3DP of foods, particularly for individuals with deficiencies or health issues, owing to various reasons. Macromolecules and other additives might be considered the main components of the recipe. Total solid concentration can vary from 5% to 50%, according to the material capability on paste consistency or gel-forming (Godoi et al. 2019). In the following part, each macromolecule will be explained in terms of their function and roles as well as aiding agents.

#### 14.3.2.1 Carbohydrates

Carbohydrates are crucial macronutrients having a high ratio in food composition. Sugar is one of the attractive ingredients, especially in 3D confectionery products. Specifically, Selective Laser Sintering (suitable for powders) is highly used in sugar powder printing with/without cocoa powder (Soni et al. 2022). Starch could be listed among typical carbohydrates for the smooth extrusion process. Low cost, gelation, and thickening ability, and high performance make starch a very good candidate for extrusion-based 3DP recipes. It enables smooth and strong features during printing according to flow and mechanical strength (Hai Alami et al. 2024). On the other side, starch has some drawbacks, such as lower resistance to heat, shear, and pressure, and aging problems during storage (Yu et al. 2023b). A denser gel matrix is formed during the extrusion owing to the increase in the level of starch molecules per volume (Demei et al. 2022). Origin of the starch type, molecular structure, amylopectin/amylose ratio, gelation temperature, concentration, etc. also have a huge impact on starch gelation and so the final quality. Corn starch suspension could be given as an extraordinary example of having shear thickening behavior under specific conditions opposite to regular types giving desired shear thinning properties. Cereal starch (Chen et al. 2022), potato starch (Yang et al. 2018), and wheat starch (Le-Bail et al. 2020) are some types of starches examined in food printing applications. Yang et al. (2018) fabricated lemon juice gels with added potato starch at 15 g/100 g for the best cylinder object manufacturing (Yang et al. 2018). Hardness and gumminess increased at higher amounts of starch because of more compact network formation. For instance, whole meal incorporation into wheat flour improved the printing process as compared to single wheat flour formulation (Zheng et al. 2021). Chen et al. (2022) reported the requirement of additional ingredients to enhance rheology. It was noted that an increase in starch concentration helps protect the organized structure of printed samples from damage caused by heat and shear forces and thereby allows for better control of digestion resistance during hot extrusion 3DP (Zhang et al. 2022). On the other hand, it might clog the nozzle, so there should be some balance in the recipe. Previous studies showed that the addition of other macromolecules such as protein, fat, and also other components like fiber, and biopolymers into starch-based inks had a positive effect on printability (Demei et al. 2022; Johansson et al. 2022; Shi et al. 2021). Bitencourt et al. (2023) worked on mineral bioaccessibility in printed and molded gels formulated with starch, modified starch by heat treatment, and κ-carrageenan in the presence of milk powder for dysphagic people. It was observed that κ-carrageenan had a higher influence on printing performance and texture than modified starch (Bitencourt et al. 2023).

Fruits, vegetables, grains, and legumes could be used as carbohydrate sources depending on goals like high/low fiber, animal source-free food, etc. Since fruit and vegetables are not suitable for printing, their extrudable versions are achieved by grinding/mashing and mixing them with flour, hydrocolloids, starch-based, or other types of additives, resulting in the enhancement of their flow behavior (Tomašević et al. 2021). Broccoli, carrots, bananas, and peas are some of the typical examples to be used in vegetable-based formulations with the addition of varying gums as texture modifiers (Yu et al. 2023b). As avocados are among high lipid sources, bananas can be seen as calcium sources. Pant et al. (2021) aimed to evaluate the printing of fresh vegetables through the use of food hydrocolloids for dysphagic patients (Pant et al. 2021). In this study, garden peas exhibited printing-friendly qualities, producing a thick and viscous consistency with high yield stress enhancing mechanical strength in printed structures. Overall, the varying starch and water content in different vegetables alters the addition of hydrocolloids, offering insights into the texture, taste, and aroma of the resulting printed forms. In essence, this underscores the distinctive printing attributes of each ingredient and how vegetable composition contributes to the sensory characteristics of the printed outcome. Fruit and vegetables not only contain carbohydrates but also different kinds of vitamins and minerals necessary for human nutrition. 3DP might serve a new service for people having difficulties in chewing and swallowing such kind of nutritious but stiff-structured foods.

Portanguen et al. (2019) reported global food loss and upcycling by-products by additive manufacturing to reduce food loss. To convert food waste or related materials into high-value-added foods, so many attempts have been made with the help of new manufacturing processes. Ahmadzadeh et al. (2023) freeze-dried imperfect broccoli and carrots after the grinding step and then converted them into ink materials with the help of wheat flour and olive oil at different ratios (Ahmadzadeh et al. 2023).

Plant-based analogs have proven to be suitable for printing applications and are actively utilized in this field owing to their structure–function relationships, unique chemical and physical features of polysaccharides and proteins giving desired rheological and printing behavior (Boehm et al. 2023). Hydrogels, 3D networks absorbing large amounts of water within their matrix, are another suitable product to be printed for personalized nutrition. They can act as a carrier system for target bioactive agents, drugs, and any compounds by different encapsulation procedures, and then they allow their controlled release under specific environmental conditions (Emir and Erunsal 2022). Different kinds of starches, gums, and polymers become the main material of gel formation. The impact of varying macronutrients on printability was reviewed by Pérez et al. (2019) so it can be recommended for further details.

#### 14.3.2.2 Proteins

Protein is another biological macromolecule composed of amino acids as basic units and important for life activities (Yu et al. 2023b). Animal or plant protein is another ink ingredient meeting the criteria for consumers in terms of nutrition, texture, and tissue framework. Milk proteins are perfect ingredients for 3DP having structure-forming capability upon gelation and then creating 3D objects by this technology. Nevertheless, their gelation mechanism might change. For instance, casein and whey are both milk proteins, but casein gels could be either obtained by selective hydrolysis of κ-casein from the surface of the casein micelles by enzyme rennet or acidification to its isoelectric pH (~4.6), or a combination of both cases (Uribe-Alvarez et al. 2021). On the other hand, whey gels could be produced by heat-set gelation, applying heat to an aqueous whey protein solution for a determined period of time and then exposing it to a cooling process (Ozel et al. 2017). Depending on their mechanisms, the printing process should be supported by heat or any post-processing steps.

Collagen and gelatin take place among the gelling agents besides their protein structure, giving desired chewiness, hardness, and viscoelastic features to the sample even at very low concentrations (ex: 1% w/v) (Demei et al. 2022). Another example might be given as milk protein concentrate, in which these particles have the function to create a hard network after aggregation (Demei et al. 2022). To increase the fluidity of the ink, whey protein or egg yolk could be utilized in recipe optimization (Anukiruthika et al. 2020). Similar to carbohydrates, some proteins might not give satisfied printing results and protein formulations could be improved by additive incorporations such as gluconolactone (Weng and Zheng 2015), carrageenan (Udo et al. 2023) or other kinds of gums.

Although animal-based proteins such as whey protein, insect protein, and egg albumin are the main sources, plant-based proteins might be preferred by specific consumers. They are good alternatives including soy protein, pea protein, etc., to mimic animal-origin-based products. Kim et al. (2021c) examined the effect of pea protein on banana blends (Kim et al. 2021c). They found that higher concentrations of protein-based raw materials improved printability, reduced the tailing effect, and kept shape retention after printing up to a level. However, it will start to affect it negatively if the limit is exceeded (Hai Alami et al. 2024).

Proteins are the well-known composition of meat, so vegan-based proteins are proposed for the production of meat analogs to replace animal origins. At that stage, muscle cells could be added into the formulation for muscle fiber production (Chao et al. 2023). Microalgae proteins (ex: *Spirulina platensis*) have begun to be so popular due to their potential as an ink material, providing a viscoelastic oil-water interface for emulsions (Guo et al. 2023). Recently, edible insects have taken great attention as a replacement for other protein sources and 3DP creates a perfect platform for their use in functional foods. Currently, edible mushrooms are gaining popularity as a substitute for animal protein, and advances in 3DP technology suggest the potential to create meat analogs from edible mushroom protein that closely mimic the taste and flavor of animal meat (Wang and Zhao 2023). This development is expected to enhance the nutritional value of edible mushroom-based foods. As people's income and consumption levels rise, and dietary preferences become more varied, it is foreseeable that edible mushrooms in meat alternatives and the broader food industry will experience promising growth prospects (Wang and Zhao 2023). Lastly, upcycling protein-rich animal by-products such as tendons, and bones by additive manufacturing can be emphasized to pay attention to the importance of sustainability concerns. 3DP serves as a perfect platform to use them in the fabrication of new products with specific characteristics (Portanguen et al. 2019). Table 14.1 summarized some of the research content and their results on meat and fish analog-related studies.

#### 14.3.2.3 Fats

The presence of fat content in the ink facilitates extrusion through the nozzle thanks to the easiness of the flow at higher fat concentrations. When considering the food industry, numerous categories prominently feature fats, including chocolate, biscuits, and dairy products. It is essential to note that each product may contain distinct fats and lipids with unique melting properties, triglyceride compositions, and other physical characteristics, leading to variations in printability attributes.

Several studies have explored the impact of different types of fats added to ink materials, investigating their influence on printability, the physicochemical properties of the final printed food, and nutritional content. For instance, Anukiruthika et al. (2020) compared the printability of egg yolk and egg white powder with rice flour blend on an extrusion printer (Anukiruthika et al. 2020). They revealed that egg yolk with higher fat and solid content, and lower protein content with a dense network of lipoprotein fractions gave better printing findings. In another study conducted by Lille et al. (2020), printed dietary fiber and protein-enriched snacks made of wholegrain rye flour and milk powder were exposed to textural and structural analyses (Lille et al. 2020). The advantages of whole-fat milk powder were listed by Lille et al. (2020) as good dispersion in water and small particle size distribution (Lille et al. 2020).

Protein-lipid interactions play another attractive role, especially in some products like cheese, hereby cheese printing. As demonstrated by Portanguen et al. (2019), molded cheese has a continuous protein phase with homogenously dispersed fat globules. Interestingly, the printing of cheese products had a notable impact on the microstructure, causing the protein phase to become discontinuous. This resulted in the loss of sphericity and volume gain in fat globules, leading to the

**Table 14.1** Examples for printed protein analogs

| Analog type and ingredients | 3DP parameters | Property | Function | Findings | Reference |
|---|---|---|---|---|---|
| Fish analog (SPI, XG, RS) | Nozzle inner diameter (μm): 200/600 extrusion flow rate (mL/min):32.56 XY-moving speed (mm/min):1200–1500 | Viscosity at low shear rate of 1 (1/s): Around $10^3$ (Pa.s), at high shear rate of 100 (1/s): Between 10–100 (Pa.s). The hardness of the sample (7.11 ± 0.87 N), printed using 200 μm filaments with a porosity of 15.86%, closely resembled that of its actual counterpart (7.04 ± 1.77 N). The springiness of steamed S20 × 3R15 food ink material matched that of the real dorsal flesh of the yellow croaker (0.58 ± 0.09). 600 μm nozzle dimater gave better printing results, too | SPI: Main protein source XG: Binder and stabilizer RS: Texture modifier | Real fish muscle fibers was successfully produced in S20X3R15 sample with the porosity of 15.86% with 200 μm filaments (similar texture to real source) | Shi et al. (2023) |
| Meat (SPI, WG, RP, canola oil) | Air heated printing | The K values of the edible inks declined as the proportion of rice protein increased, likely due to the rice proteins' capacity to disturb the formations created by soy and wheat proteins. Tanδ values for all protein paste samples remained below 1 (G′ > G″), indicating their primarily elastic characteristics. Moreover, tanδ rose with higher rice protein fractions, suggesting increased flowability of the inks. Hardness: 150–290 g springiness: 0.85–2.5 cm | This ratio was chosen according to the target protein content found in actual meat (25%) | 25% of total dry matter content was successful | Qiu et al. (2023a) |

(continued)

**Table 14.1** (continued)

| Analog type and ingredients | 3DP parameters | Property | Function | Findings | Reference |
|---|---|---|---|---|---|
| Meat (CG, SA, GM) | Coaxial nozzle-assisted, calcium curing and post cooking:Sizes of inner diameter and outer diameter of the coaxial nozzle were 1.0 mm and 1.6 mm, a nozzle height between layers of 1.8 mm, a first layer height of 1.7 mm, a nozzle speed of 20 mm/s, and an infill level of 70%, using a rectilinear infill pattern | Hardness: 20–50 N; tensile strenght: 0.2–0.6 MPa | Artificial muscle fiber production | Elastic strength of CG(1.5) GM(1.5) and CG(2.5)GM(1.5) comparable to actual meat due to GM elastic gel formation and the highest hardness with CG(2.5)GM(1.5) | Ko et al. (2021) |
| Meat (MBPI, MC, BR) | Nozzle diameter, 1.1 mm; layer height, 1.1 mm; 100% infill with one layer of vertical shell; printing speed, 30 mm/s; volume flow rate, 137 mL/h; infill pattern, concentric; and the temperature of the cartridge, the print bed, and the ambient temperature were ~ 25 °C, cooking as post process | Hardness of raw samples: 2–5 N; hardness of cooked samples:15–40 N springiness of raw samples:0.5–0.6 springiness of cooked samples:0.2–0.5; tensile strenght:7–35 kPa, elongation at break%:25–60 | Mimicking the color change of meat using plant-based materials | Increasing the quantity of xylose resulted in enhanced deformation stability of colorant-infused meat substitutes prior to cooking. This effect could be attributed to xylose's ability to elevate the shear modulus and alter texture through adjustments in interactions like protein-saccharide, protein-water, and protein-protein interactions | Wen et al. (2022) |
| Meat (SPI, cricket jel) | Nozzle size (1.1 mm), printing speed (120 mm/s), nozzle height between layers (1.1 mm), and first-layer height (1.1 mm). The object was printed in the shape of a 65 × 25 × 10 mm3 | G′ > G″; hardness: 10–160 N | *G. bimaculatus* fraction powders revealed that the pellet had the highest protein and lowest moisture content, making it is highly suitable for use as ink | 15% pellet was theoptimal fraction and concentration | Nam et al. (2023) |

| | | | | | |
|---|---|---|---|---|---|
| Meat (l-cysteine, MBPI, WG) | Layer height, 1.1 mm; volume flow rate, 1.66 mL/min; printing speed, 30 mm/s; vertical shell, 1. | Cysteine significantly softened the meat analog inks owing to the decreased yield stress when the cysteine content in the protein mixtures was increased. Yield stress: 20–100 Pa hardness of raw samples: Up to 10 N; hardness of cooked samples:Up to 100 N | Fibrous structure formation. L-cysteine improved product quality. | The cysteine-0.4% ink exhibited the highest printing resolution and structural stability | Chao et al. (2023) |
| Meat (mushroom, LD, PO, PPI, SPI, WPI, PS, MD, XG, S, BE, canola oil) | A 60 ml syringe was filled with paste and printing was carried out at room temperature with nozzle size of 1.6 mm. The layer height and line width were set closer to the nozzle size (1.65 mm). Infill rate was 100% and, printing speed:20–80 mm/s | Shear-thinning and gel-like viscoelastic behavior ($G' > G''$) with predominant elasticity ($\tan \delta < 1$). Hardness:25–215 N (commercial beef around 195 N) springiness%: 4–35 (beef around 35) | Mushroom fortification reduced hardness, stiffness, springiness, and chewiness properties of the meat analogues whereas it increased the juiciness with reasonable overall acceptance. Mushroom fortification also enhanced the nutritional value and improved release of umami amino acids. | GL was looking linty. The fortification with mushrooms positively affected the maximum printed layer. Notably, LD and PO inks brought the advantage of recycling as a result of their re-printability. All ink formulations (C, GL, LD and PO) represented similar gel-like viscoelastic behavior which contributes to a satisfactory 3DP process | Demircan et al. (2023) |

*BR* Beet red, *BE* Beetroot extract, *CG* Carrageenan, *GM* Glucomannan, *LD Lactarius deliciosus*, *MD* Maltodextrin, *MC* Methylcellulose, *MBPI* Mung bean protein isolate, *PO* Oyster, *PPI* Pea protein isolate, *PS* Potato starch, *RP* Rice protein, *RS* Rice starch, *S* Salt, *SA* Sodium alginate, *SPI* Soy protein isolate, *WG* Wheat glüten, *WPI* Wheat protein isolate, *XG* Xanthan gum, *X* Xylose emergence of interstitial fat (Le Tohic et al. 2018). Such products exhibit a soft texture without becoming sticky, as the surface fat is released during the shearing process in extrusion printing. Another study conducted by Keerthana et al. (2020) highlighted a synergistic effect between polyunsaturated fatty acids, wheat protein, and fiber, contributing to the high-performance characteristics of the printing process (Keerthana et al. 2020).

Vegan and fortified products are gaining so much popularity due to several reasons, as mentioned before, and animal-source-free recipes are attracting so much attention from most of the consumers. Incorporating lipids exceeding 5% results in lubrication within the barrel, diminishing protein denaturation. This, in turn, undermines the essential processes of structuring and fiber formation crucial for the development of extruded plant-based products. Cheese Analogue (CA) can be defined as a cheese-like product in which some components, such as milk protein and milk fat, are partially or totally substituted by vegetable-based sources, as well as other substances like acidifying agents, biopolymers, flavoring agents, etc. (Kamath et al. 2022). As reported by Kamath et al. (2022), CA could be divided into 3 main categories as dairy-based, partial dairy-based, and nondairy (plant)-based analogs. To save cost, time, and sources, some dairy sources could be substituted with others during cheese production and then called as dairy analogs. For example, whey protein concentrate could be used instead of casein due to greater protein retention and fast absorption; however it requires the addition of emulsifying salts. The reason is based on the amphiphilic nature of casein, creating an emulsifying effect of free fat during the cheese-making process, but lack of this nature on whey protein. Both could act as low-fat analogues at the same time. Some portion of fat in cheese formulation might be reduced by adding whey protein concentrate to obtain similar stability and rheological characteristics. Similar phenomena could be achieved by non-dairy alternatives. Plant-based milk, fruit pulps/extracts, and polysaccharides could be used to mimic animal-based cheese. Vegetable oils high in saturated fatty acids (SFAs) and solids under room conditions are used as a dairy fat replacer. Fat covers approximately 23–28% of CA composition, playing a role in texture and flavor. Palm oil, rice bran, soy or rapeseed oil, modified starches, and inulin could be plant-based alternatives to anhydrous milk fat, butter, or cream to some extent, mimicking their viscoelastic properties (Kamath et al. 2022).

3DP could be applied to fabricate customer-specific products such as low-fat, low-cholesterol depending on consumer demand, improving final product quality parameters. Here, nutritional quality and sustainability concerns should not be forgotten (Boehm et al. 2023). Shahbazi et al. (2021) evaluated a study about printing low-fat cheese in the presence of micro-biosurfactant (acetylated microcrystalline cellulose) showing better aftertaste, creaminess, and freeze-thaw stability (Shahbazi et al. 2021).

The lipid content of an extruded meat substitute can be elevated by integrating lipids in the form of an emulsion. Nevertheless, if a marbled appearance is preferred, any structured material needs to maintain its shape under high shear and heat conditions; otherwise, the lipid phase will only be emulsified within the structured protein network (Boehm et al. 2023). The incorporation of oleogels (gels having the capability to structure oil) is another interesting way to prevent phase separation problems and improve final quality in fat-based products. Recent chocolate studies replacing some portion of cocoa butter with oleogels or the addition of oleogels for printing improvement can be given as an example in printing applications (Huang et al. 2023). Other examples were also given in Table 14.2.

In summary, achieving the desired layer-by-layer adhesion and structural stability after printing necessitates an examination of diverse fat types, encompassing varied chain length fatty acids, unsaturation degrees, or melting behaviors. It is essential to underscore that recipe optimization is crucial for all types of ingredients. The characteristics of each macromolecule and their interactions must be carefully considered to ensure successful product fabrication with the desired nutritional content.

## 14.4 Applications of Food Printing

The food inks are grouped into natively-printable materials (confectionary, hydrogels, food purees, etc.), and non-natively printable materials (egg, rice, meat, vegetables, fruit, etc.), involving the use of unconventional ingredients (algae, insect powders, etc.) (Wilson et al. 2020; Bulut and Candoğan 2022; Ahmadzadeh et al. 2023). The suitability of the food ingredient to be prepared for 3DP performance must be evaluated. Because the food ink needs to fit through the 3DP's nozzle, authors frequently employ a trial-and-error procedure by adjusting the food's composition and viscosity properties. Generally, it is easy to extrude fluids at greater shear rates when they exhibit a shear-thinning behavior. Higher shear-thinning qualities are generally needed for 3DP materials to reduce viscosity and facilitate easy and smooth extrusion through a small-diameter nozzle (Huang et al. 2022). Yet, it must be high enough to sustain a structure that is deposited layer by layer (Liu et al. 2021b). When the shear rate increased, and the viscosity decreased, it suggests pseudoplastic fluid and shear-thinning behavior. The concentration of the food ink and the shear stress influence the viscosity (Huang et al. 2022; You et al. 2023). The internal friction that prevents food ink molecules from flowing is referred to as viscosity (Liu et al. 2021b).

The gel-like viscoelastic semi-solid behavior of food ink, which was primarily controlled in many studies, has been confirmed (Rando and Ramaioli 2021; Jo et al. 2021; Habuš et al. 2021; Armstrong et al. 2022; Du et al. 2022; Kim et al. 2022; Le et al. 2022; Letras et al. 2022; Mirazimi et al. 2022; Cai et al. 2023a; Niu et al. 2023a; Wu et al. 2023; Zhang et al. 2024; Jiang et al. 2024).

A variety of food matrices, such as confectionery, custard creams, snacks, purees, yogurt, and cheese, have been demonstrated to be compatible with the 3DP technology because they are natively printable materials. 3DP also offers advantages for individualized nutrition, such as fiber or protein-enriched products, along with the advantage of raw material diversity and customizable product design. This may include processes such as formula optimization (response surface methods), preprocessing (fermentation, etc.), and post-processing (baking, etc.) of food inks, containing non-natively printable materials (Table 14.3). It is also used in the design of

**Table 14.2** Examples for printed fat analogs

| Analog type and ingredients | Printing parameters | Property | Function | Findings | Reference |
|---|---|---|---|---|---|
| Fat (CO, GEG, KMG) | Printing speed: 20 mm/s, nozzle size: 1.95 mm, layer height: 1.5 mm, flow rate: 140%, one vertical shell with a 100% infill level, an infill pattern of concentric, printing temperature: ~25 °C | Hardness: 1–5 N tensile strength: 5–35 kPa | Adjusted the oil stability of the fat analog system before and after cooking. Due to its high solid fat content, an oil with a comparatively higher melting point (around 26.5 °C) or solid fat can enhance the stability of the product and minimize oil separation when stored at room temperature (approximately 25 °C). | Following cooking, the fat analogs experienced significant oil release. Increasing the oil content resulted in a higher cooking loss, higher oil loss, higher water loss, lower hardness, and lower tensile strength after heating. | Wen et al. (2023) |
| Fat (CO, SO, PS, L, I) | Nozzle diameter = 1.1 mm, nozzle height between layers = 2.4 mm, first layer height = 2.3 mm, infill =0% with 3 layers of vertical shells, nozzle-moving speed between layers = 40 mm/ s, first layer speed = 50 mm/s, volume flow rate = 137 mL/h, and the temperature of the cartridge, the print bed, and the ambient temperature were 25 ± 1 °C. | $G' > G''$; hardness: 1–30 N (commercial pork fat: Around 7 N, commercial beef fat: Around 42 N) | Plant-based oil inks mimicking the texture and melting behavior of traditional animal fats using 3DP have been developed. | CO-PS15 and SO-PS15 exhibited good printability and similar meltability to commercial beef and pork fats. | Wen et al. (2021) |

*CO* Coconut oil, *GEG* Glucomannan emulsion gels, *I* Inulin, *KGM* Konjac glucomannan, *L* Lecithin, *PS* Potato starch, *SO* Soybean oil

**Table 14.3** Applications of food printing

| Food analogues | Composition of food | Aim and optimum formulation | 3DP performance | References |
|---|---|---|---|---|
| Confectionary | | | | |
| Fat-reduced dark chocolate | Cocoa powder, icing sugar, soy lecithin,<br>Cacao butter or water-in-cocoa butter emulsion (gum arabic, polyglycerol polyricinoleate, golden syrup) | Substitution of cocoa butter (25%, 50%, and 75%, w/w) with water-in-cocoa butter emulsions in different water/oil ratios (2:8, 3:7, 4:6, v/v)<br>*Optimum formulation:*<br>Moderate icing sugar content and cacao powder<br>75% substitution of water-in-cocoa butter emulsion with water/oil ratio of 2:8<br>50% substitution of water-in-cocoa butter emulsion with a water/oil ratio of 3:7 | Lower melting point, and hardness via water-in-cocoa butter emulsion substitution<br>Higher printing accuracy via water-in-cocoa butter emulsion substitution | You et al. (2023) |
| Ganache, cold-extruded | Dark chocolate (57.9% cocoa), cacao powder, whipped cream | Whipped cream (20%, 30%, 40%, and 50% w/w)-based ganache with/without cacao powder in different ratios (10%, 20%, and 30%, w/w)<br>*Optimum formulation:*<br>30% whipped cream: Better appearance and emulsion stability<br>20% cacao powder: Better rheology, emulsion stability, and printing accuracy | Melting point ↓ and liquid-like behavior ↑ via whipped cream addition<br>Emulsion stability via a rise in cacao powder addition ↑ | Kim et al. (2022) |

(continued)

**Table 14.3** (continued)

| Food analogues | Composition of food | Aim and optimum formulation | 3DP performance | References |
|---|---|---|---|---|
| Dark chocolate | Unsweetened and lecithin-free dark chocolate couverture, liquid lecithin (0.05%), carob extract | Substitution of dark chocolate with carob extract in different ratios (15%, 30%, 40%, w/w) | Carob extract addition:<br>Best sensory score: 30% carob extract<br>Total aerobic mesophilic bacteria,<br>Higher printing accuracy, moisture content, and soft texture | Cikrikci Erunsal et al. (2023) |
| White chocolate with oleogels | White chocolate, monoglycerides, sucrose fatty acid ester, and hydroxypropyl methylcellulose | Individual addition of monoglycerides, sucrose fatty acid ester, and hydroxypropyl methylcellulose in different ratios (1%, 2%, and 3%, w/w)<br>*Optimum formulation:*<br>2% of sucrose fatty acid ester and monoglycerides | Faster solidification process and higher viscosity recovery via sucrose fatty acid ester and monoglycerides<br>Lower viscosity recovery via hydroxypropyl methylcellulose<br>Failed trial for hydroxypropyl methylcellulose | Huang et al. (2023) |
| Sugar/pectin jelly candy | High-methoxy pectin, citric acid, glucose, sucrose | Different ratios of pectin (10%, 12%, 14%, 16%, w/v)<br>Optimal pectin concentration: 14% | Higher elastic gel structure and solid behavior via higher pectin concentration | Huang et al. (2022) |
| Starch gummies | Corn starch, erythritol, gellan gum | Different ratios of gellan gum (0, 0.5, 1, 1.5 and 2%, w/v) | Higher elastic behavior, viscosity, hardness, rigidity, and self-supporting ability via gum addition<br>Lower water mobility via gum addition<br>Via gum addition | Niu et al. (2023b) |
| Peach gum polysaccharide in gelatin-based gummy | Peach gum, high-fructose corn syrup, gelatin, curcumin | Different ratios of peach gum (0%, 3%, 4%, 5%, 6% and 7%, w/v) | Higher printing accuracy, gelling temperature, and melting temperature, via peach gum addition<br>Lower hardness and gumminess, and reduction in the release rate of curcumin after *in-vitro* digestion in 3D structure via gelatin+peach gum | Zhou et al. (2023) |

| | | | | |
|---|---|---|---|---|
| Polyol gummies | Gelatin, water, low acyl gellan gum, liquid maltitol, erythritol, sorbitol, and xylitol | Liquid maltitol with or without erythritol, sorbitol, and xylitol addition<br>*Optimum formulation:*<br>35 g of maltitol and 30 g of erytritol | Gel network began to break down over 35 °C<br>Higher gelation temperature, gel network via polyol addition<br>Easier to shape via a rise in maltitol concentration and smoother surface via erythritol addition<br>Higher harness, gumminess, chewiness via polyol addition | Le et al. (2022) |
| Gelatin jelly | Gelatin, sugar citric acid, water | In-package cold-plasma treatment | Higher sol-gel temperature, and lower water holding capacity via cold-plasma treatment<br>Improved inferior gel formulation, elastic behavior, gel strength, and extrudability via cold plasma | In et al. (2023) |
| Agro-based | | | | |
| Purple sweet potato puree | Purple sweet potato powder, water, guar gum, xanthan gum, κ-carrageenan, and sodium alginate | Individual addition of guar gum, xanthan gum, κ-carrageenan, and sodium alginate (1.2%) | Higher gel strength, especially for κ-carrageenan and guar gum<br>Storage stability: Guar gum ˃ κ-carrageenan ˃ xanthan gum ˃ sodium alginate<br>Higher printing accuracy via hydrocolloid addition except for κ-carrageenan<br>Hardness: κ-carrageenan ˃ guar gum ˃ xanthan gum ˃ sodium alginate<br>Higher adhesiveness, cohesiveness, springiness, gumminess, and chewiness | Cai et al. (2023a) |

(continued)

**Table 14.3** (continued)

| Food analogues | Composition of food | Aim and optimum formulation | 3DP performance | References |
|---|---|---|---|---|
| Purple sweet potato puree | Purple sweet potato powder, water, hydroxypropyl methylcellulose, xanthan gum, guar gum, flaxseed gum, and κ-carrageenan | Individual addition of hydroxypropyl methylcellulose, xanthan gum, guar gum, flaxseed gum, and κ-carrageenan (5%) | Higher viscosity and lower water mobility via hydrocolloid addition<br>Best printing performance and printing accuracy: Hydroxypropyl methylcellulose<br>Poor printing performance and rough printing lines: Xanthan gum<br>Good printability, rigid but smooth structure, stability to deformation: Guar gum and flaxseed gum<br>Failed trial: κ-carrageenan<br>*In-vitro digestion of potato puree with hydrocolloids:*<br>Phenolic and anthocyanin release: Flaxseed gum $>$ xanthan gum $>$ hydroxypropyl methylcellulose ≥ guar gum<br>Flavonoid release: Hydroxypropyl methylcellulose $>$ flaxseed gum $>$ xanthan gum $>$ guar gum<br>Release of ferulic acid were higher and that of chlorogenic acid were lower | Hu et al. (2023) |
| Red ginseng puree | Red ginseng powder, water, waxy corn, tapioca, and potato starch | Individual addition of waxy corn, tapioca, and potato starch with different concentrations (7%, 10%, and 13%)<br>*Optimum formulation per 100 g paste:* 10 g potato starch | Freeze-dried samples: Higher moisture and hardness via starch addition | Jo et al. (2021) |

| | | | | |
|---|---|---|---|---|
| Strawberry snack | Strawberry pulp, water, corn and wheat starch | Individual addition of corn and wheat starch with different concentrations (10%, 15%, and 20%)<br>*Optimal starch ratio:* %15 | Water activity: ~0.95<br>Lower particle size via corn starch<br>High level of stability and minimum deviations<br>Abundant bioactive compounds with wheat starch: Total phenolics, condensed tannins, anthocyanins<br>Abundant bioactive compounds with corn starch: Hydroxycinnamic acids, flavanols, flavonoids<br>Higher FRAP and lower DPPH activity via starch<br>After 10 days of storage with vanillin or citral, no pathogenic bacteria, yeast and mold growth | Bebek Markovinović et al. (2023) |
| Food puree | Cellulose nanocrystals, water, tomato, applesauce, and spinach purees | Individual addition of tomato, applesauce, and spinach purees with different cellulose nanocrystals (2.5%, 5%, and 7.5%) | Robust gel matrix via cellulose nanocrystals<br>Easier to shape via a rise in cellulose nanocrystals | Armstrong et al. (2022) |
| Food grain-based | | | | |
| Cereal puree | SPI, wheat gluten, rice protein, canola oil, water | Different SPI: Whey gluten: Rice protein ratio (1:1:0 25, 1:1:0.4, 1:1:0.7, 1:1:1, w/w in 25 of food ink formula) | Higher apparent viscosity and lower printing accuracy via rice protein addition<br>Higher hardness and lower adhesiveness via rice protein | Qiu et al. (2023a) |
| Cereal dough | Wheat bran, pea protein, sunflower oil, any of oat or barley flour, any of sodium bicarbonate or citric acid | Pre-processed wheat bran via each ultrasound, vacuum microwave, and pulsed light | Higher shape accuracy via pre-processing of wheat bran<br>Baking loss via pre-processing: Pulsed light ˃ ultrasound ˃ vacuum microwave | Habuš et al. (2021) |

(continued)

**Table 14.3** (continued)

| Food analogues | Composition of food | Aim and optimum formulation | 3DP performance | References |
|---|---|---|---|---|
| Cereal snack | Potato flakes with gelatinized starch, wheat flour, soybean oil, and any of salt or sucrose | Post-processing via vacuum drying or microwave-vacuum drying | Smooth and greater structure stability<br>Higher drying efficiency, and lower hardness via microwave-vacuum drying compared to vacuum drying of dough without NaCl or sucrose | Liu and Zhang (2021) |
| Fiber-enriched cereal snack | Wheat flour, salt, olive oil, water, any of carrot and broccoli powder or their blend | Substitution of wheat flour with different amounts of carrot and/or broccoli powder (25%, 50%, and 75%) and the effect of baking<br>*Optimum formulation for carrot and broccoli powder per 100 g*: 50 or 75 g | Better printing accuracy via carrot powder<br>Lower hardness of baked snacks with 75% broccoli or carrot powder<br>Higher hardness with carrot powder compared to broccoli powder | Ahmadzadeh et al. (2023) |
| Gluten-free cereal snack | All-purpose corn and rice flour, salt, xanthan gum, olive oil, any of *Chlorella vulgaris* smooth or Spirulina biomass | Different amounts of *Chlorella vulgaris* smooth or spirulina biomass (5%, 10%, 15%, and 30%) | Lower printability, cohesiveness, and higher firmness and adhesiveness via higher microalgae<br>After baking, lower the hardness with spirulina<br>*Snacks with 5% microalgae biomass:* Higher total phenolic content, antioxidant activity via FRAP and DPPH | Letras et al. (2022) |
| Bread dough | Wheat starch, wheat gluten, sugar, salt, yeast, butter, water | Different starch-to-gluten ratio (89:11, 87:13, 85:15, and 83:17)<br>*Optimum gluten content after fermentation and baking:* 13% | Gas holding capacity and height after fermentation and expansion rate after baking via a higher gluten ratio | Jiang et al. (2024) |
| Microwave oven-cooked dough | Rice flour, low-gluten wheat flour, yeast, water | Different amounts of water, rice flour, yeast<br>*Optimum water content, rice flour and yeast per 100 g:* 64%, 25%, 0.36% | Lower storage and loss modulus via water content<br>Higher printing accuracy via rice flour | Du et al. (2022) |

| | | | | |
|---|---|---|---|---|
| Cookie | Wheat flour, sugar, emulsion gel (pea protein isolate and corn oil), and/or butter | Different amounts of emulsion and/or butter (6%, 12%, 18% and 24% in total batter) | Emulsion gel:<br>Lower the viscoelasticity and printability<br>Higher dimensional stability, after baking-resistance to deformation | Lim et al. (2023) |
| Fiber-enriched cookie | Wheat flour, powdered sugar, butter, milk, okara dietary fiber | Ultrasound and/or homogenization-modified okara dietary fiber addition<br>*Optimum okara fiber content per 100 g:* 6 g | Higher hardness, adhesiveness, gumminess, mechanical strength, and lower springiness via okara dietary fiber | Liu et al. (2021b) |
| Biscuit | Wholegrain cowpea sourdough and/or malted quinoa flours, sugar, sunflower oil, baking powder, water | Different ratios of wholegrain cowpea sourdough and/or malted quinoa flours, | Difference in total flavonoid contents and phenolic compounds methionine, aspartic, glutamic, iron, and phosphorus contents with conventional biscuit | Kewuyemi et al. (2023) |
| Gluten-free pizza | Sugar, salt, emulsifier, oil, yeast, water, any of gluten-free flour blend, brown rice flour, plantain flour, and quinoa flour | Comparison of wheat and gluten-free pizza dough and crust<br>*Optimum fermentation time for gluten-free dough:* 120 min | | Dey et al. (2023) |
| Noodle | Whole wheat flour, guar gum, potato peel powder, | Substitution of coarse or fine potato powder addition (20–100%)<br>*Optimum ratio of wheat flour: Fine potato peel:* 60:40 | Better appearance, higher cooking time, solid loss | Muthurajan et al. (2021) |
| Non-dairy whipping | Carrageenan, gellan gum, xanthan gum, and sugar in SPI microgel (transglutaminase and heat cross-linked), the mixture of mono-diglycerides, sucrose fatty acid esters, propylene glycol fatty acid glycerides, poly fatty acid glycerides in palm oil | Different amounts of SPI microgel (0.8–1.6%) | Lower deformation resistance via higher oil concentration and higher deformation resistance via higher protein concentration<br>Higher hardness, resilience, chewiness, printing accuracy, and stability via higher oil and protein concentration | Wu et al. (2023) |

(continued)

**Table 14.3** (continued)

| Food analogues | Composition of food | Aim and optimum formulation | 3DP performance | References |
|---|---|---|---|---|
| Tofu | Magnesium chloride, soybean milk | Different starch concentration (0–50%) | Constant hardness, adhesiveness, and gumminess at 0–30%) starch ratio<br>Water separation without starch; small bubbles with 15–30% starch addition; no water separation and bubbles with 35% starch | Arii and Nishizawa (2022) |
| Dairy and animal-based | | | | |
| Custard cream | White caster sugar, whole fat milk powder, cellulose, sodium alginate, citric acid, carrageenan, WPI, hydroxypropylated starch, water | Different ratios of WPI (0–6%), hydroxypropylated starch (0–6.0%), and carrageenan (0.05–2.0%)<br>*Optimum formulation for total WPI/*<br>*Hydroxypropylated starch content and κ-carrageenan per 100 g:* 3%–6%, and 0.5% | Lower water mobility<br>Higher influence of carrageenan on mechanical strength<br>Harder and rigid structure via κ-carrageenan | Cai et al. (2022) |
| Lemon mousse | Whipping cream, gelatin, whole pasteurized eggs, sugar, fresh lemon zest, citric acid and WPI | Different ratios of gelatin (1–2% w/w), citric acid (0.9–1.5% w/w) and WPI (8–18% w/w) | Better shape stability via higher gelatin<br>Glossier appearance via WPI<br>Higher firmness via gelatin | Chow et al. (2021) |
| Dairy-based simulant | Low-heat skim milk powder and milk protein isolate, rennet | Different pH, [$Ca^{2+}$], and temperature on the rennet gelation | Highest gel strength at pH 6.0 and temperature ramped to 40 °C followed by cooling to 15 °C | Uribe-Alvarez et al. (2021) |
| Yogurt-based gel | Commercial yogurt, gelatin, citric acid, WPI | Different ratios of WPI (0, 6, and 12%) and cooled gelatin (7.5%, 10% or 12.5% gel) | Inner filler effect of WPI<br>Higher firmness, shape stability, and resilience via higher gelatin | Riantiningtyas et al. (2021) |
| Cheese | Anjia cream cheese, emulsifying salts (sodium pyrophosphate and sodium polyphosphate), glutamine transaminase, apple polyphenol | Different ratios of apple polyphenol (0%, 0.4%, 0.8%, 1.2%, or 1.6%)<br>*Optimum apple polyphenol content:* 0.8% | Lower viscosity, extrudability, sensory overall acceptance via higher apple polyphenols | Deng et al. (2023b) |

| | | | | |
|---|---|---|---|---|
| Cheese | Anjia cream cheese, emulsifying salts (sodium pyrophosphate and sodium polyphosphate), glutamine transaminase | Post-processing via hot air drying or freeze-drying | Higher drying efficiency, casein retention rate, shape stability, and lower porosity, drying time via freeze-drying | Deng et al. (2023a) |
| Cheese | Cheddar cheese, trisodium citrate, and water | Different pH (5.4, 5.6, and 5.8), or young: Mature Cheddar cheese in different ratios at pH 5.6 (1:3, 1:1, 3:1, 4:0)<br>*Optimum formulation for young: Mature Cheddar cheese ratio:* 1:3 | Softer, gummier, and more resilient in texture at pH 5.8<br>Softer, chewier, gummier structure via young cheese<br>No significant effect of pH or cheddar maturity on printing accuracy | Ross et al. (2021) |
| Cheddar cheese | WPI (nanofibril form, treated at pH 2), cheddar cheese | Different ratios of WPI nanofibrils (0, 10, 20, 30 and 40, w/w)<br>*Optimum WPI nanofibril concentration:* 20% | Higher regulatory capacity and more homogeneous water immobilization of WPI nanofibrils than WPI<br>Lower hardness, chewiness, and water mobility via WPI and WPI nanofibrils<br>Cohesiveness: Higher via WPI nanofibrils and lower via WPI | Wang et al. (2023) |
| Chicken-based snack | Celery root, carrot, and onion boiled in freshly squeezed orange juice, cooked skinless chicken thigh meat, turmeric powder, garlic powder and salt and whole wheat flour, carrot fiber, κ-carrageenan, and olive oil | Different gelatin concentrations (0, 1, and 2%) and the effect of baking | More uniform and compact microstructure via gelatin<br>Baked: Hardness, gumminess, and chewiness via gelatin<br>Better sensory evaluation | Bulut and Candoğan (2022) |

(continued)

**Table 14.3** (continued)

| Food analogues | Composition of food | Aim and optimum formulation | 3DP performance | References |
|---|---|---|---|---|
| Chicken-based nugget | Chicken mince, refined wheat flour | Different ratio of chicken mince: Wheat flour (1:1, 2:1, and 3:1) and air-drying with deep-frying<br>*Optimum chicken meat: Wheat flour ratio:* 2:1 | Higher hardness, adhesiveness, springiness, and cohesiveness via wheat flour<br>No difference in resilience<br>*Fried nuggets:*<br>High overall acceptability, cooking loss: 27.62%, shrinkage: 10%, fat retention: 77.47% | Wilson et al. (2020) |
| Surimi | Surimi paste, potato starch | Different ratios of potato starch (3%, 6%, 9%, 12%, and 15% w/w)<br>*Optimum potato starch content:* 12% | Lower cooking loss, hardness, chewiness, and higher water-holding capacity ↑ via potato starch | Kim et al. (2021b) |
| Surimi | White chub surimi, high internal phase emulsion gel (transglutaminase-modified fish scale gelatin and fish oil, 80:20, v:v), salt | Different ratios of fish scale gelatin in emulsion gel (10–60 mg/mL)<br>*Optimum formulation:* 30 and 40 mg/mL gelatin in emulsion | Lower water holding capacity, gel strength, hardness, springiness, and chewiness via higher gelatin in emulsion | Wang et al. (2024) |
| Surimi | *Nemipterus virgatus* surimi, salt, transglutaminase | Synergistic effect of transglutaminase and different microwave power (10–60 W/g) | Better shape fidelity and lower protein solubility via heating at 40 and 50 W/g with the addition of TG<br>Microwave heating without TG:<br>Lower ionic bonds, higher hydrogen bonds, no obvious change in hydrophobic interactions and disulfide bonds | Zhao et al. (2021) |

*FRAP* Ferric reducing antioxidant power, *DPPH* 2,2-Diphenyl-1-picrylhydrazyl scavenging activity, *SPI* Soy protein isolate, *WPI* Whey protein isolate, *TG* Transglutaminase food matrices as a drug carrier or for people with special nutritional needs, such as the elderly, children, or individuals with conditions like dysphagia (Table 14.4).

### 14.4.1 Confectionary and Hydrogels

Chocolate works well as a material for 3DP since it melts at room temperature and can be piled, and sculpted while being printed, and hardened to keep its shape (Kim et al. 2022; Huang et al. 2023). The primary ingredients of chocolate are sugar, lecithin (an emulsifier) with varying proportions of cocoa mass and cocoa butter, and, in the case of milk chocolate, various amounts of milk powder. A smooth flow of chocolate at a temperature above its melting point is necessary, and once the process is complete, quick solidification is needed to preserve the intended shape. This demonstrated that the ink must be at the proper temperature for the extrusion process (Cikrikci Erunsal et al. 2023). Kim et al. (2022) recorded a decrease in melting point when chocolate is combined with whipped cream. This suggested a reaction between milk fat and cocoa butter that results in the softening. In a similar study, You et al. (2023) determined that the range of melting points in fat-reduced chocolates were 24–35 °C; 28–38 °C, and 26–35 °C at a water-to-oil ratio of 2:8, 3:7, and 4:6, respectively. Moreover, Cikrikci Erunsal et al. (2023) asserted a higher melting point via the substitution of chocolate with carob extract. Chocolate is a temperature-sensitive product and it affects sensory and rheological qualities with a melting point between 31 and 36 °C (Kim et al. 2022). Sensory factors of chocolate, such as color, shape, and texture, have an impact on how people perceive its flavor, even in highly preferred products such as chocolate. In an interesting study, Ogata et al. (2023) revealed that the 3DP of chocolate in Bouba-shaped (round) pieces was thought to be sweeter by participants than that of the Kiki-shaped (angular) pieces. Compared to traditional mold-based processes, 3DP produces chocolate layers with greater accuracy and efficiency and gives the user more creative freedom to craft layered designs in diverse shapes (Kim et al. 2022).

One of the most common chocolate sauces is ganache, which is made from chocolate and heated cream (Kim et al. 2022). Kim et al. (2022) created a chocolate emulsion with rheological characteristics (ganache) using a 3DP. Chocolate ganache with whipped cream (30%, w/w) showed the best resolution compared to the other proportions of whipped cream (20%, 40%, and 50%). In terms of rheology, emulsion stability, and printing accuracy, chocolate ganache with 20% (w/w) cocoa powder performed the best across different cocoa powder amounts (10% and 30%). The chocolate industry has faced the difficulty of developing new goods in response to specialized consumer demands, such as chocolates with fat-reduced, vegan, or sugar-free formulations (Kim et al. 2022). You et al. (2023) reported that chocolate with lowered fat and moderate icing sugar and cacao powder content had excellent printing quality with a smooth surface and acceptable 3D shape stability. Because higher icing sugar and cacao powder resulted in an excessively viscous structure. The initial printability of 3D-printed chocolate was unaffected by substituting 75%

**Table 14.4** Food simulants and drug-loaded delivery systems

| Composition of matrices | 3DP performance | IDDSI test, texture profile analysis, or drug release tests | References |
|---|---|---|---|
| Food simulants for dysphagia diet | | | |
| Shiitake mushroom powder, water, any of arabic gum, xanthan gum and κ-carrageenan gum | Higher deformation resistance capability, viscosity, mechanical strength, hardness and gumminess via xanthan gum and κ-carrageenan gum | *Optimum hydrocolloid concentration per kg puree via IDDSI test:* Arabic gum at 9 g, and xanthan gum or κ-carrageenan between 3–9 g | Liu et al. (2021a) |
| Black fungus powder, water, any of arabic gum, xanthan gum and κ-carrageenan gum | Arabic gum: Poor self-supporting performance, lower viscosity hardness and gumminess<br>Xanthan gum: Smooth and appealing surface texture with no deformation<br>κ-Carrageenan: Rough and dry surface structure with several defective points | Optimum hydrocolloid per kg puree via IDDSI test: Xanthan gum between 3–9 g | Xing et al. (2022) |
| Potato powder, soy protein acid hydrolysate, with or without powdered agar, water | Higher deformation resistance via agar addition | Hardness, cohesiveness and gumminess via agar addition ↑<br>Optimum content per kg puree via IDDSI test: Soy protein 60 g and agar 2 g | Mirazimi et al. (2022) |
| Gelatin B or whey protein isolate, xanthan gum, with or without cellulose nanocrystals | Higher viscosity and collapsed zones without cellulose nanocrystals<br>Unstable extrusion behavior over 1.5% concentration of cellulose nanocrystals | Optimum cellulose nanocrystal concentration per kg puree via IDDSI test: 5–15 g in whey protein isolate+xanthan gum; 5–10 g in gelatin+xanthan gum; all samples with 5–15 g after storage at 4 °C for 96 h or heating 50 °C for 5 min | Zhang et al. (2024) |
| Corn flour, fenugreek gum with or without flaxseed protein, water | Smooth appearance, and self-supporting ability of via fenugreek gum, improvement of printing performance and printing accuracy via appropriate concentration of flaxseed protein (15%, w/w)<br>Hardness and gumminess via flaxseed protein concentration first decrease then increased | Optimum flaxseed concentration per kg puree via IDDSI test: Only 15%, w/v with fenugreek gum (0.4%, w/v) | Niu et al. (2023a) |

| | | | |
|---|---|---|---|
| Apple powder, edible rose powder, different combination of xanthan gum and basil seed gum | Higher stiffness, self-supporting ability via gum addition<br>Apparent viscosity and yield stress lower via xanthan gum and higher via basil seed gum | Optimum xanthan gum: Basil seed gum combination for 100 g gum addition per kg puree via IDDSI test: 2:1, 1:1, 1:2, 1:0, but failure for 1:0 | Qiu et al. (2023b) |
| Honeyed red ginseng (boiling in honey), red ginseng extract, pea protein, the mixture of stevia powder, sepia color powder, and beet powder | Higher hardness, gumminess and lower springiness via pea protein | Optimum pea protein concentration per kg puree via IDDSI test: 60 and 90 g | Kim et al. (2024) |
| Native or modified cassava starch, κ-carrageenan, skim powdered milk in water | Higher firmness via κ-carrageenan<br>No obvious effect of modified starch on firmness<br>Printing quality loss via higher κ-carrageenan<br>Similar calcium, magnesium, potassium and phosphorus bioaccessibility with high level | Optimum formula via IDDSI test with texture analyzer: 0.3 g κ-carrageenan and native or modified starch while molded counterpart did not pass. | Bitencourt et al. (2023) |
| Surimi: Silver carp surimi, glycerol (0%, 5%, 10%, 15% and 20%, w/w, based on the weight of frozen surimi), spraying transglutaminase, salt, or $CaCl_2$ | Stabilized aggregation-prone intermediates, promotion of surimi myosin's hydration and prevented protein aggregation via glycerol<br>Higher printability, better texture and water distribution via dry-spraying transglutaminase compared to salt, or $CaCl_2$ | Optimum formula for surimi: 15 g glycerol and post-processing via dry-spraying of 0.5% transglutaminase<br>Suitable cooking: Electric oven, microwave, and steamer | Zhu et al. (2023) |
| Drug-loaded delivery systems | | | |
| Dark chocolate:<br>Dark chocolate (53% cocoa), corn syrup, drug: Either paracetamol (22.9 mg/g) or ibuprofen (19.6 mg/g) | A higher syrup to chocolate ratio result in too sticky ink<br>Melting point, hardness, cohesiveness and chewiness via paracetamol and ibuprofen ↓ | A significantly lower drug oral and gastric digestion release of ibuprofen compared to paracetamol | Karavasili et al. (2020) |

(continued)

**Table 14.4** (continued)

| Composition of matrices | 3DP performance | IDDSI test, texture profile analysis, or drug release tests | References |
|---|---|---|---|
| Dark chocolate:<br>Dark chocolate couverture, drug: Paracetamol (9%) | Lower hardness, breaking force, flexural strength, chewing index via drug addition | Compared to molded-chocolate:<br>Superior mass uniformity, and drug content uniformity | Chachlioutaki et al. (2022) |
| Dark chocolate:<br>Dark chocolate (72% cocoa), drug: Ibuprofen (25%) | | *In vitro* digestion: Slower release of ibuprofen in the gastric digestion were in dosages prepared at 80 °C than that at 50 °C. The release of ibuprofen via active lipase compared to inactive lipase ↑<br>*In vivo*: The serum concentration of ibuprofen in chocolate ↓ | Al-Kabariti et al. (2024) |
| Gummy:<br>Starch syrup, any of gelatin and hydroxypropyl methylcellulose, drug: Lamotrigine (10 mg) | Gummy strength depends on the gelatin concentration<br>Higher viscosity via gelatin or hydroxypropyl methylcellulose<br>Taste masking of the bitter drug with reduced sugar syrup | Higher drug release via hydroxypropyl methylcellulose<br>No obvious effect of gelatin on drug release | Tagami et al. (2021) |
| Fruit-chews:<br>Hydroxypropyl cellulose, Gelucire 48/16™, strawberry flavoring, sucralose, food grade (yellow, green, blue and red) colorings, drug: Diphenhydramine hydrochloride (2.5%) | Masking bitterness of diphenhydramine hydrochloride<br>Taste masking of the bitter drug<br>A synergistic effect of the sweetener and the strawberry flavor on sensory perception | Rapid release of diphenhydramine hydrochloride via matrix | Tabriz et al. (2021) |

*IDDSI* International Dysphagia Diet Standardization Initiative, *SPI* soy protein isolate with a 2:8 water-to-oil ratio or 50% with a 3:7 water/oil ratio with a desirable crystal polymorphism.

Cikrikci Erunsal et al. (2023) studied carob extract as an alternative to natural sugar for dark chocolate printability. One or two cycles of pre-heating were required to print the same ink composition. The product's low amount of cocoa butter and high carob extract ratio, which particularly affected the product's melting behavior, took away from its authenticity as chocolate.

White chocolate has not received as much attention as dark chocolate, because white chocolate does not contain as much cocoa powder as dark chocolate does, which has the appropriate physical attributes (like viscosity) (Huang et al. 2023). Huang et al. (2023) investigated the effects of oleogels (monoglycerides-MAG, sucrose fatty acid ester-SE, and hydroxypropyl methylcellulose-HPMC) to stabilize liquid oil to hasten the solidification of white chocolate. The viscosity recovery in the rheological analysis was less than 80% for the MAG and SE groups, and more than 80% for the HPMC group. In the bridging experiment, the white chocolate could be efficiently extruded and solidified, and the bridging distance could be increased to 10 mm when MAG and SE were supplied at a rate greater than 2%.

Gummy, candy, or jellies are also among the products that are 3D printed. In the case of a hard candy base solution, significant variations in viscosity are noted as crystallization progresses due to the dropping temperature (Kim et al. 2021a). Kim et al. (2021a) stated that at temperatures below 78 °C, the viscosity of the candy base solution rapidly changes, rendering control over the extrusion speed unattainable. As a result, to apply extrusion, the food ink's temperature must be kept at 80 °C or above. A heating controller was suggested in the 3DP to regulate the head temperature.

Traditional manufacturing of gummies has some difficulties in satisfying consumer demands. Molds are commonly employed in some creative soft candy with varied shapes of gummies, which need significant human input, denser physical handling during plating, more storage, and greater usage of plastics. The new production technique to meet the need for customized items is highly preferred (Le et al. 2022). Le et al. (2022) suggested easier shaping, a smoother surface, and higher viscosity via higher maltitol concentration and the addition of a certain amount of erythrol in the gummies. To produce a food like gummy, jelly, or candy, it is necessary to use auxiliary gelling ingredients. Food ink has made extensive use of hydrophilic colloids. According to reports, a dense gel network structure can be formed by hydrophilic colloids in food items, significantly enhancing the rheological qualities, water distribution, and textural features of printing inks (Niu et al. 2023b). Pectin (Huang et al. 2022), gelan gum (Niu et al. 2023b), gelatin with peach gum (Zhou et al. 2023), or cold-plasma-treated gelatin (In et al. 2023) were investigated as gelling materials in the literature.

Pectin gelation is a 3D network structure in which the solvent and solute are trapped within the polymer molecular chains. The molecular structure, which preserves the intermolecular interactions and the polymer molecular chain, is ultimately responsible for the circumstances needed for gelation and the characteristics of the gel (Huang et al. 2022). Huang et al. (2022) analyzed that the bases of the candies with pectin concentrations of 10% and 12% displayed a broader structure and collapsed because of the poor gelling ability of pectin and 14% was the ideal pectin concentration.

Niu et al. (2023b) recorded that as gelan gum additions increased, the material's water flow state became more constrained (dense gel network), the ink's viscosity increased, and extrusion got harder. Zhou et al. (2023) revealed that gelatin without peach gum could not be molded because of the irregular extrusion flow rate, the gel network, and poor chain density during printing. The printability of the composite gel tended to rise and then fall as the peach gum addition increased. Further, in vitro the gastric and intestinal release of curcumin in gelatin + peach gum hydrogels were reduced.

In et al. (2023) verified that the cold plasma-treated gelatin had a positive impact on the rheological characteristics of the final jelly formulation. Moreover, the application of cold plasma treatment could serve as a means of enhancing gelatin-based formulations that are not appropriate for 3DP. The jelly formulation's elastic qualities were enhanced by the addition of cold plasma-treated gelatin. The gelatin powder's non-polar functional groups (C–C bonds), which decreased the attraction between gelatin and water molecules and enhanced the material's elastic qualities.

### *14.4.2 Agro-Based Food Analogs*

Due to their high moisture content and absence of appropriate mechanisms like gelation, agglomeration, or solidification, agricultural raw materials, such as vegetables and fruits, are typically regarded as non-printable materials (Ahmadzadeh et al. 2023). However, they are a valuable food group due to their positive health benefits, such as bioactive components and dietary fibers. Agro-based food purees are combined with rheological modifiers as a thickening agent to generate food inks with lower moisture content (free water content) (Armstrong et al. 2022). For this purpose, different types of gums, sodium alginate, hydroxypropyl methylcellulose, cellulose nanocrystals, and different types of starch can be used individually for different food puree formulations (Jo et al. 2021; Armstrong et al. 2022; Bebek Markovinović et al. 2023; Cai et al. 2023a; Hu et al. 2023).

Hydrocolloids may restrict the system's ability for water molecules' mobility. The gel's physical structure boosted the amount of interstitial water bonded there, and the hydrocolloids' capacity to retain water was strong. This was explained by the way hydrocolloids and amylose molecules interacted, which altered the gel network's structure and gave the gel system good water-holding capacity (Cai et al. 2023a). For example, the purple sweet potato gel with κ-carrageenan had the maximum gel strength, followed by the samples from guar gum, xanthan gum, and sodium alginate. The interaction force between the molecules was increased and water mobility was restricted by the inclusion of hydrocolloids in purple sweet potato puree (Cai et al. 2023a). Hu et al. (2023) reported that each hydroxypropyl methylcellulose, xanthan gum, guar gum, and flaxseed gum in purple sweet potato puree provide good printable quality and stability to deformation for purple sweet potato puree, except for κ-carrageenan. Further, after in vitro digestion, glucose release was the highest via the addition of xanthan gum and flaxseed gum because they may form an impermeable shield around starch granules in purple sweet potato pure to stop enzymes from binding to the starch. Except for guar gum, all hydrocolloids increased the rate of starch digestion. The release of individual phenolics in purple sweet potato during gastrointestinal digestion and colonic fermentation was influenced by the differences in the interactions between hydrocolloids and phenolics (total phenolics, anthocyanins, and flavonoids).

According to data by Liu et al. (2021a), the addition of xanthan gum and κ-carrageenan greatly enhanced the mechanical strength of the Shiitake mushroom ink (Liu et al. 2021a) and black fungus ink (Xing et al. 2022). They create intermolecular connections via hydrogen bonds and lower the water mobility to create a network structure. The mechanical strength, viscosity, hardness, and self-supporting ability of 3D-printed structures were all reduced by the addition of arabic gum because it hindered the hydrogen bond formation in the gel network (Liu et al. 2021a; Xing et al. 2022).

Different types of starches in agro-based food inks have demonstrated significant potential. Starch is one of the hydrocolloids that has been used widely to improve the processing properties of meals. Jo et al. (2021) stated that potato starch (10 g in 100 g paste) gave the optimum shape compared to corn and tapioca starches. Because the relatively long amylose chain of potato starch encourages several hydrogen bonds between molecules. The branchless amylose's linear arrangement is aligned with the fluid's flow direction and produces minimal intermolecular interactions while the material flows. Moreover, with low shear rate conditions, food ink with potato starch is more resilient due to its comparatively high amylose content. Bebek Markovinović et al. (2023) also suggested freeze-drying as a post-processing technique for starch-based food ink formulations that would be suitable for providing a good texture. Compared to strawberry pulp with wheat starch, that with corn starch showed a noticeably higher penetrating force. The gels' hardness rose along with their starch level, making them more resilient to stress. As the starch content rises in food ink, there are more starch molecules per unit volume. Thus, it raises intermolecular hydrogen bonding and increases the compactness of network structure, and gel strength.

Armstrong et al. (2022) compared the effect of cellulose nanocrystals in tomato, spinach, and applesauce-based inks. Food inks with spinach and tomato suffered from gaps and roughness, while food ink with applesauce showed the best surface smoothness after freeze-drying (Armstrong et al. 2022). Zhang et al. (2024) reported that collapsed zones without cellulose nanocrystals in gelatin+xanthan gum or WPI + xanthan gum ink were obtained, while over a 1.5 wt% concentration of cellulose nanocrystals caused unstable extrusion behavior.

### *14.4.3 Food Grain-Based Food Analogues*

Grain-based food inks primarily contain ingredients such as natural fiber and starch, which come from the basic ingredients used. These ingredients are bran, gluten fractions (wheat) or protein fractions (pea, rice, wheat) of the grain, and grain flour (corn, oat, barley, wheat) (Habuš et al. 2021; Liu and Zhang 2021; Letras et al. 2022; Ahmadzadeh et al. 2023; Qiu et al. 2023a).

Recently, the possibility of employing 3DP to manufacture meat analogs has been explored for different types of plant proteins. Optimizing the composition and qualities of the edible inks is crucial for the effective manufacture of meat analogs using this method (Qiu et al. 2023a). Qiu et al. (2023a) created a food matrix with soy protein isolate (SPI), and wheat gluten with the addition of rice protein, which are plant-based proteins used to produce meat analogs. The rice proteins reduce the apparent viscosity of the protein mixture (SPI + wheat gluten) and impair the protein interactions between them, so this may improve the printing efficiency of the mixed protein pastes. These findings suggest that the mixed protein pastes are softer when rice protein is present, which is advantageous for the layer-by-layer construction of printed objects during the extrusion process.

Apart from the grouping of the natively or non-natively printable material, food inks for 3DP are also categorized into ready-to-eat food ink (chocolate, etc) or food inks requiring post-processing. Cookies, snacks, or pasta dough are examples of post-processed 3D-printed food (Liu and Zhang 2021). Liu and Zhang (2021) stated that the microwave-vacuum-dried cereal snacks had a significantly lower hardness than the vacuum-dried samples. This is primarily because the microwave can penetrate the material uniformly with a rapid moisture evaporation speed.

Bran addition improves the fiber and bioactive content of cereal foods. During slow 3DP, cereal foods with bran or whole-meal flour may become enzymatically dark. Bran has a high level of polyphenol oxidase activity, which catalyzes the synthesis of o-quinone in the presence of molecular oxygen and causes undesired browning (Habuš et al. 2021). Thus, Habuš et al. (2021) examined the effect of wheat bran pre-processing (ultrasound, vacuum-microwave, and pulsed light). Polyphenol oxidase activity was reduced by 77–92% by bran pre-processing without compromising printing accuracy. The inactivation of amylases was linked to the significantly greater viscosities of the wheat flour-bran mix that was produced after ultrasound treatment.

Wheat flour dough has a distinct network structure that is extensible and viscoelastic, resulting in a soft gel. Soluble dietary fibers wrap starch granules dispersed throughout the gluten network structure and efficiently absorb water. By keeping many protein molecules from becoming deeply entangled, this function inhibits the development of a coherent spatial network (Ahmadzadeh et al. 2023). According to Ahmadzadeh et al. (2023), carrot powder addition to wheat flour greatly boosted the system's ability to absorb water with higher printing accuracy because of higher fiber content. Broccoli powder needed more water to reach a particular consistency in wheat flour compared to carrot powder, and the hardness of baked broccoli snacks was lower than carrot-cereal snacks because of the higher fiber content in the broccoli. After baking, the hardness of the cereal snacks was higher with a lower amount of carrot or broccoli powder, while it was lower with a higher amount of carrot or broccoli powder. Moreover, soluble dietary fibers dilute the gluten protein in flour. The way the soluble and insoluble fibers in carrots and broccoli impact the gluten matrix explains the differences between the two vegetables (Ahmadzadeh et al. 2023).

Letras et al. (2022) reported that higher levels of microalgal biomass (*Chlorella vulgaris* and/or *Spirulina* biomass) in corn-rice flour-based snacks had a detrimental impact on the printability of the doughs. Furthermore, increased protein chain interactions limited the dough's elasticity, while the small amount of corn starch prevented any plasticizing effect. The dough without xanthan gum, even at 0.2% concentration, lacked structure and was not printable because of its very elastic behavior, which caused it to fall through the nozzle tip.

Particularly for bakery products, post-processing is necessary, because they require more customization in terms of shape and nutritional value (Muthurajan et al. 2021). Gluten content of the flour (Jiang et al. 2024), starch and gluten-free/wholegrain/multigrain flour ratio as well as wheat (Du et al. 2022; Dey et al. 2023; Kewuyemi et al. 2023), water content, yeast content (Du et al. 2022), such as Du et al. 2022), as a parameter of the product compositions should also be optimized. Determining the nutrient content and texture profile of the final product is also of great importance. The response surface technique was used to improve the dough formulation based on the textural attributes (Du et al. 2022). It's critical to use suitable formulation to control how the dough changes shape after baking (Du et al. 2022; Lim et al. 2023; Jiang et al. 2024). Jiang et al. (2024) stated that higher gluten dough viscosities were directly correlated with the emergence of tighter structures brought on by an increase in gluten aggregation and starch granules. The amounts of gluten and wheat starch affected the results of 3DP of bread dough, with the gluten content having a major effect on dimensional quality. The high gluten content strengthens the protein network and results in boosted dough stiffness. On the other side, Du et al. (2022) reported that gluten-free flour (rice flour) addition to reduce gluten makes the dough more suitable for 3DP. Due to the lack of gluten structure in rice flour, the dough's mechanical strength improved and its ability to contain air was somewhat diminished. The result of the dough's original structure being severely harmed by the overly high addition amount of rice flour (over 20%), and the rice flour having a major impact on the dough's properties. The 3D-printed bun's pores were wider and sparser than those on the traditional bun's sliced surface, making the resultant sample softer and easier to chew.

Kewuyemi et al. (2023) analyzed the nutritional profile of the 3D-printed wholegrain or multigrain biscuits. There were no differences between the traditional and 3D-printed snacks in terms of potassium, ash, fiber, protein, and amino acids (serine and threonine). Moreover, there were significant variations in the amounts of phenolic compounds, moisture, fat, carbohydrates, minerals (iron and phosphorus), amino acids (leucine, valine, and alanine), and total flavonoid content. Lim et al. (2023) suggested that emulsion gels would be a good substitute for butter in cookies. The emulsion gel in cookie dough reduced the viscoelasticity of dough and printability progressively but had better thermal and dimensional stability. Also, baked cookies had better resistance to deformation via emulsion gel, because of the better thermal stability. Liu et al. (2021b) showed that 3D-printed cookie dough with okara fiber had higher mechanical strength, apparent viscosity, and better shape retention ability with better sensory overall acceptability.

Non-dairy whip topping is a common oil-in-water emulsion. The primary components of nondairy whip topping are hydrocolloids, fat, protein, emulsifier, and sweetener (Wu et al. 2023). Wu et al. (2023) discovered that very high apparent viscosity and hardness interfered with extrusion, lower printing accuracy, and compromised the texture of the cream while declaring the stability and precision of 3D printed forms to be high. Arii and Nishizawa (2022) stated that the tofu ink with starch over 50% was unable to be assessed and The cohesiveness and brittleness were unaffected by the starch addition. Water separation may have happened in the low ratio inks containing additive starch and that water separation may be reduced by the starch.

### *14.4.4 Dairy and Animal-Based Food Analogues*

Dairy formulations preferred for 3DP applications are generally emulsion mixtures such as cream, mousse (Chow et al. 2021; Uribe-Alvarez et al. 2021; Cai et al. 2022) or cheese (Ross et al. 2021; Deng et al. 2023b; Wang et al. 2023) and gel structures such as yogurt (Riantiningtyas et al. 2021). Certain adjustments are required to enhance the qualities of flow and extrusion to produce a high-quality 3D-printed product (Bulut and Candoğan 2022). To provide changed texture for 3DP, it is common practice to incorporate hydrocolloids (thickening and gelling agents) (Chow et al. 2021; Cai et al. 2022), emulsifying agents, and enzymes (e.g., transglutaminase) (Deng et al. 2023a, b) into the formulations of typical food components that are not printable.

Custard cream is a common oil-in-water emulsion system that is used extensively in a broad range of culinary preparations (Cai et al. 2022). Cai et al. (2023b) examined that WPI and hydroxypropylated starch had less impact on the rheological and textural characteristics of custard cream than did the κ-carrageenan, which causes an entangled network and gelation. In another study by Chow et al. (2021), regardless of the dose of WPI or citric acid, lemon mousse (an aerated gel foam) stiffness dropped by almost 70%. The percentage change in resilience following printing was significantly influenced by WPI concentration, and lower WPI concentrations decreased the mousse's robustness and resilience. Although the foam structure was broken up during the extrusion process, 3D-printed mousses that were stable and aesthetically pleasing were produced compared to not-printed mousses. A higher WPI concentration improved the mousses' glossy look, silky texture, and melting mouthfeel. It lessened the lumpy, compact texture and rough surface of mousse textures linked to the gel-firming action of gelatine.

Yogurt is an acid milk gel that is created during fermentation by a starter culture of lactic acid bacteria. Riantiningtyas et al. (2021) examined the effect of inert WPI and gelatin addition to commercial yogurt. WPI served as an "inert filler" because of the restricted whey protein's ability to cross-link. Although WPI failed to unite with the yogurt matrix to create a cohesive network and weakened the yogurt gel network, it can facilitate smooth extrusion during printing.

Uribe-Alvarez et al. (2021) evaluated the rennet gelation characteristics of milk protein dispersion with 15% (w/w) protein content according to different pH, [$Ca^{2+}$], and temperature. The primary elements boosting gel formation and the final gel's strength were found to be pH, temperature, and their combination.

A 3DP can be used to print commercial processed cheese. (Ross et al. 2021; Deng et al. 2023a, b; Wang et al. 2023). As an example of a high-fat food, cheddar cheese has a distinct flavor, a high protein content, and nutritional value. Cheddar cheese has a firm structure due to its uniformly distributed fat inside a continuous protein phase (Wang et al. 2023). Ross et al. (2021) discovered that processed 3D-printed cheese with a soft, sticky consistency was produced with a higher pH (pH 5.8). Processed cheese made at higher pH levels (pH 5.7–6.1) was found to have protein particles with a "string-like" appearance, which decreased structural stability. Cohesiveness was the sole textural attribute that printing affected and poured samples were more cohesive than printed ones. In another study by Deng et al. (2023b), the apple polyphenol with a 0.8% polyphenol level in cheese ensured smooth surfaces, tightly packed layers, strong deformation resistance, and no discernible differences from the blank group sample. However, greater apple polyphenol concentrations (1.2% and 1.6%) diminish the protein gel network with a rougher shape. The same study group conducted another study on drying (hot air-drying or freeze-drying) cheese with a similar composition (Deng et al. 2023a). The sample shrinks on the outside throughout the air-drying process, but the inside moisture is not removed, which leads to fractures. The freeze-dried had low porosity of just 5.7%, strong casein retention, and good shape and color stability. The β-casein and κ-casein bands were weaker and casein retention was poorer in the AD samples because of the oxidative degradation as a result of the extended drying period and prolonged exposure to air. Additionally, the casein retention rate was greater in the freeze-dried cheese.

The protein phase will be disturbed and the size and shape of the fat globules will alter throughout the processing of processed cheese. Wang et al. (2023) showed that the protein network weakened with the WPI-nanofibrils or WPI addition, since this results in a softer texture and easier extrusion. Cheese with WPI-nanofibrils had a stronger regulatory capability than cheese with WPI. Because WPI nanofibrils can make fat globules smaller, disperse rapidly to the oil-water interface, enhance the gel microstructure's development, and quickly restore the mechanical strength necessary for deposited layers to self-support (complex network formation). This may be explained by WPI nanofibrils' unique amyloid structure and functional characteristics.

The process of 3DP meat-based items (nuggets, surimi, etc) is difficult and requires components to create filamentous structure. The capacity to produce food items that mimic the texture of meat and fish is limited by the state of 3DP technology today (Kim et al. 2021b). Because of the structural features, they are non-printable materials, so they need proper pre-processing. Meats must be homogenized to create a paste or batter, including flow enhancers that will result in an ink that can be extruded (Bulut and Candoğan 2022). Bulut and Candoğan (2022) analyzed that the addition of gelatin at 1.79% increased the complex viscosity with κ-carrageenan (as a co-gelator) in chicken-based snacks. Strong gel networks created by incorporating κ-carrageenan into gelatin gels result in unique physicochemical properties. The interactions between the double helix of κ-carrageenan and the triple helix of gelatin may be the mechanism behind this strong gel formation. (Wilson et al. 2020) blamed the denaturation of proteins and amino acids under heat for the high cooking loss (27.62%). The final product, or fried chicken nuggets, had a shrinkage percentage of 10% and a fat retention percentage of 77.47%, respectively.

Surimi is made out of deboned and minced fish meat that has been cleaned of lipids, sarcoplasmic or water-soluble proteins, and other leftovers. An intermediate fish product called surimi is ground into a paste, heated, and mixed with salt and other ingredients including fat, enzymes, and polysaccharides (Kim et al. 2021b). For the production of 3D-printed surimi, it may be necessary to add auxiliary ingredients (starch and gelatin) to the formula and apply enzymes (transglutaminase) before printing (Kim et al. 2021b; Zhao et al. 2021; Wang et al. 2024). During the heat-induced gelation process of surimi gel, transglutaminase catalyzes the transfer reaction of glutamine residues from peptide chains into the γ-carboxy amide group. This cross-links the head of the myofibrillar protein and forms non-disulfide covalent bonds (ε-(γ-Glu)-Lys), which improves the gel strength (Zhao et al. 2021). The transglutaminase-free surimi distorted after cutting, and the distortion worsened after three minutes. In surimi with transglutaminase, the interior hollow area retained a good arc form upon cutting and installation, and the structure remained stable at microwave heating at 40 and 50 W/g. When transglutaminase was added, the proteins clustered together, which forced the water out of the mixture. The protein–water interaction was reinforced and mobile water was encouraged to dwell in the surimi gel by the high-intensity microwave.

The loss of nutrients in surimi occurs during the pre-cooking and rinsing processes, particularly the high-quality polyunsaturated fatty acids (EPA, DHA, etc.) in the fish. This results in a single nutritional composition for surimi products, and the lack of lipids frequently causes the product's poor texture, which leaves it lacking in juiciness and aroma (Wang et al. 2024). Wang et al. (2024) reported that the robust structure of the emulsion gel (with fish oil) through cross-linking between fish gelatin effectively enhanced the surimi mixture's mechanical characteristics in comparison to the direct addition of oil, and improved the precision of surimi ink 3DP.

### 14.4.5 Food Simulants for Dysphagia Patients

As the population ages, the percentage of senior people experiencing difficulty chewing or swallowing food is rising quickly. Dysphagia is the term for trouble swallowing that is characterized by an abnormal delay in the movement of the food bolus during swallowing. This abnormal delay can easily lead to coughing, choking, or difficulties initiating swallowing due to food residue left in the oral cavity. 3DP offers an important opportunity in the design of nutritious food for this group (Liu et al. 2021a). Computer-controlled personalized food designs reduce material loss during processing and provide appearance-appropriate food design (Zhang et al. 2024). Overly sticky foods should not be consumed by people with dysphagia because of choking and demand more lingual effort to force food down the throat (Liu et al. 2021a). The texture of solid foods should be taken into account since, during oral processing, textural changes in solid foods happen more dramatically than in liquid foods. Understanding rheological characteristics is crucial for comprehending the swallowing process and for identifying dysphagia (Mirazimi et al. 2022).

The International Dysphagia Diet Standardization Initiative (IDDSI) test has created a practical approach for analyzing food texture and thickened fluids. Because of its simplicity and reliability, this method has been generally accepted by those reviewing diets for patients with dysphagia according to fork pressure and spoon split tests (Mirazimi et al. 2022). Standardized terminology used descriptors and metrics for texture-modified food (TMF) at various levels (3–7), providing a quick and easy way to evaluate the texture and flow of these goods at serving temperature. The IDDSI exam level 5 requires the ability to separate small soft particles solely with the force of the tongue. (Kim et al. 2024). The food must be easily mashed with minimal force on the fork, the sample holds its shape on the spoon and falls out fairly easily when the spoon is tilted or tapped, and the particles must be easily separated and pass between the tines/prongs of the fork for IDDSI to classify it as Level 5 named "Minced and Moist" (Bitencourt et al. 2023).

In a study by Liu et al. (2021a), the xanthan gum and κ-carrageenan concentration range between 0.3–0.9% in Shiitake mushroom ink passed the test, while only arabic gum at 0.9% concentration in Shiitake mushroom ink passed the test. Another study by Xing et al. (2022) revealed that the black fungus powder with arabic gum and κ-carrageenan failed the spoon tilt test, however, the sample from xanthan gum could be categorized. Zhang et al. (2024) reported that the addition of 1.5% cellulose nanocrystals into gelatin+xanthan gum failed; it was suitable for dysphagia diet formulation after storage at 4 °C. Other food simulants (gelatin or WPI and xanthan gum with cellulose nanocrystals 0.5–1.5%) were suitable even after 4 °C storage and heating to 50 °C. In a study by Qiu et al. (2023b), the spoon tilt test was not successful in the xanthan gum+basil seed gum (volumetric ratio of 1:0) in apple-edible rose blend matrix, but the other four inks were classified as suitable for dysphagia diets.

Similar studies have been conducted on protein-enriched products. Although having a lot of potential for practical application, there are few reports of 3DP dysphagia meals using proteins as modifiers (Niu et al. 2023a). When used with agar, soy protein in potato ink became harder, malleable, and resistant to deformation (Mirazimi et al. 2022). According to the IDDSI framework, soy protein hydrolysate with or without agar in potato ink samples was categorized as level 4 pure dysphagia diets (Mirazimi et al. 2022). According to Kim et al. (2024), all honeyed red ginseng puree with pea protein appeared to fit into level 5 on the fork pressure tests. However, 60 and 90 g of pea protein per kg of honeyed red ginseng puree passed all the tests. Niu et al. (2023a) explained that the only appropriate concentration of flaxseed proteins (15%, w/w) in the corn flour matrix with Fenugreek gum was suitable for dysphagia diets. On the other hand, Bitencourt et al. (2023) have not recommended milk gels (with native or modified starch) without κ-carrageenan for dysphagia. Because they tended to revert to their initial form when the fork's force was withdrawn. 3D-printed milk gels with different concentrations of κ-carrageenan (0.3–1.5%) and native or modified starch were suitable for dysphagia patients at level 5, while their molded counterparts were not (Level 6).

### *14.4.6 Drug-Loaded Delivery Systems*

Food matrices such as chocolate or hydrogels can be used to design as carriers for some drugs. The most important point is to ensure certain quality and safety standards in the production of drug dosage for specific patients (Karavasili et al. 2020). Chachlioutaki et al. (2022) stated that the 3DP of chocolate-based chewable dosage forms for paracetamol provided much better content uniformity, mass uniformity, and dose accuracy than the molded method. Because 3DP is automated, it is more effective in obtaining uniform products with lower error rates. On the other hand, Karavasili et al. (2020) analyzed that after the two-hour digestion (the oral and gastric phase) of drug-loaded dark chocolates with corn syrup, 80% of paracetamol was released within 30 min, while only 17% of ibuprofen was released after 2 h. Similarly, Al-Kabariti et al. (2024) examined that the inclusion of ibuprofen in a chocolate matrix slowed down its release compared to a reference ibuprofen tablet. Because ibuprofen has poor solubility in acidic gastric medium. Further, a delayed release of ibuprofen was obtained and slowed down with storage when the drug-loaded chocolates were prepared at 80 °C compared to those prepared at 50 °C. Moreover, when active lipase was present compared to an inactive form, the *in vitro* release of ibuprofen was faster. In vivo tests revealed that when ibuprofen was in dark chocolate formulations compared to its native form, its serum concentration was much lower.

Gummies can also be a drug-delivery matrix. In medicinal liquid formulations, hydroxypropyl methylcellulose, a water-soluble (hydrophilic) polymer forms a hydrogel as a thickening agent (Tagami et al. 2021). Tagami et al. (2021) reported stronger effects of hydroxypropyl methylcellulose on drug (lamotrigine) release and higher swelling properties compared to gelatin (no obvious effect). In a study by Tabriz et al. (2021), diphenhydramine hydrochloride, a drug, released in water was dependent on the surface area, and the synergistic effect of sweetener and strawberry flavor was improved according to the impression of sweet strawberry, fruity, and aftertaste.

Zhu et al. (2023) reported that surface crosslinking via transglutaminase with dry-spraying may improve the structure and printed form of surimi. This might also make it easier to create functional food items. 3D triangles of surimi (15 g glycerol+dry-sprayed 0.5% transglutaminase) met the requirements of IDDSI framework level 7 (easy-to-chew meals) when baked via steamer, microwave, and electric oven. Nevertheless, it was challenging to flatten and crush the 3D-printed triangles that the air fryer had fried.

## 14.5 Safety and Labeling Issues for 3D Printing of Foods

Similar to other foods, the shelf-life and safety of 3D-printed foods depend on parameters such as microbial spoilage, quality degradation, nutrient loss and the presence of toxic substances due to processing, ingredients used, migration from the package or production machines (Enfield et al. 2023). Regarding the 3D foods, the contact between the food or food ingredients with the parts of 3D printers could cause microbiological contamination and migration of leachable substances (Baiano 2022).

Powders or "food inks", can be used as 3DP materials. Powders have a long shelf-life due to their low water activity, but pastes have a higher water activity and can be susceptible to microbial growth and degradation by chemical reactions (Enfield et al. 2023). Severini et al. (2018) reported a bacterial level of 4.28 log CFU/g in their study on food inks and pointed out the need to sanitize every food contact-part of 3D printers, both in domestic/cafeteria and industrial use. Failure to properly clean the mechanical parts or cavities of food printers can result in the accumulation of harmful microbial loads (Attarin and Attaran 2020). Another microbial problem was suggested as the heating and cooling of 3D extruded foods, which makes them susceptible to microbial growth (Attarin and Attaran 2020).

The migration of toxic substances is also a safety problem related to the usage of 3D printers. In fact, there are 3D printers that can meet all the requirements, specifically built for food use. However, as these devices are quite expensive, components that can be added to 3D printers and fulfill food-related functions could be preferable as they are relatively inexpensive, but since most of these components are plastic, ultra-fine toxic particles can pass into the food during printing and can lead to adverse health effects (Sun et al. 2018).

In order to prevent both toxic substance migration and microbial growth, Baiano (2022) listed the characteristics that a 3D food printer should have in general: being safe under normal conditions of use, non-absorbent, corrosion-resistant, durable, observable in all parts, easy to clean surfaces, and having no fracture and no sharp internal angles. They also recommend coating 3D printer parts with polyurethane resins or epoxy that are food-grade to inhibit microbial growth. Kailo et al. (2023) proved that coating with epoxy resin, and using antimicrobial, and high temperature resistant filaments can improve the food safety of the 3D printers to varying degrees. The best results were reported to be obtained by epoxy resin.

It is obvious that, there is a need to expand the research on shelf life and safety of 3D foods and various processing techniques such as heat treatment, cold storage, pH control, or the addition of preservative systems for shelf-life extension of food inks (Enfield et al. 2023).

Escalante-Aburto et al. (2021) highlighted a particular biological safety issue of 3DP, which could be arise from changing eating habits and physiological rejection of traditional foods as a result of the strict consumption of 3D printed products.

In addition to food safety, the impact of 3DP on the nutritional and functional properties of foods/food ingredients and preserving these properties during 3DP are also topics to be investigated. Factors such as processing temperatures and light exposure could have a negative impact on phenolic compounds, polyunsaturated fatty acids, bioactive peptides, and other components with positive health effects (Tomašević et al. 2021). In addition, to our knowledge, no studies are available in the literature yet on the digestibility and bioavailability of 3D printed foods. Therefore, future research should address both safety and biological and nutritional effects of 3DP (Escalante-Aburto et al. 2021).

Existing regulations for 3D technology don't provide specific rules and guidelines on quality management, validation, and inspection techniques for foods produced in different areas, such as restaurants or ready-to-eat food facilities (Hai Alami et al. 2024).

Dankar et al. (2018) emphasized that it is important to develop clear regulations for 3D food production facilities and the personnel working in them (Dankar et al. 2018). However, as regulatory development has historically been slower than technological development, the applicability of existing laws to 3D foods is an issue that needs to be considered (Tran 2016). Because the potential problems such as the presence of allergens, microbial contamination, the presence of toxicants, or adulteration apply to all foods including 3D printed foods. All these problems have already been addressed and regulated by national, federal, and international mandatory norms (Baiano 2022).

Labeling can be characterized as a legal issue of particular importance for 3D-printed foods. Although it is possible to produce completely new foods with 3D, 3D printed foods are or will be imitation foods inspired by traditional foods and need to be labeled differently. In addition, unlabeled 3D printed food will be considered food fraud if the production of the traditional food requires a lower budget than the 3D printed food (Tran 2016). Although the view of 3D printed foods makes them easily noticeable from their conventional equivalents, specific labeling is strongly recommended, as is the case with GMO foods due to their potential long-term effects (Baiano 2022; Tran 2016).

3D printed foods recognized as 'novel foods' in the European Union. Based on the related regulation, the definition of a novel food represents foods that have not been substantially consumed with niyet etmek in the territory of the European Union before May 15, 1997, when the Regulation came into force, and foods that are manufactured using new production methods (European Parliement and Council 2015). In compliance with this Regulation, 'novel foods'must be safe, appropriately labeled so as not to misinstruct consumers, and not nutritionally disadvantageous if they are designed to substitute other products (Baiano 2022).

Similarly in Canada, novel foods are characterized as foods that has no history of safe use and are produced, prepared, packaged, or prepared in a manner that has not been previously implemented for that food. 3D printed foods also fall within this definition and require approval from the Health Canada before they can be placed on the market. Approvals can take between 6 months and 2 years, like EFSA approvals in the European Union (Baiano 2022).

Baiano (2022) highlighted a presumptive future scenario for 3DP as follows: 3DP could be used to produce new foods with non-food-based chemical compounds that are offered as a solution to famine or food shortages. In such a case, special legal regulations would be needed, as well as epidemiological research on the short- and long-term impacts of consuming these foods on the human body.

Another important concern about 3DP is the infringing of intellectual property, patents, and copyright laws. So much so that illegal prints are increasing with the spread of 3D prints, regardless of the area where 3D is used (Baiano 2022).

## 14.6 Conclusion

The suitability of 3DP in optimizing the physical parameters of food has been widely investigated in the literature. The rheological properties and printability (natively or non-natively materials) of food inks are important parameters in terms of 3DP performance. There are many studies in the literature on naturally printable materials such as chocolate, custard creams, purees, yogurt, etc. The use of structure-supporting materials that can form hydrogels, such as gums and polysaccharides, are common for these products. Moreover, the nutritional properties of the food should also be evaluated. Ingredients with additional nutritional value (dietary fiber, protein, nutraceuticals), or alternative food ingredients for fat-free, reduced-fat, gluten-free, and sugar-free products can also be more intensely studied. Nonetheless, studies on the bioavailability of important food components such as minerals, bioactive compounds, and proteins are limited, especially in food designs that appeal to certain special consumer groups (vegans, dysphagia patients, etc.). Additionally, in drug-carrier studies, the release and *in-vitro* digestion behavior of the drug from the material is also effective. On the other hand, some foods may need to undergo pre- or pro-processing, for example, fermentation and baking. These additional processes need to be evaluated in terms of food quality and sensory properties, and similar studies are limited in the literature. In brief, as a result, the 3DP performance of the food ink depends on (1) excellent repeatability and printing accuracy; (2) ready-to-eat matrix design; (3) additional processing characteristics; as well as the optimization of process parameters.

## References

Ahmadzadeh S, Clary T, Rosales A, Ubeyitogullari A (2023) Upcycling imperfect broccoli and carrots into healthy snacks using an innovative 3D food printing approach. Food Sci Nutr 12(1):84–93. https://doi.org/10.1002/fsn3.3820

Ainis WN, Feng R, van den Berg FWJ, Ahrné L (2023) Comparing the rheological and 3D printing behavior of pea and soy protein isolate pastes. Innovative Food Sci Emerg Technol 84:103307. https://doi.org/10.1016/j.ifset.2023.103307

Al-Kabariti AY, Arafat BT, Oriquat GA, Možná P, Jaidy H, Rehmani A, Patel K, Al-Qinna N, Alhnan MA (2024) In vitro and In vivo evaluation of dark chocolate as age-appropriate oral matrix. Eur J Pharm Sci 192:106646. https://doi.org/10.1016/j.ejps.2023.106646

Anukiruthika T, Moses JA, Anandharamakrishnan C (2020) 3D printing of egg yolk and white with rice flour blends. J Food Eng 265:109691. https://doi.org/10.1016/j.jfoodeng.2019.109691

Arii Y, Nishizawa K (2022) Tofu is a promising candidate for the development of an edible 3D-printing ink. J Food Eng 337:111249. https://doi.org/10.1016/j.jfoodeng.2022.111249

Armstrong CD, Yue L, Deng Y, Qi HJ (2022) Enabling direct ink write edible 3D printing of food purees with cellulose nanocrystals. J Food Eng 330:111086. https://doi.org/10.1016/j.jfoodeng.2022.111086

Attarin S, Attaran M (2020) Food printing: evolving technologies, challenges, opportunities, and best adoption strategies. J Int Technol Inform Manag 29(1):25–55. https://doi.org/10.58729/1941-6679.1442

Baiano A (2022) 3D printed foods: a comprehensive review on technologies, nutritional value, safety, consumer attitude, regulatory framework, and economic and sustainability issues. Food Rev Intl 38(5):986–1016. https://doi.org/10.1080/87559129.2020.1762091

Bebek Markovinović A, Putnik P, Bosiljkov T, Kostelac D, Frece J, Markov K, Žigolić A, Kaurinović J, Pavlić B, Duralija B, Zavadlav S, Bursać Kovačević D (2023) 3D printing of functional strawberry snacks: food design, texture, antioxidant bioactive compounds, and microbial stability. Antioxidants 12(2):436. https://doi.org/10.3390/antiox12020436

Bitencourt BS, Guedes JS, Saliba ASMC, Sartori AGO, Torres LCR, Amaral JEPG, Alencar SM, Maniglia BC, Augusto PED (2023) Mineral bioaccessibility in 3D printed gels based on milk/starch/κ-carrageenan for dysphagic people. Food Res Int 170:113010. https://doi.org/10.1016/j.foodres.2023.113010

Boehm MW, Nicholson RA, Baier SK (2023) Designing plant-based analogues of animal-derived foods: a review of research ranging from manufacturing to oral processing. Curr Opin Food Sci 50:100982. https://doi.org/10.1016/j.cofs.2022.100982

Bulut EG, Candoğan K (2022) Development and characterization of a 3D printed functional chicken meat based snack: optimization of process parameters and gelatin level. LWT-Food Sci Technol 154:112768. https://doi.org/10.1016/j.lwt.2021.112768

Cai Q, Zhong Y, Xu M, Huang Q, Lu X (2022) 3D printed high oil custard cream: effects of whey protein isolate, hydroxypropylated starch and carrageenan on physicochemical properties and printing performance. LWT-Food Sci Technol 156:113039. https://doi.org/10.1016/j.lwt.2021.113039

Cai L, Feng L, Nie M, Li D, Zheng T, Zhang M (2023a) Effect of different hydrocolloids on the rheological, microstructural, and 3D printing characteristics of purple sweet potato puree. Food Bioprocess Technol 16(11):2622–2634. https://doi.org/10.1007/s11947-023-03085-2

Cai Q, Zhong Y, Huang Q, Huang G, Lu X (2023b) Co-incorporation of probiotics into 3D printed custard cream with hydrophilic and hydrophobic bioactives. Food Hydrocoll 142:108809. https://doi.org/10.1016/j.foodhyd.2023.108809

Calton A, Lille M, Sozer N (2023) 3-D printed meat alternatives based on pea and single cell proteins and hydrocolloids: effect of paste formulation on process-induced fibre alignment and structural and textural properties. Food Res Int 174:113633. https://doi.org/10.1016/j.foodres.2023.113633

Chachlioutaki K, Karavasili C, Mavrokefalou EE, Gioumouxouzis CI, Ritzoulis C, Fatouros DG (2022) Quality control evaluation of paediatric chocolate-based dosage forms: 3D printing vs mold-casting method. Int J Pharm 624:121991. https://doi.org/10.1016/j.ijpharm.2022.121991

Chao C, Park HJ, Kim HW (2023) Effect of l-cysteine on functional properties and fibrous structure formation of 3D-printed meat analogs from plant-based proteins. Food Chem 439:137972. https://doi.org/10.1016/j.foodchem.2023.137972

Chen Y, McClements DJ, Peng X, Chen L, Xu Z, Meng M, Zhou X, Zhao J, Jin Z (2022) Starch as edible ink in 3D printing for food applications: a review. Crit Rev Food Sci Nutr 64(2):456–471. https://doi.org/10.1080/10408398.2022.2106546

Chow CY, Thybo CD, Sager VF, Riantiningtyas RR, Bredie WLP, Ahrné L (2021) Printability, stability and sensory properties of protein-enriched 3D-printed lemon mousse for personalised in-between meals. Food Hydrocoll 120:106943. https://doi.org/10.1016/j.foodhyd.2021.106943

Cikrikci Erunsal S, Basturk ZS, Canturkoglu I, Ozturk HI (2023) Development of 3D printed dark chocolate sweetened with carob extract. Int J Gastron Food Sci 34:100794. https://doi.org/10.1016/j.ijgfs.2023.100794

Dankar I, Haddarah A, Omar FEL, Sepulcre F, Pujolà M (2018) 3D printing technology: the new era for food customization and elaboration. Trends Food Sci Technol 75:231–242. https://doi.org/10.1016/j.tifs.2018.03.018

Demei K, Zhang M, Phuhongsung P, Mujumdar AS (2022) 3D food printing: controlling characteristics and improving technological effect during food processing. Food Res Int 156:111120. https://doi.org/10.1016/j.foodres.2022.111120

Demircan E, Aydar EF, Mertdinç Z, Kasapoğlu KN, Özçelik B (2023) 3D printable vegan plant-based meat analogue: fortification with three different mushrooms, investigation of printability, and characterization. Food Res Int 173(1):113259. https://doi.org/10.1016/j.foodres.2023.113259

Deng Y, Cheng K, Shao X, Dong X, Jiang H, Xiao G (2023a) Effects of drying method on the stability and quality of post-processing of 3D-printed processed cheese. Dry Technol 41(6):1060–1067. https://doi.org/10.1080/07373937.2022.2126494

Deng Y, Zhao G, Cheng K, Shi C, Xiao G (2023b) Effect of apple polyphenols on the antioxidant activity and structure of three-dimensional printed processed cheese. Food Secur 12(8):1731. https://doi.org/10.3390/foods12081731

Derossi A, Paolillo M, Caporizzi R, Severini C (2020) Extending the 3D food printing tests at high speed. Material deposition and effect of non-printing movements on the final quality of printed structures. J Food Eng 275:109865. https://doi.org/10.1016/j.jfoodeng.2019.109865

Dey S, Maurya C, Hettiarachchy N, Seo HS, Zhou W (2023) Textural characteristics and color analyses of 3D printed gluten-free pizza dough and crust. J Food Sci Technol 60(2):453–463. https://doi.org/10.1007/s13197-022-05596-w

Du Y, Zhang M, Mujumdar AS, Phuhongsung P, Yang C (2022) Effect of addition of rice flour and yeast on improving 3D printing of fermented dough. J Food Proc Preserv 46(11):e17016. https://doi.org/10.1111/jfpp.17016

Emir AA, Erunsal SC (2022) Impact of oleuropein on LCD-based stereolithography-assisted fabrication of 3D printed PEGDMA hydrogels. Eur Polym J 180:111592. https://doi.org/10.1016/j.eurpolymj.2022.111592

Enfield RE, Pandya JK, Lu J, McClements DJ, Kinchla AJ (2023) The future of 3D food printing: opportunities for space applications. Crit Rev Food Sci Nutr 63(29):10079–10092. https://doi.org/10.1080/10408398.2022.2077299

Escalante-Aburto A, Trujillo-de Santiago G, Álvarez MM, Chuck-Hernández C (2021) Advances and prospective applications of 3D food printing for health improvement and personalized nutrition. Compr Rev Food Sci Food Saf 20(6):5722–5741. https://doi.org/10.1111/1541-4337.12849

European Parliament and Council (2015) Regulation (EU) 2015/2283 of the European Parliament and of the Council of 25 November 2015 on novel foods, amending regulation (EU) no 1169/2011 of the European Parliament and of the Council and repealing regulation (EC) no

258/97 of the European Parliament and of the Council and Commission Regulation (EC) no 1852/2001. Off J Eur Union L327:1–22

Godoi FC, Prakash S, Bhandari BR (2016) 3d printing technologies applied for food design: status and prospects. J Food Eng 179:44–54. https://doi.org/10.1016/j.jfoodeng.2016.01.025

Godoi FC, Bhandari BR, Prakash S, Zhang M (2019) Fundamentals of 3D food printing and applications. Academic Press. https://doi.org/10.1016/C2017-0-01591-4

Guo J, Gu X, Du L, Meng Z (2023) Spirulina platensis protein nanoparticle-based bigels: dual stabilization, phase inversion, and 3D printing. Food Hydrocoll 135:108160. https://doi.org/10.1016/j.foodhyd.2022.108160

Habuš M, Golubić P, Vukušić Pavičić T, Čukelj Mustač N, Voučko B, Herceg Z, Ćurić D, Novotni D (2021) Influence of flour type, dough acidity, printing temperature and bran preprocessing on Browning and 3D printing performance of snacks. Food Bioprocess Technol 14(12):2365–2379. https://doi.org/10.1007/s11947-021-02732-w

Hai Alami A, Ghani Olabi A, Khuri S, Aljaghoub H, Alasad S, Ramadan M, Ali Abdelkareem M (2024) 3D printing in the food industry: recent progress and role in achieving sustainable development goals. Ain Shams Eng J 15(2):102386. https://doi.org/10.1016/j.asej.2023.102386

Holland S, Foster T, MacNaughtan W, Tuck C (2018) Design and characterisation of food grade powders and inks for microstructure control using 3D printing. J Food Eng 220:12–19. https://doi.org/10.1016/j.jfoodeng.2017.06.008

Hu H, Guo J, Huang Z, Huang Q, Lu X (2023) Effect of different hydrocolloids on the physicochemical, printing, and digestion properties of 3D printed purple sweet potato. Food Hydrocoll 144:109058. https://doi.org/10.1016/j.foodhyd.2023.109058

Huang JHR, Wu CY, Chan HM, Ciou JY (2022) Printing parameters of sugar/pectin jelly candy and application by using a decision tree in a hot-extrusion 3D printing system. Sustain For 14(18):11618. https://doi.org/10.3390/su141811618

Huang JHR, Lim GGCW, Su CHJ, Ciou JY (2023) Improvement of 3D white chocolate printing molding effect with oleogels. Heliyon 9(9):e19165. https://doi.org/10.1016/j.heliyon.2023.e19165

In J, Im S, Min SC (2023) Cold plasma treatment to improve jelly production using a fused deposition modeling 3D printer. Food Sci Biotechnol 32(1):39–46. https://doi.org/10.1007/s10068-022-01170-4

ISO. (2021). *ISO/ASTM 52900:2021: Additive manufacturing—General principles, fundamentals and vocabulary*, Vol. 2

Jiang Q, Wei X, Liu Q, Zhang T, Chen Q, Yu X, Jiang H (2024) Rheo-fermentation properties of bread dough with different gluten contents processed by 3D printing. Food Chem 433:137318. https://doi.org/10.1016/j.foodchem.2023.137318

Jin Z, Li Y, Yu K, Liu L, Fu J, Yao X, Zhang A, He Y (2021) 3D printing of physical organ models: recent developments and challenges. Adv Sci 8(17):2101394. https://doi.org/10.1002/advs.202101394

Jo GH, Lim WS, Kim HW, Park HJ (2021) Post-processing and printability evaluation of red ginseng snacks for three-dimensional (3D) printing 3D printing of snacks according to starch type. Food Biosci 42:101094. https://doi.org/10.1016/j.fbio.2021.101094

Johansson M, Nilsson K, Knab F, Langton M (2022) Faba bean fractions for 3D printing of protein-, starch- and fibre-rich foods. Processes 10(3):466. https://doi.org/10.3390/pr10030466

Kailo GG, Gáspár I, Abdisa KB, Pajčin I, Vlajkov V, Jokić A, Cvetković D, Grahovac J, Koris A (2023) Food safety-based evaluation of 3D printed objects. Analecta Technica Szegedinensia 17(4):1–9. https://doi.org/10.14232/analecta.2023.4.1-9

Kamath R, Basak S, Gokhale J (2022) Recent trends in the development of healthy and functional cheese analogues-a review. LWT-Food Sci Technolo 155:112991. https://doi.org/10.1016/j.lwt.2021.112991

Karavasili C, Gkaragkounis A, Moschakis T, Ritzoulis C, Fatouros DG (2020) Pediatric-friendly chocolate-based dosage forms for the oral administration of both hydrophilic and lipophilic drugs fabricated with extrusion-based 3D printing. Eur J Pharm Sci 147:105291. https://doi.org/10.1016/j.ejps.2020.105291

Keerthana K, Anukiruthika T, Moses JA, Anandharamakrishnan C (2020) Development of fiber-enriched 3D printed snacks from alternative foods: a study on button mushroom. J Food Eng 287:110116. https://doi.org/10.1016/j.jfoodeng.2020.110116

Kewuyemi YO, Kesa H, Meijboom R, Alimi OA, Adebo OA (2023) Comparison of nutritional quality, phenolic compounds, and antioxidant activity of conventional and 3D printed biscuits from wholegrain and multigrain flours. Innov Food Sci Emerg Technol 83:103243. https://doi.org/10.1016/j.ifset.2022.103243

Kim HB, Kim SW, Ko JB, Yang YJ, Oh GY, Kim YW, Kim HC (2021a) Optimization of extrusion process for 3D hard candy manufacture. J Manuf Process 71:580–588. https://doi.org/10.1016/j.jmapro.2021.09.035

Kim SM, Kim HW, Park HJ (2021b) Preparation and characterization of surimi-based imitation crab meat using coaxial extrusion three-dimensional food printing. Innov Food Sci Emerg Technol 71:102711. https://doi.org/10.1016/j.ifset.2021.102711

Kim Y, Kim HW, Park HJ (2021c) Effect of pea protein isolate incorporation on 3D printing performance and tailing effect of banana paste. LWT-Food Sci and Technol 150:111916. https://doi.org/10.1016/j.lwt.2021.111916

Kim SM, Woo JH, Kim HW, Park HJ (2022) Formulation and evaluation of cold-extruded chocolate ganache for three-dimensional food printing. J Food Eng 314:110785. https://doi.org/10.1016/j.jfoodeng.2021.110785

Kim J, Kim JS, Lim JH, Moon KD (2024) Effects of isolated pea protein on honeyed red ginseng manufactured by 3D printing for patients with dysphagia. LWT-Food Sci and Technol 191:115570. https://doi.org/10.1016/j.lwt.2023.115570

Kjar A, Huang Y (2019) Application of micro-scale 3D printing in pharmaceutics. Pharmaceutics 11(8):390. https://doi.org/10.3390/pharmaceutics11080390

Ko HJ, Wen Y, Choi JH, Park BR, Kim HW, Park HJ (2021) Meat analog production through artificial muscle fiber insertion using coaxial nozzle-assisted three-dimensional food printing. Food Hydrocoll 120:106898. https://doi.org/10.1016/j.foodhyd.2021.106898

Kruth J-P, Levy G, Klocke F, Childs THC (2007) Consolidation phenomena in laser and powder-bed based layered manufacturing. CIRP Ann 56(2):730–759. https://doi.org/10.1016/j.cirp.2007.10.004

Le Tohic C, O'Sullivan JJ, Drapala KP, Chartrin V, Chan T, Morrison AP, Kerry JP, Kelly AL (2018) Effect of 3D printing on the structure and textural properties of processed cheese. J Food Eng 220:56–64. https://doi.org/10.1016/j.jfoodeng.2017.02.003

Le H, Wang X, Wei Y, Zhao Y, Zhang J, Zhang L (2022) Making polyol gummies by 3D printing: effect of polyols on 3D printing characteristics. Food Secur 11(6):874. https://doi.org/10.3390/foods11060874

Le-Bail A, Maniglia BC, Le-Bail P (2020) Recent advances and future perspective in additive manufacturing of foods based on 3D printing. Curr Opin Food Sci 35:54–64. https://doi.org/10.1016/j.cofs.2020.01.009

Letras P, Oliveira S, Varela J, Nunes MC, Raymundo A (2022) 3D printed gluten-free cereal snack with incorporation of Spirulina (Arthrospira platensis) and/or Chlorella vulgaris. Algal Res 68:102863. https://doi.org/10.1016/j.algal.2022.102863

Lille M, Nurmela A, Nordlund E, Metsä-Kortelainen S, Sozer N (2018) Applicability of protein and fiber-rich food materials in extrusion-based 3D printing. J Food Eng 220:20–27. https://doi.org/10.1016/j.jfoodeng.2017.04.034

Lille M, Kortekangas A, Heiniö RL, Sozer N (2020) Structural and textural characteristics of 3D-printed protein- and dietary fibre-rich snacks made of milk powder and wholegrain rye flour. Food Secur 9(11):1527. https://doi.org/10.3390/foods9111527

Lim WS, Lim N, Kim HW, Park HJ (2023) Effect of emulsion gel as butter substitute on the dimensional stability and nutritional profile of 3D printed cookies. Food Biosci 56:103207. https://doi.org/10.1016/j.fbio.2023.103207

Lipton JI, Cutler M, Nigl F, Cohen D, Lipson H (2015) Additive manufacturing for the food industry. Trends Food Sci Technol 43(1):114–123. https://doi.org/10.1016/j.tifs.2015.02.004

Liu Z, Zhang M (2021) Texture properties of microwave post-processed 3D printed potato snack with different ingredients and infill structure. Future Foods 3:100017. https://doi.org/10.1016/j.fufo.2021.100017

Liu Z, Zhang M, Bhandari B, Wang Y (2017) 3D printing: printing precision and application in food sector. Trends Food Sci Technol 69:83–94. https://doi.org/10.1016/j.tifs.2017.08.018

Liu Z, Bhandari B, Guo C, Zheng W, Cao S, Lu H, Mo H, Li H (2021a) 3D printing of shiitake mushroom incorporated with gums as dysphagia diet. Food Secur 10(9):2189. https://doi.org/10.3390/foods10092189

Liu Y, Yi S, Ye T, Leng Y, Alomgir Hossen M, Sameen DE, Dai J, Li S, Qin W (2021b) Effects of ultrasonic treatment and homogenization on physicochemical properties of okara dietary fibers for 3D printing cookies. Ultrason Sonochem 77:105693. https://doi.org/10.1016/j.ultsonch.2021.105693

Mirazimi F, Saldo J, Sepulcre F, Gràcia A, Pujola M (2022) Enriched puree potato with soy protein for dysphagia patients by using 3D printing. Food Front 3(4):706–715. https://doi.org/10.1002/fft2.149

Miyanaji H, Momenzadeh N, Yang L (2018) Effect of printing speed on quality of printed parts in binder jetting process. Addit Manuf 20:1–10. https://doi.org/10.1016/j.addma.2017.12.008

Muthurajan M, Veeramani A, Rahul T, Gupta RK, Anukiruthika T, Moses JA, Anandharamakrishnan C (2021) Valorization of food industry waste streams using 3D food printing: a study on noodles prepared from potato peel waste. Food Bioprocess Technol 14(10):1817–1834. https://doi.org/10.1007/s11947-021-02675-2

Nam HK, Kang TW, Kim IW, Choi RY, Kim HW, Park HJ (2023) Physicochemical properties of cricket (Gryllus bimaculatus) gel fraction with soy protein isolate for 3D printing-based meat analogue. Food Biosci 53:102772. https://doi.org/10.1016/j.fbio.2023.102772

Niu D, Zhang M, Mujumdar AS, Li J (2023a) Investigation of 3D printing of toddler foods with special shape and function based on fenugreek gum and flaxseed protein. Int J Biol Macromol 253:127203. https://doi.org/10.1016/j.ijbiomac.2023.127203

Niu D, Zhang M, Tang T, Mujumdar AS, Li J (2023b) Investigation of 3D printing of children starch gummies with precise and special shape based on change of model parameters. J Food Eng 356:111568. https://doi.org/10.1016/j.jfoodeng.2023.111568

Ogata K, Gakumi R, Hashimoto A, Ushiku Y, Yoshida S (2023) The influence of Bouba- and Kiki-like shape on perceived taste of chocolate pieces. Front Psychol 14:1170674. https://doi.org/10.3389/fpsyg.2023.1170674

Ozel B, Cikrikci S, Aydin O, Oztop MH (2017) Polysaccharide blended whey protein isolate-(WPI) hydrogels: a physicochemical and controlled release study. Food Hydrocoll 71:35–46. https://doi.org/10.1016/j.foodhyd.2017.04.031

Pant A, Lee AY, Karyappa R, Lee CP, An J, Hashimoto M, Tan U-X, Gladys W, Chua CK, Zhang Y (2021) 3D food printing of fresh vegetables using food hydrocolloids for dysphagic patients. Food Hydrocoll 114:106546. https://doi.org/10.1016/j.foodhyd.2020.106546

Pérez B, Nykvist H, Brøgger AF, Larsen MB, Falkeborg MF (2019) Impact of macronutrients printability and 3D-printer parameters on 3D-food printing: a review. Food Chem 287:249–257. https://doi.org/10.1016/j.foodchem.2019.02.090

Portanguen S, Tournayre P, Sicard J, Astruc T, Mirade PS (2019) Toward the design of functional foods and biobased products by 3D printing: a review. Trends Food Sci Technol 86:188–198. https://doi.org/10.1016/j.tifs.2019.02.023

Pulatsu E, Lin M (2021) A review on customizing edible food materials into 3D printable inks: approaches and strategies. Trends Food Sci Technol 107:68–77. https://doi.org/10.1016/j.tifs.2020.11.023

Qiu Y, McClements DJ, Chen J, Li C, Liu C, Dai T (2023a) Construction of 3D printed meat analogs from plant-based proteins: improving the printing performance of soy protein- and gluten-based pastes facilitated by rice protein. Food Res Int 167:112635. https://doi.org/10.1016/j.foodres.2023.112635

Qiu L, Zhang M, Bhandari B, Chitrakar B, Chang L (2023b) Investigation of 3D printing of apple and edible rose blends as a dysphagia food. Food Hydrocoll 135:108184. https://doi.org/10.1016/j.foodhyd.2022.108184

Rando P, Ramaioli M (2021) Food 3D printing: effect of heat transfer on print stability of chocolate. J Food Eng 294:110415. https://doi.org/10.1016/j.jfoodeng.2020.110415
Riantiningtyas RR, Sager VF, Chow CY, Thybo CD, Bredie WLP, Ahrné L (2021) 3D printing of a high protein yoghurt-based gel: effect of protein enrichment and gelatine on physical and sensory properties. Food Res Int 147:110517. https://doi.org/10.1016/j.foodres.2021.110517
Ross MM, Crowley SV, Crotty S, Oliveira J, Morrison AP, Kelly AL (2021) Parameters affecting the printability of 3D-printed processed cheese. Innov Food Sci Emerg Technol 72:102730. https://doi.org/10.1016/j.ifset.2021.102730
Severini C, Derossi A, Ricci I, Caporizzi R, Fiore A (2018) Printing a blend of fruit and vegetables. New advances on critical variables and shelf life of 3D edible objects. J Food Eng 220:89–100. https://doi.org/10.1016/j.jfoodeng.2017.08.025
Shahbazi M, Jäger H, Ettelaie R (2021) Application of Pickering emulsions in 3D printing of personalized nutrition. Part II: functional properties of reduced-fat 3D printed cheese analogues. Colloids Surf A Physicochem Eng Asp 624:126760. https://doi.org/10.1016/j.colsurfa.2021.126760
Shi Y, Zhang M, Bhandari B (2021) Effect of addition of beeswax based oleogel on 3D printing of potato starch-protein system. Food Struct 27:100176. https://doi.org/10.1016/j.foostr.2021.100176
Shi H, Li J, Xu E, Yang H, Liu D, Yin J (2023) Microscale 3D printing of fish analogues using soy protein food ink. J Food Eng 347:111436. https://doi.org/10.1016/j.jfoodeng.2023.111436
Soni R, Ponappa K, Tandon P (2022) A review on customized food fabrication process using food layered manufacturing. LWT-Food Sci Technol 161:113411. https://doi.org/10.1016/j.lwt.2022.113411
Sun J, Zhou W, Yan L, Huang D, Lin L (2018) Extrusion-based food printing for digitalized food design and nutrition control. J Food Eng 220:1–11. https://doi.org/10.1016/j.jfoodeng.2017.02.028
Tabriz AG, Fullbrook DHG, Vilain L, Derrar Y, Nandi U, Grau C, Morales A, Hooper G, Hiezl Z, Douroumis D (2021) Personalised tasted masked chewable 3d printed fruit-chews for paediatric patients. Pharmaceutics 13(8):1301. https://doi.org/10.3390/pharmaceutics13081301
Tagami T, Ito E, Kida R, Hirose K, Noda T, Ozeki T (2021) 3D printing of gummy drug formulations composed of gelatin and an HPMC-based hydrogel for pediatric use. Int J Pharm 594:120118. https://doi.org/10.1016/j.ijpharm.2020.120118
Tay JU, Zhou C, Lee HW, Lu Y, Huang D (2023) 3D printing of salmon fillet mimic: imparting printability via high-pressure homogenization and post-printing texturization via transglutaminase. Food Hydrocoll 140:108564. https://doi.org/10.1016/j.foodhyd.2023.108564
Tomašević I, Putnik P, Valjak F, Pavlić B, Šojić B, Bebek Markovinović A, Bursać Kovačević D (2021) 3D printing as novel tool for fruit-based functional food production. Curr Opin Food Sci 41:138–145. https://doi.org/10.1016/j.cofs.2021.03.015
Tran JL (2016) 3D-printed food. Minn J Law Sci Technol 17:855
Udo T, Mummaleti G, Mohan A, Singh RK, Kong F (2023) Current and emerging applications of carrageenan in the food industry. Food Res Int 173(2):113369. https://doi.org/10.1016/j.foodres.2023.113369
Uribe-Alvarez R, O'Shea N, Murphy CP, Coleman-Vaughan C, Guinee TP (2021) Evaluation of rennet-induced gelation under different conditions as a potential method for 3D food printing of dairy-based high-protein formulations. Food Hydrocoll 114:106542. https://doi.org/10.1016/j.foodhyd.2020.106542
Wang M, Zhao R (2023) A review on nutritional advantages of edible mushrooms and its industrialization development situation in protein meat analogues. J Future Food 3(1):1–7. https://doi.org/10.1016/j.jfutfo.2022.09.001
Wang D, Guo J, Wang Y, Yang Y, Jiang B, Li D, Feng Z, Liu C (2023) Whey protein isolate nanofibrils as emulsifying agent to improve printability of Cheddar cheese for 3D printing. Food Hydrocoll 142:108807. https://doi.org/10.1016/j.foodhyd.2023.108807
Wang Y, Hu D, Yue W, You J, Yin T, Rong J, Liu R, Xiong S, Hu Y (2024) High internal phase emulsions of ω-3 fatty acids stabilized by fish scale gelatin particles: application to lipid-

enhanced surimi 3D printing. Food Hydrocoll 146:109254. https://doi.org/10.1016/j.foodhyd.2023.109254

Waseem M, Tahir AU, Majeed Y (2023) Printing the future of food: the physics perspective on 3D food printing. Food Phys 1:100003. https://doi.org/10.1016/j.foodp.2023.100003

Wen Y, Che QT, Kim HW, Park HJ (2021) Potato starch altered the rheological, printing, and melting properties of 3D-printable fat analogs based on inulin emulsion-filled gels. Carbohydr Polym 269:118285. https://doi.org/10.1016/j.carbpol.2021.118285

Wen Y, Kim HW, Park HJ (2022) Effect of xylose on rheological, printing, color, texture, and microstructure characteristics of 3D-printable colorant-containing meat analogs based on mung bean protein. Food Res Int 160:111704. https://doi.org/10.1016/j.foodres.2022.111704

Wen Y, Cho B, Park HJ, Kim HW (2023) Development of 3D-printable plant-based fat analogs utilizing coconut oil and glucomannan emulsion gels. Food Biosci 56:103255. https://doi.org/10.1016/j.fbio.2023.103255

Weng W, Zheng W (2015) Effect of setting temperature on glucono-δ-lactone-induced gelation of silver carp surimi. J Sci Food Agric 95(7):1379–1570. https://doi.org/10.1002/jsfa.6857

Wilms P, Daffner K, Kern C, Gras SL, Schutyser MAI, Kohlus R (2021) Formulation engineering of food systems for 3D-printing applications—a review. Food Res Int 148:110585. https://doi.org/10.1016/j.foodres.2021.110585

Wilson A, Anukiruthika T, Moses JA, Anandharamakrishnan C (2020) Customized shapes for chicken meat–based products: feasibility study on 3D-printed nuggets. Food Bioprocess Technol 13(11):1968–1983. https://doi.org/10.1007/s11947-020-02537-3

Wu C, Liu Z, Zhi L, Jiao B, Hu H, Ma X, Zhu J, Pignitter M, Wang Q, Shi A (2023) Effect of oil content and protein particles concentration on non-dairy whip topping based on Pickering emulsion system and their 3D printing properties. Food Hydrocoll 144:109012. https://doi.org/10.1016/j.foodhyd.2023.109012

Xing X, Chitrakar B, Hati S, Xie S, Li H, Li C, Liu Z, Mo H (2022) Development of black fungus-based 3D printed foods as dysphagia diet: effect of gums incorporation. Food Hydrocoll 123:107173. https://doi.org/10.1016/j.foodhyd.2021.107173

Yang F, Zhang M, Bhandari B, Liu Y (2018) Investigation on lemon juice gel as food material for 3D printing and optimization of printing parameters. LWT-Food Sci Technol 87:67–76. https://doi.org/10.1016/j.lwt.2017.08.054

You S, Huang Q, Lu X (2023) Development of fat-reduced 3D printed chocolate by substituting cocoa butter with water-in-oil emulsions. Food Hydrocoll 135:108114. https://doi.org/10.1016/j.foodhyd.2022.108114

Yu J, Li D, Wang L, Wang Y (2023a) Improving freeze-thaw stability and 3D printing performance of soy protein isolate emulsion gel inks by guar & xanthan gums. Food Hydrocoll 136:108293. https://doi.org/10.1016/j.foodhyd.2022.108293

Yu Q, Zhang M, Bhandari B, Li J (2023b) Future perspective of additive manufacturing of food for children. Trends Food Sci Technol 136:120–134. https://doi.org/10.1016/j.tifs.2023.04.009

Zhang Z, Zheng B, Tang Y, Chen L (2022) Starch concentration is an important factor for controlling its digestibility during hot-extrusion 3D printing. Food Chem 379:132180. https://doi.org/10.1016/j.foodchem.2022.132180

Zhang C, Wang CS, Girard M, Therriault D, Heuzey MC (2024) 3D printed protein/polysaccharide food simulant for dysphagia diet: impact of cellulose nanocrystals. Food Hydrocoll 148:109455. https://doi.org/10.1016/j.foodhyd.2023.109455

Zhao Z, Wang Q, Yan B, Gao W, Jiao X, Huang J, Zhao J, Zhang H, Chen W, Fan D (2021) Synergistic effect of microwave 3D print and transglutaminase on the self-gelation of surimi during printing. Innov Food Sci Emerg Technol 67:102546. https://doi.org/10.1016/j.ifset.2020.102546

Zheng L, Liu J, Liu R, Xing Y, Jiang H (2021) 3D printing performance of gels from wheat starch, flour and whole meal. Food Chem 356:129546. https://doi.org/10.1016/j.foodchem.2021.129546

Zhou L, Meng FB, Li YC, Shi XD, Yang YW, Wang M (2023) Effect of peach gum polysaccharide on the rheological and 3D printing properties of gelatin-based functional gummy candy. Int J Biol Macromol 253:127186. https://doi.org/10.1016/j.ijbiomac.2023.127186

Zhu S, Stieger MA, van der Goot AJ, Schutyser MAI (2019) Extrusion-based 3D printing of food pastes: correlating rheological properties with printing behaviour. Innov Food Sci Emerg Technol 58:102214. https://doi.org/10.1016/j.ifset.2019.102214

Zhu J, Cheng Y, Ouyang Z, Yang Y, Ma L, Wang H, Zhang Y (2023) 3D printing surimi enhanced by surface crosslinking based on dry-spraying transglutaminase, and its application in dysphagia diets. Food Hydrocoll 140:108600. https://doi.org/10.1016/j.foodhyd.2023.108600

# Chapter 15
# Packaging Aspects of Food Analogues

**Hatice Kavuncuoglu**

## 15.1 Introduction

The food industry is undergoing constant evolution, and a significant aspect of this transformation is the emergence of food analogues. Food analogues, also referred to as food substitutes or alternatives, are meticulously formulated products designed to closely replicate the sensory characteristics, nutritional profiles, and functional attributes of conventional food items (Del Nobile and Conte 2013). These innovative products cater to a wide range of consumer preferences and dietary requirements, including plant-based diets, lactose intolerance, gluten sensitivity, and ethical considerations such as sustainability and animal welfare.

In the continually evolving realm of the food industry, the role of packaging emerges as an essential component that warrants comprehensive investigation and scholarly examination. Packaging goes beyond its traditional function of containment and safeguarding, serving as a multifaceted interface that profoundly influences the tangible attributes of food analogues, as well as their perceptual reception, consumer acceptance, and strategic market positioning (Ma et al. 2022). The packaging of food analogues requires special methods and technologies to meet the requirements of these products, in terms of selection of appropriate packaging materials, techniques to extend shelf life and maintenance of quality and safety (Abdelrazek et al. 2022).

Additionally, addressing environmental concerns regarding packaging in the context of food analogues requires scientific research. The packaging industry places great emphasis on sustainability and regulatory compliance. (Shaikh et al.

H. Kavuncuoglu (✉)
Department of Gastronomy and Culinary Arts, School of Akcakoca Tourism and Hotel Management, Duzce University, Duzce, Turkey
e-mail: haticekavuncuoglu@duzce.edu.tr



Ö. P. CAN et al. (eds.), *Food Analogues: Emerging Methods and Challenges*,
https://doi.org/10.1007/978-3-031-69872-9_15

2021). Therefore, for packaging solutions that will address environmental concerns and meet legal obligations, it is necessary to correctly understand the connection between these two concepts and work accordingly.

Innovations in food analog packaging have become a focus of attention for research and development activities. AP and IP methods have emerged as promising approaches to provide the desired features in food analogue packaging (Realini and Marcos 2014). AP systems interact with the food analogue to extend its shelf life, while intelligent packaging systems provide information about the condition of the wrapped food analogue, such as freshness and storage conditions (Muller and Schmid 2019). In addition, studies such as packaging designs that will provide convenience have increased in order to meet the increasing demands and needs of consumers (Biji et al. 2015).

These packaging strategies significantly impact the market success of food analogues. Research and findings from the food packaging industry concretely reveal this fact (Silayoi and Speece 2007). By analyzing and applying these methods and technologies, stakeholders in the food sector can increase place of food analogues in industry and their consumer appeal.

As the field of food analogue packaging continues to evolve, it is crucial to identify promising research trajectories and future directions. This includes exploring emerging technologies, addressing sustainability challenges, and considering the evolving needs and expectations of consumers. By providing a comprehensive guide to the packaging aspects of food analogues, this book chapter aims to contribute to the scholarly inquiry and practical knowledge in this nascent sector. It caters to the intellectual needs of academics, industry practitioners, and stakeholders vested in the ever-evolving panorama of food analogue packaging. Major components of food analogues packaging were given in Table 15.1.

**Table 15.1** Major components of food analogues packaging

| Component | Description |
|---|---|
| Packaging material | Material used for the packaging, e.g., plastic, paper, or metal |
| Barrier properties | The ability of the packaging to prevent gas and moisture permeation |
| Shelf life | The duration during which the product remains safe and of good quality |
| Environmental impact | Assessment of the packaging's ecological footprint |
| Sensory attributes | How the packaging affects the sensory perception of the food |
| Labeling and information | Information provided on the packaging, including nutritional details |
| Safety regulations | Compliance with regulations and standards related to food packaging |
| Cost and pricing | Evaluation of the economic feasibility of the packaging material |
| Packaging design | Aesthetics, functionality, and user-friendliness of the packaging |
| Sustainability | Consideration of the package's impact on social and environmental aspects |

## 15.2 Packaging Materials for Food Analogues

Selecting appropriate packaging materials is critical for preserve from spoilage of food analogues. The food industry relies on a variety of packaging materials to shield these products from external factors such as moisture, oxygen, light, and microbial contamination (Smith 1993). In the selection of packaging material, factors such as the composition of the food analogue, its sensory and structural properties, and the desired shelf life of the product are taken into account (Wilbey 2005).

The first technology that stands out to provide the desired features in the packaging of food analogues is active packaging. Active packaging materials contain active ingredients such as antioxidant or/and antimicrobial substances that can help ensure quality and safety of the packaged food throughout its shelf life. (Vasile and Baican 2021).

Furthermore, intelligent packaging offers a significant innovation in the packaging of food analogues. This type of packaging incorporates advanced technologies such as sensors and indicators to supply real-time information such as the condition, freshness and integrity of packaged food analogue. These intelligent packaging systems allow consumers to stay informed about the products they choose and consume, ultimately contributing to the optimization of food quality and safety. Intelligent packaging provides transparency in the food supply chain by acting as a bridge between producers and consumers.

In addition to active and intelligent packaging, additional methods are available to ensure the quality and safety of food analogues throughout the chain from production to consumption. $O_2$ and $CO_2$ scavengers and aroma absorbers can be evaluated in this context. These advanced tools are aimed at extending the shelf life of food analogues by effectively minimized the adverse effects of oxygen and undesirable aromas.

The food sector is progressively attaching importance to sustainability and environmentally friendly practices. In this context, sustainable packaging solutions such as biodegradable and edible materials are gaining importance. These materials are designed to fit seamlessly to prevent packaging materials from causing ecological pollution and reduce environmental concerns. In addition to its environmentally friendly qualities, it also promises great promise in preserving the safety and quality of food analogues due to its sustainability.

To sum up, the selection of packaging materials, taking into account technologies such as active and intelligent packaging, besides the sustainable alternatives, plays an important role in preserving the quality and safety of food analogues. The dynamic environment of packaging materials and technologies requires constant research and innovation for the food industry to meet the ever-evolving requests of consumers.

## *15.2.1 Active Packaging (AP)*

Active packaging (AP) systems are designed to release or absorb active ingredients into the food or its environment. These systems are regulated by European Regulation 450/2009 (European Commission 2009). These systems primarily aim to maintain the safety of packaged food analogues throughout their shelf life and to improve their general condition (Kline 2001). AP typically falls into two main categories: active absorber-release systems. The first aims to remove undesirable elements such as oxygen, carbon dioxide, moisture, ethylene or odor from the food analogue or its immediate environment. In contrast, the latter involves the incorporation of specific compounds such as antimicrobials, antioxidants, carbon dioxide, ethylene, or ethanol into the packaged food or its environment (Yildirim et al. 2018).

AP systems allow the expected benefit of active ingredients to be obtained through the packaging system instead of adding them directly to the food analogue. This approach allows the use of less of the required active substance and the reduction of the interaction of these substances with the food analogue. For this reason, it is emphasized that incorporating active ingredients into food systems through packaging is a more effective method (Gómez-Estaca et al. 2014). Nanotechnology has attracted great attention in the field of AP because of its potential to advance the desired properties of food packaging materials such as mechanical and barrier properties. Nanomaterials show completely different and superior properties due to the increased surface area/volume, giving the material the desired advantages.

Nanotechnology offers promising opportunities for improving the characteristics of materials used in food packaging and enhancing their functionality. By integrating active substances into the packaging system, the amount of active substances required can be reduced, and potential interactions with food components can be minimized. This approach provides a more effective and efficient method for delivering active substances to food systems. AP systems are summarized in Table 15.2.

### 15.2.1.1 Antioxidant Packaging

Antioxidants have become very important in both research and development activities and industry applications due to their significant effects in increasing the stability of food products susceptible to oxidative degradation. Oxidative spoilage not

**Table 15.2** Active packaging system

| |
|---|
| **Antioxidant packaging**<br>Focused on increasing the stability of food analogues susceptible to oxidative degradation through the integration of antioxidants into a polymer matrix |
| **Antimicrobial packaging**<br>Involves inhibiting the growth of pathogenic microorganisms and the gradual release of antimicrobial substances in packaging materials |
| **Other active packaging systems**<br>Ethylene absorbers—Moisture absorbers—Carbon dioxide scavengers and emitters—Odor and taste absorbers |

only causes loss of quality in food, but also negatively affects its shelf life. Antioxidants are added to the formulations in order to preserve the quality parameters of food analogues such as color and taste. Additionally, it may even eliminate the need for some food additives (Gómez-Estaca et al. 2014).

Antioxidant packaging represents an innovative strategy to prevent or delay oxidative degradation and extend the shelf life of food analogues. This method involves the integration of antioxidants into a polymer matrix; thus, a relatively smaller amount of active ingredient is required compared to direct addition to the formulation of the food analogue. This approach eliminates the need for complex processes such as mixing, dipping or spraying. It also facilitates a controlled and regular release of active ingredients into the food analogue (Bolumar et al. 2011). This innovative arrangement not only ensures maximum efficiency, but also places a strong emphasis on safety, making this technology attractive (Almasi et al. 2021).

It is necessary to emphasize the importance of compatibility when it comes to the selection of antioxidant substances included in AP systems. These systems can contain a wide variety of synthetic and natural antioxidant compounds known for their ability to increase the oxidative stability of food products. When making these choices, it is crucial to take into account factors such as the composition of the packaged food analogue, storage conditions, and the synergy between the selected antioxidant compound and the packaging material. The effective distribution of these antioxidants is affected by the compatibility between the antioxidant used and the packaging material.

Additionally, the controlled release of these antioxidants into food or the environment is vital. Controlled-release packaging is a new approach that uses packaging containing varying concentrations of active compounds around a food analogue to extend shelf life. These concentration profiles change over time, providing controlled release rates of active compounds for food quality and safety (Moshe et al. 2016). This technique serves as a delivery system that enables effective use of the initial concentration of antioxidants to extend shelf life during storage and distribution, making it a valuable innovation in functional packaging. In addition, the specific type of antioxidant chosen must match the specific properties of the food analogue in question. For example, nonpolar antioxidants are more suitable for foods with high lipid content, while polar antioxidants may be preferred for other types of foods. By carefully considering these aspects it is possible to alleviate the 'antioxidant paradox', making the nature of the packaged product a very important factor in the selection process (Dobrucka and Przekop 2019).

The possible harms of synthetic antioxidants on human health have limited their use and directed research towards natural antioxidants (Echevarría et al. 2015; Silva et al. 2014; Ganiari et al. 2017). Natural antioxidant compounds and extracts obtained from sources such as rosemary, grape seed, black cumin, green tea and herbs and spices such as cinnamon, lemongrass, ginger and thyme can be used safely and effectively instead of synthetic antioxidants in AP (Silva et al. 2014; Amorati et al. 2013).

However, an important aspect that limits the utilize of natural antioxidants is that they require larger amounts to achieve the desired antioxidant activity (Tian et al.

2013). Despite such disadvantages, the utilize of by-products from the food sector as potential antioxidant sources for food packaging has been investigated. In a study, LDPE films coated with extracts from brewing waste were successfully used to package beef, achieved to a remarkable 80% reduction in lipid oxidation in cold storage (Barbosa-Pereira et al. 2014). Additionally, barley husk, another promising byproduct from the beer industry, was found effective in slowing lipid hydrolysis and delaying oxidative degradation in blue shark muscle (Pereira de Abreu et al. 2011).

Antioxidants can be integrated into packaging systems in a variety of ways, including bags, labels, surface coatings, immobilization on packaging materials, incorporation into polymer matrices, and the development of multilayer films.

#### 15.2.1.2 Antimicrobial Packaging

In recent years, another noteworthy approach that has garnered substantial attention is antimicrobial packaging, which falls within the purview of active and controlled release packaging. Antimicrobial packaging entails the integration of packaging materials with antimicrobial agents designed to be gradually released over time, thereby targeting and eradicating pathogenic microorganisms that pose a threat to food products (Sharma et al. 2017). This technology exhibits significant promise in extending the shelf life of food products by effectively restraining microbial proliferation.

Antimicrobial food packaging systems are meticulously engineered to counteract the proliferation of pathogenic microorganisms within food products, thus serving as a robust deterrent to food spoilage. Among the various active components under extensive research, antimicrobial agents assume paramount importance, given their pivotal role in suppressing the growth of both pathogenic and deteriorative microorganisms, which stand as fundamental contributors to food losses (Ahmed et al. 2017).

Metals, including Ag, Cu, Au, and Pt, along with metal oxides like $TiO_2$, and a wide array of essential oils such as rosemary, clove, thyme, tea tree, lemongrass, and basil, have been harnessed for their antimicrobial properties for millennia and continue to be some of the most extensively employed antimicrobial agents. Furthermore, biopolymers such as peptides and proteins (e.g., nisin and lactoferrin), enzymes (e.g., lysozyme), and polysaccharides such as chitosan, are subjects of active research and application due to their firmly established antimicrobial properties.

Furthermore, synthetic antimicrobial agents, including quaternary ammonium salts, and organic acids like benzoic, acetic, and propionic acid, have diverse applications in several domains (Aziz and Karboune 2018). These organic acids function as robust antimicrobial agents by disrupting membrane transport and permeability and reducing intracellular pH. In many cases, ascorbic acid and citric acid are employed in fresh produce and fruit juices. Packaging films that contain organic acids must come into direct contact with food materials to release active compounds onto the food surface.

Antimicrobial food packaging materials function by extending the lag phase, shortening log phase, or reducing the microbial count, making antimicrobial AP a highly promising approach. Typically, antimicrobial agents act by disrupting cell walls or membranes. There are three different methods for incorporating antimicrobial agents into the system: release, absorption, and immobilization (Lee et al. 2015). In the release method, antimicrobial agents migrate into the food or the headspace of the package to inhibit microbial growth. The absorption method eliminates factors that promote microbial growth in food. In contrast, immobilization prevents deterioration upon contact with the surface and is more suitable for liquid foods than solid foods.

#### 15.2.1.3 Other Active Packagins Systems

Beyond the previously mentioned strategies, there is a diverse category of AP focused on ethylene absorbers. Ehtylene, a plant hormone, plays a crucial role in respiration and ripening of fruits, vegetables, and flowers. It's also a highly reactive compound, involved in chemical breakdown, adsorption, and modification reactions. Excessive ethylene production in the presence of various fruits and vegetables can hasten deterioration during ripening, accelerating mechanical or chemical damage, and increasing the risk of bacterial or fungal infections (Lee et al. 2015). Ethylene absorbers function by reducing ethylene production, typically by blocking its binding sites. They are frequently used to retard the ripening process of fresh fruits and vegetables, which is essential for international trade. Common ethylene absorbers include potassium permanganate, zeolite, and activated carbon (Álvarez-Hernández et al. 2018).

Another significant category within AP involves liquid or moisture absorbers. Managing the presence of liquid or moisture in food packaging is crucial for maintaining product quality and safety. Meat and fish products often face challenges related to liquid leakage (such as water, blood, or other fluids). Excessive moisture within the packaging can promote microbial growth, cause dry and crispy products like pasta, cookies, and biscuits to become soft, and lead to clumping. It can also accelerate the oxidation of fats. These AP solutions primarily function through an absorption process that, in some cases, eliminates excess liquids and, in others, regulates the relative humidity within the package. To achieve these goals, highly hygroscopic and dehydrating agents come into play.

In addition to the previously discussed strategies, AP for food analogues includes various innovative techniques and solutions that cater to different needs and challenges. Among these, carbon dioxide ($CO_2$) scavengers and emitters also stand out for food analogues (Sivertsvik et al. 2002). $CO_2$ scavengers are designed to reduce undesirable levels of carbon dioxide in the packaging atmosphere, preventing undesirable consequences such as off-flavors and rheological changes. They are particularly useful for products that require a modified atmosphere to extend shelf life, such as baked goods and fresh produce. These AP solutions help maintain the desired sensory attributes and overall quality of food analogues by balancing carbon dioxide levels within the package (Realini and Marcos 2014). Conversely, $CO_2$ emitters can be used in situations where carbon dioxide can contribute to product quality. For example, controlled $CO_2$ release in packaging fresh-cut produce helps maintain the desired firmness and color of the product. Sensory properties and overall quality of food analogues can be preserved by using AP solutions that regulate carbon dioxide levels (Yildirim et al. 2018). The development of AP technologies, including $CO_2$ scavengers and diffusers, has been developed to meet changes in retail practices and consumer demands for safer, higher quality foods with longer shelf life (Zhang et al. 2020). These innovative packaging techniques ensure effective preservation of the food analogue throughout its shelf life.

Additionally, odor and taste absorbers are other approaches that can be successfully applied to preserve the desired sensory properties of food analogues. Aroma and taste are among the important factors affecting consumer preferences and product acceptance (Small et al. 2004). These AP solutions can adsorb volatile compounds responsible for unpleasant odors or undesirable tastes (Becerril et al. 2020). These absorbents help preserve the original aroma and taste of food analogues by preventing the migration of these compounds in the packaging.

Undoubtedly, food analogue packaging requires additional considerations to preserve sensory attributes such as aroma and taste, playing an important role in consumer preferences. These approaches, which include materials such as activated carbon, zeolites or clay minerals, are excellent at adsorption of volatile compounds responsible for unpleasant odors or undesirable tastes (Becerril et al. 2020). By reducing the migration of these compounds within the packaging, these absorbents have proven effective in preserving the original aroma and taste of food analogues. This not only ensures sensory consistency but also supports consumer satisfaction, highlighting the importance of incorporating such technologies into food analog packaging applications.

The incorporation of these advanced AP strategies into food analog packaging systems is a dynamic area of research and application. The ability to regulate the atmosphere inside the packaging to meet the specific requirements of different food analogues ensures that product quality is maintained, shelf life is extended and consumer satisfaction is increased. Therefore, carbon dioxide scavengers/diffusers and taste/odor absorbers are important alternatives to meet the evolving demands of the market and consumer expectations. These versatile technologies enable tailored packaging solutions that meet the diverse needs of food analogues, contributing to the growth and sustainability of the food industry.

### *15.2.2 Intelligent Packaging (IP)*

Intelligent packaging (IP) systems refer to systems that can detect environmental changes and communicate these changes to consumers (Muller and Schmid 2019). These systems can provide real-time data and enable monitoring of the storage conditions of the food analogue throughout the supply chain. While AP aims to extend the shelf life of the product, the main purpose of IP is to monitor quality and safety of food analogue. In other words, AP is the system that acts effectively, and IP is the system that detects and transmits information (Hogan and Kerry 2008).

Nanotechnology has also highlighted its potential in the field of IP with applications such as the use of sensors with advanced sensor capabilities (Sharma et al. 2017). This technology enables the packaging to interact with internal and external factors, providing real-time information about the food analogue. In a study, investigated the development of IP with whey protein isolate/polyaniline composites, which exhibit properties suitable for IP applications (Oliveira et al. 2019). This technology enables the packaging to interact with internal and external factors, providing real-time information about the food analogue. In another study, investigated the development of IP with whey protein isolate/polyaniline composites, which exhibit properties suitable for IP applications (Wang et al. 2022).

Interest in IP systems has increased due to the rising use of active ingredients in food packaging; this suppose a tool to monitor the expected effect. On the other hand, the high complexity required to reach a reliable and efficient smart system was an obstacle to the development of smart devices (Malhotra et al. 2015). In addition, the improvement of such devices requires interdisciplinary work such as food science, materials science and chemical and electrochemical engineering, which makes the studies more complex. (Garba 2023). IP systems are summarized in Table 15.3.

#### 15.2.2.1 Indicators

Indicators that determine the presence and concentration of a substance and the reaction between different substances are among the most important components of IP systems. Typically the information they provide is qualitative or semi-quantitative; Their main purpose is to increase consumer awareness and confidence in food products. They are placed inside or outside the packaging, depending on the type of indicator (Ghaani et al. 2016).

**Table 15.3** Intelligent packaging systems

| |
|---|
| **Indicators** |
| • Temperature/Gas/Freshness indicators |
| **Data carriers** |
| • RFID |
| • Barcodes |
| **Sensors** |
| • Gas sensors |
| • Biosensors |
| **Other intelligent packaging systems** |
| • Doneness indicators |
| • Thermochromic ink |

Temperature, gas, and freshness indicators hold pivotal roles within the food packaging industry, significantly influencing consumer perceptions of product quality and safety. These indicators convey valuable information, often in a qualitative or semi-quantitative manner, enabling consumers to make informed choices.

#### 15.2.2.1.1 Temperature Indicators (TI)

Temperature indicators (TI) are instrumental for monitoring and communicating the temperature history of products throughout the supply chain. Typically, they rely on color changes or other visual cues to indicate temperature variations. For instance, in the case of perishable items like dairy products or ready-to-eat meals, temperature indicators serve as a critical tool for consumers. They offer a quick assessment of whether a product has been exposed to temperature conditions that may compromise its quality and safety. In essence, temperature indicators empower consumers to evaluate the cold chain integrity of products.

TI can be categorized into two main types: simple temperature indicators and time-temperature integrators (TTIs) as discussed by Ahvenainen and Hurme (1997). Simple temperature indicators provide information regarding whether a product has been exposed to temperatures exceeding or falling below a critical threshold, thereby indicating potential impacts on microorganism survival and protein denaturation during processes such as freezing or defrosting. In contrast, TTIs, also referred to as integrators, are specifically engineered to monitor temperature variations over time across the food supply chain. They serve as early warning systems, alerting to any unfavorable deviations from a critical reference value. These indicators operate through various mechanisms, including mechanical, chemical, electrochemical, enzymatic, or microbiological changes, which manifest as visible signs such as mechanical deformation, color changes, or color migration (Ahvenainen 2003).

Time-temperature integrators (TTIs) have garnered considerable attention owing to their pivotal role in assessing the kinetics of physical and chemical degradation over time and temperature (Biegańska et al. 2014). They offer valuable insights into the temperature history of wrapped food, aiding in the prevention of potential misuse or mishandling. TTIs are user-friendly devices that consumers readily grasp, as they serve as direct indicators of food quality at specific temperature thresholds, as noted by Pereira et al. (2015). Typically, TTIs are small, self-adhesive labels that can be attached to food packaging to help consumers track freshness.

#### 15.2.2.1.2 Gas Indicators

Gas indicators are particularly relevant in the context of modified atmosphere packaging (MAP), where the composition of gases within a package is pivotal in determining product freshness and shelf-life. These indicators continuously monitor and communicate the presence and concentration of specific gases inside the packaging environment. Notably, gas indicators play a significant role in the preservation of products like fruits, vegetables, and baked goods. By providing insight into the internal atmosphere of the package, these indicators enable consumers to choose products with confidence, knowing that the gas composition has been optimized for freshness preservation (Mills 2005).

##### 15.2.2.1.3 Freshness Indicators

Freshness indicators are designed to keep consumers informed about a product's state of freshness and its remaining shelf-life. These indicators are of utmost importance for products such as seafood, meat, and bakery items. They offer consumers a reliable means of assessing the product's condition inside the package (O'Grady & Kerry 2008). This knowledge empowers consumers to make purchase decisions aligned with their preferences for immediate consumption or more extended storage. The freshness indicator can inform the consumer in an easily perceptible way, such as color change.

The integration of temperature, gas, and freshness indicators into food packaging aligns with the overarching mission of enhancing food quality and safety. These indicators serve as essential tools, providing consumers with vital insights into product condition and quality, ultimately boosting consumer confidence and satisfaction in the food products they select.

#### 15.2.2.2 Data Carriers

Data carriers, often called automatic identification devices, help make the flow of information within the supply chain more efficient (Manthou and Vlachopoulou 2001). The purpose of these devices is not to directly evaluate food quality; The main goal is to focus on automating processes, ensuring traceability, preventing theft and combating adulteration. For food analogues, barcode labels and a Radio Frequency Identification (RFID) labels are among the IP technologies that carry data and provide convenience. When correctly integrated into the packaging material, these devices facilitate the smooth functioning of supply chains, facilitating precise and timely delivery of food analogues to consumers (Sarac et al. 2010). RFID tags, consisting of a tag, antenna and a middleware, are the most advanced example of data carrier devices. With RFID technologies, many different codes can be stored on the tag and information can be transmitted even from long distances (Müller and Schmid 2019). Thus providing the perfect solution for automatic product identification and traceability. RFID technology comprise two types of tags: active tags and passive tags, depending on whether they are connected to a battery.

Barcodes are inexpensive, easy to use, provide tracking of product information, manage inventory, and facilitate traceability (Drobnik 2015). They can be examined in two classes: one-dimensional and two-dimensional barcodes. A One-dimensional barcode consists of parallel spaces and bars. The laser beam developed for this purpose reads the barcode by measuring the time required to scan the dark bars and bright areas. Two-dimensional barcodes are a combination of points and spaces arranged in an array or matrix and offer greater memory capacity. QR (quick response) codes preferred by this industry and consumers are examples of 2D barcodes.

#### 15.2.2.3 Sensors

Sensors contribute significantly to packaging technologies by providing significant developments in the field of IP. The sensor is a complex device that includes control and processing electronics, a complex interconnection network, and software. Among the sensor types, gas sensors and biosensors stand out as key players, ensuring the integrity of packaged food analogues.

Gas sensors are designed to respond quantitatively and reversibly to specific gases within the packaging. Gas sensors provide real-time monitoring of gas levels, allowing precise control and optimization of the packaging environment to maintain product quality. On the other hand, these systems bring some disadvantages such as cross-sensitivity to carbon dioxide and hydrogen sulfide. To prevent these, studies have focused mostly on $O_2$ and $CO_2$ sensors. The development of smart sensors to measure oxygen, which is one of the most important factors limiting the shelf life of a food analogue, has been the main research topic of scientific studies (Kerry et al. 2006).

Biosensors, on the other hand, use biological components such as enzymes, antibodies, and antigens to detect specific substances in food analogues. They can provide valuable information about the presence or absence of compounds of interest to detect, such as pathogens, allergens or pollutants. Biosensors offer high sensitivity and specificity, making them effective tools in ensuring food safety and quality.

Integration of gas sensors and biosensors into food packaging systems enables continuous monitoring of critical parameters such as gas composition or pathogen presence, which directly affect the quality and safety of food analogues. This real-time monitoring is attractive to producers and consumers as it allows to detect any deviations from desired conditions (Drago et al. 2020).

In summary, gas sensors and biosensors are integral components of IP systems. Gas sensors enable real-time monitoring of gas levels, optimizing the packaging environment for freshness and shelf-life. Biosensors provide sensitive and specific detection of specific substances, contributing to food safety. The integration of these sensors into packaging systems enhances the ability to monitor and maintain the quality and safety of packaged food products.

#### 15.2.2.4 Other Intelligent Packaging Systems

IP systems encompass a range of devices that go beyond the commonly used technologies. Doneness indicators and thermochromic ink convenience-enhancing systems are two examples of devices that can be used to improve the convenience and safety of food preparation. Thermochromic inks, for instance, are printed on packaging materials, such as shrink sleeves on beverage cans, and undergo a color change within a specific pre-set temperature range that indicates the optimal temperature for food consumption (Prikh & Jancke 2008). Some thermochromic inks even display short messages, like "ready to serve," alongside the color change. Similarly, doneness indicators inform consumers when heated food has reached the desired level of doneness. However, a difficulty with doneness indicators is that the color change cannot be clearly observed if the oven is off (Robertson 2016).

Another class of intelligent devices is dedicated to addressing concerns related to tampering, counterfeiting, and theft. These systems, which are not yet properly used in the food sector, have the potential to reduce the economic challenges associated with such threats. As an instance of theft prevention, electronic article surveillance systems, which employ electronic tagging, are being explored. On the other hand, anti-counterfeiting and anti-tampering devices encompass a range of technologies, including holograms, thermochromic inks, micro-tags, tear labels, and tapes (Han et al. 2005).

When applying an IP system to a food analogue, it must be evaluated whether it is suitable for the product. Although IP systems bring a great innovation to the packaging industry, they cannot be considered reliable on their own. Malfunction or misuse of the system used may cause product loss. Due to the complex structure and composition of food analogues, the spoilage factor may be multiple and monitoring a single factor may mislead the consumer. To prevent these situations, systems can be integrated.

In summary, unlike traditional technologies, IP systems ensure the information about the quality and safety of the food analogue in which it is packaged throughout the supply chain. Although their high cost limits their usage area because they require advanced technologies, they continue to be a focus of interest for scientific research and industry.

## 15.3 Enhancing Food Shelf-Life: Innovations in Packaging Systems

Food packaging is the most important aspect to protect against deterioration of food analogues. One of the prominent areas of innovation in recent years is the development of antimicrobial and/or antioxidant packaging systems that provide controlled release. These systems will ensure the release of active ingredients at targeted levels so that the packaged food analogue will not deteriorate during its shelf life. In addition, nanotechnology is a promising approach for food packaging that offers new possibilities to enhance the quality and safety of food analogues. Nanotechnology-based packaging materials attract attention with their superior barrier properties, thermal and mechanical properties (Duncan 2011). These materials interact with factors related to the food or its environment, such as humidity and temperature, to trigger actions that extend shelf life and preserve food safety (Sharma et al. 2017).

Improving the barrier properties of packaging materials by incorporating fillers into a polymer matrix is one of the notable applications of nanotechnology. This effect occurs through the creation of a complex gas diffusion path that requires gas molecules to circulate around impermeable inorganic crystals, effectively widening the average diffusion path of gases within the film (Choudalakis and Gotsis 2009).

Taking into account possible interactions between packaging materials and food analogues, precautions should be taken if necessary. The transfer of substances from packaging materials to food is significantly affected by the properties of the packaged product such as pH or storage conditions such as temperature. (Marsh and Bugusu 2007). Therefore, in order to avoid possible health risks and comply with legal regulations, care should be taken to ensure the migration of the materials contained in the packaging material into the food analogue.

Recently, numerous researchers have made notable advancements in packaging materials for food products. Duncan (2011) reviewed various applications of nanomaterials in food packaging and safety, including nanocomposites, silver nanoparticles, and nanosensors to detect relevant analytes and foodborne pathogens. Kamarudin et al. (2022) emphasized that antimicrobial properties are gained by incorporating biodegradable polymer composites into the packaging. Muller & Schmid (2019) provided a brief overview of IP in the food sector, emphasizing its potential to improve product safety and reduce environmental impact. Marsh and Bugusu (2007) discussed concepts, materials, and environmental issues related to food packaging. In addition to the functional advantages of plastics, the concerns brought about by their increasing use were emphasized. Yousefi et al. (2019) reviewed smart sensing technologies for monitoring food quality in IP. Topuz and Uyar (2020) discussed the development of antioxidant, antibacterial, and antifungal electrospun nanofibers for food packaging applications. Sharma et al. (2017) highlighted the potential of nanotechnology in food packaging, including the use of copper nanoparticles for antimicrobial activity. Malhotra et al. (2015) discussed the potential and pitfalls of antimicrobial food packaging, including the development of resistance among microorganisms. Yildirim et al. (2018) reviewed AP applications for food, considering consumer demands for safer and higher-quality foods. Yan et al. (2022) explored innovative food packaging explored food quality and safety and consumer perspectives, including the potential risks and costs associated with certain packaging technologies.

Selection of the correct packaging materials in which food analogues will be packaged is very important to ensure their quality and safety. AP and IP systems and the integration of nanotechnology into these systems offer innovative solutions that increase shelf life, product safety and preserve sensory properties. In addition, the interaction between the packaging material and the food analogue in the new systems developed should be carefully examined and necessary precautions should be taken.

Prolonging the shelf life of food products is of utmost importance, with a direct impact on their overall quality and safety. Numerous strategies and technologies have been developed for this purpose, with AP being a prominent approach. AP involves the purposeful integration of elements that either release or absorb substances into or from the packaged food or its surrounding environment. The primary objective is to enhance food preservation and extend shelf life, achieved through diverse methods such as temperature regulation, oxygen removal, moisture control, and the incorporation of chemicals like salt, sugar, carbon dioxide, or natural acids, either individually or in combination with effective packaging (Suppakul et al. 2003). Advancements in this field play a significant role in improving food quality, safety, and sustainability (Restuccia et al. 2010; Robertson 2016).

Food packaging technologies are constantly evolving due to increasing awareness and demand for better quality and safer foods. AP and IP systems, in which nanotechnology is effectively used, have made a significant contribution to extending the shelf life of food analogues. Increasing consumer preferences for natural and minimally processed preserved foods have triggered exploration of new materials, technologies and approaches that can promote sustainable food packaging practices.

## 15.4 Aroma and Flavor Preservation in Food Analogue Packaging

Preserving aroma and flavor in food analogues requires the combination of various strategies and technologies. MAP is an effective approach to maintain the desired quality in this context. MAP is a method that involves changing the composition of the atmosphere surrounding food and is widely used to preserve the aroma and flavor profiles of freshly cut fruits and vegetables (Zhuang (2011). MAP can minimize the loss of volatile compounds responsible for aroma and taste by controlling the levels of important gases such as oxygen, carbon dioxide and ethylene.

Food analogues often consist of a combination of plant-based ingredients and specialized formulations to mimic the taste and texture of traditional animal-based foods. These novel food products may contain delicate aroma compounds that are susceptible to degradation over time. Thus, in the realm of food analogs, the preservation of aroma and flavor is not only vital for ensuring product quality but also for maintaining consumer acceptance. As such, research into tailored MAP strategies, AP materials, and controlled storage conditions specifically designed for food analogs is crucial. These considerations extend to the choice of packaging materials, gas composition, and storage temperature, among other factors. By leveraging technology and expertise in food science and packaging engineering, it becomes possible to enhance the aroma and flavor stability of food analogs, thereby meeting the growing consumer demand for high-quality, plant-based alternatives (Kader 2008).

Encapsulation is another technique that can be used to preserve flavor in food products during storage. Encapsulation ensures that volatile compounds are trapped within a protective matrix, preventing their interaction with other food components and minimizing flavor loss (Madene et al. 2006). This approach can preserve the aroma of the food analogue throughout its shelf life and, thanks to controlled release, ensure the release of the desired flavor and aroma over time. Various encapsulation methods have been developed to encapsulate aromas, such as spray drying, coacervation, and extrusion (Barnett et al. 2019). In the context of food analogues, encapsulation has proven to be a particularly promising avenue. One of the notable advantages of encapsulation is its ability to extend the shelf life of flavors and facilitate controlled release, thus ensuring the release of the desired flavor and aroma at the optimal time. As the demand for high-quality food analogues continues to grow, ongoing research and innovation in encapsulation methods is increasing (Bell et al. 2018).

There has been significant research in the field of food packaging, particularly in the area of aroma preservation (Sivertsvik et al. 2002). Biopolymers, such as peptides and proteins, enzymes, and polysaccharides, have been actively studied and implemented as active agents due to their well-established antimicrobial activities (Jariyasakoolroj et al. 2020). These active agents can help inhibit the growth of spoilage microorganisms, which can contribute to the degradation of aroma compounds in food products.

Due to ecological concerns, it has become important to develop biodegradable and sustainable packaging materials to preserve aroma. Bioplastics derived from renewable resources such as starch, cellulose and polylactic acid (PLA) are attracting attention as potential alternatives to traditional packaging materials. Although these materials do not provide as effective a barrier as plastics to preserve the aroma and taste of food products, they offer the advantage of being environmentally friendly.

In addition to the selection of appropriate packaging materials, the incorporation of active agents and the optimization of packaging design are crucial for aroma preservation. For example, the use of AP systems that release aroma compounds or absorb undesirable compounds can help maintain the desired aroma profile of food products. The selection of suitable packaging materials with specific barrier properties can also play a significant role in preventing the loss of aroma compounds during storage and transportation (Reynier et al. 2004). Barrier materials also play a crucial role in aroma and flavor preservation. Ethylene vinyl alcohol copolymer (EVOH) is one such material known for its excellent barrier properties against gases and organic substances, including aromas and flavors. By acting as a functional barrier, EVOH can prevent the migration of volatile compounds and maintain the desired sensory characteristics of the food.

Additionally, the addition of flavor enhancers can increase the perception of aroma and flavor in food products. Flavor enhancers are compounds that, when added at low concentrations, increase the taste and aroma of food, making it more attractive to the consumer (Vasilaki et al. 2022). Monosodium glutamate, which increases the umami taste, and herbs and spices, which add complexity and depth to the flavor profile of food, are flavor enhancers that can be used for this purpose.

In summary, the preservation of aroma and flavor in food packaging necessitates the utilization of a range of strategies and technological solutions. MAP, encapsulation, the use of barrier materials such as EVOH, and the addition of flavor enhancers are all effective approaches to enhance aroma and flavor retention in packaged foods. By employing these techniques, food manufacturers can ensure that the sensory qualities of their products are maintained throughout the shelf life, providing consumers with a satisfying and enjoyable eating experience. Further research and development in this field are necessary to explore new materials, technologies, and approaches that can further enhance aroma and flavor preservation in food analogue packaging.

## 15.5 Sustainable Packaging Solutions for Food Analogues

The necessity of sustainable packaging materials for food analogues has witnessed a significant rise in recent years, driven by increasing awareness of environmental sustainability and the imperative to reduce environmental pollution caused by packaging materials. Based on this, various strategies and technologies have been formulated to evolve sustainable packaging methodologies.

One approach is the utilize of biodegradable and compostable materials for food analogue packaging. Biodegradable materials, such as bioplastics derived from renewable resources, offer an alternative to traditional petroleum-based plastics that take a long time to degrade in the environment (Kool et al. 2010). These materials have the ability to naturally break down over time, reducing the accumulation of plastic waste and minimizing the environmental footprint associated with packaging. By adopting biodegradable packaging solutions, the food analogue industry can conduce to the reduction of plastic pollution and promote a more sustainable approach to packaging. However, it is important to consider the specific requirements of food analogues, such as barrier properties and shelf life, when selecting suitable biodegradable materials. Further research and development are needed to optimize the performance and functionality of biodegradable packaging materials for food analogues, ensuring both environmental sustainability and product quality (Arrieta et al. 2017).

Another effective strategy for achieving sustainable packaging for food analogues involves the incorporation of recyclable materials. Recycling plays an important role in reducing raw material consumption and the energy required to produce packaging materials. Adopting designs that will facilitate the recycling and reuse of packaging waste and compatibility with existing recycling processes will facilitate the process (Mohr et al. 2010). This focus on recyclable packaging is not only compatible with environmental protection efforts, but also promotes a circular economy in the food analogue industry, contributing to the responsible management of resources and reducing waste in the packaging lifecycle.

The concept of lightweight packaging has also begun to gain attention as a sustainable solution. Lightweight packaging means reducing the amount of material used in packaging without compromising the functionality and protective properties of the packaging. This approach not only reduces the amount of waste produced, but also reduces transportation costs and energy consumption throughout the supply chain (Fulop et al. 2018).

In addition to material choices, innovative packaging designs can also contribute to sustainability. For example, thanks to AP systems containing oxygen scavengers or moisture absorbers, the shelf life of food analogues can be extended, which helps reduce food waste and the amount of packaging used (Somerville and Casey 2010). Similarly, IP systems that utilize sensors and indicators can provide real-time information about the freshness and quality of the product, enabling consumers to make informed decisions and reduce food waste (Geier et al. 2010).

In the context of promoting sustainable packaging solutions, collaboration between stakeholders in the packaging industry is of critical importance. This collaboration aims to develop and implement sustainable packaging practices between packaging manufacturers, food producers, retailers and consumers. Additionally, government regulations and incentives can significantly influence this transformation process by supporting innovation and investment in sustainable packaging solutions and providing an enabling environment (Ernst et al. 2006).

To summarize, sustainable packaging solutions for food alternatives are critical to reducing the environmental impact of packaging materials and promoting a more sustainable food system. Factors such as biodegradable and compostable materials, recyclable packaging, lightweight designs and innovative packaging technologies play an important role in achieving sustainability goals. Additionally, collaboration between stakeholders and supportive government policies play a key role in supporting the adoption of sustainable packaging practices. By adopting these strategies, the food industry has the potential to contribute to the construction of a more sustainable future while preserving the safety and quality of food analogues.

## 15.6 Regulatory Compliance and Labeling for Food Analogues Packaging

Packaging of food analogues is an extremely important process with the aim of protecting the health and safety of consumers and ensuring the transparency of food products. This process involves a number of critical factors that require strict compliance with appropriate regulations and labeling requirements. The obligation to provide consumers with accurate and complete information about products is a fundamental component of packaging and labeling processes.

An important regulation within the European Union is Regulation (EU) 1169/2011 on the provision of food information to consumers. This regulation requires that packaging of food products provide clear and understandable information about the product, from the ingredients they contain to allergens, nutritional values and special storage or use instructions (Vermeulen et al. 2012). This regulation aims to provide consumers with comprehensive information about the food analogues they purchase and consume.

In addition to general food labeling regulations, there are specific regulations for certain types of food analogues. For example, if a food analogue product contains genetically modified organisms (GMOs), these products may be subject to labeling requirements specific to GMO foods and feeds covered by Regulation (EU) No 1829/2003 (Vermeulen et al. 2012). The regulation in question aims to help consumers make informed choices by requiring food products containing GMOs to be labeled clearly and understandably to consumers.

There are regulations that deal with specific packaging materials or substances used in the packaging of food analogue products. For example, regulations such as Regulation (EC) No 1935/2004 and Regulation (EU) No 10/2011 address food packaging, including food contact materials and articles defines the requirements for materials (Biji et al. 2015). These regulations aim to ensure that packaging materials prevent the migration of substances harmful to food and are safe for the consumer. Such regulations aim to maximize food safety in the packaging process of food analogue products. The suitability of packaging materials and substances for food aims to protect the quality of the products and consumer health.

Compliance with these regulations is critical for food analog product manufacturers and packaging suppliers because it requires them to ensure that their products both comply with regulatory requirements and maintain consumer confidence. Detection of non-compliance may result in serious legal consequences and damage the brand's reputation. Therefore, it is extremely important for companies to stay informed about current regulations and ensure that their packaging and labeling practices are in full compliance with these requirements. Therefore, manufacturers need to view these regulations as a guide to not only meet regulatory requirements but also improve the quality and safety of food analogue products.

To summarize, regulatory compliance and labeling in the packaging processes of food analogue products are of critical importance that cannot be ignored. Regulations such play a vital role in ensuring that consumers have access to accurate and transparent information about food products. These serve as essential tools for food analogue product manufacturers and packaging companies to meet requirements. These regulations aim to minimize potential risks arising from the contact of packaging materials and substances with food and to ensure consumer confidence. Compliance with these regulations is an essential requirement for food analog product manufacturers. Ensuring ongoing compliance not only meets regulatory requirements, but is also crucial to maintaining consumer trust and maintaining the security and transparency of products offered.

## 15.7 Case Studies: Effective Packaging Strategies for Food Analogues

Effective strategies in the packaging processes of food analog products play a strategic role in ensuring product quality, safety and consumer satisfaction. The unique challenges associated with packaging food analogue products have necessitated the development of various innovative approaches in this field. An important aspect of effective packaging strategies to highlight is AP.

AP systems incorporate components that interact with the packaged food to extend shelf life, improve safety, and enhance sensory properties (Biji et al. 2015). For instance, the inclusion of oxygen scavengers within the packaging serves to mitigate oxidative reactions and uphold the food's freshness. Likewise, antimicrobial agents prove effective in restraining the proliferation of spoilage microorganisms, thereby fortifying food safety.

Sustainable packaging solutions are also gaining prominence in the food industry. These strategies aim to minimize the environmental impact of packaging materials and reduce waste. Biodegradable and compostable materials, such as bioplastics derived from renewable resources, are being explored as alternatives to traditional petroleum-based plastics (Guillard et al. 2018). Additionally, lightweight packaging designs are being developed to reduce material usage and transportation costs, while maintaining the necessary protective properties (Licciardello 2017). This progression aligns with the industry's commitment to sustainable and environmentally responsible practices.

Innovations in nanotechnology have made significant contributions to effective packaging strategies for food analogues. Nanomaterials can improve the barrier properties of packaging materials by protecting food from undesirable factors such as moisture and oxygen (Mlalila et al. 2016). Additionally, nanosensors can be integrated into packaging to monitor the quality and safety of food; Real-time information can be obtained about factors such as temperature, humidity, and microbial activity. Consumer perception and behavior play a critical role in the development of effective packaging strategies. Understanding consumer preferences and concerns can provide important guidance in shaping packaging design and communication strategies. Studies have revealed that consumers place great value on useful, informative and environmentally friendly packaging (Lindh et al. 2016). Therefore, packaging that provides convenience, provides clear information about the product, and is also compatible with sustainable practices has the potential to increase consumer satisfaction.

Regulatory compliance and labeling requirements form a key component of effective packaging strategies for food analogues. Compliance with regulations ensures that packaging materials are safe during contact with food and ensures that accurate information is presented to consumers (Han et al. 2018). In particular, accurate labeling of allergens, nutritional information and storage instructions is of critical importance for consumer safety and informed decision-making. Strict adherence to regulations is essential when placing food analogue products on the market in order to protect consumer health and provide safe food.

Therefore, effective packaging strategies for food analogues include various approaches such as AP and IP, effective use of nanotechnology, sustainable solutions, consumer-centered design and regulatory compliance. By implementing these strategies, food manufacturers can ensure the quality, safety and sustainability of their products and meet consumer expectations.

## 15.8 Future Trends in Food Analogue Packaging

Future trends regarding the packaging of food analogue products are taking into account innovation, sustainability, and a focus on consumer preferences. As the food industry is constantly evolving, the vital role of packaging in ensuring product quality, safety and ease of use cannot be ignored. There are a number of important trends that are expected to shape the future of food analogue product packaging.

Sustainability is one of today's prominent trends and is also taken into account in food analogue product packaging. Consumers and regulatory bodies demand packaging solutions that contribute to minimizing environmental impact and supporting the circular economy (Guillard et al. 2018). In this context, biodegradable and compostable materials have attracted attention as alternatives to traditional petroleum-based plastics (Chen et al. 2019). Additionally, factors such as lightening packaging designs, maintaining protective properties and reducing material use form part of strategies developed to both reduce environmental impact and reduce transportation costs (Guillard et al. 2018). These approaches can contribute to the adoption of sustainable packaging, leading to a more environmentally and economically sustainable future.

Another important emerging trend is the combination of AP and IP technologies. AP systems have the potential to extend the shelf life and increase the safety of food analogue products by supporting components such as oxygen scavengers and antimicrobial agents (Chen et al. 2019). The AP can limit the growth of harmful microorganisms while preserving the freshness of food analogue products. On the other hand, IP aims to make food packaging more functional. This approach integrates components such as sensors, indicators, and RFID technology to monitor the condition and quality of packaged foods (Enescu et al. 2019). These technologies play an important role in improving food quality control as well as enriching the consumer experience. The combined use of these two systems allows both to extend the shelf life of the food analogue through packaging and to monitor the quality and safety of the product at every stage of the supply chain.

Recent advances in nanotechnology are important in shaping the future of food analogue product packaging. The use of nanomaterials can improve the barrier properties of packaging materials, improving their ability to protect against moisture, oxygen and other external factors. In addition, integrating nanosensors into packaging, monitoring product freshness, detecting spoilage processes and enables applications such as increasing product safety (Eleftheriadou et al. 2017). Therefore, more effective use of nanotechnology in the packaging industry will contribute greatly to the development of food analogue packaging in the future.

Personalization and customization are one of the important trends that are rising in the food industry and are also effective in the field of packaging. Custom packaging designs and formats can respond to individual consumer preferences. For example, portion-controlled packaging or packaging with resealable features can both provide convenience to consumers and help minimize food waste (Han et al. 2018). This type of packaging allows consumers to use products more efficiently.

Additionally, interactive packaging through technologies such as augmented reality or QR codes can engage consumers. This type of innovative packaging offers the opportunity to interact with products and learn more. It can help consumers learn more about products and support their purchasing decisions. As a result, personalization and customization emphasize the importance of customer-focused solutions in the field of food packaging, and these trends are expected to grow further in the coming period (Vasile and Baican 2021).

As a result, future trends in food analog product packaging can be expressed as elements such as sustainability, AP and IP, nanotechnology, personalization and digitalization. These trends aim to address environmental concerns, improve product quality and safety, improve consumer experience, and optimize supply chain efficiency. By focusing on these trends, the food industry can meet changing consumer demands, reduce environmental impact and ensure the success of food-like products in the market. Following these trends and adopting these innovative approaches has the potential to provide more sustainable, safe and effective solutions in the field of packaging of food analogue products in the future. Additionally, adopting these approaches as part of a broader transformation in the food industry could help the sector respond more effectively to its future challenges.

## 15.9 Conclusion

This book chapter has provided a comprehensive overview of the packaging requirements of food analogues. Selection of appropriate packaging materials and technologies is of great importance in the process of these products from production to consumption. The use of biodegradable and edible packaging materials provides a sustainable alternative to traditional packaging methods, effectively reducing the environmental damage caused by food packaging. Starch-based films and edible coatings, in particular, show the potential to preserve the quality and safety of food analogues. Additionally, packaging technologies such as MAP play an important role in extending the shelf life of food analogues, regulating gas composition, and preserving the sensory and nutritional properties of products. IP systems with sensors and indicators provide real-time information about the quality and freshness of food analogues. This technology helps consumers make informed choices about the products they choose and consume. Monitoring additives passing from packaging materials into food products plays a critical role in ensuring the safety of food analogues. Migration and potential health risks of substances such as plasticizers and nanometals were highlighted. Further research is needed to better analyze and prevent the migration of these additives. In general, the packaging requirements of food analogues must be carefully addressed to meet consumers' demands for sustainable, safe and high-quality products. Future research should focus on optimizing packaging solutions, addressing additive migration challenges, and developing more innovative packaging technologies for the preservation and presentation of food analogues.

## References


Abdelrazek SG, Abou Taleb EMA, Mahmoud AS, Hamouda T (2022) Utilization of polylactic acid (PLA) in textile food packaging: a review. Egypt J Chem 65(3):725–738. https://doi.org/10.21608/ejchem.2021.92005.4368

Ahmed J, Arfat YA, Al-Attar H, Auras R, Ejaz M (2017) Rheological, structural, ultraviolet protection and oxygen barrier properties of linear low- density polyethylene films reinforced with zinc oxide (ZnO) nanoparticles. Food Packag Shelf Life 13:20–26. https://doi.org/10.1016/j.fpsl.2017.04.005

Ahvenainen R (Ed) (2003) Related titles from Woodhead's food science, technology and nutrition list. In: Novel food packaging techniques. Woodhead Publishing, p ii. https://doi.org/10.1016/B978-1-85573-675-7.50001-8

Ahvenainen R, Hurme E (1997) Active and smart packaging for meeting consumer demands for quality and safety. Food Additives & Contaminants 14(6–7):753–763. https://doi.org/10.1080/02652039709374586

Almasi H, Jahanbakhsh Oskouie M, Saleh A (2021) A review on techniques utilized for design of controlled release food active packaging. Crit Rev Food Sci Nutr 61(15):2601–2621. https://doi.org/10.1080/10408398.2020.1783199

Álvarez-Hernández MH, Artés-Hernández F, Ávalos-Belmontes F, Castillo-Campohermoso MA, Contreras-Esquivel JC, Ventura-Sobrevilla JM, Martínez-Hernández GB (2018) Current scenario of adsorbent materials used in ethylene scavenging systems to extend fruit and vegetable postharvest life. Food Bioprocess Technol 11(3):511–525. https://doi.org/10.1007/s11947-018-2076-7

Amorati R, Foti MC, Valgimigli L (2013) Antioxidant activity of essential oils. J Agric Food Chem 61(46):10835–10847. https://doi.org/10.1021/jf403496k

Arrieta MP, Samper MD, Aldas M, López J (2017) On the use of PLA-PHB blends for sustainable food packaging applications. Materials 10(9):Article 9. https://doi.org/10.3390/ma10091008

Aziz M, Karboune S (2018) Natural antimicrobial/antioxidant agents in meat and poultry products as well as fruits and vegetables: a review. Crit Rev Food Sci Nutr 58(3):486–511. https://doi.org/10.1080/10408398.2016.1194256

Barbosa-Pereira L, Aurrekoetxea GP, Angulo I, Paseiro-Losada P, Cruz JM (2014) Development of new active packaging films coated with natural phenolic compounds to improve the oxidative stability of beef. Meat Sci 97(2):249–254. https://doi.org/10.1016/j.meatsci.2014.02.006

Barnett SM, Sablani SS, Tang J, Ross CF (2019) Utilizing herbs and microwave-assisted thermal sterilization to enhance saltiness perception in a chicken pasta meal. J Food Sci 84(8):2313–2324. https://doi.org/10.1111/1750-3841.14736

Becerril R, Nerín C, Silva F (2020) Encapsulation systems for antimicrobial food packaging components: an update. Molecules *25*(5), Article 5. https://doi.org/10.3390/molecules25051134

Bell L, Oloyede OO, Lignou S, Wagstaff C, Methven L (2018) Taste and flavor perceptions of Glucosinolates, Isothiocyanates, and related compounds. Mol Nutr Food Res 62(18):1700990. https://doi.org/10.1002/mnfr.201700990

Biegańska M, Gwiazdowska D, Kozak W, Kluczyńska K (2014) The use of TTI indicators for quality monitoring of freshly squeezed juices. The International Forum on Agri-Food Logistics II Domestic Scientific Conference Agrologistyka Poland

Biji KB, Ravishankar CN, Mohan CO, Srinivasa Gopal TK (2015) Smart packaging systems for food applications: a review. J Food Sci Technol 52(10):6125–6135. https://doi.org/10.1007/s13197-015-1766-7

Bolumar T, Andersen ML, Orlien V (2011) Antioxidant active packaging for chicken meat processed by high pressure treatment. Food Chem 129(4):1406–1412. https://doi.org/10.1016/j.foodchem.2011.05.082

Chen H, Wang J, Cheng Y, Wang C, Liu H, Bian H, Pan Y, Sun J, Han W (2019) Application of protein-based films and coatings for food packaging: a review. Polymers 11(12), Article 12. https://doi.org/10.3390/polym11122039

Choudalakis G, Gotsis AD (2009) Permeability of polymer/clay nanocomposites: a review. Eur Polym J 45(4):967–984. https://doi.org/10.1016/j.eurpolymj.2009.01.027
Commission Regulation (EC) No 450/2009 of 29 May 2009 on active and intelligent materials and articles intended to come into contact with food Text with EEA relevance. (n.d.)
de Oliveira ACS, Ugucioni JC, da Rocha RA, Borges SV (2019) Development of whey protein isolate/polyaniline smart packaging: morphological, structural, thermal, and electrical properties. J Appl Polym Sci 136(14):47316. https://doi.org/10.1002/app.47316
Del Nobile MA, Conte A (2013) Packaging for food preservation. Springer. https://doi.org/10.1007/978-1-4614-7684-9
Dobrucka R, Przekop R (2019) New perspectives in active and intelligent food packaging. J Food Proc Preserv 43(11):e14194. https://doi.org/10.1111/jfpp.14194
Drago E, Campardelli R, Pettinato M, Perego P (2020) Innovations in smart packaging concepts for food: an extensive review. *Foods 9*(11), Article 11. https://doi.org/10.3390/foods9111628
Drobnik O (2015) Barcodes with iOS. https://www.simonandschuster.com/books/Barcodes-with-iOS/Oliver-Drobnik/9781617292156
Duncan TV (2011) Applications of nanotechnology in food packaging and food safety: barrier materials, antimicrobials and sensors. J Colloid Interface Sci 363(1):1–24. https://doi.org/10.1016/j.jcis.2011.07.017
Eleftheriadou M, Pyrgiotakis G, Demokritou P (2017) Nanotechnology to the rescue: using nano-enabled approaches in microbiological food safety and quality. Curr Opin Biotechnol 44:87–93. https://doi.org/10.1016/j.copbio.2016.11.012
Enescu D, Cerqueira MA, Fucinos P, Pastrana LM (2019) Recent advances and challenges on applications of nanotechnology in food packaging. A literature review. Food Chem Toxicol 134:110814. https://doi.org/10.1016/j.fct.2019.110814
Ernst M, Pine DS, Hardin M (2006) Triadic model of the neurobiology of motivated behavior in adolescence. Psychol Med 36(3):299–312. https://doi.org/10.1017/S0033291705005891
Fulop T, Witkowski JM, Bourgade K, Khalil A, Zerif E, Larbi A, Hirokawa K, Pawelec G, Bocti C, Lacombe G, Dupuis G, Frost EH (2018) Can an infection hypothesis explain the Beta amyloid hypothesis of Alzheimer's disease? Front Aging Neurosci 10. https://doi.org/10.3389/fnagi.2018.00224
Ganiari S, Choulitoudi E, Oreopoulou V (2017) Edible and active films and coatings as carriers of natural antioxidants for lipid food. Trends Food Sci Technol 68:70–82. https://doi.org/10.1016/j.tifs.2017.08.009
Garba AI (2023) Food preservation packaging. In: Food processing and packaging technologies—recent advances. Intech Open. https://doi.org/10.5772/intechopen.110043
Geier CF, Terwilliger R, Teslovich T, Velanova K, Luna B (2010) Immaturities in reward processing and its influence on inhibitory control in adolescence. Cereb Cortex 20(7):1613–1629. https://doi.org/10.1093/cercor/bhp225
Ghaani M, Cozzolino CA, Castelli G, Farris S (2016) An overview of the intelligent packaging technologies in the food sector. Trends Food Sci Technol 51:1–11. https://doi.org/10.1016/j.tifs.2016.02.008
Gómez-Estaca J, López-de-Dicastillo C, Hernández-Muñoz P, Catalá R, Gavara R (2014) Advances in antioxidant active food packaging. Trends Food Sci Technol 35(1):42–51. https://doi.org/10.1016/j.tifs.2013.10.008
Guillard V, Gaucel S, Fornaciari C, Angellier-Coussy H, Buche P, Gontard N (2018) The next generation of sustainable food packaging to preserve our environment in a circular economy context. Front Nutr 5. https://doi.org/10.3389/fnut.2018.00121
Han JH, Ho CHL, Rodrigues ET (2005) 9—Intelligent packaging. In: Han JH (ed) Innovations in food packaging. Academic, pp 138–155. https://doi.org/10.1016/B978-012311632-1/50041-3
Han J-W, Ruiz-Garcia L, Qian J-P, Yang X-T (2018) Food packaging: a comprehensive review and future trends. Compr Rev Food Sci Food Saf 17(4):860–877. https://doi.org/10.1111/1541-4337.12343

Hogan SA, Kerry JP (2008) Smart packaging of meat and poultry products. In: Smart packaging technologies for fast moving consumer goods. John Wiley & Sons, Ltd., pp 33–59. https://doi.org/10.1002/9780470753699.ch3

Jariyasakoolroj P, Leelaphiwat P, Harnkarnsujarit N (2020) Advances in research and development of bioplastic for food packaging. J Sci Food Agric 100(14):5032–5045. https://doi.org/10.1002/jsfa.9497

Kader AA (2008) Flavor quality of fruits and vegetables. J Sci Food Agric 88(11):1863–1868. https://doi.org/10.1002/jsfa.3293

Kamarudin SH, Rayung M, Abu F, Ahmad S, Fadil F, Karim AA, Norizan MN, Sarifuddin N, Mat Desa MSZ, Mohd Basri MS, Samsudin H, Abdullah LC (2022) A review on antimicrobial packaging from biodegradable polymer composites. Polymers *14*(1), Article 1. https://doi.org/10.3390/polym14010174

Kerry JP, O'Grady MN, Hogan SA (2006) Past, current and potential utilisation of active and intelligent packaging systems for meat and muscle-based products: a review. Meat Sci 74(1):113–130. https://doi.org/10.1016/j.meatsci.2006.04.024

Kline ALB, Strupinsky EP, Lauri R (2001) Active packaging for food applications. CRC Press. https://doi.org/10.1201/9780367801311

Kool W, McGuire JT, Rosen ZB, Botvinick MM (2010) Decision making and the avoidance of cognitive demand. J Exp Psychol Gen 139(4):665–682. https://doi.org/10.1037/a0020198

Lee SY, Lee SJ, Choi DS, Hur SJ (2015) Current topics in active and intelligent food packaging for preservation of fresh foods. J Sci Food Agric 95(14):2799–2810. https://doi.org/10.1002/jsfa.7218

Licciardello F (2017) Packaging, blessing in disguise. Review on its diverse contribution to food sustainability. Trends Food Sci Technol 65:32–39. https://doi.org/10.1016/j.tifs.2017.05.003

Lindh H, Olsson A, Williams H (2016) Consumer perceptions of food packaging: contributing to or counteracting environmentally sustainable development? Packag Technol Sci 29(1):3–23. https://doi.org/10.1002/pts.2184

Ma Y, Yang W, Xia Y, Xue W, Wu H, Li Z, Zhang F, Qiu B, Fu C (2022) Properties and applications of intelligent packaging indicators for food spoilage. Membranes 12(5):Article 5. https://doi.org/10.3390/membranes12050477

Madene A, Jacquot M, Scher J, Desobry S (2006) Flavour encapsulation and controlled release—a review. Int J Food Sci Technol 41(1):1–21. https://doi.org/10.1111/j.1365-2621.2005.00980.x

Malhotra B, Keshwani A, Kharkwal H (2015) Antimicrobial food packaging: potential and pitfalls. Front Microbiol 6. https://doi.org/10.3389/fmicb.2015.00611

Manthou V, Vlachopoulou M (2001) Bar-code technology for inventory and marketing management systems: a model for its development and implementation. Int J Prod Econ 71(1):157–164. https://doi.org/10.1016/S0925-5273(00)00115-8

Marsh K, Bugusu B (2007) Food packaging—roles, materials, and environmental issues. J Food Sci 72(3):R39–R55. https://doi.org/10.1111/j.1750-3841.2007.00301.x

Mills A (2005) Oxygen indicators and intelligent inks for packaging food. Chem Soc Rev 34(12):1003–1011. https://doi.org/10.1039/B503997P

Mlalila N, Kadam DM, Swai H, Hilonga A (2016) Transformation of food packaging from passive to innovative via nanotechnology: concepts and critiques. J Food Sci Technol 53(9):3395–3407. https://doi.org/10.1007/s13197-016-2325-6

Mohr PNC, Biele G, Heekeren HR (2010) Neural processing of risk. J Neurosci 30(19):6613–6619. https://doi.org/10.1523/JNEUROSCI.0003-10.2010

Moshe I, Weizman O, Natan M, Jacobi G, Banin E, Dotan A, Ophir A (2016) Multiphase thermoplastic hybrid for controlled release of antimicrobial essential oils in active packaging film. Polym Adv Technol 27(11):1476–1483. https://doi.org/10.1002/pat.3817

Müller P, Schmid M (2019) Intelligent packaging in the food sector: a brief overview. Food Secur 8(1), Article 1. https://doi.org/10.3390/foods8010016

Nieva-Echevarría B, Manzanos MJ, Goicoechea E, Guillén MD (2015) 2,6-Di-Tert-butyl-Hydroxytoluene and its metabolites in foods. Compr Rev Food Sci Food Saf 14(1):67–80. https://doi.org/10.1111/1541-4337.12121

O'Grady MN, Kerry JP (2008) Smart packaging technologies and their application in conventional meat packaging systems. In: Toldrá F (ed) Meat biotechnology. Springer, New York, pp 425–451. ISBN 978-0-387-79381-8

Pereira de Abreu DA, Paseiro Losada P, Maroto J, Cruz JM (2011) Natural antioxidant active packaging film and its effect on lipid damage in frozen blue shark (Prionace glauca). Innovative Food Sci Emerg Technol 12(1):50–55. https://doi.org/10.1016/j.ifset.2010.12.006

Pereira VA, de Arruda INQ, Stefani R (2015) Active chitosan/PVA films with anthocyanins from Brassica oleraceae (Red Cabbage) as Time–Temperature Indicators for application in intelligent food packaging. Food Hydrocoll 43:180–188. https://doi.org/10.1016/j.foodhyd.2014.05.014

Prikh D, Jancke G (2008) Localization and segmentation of A 2D high capacity color barcode. In: Applications of Computer Vision, 2008. WACV 2008. IEEE Workshop. https://doi.org/10.1109/WACV.2008.4544033

Realini CE, Marcos B (2014) Active and intelligent packaging systems for a modern society. Meat Sci 98(3):404–419. https://doi.org/10.1016/j.meatsci.2014.06.031

Restuccia D, Spizzirri UG, Parisi OI, Cirillo G, Curcio M, Iemma F, Puoci F, Vinci G, Picci N (2010) New EU regulation aspects and global market of active and intelligent packaging for food industry applications. Food Control 21(11):1425–1435. https://doi.org/10.1016/j.foodcont.2010.04.028

Robertson GL (2016) Food packaging: principles and practice, 3rd edn. CRC Press. https://doi.org/10.1201/b21347

Sanches-Silva A, Costa D, Albuquerque TG, Buonocore GG, Ramos F, Castilho MC, Machado AV, Costa HS (2014) Trends in the use of natural antioxidants in active food packaging: a review. Food Additiv Contam Part A 31(3):374–395. https://doi.org/10.1080/19440049.2013.879215

Sarac A, Absi N, Dauzère-Pérès S (2010) A literature review on the impact of RFID technologies on supply chain management. Int J Prod Econ 128(1):77–95. https://doi.org/10.1016/j.ijpe.2010.07.039

Shaikh S, Yaqoob M, Aggarwal P (2021) An overview of biodegradable packaging in food industry. Curr Res Food Sci 4:503–520. https://doi.org/10.1016/j.crfs.2021.07.005

Sharma C, Dhiman R, Rokana N, Panwar H (2017) Nanotechnology: an untapped resource for food packaging. Front Microbiol 8. https://doi.org/10.3389/fmicb.2017.01735

Silayoi P, Speece M (2007) The importance of packaging attributes: a conjoint analysis approach. Eur J Mark 41(11/12):1495–1517. https://doi.org/10.1108/03090560710821279

Sivertsvik M, Jeksrud WK, Rosnes JT (2002) A review of modified atmosphere packaging of fish and fishery products—significance of microbial growth, activities and safety. Int J Food Sci Technol 37(2):107–127. https://doi.org/10.1046/j.1365-2621.2002.00548.x

Small DM, Voss J, Mak YE, Simmons KB, Parrish T, Gitelman D (2004) Experience-dependent neural integration of taste and smell in the human brain. J Neurophysiol 92(3):1892–1903. https://doi.org/10.1152/jn.00050.2004

Smith JP (1993) Food packaging—principles and practice: edited by G. L. Robertson, Marcel Dekker, Inc., New York, 1993. pp. 667, price US$165·00. ISBN 0-8247-8749-9. Food Res Int 26(3):233–234. https://doi.org/10.1016/0963-9969(93)90058-Q

Somerville LH, Casey B (2010) Developmental neurobiology of cognitive control and motivational systems. Curr Opin Neurobiol 20(2):236–241. https://doi.org/10.1016/j.conb.2010.01.006

Suppakul P, Miltz J, Sonneveld K, Bigger S, w. (2003) Active packaging technologies with an emphasis on antimicrobial packaging and its applications. J Food Sci 68(2):408–420. https://doi.org/10.1111/j.1365-2621.2003.tb05687.x

Tian F, Decker EA, Goddard JM (2013) Controlling lipid oxidation of food by active packaging technologies. Food Funct 4(5):669–680. https://doi.org/10.1039/C3FO30360H

Vasilaki A, Panagiotopoulou E, Koupantsis T, Katsanidis E, Mourtzinos I (2022) Recent insights in flavor-enhancers: definition, mechanism of action, taste-enhancing ingredients, analytical techniques and the potential of utilization. Crit Rev Food Sci Nutr 62(32):9036–9052. https://doi.org/10.1080/10408398.2021.1939264

Vasile C, Baican M (2021) Progresses in food packaging, food quality, and safety—controlled-release antioxidant and/or antimicrobial packaging. Molecules 26(5), Article 5. https://doi.org/10.3390/molecules26051263

Vermeulen SJ, Campbell BM, Ingram JSI (2012) Climate change and food systems. Annu Rev Environ Resour 37(1):195–222. https://doi.org/10.1146/annurev-environ-020411-130608

Wang G, Cheng J, Ma J, Huang S, Zhang T, Li Y, Li H (2022) Preparation and application of an olfactory visualization freshness sensor array based on microfluid paper-based chip. J Food Proc Preserv 46(6):e16530. https://doi.org/10.1111/jfpp.16530

Wilbey RA (2005) Food packaging technology. Int J Dairy Technol 58(2):125–125. https://doi.org/10.1111/j.1471-0307.2005.00157.x

Yan MR, Hsieh S, Ricacho N (2022) Innovative food packaging, food quality and safety, and consumer perspectives. PRO 10(4):Article 4. https://doi.org/10.3390/pr10040747

Yousefi H, Su HM, İmani SM, Alkhaldi K, Filipe CDM, Didar TF (2019) Intelligent food packaging: A review of smart sensing technologies for monitoring food quality. 4(4):808–821. https://doi.org/10.1021/acssensors.9b00440

Topuz F, Uyar T (2020) Antioxidant, antibacterial and antifungal electrospun nanofibers for food packaging applications. Food Research International 130:108927. https://doi.org/10.1016/j.foodres.2019.108927

Yildirim S, Röcker B, Pettersen MK, Nilsen-Nygaard J, Ayhan Z, Rutkaite R, Radusin T, Suminska P, Marcos B, Coma V (2018) Active packaging applications for food. Compr Rev Food Sci Food Saf 17(1):165–199. https://doi.org/10.1111/1541-4337.12322

Zhang C, Li Y, Wang P, Zhang H (2020) Electrospinning of nanofibers: potentials and perspectives for active food packaging. Compr Rev Food Sci Food Saf 19(2):479–502. https://doi.org/10.1111/1541-4337.12536

Zhuang H (2011) Modified atmosphere packaging for fresh-cut fruits and vegetables. In Modified atmosphere packaging for fresh-cut fruits and vegetables. John Wiley & Sons, Ltd., pp 1–7. https://doi.org/10.1002/9780470959145.ch1

# Chapter 16
# Regulatory Aspects of Food Analogues in Food Sciences. Food Analogues in the European Union: A Legal Perspective

**Valeria Paganizza**

## 16.1 Introduction

In recent years, the interest in alternative proteins and foods which could replace commonly used products of animal origin has been increasing. In 2020, the Directorate-General for Research and Innovation of the European Union published the document «FOOD 2030 Pathways for Action—Alternative proteins and dietary shift» (European Commission, DG R&I 2020), which underlined the positive aspects of a change in nutritional habits. Starting with the contribution to the containment of global warming, the short leaflet described the bright sides of dietary modification in nutrition, health, climate change, sustainability, circularity, resource efficiency, innovation and effects on communities.

Alternative proteins are therefore seen as one of the main drivers of food market development for the next decade: they could represent one of the sustainable solutions to feed a growing population as well as the answer to the increasing demand for replacing products of animal origin, in connection with the expanding popularity of some dietary patterns, like veganism, vegetarianism or flexitarianims (CBI 2023).

If the theme of food analogues is collecting mounting interest among the experts in the technology and the food processing sectors, then legal scholars are also prompted to look into the new challenges that the development of new foods entails.

The purpose of this chapter is therefore to sketch a picture of the regulatory framework for food analogues, trying to pinpoint possible criticalities of the applicable provisions.

V. Paganizza (✉)
Department of Law, Politics and International Studies, Department of Excellence 2023-2027, financed through funds of the Italian Ministry of University and Research, University of Parma, Parma, Italy
e-mail: valeria.paganizza@unipr.it



Ö. P. CAN et al. (eds.), *Food Analogues: Emerging Methods and Challenges*,
https://doi.org/10.1007/978-3-031-69872-9_16

Though recognizing that food analogues do not include only alternative proteins, the investigation will mainly focus on these components due to the increasing consideration that new protein sources are receiving.

Starting to look for a definition of *food analogues*, the chapter will shift from the Codex Alimentarius work on the topic to the European Union's regulatory framework. Due to the close dependency between the old novel food regulation and the new one, it will offer an overview of both the acts and of the importance of unified legislation on this theme. In this regard, the chapter will also consider the most recent effects of the contemporary sovereigntist approaches of some Member States towards innovation and some collateral aspects of food analogues marketing. Attention will be given, in particular, to the use of meat-sounding names and to the fierce opposition to this practice, expressed at a political level by the European Parliament's questions to the European Commission.

The theme of nutrition claims and the need to adequately inform consumers will be finally discussed.

## 16.2 Food Analogues: A Lack of Legal Definition

When engaging with legal studies, it is imperative for researchers to commence their study by establishing a clear and comprehensive definition of the specific area under examination. Hence, since the objective of this chapter is to elucidate the regulatory framework governing food analogues inside the European Union, the initial stage entails comprehending the precise meaning of the phrase *food analogue*. Neither a quick nor an advanced inquiry conducted using the European Union legal search engine (Eur-lex) yielded any results, meaning that the expression has never been used in one of the acts of the EU Institutions. A definition of *food analogue* must therefore be sought beyond the confines of legislative provisions, in the natural sciences sector, where it is instead quite recurrent, even if it might refer to different concepts. It might be used, for instance, to describe substances and products characterized by properties that are similar to those of real foods (Carson et al. 2004), thus entailing the non-food nature of such items. The most common (and modern) definition refers, however, to a food that can be thought of as an alternative to or a substitute for a food that is typically consumed, such as plant-based proteins as replacements for meat or dairy products (Lee et al. 2023; Seung Yun Lee et al. 2023). If we consider the Codex General Standard for Food Additives, specifically the part of food categories and relevant details, when referring to *Dairy products and analogues*, we could see that the word *analogues* refers to the whole or partial replacement of the fatty part of milk by vegetable fats or oils. This is due to the fact that these foods are designed and manufactured to imitate the replaced product. Such a replication may rely on visual, organoleptic, and content analogies (Bohrer 2019; Keefe 2018; Kumar et al. 2017; Joshi and Kumar 2015). The chapter will take into account the second definition: it currently proves to be the most common, due to the resounding success that this second type of *food alternatives* is attaining and because of the legal issues that its development is implying.

Food, however, has a very heterogeneous nature, since it includes products that have commonly been used as well as new ones; it comprises food with specific technological purposes such as additives, colours and enzymes; it contains food with specific nutritional properties for certain categories of people or intended to simply supplement the diet. This means that when considering food analogues with the purpose of understanding the applicable provisions, we must take into account the specific category of food and the vertical provisions that might integrate or replace the applicable horizontal legislation.

There are then some new products that could hardly fit into the definition. An example could be meat obtained from cell cultivation. Once the technological process permits to produce a perfect *alternative* product that also meets the aesthetic, organoleptic and chemical characteristics of *traditional* meat, a doubt about the qualification of the former may arise. The obtained product will not be *analogue* to meat: it will be meat itself, composed of cultured cells grown on a bioreactor, exactly as it was produced by slaughtering an animal. As we will say in paragraph no 8 there will be just one difference: the name *meat* will not apply.

Specific considerations should be drawn for the further emerging category constituted by 3D printed food: some of these products will be perfectly identical, in composition, to traditionally processed food, though having been obtained with an additive manufacturing process. This is the case of a 3D pasta obtained with customary ingredients. We cannot talk about *food analogues* but, once again, we need to consider them identical. Some of them, instead, could be used to produce functional food, incorporating bioactive compounds or other beneficial substances in order to shape *personalized* food (Tomaševic et al. 2021), which could replace the *non-personalized* form of the same product: printed food will be characterized by customized properties but will also be produced in order to obtain an alternative result.

There might then be some other products for which the classification can be controversial: this is the case of insects. Their use in human nutrition is being promoted to replace protein intake from *traditional* animals. Whole insects cannot be included in the category of *food analogues* since they do not replicate common food under aesthetic and organoleptic characteristics. As for products obtained from insects, they might be used to produce food *analogues* (let's think to insect balls or steak). Up till now, however, at least in the European Union, the market will be characterized only by traditional food, supplemented with insect proteins (in the form of frozen, paste, dried and powder insects), without any whole replacement. Just a few products had been authorized: Alphitobius diaperinus *larvae* (lesser mealworm), frozen, reduced to paste, dried or as powder; Acheta domesticus (house cricket) in a partially defatted powder; *Locusta migratoria* (migratory locust) and *Tenebrio molitor* larva (yellow mealworm), both in frozen, dried or powder forms. Once again, such products do not fit the definition that represented the starting point of this chapter, even if some of the reasoning we will draw in paragraph 5 applies.

## 16.3 Food Analogues at an International Level

In 2022, the Food and Agriculture Organization of the United Nations (FAO) published a report on the Future of Food Safety (FAO 2022). Two of the report sections focus, respectively, on plant-based alternatives (section 4.3) and on cell-based food production (section 4.5), seen as possible answers to the reduction of natural resources and the negative environmental impacts deriving from food production. According to the report, since these alternatives are intended to be used as food, they need to comply with food safety requirements to exclude possible risks to human health (Foreword section).

Up till now the Codex Alimentarius has not issued any texts (standards, codes of practice or guidelines) specifically addressed to food analogues. However, the General guidelines for the utilization of vegetable protein products (VPP) in foods, which date back to 1989 (Codex Alimentarius 1989, CXG 4-1989) and are particularly focused on protein quality, encompass the partial or complete substitution of animal protein in food. Besides some indications on the need to inform consumers with clear reference to the presence of these proteins, the Guidelines cover the testing activities for the safety and nutritional quality of vegetable protein products.

The Committees' work is still in progress. The 81st Session of the Executive Committee of the Codex Alimentarius Commission (CCEXEC81) agreed on the establishment of a subcommittee to address the issues emerged in the mentioned FAO document on new food sources and production systems: the decision was confirmed during the 44th Session of the Codex Alimentarius Commission (Codex Alimentarius Commission 2021, CAC44). To this end, the Codex Alimentarius Commission's Secretariat requested the Codex Contact Points and the Contact Points of international organizations having observer status with Codex to share information on the potential issues, regulatory matters, labelling, nutritional, food safety, and quality aspects, including, but not limited to, cultivated meat, seafood, dairy products, fermentation-derived ingredients, plant-based protein alternatives, seaweed, edible insects, 3-D printed foods, microalgae (Codex Alimentarius Commission's Secretariat 2022).

In 2023, alternative proteins were also the subject of debate within the Codex Committee on Nutrition and Foods for Special Dietary Uses (Codex Committee on Nutrition and Foods for Special Dietary Uses (CCNFSDU) 2023; Codex Alimentarius Commission 2023a, b). Following the 2020 decision to collect proposals for new work, several requests were submitted. In particular, the United States and Canada proposed to elaborate Guidelines, including general principles, for the nutritional composition of foods and beverages made from plant-based and other alternative protein sources (excluding animal based or animal-cells proteins). The proposal recognized the growing success of plant-based diet patterns due also to their encouragement in dietary guidance recommendations (like the Dietary Guidelines for Americans 2020–2025 or the Eatwell Guide 2016 for the EU) and of other *new* protein products obtained from fungi, insects and fermentation-derived proteins, including products that imitate animal-based products in the way in which they are used, represented, or in their aspect (Codex Committee on Nutrition and Foods for Special Dietary Uses (CCNFSDU) 2023, CX/NFSDU 23/43/7—Introduction, p. 32). Meanwhile, the United States and Canada, starting from studies and reviews (Bohrer 2019) on the nutritional composition of these food analogues, emphasized that some concerns could arise from the high content of some nutrients that should instead require a limited intake, like sodium, saturated fats and sugars. The food industry's response to the increasing demand for food analogues and the fact that a lack of dedicated regulations could result in health and trade implications led the proposers to urge the adoption of nutritional guidelines for such products.

The United States and Canada's proposal received some comments, with a request to the CODEX Committee on Nutrition and Foods for Special Dietary Uses to refrain from supporting the text (CCNFSDU 2023). Concerns were raised on the identification of food to which the proposal would apply as well as on the starting assumption according to which consumers would buy food analogues only to replace food of animal origin, while they could actually choose such products just to integrate their diets without substituting them. Moreover, differentiating food composition, rather than standardize it by means of guidelines, would offer consumers a wider food choice and would also allow the future development of new products. Establishing guidelines, on the contrary, could result in a limit on the expansion of plant-based food (Good Food Institute recommendation, CCNFSDU 2023).

Another major debate paid attention to the technical processes undergone by plant-based products and other alternative protein sources. With specific reference to insects, moreover, potential allergenic reactions and parasites were included among the alarming elements to be carefully considered. Other remarks focused instead on the essential nutrients, suggesting an amendment of the Principles for the Addition of Essential Nutrients to Foods (CXG 9-1987).

On the United States and Canada's proposal, the Representative of the International Organization for Standardization (ISO) within Codex Alimentarius added that the work of the organization on this topic, which would lead to a draft standard by the end of 2023 and to a finalization in 2024, will entail the identification of what could be defined as *plant-based food* and of the ingredients that could be used, while excluding the nutritional composition of plant-based products (Codex Alimentarius Commission 2023, point 107).

One Member suggested that plant-based alternatives should fall under the scope of *new food sources and production systems* (NFPS), discussed during the Codex Alimentarius Commission meetings in 2021 (CAC44) and 2022 (CAC45). During the two conventions, Members were engaged in discussing *new foods*: Products that fall into this category have not been widely consumed as food. This could depend on the fact that they have only recently entered the global retail market as a result of technological advancements or because their consumption has traditionally been limited to particular regions throughout the world (CX/CAC 21/44/15 Add.1 2021). From the Codex Commission's work in 2021, the need for international standards and guidance emerged due to the growing importance and constant development of such new foods. The cross-cutting nature of these products, however, showed that the current Committees' structure cannot efficiently answer the evolving tasks (CX/CAC 21/44/15 Add.1 2021). During the 2022 meeting, therefore, Members were encouraged to submit proposals with reference to the new food sources and production systems, using the existing Codex mechanism and to share possible issues that the Codex structure and procedures could not address. During the meeting held in November 2023, the Codex Alimentarius Commission concluded that the Codex working structures currently in place are sufficient to deal with any additional work on new sources and production systems and invited members to submit discussion papers or new work proposals (CAC46 Report, para. 206).

As mentioned in paragraph 1 and as we will see also in paragraphs 5, 8 and 9, if some food analogues could definitely be included in the category of *new food sources and production systems*, some products have actually been used for decades and the only criticality they might carry around is linked to the need to inform consumers about the food characteristics, with particular reference to the quality protein of plant-based products and other alternative sources.

## 16.4 Food Analogues in the European Union: A Protein Strategy

If the Codex Alimentarius' work on food analogues is in progress, at an EU level the attention is similarly high, and it is mainly based on alternative proteins. The reasons underlying the European interest are the consequences of an integrated approach built on several needs. The Russia-Ukraine conflict has been threatening EU food security and has shown the strong dependence of the EU agrifood sector on imports of raw materials, mainly cereals, to be used as protein sources in animal farming and the need to diversify food sources. The direction towards sources which could replace *traditional* food has actually already been suggested by the European Commission in its 2018 report on the development of plant proteins in the European Union, COM (2018) 757 final. The report underlined how the EU demand for plant proteins amounted to around 27 million tons of crude protein, even if most of the absorption depended on animal feed. The document emphasized the market change for meat and dairy alternatives; if it could previously be considered a niche market, already in 2018, plant-based food analogues could be included in the general segments of the market, where also major retailers were active with their brands. According to the Commission's report, meat and dairy alternatives had counted annual growth rates of 14% and 11%, respectively, being consumed not only by vegans or vegetarians, but also by the so-called flexitarians, meaning people who mainly privilege plant-based protein but also include some animal proteins (section 3.3 of the report).

The greener path, first indicated rather than designed by the European Commission, found one of its major enhancements with the European Green Deal COM (2019) 640: the push towards sustainability could not neglect the food production system. The Farm to Fork strategy, anticipated in the Green Deal and addressed to the achievement of a circular economy, was also intended to launch activities for innovation in food and feed products (Green Deal, Section 2.1.6) and to enhance plant-based food analogues. According to the Farm to Fork strategy, increasing the availability and source of alternative proteins (which include not only plant sources, but also microbial and marine proteins, as well as proteins from insects and meat substitutes) will be a key element for future research to enable the transition towards a more sustainable food system [point 3.1 COM (2020) 381 final].

On October 19, 2023, the European Parliament passed a resolution on an EU protein strategy [2023/2015/INI]. It demanded that the Commission promptly deliver an exhaustive protein strategy that would cover the sustainable production and consumption of proteins in the EU, particularly those derived from plants and animals. The strategy should develop on six points: vision for strategic and sustainable EU protein production and trade flows that fits needs and demands; an action plan for increased EU plant-based protein production and consumption; better conditions for more sustainable production of both plant- and animal-based protein in the EU; the development of sustainable protein systems for food and feed; a holistic approach that necessarily includes farmers along with the whole food value chain, taking into account the principles of the circular economy; concrete science-based policy actions for development, innovation and research on sustainably produced proteins [European Parliament's resolution, point 2, letters a)-f)].

The call on the Commission is certainly appropriate, considering that, by the end of 2023, the Institution had planned the publication of a proposal for an act on sustainable food systems. Even though the expected document was not mentioned during the speech of the EU Commission's President to the European Parliament in September 2023, the European Parliament's pressure could contribute to including provisions on food analogues in the future proposal.

## 16.5 Food Analogues as Novel Foods: A Regulatory Framework and Some Issues for Marketing

Lacking dedicated provisions, the current general food legislative framework applies. Starting from Regulation (EC) no 178/2002 which is the general food law, containing the main definitions and principles applicable to the food sector, to Regulation (EC) no 852/2004 on food hygiene, to Regulation (EU) 2017/625 on official controls, just to give a few examples, the sector of food analogues is subject to all the horizontal food provisions of the European Union.

Nevertheless, food analogues may be mere new cooking options obtained with ingredients that are commonly consumed as food or may instead be the result of the use of new raw materials or of new processing technologies or methods. While the first occurrence does not entail any peculiar obstacle to the placing on the market (with the exception of some issues related to the name that can be used for the products, as we will say in paragraph 8), the second case is covered, in the European Union, by a specific regulatory framework on novel foods, originally set in 1997 and last substantially revised in 2015.

### *16.5.1 The Old Novel Food Regulation*

More than twenty years have passed (more than thirty if we consider the date of the proposal), since the Community legislator decided to adopt a discipline for *novel foods*, with the intention to ensure the free movement of goods, the smooth functioning of the internal market and competition, on the one hand, and human health on the other (Costato 1997). The primary commitment was to ensure that foods obtained through the use of new technologies were placed on the market only after a thorough assessment of their safety, as required precisely by the novelty of the products. Although, over time, the scope of application of the first Regulation on new foods has been adjusted to include any substance that had not been significantly consumed as a food by May 15, 1997, it is undoubted that the original intention of the EC institutions was to require an authorization procedure only for foods that were *new* from a technological point of view. There are many hints in this respect. Firstly, the difference between the original 1992 proposal, the amended 1993 proposal and the final text of Regulation (EC) No 258/97, in its earlier version. Annex I to the 1992 Commission proposal explicitly listed the products to which the new regulation should apply: in addition to three classes of foods subjected to unused modifications or processes, there was a generic class containing any product produced from an organism that did not have a recognized past of use as food or consisting of that organism or containing it. Instead, this category disappeared from the 1993 proposal, which limited the scope of application of the Regulation to the placing on the market of food products or food ingredients not yet used to a significant extent for human consumption or produced by processes involving significant changes in composition, nutritional value and use, falling within at least one of the categories identified in Article 1(2) of the proposal. Looking at the food classes, it is clear that the new proposal was intended to regulate foods undergoing innovative processes or modifications: the first two categories were dedicated to genetically modified organisms (GMOs), the third to food products or ingredients with a new or deliberately modified primary molecular structure, single-cell proteins, food or food ingredients undergoing manufacturing processes that had not yet been used in food production or, if used, had not been placed on the market and for which the process of manufacture had resulted in significant changes in the composition or structure of the finished product, such as to affect the nutritional value, digestibility, metabolism or the content of undesirable substances in the finished food. However, the troubled course of the Regulation led only in 1997 to the adoption of a text significantly different from the first proposals, also in relation to the identification of food classes to be considered *new*, while maintaining, with few exceptions, the *technological* concept that inspired the Commission's idea.

A second indication that confirms the assumption that the authorization procedure, in the original intentions of the institutions, should be directed at foods obtained or otherwise subject to the use of technologies, is the same provision of a double condition for the qualification of a food as new: the product or the ingredient, not being consumed as a food before May 15, 1997 (date coinciding with the entry into force of the Regulation) and the simultaneous belonging to one of the classes listed in the same act. If, in fact, the Regulation had intended to authorize any food not consumed significantly in that period, then the Community would have had no reason to request its inclusion in one of the classes, nor would it have any reason to list them.

A third element that suggests a reading like the one mentioned above is the extreme attention paid to GMOs, organisms of such prominent relevance that they were subsequently excluded from the text of the Regulation in order to obtain a specific discipline.

Fourthly, the provision of a simplified notification procedure for *substantially equivalent* foodstuffs, which was not included in the Commission's original proposal, but was introduced only later, could also be considered a corollary to the aforementioned intention: where the application of new technologies would have resulted in a food similar to another already existing food, in terms of composition, nutritional value, metabolism, use and content of undesirable substances, the safety check would have led to a less costly *iter*.

The practical application of Regulation (EC) No 258/97 soon revealed the limits of the Act. Firstly, as it has been mentioned, the specific nature of GMOs and the inopportune inclusion of these products in the scope of the simplified notification procedure suggested that a specific regulation had to be devoted to them, with the resulting deletion of the relevant categories from the list of the Regulation.

A second critical element of the previous act was the absence of a simplified procedure for food traditionally consumed in third countries. Concerns, in other words, were expressed about the need to submit to authorization procedures foodstuffs that were used significantly in third countries but that were unknown to the market and the use in the European Union.

One additional vulnerability of the authorization system pertained to the initial phase being limited to a national scope. This could have led to variations in authorization procedures among Member States, as competent authorities may have adopted divergent approaches to ensuring food safety.

Ultimately, it became evident that the classification into classes did not align with the diverse expectations that were observed in practice. What was the classification for food supplements, for substances of mineral origin or for insects in the context being discussed? Furthermore, could a food product composed of substances that were deemed safe be considered exempt from the regulatory framework governing novel foods? The Court of Justice provided some responses to these questions, sometimes in an unsatisfactory manner.

### *16.5.2 The New Novel Food Regulation*

In addition to the aforementioned key features, which were partially shaped by the case-law of the Court of Justice, the necessity for a "modernization" of the 1997 Regulation led to an adjustment in accordance with the principles and definitions outlined in Regulation (EC) No 178/2002. After experiencing a disruption in early 2008 (Gonzalez Vaqué 2008; Bonora 2016), which was further prolonged due to the inability to reach a consensus in the Conciliation Committee in 2011, the process was recommenced in 2013 (Ballke 2014). Consequently, Regulation (EU) 2015/2283 was enacted in 2015 with the primary objective of modernizing and simplifying the authorization procedures (fourth recital). Additionally, the new regulation sought to provide clarity regarding the definition of novel food (sixth recital), as well as to review, clarify, and update the existing categories (eighth recital). Furthermore, it aimed to encompass substances that were exclusively used as supplements but are now intended to be consumed as regular food, while also introducing a simplified procedure for traditional foods originating from third countries.

The exclusions from the scope of the new Regulation are the same as in the old act: genetically modified foods, foods utilized as additives, flavors, enzymes, and extraction solvents (all subject to other authorization procedures). The conceptualization of *novel food* undergoes slight modifications. Regulation (EC) No 258/97 referred to *foodstuffs* and *food ingredients*. However, the new Act has been revised to align with Regulation No 178/2002, which provided the first harmonized definition of *food*, for better consistency. In this regard, it defines *novel food* as any *food* that was not extensively consumed by humans in the European Union before May 15, 1997 (cut-off date), regardless of the accession date of Member States. Additionally, to be considered as *novel, food* must fall into at least one of the listed categories. As expected, the categories have been either added, updated, or newly inserted: (i) food with a new or intentionally modified molecular structure. The structure was not used in any food within the Union, nor was it used in any food before May 15, 1997. The second, third and fourth categories cover food consisting of, isolated from or produced from, respectively, microorganisms, fungi or algae (ii), material of mineral origin (iii), plants or their parts (iv). The fourth category does not, however, include foods with a history of safe use as food within the Union, provided that two further alternative conditions are met. The plant or variety of the same species that constitute the food or from which the food is isolated or produced must have been obtained by traditional propagating practices (used for food production within the Union before May 15, 1997). In the alternative, the plant or varieties could have been obtained by non-traditional propagating practices not used for food production within the Union before May 15, 1997, provided that no significant changes were detected in the composition or structure of the food that could impact the nutritional value, the metabolism or the level of undesirable substances. The fifth category includes food consisting of animals or their parts (or isolated from or produced from them). Once again, an exception is made for animals obtained by traditional breeding practices used for food production within the Union before

May 15, 1997, provided that the product has a history of safe use (as food) within the Union (v). The sixth category covers food consisting of, isolated from or produced from cell culture or tissue culture derived from animals, plants, microorganisms, fungi or algae (vi). Changes in the composition or structure of a food that could impact the nutritional value, metabolism or level of undesirable substances are also the limits for the seventh category, which includes food resulting from a production process not used for food production within the Union before May 15, 1997 (vii). The eighth category covers food consisting of engineered nanomaterials (viii) while the ninth class includes vitamins, minerals and other substances to be used in or as food supplements (Directive 2002/46/EC), in fortified food under Regulation (EC) No 1925/2006 or in food under Regulation (EU) No 609/2013. To be considered novel food, however, these vitamins, minerals and other substances have to be obtained by means of a production process not used for food production within the Union before May 15, 1997 or they must contain or consist of engineered nanomaterials (ix). The last category includes substances that were used exclusively in food supplements within the Union before May 15,1997, but that are now intended as *common* foods.

With regards to food analogues that were not used prior to 1997, several categories could apply, depending on the nature and the production process: they might be products obtained from plants, from microorganisms, fungi or algae or from cell cultivation. If we scroll through the European Union list of authorized and notified novel foods, we will identify several products that are intended to replace, wholly or partly, some food or to be used in food analogues: N-Acetyl-D-neuraminic acid and 2′-Fucosyllactose, Lacto-N-neotetraose can be used in dairy analogues, including beverage whiteners; *Acheta domesticus* (house cricket) partially defatted powder can be used in meat analogues; the frozen, paste, dried and powder forms of *Alphitobius diaperinus* larvae (lesser mealworm), pea and rice protein fermented by *Lentinula edodes* (Shiitake mushroom) mycelia, Vitamin D2 mushroom powder can instead serve in both meat and dairy analogues; *Allanblackia* seed oil can be used in mixtures of vegetable oils and milk included in the category of dairy analogues; Antarctic Krill oil from *Euphausia superba*, *Buglossoides arvensis* seed oil, Chia seeds (*Salvia hispanica*), oil extracted from squids can also be used in dairy analogues while the *Schizochytrium sp.* oil rich in DHA and EPA could serve in fish, meat and dairy analogues.

One of the notable changes introduced by Regulation (EU) 2015/2283 is the elimination of the simplified process for foods that are deemed substantially equivalent to other products already present in the EU market. Instead, a notification procedure is now in place for traditional foods originating from third countries. These traditional foods, which exclude those falling under classes i), iii), vii), viii), ix), and x) as previously described, must demonstrate their safety through data on their composition and evidence of their long-standing consumption in the regular diets of a significant population, in at least one third country, for a minimum period of 25 years prior to notification.

The introduction of this new specific procedure is purportedly a response to the stringent requirements of the previous regulation, which required a prior authorization even for foods with a well-established record of safety, solely based on their origin from third countries.

With reference to the new simplified procedure, uncertainty may arise regarding the definition of *food derived from primary production*. One might doubt whether the expression should refer to unprocessed food directly sourced from agriculture or include any food initially obtained from primary production, regardless of subsequent processing. The second interpretation would, however, require that the notification procedure encompass both the food itself and its manufacturing process, to assess food safety, taking into account any possible changes that the food might undergo. The notification procedure ends with an EU Commission's act, which authorizes the notified product, contingent upon the absence of objections raised by the Member States or the European Food Safety Authority.

The new regulation has implemented a structured approach for both the notification procedure for a traditional food from a third country and the *ordinary* authorization procedure, which were previously characterized by a national phase. This structured approach has established Union unitary routes, thereby eliminating the previous national stage. By adopting this approach, the potential for varying perspectives among national authorities is reduced, though not entirely eradicated.

It is worth noting that Regulation (EU) 2015/2283 outlines a preliminary consultation procedure in Article 4. Currently implemented at a national level, this procedure resolves uncertainties about classifying a food as *novel* under the regulation. In other words, it helps determine if a food should undergo one of the procedures of the regulation. The consultation process, as outlined in Article 4(3) of the novel food regulation and Article 6(2) of Implementing Regulation (EU) 2018/456, indicates that the member states' engagement with other nations and the Commission is discretionary.

Moreover, the Regulation grants the Commission the competence to determine the classification of a product as novel, either autonomously or upon the request of a Member State, by means of implementing acts (Article 5).

Another innovation brought about by Regulation (EU) 2015/2283 was the implementation of the European list of novel foods. Both under the previous regulatory framework and the current one, a novel food catalogue is available on an EU official webpage. This catalogue facilitates the verification of the classification of specific foods as either *novel*, *not novel* or exclusively recognized for supplement purposes. The European list of novel foods is instead a comprehensive and regularly updated compilation, officially laid down by Implementing Regulation (EU) 2017/2470. This inventory encompasses all novel foods that have been authorized for consumption within the European Union. It includes pertinent information such as the stipulated usage conditions, any supplementary labelling obligations, and any additional requirements mandated by the authorization process. These details are presented in Table 1 of the Annex to the Regulation. Furthermore, Table 2 provides specifications for the various listed products. A food product that is intended to be included in the European Union list and made available for sale must satisfy three criteria, which are akin to the three conditions specified in Article 3 of Regulation (EC) no 258/97: first, the food does not pose any safety concerns for human health. Then, the intended use of the food does not mislead consumers, especially when the product is intended as a substitute for another food and there is a notable alteration in its nutritional composition. Additionally, if the food is meant to substitute another food, it should not vary in a way that would make its regular consumption nutritionally disadvantageous for the consumer.

The authorization for the introduction of a novel food into the market does not exhaust all the requirements outlined in Regulation (EU) 2015/2283. On the one hand, like other types of goods that are subject to different authorization procedures, such as genetically modified foods, the Commission has the authority to impose post-market monitoring requirements for food safety purposes, following consultation with the European Food Safety Authority.

However, it is the responsibility of the food business operator who has placed the new food product on the market to notify the Commission about any newly discovered scientific or technical information that could potentially impact the safety assessment of the novel food. This information will then be made available to the member states. Additionally, food business operators must inform the Commission about any prohibitions or restrictions that a third country has imposed on the food being marketed (Article 25).

In this regard, Regulation (EU) 2015/2283 aligns with the provisions concerning emergency measures that the Commission has the authority to implement, as outlined in Article 53 of Regulation (EC) no 178/2002. These emergency measures are intended to address situations where there is a significant threat to human health, animal health, or the environment and where national measures are insufficient to effectively mitigate the risks.

The novel food regulation effectively addresses the dual objectives of safeguarding consumer health and interests and promoting fair competition. Simultaneously, it also aims to protect the substantial investments made by food business operators in gathering essential data, often through commissioned studies, required for the authorization process. The regulation on novel foods has two methods that facilitate the examination of these assumptions.

When seeking authorization for a new food or notifying a traditional food from a third country, the food business operator can make a specific request to the Commission. This application pertains to the confidential treatment of certain information, as explicitly specified by the operator. Information excluded from confidentiality includes the operator's name and address, the name and description of the food, the proposed conditions of use, the summary of submitted studies, the results of safety studies conducted on the product, the methods of analysis, and any prohibitions or restrictions imposed by third countries in relation to the food. The purpose of this confidential treatment is to prevent the disclosure of information that could potentially harm competition. After receiving notification from the Commission of its stance, the applicant has the option to either withdraw the application, in which case the material will remain confidential, or choose to retain the record. In this particular scenario, similar to the one involving non-response, the responsibility of determining which information can be classified as confidential lies with the Commission.

The concept of confidentiality, as outlined in Article 23 of Regulation (EU) 2015/2283, pertains to the non-disclosure of information to the public. It is important to note that the acknowledgment of confidentiality does not hinder the sharing of information between the Commission, the European Food Safety Authority, and the Member States. This exchange of information is necessary for the effective implementation of the procedures and obligations outlined in the Regulation.

The second instrument pertains to the opportunity provided to the applicant to seek protection for novel evidence or newly discovered scientific data that are presented in support of the application. The provision of such protection will serve to prohibit the utilization of said proof and data in support of a subsequent application within a span of five years following the authorization of the new food product, unless, naturally, the applicant who has obtained the protection has provided their explicit approval. The Commission will offer protection under three circumstances: if the applicant indicated the new evidence or scientific data as protected by industrial property at the time of application submission; if the applicant holds an exclusive right of reference to evidence or scientific data protected by industrial property at the time of application submission; and if the evaluation of the food by the European Food Safety Authority and subsequent authorization by the Commission would not have been possible without the presence of such evidence and scientific data protected by industrial property.

The extent of the protection is significant. With the old regulation, the marketing authorization was granted through a decision that only applied to the applicant. However, under the new provisions, authorization is granted through a general implementing regulation, an act that has general and direct applicability. Under Regulation (EU) 2015/2283, once the authorization has been obtained, it is theoretically possible for anyone, provided that they comply with the conditions outlined in the authorizing implementing regulations, to introduce the product into the market. In contrast, in cases where the protection of scientific data is recognized, the authorizing act will include a statement that specifies that the authorization is limited to the applicant. This condition is justified by the fact that entry into the EU list of novel foods is strictly dependent on the aforementioned evidence and protected scientific data.

There are, however, no inherent barriers that prevent future entities from acquiring authorization for the same food by independently presenting their own scientific data and evidence.

Regulation (EU) 2019/1381 on transparency and sustainability of risk analysis in the food chain has amended the procedure governing the authorization of novel foods and the notification process for traditional foods originating from third countries. The most significant transformation depended on Article 32ter of Regulation (EC) no 178/2002, as introduced by the Transparency Regulation, which raises some doubts regarding the requirements it imposes. Specifically, it mandates that any operator seeking authorization or notification, including those outlined in Regulations (EU) 2015/2283, must promptly inform the European Food Safety Authority about the *title and scope of all studies* commissioned or conducted to support the application or notification. Additionally, operators must provide information about the laboratory or facility responsible for conducting the tests associated with the study, as well as the start and end dates of the research. The aforementioned requirements will also be applicable to the laboratories and other facilities responsible for conducting tests within the Union. These entities are obligated to promptly inform the European Food Safety Authority of the relevant details of all studies commissioned by the operators and conducted by the laboratories and facilities involved in performing tests to support an application or notification. This includes providing information on the title and scope of the studies, the expected start and end dates and the identity of the economic operator who commissioned the study. All the collected data and studies will be incorporated into a database established and managed by the European Food Safety Authority.

The amendments to the Regulation convey a perception of an excess of transparency. The procedure does not require just the involvement of the food business operators that have invested in research (and might not be willing to see information disclosed) but affect also the laboratories responsible for conducting the studies, thus making public the relationships between food business operators and research or analysis centers. These provisions do not seem to offer adequate safeguards for confidentiality and intellectual property rights. As stated in recital 28 of the Transparency Regulation, the European Union legislator aims to ensure general public access to the scientific basis (data and information) that represent the foundation for authorizations, approvals or the applications for other scientific output and requires business operators to actively contribute to the purpose with a timely notification. Simultaneously, the aforementioned recital declares that no prejudice to provisions safeguarding innovators' investments should be caused in gathering the information and data supporting relevant applications or notifications. The consideration of the two conflicting requirements and the adoption of the *practical arrangements* for the execution of future Article 32b are of great importance, particularly in relation to the European Food Safety Authority's role in this matter.

## 16.6 The Importance of a Unified Framework Against Nationalisms

Since 1997, the theme of novel food has been regulated with a *Regulation*. The use of this specific type of act is of paramount importance. In the European Economic Community, the first category of acts used for the food sector were directives. A directive is an act that sets the objectives that member states have to achieve but leaves national legislative institutions free to decide which form and methods to use to reach those results (Article 288 of the Treaty on the Functioning of the European Union, TFEU). Directives were the fundamental tool for the approximation of the legislation of Member States since they could match the need for harmonization, obtained in setting the objectives, with the peculiarities and nationalistic attitude of member states, which were still allowed to adapt their legislation to reach the targets. Requiring a national implementing act, directives could however lead to some discrepancies among member states. Once national provisions had been approximated, the following step of *unification* could be covered with a regulation, which is a binding act characterized by general and direct applicability, as it is, in all member states (Article 288 TFEU). When a subject is encompassed by a regulation, there should not be any space for national measures, unless expressly stated by the act itself.

When talking about food analogues, in the case of products falling under the novel food regulation, their placing on the market is *only* subject to the authorization procedure and the compliance with the authorizing implementing regulation of the European Commission. Nevertheless, in 2023, some member states tried to bypass the path set by Regulation (EU) 2015/2283, proposing national acts intended to introduce further requirements or prohibitions for certain types of products.

According to Directive (EU) 2015/1535, member states must immediately communicate to the Commission any draft technical provision; the obligation should avoid the creation of barriers to trade between countries. The summary of the notified act will be published by the European Commission, in order to collect any possible comments from interested parties. From the date of the notification, the standstill period will run; the notifying member could not pass the final act before the ending of such a period.

With notification no. 2023/0109/HU, on March 16, 2023, Hungary informed the Commission of its intention to approve a Decree amending Ministry of Agriculture Decree No 36/2014 of December 17, 2014 on food information. The new provisions mandated that food containing insect proteins must include a warning about the presence of such proteins. The indication had to be written in at least the same font size as that used for the name of the product. Food containing insects had moreover to be sold on separate shelves from other foods.

With notification n. 2023/111/RO, on March 17, 2023, Romania informed the Commission of an Order that asked economic operators to provide consumers with clear and complete information on foods containing insects that had been authorized as novel foods, namely yellow mealworm, *Locusta migratoria*, house cricket and *Alphitobius diaperinus* larvae (lesser mealworm). The Order would have extended to any other insect species included in the Union list of novel foods. The Romanian act would have covered insects sold both in conventional markets and by means of distance sales, as well as food to be consumed on the spot or at home.

On March 22, 2023, the Italian Standing Committee State-Regions-Autonomous Provinces agreed upon the text of four interministerial decrees requiring additional mandatory particulars for placing on the market the four authorized insects of the European Union.

With notification n. 2023/0469/IT, on July 27, 2023, Italy informed the Commission of a proposal for a decree laying down provisions relating to the prohibition of the production and placing on the market of food and feed consisting of, isolated from or produced from cellular crops or tissues derived from vertebrate animals and of the ban on the designation of meat for processed products containing plant proteins. The reasons underlying the proposal were identified in the need to protect human health and consumers' interests and on the protection of the agri-food heritage, meant as a set of products that are expressions of the process of socio-economic and cultural Italian development, of strategic importance for the national interest.

There is no need to say that the national acts notified to the Commission raised criticism from states, associations, scholars and other interested partners (as the feedback on the TRIS platform showed). Once the European Union covers a certain subject with a regulation, as mentioned above, member states do not have any room for further national legislation. In the cases at issue, now Regulation (EU) 2015/2283 entirely encompasses insects. Furthermore, the four species identified both in the Romanian and Italian acts have been authorized with implementing regulations that also lay down provisions on the labelling of new products, to inform consumers of possible allergies resulting from the consumption of these insects.

With reference to foods obtained from cultivation of cells animal origin, up till now the European Commission has not authorized any products yet.

As can be easily inferred from the invoked reasons for its approval, the Italian project is groundless. As for the protection of human health, the whole European Union legislation is committed to defending it. The regulatory framework for novel foods is an example of such a pledge. If the free movement of goods has been one of the fundamental principles since the first steps of the European Economic Community, the protection of human health is one of the primary exceptions that could be opposed to the full achievement of that freedom. If generally food does not require a pre-market authorization, novel food is subject to an in-depth risk assessment to evaluate both their safety and their attitude not to mislead consumers. If the novel food is authorized, its safety has already been assessed by means of a European Union procedure, which usually also involves the European Food Safety Authority and where States can intervene. A national decree overstepping a unified procedure is clearly inconsistent with the principle of EU law prevalence.

The same reasoning applies to the protection of consumers' interests that the proposed decree claims to ensure. As we said in paragraph 5, one of the conditions that the European Union requires for a novel food to be included in the Union list is that the product does not mislead consumers and does not create disadvantages when consumed. A decree that states that provisions are needed to protect consumers' interests (while limiting the possibility for consumers to choose) implies a failure of the existing supranational legal framework and seems to infringe EU law as far as it overlaps with it.

Besides lacking legal grounding, the Italian proposed decree, therefore, does not have even a logical reason: in the European Union, food business operators are not allowed to place on the market novel food without having obtained an authorization. Transposing the statement in the specific case, this means that in the European Union no cell-cultivated meat is currently sold or can be placed on the market.

The declared protection of agri-food heritage as an asset of strategic importance for the national interest is a further element that conflicts with EU law if it becomes a tool to extend an exception to the free movement of goods beyond what has been allowed by the regulation. It is a clear exhibition of the alleged sovereignisms that are undermining the European project while instilling doubts about the effectiveness of the existing set of regulations. It is not clear how the possibility for consumers to choose products obtained by cell cultivation could threaten the agri-food heritage: it is just a further food option given to consumers. People willing to buy and consume products related to the culture of that country will have the possibility to do so: novel foods are not intended to wholly replace existing products but to offer an alternative to consumers. If adequate information is available (and the European Union is definitely committed in this regard), there will not be any reason to prohibit the production and placing on the market of cell-based food.

## 16.7 Collateral Themes: Land Planning, Legal Qualifications and Taxation

The production and placing on the market of some novel food products used as or in food analogues might affect collateral themes or give rise to further legal questions that could be solved only at a national level, depending on provisions that are not harmonized in the European Union.

This chapter is not the most suitable place to discuss elements that might vary from state to state; we will therefore provide just a quick overview of the collateral themes that could be the object of future sectoral investigations.

One of the main topics that scholars use to emphasize when talking about food analogues, referring to either plant-based products, insects, or food obtained from cell cultivation, is the opportunity to reduce soil consumption in primary production and the negative environmental impact (Polyak et al. 2023). Without prejudice for the need for a life cycle assessment that takes into account all the different variables that can affect the results, food analogues are considered to be more environmentally sustainable than *traditional* food of animal origin (Nelson et al. 2016). In its Communication Farm to Fork Strategy, the European Union Commission acknowledged that adopting a diet that is more plant-based, with a reduction in processed and red meat and a greater proportion of vegetables and fruits, will not only decrease the likelihood of contracting life-threatening diseases, but it will also reduce the impact that the food system has on the environment (European Commission 2021). On the one hand, replacing part of the proteins of animal origin with plant-based alternatives could lead to a decrease in animal farming and required feed. On the other hand, cell-based options could be produced with a limited use of agricultural raw materials.

The potential beneficial effects on the environment deriving from an increased use of plant-based proteins are also recognized, though with a milder tone, by the European Parliament in its Resolution on a European Protein Strategy (European Parliament 2023, points 24 et seq).

The future shift towards a more plant-based diet and food analogues could also affect land planning, in a reverse process compared to the path that agriculture has been walking in the last decades (Strapasson et al. 2020): if food analogues could determine a decrease in the use of soil, land could, at least to some extent, be converted to forestation or reforestation.

This consideration brings to attention the issue of the qualification of some food analogue production activities. In the case of insects, for instance, some reasoning should be developed to understand if the activity is an agricultural activity or not. In Italy, for instance, despite the clear qualification of animal farming as *agriculture*, under Article 2135 of the civil code, from a fiscal point of view, the classification might be controversial.

Also, the VAT rate for food analogues, in some countries, is different compared to that of the equivalent product (mainly of animal origin). In Italy, for instance, while milk has a VAT rate of 4% or 10% depending on the durability (fresh or pasteurized), the rate for plant-based drinks is 22%.

## 16.8 Naming "Non-Novel" Food Analogues

Under Regulation (EU) no 1169/2011, one of the mandatory particulars that should always be present when labelling or presenting the food intended to consumers is its name, which might be its legal, customary or descriptive name. The legal name of the food is the name prescribed by law, at any level. It might be a name prescribed either by the EU or by the provisions of the member state where the food is sold at the retail stage. Some examples of legal name, for EU law, are *milk* [Regulation (EU) no 1308/2013]; *extra virgin olive oil* [Regulation (EU) no 1308/2013, Annex VII, part VIII)]; *chocolate* [Directive 2000/36/EC, Annex I, A.3]. When a legal name does not exist, then the customary name must be used; it identifies the word or words that consumers use to call that product in the Member State where it is sold, and does not require further explanation. An example of a customary name is the word *cookies*. If even the customary name does not exist for that food, then the *descriptive name* must be used. The descriptive name consists of a description of the product, which might be accompanied by information on its use. The name should allow consumers to clearly understand the nature of the food and recognize it among other products without any possible confusion.

Annex VI to Regulation (EU) no 1169/2011 lays down further requirements for the food name, such as the obligation to specify the physical condition of the food or the specific treatment undergone, when omitting it could mislead the purchaser.

When a food analogue or one of its ingredients are authorized or notified as novel foods, in the European Union, the Commission Implementing Regulation of authorization includes the name of the product that must be used; that name will become the *legal name*, recorded in the European Union list of novel foods and subject to the conditions of use detailed in Regulation (EU) 2017/2470.

When a food analogue does not require authorization, some criticalities might occur. As we mentioned several times, usually food analogues are designed, produced and presented as alternatives to commonly used foods, generally of animal origin. The Codex Alimentarius General Guidelines for utilizing vegetable protein products (VPP) in foods state that if a food with proteins entirely derived from vegetables mimics an animal product, it should be named after the VPP, along with descriptive words for its flavor or characteristics (Codex Alimentarius 1989, CXG 4-1989). In the European Union, several names of meat or dairy products are defined by regulations, thus becoming *legal names*. Such names cannot be used to identify products of non-animal origin or any other product that does not comply with the requirements of the Regulation. So, for instance, according to Annex VII, part III, 1, milk can be used only for the product that is normally secreted by the mammary glands of mammals and that is obtained from milking, without any addition or extraction. The name *milk* cannot therefore be used for plant-based beverages, like soy drinks or rice drinks. This was also confirmed by the Court of Justice in C-422/16 which stated that the mentioned provisions preclude «the term 'milk' and the designations reserved exclusively for milk products from being used to designate a purely plant based product in marketing or advertising, even if those terms are expanded upon by clarifying or descriptive terms indicating the plant origin of the product at issue» (point 53 of the Judgment). The only exception is for products listed in Annex I to Commission Decision 2010/791/EU of December 20, 2010: the reference is to a few products that are well known by consumers thanks to their traditional use or because those names describe the distinctive quality of the food (Decision 2010/791/EU, recital 2). Member states had to notify the Commission of the products they identified as matching the mentioned criteria. This is the reason why in the European Union market coconut milk or cocoa butter maintained their *dairy-like* names.

Some food analogues lack a name defined by the EU or national law and do not fall into one of the foods listed in Decision 2010/791/EU: if, on the one hand, they could be presented to consumers with customary or descriptive names, on the other hand, debate on their names is still quite hot. When they are presented as an alternative to meat products, for instance, they are designed and shaped with forms that can recall such products: legume balls, veggie nuggets, plant-based burgers or cutlets are some of the products that are currently placed on the EU market.

Over the years, the European Parliament raised several parliamentary questions about the meat-sounding nature of the names used and deferred them to the European Commission. Just to make an example (which is actually emblematic of the frequent Parliament's inquiries on the theme), with question for written answer E-004966/2020 the European Parliament asked if the Commission was going to adopt rules to clearly distinguish between products with animal origin and plant-based products and if it was going to prohibit the use of meat names for vegetarian and vegan products. The Commission's answer, replicated as a reply to all the similar Parliament's questions, considered EU legislation (and in particular Regulation (EU) no 1169/2011) adequate to protect both consumers and competition. In 2022, the European Parliament rejected the attempt to introduce, during the works on the Strategic Plans Regulation (COM(2018)0394), a prohibition for plant-based products to use names that are often linked with or usually reserved for items containing meat (e.g. plant-based burgers, sausages or steaks) (question for written answer E-002681/2022).

Associations that support the request for such a prohibition usually base their reasoning on the likelihood that vegetarian or vegan foods bearing meat-sounding names will mislead consumers. The risk is, however, totally absent. On the one hand, people who usually purchase meat products and are not interested in food analogues know exactly what products they are looking for, where they could find them, and how they are labeled and presented. They could not be misled by a labeling declaring that balls, cutlets or burgers are veggie/plant-based/vegetarian or similar. On the other hand, consumers who are vegan, vegetarian or flexitarian are usually well informed about the products suitable for their diet and very careful about their consumption choices: they will not be misled by the name. Finally, food business operators placing on the market food analogues have as their primary interest the differentiation from *common* food: they will present and promote their products using their peculiarity as a point of strength. Claims, colors, positioning will generally play on the differences, emphasizing the *alternative* aspects.

## 16.9 Food Analogues and Nutrition Claims

In the European Union, nutrition claims (as well as health claims) are regulated under the provisions of Regulation (EC) no 1924/2006. Food analogues, being alternatives to common foods, are often sold with some nutrition claims, emphasizing the high protein content or the low-fat formulation. The Annex to Regulation (EC) no 1924/2006 lists all the authorized nutrition claims and the specifications for their use while the general conditions are described in Article 5. Claims are authorized

- If it has been demonstrated that the presence, absence, or reduced quantity of a nutrient or other substance in a food or category of food has been proved to have a favorable effect on the body's physiological or nutritional functioning, as established by generally recognized scientific evidence.
- If the nutrient or other substance related to the claim is present in a significant quantity. The measure that specifies if the amount is significant must be defined by EU provisions or determined according to generally accepted scientific evidence that proves the connection with the nutritional or physiological effect. Some claims could, however, emphasize the absence or the reduced quantity of certain substances. Generally accepted scientific evidence will again represent the basis for the used claim.
- If bioavailability is proven. In other words, the nutrients or other substances related to the claim are in a form that is available to be used by the body.
- The reasonable quantity of food that the consumer is expected to have entails a sufficient intake of substances with nutritional or physiological effects to substantiate the claims according to generally accepted scientific evidence.
- Products comply with the requirements of the regulation on nutrition and claims.

The bioavailability and nutritional impact of dietary protein depend not only on the quantity but also on their quality (Bennett et al. 2023) and, with reference to food analogues, scientific literature has stressed the importance of investigating both protein quality and digestibility (Cutroneo et al. 2023; Rojas et al. 2022). If research on these aspects is still going on, the European Legislature has not amended the Regulation. The result is that food analogues can bear the nutrition claim of high protein contents based just on their quantity. Consumers are not provided with the necessary tools to truly understand the food content.

A further aspect of the claims regulation that might be of interest when talking about food is the possibility to use comparative claims, provided that the conditions listed in Article 9 are met. Compared food must belong to the same category; the comparison shall take into consideration a range of foods of that category with a composition that does not allow those products to bear a claim; the comparison refers to the same quantity of food; the difference in quantity of a nutrient or of energy should be stated.

Given all this, could food analogues be considered in the same category as their alternative products? In other words, could, for instance, a plant-based food analogue be compared to its alternative of animal origin? According to the «Guidance on the implementation of regulation n° 1924/2006 on nutrition and health claims made on foods. Conclusions of the Standing Committee on the Food Chain and Animal Health» of December 14, 2007, comparisons could only be allowed between foods that are similar in terms of nutritional content and that belong to a group. If we limit the reasoning to these words, food analogues and the products they are intended to replace should fit in different groups that might be certainly similar as for nutritional content (though with the expressed concerns we mentioned above on protein quality), but that are not comparable as for nature. Nevertheless, the Guidelines, using as an example butter and margarine, focus on *alternative* foods, allowing comparison for such products, provided that they are clearly identified to the final consumers (recital 21 of the Regulation).

## 16.10 Conclusions

In the previous paragraphs, starting from the lack of definition of *food analogues*, we tried to draw a picture of their regulatory framework within the European Union, taking stock of the work that also international organizations are carrying out on this subject. We mentioned that food analogues could be placed on the market without any authorization or notification, to the extent that they are not new, while they require pre-market approval when they had not been consumed as food before May 15, 1997 and provided that they belong to one of the categories listed by the novel food regulation. We also considered that while implementing regulations that authorize novel foods clearly indicate the name that shall be used for their marketing, other products are subject to criticism, when using meat-sounding products, on the grounds that they might mislead consumers.

We explained that, to the extent that the name is not reserved for certain products, the words could be used, provided that consumers are informed of the true nature of those products.

This does not entail that purchasers are absolutely protected when choosing food analogue: no provisions exist on the need to inform consumers of the difference in quality of some vegetable proteins when compared to those of animal origin. The regulation on claims, as interpreted by the guidelines of 2007, also seems to allow the comparison between food analogues and *replaced* foods, provided that consumers are informed of the terms of comparison.

Some doubts could, however, arise about the adequacy of the system of information applicable to food analogues: providing consumers with all the details that could be useful to make a thoughtful decision should be the first step for a truly transparent growth of food analogues on the market.

In this regard, the European Union (but also the work carried out at an international level) should keep going on with an increasing unification (rather than mere harmonization) covering the critical points still open. A firm and clear answer should be provided, in particular, against all the different attempts of destabilization enacted by those Member States that, invoking sovereignty and cultural heritage and spreading ungrounded rumors on food safety, try to fragment the applicable legislative framework, taking advantage of incorrect news. Once again, information, a truthful and science-based information, proves to be the best tool to balance tradition and innovation, in compliance with consumers' rights.

## References

Ballke C (2014) The new novel food regulation—reform 2.0. European Food and Feed Law Review 5:285–292

Bennett L, Piper M, Ernst A, Wu T (2023) ePROB: a new tool for defining protein quality. Food Australia 75(1):9–21. https://search.informit.org/doi/10.3316/informit.858843876174869

Bohrer BM (2019) An investigation of the formulation and nutritional composition of modern meat analogue products. Food Sci Human Wellness 8(4):320–329. https://doi.org/10.1016/j.fshw.2019.11.006

Bonora G (2016) I Novel Foods nel Reg. (UE) n. 2015/2283 e gli insetti: una possibile evoluzione dei costumi alimentari? Rivista di diritto alimentare 10(1):42–54

Carson JK, Lovatt SH, Tanner DJ, Cleland AC (2004) Experimental measurements of the effective thermal conductivity of a pseudo-porous food analogue over a range of porosities and mean pore sizes. J Food Eng 63(1):87–95. https://doi.org/10.1016/S0260-8774(03)00286-3

Centre for the Promotion of Imports from developing countries (CBI) (2023) Which trends offer opportunities or pose threats on the European processed fruit and vegetables market? https://www.cbi.eu/market-information/processed-fruit-vegetables-edible-nuts/trends

Codex Alimentarius (1989) General guidelines for the utilization of vegetable protein products (VPP) in foods (CXG 4-1989). https://www.fao.org/fao-who-codexalimentarius/sh-proxy/en/?lnk=1&url=https%253A%252F%252Fworkspace.fao.org%252Fsites%252Fcodex%252FStandards%252FCXG%2B4-1989%252FCXG_004e.pdf

Codex Alimentarius Commission (2021) New food sources and production systems: need for codex attention and guidance? (CX/CAC 21/44/15 Add.1). https://www.fao.org/fao-who-

codexalimentarius/sh-proxy/en/?lnk=1&url=https%253A%252F%252Fworkspace.fao.org%252Fsites%252Fcodex%252FMeetings%252FCX-701-44%252FWorking%2BDocuments%252Fcac44_15.Add.1e.pdf
Codex Alimentarius Commission (2023a) Report of the forty-fifth Session (REP22/CAC). https://www.fao.org/fao-who-codexalimentarius/sh-proxy/en/?lnk=1&url=https%253A%252F%252Fworkspace.fao.org%252Fsites%252Fcodex%252FMeetings%252FCX-701-45%252FFinal%252520Report%252520CAC45%252FCompiled%2BREP22_CAC.pdf
Codex Alimentarius Commission (2023b) Report of the forty-third session of the Codex Committee on nutrition and foods for special dietary uses. https://www.fao.org/fao-who-codexalimentarius/sh-proxy/en/?lnk=1&url=https%253A%252F%252Fworkspace.fao.org%252Fsites%252Fcodex%252FMeetings%252FCX-720-43%252FFinal%252520Report%252FREP23_NFSDUe.pdf
Codex Alimentarius Commission's Secretariat (2022) Request for Information on New Food Sources and production systems; Need for Codex guidance and attention to inform the CCEXEC sub-committee on working on this topic. https://www.fao.org/fao-who-codexalimentarius/sh-proxy/en/?lnk=1&url=https%253A%252F%252Fworkspace.fao.org%252Fsites%252Fcodex%252FCircular%252520Letters%252FCL%2525202022-06%252Fcl22_06e.pdf
Codex Committee on Nutrition and Foods for Special Dietary Uses (CCNFSDU) (2023) Agenda Item 7. Report Prioritization mechanism/emerging issues or new work proposals. https://www.fao.org/fao-who-codexalimentarius/sh-proxy/pt/?lnk=1&url=https%253A%252F%252Fworkspace.fao.org%252Fsites%252Fcodex%252FMeetings%252FCX-720-43%252Fdocuments%252Fnf43_07e.pdf
Costato L (1997) Organismo biologicamente modificati e novel foods. Riv Dirit Agrar 76(I):137–164
Cutroneo S, Prandi B, Faccini A, Pellegrini N, Sforza S, Tedeschi T (2023) Comparison of protein quality and digestibility between plant-based and meat-based burgers. Food Res Int (Ottawa, Ont) 172:113183. https://doi.org/10.1016/j.foodres.2023.113183
European Commission (2018) Report from the Commission to the Council and the European Parliament on the development of plant proteins in the European Union, COM/2018/757 final
European Commission (2019) Communication from the Commission The European Green Deal, COM (2019) 640 final
European Commission (2020) Directorate general. Food 2030 pathways for action. Alternative proteins and dietary shift. Luxembourg. https://op.europa.eu/en/publication-detail/-/publication/2a3f49eb-2564-11eb-9d7e-01aa75ed71a1/language-en/format-PDF/source-295653209
European Commission (2021) Communication from the Commission to the European Parliament, the Council, the European Economic and Social Committee and the Committee of the Regions A Farm to Fork Strategy for a fair, healthy and environmentally-friendly food system, COM/2020/381 final
European Parliament (2023) Resolution of 19 October 2023 European protein strategy (2023/2015(INI))
FAO (2022) Thinking about the future of food safety—a foresight report. Rome https://doi.org/10.4060/cb8667en
González Vaqué L (2008) Modificación del régimen comunitario relativo a los Novel Foods: la Propuesta de la Comisión. Rivista di diritto alimentare 2(1):38–41
Joshi VK, Kumar S (2015) Meat analogues: plant based alternatives to meat products—a review. Int J Food Ferment Tech 5:107–119. https://doi.org/10.5958/2277-9396.2016.00001.5
Keefe LM (2018) #FakeMeat: how big a deal will animal meat analogs ultimately be? Anim Front 8:30–37. https://doi.org/10.1093/af/vfy011
Kumar P, Chatli MK, Mehta N, Singh P, Malav OP, Verma AK (2017) Meat analogues: health promising sustainable meat substitutes. Crit Rev Food Sci Tech 57(5):923–932. https://doi.org/10.1080/10408398.2014.939739

Lee JJ, Srebot S, Ahmed M, Mulligan C, Hu G, L'Abbé MR (2003) Nutritional quality and price of plant-based dairy and meat analogs in the Canadian food supply system. J Food Sci 88(8):3594–3606. https://doi.org/10.1111/1750-3841.16691

Nelson ME, Hamm MW, Hu FB, Abrams SA, Griffin TS (2016) Alignment of healthy dietary patterns and environmental sustainability: a systematic review. Adv Nutr 7(6):1005–1025. https://doi.org/10.3945/an.116.012567

Polyak E, Breitenbach Z, Frank E, Mate O, Figler M, Zsalig D, Simon K, Szijarto M, Szabo Z (2023) Food and sustainability: is it a matter of choice? Sustain For 15(9):7191. https://doi.org/10.3390/su15097191

Rojas Conzuelo Z, Robyr R, Kopf-Bolanz KA (2022) Optimization of protein quality of plant-based foods through digitalized product development. Front Nutr 9:902565. https://doi.org/10.3389/fnut.2022.902565

Seung Yun Lee, Lee DY, Jeong JW, Kim JH, Yun SH, Mariano E Jr, Lee J, Park S, Jo C, Hur SJ (2023) Current technologies, regulation, and future perspective of animal product analogs. A review. Anim Biosci 36(10):1465–1487. https://doi.org/10.5713/ab.23.0029

Strapasson A, Woods J, Meessen J, Mwabonje O, Baudry G, Mbuk K (2020) EU land use futures: modelling food, bioenergy and carbon dynamics. Energ Strat Rev 31:100545. https://doi.org/10.1016/j.esr.2020.100545

Tomašević I, Putnik P, Valjak F, Pavlić B, Šojić B, Bebek Markovinović A, Bursać Kovačević D (2021) 3D printing as novel tool for fruit-based functional food production. Curr Opin Food Sci 41:138–145. https://doi.org/10.1016/j.cofs.2021.03.015

# Chapter 17
# Environmental Impacts/Benefits/Risk of Food Analogues

**Selin Yabaci Karaoğlan and Merve Darici**

## 17.1 Introduction

Food analogues, emerging as pivotal components of the modern food revolution, are products derived from plant or synthetic sources that serve as substitutes for conventional foods or additives. They aim to replicate key properties such as flavor and texture (Joshi and Kumar 2015; Bachmann 2001; Kyriakopoulou et al. 2019; Girija et al. 2021; Kazir and Livney 2021) while also addressing nutritional concerns like cholesterol and fat content (Girija et al. 2021).

These alternatives hold particular appeal for individuals following vegetarian or vegan diets, those with allergies, ethical concerns, or a commitment to ecological food production practices (Ismail et al. 2020; Kołodziejczak et al. 2021). The evolution of eating habits, driven by environmental sensitivities, ethical considerations regarding animal rights, and the quest for healthier lifestyles, often necessitates a departure from traditional eating patterns. The challenges of such transitions are manifold, encompassing not just the stress of altering habits but also navigating the ecological and social ramifications of dietary changes (Bhuva et al. 2021). Herein lies the significance of food analogues: they offer a means to maintain familiar dietary choices amidst these shifting paradigms, enabling the enjoyment of traditional flavors with a reduced environmental footprint. Notably, plant-based burgers have garnered attention for their ability to mimic beef while utilizing fewer natural

S. Y. Karaoğlan (✉)
Department of Food Engineering, Faculty of Engineering, Adana Alparslan Turkes Science and Technology University, Adana, Türkiye
e-mail: syabaci@atu.edu.tr

M. Darici
Department of Food Engineering, Faculty of Engineering, Cukurova University, Adana, Türkiye
e-mail: mdarici@cu.edu.tr

 
Ö. P. CAN et al. (eds.), *Food Analogues: Emerging Methods and Challenges*,
https://doi.org/10.1007/978-3-031-69872-9_17

resources and mitigating greenhouse gas (GHG) emissions. Amidst a rapidly evolving food industry landscape, which imposes significant environmental burdens, the identification of high-impact foods becomes crucial. Studies highlight dairy and meat products, especially beef-pork and cheese, as major contributors to environmental degradation (Notarnicola et al. 2017; Rosi et al. 2017; Van Der Weele et al. 2019). The pursuit of sustainable diets, emphasizing decreased the processed foods and increased plant-based food consumption, aligns with global efforts to lessen the ecological impacts of food production (Alsaffar 2016).

The challenges of food production are exacerbated by an ever-increasing global population and escalating nutritional demands, pushing the limits of our natural resources. Animal agriculture, in particular, exerts substantial pressure on water resources, while crop production demonstrates a relatively more efficient use of water (Ismail et al. 2020; Alexander et al. 2017). Agricultural expansion contribute to deforestation and biodiversity loss, posing significant threats to environmental sustainability (Zhuang et al. 2022). Concurrently, the food industry grapples with food waste, contrasting starkly with the ongoing issue of food insecurity affecting millions globally. Additionally, the environmental repercussions of food waste, energy-intensive food production processes, and the chemical usage in agriculture underscore the urgency of embracing sustainable food practices (Zhuang et al. 2022; Alexander et al. 2017; Viana et al. 2023). In this context, food analogues emerge as innovative solutions capable of mitigating environmental concerns associated with traditional food production (Smetana et al. 2015). By leveraging advanced technologies such as cultured/lab-based meat production, food analogues may drop the environmental impact of animal husbandry (Tuomisto and Teixeira De Mattos 2011; Smetana et al. 2015). Figure 17.1 shows the introduction of food analogues to the market. As consumer preferences evolve, food analogues increasingly become integral to our diets, necessitating a careful balance of taste, nutritional value, and sustainability (Alexander et al. 2017). However, the ascent of food analogues invites critical examination of their potential limitations, including nutritional adequacy and ecological impacts (Kołodziejczak et al. 2021).

In general, foods from plant sources have been found to have a less impact than foods from animal-sourced, environmentally (Clark et al. 2022). Nevertheless, the rapidly evolving food industry, a significant contributor to environmental strain,

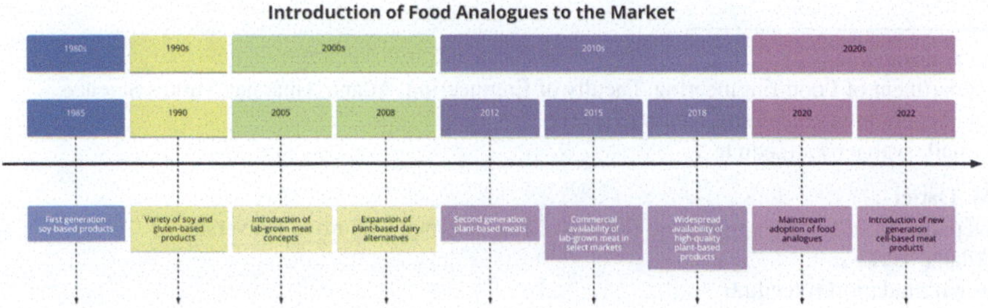


**Fig. 17.1** Introduction of food analogues to the market

necessitates a critical examination of food analogues. While these alternatives can mitigate some environmental concerns, they are not devoid of challenges (Springmann et al. 2018). Some certain ingredients like legumes and oilseeds may be constrained by antinutrients in meat analogues (Joshi and Kumar 2015). The production processes for some plant-based ingredients might result in notable alterations in land usage, monoculture farming practices and increased chemical inputs, all of which carry ecological ramifications (Fu et al. 2023; Kołodziejczak et al. 2021). Moreover, the high energy requirements involved in producing cultured meat may diminish some of the carbon emissions savings.

This chapter aims to provide a balanced exploration of food analogues, examining their production methods, ecological implications, and their role in shaping a more eco-friendly food system. Acknowledging both the advantages and the challenges of food analogues, helps a better understand their capacity to participate to a more sustainable and eco-friendly food future.

## 17.2 The Main Categories and Production Methods of Food Analogues

The introduction of food analogues signifies an important change in traditional dietary habits, as they present environmentally friendly and sustainable replacements for typical animal-based products. Figure 17.2 shows the main categories of food analogues. The purpose of these various analogues is to replicate the nutritional and sensory attributes of meats (for beef-cattle, poultry or fish analogues) generally based on plants, lab-grown/cultured or insects, dairy (analogues for milk, cream, spreads or cheeses) and eggs (plant-based gels like aquafaba, gel of chia seeds and plant-based flours-proteins like pea, potato, rice) (Viana et al. 2023). The main groups of food analogues are categorized in this article, which facilitates an in-depth assessment of their environmental impacts.

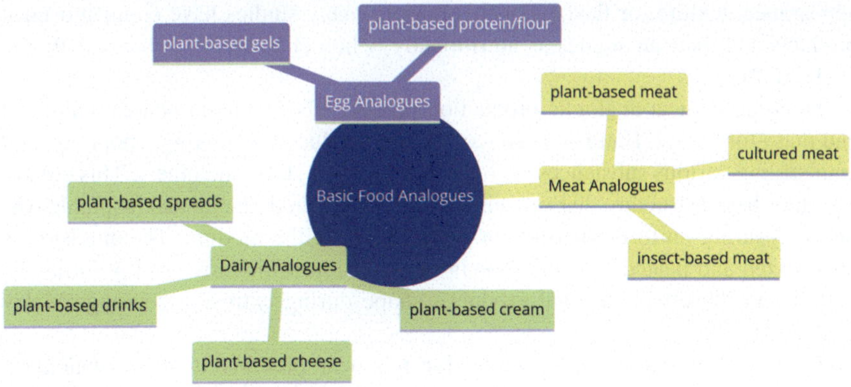


**Fig. 17.2** The main categories of food analogues

### 17.2.1 Meat Analogues

Meat analogues, also can be called as meat alternatives-substitutes aim to mimic the aroma, taste, texture, and the composition of nutritions of actual animal meat (Li et al. 2022b). These substitutes are typically created utilizing plant-based proteins derived from legumes such as soybeans, oil seeds, as well as grains like wheat, barley, rye, or filamentous fungus (Marcone 1999; Kang et al. 2007; Sun et al. 2008; Denny et al. 2008; Asgar et al. 2010). The food industry is particularly interested in meat analogues which is plant originated as replacement for traditional meat products. These analogues have the potential to imitate the flavour, texture, and nutritional value of actual meat, while also having beneficial impacts on the human health, environment, and animal welfare (Kyriakopoulou et al. 2021; Yang et al. 2023). Plant proteins constitute the fundamental components in plant-based meat alternatives, providing the structure of a meat-like fibrous tissue by breaking into and reassembling proteins (Zhang et al. 2022; Yang et al. 2023). Furthermore, oil, colorant, adhesive, and other non-animal components of food can be utilized to alter and adjust the shape, color, aroma, and taste to closely mimic genuine animal meat (Wild et al. 2014; Du et al. 2023; Zheng et al. 2023).

The function of protein is an indicator for the general characteristics of actual meat, such as their appearance, mouthfeel and texture. The aspect is essential as no other food proteins can replicate the functional characteristics of meat proteins (Xiong 2004). The primary protein functions found in processed meats include gelation and its linked qualities such as meat particle binding- adhesion, water-holding capacity and emulsification. Gelatin is considered a distinctive protein utilized in the food sector, with a variety of purposes across multiple sectors (Karim and Bhat 2008). Although the dietary benefit of meat products mostly arises from their higher protein-based biological values as well as a high level of vitamins and minerals, meat analogues are becoming more important due to the economic reasons and positive impacts on human health, the environment and animal welfare (Asgar et al. 2010; Yang et al. 2023). Moreover, consumers' preferences for meat analogues are primarily driven by their nutritional content, rather than just their appearance, texture, or flavor. For this reason, many studies have compared meat products and their analogues as nutritionally (Zhou et al. 2021; Bohrer 2019; Xie et al. 2022).

The typical meat analogue production process has two main phases: emulsion and mass formation. Emulsions are commonly produced by mixing, chopping, and emulsifying various substances such as proteins, salts, fats, and others. This creates a protein-based structure that surrounds and contains both the fat and insoluble particles, ensuring uniform distribution and stability of the mixture. The emulsion is pressurized and heated, causing protein denaturation, which irreversibly forms the matrix and preserves the final product's shape during extrusion cooking (Malav et al. 2015).

There are numerous applications for nonmeat proteins in meat analogues. Figure 17.3 is an overview of significant protein sources.

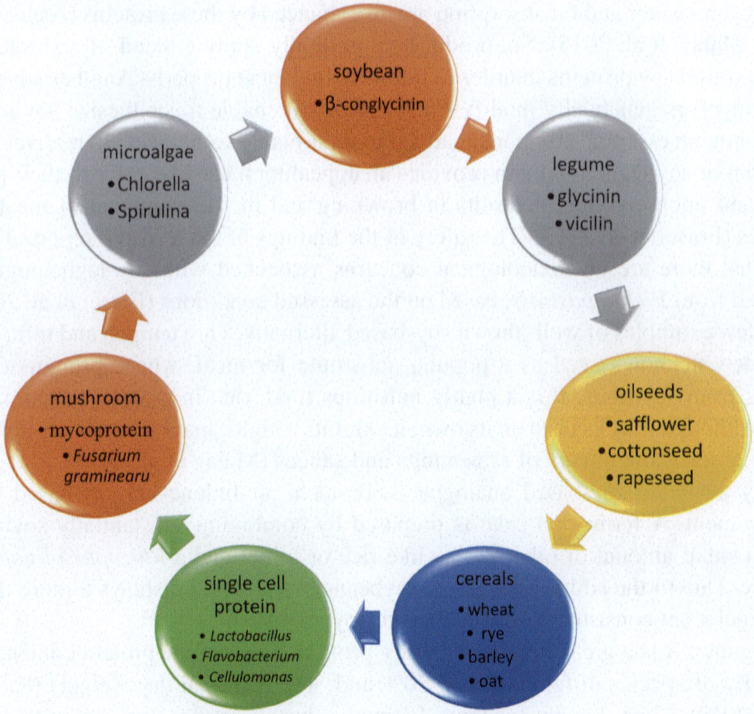


**Fig. 17.3** Major meat analogue protein sources

#### 17.2.1.1 Major Protein Sources for Meat Analogues

Soybean: It provides a valuable and highly capable plant-based protein source for human consumption. Soybeans display substantial variations in dietary composition. On average, they comprise 35–40% protein, 15–20% fat, 30% carbohydrates, and ensure an optimal proportion of amino acids with 10–30% water content (Golbitz and Jordan 2006). Additionally, soybeans are abundant in iron, zinc, calcium, zinc, fiber and B vitamins. Based on their protein content, soy-protein products can be classified into three main categories: soy flour, which includes 50% protein on a moisture-free basis; isolated soy proteins, which include 90% protein on a moisture-free basis; and soy protein concentrates, which include 70% protein on a moisture-free basis (Asgar et al. 2010; Malav et al. 2015). Additionally, soy-protein products can also be characterized as either textured or non-textured. Textured soy proteins consist of soy flours and soy concentrates, which are utilized to enhance texture and mouthfeel, improve appearance, and water binding capacity. When hydrated, they develop a meat-like texture and feature an amino acid profile that offers a high protein content comparable to that found in actual meat. Otherwise, non-textured proteins, such as soy isolated soy proteins and protein concentrates, may be applied for integration. Protein fortification, emulsification which impacts texture, and water and fat absorption are all enhanced by these proteins (Asgar et al. 2010; Malav et al. 2015). So, producers commonly apply a blend of textured and non-textured soy proteins in order to mimic actual meat properly. Another advancement involves genetically modify *Pichia pastoris*, enable to synthesize soy leghemoglobin, an essential component added to their plant-based meat alternatives. The addition of soy leghemoglobin provides an appealing meat-like color to their products, and upon cooking, it results in browning and the development of meat-like aromas (Fraser et al. 2017). The safety of the findings of the research reported indicate that there are no toxicological concerns associated with soy leghemoglobin sourced from *Pichia pastoris*, based on the assessed conditions (Fraser et al. 2018).

A few examples of well-known soy-based alternatives are tempeh and tofu. Tofu is widely acknowledged as a popular substitute for meat, whose protein source comes from soybeans. It is a highly nutritious food, rich in protein, calcium, and iron. Although it lacks taste on its own, it exhibits a high capacity for absorption and easily absorbs the flavors of seasonings and sauces (Malav et al. 2015). The other widely available soy-based analogue is Tempeh, an Indonesian soy-based meat replacement. A fermented cake is prepared by combining substantially soybeans with a small amount of other grains like rice or millet and a *Rhizopus oligoporus* culture. Due to the addition of entire soybeans, the product displays a more dense and significant consistency (Shurtleff and Aoyagi 1989).

Legume: It is a great source of dietary protein. The legume protein content in a majority of species differs between 20% and 30% (on total dry weight) (Riascos et al. 2010). Then, known for their distinctive beany aroma, they are used as the primary raw component in the manufacturing of plant-based meat alternatives. But their odor is undesirable and it might influence the taste of the product (Roland et al. 2017). For this reason, the beany aroma could be concealed and substituted with a flavor of meat by including natural seasonings and aromas associated with meat and savory odor (Fu et al. 2023; Li and Li 2020). The utilization of legume proteins as components of industrially produced food has significantly grown due to many factors such as enhancing nutritional quality, improving texture, optimizing functional qualities, and economic considerations. They hold a significant role in the nutrition of individuals, particularly in the dietary patterns of those with low incomes (Asgar et al. 2010; Malav et al. 2015).

Oilseed Protein: Storage proteins, which consist of enzymes, are the proteins found in oilseeds. These storage proteins are abundant and commonly found in oilseeds. Safflower, cottonseed, and rapeseed contain between 13% and 17% protein, 23% protein, and 25% protein, respectively (Asgar et al. 2010; Malav et al. 2015).

Cereal Protein: It is possible that cereals may be implemented in various forms, including granules, grain flour, or seeds. These may be derived from maize, wheat, rice, barley, oats, or rye. The protein compositions of these cereals, expressed as a percentage of dry matter, are as follows: barley (7–14.6%), wheat (8–17.5%), oats (8.7–16%), and rye (7–14%) (Asgar et al. 2010; Malav et al. 2015). The proteins found in cereals can be classified into the following categories: globulins, gliadins, glutelins, and albumins (Guerrieri 2004). Especially wheat gluten offers an extensive range of plant-base protein constituents that can be integrated into products as meat substitutes and meat extenders (high content of protein) in order to mimic actual meat (Orcutt et al. 2006). The production of meat extenders can involve combining gluten with soy concentrate or flour containing a high protein content. In order to generate an assortment of meat substitutes, gluten that is hydrated can be processed (as texturized), shredded, or extruded into the fibers (Asgar et al. 2010, Malav et al. 2015). One of the example for wheat based analogues is Seitan, frequently referred to as "wheat meat", is another effective substitute to meat for vegetarians. It is composed of wheat flour and formed by rinsing the flour dough until the starch disintegrates, leading to a dense and elastic mass of protein-containing gluten (Malav et al. 2015).

Mushroom: They typically have a fibrous texture that improves their chewiness. Their cell membranes have a high concentration of polyunsaturated fatty acids, and the hyphae wall contains a significant amount of dietary fibers. Additionally, their cytoplasm contains higher concentrations of protein (Singh and Sit 2022). Mushrooms provide a complete source of protein, minerals, and vitamins necessary for optimal nutrition. *Fusarium graminearu*, an edible filamentous fungus, was the initial choice for food products such as burger patties and sausages, serving as an alternative for meat. The amino acids and sugars that compose fungi improve the taste of food products, adding sweetness and umami flavors that mimic the taste of meat (Oosone et al. 2020; Li et al. 2022b). Mycoprotein-based meat alternatives are prepared by combining fermented mushrooms with eggs, spices, and flavorings. It is frequently used as a replacement for both beef and chicken (Singh and Sit 2022).

Fungal proteins (mycoprotein) have several advantages. Firstly, their distinct mushroom flavor is quite desirable. Additionally, they may be cooked to get a more succulent and firm texture, similar to the texture of meat. By including spices and colorants, they can approximate the taste and appearance of fish, chicken, or ham. Moreover, fungal proteins provide health and nutrition benefits that above those of bacterial and vegetable proteins (Rousta et al. 2021; Singh and Sit 2022). Therefore, a meat analogue could be created by combining edible mushrooms, gelling substances (polysaccharides), texturing ingredients (proteins), and seasoning. Replacing texturized soy protein with mushrooms enhances the sensory experience of meat analogues by improving their taste and aroma as well as their consumer acceptance (Kumar and Kumar 2011; Kumar et al. 2011).

Microalgae: Microalgae have the ability to improve the nutritional value of typical food product, which can have a good impact on the health of both human and animal (Spolaore et al. 2006). Their composition includes various fatty acids (long-chain polyunsaturated), carotenoids, proteins, phenolic compounds vitamins, and minerals (Plaza et al. 2009). The significant protein concentration present in numerous species of microalgae makes them a compelling option as an alternative protein source (Iwamoto 2003; Pulz and Gross 2004; Soletto et al. 2005). Since the cells have the ability to produce all amino acids, they are able to provide the essential protein (Guil-Guerrero et al. 2004). Furthermore, microalgae display a high level of overall digestibility, hence allowing for accessible use in food products (Becker 2004). The commercially significant species include *Spirulina, Dunaliella, Chlorella, Haematococcus, Botyococcus, Chaetoceros, Phaeodactylum,*

*Crypthecodinium, Isochrysis, Porphyridium, Nannochloris, Nitzschia, Tetraselmis, Speleothem* and *Schizochytrium*. At the present market is mostly dominated by products derived from *Chlorella* and *Spirulina* (Singh and Sit 2022; Sathasivam et al. 2019). *Spirulina maxima* contains around 60–71% protein, 13–16% carbohydrates, and 6–7% lipids in terms of dry matter. Also, *Chlorella vulgaris* has around 51–58% protein, 12–17% carbohydrates, and 14–22% lipids in terms of dry matter. Generally, microalgae cells contain lipids ranging from 1% to 40% (on dry weight); however, under particular circumstances, it is possible to represent as much as 85% (on dry weight) (Aly et al. 2023). Meanwhile, microalgae are a rich source of almost all necessary vitamins, including A, B1, B2, B6, B12, C, E, nicotinate, folic acid, pantothenic acid, and biotin. An additional characteristic that contributes to the commercial worth of microalgae is their substantial pigment content, which includes carotenoids (approximately 0.1–0.2% of dry matter and chlorophyll (approximately 0.5–1% by dry matter). Additionally, β-carotene, specifically, can be detected in *Dunaliella salina* at levels as high as 14% of dry matter (Spolaore et al. 2006; Becker 2004).

Microalgae can be produced in multiple forms, including tablets, capsules, and liquids. These components are diverse and are able to be included in a diverse variety of products such as desserts, chewing gum, meals, noodles, wine, cereals and food analogues (Spolaore et al. 2006; Singh and Sit 2022). According to specific research, the processing of *Spirulina* leads to the production of products that have a lack of fibrous components. Therefore, it is suggested to combine *Spirulina* biomass with other plant based proteins that have the ability to create texture (Palanisamy et al. 2019; Grahl et al. 2018).

Single cell protein (SCP): It implies microbial biomass that is utilized as a component in food. SCP may refer to both the isolated cell protein and the entire cell material. Substances such as carbonhydrates, methane, methanol, cellulose alkanes are evaluated for their possible use as SCP precursors. Various yeast species, including *Candida tropicalis, Candida oleophila*, and *Saccharomycopsis lipolytica*, are capable of catabolizing alkanes. Methanol is the most essential component in the SCP's generation. Fungi (such as *Gliocladium delinquescens* and *Trichoderma lignorum*), yeast (such as *Candida boidinii, Hansenula capsulate, Pichia haplophila* and *Torulopsis glabrata*), and bacteria (such as *Pseudomonas, Bacillus* and *Rhodopseudomonas*), can all be utilized in the production of SCP from the precursors of methanol (Crueger and Crueger 1990). Moreover, bacteria's rapid growth and brief generation cycle contribute to their being well-suited to generating single cell protein (Suman et al. 2015). The bacteria have the ability to develop on various raw substances, including carbohydrates like starch and sugar, as well as hydrocarbons like methane and petroleum components. Hence, a wide range of bacteria may be utilized to generate substantial amounts of single-cell proteins. The microorganisms pointed out are *Lactobacillus* spp., *Flavobacterium* spp., *Cellulomonas* spp., *Rhodopseudomonas capsulate, Methylomonas methylotrophus, Thermomonospora fusca*, and *Pseudomonas fluorescens* (Dharumadurai et al. 2011), *Acromobacter delvaevate, Bacillus megaterium, Bacillus subtilis, Aeromonas hydrophila, Methylophilus methylotrophus, and Acinetobacter calcoacenticus* (Gomashe et al. 2014).

#### 17.2.1.2 Production Process

##### 17.2.1.2.1 Processing Technology for the Production of Plant-Based Meat Analogues

The producing of meat analogues or alternatives involves the use of high-quality protein sources. These sources undergo various techniques to extract protein components, which are then transformed into texturized intermediary products. These intermediary products are used to create meat substitutes, which are then distributed to consumers (Pearson 2012; Singh and Sit 2022).

The flavor and fibrous structure of meat analogues are significantly linked to the ingredients and processing techniques (Guo et al. 2020). In addition to proteins, non-protein additives have also a significant impact on the solidification and flavor of meat analogues (Palanisamy et al. 2019; Boukid 2021). When multiphase mixtures, such as a combination of protein and polysaccharides, are texturized, the resultant substance is anisotropic. In contrast, pure proteins create an isotropic substance (Wi et al. 2020; Grabowska et al. 2014). Therefore, texturization is an essential process as consumers expected a product that closely resembles the texture, mouthfeel, aroma, taste and nutritional content of meat products (Palanisamy et al. 2019, Boukid 2021). Textured proteins are created when plant-based proteins undergo a number of changes that transform their basic globular structure into a linear structure, thus mimicking the fibrous texture observed in meat muscle.This is achieved by several processing techniques of meat analogue producing methods which are consist of extrusion, freeze structuring, shear induced structuring, hydrocolloids merging with plant protein, wet-spinning, electrospinning, bioprinting, cell culturing, mycoprotein. These production technologies have various related advantages and disadvantages (Boukid 2021; Singh and Sit 2022).

*Extrusion*: It maintains a widely applied process despite being the oldest patented (method for protein texturization (Wenger et al., 1977; Osterhaus 1978). Initially, a substantial number of extrusion methods were created for soy protein over the years. However, current research has expanded the number of evaluations to include a broader range of sources of protein, including *Cannabis sativa*, pea, and microalgae (Caporgno et al. 2020; Guo et al. 2020; Zahari et al. 2020). Techniques that have significant productivity, adaptability, energy efficiency, and cost-effectiveness further contribute to their continued prevalence. Furthermore, extrusion deactivates hydrolytic enzymes such as, peroxidases, lipoxidases and lipoxygenases, and it also induces the breakdown of heat-sensitive substances such as hemagglutinins and trypsin inhibitors, consequently enhancing the ability of proteins to digest (Kamau et al. 2020; Boukid 2021).

Regarding extrusion, there are two distinct techniques: low-moisture and high-moisture extrusion (Boukid 2021). The process of applying the low-moisture extrusion technique, whose moisture content is below 35%, meat analogues develop a spongy texture that needs to be further hydrated in order to get a texture similar to actual meat (Osen et al. 2014). Nevertheless, by controlling the product's expansion better, high moisture extrusion (40–80%) allowed for improved emulsification of fat. The process of gelation of proteins was employed, leading to the reorganization and formation of particles, ultimately culminating in a fibrous meat-like structure (Alam et al. 2016; Pietsch et al. 2019; Kamau et al. 2020). Additionally to melting and polymerizing proteins, extrusion at high temperatures (140–180 °C) can cause color changes as a result of the Maillard reaction, caramelization, hydrolysis, and pigment degradation (Guo et al. 2020; Zahari et al. 2020).

*Freeze structuring*: The process of freeze structuring involves creating a unique fibrous form by freezing a protein emulsion (Dekkers et al. 2018). The development of a fibrous structure in plant protein depends significantly on factors involving the source and features of the protein (such as water absorption, gelation, and solubility), and the conditions of freezing and drying (like the temperature and time). In this procedure, proteins are mixed with other substances to achieve a homogeneous emulsion. The resulting mixture was shaped, exposed to freezing to create layers of ice crystals, and then dehydrated via methods such as steaming, baking, or frying (Yuliarti et al. 2021). Subjecting a substance to high temperatures enables the creation of a fibrous based texture structure that is permanently insoluble, while also preventing the ice crystals from melting (Boukid 2021). The textural characteristics of proteins can be adjusted by altering the conditions of freezing, which include freezing rate, pH, solids content, surface impacts, heat exchange impact, degree of confinement, and pressure impact (Boyer and Middendorf 1971). Plant-based proteins that include pea, soy, wheat, and mushroom proteins are capable of generating fibrous structures in meat analogues by the freeze-structuring process (Joshi and Kumar 2015).

*Hydrocolloids merging with plant protein:* Combining plant proteins and hydrocolloids is another method, covered by patent as well, for producing meat (Mehran et al. 2013). A sustained emulsion and a solution with colloidal particles were produced by shearing a mixture of water, a vegetable lipid containing a protein sourced from lupine, pea, potato, and hydrocolloids such as sodium alginate and methylcellulose. Casein, which can solidify with positively charged particles, was added to the emulsion particles to capture the anisotropic structures and thus, trigger the fiber creation process. The structure of the fibrous substance can be adjusted by varying the level of hydrocolloids and divalent metal cations, which are required to facilitate the precipitation and formation of micellar casein (Boukid 2021).

*Shear induced structuring*: This process has two types of apparatus with distinct shapes, the first cone-on-cone device and the other cylinder-in-cylinder device, that serve the purpose of ensuring the formation of protein structures under the impact of shearing (Krintiras et al. 2015; Dekkers et al. 2016). The cone-in-cone apparatus is specifically built to position the product within the gap between two cones. The lower cone rotates in opposition to the upper cone, which keeps still, allowing the product to undergo heating from both sides using an oil bath. This process occurs under high shear stress and at temperatures ranging from 95 to 140 °C (Schreuders et al. 2019; Dekkers et al. 2016; Boukid 2021). Alternatively, the cylinder-in-cylinder device involves placing a product between a stationary outer cylinder (which remains still) and a moving inner cylinder, resulting in a shearing motion similar to that seen in the cone-in-cone device (Krintiras et al. 2014; Krintiras et al.

2015). A critical parameter is the temperature; for instance, shearing at a temperature of 140 °C produced a firm texture that was anisotropic, whereas low temperatures led to the formation of a layered structure (Grabowska et al. 2014). Advantages of high-temperature-induced shearing in plant protein texturization have been demonstrated to be a simple and cost-effective process (Estrada et al. 2018; Boukid 2021). By applying shear-induced structuring instead of extrusion, precise fiber patterns are created due to the synergistic effect of shear and heating (Krintiras et al. 2014; Singh and Sit 2022).

*Wet spinning*: It has been used for a significant period of time to produce fibrous protein-based products (Tolstoguzov 1988; Yang and Olsen 1970; Rampon et al. 1999). The wet-spinning procedure involved the application of an alkaline protein mixture that was inserted into a spinnerette and subsequently placed in an acid bath to cause precipitation and coagulation (Huang et al. 1995). The particles that are generated, with a thickness of approximately 20 μm, can be gathered and elongated to align the structure of the fibers (Yang and Olsen 1970). Yet, this procedure involves the use of proteins, high concentrations of salt, acidic conditions, and additives (Krintiras et al. 2015; Liu et al. 2017). Further, significant amounts of waste substance occur during this procedure, including wastewater flows resulting from the coagulation and washing stages (Grabowska et al. 2014; Boukid 2021).

*Electrospinning:* This process is a cost-efficient and sustainable technique to create very thin fibrils (Wongkanya et al. 2017). Electrospinning is the application of a powerful electric field to a polymer solution using a hollow needle or spinneret. A polymer solution that is electrically charged moves towards a collector once the applied electrical forces are in excess of the solution's surface tension (Kutzli et al. 2020a). During their moves, the solvent quickly evaporates and the jets elongate and curve into very thin dry fibrils of around 100 nm (Nieuwland et al. 2014; Kutzli et al. 2019). The electrospinning process is influenced by several key parameters, including the type, structure and concentration of the polymer; and the surrounding temperature and humidity; electrical conductivity, surface tension, and viscosity of the solvent (Kutzli et al. 2020a). In order to facilitate the formation of a linked system, polymers require a high level of solubility and concentration, which promotes sufficient molecular overlap. Additionally, the solution must possess appropriate levels of conductivity, viscosity, and surface tension. Proteins are often challenging to achieve electrospin. Because of that, an ideal strategy for improving the effectiveness of the method involves the combination of spinnable polymers, like maltodextrin or cellulose, with plant proteins, in particular pea and soy proteins (Kutzli et al. 2020b).

*3D printing*: The utilization of 3D printing technology was employed to mimic the meat substitutes composed of plant proteins (Boukid 2021). It is a prominent form of additive manufacturing (AM), combining techniques such as photographic and stereolithographic processes (Severini et al. 2018). Multiple approaches which involve inkjet printing, extrusion, binder jetting, and bioprinting may be applied to the 3D printing of meat substitutes. Among the several processes being investigated for 3D printing of meat-based products through specific designs, the extrusion process has the greatest potential (Singh and Sit 2022). Meat analogues are printed through 3D technology with extrusion process to compose fibrous structures (Singh and Sit 2022). This approach is founded on the extrusion of a paste composed of plant proteins with other substances such as water, lipids, and polysaccharides via a narrow nozzle to construct blocks with several layers (Godoi et al. 2016). In addition, extrusion is usually associated with the application of semisolid pastes, including dough, chocolate, or meat purees (Boukid 2021; Singh and Sit 2022). To achieve the ideal structure, the paste's viscosity is essential and several rheological modifiers are used to get the rheological characteristics that are required. The developed meat substitute undergoes a subsequent maturation stage in a bioreactor in order to preserve its structural stability (Godoi et al. 2016; Voon et al. 2019; Boukid 2021). This technology allows for the creation of substances that possess a texture resembling muscle fibers and have customized nutritional composition. However, the primary disadvantages of the technique are linked to the high processing expenses, the intricate spatial arrangement, and the lengthy maturation period (Lupton and Turner 2018).

*Cultured meat*: It is produced by utilizing animal-derived stem cells and then cultivating cells in a controlled in vitro growth medium (Singh and Sit 2022). This was achieved by the assistance of simulated oxygenation and nutrient supplementation. Since it shares biological characteristics of meat, it is not derived from any animal in order to prioritize animal welfare. This approach, which is referred to as "cell culturing" uses biotechnological approach for generating meat. Cell lines are typically acquired through either genetic engineering or altering genes, or by causing spontaneous mutations in primary animal cells that have been adjusted (Ramboer et al. 2014). Cells are cultured using specially formulated liquid mediums that create the necessary conditions for growth of tissues. The technique requires specific nutrients, including, chicken embryo extract, horse serum, and collagen (Burattini et al. 2004; Chiron et al. 2012). Cells involve a structure known as a scaffold to extend and establish a base that creates a tissue, such as a muscle (Stephens et al. 2018). Then, muscle contractions generate tension in the base structure, which enhances the growth of muscles and protein synthesis. The process of collecting the muscle fiber requires a duration of 3 weeks (Boukid 2021; Boonen et al. 2010). Moreover, progress in this specific area includes the improvement of cultured meat is important in order to reduce the environmental impacts of meat analogues which is vary significantly based on the agricultural practices applied to crop cultivation, the processing technology utilized, and the scale of production. Figure 17.4 shows the cultured meat production in representative order (Generated by OpenAI's DALL·E, 2024).

*Mycoproteins*: Various meat replacements were created in conjunction with *Fusarium venenatum*. The fungus is generated in bioreactors using a continuous fermentation process. Temperatures and pH levels are critical in the growth of fungi, consequently they require being precisely regulated within the bioreactor (Dekkers et al. 2018). In order to enhance the release of RNA from the cells during fermentation, it is necessary to apply it to heat treatment, which will breakdown it into monomers. Heating and centrifugation the remaining biomass yield a paste-like substance containing 20% in solids (Wiebe 2004). In spite of the centrifugation stage,

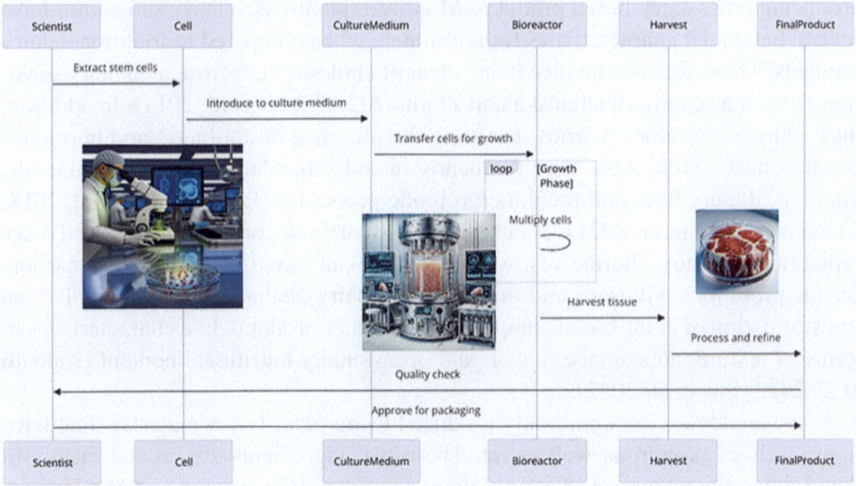


**Fig. 17.4** Production process of cultured meat

filamentous fungi lacks structure and hence require many more processing steps, including forming, steaming, chilling, and texturizing for fibrous form (Singh and Sit 2022; Smetana et al. 2015).

### *17.2.2 Dairy Analogues*

Dairy products have long been a significant and remaining part of many people's diet, serving not only as a source of pleasure but also to fulfill nutritional requirements and provide particular nutritional benefits, such as probiotic and mineral consumption (Tamang et al. 2016; Pua et al. 2022). Notwithstanding the current widespread acceptance of standard cow dairy products, consumers are exhibiting a proactive tendency towards exploring replacements because of possible long-term negative consequences on animal welfare, the environment, and human health (Mendly-Zambo et al. 2021). Consequently, there has been an increasing demand in plant-based replacements known as dairy analogues.

Dairy analogues are intended to reproduce the sensory and physical attributes of milk obtained from animals and its byproducts, such as yogurt and cheese, by mimicking the composition derived from plant sources and employing suitable technologies to achieve a comparable texture and flavor (Lima et al. 2022). Several plant-based fundamentals have been utilized to generate dairy analogues (Pua et al. 2022). Almonds, soy, oats, and coconut are the most commonly utilized plant-based sources.

Dairy analogues are produced through the following steps: cultivation of these crops, processing into a liquid that mimics milk, and further processing for producing other dairy-based products. Moreover, plant-based dairy substitutes have several beneficial characteristics for consumers when compared to traditional dairy products. These features involve being clear of cholesterol, lactose, and dairy sensitivities such as casein (Cichońska and Ziarno 2021; Yadav et al. 2017). In addition, they address consumer worries regarding the presence of antibiotic and hormones residues in the milk. Also, they frequently include abundant amounts of minerals, vitamins, dietary fiber and prebiotic/probiotic properties (Roselló-Soto et al. 2018; Cichońska and Ziarno 2021). There are two primary categories of plant-based dairy replacements: dairy alternatives, which utilize plant-based components as carriers for the probiotics, vitamins and minerals; and dairy analogues, which involve the transformation of plant-based components into dairy products-like characteristics in terms of texture, appearance, flavor, and occasionally nutritional content (Kamath et al. 2022; Pua et al. 2022).

Dairy analogues are commonly produced using plant-based materials that have high levels of protein as well as fat. These two ingredients are crucial factors in determining the texture and flavor of dairy analogues (Grossmann and Mcclements 2021). Proteins have a crucial role in several physico-chemical qualities that are important for dairy products. These features include the ability to water-absorption capacity, gelation/gel-transition, and contribute to the development of aroma precursors and aroma and taste compounds. Fats, however, have an impact on both the mechanical properties of the products, such as strength, hardness, ductility, and stiffness, and the sensory characteristics of the products, such as texture, aroma, taste, and the ability to be a flavor carrier (Jeske et al. 2018; Grossmann and Mcclements 2021; Pua et al. 2022). Therefore, the majority of the ingredients utilized in the manufacturing of dairy analogues could be grouped into four botanical classifications: "legumes", "grains", "nuts, drupes and seeds", and "tubers" (Souza et al. 2015; Chandrasekara and Josheph Kumar 2016; Petrova and Petrov 2020; Cichońska and Ziarno 2021). The classification can be seen in Fig. 17.5. Additionally, the majority of these ingredients possess the advantage of being rich in nutrients (Pua et al. 2022; Lima et al. 2022).

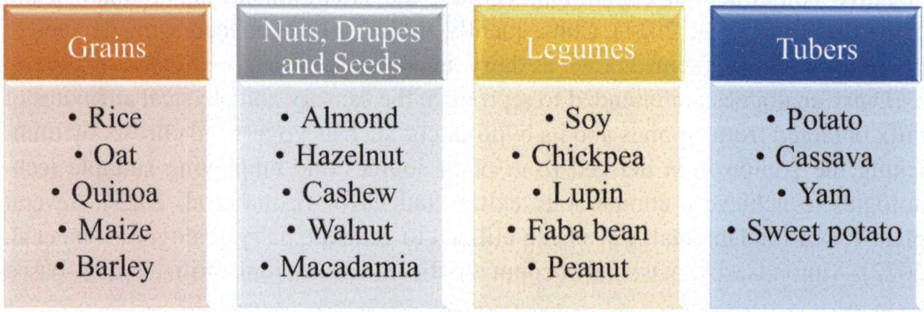


**Fig. 17.5** Major plant-based ingredients

#### 17.2.2.1 Plant-Based Milks/Drinks

It is one of the most known and vital dairy analogues in the vegan food industry since they serve as a fundamental ingredient in many dairy alternatives products, including plant-based yogurt, cheese, butter, and ice cream. Furthermore, those with lactose intolerance or a milk allergy also require plant-based milk alternatives (Aydar et al. 2020).

The production techniques of plant-based drinks exhibit variations dependent on the particular raw components applied. Nevertheless, the manufacturing of plant-based beverages involves comparable methodologies. These procedures could include roasting, peeling, soaking in water, blanching, dry grinding/wet milling, filtration, fortification, homogenization, sterilization, and packaging. Production may involve a few or multiple of these steps (Aydar et al. 2020). Firstly, it is necessary to determine whether or not to remove the outer layer of the preferred raw material. The product could be available in both pre-peeled and un-peeled forms, as well as in dried or fresh forms. In the circumstance that the primary ingredient is obtained intact along with its fresh form, it is suggested to submerge it in heated water in order to subsequently remove the outer layer (Manzoor et al. 2017). Following the removal of the outer layer, the preparation must undergo a drying process. Alternatively, if the food is already in dried form, it continues immediately to a stage of dry roasting or grinding (Kohli et al. 2017; Aydar et al. 2020). Extracts of plant substances are typically dissolved in water and subsequently extracted in order to achieve further homogenization. Thus, concise descriptions of the procedures involved in the production of plant-based drinks are provided below.

**Roasting**: The process of roasting is mainly used in the production of peanut, sesame, and hazelnut drinks to enhance the stability of the emulsion, improve the protein solubility, and prevent bitterness. However, it can also lead to a decrease in acidity, total solids, protein, and fat (Ahmadian-Kouchaksaraei et al. 2014; Chavan et al. 2018; Ilyasoglu and Yilmaz 2019; Zaaboul et al. 2019).

**Peeling**: The peeling process eliminates contaminants from the skin's surface and bitterness. Acidic solution such as citric acid or alkaline solutions like sodium hydroxide are used in the peeling process. The process could take from 18 to 24 h, depending on the selected raw material. Following that, a rinse should be conducted to eliminate any residual alkaline or acidic solution. One drawback is the requirement to use inorganic chemical compounds, resulting in higher water consumption and increased wastewater generation during process (Chen et al. 2014; Cui et al. 2013; Kizzie-Hayford et al. 2016; Kluczkovski et al. 2017; Maghsoudlou et al. 2016; Quasem et al. 2009).

**Soaking in water**: Through soaking in water, nuts and cereals gain hydration and become softer. Generally, it is utilized for rice, almonds, hazelnuts, peanuts, soybeans, and sesame (Aboulfazli et al. 2016; Alozie Yetunde and Udofia 2015; Bernat et al. 2014, 2015; Chavan et al. 2018; Kizzie-Hayford et al. 2016; Kohli et al. 2017; Padma et al. 2018; Sethi et al. 2016). This process gives an advantage to toxin reduction through release in water and decreases the processing time for other processes such as blanching as a result of the softening (Padma et al. 2018). On the downside, the soaking process requires up to 24 h and increases water consumption (Lima et al. 2022).

**Blanching**: It is a process used to decrease microbial load and deactivate enzymes like lipase and lipoxygenase, which are responsible for the beany flavor in soymilk alternatives (Pardeshi et al. 2014; Seow and Gwee 1997; Sethi et al. 2016). This procedure is often utilized for the preparation of soybeans, almonds, peanuts, coconut, quinoa, rice, and sesame (Bolarinwa et al. 2018; Diarra et al. 2005; Kohli et al. 2017; Kundu et al. 2018; Maghsoudlou et al. 2016; Maria and Victoria 2018; Padma et al. 2018; Pardeshi et al. 2014; Pineli et al. 2015; Quasem et al. 2009). Furthermore, blanching with a steam approach can be employed rather than soaking to enhance the total solids and yield of proteins (Quasem et al. 2009). The blanching process causes an increase in wastewater (Lima et al. 2022).

**Wet milling/Dry grinding**: Water must be integrated into the fundamental material throughout the grinding procedure for wet milling. A number of variables influence the end product, including milling type, milling temperature, water amount, rate of feed, and pH (Seow and Gwee 1997). Frequently prepared using this method are almonds, peanuts, hazelnuts, walnuts, cashews, coconut and hemp seeds (Aidoo et al. 2012; Cui et al. 2013; Dhakal et al. 2016; Kizzie-Hayford et al. 2016; Maleki et al. 2015; Manzoor et al. 2017; Pardeshi et al. 2014; Seow and Gwee 1997; Wang et al. 2018; Zaaboul et al. 2019).

Consumption of water increases as a result of the wet milling process (Lima et al. 2022). Otherwise, dry grinding is another option (Mäkinen et al. 2016). While not preferred for use in the production of plant-based driks, a patented option to almond milk is produced by dry milling roasted almonds (Berger et al. 1997).

**Filtration:** The aim of the filtration process is to distinguish the cake from the liquid component of the ground raw material. An extensive range of filtering materials could be employed, including muslin cloth, double-layered cheesecloth, and filter paper in various sizes (Ahmadian-Kouchaksaraei et al. 2014; Balogun et al. 2016; Bolarinwa et al. 2018; Chen et al. 2014; codina-torrella et al. 2017; Cui et al. 2013; Dhakal et al. 2016; Ermiş et al. 2018; Ilyasoglu and Yilmaz 2019; Maghsoudlou et al. 2016; Manzoor et al. 2017; Padma et al. 2018; Pardeshi et al. 2014).

**Addition the ingredients**: Some additives are utilized during the manufacturing process of plant-based milk in order to improve the final product's physical, chemical, and sensory attributes. Examples include locust bean gums, gellan, and sunflower lecithin, which enhance the stability of final beverages. Xanthan gum functions as a common thickener and stabilizer (Padma et al. 2018; Bernat et al. 2014). Ascorbic acid is incorporated for oxidation prevention (Kohli et al. 2017). By utilizing the α-amylase enzyme, gelatinization can be prevented throughout the thermal treatment process (codina-torrella et al. 2017). Olive oil and sunflower oil are used to improve the product's silky appearance (Lima et al. 2022). Sweeteners like sucrose, fructose or glucose syrup, and salt can be used for improving the aroma and taste of the drinks. Additionally, certain flavoring compounds, including cocoa and vanilla, may be utilized (Manzoor et al. 2017; Quasem et al. 2009; Pineli et al. 2015). As a drawback, certain components could trigger allergic reactions and further raise the overall cost of the end products (Lima et al. 2022).

**Fortification and enrichment**: Consumers that select plant-based milk alternatives consider vitamin, mineral and the protein composition of the product to be important considerations. Various substances are added during the production to enhance the nutritional qualities and sensory characteristics of the product (Aydar et al. 2020). An example of enhancing protein intake is by including lentils (Jeske et al. 2019; Sethi et al. 2016). Vitamins such as A, B, D, and E and calcium are included in order to enhance the nutritional value (Sethi et al. 2016; Lima et al. 2022).

**Homogenization**: It is used to improve the physical consistency of the plant-based drink. Homogenization is being conducted through low-, high-, and ultra-high-pressure approaches (Aydar et al. 2020). Several studies have shown that the clarity and stability of milk substitutes (plant-based drinks) improve with increasing homogenization pressure (Bernat et al. 2015; Briviba et al. 2016; Zaaboul et al. 2019).

**Sterilization/Non-thermal approach**: Termal treatment could be performed via pasteurization, sterilization, or ultra-high sterilization to improve the shelf life (Khuenpet et al. 2016; Lee and Rhee 2003). Conversely, sterilization adversely impact the nutritional and sensory qualities of drinks (Lima et al. 2022).

Since various new non-thermal methods have been evaluated for their ability to inactivate microorganisms (Munekata et al. 2020), it has been demonstrated that they also contribute to both flavor and texture (Pua et al. 2022). Consequently, various techniques have been implemented in plant-based milk/drinks applications, including ultrasonication (Iswarin and Permadi 2012; Gharibzahedi and Smith 2020), high hydrostatic pressure (HHP) (Sim et al. 2021; Wang et al. 2021), and pulsed electric field (PEF) treatment (Wang et al. 2021; Liu et al. 2019).

Consequently, plant-based drinks present the advantage of being perceived as more sustainable. But the potential downside of plant-based drinks is their lack of flavor. In consequence of this, there is concern over the inclusion of additional sugars and sweetening agents. But fermentation could have the potential to improve the nutritional and sensory quality of plant-based drinks. In contrast, as will be explained in more detail in subsequent sections, almond drinks have a greater ecological adverse effect than milk because of the consequences of irrigation (Silva et al. 2020; Mccarthy et al. 2017).

#### 17.2.2.2 Fermented Dairy Analogues

Common enhancements to plant-based dairy products consist of modifications that need to improve flavor, texture, nutritional composition, and food safety (Tamang et al. 2016). Given the prevalence of off-flavors in several plant-based formulations, it becomes challenging to create a product that closely resembles dairy analogues without the frequent addition of processing additives and flavorings. Fermentation could be a crucial step in creating dairy-like flavor and texture in plant-based dairy analogues. By integrating fermentation into the process, it is achievable to improve the sensory characteristics and nutrient values of plant-based dairy analogues that fulfill consumer expectations (Pua et al. 2022). Thus, several plant-based components could be used to produce fermented dairy analogues. Therefore, fermentation enhances sensory characteristics by masking or eliminating undesired odors in plant materials, increasing acid production, and stimulating the production of exopolysaccharides (EPS) that improve the plant matrix and mimic the texture of dairy products (Grossmann and Mcclements 2021; Yépez et al. 2019). Furthermore, it enhances the stability and shelf life of the analogues by acidifying, developing antimicrobial substances, and surpassing unwanted microorganisms (Rasika et al. 2021; Aryana and Olson 2017). Various starter cultures, including lactic acid bacteria (LAB), yeast, and fungus, are applied in fermentation to develop plant-based dairy alternatives. LAB already play important role in dairy fermentation for creating flavors, affecting textures, and preserving foods. Moreover, fungi and yeasts are also typically utilized as starting cultures in dairy fermentations. Especially while involved in cheese fermentation, yeast and fungi have several distinctive roles, such as developing the characteristics of surface-ripened and mold-ripened cheeses (Pua et al. 2022). Consequently, multiple studies have further demonstrated the significance of LABs, yeasts, fungi, and particularly their mixtures, in the production of plant-based dairy alternatives (Tangyu et al. 2021; Ogundipe et al. 2021; Aydar et al. 2021; Masiá et al. 2022; Ben-Harb et al. 2020; Li et al. 2020). A brief introduction of fermented dairy alternatives, such as cheese and yogurt alternatives, is given below.

**Yoghurt Analogues**: Yoghurt analogues are various fermented dairy alternatives that come in many forms, such as set, stirred, and flavored, based on their microbial composition. The conventional method for producing yogurt involves the fermentation of milk using two distinct cultures: *Streptococcus thermophilus* and *Lactobacillus delbrueckii* subsp. *bulgaricus* (Herve-Jimenez et al. 2009; Kaneko et al. 2014). Additionally, yoghurt fermentation has begun to incorporate LAB cultures, including the cultures of *Lactobacillus acidophilus, L. casei*, and *Bifidobacterium* spp. (Aryana and Olson 2017). Lactose is converted into lactic acid during fermentation for the purpose of producing dairy yogurt with a characteristic sour taste and a dairy and buttermilk aroma (Dan et al. 2017). Therefore, there is a significant amount of study about the application of these particular lactic acid bacteria (*Streptococcus thermophilus* or *Lactobacillus delbrueckii* subsp. *bulgaricus*) in order to produce plant-based yoghurt alternatives (Khrundin et al. 2021; Ogundipe et al. 2021; Yang et al. 2021; Aydar et al. 2021; Pontonio et al. 2020). Previous studies have demonstrated that yoghurt cultures can stimulate the development of yoghurt-like texture, generate pleasant, creamy odors similar to actual dairy yoghurt, and also mask or minimize the unpleasant odors, like beany notes, depending on the conditions (Pandey and Mishra 2015; Ogundipe et al. 2021; Yang et al. 2021; Khrundin et al. 2021; Aydar et al. 2021; Brückner-Gühmann et al. 2019). Furthermore, additional studies have demonstrated that non-dairy strains, including

*L. plantarum* and *L. brevis*, are also able to generating an aroma similar of yogurt (Hickisch et al. 2016; Pontonio et al. 2020), in addition to improving acidity and reducing earthy flavors (Luana et al. 2014). The LAB fermentation is also causing the acid gelation of casein, which creates the thick, firm, and uniform gel that is distinctive of dairy yogurts (Boeck et al. 2021; Aryana and Olson 2017). The lack of casein in plant-based materials and the undesired plant-based aroma are significant obstacles to creating firm, uniform, and dairy-like yogurt alternatives (Boeck et al. 2021; Hickisch et al. 2016; Lopusiewicz et al. 2020). In conclusion, commercially accessible fermented plant-based yoghurt alternatives are produced primarily from cashew, almond, and coconut combined with traditional yoghurt culture along with additives like starch, gums, and pectin (Pua et al. 2022).

**Cheese Analogues**: Cheese analogues are produced through a combination of plant-based protein sources and possibly fat, along with closely mimics the texture and flavor of actual cheese. It is important to create an appropriate processing system that can blend these components to achieve the desired texture and flavor (Bachmann 2001). Various formulas of cheese replacements available. One formulation includes caseinates and vegetable oils, while another formulation is dairy-free and utilizes only plant-based components (Shaw 1984; Kiely et al. 1991). Cheese analogues could be completely vegan, excluding milk and other dairy products, or substantially apply dairy products such casein or butter oil (Bachmann 2001). Cheese alternatives include plant-based protein, such as peanut or soybean proteins and hydrogenated vegetable oils such as palm or soybean (Bachmann 2001).

Generally, in order to produce conventional cheese, starter cultures must be added to milk to acidify the matrix through the metabolism of lactose to lactate; this results in the formation of coagulation due to the caseins within the milk (Mcsweeney 2004). The composition of starter cultures differs depending on the specific cheese type. Different culture combinations of starter cultures such as *Str. lactis, Lb. casei, Lb. plantarum*, and *Lb. brevis* could be utilized in order to develop different type of cheese. Starter cultures containing bacterial mixtures can influence acid formation rates (slow, medium, and fast), thus affecting the cheese texture (such as hard, semi-hard, or soft cheese) (Mcsweeney 2004). Additionally, cultures grow in the cheese matrix during the ripening stage, metabolizing different substrates to create gas, color, or various aroma compounds in cheese (Blaya et al. 2018; Irlinger et al. 2017; Pua et al. 2022).

Conversely, plant-based proteins do not have the same coagulation or gelation characteristics as casein micelles. So, adding LAB cultures to soy-based analogues (Matias et al. 2014), as well as a pea-based-olive oil mixture (Masiá et al. 2022) made the gels harder. Product composition, including the use of vegetable fats and gums, has also significantly influenced the textural persistence of the gels (Pua et al. 2022). The important influence of processing and extraction parameters on the textural characteristics of cheese analogues has been highlighted in recent research (Mefleh et al. 2022; Ben-Harb et al. 2020; Jeewanthi and Paik 2018).

Furthermore, fungi, more specifically *G. candidum*, have been utilized as additional cultures in order to produce smooth cheese analogues with a resemblance to Camembert and Reblochon. *G. candidum* is valued due to its proteolytic and lipolytic activities, production of volatile sulfur compounds, and ability to increase acidity. A camembert cheese analogue was created by fermenting flaxseed oil cake using LAB starters (*Lactococcus lactis, Lactococcus lactis subsp. cremoris*, and *Streptococcus thermophilus*) and fungal starters (*Penicillium camemberti* and *Geotrichum candidum*), which indicated that *Geotrichum candidum* maintains a distinct soft and chewy texture in analogues (Lopusiewicz et al. 2020). Additionally, research on soy-based soft cheese alternatives discovered that using *Geotrichum candidum* led to a softer and firmer texture in comparison to using just LAB starters (*Lactabacillus bulgaricus, Streptococcus thermophilus*). The difference was associated with the considerable degradation of fat and protein in the analogues that contained *Geotrichum candidum* (Irlinger et al. 2017). Fermented tofu, known for a considerable period, has also been utilized as a replacement due to its strong flavor consisting of mold-ripened cheeses such as Camembert or Roquefort. Tofu cubes are inoculated with a particular mold and matureted in a warm atmosphere in order to facilitate the growth of the mycelium for the production of this alternative cheese (Alcorta et al. 2021).

### *17.2.3 Egg Analogues*

Egg analogues are composed of plant-based proteins that accumulate and spread within a comparable temperature interval of 63–93 °C as egg proteins in order to create a similar texture to the egg properties (Mcclements and Grossmann 2021; Lima et al. 2022). Egg replacements aim to mimic the essential characteristics of egg protein, including coagulating or gelling, emulsifying, and foaming features (Woodward 1990), for culinary purposes, thus benefiting consumers adhering to a vegan diet or those with egg allergies (Alcorta et al. 2021). Proteins in egg yolk undergo denaturation when exposed to heat, alkali, and salt, leading to interactions involving protein-lipid and protein-protein, resulting in the creation of gels (Zhao et al. 2021). Hence, there has been a focus on plant-based proteins capable of exhibiting these characteristics. Various plant-based proteins, including pea, chickpea, bean, soybean, and sunflower, may create gels when exposed to temperatures higher than their thermal denaturation points. Nevertheless, precise adjustment of salt concentrations, protein structures, protein concentrations, pH levels, and temperatures is necessary. Various flavorings, including sugars, salts, herbs, or spices, can be incorporated in order to obtain an egg-like odor and taste (Mcclements and Grossmann 2021; Lima et al. 2022).

Substituting egg protein using soy and pea protein has been researched to be effective since both soy and pea protein contain qualities comparable to real eggs; however, pea protein exhibits worse gelling capabilities. The protein digestibility-corrected amino acid score (PDCAAS) also showed that soy protein exhibited roughly the same quality as egg protein. In contrast, pea protein was determined to be insufficient (Söderberg 2013). Therefore, pea protein may be considered fulfilled by mixing it with other plant-based protein sources that have complementary amino acid compositions. Conversely, an important point to consider for the manufacturer is that soy and pea proteins have unique aromas that are typically identified as beany or green odors and therefore may not appeal to consumers (Söderberg 2013; Alcorta et al. 2021).

Soy milk combines with stabilizers such as xanthan gum, guar gum, and arabic gum in order to create vegan mayonnaise without an actual egg (Nikzade et al. 2012; Ali and El Said 2020). Various plant-based proteins such as soy, pea, sunflower, wheat, white lupin, faba bean, and tomato seed were effectively evaluated for their capacity for stabilizing oil-in-water emulsions while replicating the emulsifying properties of egg proteins (Ali and El Said 2020; Alcorta et al. 2021). Conversely, the nutritional value of egg alternatives has also been taken into account, so legumes with high protein contents might provide similar protein content (Alcorta et al. 2021).

## 17.3 Environmental Impacts and Benefits of Food Analogues

The shift towards food analogues is often motivated by the potential for reduced environmental impacts. The world we live in provides space to feed our crowded population with limited fertile lands. According to the literature, ensuring global food security will pose growing challenges, with projections indicating a potential global population surpassing 9 billion by 2050 (Mulia and Doi 2019). So as to adequately meet global food demands by 2050, agricultural production must increase by approximately 60%, while livestock and dairy production must increase by 70% (Alexandratos and Bruinsma 2012). Livestock provides almost 15% of dietary energy and 25% of all dietary protein and poses a significant burden on nature (Van Huis 2013; Alexander et al. 2017; Doi and Mulia 2021). The ecological burden of foods may vary greatly depending on the product type, methods for production, transportation- storage conditions, and waste management. In the assessment of environmental impact, the individual impact of foods on the environment are evaluated in the process from production to consumption. These effects can be examined in different dimensions, such as water and land use, GHG emissions, energy consumption and pressures on biodiversity, or more generally, the carbon footprint (Fig. 17.6). In this section, the comparative environmental effects of basic food analogues, whose production methods were mainly specified in the previous section, will be examined.

### *17.3.1 Resource Use*

In today's world, where arable land is decreasing and environmental challenges are increasing, the efficiency of land and water use is a critical factor in sustainability and food security (Gebrehiwot et al. 2022; Tramberend et al. 2019). Animal husbandry has a big impact on land and water resources (Zhang et al. 2019; Karlsson Potter et al. 2020). About 35% of cropland and 20% of clean water is dedicated to the feeding of cattle and animal rearing (Bhuva et al. 2021). While Table 17.1 shows

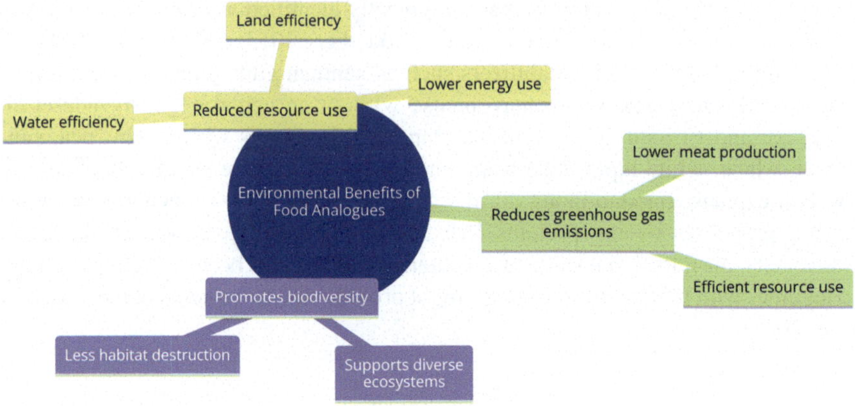


**Fig. 17.6** Environmental benefits of food analogues

**Table 17.1** Water use data for 1 kg product

| | Food product | Water use m³/kg | References |
|---|---|---|---|
| Meat (bone-free) | Beef | 15.4 | Kassem et al. (2022) and Karlsson Potter et al. (2020) |
| | Poultry-chicken | 2–3 | Kassem et al. (2022) and Karlsson Potter et al. (2020) |
| | Pork | 5–6 | Karlsson Potter et al. (2020) |
| | Fish | 0.4–1.9 | Guzmán-Luna et al. (2021) |
| Meat analogous including fish analogues | Plant based (include tofu-tempeh) | 0.03–1.6 | Desiderio et al. (2023), Heller and Keoleian (2018), Khan et al. (2019), Goldstein et al. (2017), Dettling et al. (2016), Santo et al. (2020), and Karlsson Potter et al. (2020) |
| | Lab-based/cultured meat-mycoprotein based | 0.4–0.5 | Tuomisto and Teixeira de Mattos (2011) |
| | Insect | 0.62 | Smetana et al. (2021b) |
| Dairy | Milk | 5.68 | Winans et al. (2020) |
| | Cheese | 6–17 | Berlin (2002) |
| | Plant based drink | 0.1–1.2 | Karlsson Potter et al. (2020) |
| | Coconut drink | 4.1–5.4 | Karlsson Potter et al. (2020) |
| | Plant based cream | 0.64 | Carlsson Kanyama et al. (2021) and Potter and Röös (2021) |
| | Plant based cheese | 0.03 | Carlsson Kanyama et al. (2021) and Potter and Röös (2021) |
| Egg | Egg | 0.5–0.6 | Xing et al. (2019) |
| Egg analogue | Pea protein flour | 1 | Carlsson Kanyama et al. (2021) |

**Table 17.2** Land use data for 1 kg product

| | Food product | Land use m²/kg | References |
|---|---|---|---|
| Meat (bone-free) | Beef | 223.63 | Mulia and Doi (2019) |
| | Lamb | 231.18 | Mulia and Doi (2019) |
| | Poultry-chicken | 5–17.4 | Mulia and Doi (2019) and Nijdam et al. (2012) |
| | Pork | 8–28.59 | Mulia and Doi (2019) and Nijdam et al. (2012) |
| | Fish from aquaculture | 1.2–10.7 | Guzmán-Luna et al. (2021) and Nijdam et al. (2012) |
| | Fish[a] | 0 | Mulia and Doi (2019) |
| | Plant based | 0.4–3 | Karlsson Potter et al. (2020) and Nijdam et al. (2012) |
| Meat analogous including fish analogues[a] | Mixed with egg | 2.4–3.4 | Karlsson Potter et al. (2020) |
| | Lab-based/ cultured meat | 0.1–0.2 | Tuomisto and Teixeira De Mattos (2011) |
| | Insect | 3.56 | Mulia and Doi (2019) |
| Dairy | Milk | 2.9 | Röös et al. (2018) |
| | Cheese | 6–17 | Nijdam et al. (2012) |
| Dairy Analogues (plant based) | Drinks (Milk-like) | 0.6–1.7 | Carlsson Kanyama et al. (2021) and Röös et al. (2018) |
| | Cream-like | 0.58 | Carlsson Kanyama et al. (2021) |
| | Yoghurt-like | 0.42 | Carlsson Kanyama et al. (2021) |
| | Spreads/cheese like | 0.33 | Carlsson Kanyama et al. (2021) |
| | Tofu/tempeh | 1.8–3.5 | Karlsson Potter et al. (2020) |
| Egg | Egg | 4–8.7 | Copley et al. (2023) and Nijdam et al. (2012) |
| Egg analogue | Pea protein flour | 4.9 | Karlsson Potter et al. (2020) |

[a]Fish farming does not require land, unlike other types of livestock farming. However, its environmental impact depends on the fishing or specific production methods used (Clark and Tilman 2017)

the water uses, Table 17.2 shows the land use data for 1 kg of each product, respectively.

#### 17.3.1.1 Water Use

Water shortage is worsening across arid/semi-arid parts of India, Northeastern China, Pakistan, North Africa, and the Middle East, which threatens economic progress and food security (Tramberend et al. 2019). It's estimated that by 2025, approximately 65% of the global populace will encounter water stress, and by the year 2050, severe water scarcity is expected to affect 3–5 billion individuals worldwide (Bhuva et al. 2021). Freshwater usage for household purposes accounts for roughly 10%, while agriculture consumes about 66–70% of water resources

(Karlsson Potter et al. 2020), a notable portion of this is allocated to the production of livestock (Bhuva et al. 2021). Organic systems cause more eutrophication, even (Clark and Tilman 2017).

Most of the food analogues are made from crops like soy, wheat and peas which require significantly less water to grow than the feed crops for livestock (Hoekstra 2012; Smetana et al. 2015; Mekonnen and Gerbens-Leenes 2020). The production processes for these analogues are also less water-intensive (Hoekstra 2012). It's understandable that certain nuts, like almonds, coconuts, and cashews, require more soil and water resources than other plant products due to their low yield/hectare. For instance, for 1 kg almond production in the USA needs an irrigation of approximately 4,3 $m^3$ water. Nonetheless, the water consumption for almond milk production is significantly less, as it contains only 2–3% almonds (Winans et al. 2020). Despite these factors, these products environmental impact is still considerably lower than animal-derived counterparts (Karlsson Potter et al. 2020). Although still in the initial steps of commercialization, lab-grown meats promise to drastically reduce water usage. These products are developed from animal cells in bioreactors, eliminating the need for feed cultivation and significantly reducing water consumption. Early studies suggest that cultured meat could reduce water usage by up to 90% than production of traditional meat (beef) (Tuomisto and Teixeira De Mattos 2011).

Insect-based food alternatives represent an innovative and sustainable addition to the conversation about reducing the water usage of our food systems. Insects, as a food source, offer a highly efficient alternative to traditional livestock and plant-based analogues, with a significantly lower environmental footprint, including reduced water usage (Alexander et al. 2017). Insect farming stands out for its minimal water requirements. Insects naturally require far less water than traditional livestock, partly because they are ectothermic (cold-blooded), meaning they do not need to consume water to maintain body heat (Alexander et al. 2017). Additionally, insects demonstrate remarkable efficiency in converting feed into protein, leading to less water being used for their feed compared to the feed for cattle, pigs, or chickens. For instance, producing 1 kg of crickets can use less than 1000 l of water, which is a small portion of the water required for the same amount of beef, highlighting the substantial water savings that can be achieved (Smetana et al. 2021b).

The comparison of average water use of certain foods and their analogues can be seen in Fig. 17.7 ($m^3$/kg). The figure is created from the average data of water use table (Table 17.1).

#### 17.3.1.2 Land Use

Large pasture areas are required for grazing and growing forage crops (Gerbens-Leenes and Nonhebel 2005). This not only contributes to deforestation and habitat destruction in our increasingly crowded world with limited resources, but also leads to a significant carbon footprint due the use of fertilizers in feed production and emission of methane from cattle. According to De Sy et al. (2015) data (1990–2005)

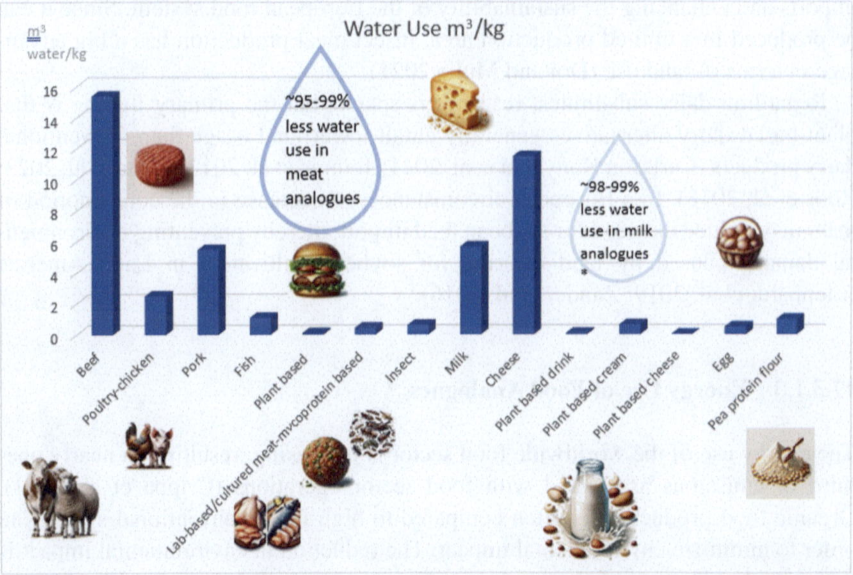


**Fig. 17.7** Average water use of certain foods and their analogues ($m^3$/kg). *Except almond milk

livestock used to contribute more than 85% of deforestation in South America. Certified food products even have higher land footprint (Bellassen et al. 2021; Clark and Tilman 2017).

Food analogues, particularly those derived from plants, utilize less land compared to conventional animal agriculture (Fresán et al. 2019b; Hoekstra 2012). By optimizing the use of farming land, this approach minimizes the necessity for deforestation, thereby safeguarding natural habitats and preserving biodiversity. The vast of comparisonal studies about land efficiency is focused on different meat product and their alternatives (Pilorgé et al. 2021; Van Der Weele et al. 2019; Mulia and Doi 2019; Xu et al. 2021). Meat analogues, including those created from algae proteins, legumes, and oil seeds, are being investigated as eco-friendly replacements for conventional meat, owing to their reduced environmental impact (Bhuva et al. 2021; Joshi and Kumar 2015; Bakhsh et al. 2023). As stated by Alexander et al. (2017) reducing meat consumption will have a greatest effect on existing land use footprint of food production. But also they are adding that replacing only beef or lamb with any of the other alternatives in the work offers the possibility of considerable sustainability adventages. Poultry meat, milks and eggs were also declareted as have land usage benefits than beef and lamb (Nijdam et al. 2012). As per Pilorgé et al. (2021), if the EU were to fully substitute meat with plant-based alternatives, it would require approximately 12% of arable land for producing "pea meat". By replacing 25% of meat consumption, the EU could generate food proteins without expanding cultivated areas, thus averting the need for importing soybeans and corn for feed. The authors also found that "pea meat," as well as other vegetable proteins suitable for consumption, play a crucial role in reducing deforestation caused by imports and enhancing the sustainability of the European food system. Since it can be produced in a limited production area, insect meat production has a big advantage in terms of land use (Doi and Mulia 2021).

Regarding dairy substitutes, research is sparse, but the primary finding is that plant-based dairy alternatives generally entail lower land usage than conventional dairy products (Carlsson Kanyama et al. 2021; Temme et al. 2013; Costa et al. 2023; Röös et al. 2018). In a European circumstances, a decrease in the consumption of meat may reduce the need for soybean feed import, thereby preventing environmental damage caused by land clearing for soybean cultivation in Latin America (Lienhardt et al. 2019; Zander et al. 2016).

#### 17.3.1.3 Energy Use of Food Analogues

The energy use of the worldwide food sector is increasing, resulting in nearly one-third of emissions associated with food sector operations (Crippa et al. 2021). Organic food production is often compared to high-input conventional systems in order to minimize environmental impact. The reduction in environmental impact is related to the changing of chemical inputs using ecological alternatives, like manure, or utilizing ecosystem services like pest control (Azadi et al. 2011). The energy required to produce food analogues varies widely. Plant-based alternatives often have lower energy requirements in cultivation and processing than traditional meat production (plant-based meats use 46% less energy- (Desiderio et al. 2023)), which involves energy-intensive stages like feed production, animal rearing, and refrigeration. Cultured meat, while promising reduced land and water use, currently faces challenges related to high energy consumption for cell cultivation and growth. According to literature, the energy needs for cultivating meat are lower than those for beef, lamb, and pork, but higher than those for poultry (Tuomisto and Teixeira De Mattos 2011).

### *17.3.2 Greenhouse Gas Emission (GHG)*

GHG are recognized as a significant issue concerning climate change. It takes into account gases including methane ($CH_4$), carbon dioxide ($CO_2$), and nitrous oxide ($N_2O$) which are fluorinated and capture heat within the earth's atmosphere, hence creating the greenhouse effect (Crippa et al. 2021; Nette et al. 2016). Human activities have risen the concentration of such gases by large amounts and intensified the greenhouse effect causing global warming. In this regard, understanding GHG emissions is a very important tool for chalking out strategies to mitigate the climate change impacts and to save the environment for the coming generations. The food system, spanning from beginning to end-consumption, is responsible for approximately 34% of total GHG emission, with agriculture and land use/land-use change activities participating the majority, comprising 71% of this figure, globally. The remaining emissions stem from supply chain operations like industrial processes, retail, fuel production, packaging transportation, consumption, and waste management (Crippa et al. 2021; Mrówczyńska-Kamińska et al. 2021; Tubiello et al. 2021). Globally food-system emissions totaled 18 gigatons of $CO_2$ equivalent annually in 2015 (Crippa et al. 2021). Organic farming leads to more eutrophication and uses less energy, yet results in GHG emissions comparable to conventional farming methods. Similarly, beef from grass-fed cattle demands more land while producing GHG emissions on par with grain-fed beef. On the other hand, low-input aquaculture and fisheries that do not engage in trawling emit significantly fewer GHGs than trawling operations. Moreover, across various environmental metrics and nutritional measures, plant-based products consistently exhibit the least environmental impact (Shanmugam et al. 2023). Eggs, dairy, pork, poultry, sustainable fisheries, and aquaculture with minimal water exchange have moderate environmental effects, while the environmental effect of ruminant meat is roughly 100 times greater than that of plant-based foods (Clark and Tilman 2017). Elements affecting the GHG concentration in food supply chains consist of factors like cereal yield, nitrogen fertilizer usage, and the Human Development Index (Mrówczyńska-Kamińska et al. 2021). All these studies show the importance of implementing sustainable methods in both food production and consumption to lower the environmental footprint of global food systems. Because livestock system is a major emitter of methane and nitrous oxide, which have a greater potential for warming comparing carbon dioxide, one of the most significant environmental benefits of food analogues is their potential to reduce GHG emissions. Figure 17.8 shows the average GHG reduction potentials of certain food analogues (Data is taken from the Table 17.3 which shows the GHG emissions of certain food and their analogues).

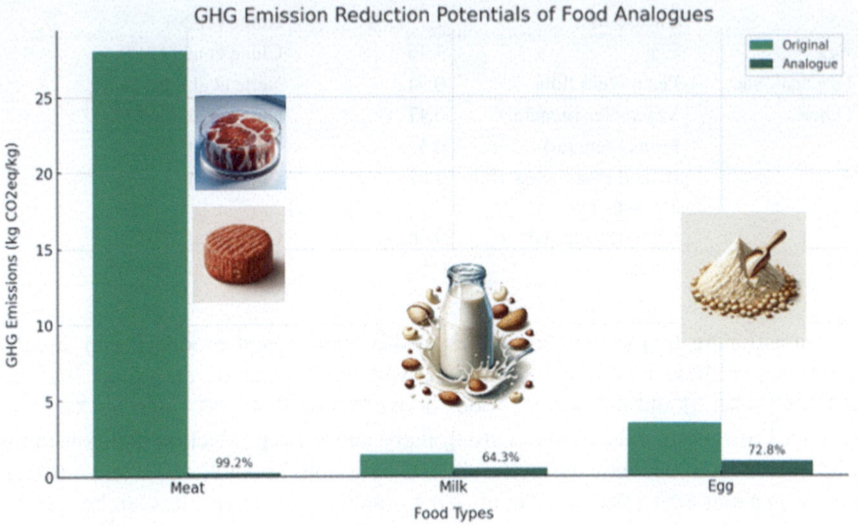


**Fig. 17.8** Average GHG reduction potentials of certain food analogues

**Table 17.3** GHG emissions of certain food and their analogues

| | Food product | GHG emissions kg $CO_2$eq/kg | References |
|---|---|---|---|
| Meat (bone-free) | Ruminant meat (lamb-beef) | 28–29 | Nette et al. (2016), Aan den Toorn et al. (2018), and Clune et al. (2017) |
| | Poultry-chicken | 4.12 | Clune et al. (2017) |
| | Pork | 5.85 | Clune et al. (2017) and Aan den Toorn et al. (2018) |
| Fisheries | Prawns/shrimp | 14.85 | Clune et al. (2017) |
| | Fish: All species combined | 4.41 | Clune et al. (2017) |
| Meat analogous including fish* | Plant-based (wheat, soy or nut based found same) | 0.21 | Fresán et al. (2019b) |
| | Plant-based (wheat/soy) | 0.23 | Fresán et al. (2019b) |
| | Meat analogue in presence egg | 0.27 | Fresán et al. (2019b) |
| | Lab-based/cultured meat | 1.9–2.24 | Tuomisto and Teixeira de Mattos (2011) |
| | Insect base meat | 2.3–3.1 | Smetana et al. (2021b) |
| Dairy | Cow Milk | 1.4 | Clune et al. (2017) and Aan den Toorn et al. (2018) |
| | Cheese | 8.8 | Clune et al. (2017) and Nette et al. (2016) |
| | Butter | 11.52 | Clune et al. (2017) |
| Dairy Analogues (plant based) | Drinks (Milk-like) | 0.53–0.99 | Clune et al. (2017) and Carlsson Kanyama et al. (2021) |
| | Yoghurt-like | 0.80 | Kolbe (2020) |
| | Cheese-like | 1.7–1.78 | Kolbe (2020) |
| Egg | Egg | 3.46 | Clune et al. (2017) |
| Egg analogue | Pea protein flour | 0.94 | Nette et al. (2016) |
| Others | Vegetables (general) | 0.47 | Clune et al. (2017) |
| | Fruits (general) | 0.5 | Clune et al. (2017) |
| | Heated greenhouse fruit and vegetable | 2.81 | Clune et al. (2017) |
| | Legumes and pulses | 0.66 | Clune et al. (2017) |
| | Cereals | 0.53 | Clune et al. (2017) |

Transitioning to food analogues, especially plant-based products, can significantly lower these emissions. Animal-derived foods generate emissions that are notably higher, doubling those of foods derived from plants, as noted by Xu et al. (2021). Projections indicate that cutting dairy and meat production and consumption in half could decrease agricultural GHG emissions about 25–40% and nitrogen emission about 40% (Westhoek et al. 2014; Zander et al. 2016). Researches on the GHG emissions of analogue foods have mostly focused on meat analogues to understand their sustainability and environmental impact in comparison to products from animal origined (Alexander et al. 2017; Desiderio et al. 2023). Meat alternatives typically generate emissions that are the same as or lower than those from pork and poultry (Copley et al. 2023; Smetana et al. 2015; Tuomisto and Teixeira De Mattos 2011).

In one study mentioned, no significant difference was made among the meat analogues with the differing main protein sources (wheat, soy, wheat and soy, nuts). On the other hand, egg-based products resulted in much greater GHG. This study concludes that for environmental gains, one should go for whole plant-based meat analogues because their nutritional composition was found quite similar across different analogues (Fresán et al. 2019b). Another did a life cycle assessment (LCA) (partial) in quantifying the emissions of GHG across the wide range of meat analogue products. They found that most of the GHG footprint was relatively low, with mean emissions attributed to the manufacturing processes. This is hence evidence of meat analogues as the 'lower carbon' alternative to traditional meats where there are fluctuations of GHG emissions between food product and location of production (Mejia et al. 2019). Reduced meat consumption and a shift to plant origined diet which include meat analogues consumption can bring about large reductions in dietary GHG emissions. This further supports environmental benefits associated with food analogues (Scarborough et al. 2014).

Cultured meat also presents an opportunity for emission reduction, though its current energy-intensive production processes may limit its immediate benefits (Blonk et al. 2008). Poultry meat, eggs and milk was also declareted as have GHG emission benefits than beef and lamb (Nijdam et al. 2012; Carlsson Kanyama et al. 2021). The research available through the search results primarily focuses on the GHG emissions associated with traditional fisheries (Gephart et al. 2021; Parker et al. 2018) rather than directly addressing fishery analogues, such as plant-based or lab-grown seafood alternatives. However, insights from these studies on traditional fisheries highlight the environmental impacts of conventional fish harvesting and provide a useful context for understanding the potential benefits of fishery analogues in terms of GHG emissions. A study focused on the global fishing fleet's fuel use and GHG emissions from 1990 to 2011 revealed that, by 2011, the fisheries sector had consumed 40 billion liters of fuel, leading to the production of 179 million tonnes of $CO_2$-equivalent greenhouse gases. This constituted 4% of the total GHG emission from worldwide food production. The research pointed out that emissions attributed to the fishing industry increased by 28% over this period, primarily driven by a rise in fuel use, especially in the harvesting of crustaceans (Parker et al. 2018). Another research comparing the environmental footprints of different types of blue foods revealed that farmed bivalves and seaweeds produced the lowest environmental stressors, including GHG emissions, compared to other seafood options. Among farmed finfish and crustaceans, certain species like carps had lower GHG emissions but higher water use, while farmed salmon and trout used the least land and water. This study highlights the diversity in environmental impacts across different types of seafood and points to the potential for certain types of aquaculture to offer more sustainable alternatives (Gephart et al. 2021). These findings suggest that while traditional fisheries, especially those that are fuel-intensive, contribute significantly to GHG emissions, there are opportunities within the sector to reduce these impacts through sustainable practices and alternative seafood production methods, such as plant-based or lab-grown fishery analogues. The potential for fishery analogues to reduce GHG emissions lies in their ability to bypass the environmental impacts associated with traditional capture fisheries and aquaculture, including fuel use for fishing vessels and feed production for farmed fish. By providing a more environmentally friendly alternative, fishery analogues could contribute to reducing the climate footprint associated with seafood production.

The information about GHG emission of dairy analogues are very limited (Lima et al. 2022; Carlsson Kanyama et al. 2021). The existing research and reviews mainly concentrate on GHG emissions related to conventional dairy farming, covering the efficiency of production and the ecological effects associated with dairy cattle and the management of dairy farms. Conventional dairy farming is acknowledged as a significant source of GHG emissions, primarily caused by the production and processing of feed, digestive processes in livestock, and the handling of manure (Rotz et al. 2010; Rotz 2018; Gerber et al. 2011). The pollution that comes from dairy systems is usually divided into milk (primary product), meat, and excess calves (co-products) (Costa et al. 2023). While the documents reviewed provided limited details on the environmental impacts of dairy substitutes, the broader body of research consistently shows that plant-based dairy alternatives typically produce fewer GHGs compared to conventional dairy products (Nette et al. 2016). This decline in GHG emissions is associated with the generally lower resource needs for plant cultivation, the absence of enteric fermentation processes that release methane, and potentially more efficient feed-to-product conversion rates (Aan Den Toorn et al. 2018). Nevertheless, the exact environmental footprint varies depending on the specific dairy alternative, its production methods, and the logistics of its supply chain. According to a study, changing traditional milk with soy-based drink alone won't substantially reduce GHG emissions unless it's coupled with a significant decrease in beef consumption (Costa et al. 2023). This occurs due to the shift to less efficient suckler-beef systems from dairy-beef production. However, replacing cow's milk can contribute to GHG reduction if the land is freed up because of reforestation. Research indicates that milk's carbon footprint, at 1.4 kg $CO_2$-equivalents/kg, is lower than that of meat products. However, alternatives derived from plants and yeast have an even smaller footprint, ranging from 0.4–0.9 kg $CO_2$ quivalents/kg (Clune et al. 2017; Steer 2015). According to these data swapping meat and dairy products with their alternatives may lead to a decline in GHG emissions (Aan Den Toorn et al. 2018).

Over the last five decades, the ecological impact of the U.S. egg sector has markedly diminished, thanks to advancements in how feed is utilized and composed, as well as better handling of manure (Pelletier et al. 2013). Despite these improvements, the contribution of egg consumption to GHG emissions remains considerable, exhibiting a broad spectrum of emissions levels among various food products (Carlsson-Kanyama and González 2009). But, it should be added that studies show that egg production has lower GHG emissions than meat production. While the search did not yield specific studies on the GHG emissions of egg substitutes, considering the environmental challenges associated with traditional egg production, especially related to feed production and manure management, egg substitutes made from plant-based ingredients could potentially offer lower GHG emissions (Cleveland et al. 2021). Ingredients such as pea protein have been suggested as alternatives in food products to reduce environmental impacts. A study on pasta products found that substituting animal components with plant-based ingredients, like peas, could reduce GHG emissions, suggesting a similar potential for egg substitutes (Nette et al. 2016). Optimizing feed production and manure management in egg production has led to reduced environmental footprints, suggesting that egg substitutes could further mitigate environmental impacts, especially if made from low-impact plant-based ingredients.

Finally, it can be said that, food analogues, especially those that are fully plant-based, have the potential to act as an eco-friendly alternative to traditional meat products and, therefore, entail less GHG emissions and other possible environmental benefits. In this regard, such a measure could therefore be an important one in the light of mitigating climate change impacts. In its 2019 report, the UN Climate Panel indicates that adopting a diet low in or devoid of meat and dairy components can result in significant decreases in GHG emissions and land utilization (Debonne 2019).

### *17.3.3 Waste Management*

It should be notted that not only meat product consumption but also food wastes are a big problem. Each year, around 7.5% of dairy and 14.5% of meat products are wasted during retail-final consumption (Aan Den Toorn et al. 2018). Inadequate regulations on food waste recycling and insufficient attention to practical utilization of compost and digestate may hinder attempts to decrease food waste and foster a circular bioeconomy. Annually, global food waste represents 28% of arable land and 24% of global crop water usage (FAO 2013). Even in there meat products has bigger fottprints than plantbased wastes (Li et al. 2022a). The production of food analogues generates different types of waste compared to conventional animal agriculture. While livestock farming produces large amounts of organic waste, including manure and methane, the waste from food analogue production is often less environmentally damaging (Mcclements and Grossmann 2021). Proper management and utilization of agricultural by-products in food analogue production can further mitigate waste-related impacts (Plazzotta and Manzocco 2019; Aschemann-Witzel et al. 2015; Bajželj et al. 2020).

### *17.3.4 Biodiversity and Ecological Balance*

By reducing the demand for land-intensive animal agriculture, food analogues can contribute to biodiversity conservation and the restoration of ecological balance. Conventional livestock farming is a leading cause of deforestation and habitat destruction, driving biodiversity loss (Machovina et al. 2015; Alkemade et al. 2013). The food analogues lower land footprint offers a pathway to preserve natural habitats and protect species (Alexander et al. 2017; Costa et al. 2023). As it is shown from the studies, plant-based foods display the smallest environmental footprints, whereas ruminant meat has impacts roughly 100 times higher than those of plant-based foods. Dairy, eggs, poultry, pork, non-trawling fisheries, and non-recirculating aquaculture demonstrate average impacts (Clark and Tilman 2017). As we grapple with the complexities of feeding a growing global population, we seek sustainable sources of protein. It becomes clear that no single protein source is the silver bullet for environmental sustainability. Instead, a harmonious integration of plant-based, lab-grown, and insect-derived proteins could foster a more resilient and sustainable food system. Ways to reduce the environmental burden of foods include adopting sustainable farming methods, consuming local and seasonal produce, preventing food waste, and promoting a plant origined diet. Consumers have the potential to make a meaningful impact in this process through their shopping and eating behaviors.

## 17.3.5 *Case Studies Food Analogue Production*

### 17.3.5.1 Beyond Meat and Impossible Foods

Both companies have made significant strides in lowering the environmental trace of their plant-based meat products through life cycle assessments and continuous improvement in their production processes (Heller and Keoleian 2018). According to studies consumers can decrease water use, land use and GHG emissions between

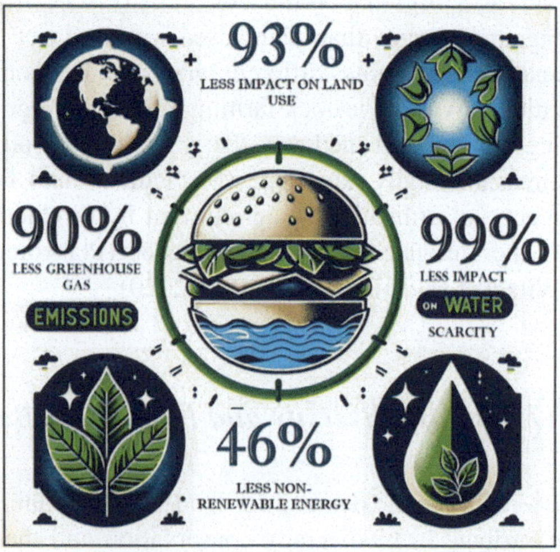


**Fig. 17.9** Beyond burger's life cycle analysis results

87% and 96% by changing a regular beef patty with a Beyond (Heller and Keoleian 2018) or Impossible burger (Kustar and Patino-Echeverri 2021). Figure 17.9 shows the key points of the life cycle analysis results of beyond burger.

Impossible Foods has been pioneering the development of plant-based meat products, with a strong emphasis on sustainability and reducing the environmental footprint associated with meat consumption. Their key innovation, the use of heme (a molecule contains iron) derived from genetically modified yeast, has allowed them to create a meat analogue which replicates the texture and taste of beef more closely than previous alternatives (Ransom 2021; Sarkar 2021). But the long-term health effects of consuming these products remain largely unknown. There are various discussions about the long-term effects of genetically modified organisms (GMO) products used in these products (Sarkar 2021; Perro 2019).

#### 17.3.5.2 Memphis Meats

As a pioneer in cultured meat (cell-based meat and poultry), Memphis Meats is working on reducing the cost and increasing the efficiency of cultured meat production, aiming to make it a viable alternative to conventional meat (Ransom 2021). Cultivated meat has the potential to result in a reduction of up to 96% in water usage, 99% in land usage and up to 90% in GHG emission (Hoogenkamp 2018).

## 17.4 Risks and Challenges

While food analogues offer promising environmental and health benefits, their production and consumption also pose unique challenges and risks. Understanding these factors is crucial for mitigating negative impacts and enhancing the

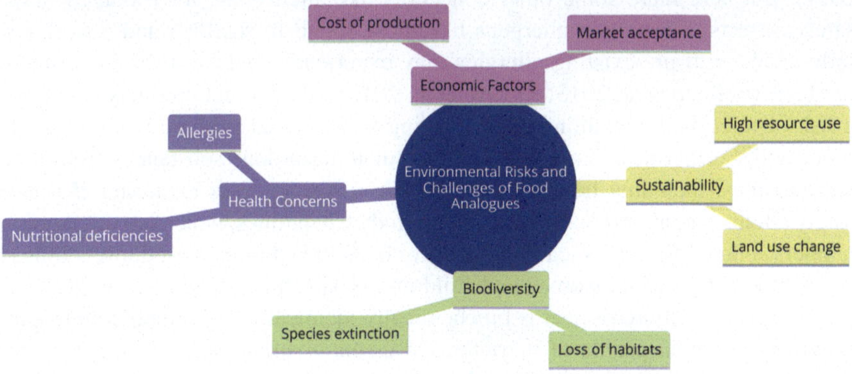


**Fig. 17.10** Environmental risks and challanges of food analogues

sustainability of these alternatives. Figure 17.10 visualizes the environmental risks and challenges of food analogues.

### 17.4.1 Health Concerns

Plant-based analogues are often lauded for their lower saturated fat content and reduced calorie count. Nevertheless, they might be deficient in vital nutrients typically found in natural animal products, including essential amino acids, vitamin B-12, iron, and omega-3 fatty acids (Pingali et al. 2023; Costa-Catala et al. 2023). Specific vitamins and minerals, such as vitamin B12, iron, and calcium, are more readily available or bioavailable in animal-derived products. There is a large nutritional differences between grass-fed meat and plant-based meat analogue (Van Vliet et al. 2021). The nutrient bioavailability in plant-based analogues can also be lower, potentially leading to nutritional deficiencies if not properly managed (Craig and Mangels 2009). Ensuring that food analogues provide adequate nutritional value requires careful formulation and sometimes supplementation of essential nutrients (Pingali et al. 2023).

To simulate the texture and flavour of dairy and meat products, plant-based analogues frequently contain high levels of additives, including salt, sugar, and artificial flavorings. Regular intake of these additives may heighten the risk of health conditions for instance cardiovascular disease, type 2 diabetes, and hypertension (Fardet and Rock 2019; Micha et al. 2017).

When we look at the insect-based products, we are facing with a high potencial about nutritional value. Insect-derived products are abundant in protein and encompass all essential amino acids, thus serving as a complete protein source akin to traditional meat. Comparisons of nutritional values have shown that both insects and meats are nutrient-rich, although their contents vary. They also offer high levels of vitamins-minerals, like vitamins B, zinc, and iron (Smetana et al. 2021a; Orkusz 2021). But here there some other concenrns like allergy and microbial/chemical safety. Insects can contain allergens like those found in shellfish and dust mites. Individuals with pre-existing allergies may experience adverse reactions to these products (Belluco et al. 2013; Ribeiro et al. 2018). Also, insect products may carry a risk of microbial contamination, including bacteria and viruses, if not properly processed. Furthermore, insects may accumulate hazardous substances from their surroundings, including heavy metals, which raises concerns regarding chemical safety (Belluco et al. 2013). In an outbreak study Chapulines, which are dried grasshoppers sourced from Oaxaca-Mexico, have been linked to increased levels of lead in the blood of pregnant women and children in California (Handley et al. 2007).

Lab-grown (cultured) meat is biochemically identical to its animal counterpart, providing a similar profile of nutrients and essential amino acids. Since it is produced in controlled environments, it has the potential for enhanced nutritional profiles, such as reduced saturated fat levels (Treich 2021).While lab-grown/cultured food analogues promise a sustainable and ethical alternative to traditional meat, the long-term health effects of consuming these products remain largely unknown. Safety assessments are crucial to ensure that these novel foods do not pose unforeseen health risks (Post 2012).

### 17.4.2 Sustainability, Resource Use

Although food analogues are typically more environmentally sustainable than their animal-based equivalents, the procurement of ingredients can still present environmental obstacles. Some food analogues, especially highly processed plant-based products, and lab-grown meats, require energy-intensive production processes. These processes can include the use of bioreactors for cultivating meat or the extensive processing needed to convert plants into the products that simulate the texture and taste of actual meats. Therefore, they may result in environmental impacts and water consumption similar to those of unprocessed or minimally processed animal-derived products (Fresán et al. 2019a). Although generally more resource-efficient than traditional livestock farming, the cultivation of crops like soy, peas, and almonds—common ingredients in food analogues—can still be resource-intensive (Aan Den Toorn et al. 2018). For instance, almond cultivation, a key ingredient in dairy alternatives, demands significant water usage, highlighting the need for sustainable water management practices in the supply chain of food analogues (Winans et al. 2020). The cultivation of crops like soy, which is a common ingredient in many plant-based analogues, has been linked to deforestation and habitat destruction (Sauer 2018). Promoting sustainable crop production and sourcing is essential to mitigate these impacts.

### 17.4.3 Biodiversity

The agricultural practices associated with growing crops for food analogues can impact biodiversity. The increased demand for specific crops to produce food analogues, such as soy, peas, and almonds, could lead to monoculture farming practices. Monoculture farming practices have been widely adopted due to their operational efficiencies and economic benefits. Monocultures are detrimental to soil health, increase dependency on chemical fertilizers and pesticides, and reduce biodiversity (Grant 2007). Natural ecosystems are composed of a wide variety of plants, animals, and microorganisms, which interact in complex ways to maintain ecological balance. By replacing these ecosystems with single-crop systems, monocultures disrupt these interactions, leading to the loss of species and genetic diversity. This loss not only affects the immediate environment but also has broader implications for global biodiversity (Altieri 1999; Tilman et al. 2002; Foley et al. 2011). Soil health is crucial for sustainable agriculture, yet monoculture practices contribute to soil degradation in several ways (Tscharntke et al. 2012). Repeatedly growing the same crop leads to the depletion of specific nutrients, resulting in soil fertility loss. Additionally, monoculture fields are often more susceptible to erosion due to the lack of vegetative variety to anchor the soil. Over time, this can lead to significant declines in soil quality, making land less productive and reducing its capacity to support agriculture (Gomiero et al. 2011). Monocultures create an environment where pests and diseases can easily spread and become established. With a single crop type as the host, pests and pathogens that specialize in that crop can thrive, leading to outbreaks that can devastate yields. This increased vulnerability often leads to a greater reliance on chemical pesticides and fungicides, which may worsen environmental consequences (Tilman et al. 2002). The extensive application of herbicides-pesticides, and chemical fertilizers in monoculture farming poses a risk of water source contamination. Runoff from fields can carry these chemicals into rivers, lakes, and groundwater, harming aquatic ecosystems and affecting water quality. This pollution not only impacts wildlife but also poses risks to human health, particularly in communities reliant on these water sources for drinking and recreation (Gomiero et al. 2011). Adopting sustainable agricultural practices, like crop rotation, polyculture systems, and organic fertilizer usage, provides feasible alternatives that can reduce these environmental effects. By adopting more diverse farming practices, it is possible to achieve a balance between agricultural productivity and environmental health.

The production of food analogues sometimes involves the use of GMOs to increase yield, improve nutritional content, or enhance other desirable qualities. While GMOs can contribute to food security and agricultural efficiency, their environmental impact, including effects on non-target species and the potential for gene transfer, remains a subject of debate (Sarkar 2021, Perro 2019).

### *17.4.4 Economic Factors, Consumer Acceptance, Wastes*

The regulatory landscape for food analogues, especially for novel products like cultured meat, is still evolving. Issues around labeling, safety standards, and market approval pose significant hurdles for new entrants. Currently, many food analogues, especially novel products like cultured meat, are more expensive than traditional animal-based foods. This cost barrier can limit accessibility and adoption, particularly in lower-income populations (Aan Den Toorn et al. 2018). Acceptance of food analogues by consumers is critical for their success. Taste, texture, cultural preferences, and food traditions play significant roles in the adoption of these products. Food is deeply intertwined with culture and tradition. Introducing food analogues as replacements for traditional foods can face resistance, necessitating thoughtful engagement and education efforts. Plant-based, lab-grown or edible insects are seen as promising alternatives, but their palatability and consumer acceptance remain challenges for now (Lee et al. 2020; Anusha Siddiqui et al. 2023; Alexander et al. 2017; Roy et al. 2021). Although it is thought to be difficult today, society's perspective may change as analogue products become more common in the future.

Research indicates that consumers tend to become more accepting of unknown meat alternatives with repeated exposure, unless the first impression are unpleasant (Hoek et al. 2013). The sustainability and resource usage challenges of food analogues are significant but not insurmountable. Through a combination of technological innovation, policy support, and shifts in consumer behavior, it is possible to mitigate these challenges and move towards a more environmental food system.

## 17.5 Conclusions

The environmental benefits of food analogues are clear and compelling. By significantly reducing GHG emissions, conserving water and land resources, and mitigating pollution and biodiversity loss, food analogues offer a viable pathway toward reducing the ecological footprint of our diets. However, the transition to food analogues is not without its challenges. Nutritional adequacy, economic and social implications, consumer acceptance, and sustainability of ingredient sourcing and bodiversity are critical issues that require thoughtful consideration and action. The future perspectives and innovations discussed highlight the ongoing efforts to address these challenges, from technological advancements improving the sensory and nutritional qualities of food analogues to policy frameworks that support their development and integration into the food system. Achieving the full environmental and health potential of food analogues necessitates a collaborative effort among multiple stakeholders. Producers, researchers, policymakers, and consumers all play pivotal roles in shaping the future of food. Producers must continue to innovate and enhance the sustainability and nutritional value of their products. Researchers are tasked with advancing our understanding of the impacts and benefits of food analogues. Policymakers should create supportive regulatory environments that facilitate the growth of sustainable food options. Finally, consumers hold the power to drive change through their food choices, demanding more sustainable and ethical products.

As we anticipate the future, the direction of food analogues will certainly be shaped by global trends in population growth, climate change, and shifting dietary patterns. The continued evolution of food technologies, along with increasing consumer awareness of environmental and health issues, positions food analogues as a key component of sustainable diets. However, the successful integration of food analogues into global food systems will require not only innovation but also a commitment to addressing the complex web of environmental, nutritional, and social factors that influence food production and consumption. Food analogues stand at the forefront of a revolution in how we produce-consume food, offering hope for a more equitable and sustainable world. As we embrace these changes, it is imperative to remain vigilant, ensuring that the benefits are realized for all and that the risks are carefully managed. The journey of food analogues is not merely about replacing animal-based foods but about reimagining the possibilities of the food system to nourish humanity while protecting our planet.

## References

Aan Den Toorn S, Tziva M, Van Den Broek M, Negro S, Hekkert M, Worrell E (2018). Climate innovations in meat and dairy. Reinvent Project NR, 730053

Aboulfazli F, Shori AB, Baba AS (2016) Effects of the replacement of cow milk with vegetable milk on probiotics and nutritional profile of fermented ice cream. LWT 70:261–270

Ahmadian-Kouchaksaraei Z, Varidi M, Varidi MJ, Pourazarang H (2014) Influence of processing conditions on the physicochemical and sensory properties of sesame milk: a novel nutritional beverage. LWT-Food Sci Technol 57:299–305

Aidoo H, Sakyi-Dawson E, Abbey L, Tano-Debrah K, Saalia FK (2012) Optimisation of chocolate formulation using dehydrated peanut–cowpea milk to replace dairy milk. J Sci Food Agric 92:224–231

Alam M, Kaur J, Khaira H, Gupta K (2016) Extrusion and extruded products: changes in quality attributes as affected by extrusion process parameters: a review. Crit Rev Food Sci Nutr 56:445–473

Alcorta A, Porta A, Tárrega A, Alvarez MD, Vaquero MP (2021) Foods for plant-based diets: challenges and innovations. Food Secur 10:293

Alexander P, Brown C, Arneth A, Dias C, Finnigan J, Moran D, Rounsevell MD (2017) Could consumption of insects, cultured meat or imitation meat reduce global agricultural land use? Glob Food Sec 15:22–32

Alexandratos N, Bruinsma J (2012) World agriculture towards 2030/2050: the 2012 revision

Ali MR, El Said RM (2020) Assessment of the potential of Arabic gum as an antimicrobial and antioxidant agent in developing vegan "egg-free" mayonnaise. J Food Saf 40:E12771

Alkemade R, Reid RS, Van Den Berg M, De Leeuw J, Jeuken M (2013) Assessing the impacts of livestock production on biodiversity in rangeland ecosystems. Proc Natl Acad Sci 110:20900–20905

Alozie Yetunde E, Udofia US (2015) Nutritional and sensory properties of almond (Prunus Amygdalu Var. Dulcis) seed milk. World J Dairy Food Sci 10:117–121

Alsaffar AA (2016) Sustainable diets: the interaction between food industry, nutrition, health and the environment. Food Sci Technol Int 22:102–111

Altieri MA (1999) The ecological role of biodiversity in agroecosystems. In: Invertebrate biodiversity as bioindicators of sustainable landscapes. Elsevier

Aly HF, Rizk MZ, Shams SGE (2023) Microalgae in food and feed: Safety and toxicological aspects. Handbook of Food and Feed from Microalgae: Production, Application, Regulation, and Sustainability, 549–565. https://doi.org/10.1016/B978-0-323-99196-4.00004-8

Anusha Siddiqui S, Bahmid NA, Mahmud CM, Boukid F, Lamri M, Gagaoua M (2023) Consumer acceptability of plant-, seaweed-, and insect-based foods as alternatives to meat: a critical compilation of a decade of research. Crit Rev Food Sci Nutr 63:6630–6651

Aryana KJ, Olson DW (2017) A 100-year review: yogurt and other cultured dairy products. J Dairy Sci 100:9987–10013

Aschemann-Witzel J, De Hooge I, Amani P, Bech-Larsen T, Oostindjer M (2015) Consumer-related food waste: causes and potential for action. Sustain For 7:6457–6477

Asgar MA, Fazilah A, Huda N, Bhat R, Karim A (2010) Nonmeat protein alternatives as meat extenders and meat Analogs. Compr Rev Food Sci Food Saf 9:513–529

Aydar EF, Tutuncu S, Ozcelik B (2020) Plant-based Milk substitutes: bioactive compounds, conventional and novel processes, bioavailability studies, and health effects. J Funct Foods 70:103975

Aydar AY, Mataracı CE, Sağlam TB (2021) Development and Modeling of a novel plant-based yoghurt produced by Jerusalem artichoke and almond milk using L-optimal mixture design. J Food Meas Charact 15:3079–3087

Azadi H, Schoonbeek S, Mahmoudi H, Derudder B, De Maeyer P, Witlox F (2011) Organic agriculture and sustainable food production system: main potentials. Agric Ecosyst Environ 144:92–94

Bachmann H-P (2001) Cheese analogues: a review. Int Dairy J 11:505–515
Bajželj B, Quested TE, Röös E, Swannell RP (2020) The role of reducing food waste for resilient food systems. Ecosyst Serv 45:101140
Bakhsh A, Park J, Baritugo KA, Kim B, Sil Moon S, Rahman A, Park S (2023) A holistic approach toward development of plant-based meat alternatives through incorporation of novel microalgae-based ingredients. Front Nutr 10:1110613
Balogun MA, Kolawole FL, Joseph J, Adebisi TT, Ogunleye OT (2016) Effect of fortification of fresh cow milk with coconut milk on the proximate composition and yield of Warankashi, a traditional cheese. Croatian J Food Sci Technol 8:10–14
Becker W (2004) Microalgae in human and animal nutrition. In: Richmond A (ed) Handbook of microalgal culture. Blackwell Publishing, Oxford
Bellassen V, Drut M, Antonioli F, Brečić R, Donati M, Ferrer-Pérez H, Gauvrit L, Hoang V, Knutsen Steinnes K, Lilavanichakul A (2021) The carbon and land footprint of certified food products. J Agric Food Ind Organ 19:113–126
Belluco S, Losasso C, Maggioletti M, Alonzi CC, Paoletti MG, Ricci A (2013) Edible insects in a food safety and nutritional perspective: a critical review. Compr Rev Food Sci Food Saf 12:296–313
Ben-Harb S, Irlinger F, Saint-Eve A, Panouillé M, Souchon I, Bonnarme P (2020) Versatility of microbial consortia and sensory properties induced by the composition of different Milk and pea protein-based gels. LWT 118:108720
Berger J, Bravay G, Berger M (1997) Almond milk preparation process and products obtained. United States Patent Application
Berlin J (2002) Environmental life cycle assessment (LCA) of Swedish semi-hard cheese. Int Dairy J 12:939–953
Bernat N, Cháfer M, Chiralt A, González-Martínez C (2014) Hazelnut milk fermentation using probiotic Lactobacillus rhamnosus GG and inulin. Int J Food Sci Technol 49:2553–2562
Bernat N, Cháfer M, Chiralt A, Laparra JM, González-Martínez C (2015) Almond milk fermented with different potentially probiotic bacteria improves iron uptake by intestinal epithelial (Caco-2) cells. Int J Food Stud 4
Bhuva V, Morya S, Borah A (2021) A review on meat analogue: is this time to see the algal proteins as a sustainable substitute for the meat proteins. Pharma Innov J 10:1160–1168
Blaya J, Barzideh Z, Lapointe G (2018) Symposium review: interaction of starter cultures and nonstarter lactic acid bacteria in the cheese environment. J Dairy Sci 101:3611–3629
Blonk H, Kool A, Luske B, De Waart S, Blonk Milieuadvies G, Vegetariërsbond N (2008) Environmental effects of protein-rich food products in The Netherlands: consequences of animal protein substitutes. Blonk Consulktants, Gouda
Boeck T, Sahin AW, Zannini E, Arendt EK (2021) Nutritional properties and health aspects of pulses and their use in plant-based yogurt alternatives. Compr Rev Food Sci Food Saf 20:3858–3880
Bohrer BM (2019) An investigation of the formulation and nutritional composition of modern meat analogue products. Food Sci Human Wellness 8:320–329
Bolarinwa I, Aruna T, Adejuyitan J, Akintayo O, Lawal O (2018) Development and quality evaluation of soy-walnut milk drinks. Int Food Res J 25:2033–2041
Boonen KJ, Langelaan ML, Polak RB, Van Der Schaft DW, Baaijens FP, Post MJ (2010) Effects of a combined mechanical stimulation protocol: value for skeletal muscle tissue engineering. J Biomech 43:1514–1521
Boukid F (2021) Plant-based meat analogues: from niche to mainstream. Eur Food Res Technol 247:297–308
Boyer, R. A. & Middendorf, J. E. 1971. Method of producing a meat simulating textured food product
Briviba K, Gräf V, Walz E, Guamis B, Butz P (2016) Ultra high pressure homogenization of almond milk: physico-chemical and physiological effects. Food Chem 192:82–89

Brückner-Gühmann M, Banovic M, Drusch S (2019) Towards an increased plant protein intake: rheological properties, sensory perception and consumer acceptability of lactic acid fermented, oat-based gels. Food Hydrocoll 96:201–208

Burattini S, Ferri P, Battistelli M, Curci R, Luchetti F, Falcieri E (2004) C2C12 murine myoblasts as a model of skeletal muscle development: morpho-functional characterization. Eur J Histochem 48:223–234

Caporgno MP, Böcker L, Müssner C, Stirnemann E, Haberkorn I, Adelmann H, Handschin S, Windhab EJ, Mathys A (2020) Extruded meat analogues based on yellow, heterotrophically cultivated Auxenochlorella protothecoides microalgae. Innovative Food Sci Emerg Technol 59:102275

Carlsson Kanyama A, Hedin B, Katzeff C (2021) Differences in environmental impact between plant-based alternatives to dairy and dairy products: a systematic literature review. Sustain For 13:12599

Carlsson-Kanyama A, González AD (2009) Potential contributions of food consumption patterns to climate change. Am J Clin Nutr 89:1704S–1709S

Chandrasekara A, Josheph Kumar T (2016) Roots and tuber crops as functional foods: a review on phytochemical constituents and their potential health benefits. Int J Food Sci 2016:3631647

Chavan M, Gat Y, Harmalkar M, Waghmare R (2018) Development of non-dairy fermented probiotic drink based on germinated and Ungerminated cereals and legume. LWT 91:339–344

Chen Y, Lu Y, Yu A, Kong X, Hua Y (2014) Stable mixed beverage is produced from walnut milk and raw soymilk by homogenization with subsequent heating. Food Sci Technol Res 20:583–591

Chiron S, Tomczak C, Duperray A, Laine J, Bonne G, Eder A, Hansen A, Eschenhagen T, Verdier C, Coirault C (2012) Dynamic interplay between human myoblasts and 3D fibrin-based matrix. PLoS One 7:E36173

Cichońska P, Ziarno M (2021) Legumes and legume-based beverages fermented with lactic acid bacteria as a potential carrier of probiotics and prebiotics. Microorganisms 10:91

Clark M, Tilman D (2017) Comparative analysis of environmental impacts of agricultural production systems, agricultural input efficiency, and food choice. Environ Res Lett 12:064016

Clark M, Springmann M, Rayner M, Scarborough P, Hill J, Tilman D, Macdiarmid JI, Fanzo J, Bandy L, Harrington RA (2022) Estimating the environmental impacts of 57,000 food products. Proc Natl Acad Sci 119:E2120584119

Cleveland DA, Gee Q, Horn A, Weichert L, Blancho M (2021) How many chickens does it take to make an egg? Animal welfare and environmental benefits of replacing eggs with plant foods at the University of California, and beyond. Agric Hum Values 38:157–174

Clune S, Crossin E, Verghese K (2017) systematic review of greenhouse gas emissions for different fresh food categories. j clean prod 140:766–783

codina-torrella i, guamis b, ferragut v, trujillo a (2017) potential application of ultra-high pressure homogenization in the physico-chemical stabilization of tiger nuts' milk beverage. Innovative Food Sci Emerg Technol 40:42–51

Copley M, Wiedemann S, Mcgahan E (2023) Environmental impacts of the Australian poultry industry. 2. Egg production. Anim Prod Sci 63:505–521

Costa MP, Saget S, Zimmermann B, Petig E, Angenendt E, Rees RM, Chadwick D, Gibbons J, Shrestha S, Williams M (2023) Environmental and land use consequences of replacing milk and beef with plant-based alternatives. J Clean Prod 424:138826

Costa-Catala J, Toro-Funes N, Comas-Basté O, Hernández-Macias S, Sánchez-Pérez S, Latorre-Moratalla ML, Veciana-Nogués MT, Castell-Garralda V, Vidal-Carou MC (2023) Comparative assessment of the nutritional profile of meat products and their plant-based analogues. Nutrients 15:2807

Craig WJ, Mangels AR (2009) Position of the American Dietetic Association: vegetarian diets. J Am Diet Assoc 109:1266

Crippa M, Solazzo E, Guizzardi D, Monforti-Ferrario F, Tubiello FN, Leip A (2021) Food systems are responsible for a third of global anthropogenic GHG emissions. Nature Food 2:198–209

Crueger W, Crueger A (1990) Biotechnology: a textbook of industrial microbiology. Single Cell protein

Cui X-H, Chen S-J, Wang Y, Han J-R (2013) Fermentation conditions of walnut Milk beverage inoculated with Kefir grains. LWT-Food Sci Technol 50:349–352

Dan T, Wang D, Jin R, Zhang H, Zhou T, Sun T (2017) Characterization of volatile compounds in fermented milk using solid-phase microextraction methods coupled with gas chromatography-mass spectrometry. J Dairy Sci 100:2488–2500

De Sy V, Herold M, Achard F, Beuchle R, Clevers J, Lindquist E, Verchot L (2015) Land use patterns and related carbon losses following deforestation in South America. Environ Res Lett 10:124004

Debonne N (2019) Climate change and land: An IPCC special report on climate change, desertification, land degradation, sustainable land management, food security, and greenhouse gas fluxes in terrestrial ecosystems: Chapter 1: Framing and context

Dekkers BL, Nikiforidis CV, Van Der Goot AJ (2016) Shear-induced fibrous structure formation from a pectin/SPI blend. Innovative Food Sci Emerg Technol 36:193–200

Dekkers BL, Boom RM, Van Der Goot AJ (2018) Structuring processes for meat analogues. Trends Food Sci Technol 81:25–36

Denny A, Aisbitt B, Lunn J (2008) Mycoprotein and health. Nutr Bull 33:298–310

Desiderio E, Shanmugam K, Östergren K (2023) Plant based meat alternative, from cradle to company-gate: a case study uncovering the environmental impact of the Swedish pea protein value chain. J Clean Prod 418:138173

Dettling J, Tu Q, Faist M, Delduce A, Mandlebaum S (2016) A comparative life cycle assessment of plant-based foods and meat foods. Quantis USA, Boston

Dhakal S, Giusti MM, Balasubramaniam V (2016) Effect of high pressure processing on dispersive and aggregative properties of almond milk. J Sci Food Agric 96:3821–3830

Dharumadurai D, Subramaniyan L, Subhasish S, Nooruddin T, Annamalai P (2011) Production of single cell protein from pineapple waste using yeast. Innov Rom Food Biotechnol:26–32

Diarra K, Nong ZG, Jie C (2005) Peanut milk and Peanut milk based products production: a review. Crit Rev Food Sci Nutr 45:405–423

Doi H, Mulia RN (2021) Future land use for insect meat production among countries: a global classification. Front Nutr 8:661056

Du M, Zhang Y, Zhao Y, Fang Y (2023) Agarose/Konjac Glucomannan double network hydrogels to mimic the texture of beef tripe. Food Hydrocoll 135:108173

Ermiş E, Güneş R, Inci Z, Çağlar MY, Yilmaz MT (2018) Characterization of hazelnut Milk fermented by lactobacillus Delbrueckii Subsp. Bulgaricus and Streptococcus thermophilus. Gıda 43:677–686

Estrada PD, Berton-Carabin CC, Schlangen M et al (2018) Protein oxidation in plant protein-based fibrous products: effects of encapsulated iron and process conditions. J Agric Food Chem 66:11105–11112. https://doi.org/10.1021/acs.jafc.8b02844

FAO (2013) Food wastage footprint: impacts on natural resources: summary report. Food Wastage Footprint. Food & Agriculture Organization of the UN (FAO)

Fardet A, Rock E (2019) Ultra-processed foods: a new holistic paradigm? Trends Food Sci Technol 93:174–184

Foley JA, Ramankutty N, Brauman KA, Cassidy ES, Gerber JS, Johnston M, Mueller ND, O'connell C, Ray DK, West PC (2011) Solutions for a cultivated planet. Nature 478:337–342

Fraser R, Davis SC, Brown POR (2017). Secretion of heme-containing polypeptides. Google Patents

Fraser RZ, Shitut M, Agrawal P, Mendes O, Klapholz S (2018) Safety evaluation of soy leghemoglobin protein preparation derived from Pichia pastoris, intended for use as a flavor catalyst in plant-based meat. Int J Toxicol 37:241–262

Fresán U, Marrin D, Mejia MA, Sabaté J (2019a) Water footprint of meat analogs: selected indicators according to life cycle assessment. Water 11:728

Fresán U, Mejia MA, Craig WJ, Jaceldo-Siegl K, Sabaté J (2019b) Meat Analogs from different protein sources: a comparison of their sustainability and nutritional content. Sustain For 11:3231

Fu J, Sun C, Chang Y, Li S, Zhang Y, Fang Y (2023) Structure analysis and quality evaluation of plant-based meat Analogs. J Texture Stud 54:383–393

Gebrehiwot AA, Hashemi-Beni L, Kurkalova LA, Liang CL, Jha MK (2022) Using ABM to study the potential of land use change for mitigation of food deserts. Sustain For 14:9715

Gephart JA, Henriksson PJ, Parker RW, Shepon A, Gorospe KD, Bergman K, Eshel G, Golden CD, Halpern BS, Hornborg S (2021) Environmental performance of blue foods. Nature 597:360–365

Gerbens-Leenes W, Nonhebel S (2005) Food and land use. The influence of consumption patterns on the use of agricultural resources. Appetite 45:24–31

Gerber P, Vellinga T, Opio C, Steinfeld H (2011) Productivity gains and greenhouse gas emissions intensity in dairy systems. Livest Sci 139:100–108

Gharibzahedi SMT, Smith B (2020) The functional modification of legume proteins by ultrasonication: a review. Trends Food Sci Technol 98:107–116

Girija J, Kamalasundari S, Hemalatha G, Umamaheswari T (2021) Production methodologies of meat analogues: a review. J Agric Eng 58:137–148

Godoi FC, Prakash S, Bhandari BR (2016) 3d printing technologies applied for food design: status and prospects. J Food Eng 179:44–54

Golbitz P, Jordan J (2006) Soyfoods: market and products. In: Soy applications in food, pp 1–21

Goldstein B, Moses R, Sammons N, Birkved M (2017) Potential to curb the environmental burdens of American beef consumption using a novel plant-based beef substitute. PLoS One 12:E0189029

Gomashe A, Pounikar M, Gulhane P (2014) Liquid whey: a potential substrate for single cell protein production from bacillus subtilis NCIM 2010. Int J Life Sci 2:119–123

Gomiero T, Pimentel D, Paoletti MG (2011) Environmental impact of different agricultural management practices: conventional vs. organic agriculture. Crit Rev Plant Sci 30:95–124

Grabowska KJ, Tekidou S, Boom RM, Van Der Goot A-J (2014) Shear structuring as a new method to make anisotropic structures from soy–gluten blends. Food Res Int 64:743–751

Grahl S, Palanisamy M, Strack M, Meier-Dinkel L, Toepfl S, Mörlein D (2018) Towards more sustainable meat alternatives: how technical parameters affect the sensory properties of extrusion products derived from soy and algae. J Clean Prod 198:962–971

Grant SM (2007) The importance of biodiversity in crop sustainability: a look at monoculture. J Hunger Environ Nutr 1:101–109

Grossmann L, Mcclements DJ (2021) The science of plant-based foods: approaches to create nutritious and sustainable plant-based cheese analogs. Trends Food Sci Technol 118:207–229

Guerrieri N (2004) Cereal Proteins. In: Proteins in food processing. Woodhead Publishing Limited, London

Guil-Guerrero J, Navarro-Juárez R, López-Martınez J, Campra-Madrid P, Rebolloso-Fuentes M (2004) Functional properties of the biomass of three microalgal species. J Food Eng 65:511–517

Guo Z, Teng F, Huang Z, Lv B, Lv X, Babich O, Yu W, Li Y, Wang Z, Jiang L (2020) Effects of material characteristics on the structural characteristics and flavor substances retention of meat analogs. Food Hydrocoll 105:105752

Guzmán-Luna P, Gerbens-Leenes P, Vaca-Jiménez S (2021) The water, energy, and land footprint of tilapia aquaculture in Mexico, a comparison of the footprints of fish and meat. Resour Conserv Recycl 165:105224

Handley MA, Hall C, Sanford E, Diaz E, Gonzalez-Mendez E, Drace K, Wilson R, Villalobos M, Croughan M (2007) Globalization, binational communities, and imported food risks: results of an outbreak investigation of lead poisoning in Monterey County, California. Am J Public Health 97:900–906

Heller, M. C. & Keoleian, G. A. 2018. Beyond meat's beyond burger life cycle assessment: a detailed comparison between

Herve-Jimenez L, Guillouard I, Guedon E, Boudebbouze S, Hols P, Monnet V, Maguin E, Rul F (2009) Postgenomic analysis of Streptococcus thermophilus cocultivated in milk with Lactobacillus delbrueckii Subsp. Bulgaricus: involvement of nitrogen, purine, and iron metabolism. Appl Environ Microbiol 75:2062–2073

Hickisch A, Beer R, Vogel RF, Toelstede S (2016) Influence of lupin-based milk alternative heat treatment and exopolysaccharide-producing lactic acid bacteria on the physical characteristics of lupin-based yogurt alternatives. Food Res Int 84:180–188

Hoek AC, Elzerman JE, Hageman R, Kok FJ, Luning PA, De Graaf C (2013) Are meat substitutes liked better over time? A repeated in-home use test with meat substitutes or meat in meals. Food Qual Prefer 28:253–263

Hoekstra AY (2012) The hidden water resource use behind meat and dairy. Anim Front 2:3–8

Hoogenkamp B (2018) Clean cultured meat for today's future. Fleischwirtschaft International 2:42–46

Huang H, Hammond E, Reitmeier C, Myers D (1995) Properties of fibers produced from soy protein isolate by extrusion and wet-spinning. J Am Oil Chem Soc 72:1453–1460

Ilyasoglu H, Yilmaz F (2019) Preliminary investigation of yoghurt enriched with hazelnut Milk. Int Food Res J

Irlinger F, Hélinck S, Jany JL (2017) Secondary and adjunct cultures. In: Cheese. Elsevier

Ismail I, Hwang Y-H, Joo S-T (2020) Meat analog as future food: a review. J Anim Sci Technol 62:111

Iswarin SJ, Permadi B (2012) Coconut milk's fat breaking by means of ultrasound. Int J Basic Appl Sci 12:1–5

Iwamoto H (2003) Industrial production of microalgal cell-mass and secondary products-major industrial species: chlorella. In: Handbook of microalgal culture: biotechnology and applied Phycology, pp 253–263

Jeewanthi RKC, Paik H-D (2018) Modifications of nutritional, structural, and sensory characteristics of non-dairy soy cheese analogs to improve their quality attributes. J Food Sci Technol 55:4384–4394

Jeske S, Zannini E, Arendt EK (2018) Past, present and future: the strength of plant-based dairy substitutes based on gluten-free raw materials. Food Res Int 110:42–51

Jeske S, Bez J, Arendt EK, Zannini E (2019) Formation, stability, and sensory characteristics of a lentil-based milk substitute as affected by homogenisation and pasteurisation. Eur Food Res Technol 245:1519–1531

Joshi V, Kumar S (2015) Meat analogues: plant based alternatives to meat products – a review. Int J Food Ferment Technol 5:107–119

Kamath R, Basak S, Gokhale J (2022) Recent trends in the development of healthy and functional cheese analogues – a review. LWT 155:112991

Kamau EH, Nkhata SG, Ayua EO (2020) Extrusion and nixtamalization conditions influence the magnitude of change in the nutrients and bioactive components of cereals and legumes. Food Sci Nutr 8:1753–1765

Kaneko D, Igarashi T, Aoyama K (2014) Reduction of the off-flavor volatile generated by the yogurt starter culture including Streptococcus thermophilus and Lactobacillus delbrueckii Subsp. Bulgaricus in soymilk. J Agric Food Chem 62:1658–1663

Kang I-H, Srivastava P, Ozias-Akins P, Gallo M (2007) Temporal and spatial expression of the major allergens in developing and germinating peanut seed. Plant Physiol 144:836–845

Karim AA, Bhat R (2008) Gelatin alternatives for the food industry: recent developments, challenges and prospects. Trends Food Sci Technol 19:644–656

Karlsson Potter, H., Lundmark, L. & Röös, E. 2020. Environmental impact of plant-based foods–data collection for the development of a consumer guide for plant-based foods

Kassem Y, Gökçekuş H, Kouzehgarani JK (2022) Water footprint assessment of animal-based and plant-based products in Iran. Future Technol 1:18–24

Kazir M, Livney YD (2021) Plant-based seafood analogs. Molecules 26:1559

Khan S, Dettling J, Loyola C, Hester J, Moses R (2019) Environmental life cycle analysis: impossible burger 2.0. Quantis, Boston

Khrundin D, Ponomarev V, Yunusov E, Ezhkova G (2021) The use of plant proteins in the technology of fermented dairy-free products. In: IOP conference series: earth and environmental science. Iop Publishing, 012066

Khuenpet K, Jittanit W, Hongha N, Pairojkul S (2016) UHT skim coconut milk production and its quality. In: SHS web of conferences. EDP Sciences, 03002

Kiely L, Mcconnell S, Kindstedt P (1991) Observations on the melting behavior of imitation mozzarella cheese. J Dairy Sci 74:3568–3572

Kizzie-Hayford N, Jaros D, Zahn S, Rohm H (2016) Effects of protein enrichment on the microbiological, physicochemical and sensory properties of fermented tiger nut Milk. LWT 74:319–324

Kluczkovski A, Lima N, Oliveira MK (2017) Brazil nut powdered milk properties. J Food Process Preserv 41:E13147

Kohli D, Kumar S, Upadhyay S, Mishra R (2017) Preservation and processing of soymilk: a review. Int J Food Sci Nutr 2:66–70

Kolbe K (2020) Mitigating climate change through diet choice: costs and CO2 emissions of different cookery book-based dietary options in Germany. Adv Clim Chang Res 11:392–400

Kołodziejczak K, Onopiuk A, Szpicer A, Poltorak A (2021) Meat analogues in the perspective of recent scientific research: a review. Food Secur 11:105

Krintiras GA, Göbel J, Bouwman WG, Van Der Goot AJ, Stefanidis GD (2014) On characterization of anisotropic plant protein structures. Food Funct 5:3233–3240

Krintiras GA, Göbel J, Van Der Goot AJ, Stefanidis GD (2015) Production of structured soy-based meat analogues using simple shear and heat in a Couette cell. J Food Eng 160:34–41

Kumar P, Kumar R (2011) Product profile comparison of analogue meat nuggets versus chicken nuggets. Fleischwirtschaft Int J Meat Prod Meat Process:72–74

Kumar P, Sharma B, Kumar R (2011) Optimization of the level of mushroom in analogue meat nuggets. J Meat Sci 7:53–55

Kundu P, Dhankhar J, Sharma A (2018) Development of non dairy milk alternative using soymilk and almond milk. Curr Res Nutr Food Sci J 6:203–210

Kustar A, Patino-Echeverri D (2021) A review of environmental life cycle assessments of diets: plant-based solutions are truly sustainable, even in the form of fast foods. Sustain For 13:9926

Kutzli I, Gibis M, Baier SK, Weiss J (2019) Electrospinning of whey and soy protein mixed with maltodextrin–influence of protein type and ratio on the production and morphology of fibers. Food Hydrocoll 93:206–214

Kutzli I, Beljo D, Gibis M, Baier SK, Weiss J (2020a) Effect of maltodextrin dextrose equivalent on electrospinnability and glycation reaction of blends with pea protein isolate. Food Biophys 15:206–215

Kutzli I, Griener D, Gibis M, Grossmann L, Baier SK, Weiss J (2020b) Improvement of emulsifying behavior of pea proteins as plant-based emulsifiers via maillard-induced glycation in electrospun pea protein–maltodextrin fibers. Food Funct 11:4049–4056

Kyriakopoulou K, Dekkers B, Van Der Goot AJ (2019) Plant-based meat analogues. In: Sustainable meat production and processing. Elsevier

Kyriakopoulou K, Keppler JK, Van Der Goot AJ (2021) Functionality of ingredients and additives in plant-based meat analogues. Food Secur 10:600

Lee S-W, Rhee C (2003) Processing suitability of a rice and pine nut (Pinus Koraiensis) beverage. Food Hydrocoll 17:379–385

Lee HJ, Yong HI, Kim M, Choi Y-S, Jo C (2020) Status of meat alternatives and their potential role in the future meat market—a review. Asian Australas J Anim Sci 33:1533

Li X, Li J (2020) The flavor of plant-based meat analogues. Cereal Foods World 65:0040

Li Y, Zhang X, Yang JJ, Ma XY, Jia XD, Du P, Li AL (2020) Influence of the addition of Geotrichum candidum on the microbial, chemical, textural, and sensory features of soft soy cheese. J Food Process Preserv 44:E14823

Li M, Zhuo L, Liu Y, Yan C, Wu P (2022a) Spatiotemporal heterogeneities in water and land appropriations related to food losses and waste in China. Environ Res Lett 17:054020

Li X, Manickavasagan A, Lim L-T, Ali A (2022b) Plant-based protein meat analogues. Springer, Plant Protein Foods

Lienhardt T, Black K, Saget S, Costa MP, Chadwick D, Rees RM, Williams M, Spillane C, Iannetta PM, Walker G (2019) Just the tonic! Legume biorefining for alcohol has the potential to reduce Europe's protein deficit and mitigate climate change. Environ Int 130:104870

Lima M, Costa R, Rodrigues I, Lameiras J, Botelho G (2022) A narrative review of alternative protein sources: highlights on meat, fish, egg and dairy analogues. Foods 11:2053

Liu P, Xu H, Zhao Y, Yang Y (2017) Rheological properties of soy protein isolate solution for fibers and films. Food Hydrocoll 64:149–156

Liu G-C, Xiao-Yun G, Jiang Y, Yi-Qing P, Xiang F, Huang Z-J, Di T (2019) Research and application of mass spectrometry with low energy electron impact ionization source. Chin J Anal Chem 47:686–694

Lopusiewicz L, Drozlowska E, Siedlecka P, Mezynska M, Bartkowiak A (2020) Preparation and characterization of novel flaxseed oil cake yogurt-like plant milk fortified with inulin. J Food Nutr Res 59

Luana N, Rossana C, Curiel JA, Kaisa P, Marco G, Rizzello CG (2014) Manufacture and characterization of a yogurt-like beverage made with oat flakes fermented by selected lactic acid bacteria. Int J Food Microbiol 185:17–26

Lupton D, Turner B (2018) Food of the future? Consumer responses to the idea of 3D-printed meat and insect-based foods. Food Foodways 26:269–289

Machovina B, Feeley KJ, Ripple WJ (2015) Biodiversity conservation: the key is reducing meat consumption. Sci Total Environ 536:419–431

Maghsoudlou Y, Alami M, Mashkour M, Shahraki MH (2016) Optimization of ultrasound-assisted stabilization and formulation of almond milk. J Food Process Preserv 40:828–839

Mäkinen OE, Wanhalinna V, Zannini E, Arendt EK (2016) Foods for special dietary needs: non-dairy plant-based milk substitutes and fermented dairy-type products. Crit Rev Food Sci Nutr 56:339–349

Malav O, Talukder S, Gokulakrishnan P, Chand S (2015) Meat analog: a review. Crit Rev Food Sci Nutr 55:1241–1245

Maleki N, Khodaiyan F, Mousavi SM (2015) Antioxidant activity of fermented hazelnut milk. Food Sci Biotechnol 24:107–115

Manzoor MF, Manzoor A, Siddique R, Ahmad N (2017) Nutritional and sensory properties of cashew seed (Anacardium occidentale) milk. Mod Concepts Dev Agron 1:1–4

Marcone MF (1999) Biochemical and biophysical properties of plant storage proteins: a current understanding with emphasis on 11S seed globulins. Food Res Int 32:79–92

Maria MF, Victoria AT (2018) Influence of processing treatments on quality of vegetable milk from almond (Terminalia catappa) kernels. Acta Sci Nutr Health 2:37–42

Masiá C, Jensen PE, Petersen IL, Buldo P (2022) Design of a functional pea protein matrix for fermented plant-based cheese. Food Secur 11:178

Matias NS, Bedani R, Castro IA, Saad SM (2014) A probiotic soy-based innovative product as an alternative to petit-suisse cheese. LWT-Food Sci Technol 59:411–417

Mccarthy K, Parker M, Ameerally A, Drake S, Drake M (2017) Drivers of choice for fluid milk versus plant-based alternatives: what are consumer perceptions of fluid milk? J Dairy Sci 100:6125–6138

Mcclements DJ, Grossmann L (2021) The science of plant-based foods: constructing next-generation meat, fish, milk, and egg analogs. Compr Rev Food Sci Food Saf 20:4049–4100

Mcsweeney PL (2004) Biochemistry of cheese ripening. Int J Dairy Technol 57:127–144

Mefleh M, Pasqualone A, Caponio F, Faccia M (2022) Legumes as basic ingredients in the production of dairy-free cheese alternatives: a review. J Sci Food Agric 102:8–18

Mehran, R., Andreas, G. & Wolfgang, M. 2013. Method for producing meat substitute products

Mejia M, Fresán U, Harwatt H, Oda K, Uriegas-Mejia G, Sabaté J (2019) Life cycle assessment of the production of a large variety of meat analogs by three diverse factories. J Hunger Environ Nutr

Mekonnen MM, Gerbens-Leenes W (2020) The water footprint of global food production. Water 12:2696

Mendly-Zambo Z, Powell LJ, Newman LL (2021) Dairy 3.0: cellular agriculture and the future of milk. Food, Cult Soc 24:675–693

Micha R, Peñalvo JL, Cudhea F, Imamura F, Rehm CD, Mozaffarian D (2017) Association between dietary factors and mortality from heart disease, stroke, and type 2 diabetes in the United States. JAMA 317:912–924

Mrówczyńska-Kamińska A, Bajan B, Pawłowski KP, Genstwa N, Zmyślona J (2021) Greenhouse gas emissions intensity of food production systems and its determinants. PLoS One 16:E0250995

Mulia RN, Doi H (2019) Global simulation of insect meat production under climate change. Front Sustain Food Syst 3:91

Munekata PE, Domínguez R, Budaraju S, Roselló-Soto E, Barba FJ, Mallikarjunan K, Roohinejad S, Lorenzo JM (2020) Effect of innovative food processing technologies on the physicochemical and nutritional properties and quality of non-dairy plant-based beverages. Food Secur 9:288

Nette A, Wolf P, Schlüter O, Meyer-Aurich A (2016) A comparison of carbon footprint and production cost of different pasta products based on whole egg and pea flour. Food Secur 5:17

Nieuwland M, Geerdink P, Brier P, Van Den Eijnden P, Henket JT, Langelaan ML, Stroeks N, Van Deventer HC, Martin AH (2014) Reprint of "food-grade electrospinning of proteins". Innovative Food Sci Emerg Technol 24:138–144

Nijdam D, Rood T, Westhoek H (2012) The price of protein: review of land use and carbon footprints from life cycle assessments of animal food products and their substitutes. Food Policy 37:760–770

Nikzade V, Tehrani MM, Saadatmand-Tarzjan M (2012) Optimization of low-cholesterol–low-fat mayonnaise formulation: effect of using soy milk and some stabilizer by a mixture design approach. Food Hydrocoll 28:344–352

Notarnicola B, Tassielli G, Renzulli PA, Castellani V, Sala S (2017) Environmental impacts of food consumption in Europe. J Clean Prod 140:753–765

Ogundipe OO, Fasogbon BM, Ogundipe FO, Oredope O, Amaezenanbu RU (2021) Nutritional composition of non-dairy yogurt from sprouted tigernut tubers. J Food Process Preserv 45:E15884

Oosone S, Kashiwaba A, Yanagihara N, Yoshikawa J, Kashiwagi Y, Maehashi K (2020) The role of Amylolytic and proteolytic enzyme activities of vegetables, fruits, and edible fungi in flavor enhancement during cooking. Int J Gastron Food Sci 22:100264

Orcutt M, Mcmindes M, Chu H, Mueller I, Bater B, Orcutt A (2006) Textured soy protein utilization in meat and meat analog products. In: Soy applications in food, pp 155–184

Orkusz A (2021) Edible insects versus meat—nutritional comparison: knowledge of their composition is the key to good health. Nutrients 13:1207

Osen R, Toelstede S, Wild F, Eisner P, Schweiggert-Weisz U (2014) High moisture extrusion cooking of pea protein isolates: raw material characteristics, extruder responses, and texture properties. J Food Eng 127:67–74

Osterhaus EJ (1978) Double-extrusion apparatus for producing dense, uniformly layered vegetable protein meat analogue. Google Patents

Padma M, Jagannadarao P, Edukondalu L, Ravibabu G, Aparna K (2018) Physico-chemical analysis of milk prepared from broken Rice. Int J Curr Microbiol App Sci 7:426–428

Palanisamy M, Töpfl S, Berger RG, Hertel C (2019) Physico-chemical and nutritional properties of meat analogues based on spirulina/lupin protein mixtures. Eur Food Res Technol 245:1889–1898

Pandey SM, Mishra H (2015) Optimization of the prebiotic & probiotic concentration and incubation temperature for the preparation of synbiotic soy yoghurt using response surface methodology. LWT-Food Sci Technol 62:458–467
Pardeshi I, Murumkar R, Tayade P (2014) Optimization of process for spray drying of soymilk and sprouted soybean milk. J Grain Process Storage 1:13–20
Parker RW, Blanchard JL, Gardner C, Green BS, Hartmann K, Tyedmers PH, Watson RA (2018) Fuel use and greenhouse gas emissions of world fisheries. Nat Clim Chang 8:333–337
Pearson AM (2012) Muscle and meat biochemistry. Elsevier
Pelletier N, Ibarburu M, Xin H (2013) A carbon footprint analysis of egg production and processing supply chains in the Midwestern United States. J Clean Prod 54:108–114
Perro M (2019) Rat feeding study suggests the impossible burger may not be safe to eat. GmoscienceOrg [Online]. Available: https://Gmoscience.Org/2019/06/25/Rat-Feeding-Studies-Suggest-The-Impossible-Burger-May-Not-Be-Safe-To-Eat/?Fbclid=Iwar0ulka2qo8a-Zs3lsksn31j94cxc9nddaqzuutyict0qg3ne_Uq8YGfKE8&Amp=1
Petrova P, Petrov K (2020) Lactic acid fermentation of cereals and pseudocereals: ancient nutritional biotechnologies with modern applications. Nutrients 12:1118
Pietsch VL, Bühler JM, Karbstein HP, Emin MA (2019) High moisture extrusion of soy protein concentrate: influence of thermomechanical treatment on protein-protein interactions and rheological properties. J Food Eng 251:11–18
Pilorgé E, Kezeya B, Stauss W, Muel F, Mergenthaler M (2021) Pea and rapeseed acreage and land use for plant-based meat alternatives in the EU. OCL 28:54
Pineli LDLDO, Botelho RB, Zandonadi RP, Solorzano JL, De Oliveira GT, Reis CEG, Teixeira DDS (2015) Low Glycemic Index and increased protein content in a novel quinoa milk. LWT-Food Sci Technol 63:1261–1267
Pingali P, Boiteau J, Choudhry A, Hall A (2023) Making meat and milk from plants: a review of plant-based food for human and planetary health. World Dev 170:106316
Plaza M, Herrero M, Cifuentes A, Ibanez E (2009) Innovative natural functional ingredients from microalgae. J Agric Food Chem 57:7159–7170
Plazzotta S, Manzocco L (2019) Food waste valorization. Saving food. Elsevier
Pontonio E, Raho S, Dingeo C, Centrone D, Carofiglio VE, Rizzello CG (2020) Nutritional, functional, and technological characterization of a novel gluten-and lactose-free yogurt-style snack produced with selected lactic acid bacteria and leguminosae flours. Front Microbiol 11:1664
Post MJ (2012) Cultured meat from stem cells: challenges and prospects. Meat Sci 92:297–301
Potter HK, Röös E (2021) Multi-criteria evaluation of plant-based foods–use of environmental footprint and LCA data for consumer guidance. J Clean Prod 280:124721
Pua A, Tang VCY, Goh RMV, Sun J, Lassabliere B, Liu SQ (2022) Ingredients, processing, and fermentation: addressing the organoleptic boundaries of plant-based dairy analogues. Food Secur 11:875
Pulz O, Gross W (2004) Valuable products from biotechnology of microalgae. Appl Microbiol Biotechnol 65:635–648
Quasem JM, Mazahreh AS, Abu-Alruz K (2009) Development of vegetable based milk from decorticated sesame (Sesamum indicum). Am J Appl Sci 6:888
Ramboer E, De Craene B, De Kock J, Vanhaecke T, Berx G, Rogiers V, Vinken M (2014) Strategies for immortalization of primary hepatocytes. J Hepatol 61:925–943
Rampon V, Robert P, Nicolas N, Dufour E (1999) Protein structure and network orientation in edible films prepared by spinning process. J Food Sci 64:313–316
Ransom E (2021) Impossible solutions: competing values in marketing alternative proteins for sustainable food systems. J Rural Stud 86:694–701
Rasika D, Vidanarachchi JK, Luiz SF, Azeredo DRP, Cruz AG, Ranadheera CS (2021) Probiotic delivery through non-dairy plant-based food matrices. Agriculture 11:599
Riascos JJ, Weissinger AK, Weissinger SM, Burks AW (2010) Hypoallergenic legume crops and food allergy: factors affecting feasibility and risk. J Agric Food Chem 58:20–27

Ribeiro JC, Cunha LM, Sousa-Pinto B, Fonseca J (2018) Allergic risks of consuming edible insects: a systematic review. Mol Nutr Food Res 62:1700030
Roland WS, Pouvreau L, Curran J, Van De Velde F, De Kok PM (2017) Flavor aspects of pulse ingredients. Cereal Chem 94:58–65
Röös, E., Garnett, T., Watz, V. & Sjörs, C. 2018. The role of dairy and plant based dairy alternatives in sustainable diets
Roselló-Soto E, Garcia C, Fessard A, Barba FJ, Munekata PE, Lorenzo JM, Remize F (2018) Nutritional and microbiological quality of tiger nut tubers (Cyperus esculentus), derived plant-based and lactic fermented beverages. Fermentation 5:3
Rosi A, Mena P, Pellegrini N, Turroni S, Neviani E, Ferrocino I, Di Cagno R, Ruini L, Ciati R, Angelino D (2017) Environmental impact of omnivorous, ovo-lacto-vegetarian, and vegan diet. Sci Rep 7:6105
Rotz CA (2018) Modeling greenhouse gas emissions from dairy farms. J Dairy Sci 101:6675–6690
Rotz C, Montes F, Chianese D (2010) The carbon footprint of dairy production systems through partial life cycle assessment. J Dairy Sci 93:1266–1282
Rousta N, Hellwig C, Wainaina S, Lukitawesa L, Agnihotri S, Rousta K, Taherzadeh MJ (2021) Filamentous fungus Aspergillus oryzae for food: from submerged cultivation to fungal burgers and their sensory evaluation—a pilot study. Food Secur 10:2774
Roy B, Hagappa A, Ramalingam YD, Mahalingam N (2021) A review on lab-grown meat: advantages and disadvantages. Quest Int J Med Health Sci 4:19–24
Santo RE, Kim BF, Nachman KE (2020) Questions and concerns re: blue water footprints reported in "water footprint of meat analogs: selected indicators according to life cycle assessment". Water 12:1270
Sarkar S (2021) The impossible burger 3
Sathasivam R, Radhakrishnan R, Hashem A, Abd-Allah EF (2019) Microalgae metabolites: a rich source for food and medicine. Saudi J Biol Sci 26:709–722
Sauer S (2018) Soy expansion into the agricultural frontiers of the Brazilian Amazon: the agribusiness economy and its social and environmental conflicts. Land Use Policy 79:326–338
Scarborough P, Appleby PN, Mizdrak A, Briggs AD, Travis RC, Bradbury KE, Key TJ (2014) Dietary greenhouse gas emissions of meat-eaters, fish-eaters, vegetarians and vegans in the UK. Clim Chang 125:179–192
Schreuders FK, Dekkers BL, Bodnár I, Erni P, Boom RM, Van Der Goot AJ (2019) Comparing structuring potential of pea and soy protein with gluten for meat analogue preparation. J Food Eng 261:32–39
Seow CC, Gwee CN (1997) Coconut milk: chemistry and technology. Int J Food Sci Technol 32:189–201
Sethi S, Tyagi SK, Anurag RK (2016) Plant-based Milk alternatives an emerging segment of functional beverages: a review. J Food Sci Technol 53:3408–3423
Severini C, Derossi A, Ricci I, Caporizzi R, Fiore A (2018) Printing a blend of fruit and vegetables. New advances on critical variables and shelf life of 3D edible objects. J Food Eng 220:89–100
Shanmugam K, Bryngelsson S, Östergren K, Hallström E (2023) Climate impact of plant-based meat analogues: a review of life cycle assessments. Sustain Prod Consump 36:328–337
Shaw M (1984) Cheese substitutes: threat or opportunity?
Shurtleff W, Aoyagi A (1989) Bibliography of tempeh and tempeh products: 1416 references from 1815 to 1989
Silva AR, Silva MM, Ribeiro BD (2020) Health issues and technological aspects of plant-based alternative milk. Food Res Int 131:108972
Sim SYJ, Srv A, Chiang JH, Henry CJ (2021) Plant proteins for future foods: a roadmap. Food Secur 10:1967
Singh A, Sit N (2022) Meat analogues: types, methods of production and their effect on attributes of developed meat analogues. Food Bioprocess Technol 15:2664–2682
Smetana S, Mathys A, Knoch A, Heinz V (2015) Meat alternatives: life cycle assessment of most known meat substitutes. Int J Life Cycle Assess 20:1254–1267

Smetana S, Profeta A, Voigt R, Kircher C, Heinz V (2021a) Meat substitution in burgers: nutritional scoring, sensorial testing, and life cycle assessment. Future Foods 4:100042

Smetana S, Spykman R, Heinz V (2021b) Environmental aspects of insect mass production. J Insects Food Feed 7:553–571

Söderberg, J. 2013. Functional properties of legume proteins compared to egg proteins and their potential as egg replacers in vegan food

Soletto D, Binaghi L, Lodi A, Carvalho J, Converti A (2005) Batch and fed-batch cultivations of Spirulina platensis using ammonium sulphate and urea as nitrogen sources. Aquaculture 243:217–224

Souza RG, Gomes AC, Naves MM, Mota JF (2015) Nuts and legume seeds for cardiovascular risk reduction: scientific evidence and mechanisms of action. Nutr Rev 73:335–347

Spolaore P, Joannis-Cassan C, Duran E, Isambert A (2006) Commercial applications of microalgae. J Biosci Bioeng 101:87–96

Springmann M, Clark M, Mason-D'croz D, Wiebe K, Bodirsky BL, Lassaletta L, De Vries W, Vermeulen SJ, Herrero M, Carlson KM (2018) Options for keeping the food system within environmental limits. Nature 562:519–525

Steer M (2015) A comparison of land, water and energy use between conventional and yeast-derived dairy products: An initial analysis. Report prepared for Perfect Day. University of the West of England, Bristol

Stephens N, Di Silvio L, Dunsford I, Ellis M, Glencross A, Sexton A (2018) Bringing cultured meat to market: technical, socio-political, and regulatory challenges in cellular agriculture. Trends Food Sci Technol 78:155–166

Suman G, Nupur M, Anuradha S, Pradeep B (2015) Single cell protein production: a review. Int J Curr Microbiol App Sci 4:251–262

Sun P, Li D, Li Z, Dong B, Wang F (2008) Effects of glycinin on Ige-mediated increase of mast cell numbers and histamine release in the small intestine. J Nutr Biochem 19:627–633

Tamang JP, Shin D-H, Jung S-J, Chae S-W (2016) Functional properties of microorganisms in fermented foods. Front Microbiol 7:578

Tangyu M, Fritz M, Aragao-Börner R, Ye L, Bogicevic B, Bolten CJ, Wittmann C (2021) Genome-based selection and application of food-grade microbes for chickpea milk fermentation towards increased L-lysine content, elimination of indigestible sugars, and improved flavour. Microb Cell Factories 20:1–21

Temme EH, Van Der Voet H, Thissen JT, Verkaik-Kloosterman J, Van Donkersgoed G, Nonhebel S (2013) Replacement of meat and dairy by plant-derived foods: estimated effects on land use, iron and SFA intakes in young Dutch adult females. Public Health Nutr 16:1900–1907

Tilman D, Cassman KG, Matson PA, Naylor R, Polasky S (2002) Agricultural sustainability and intensive production practices. Nature 418:671–677

Tolstoguzov V (1988) Creation of fibrous structures by spinneretless spinning. In: Mitchell JR, Blanshard JMV (eds) Food structure: creation and evaluation. Woodhead Publishing Limited, Cambridge, pp 181–196

Tramberend S, Fischer G, Bruckner M, Van Velthuizen H (2019) Our common cropland: quantifying global agricultural land use from a consumption perspective. Ecol Econ 157:332–341

Treich N (2021) Cultured meat: promises and challenges. Environ Resour Econ 79:33–61

Tscharntke T, Clough Y, Wanger TC, Jackson L, Motzke I, Perfecto I, Vandermeer J, Whitbread A (2012) Global food security, biodiversity conservation and the future of agricultural intensification. Biol Conserv 151:53–59

Tubiello FN, Rosenzweig C, Conchedda G, Karl K, Gütschow J, Xueyao P, Obli-Laryea G, Wanner N, Qiu SY, De Barros J (2021) Greenhouse gas emissions from food systems: building the evidence base. Environ Res Lett 16:065007

Tuomisto HL, Teixeira De Mattos MJ (2011) Environmental impacts of cultured meat production. Environ Sci Technol 45:6117–6123

Van Der Weele C, Feindt P, Van Der Goot AJ, Van Mierlo B, Van Boekel M (2019) Meat alternatives: An integrative comparison. Trends Food Sci Technol 88:505–512

Van Huis A (2013) Potential of insects as food and feed in assuring food security. Annu Rev Entomol 58:563–583
Van Vliet S, Bain JR, Muehlbauer MJ, Provenza FD, Kronberg SL, Pieper CF, Huffman KM (2021) A metabolomics comparison of plant-based meat and grass-fed meat indicates large nutritional differences despite comparable nutrition facts panels. Sci Rep 11:13828
Viana A, Ethur EM, De Freitas EM, Hoehne L (2023) Chicken eggs substitute using vegetable origin – a review. Food Bioprocess Technol:1–16
Voon SL, An J, Wong G, Zhang Y, Chua CK (2019) 3D food printing: a categorised review of inks and their development. Virtual Phys Prototyp 14:203–218
Wang Q, Jiang J, Xiong YL (2018) High pressure homogenization combined with Ph shift treatment: a process to produce physically and oxidatively stable hemp milk. Food Res Int 106:487–494
Wang C, Yin H, Zhao Y, Zheng Y, Xu X, Yue J (2021) Optimization of high hydrostatic pressure treatments on soybean protein isolate to improve its functionality and evaluation of its application in yogurt. Food Secur 10:667
Wenger LVG, Osterhaus EJ, Smith OB (1977) Dense, uniformly layered vegetable protein meat analogue. Google Patents
Westhoek H, Lesschen JP, Rood T, Wagner S, De Marco A, Murphy-Bokern D, Leip A, Van Grinsven H, Sutton MA, Oenema O (2014) Food choices, health and environment: effects of cutting Europe's meat and dairy intake. Glob Environ Chang 26:196–205
Wi G, Bae J, Kim H, Cho Y, Choi M-J (2020) Evaluation of the physicochemical and structural properties and the sensory characteristics of meat analogues prepared with various non-animal based liquid additives. Food Secur 9:461
Wiebe MG (2004) Quorntm Myco-protein-overview of a successful fungal product. Mycologist 18:17–20
Wild F, Czerny M, Janssen AM, Kole A, Zunabovic M, Domig KJ (2014) The evolution of a plant-based alternative to meat. Agro Food Industry Hi Tech 25:45–49
Winans KS, Macadam-Somer I, Kendall A, Geyer R, Marvinney E (2020) Life cycle assessment of California unsweetened almond Milk. Int J Life Cycle Assess 25:577–587
Wongkanya R, Chuysinuan P, Pengsuk C, Techasakul S, Lirdprapamongkol K, Svasti J, Nooeaid P (2017) Electrospinning of alginate/soy protein isolated nanofibers and their release characteristics for biomedical applications. J Sci Adv Mater Dev 2:309–316
Woodward SA (1990) Egg protein gels. In: Food gels. Springer
Xie Y, Cai L, Zhao D, Liu H, Xu X, Zhou G, Li C (2022) Real meat and plant-based meat analogues have different in vitro protein digestibility properties. Food Chem 387:132917
Xing H, Zheng W, Li B, Liu Z, Zhang Y (2019) Water footprint assessment of eggs in a parent-stock layer breeder farm. Water 11:2546
Xiong Y (2004) Muscle Protein. In: Proteins in food processing. Woodhead Publishing Limited, London
Xu X, Sharma P, Shu S, Lin T-S, Ciais P, Tubiello FN, Smith P, Campbell N, Jain AK (2021) Global greenhouse gas emissions from animal-based foods are twice those of plant-based foods. Nat Food 2:724–732
Yadav, D. N., Bansal, S., Jaiswal, A. K. & Singh, R. 2017. Plant based dairy analogues: An emerging food
Yang JH, Olsen RA (1970) Meat analogs having the fiber structure of meat
Yang M, Li N, Tong L, Fan B, Wang L, Wang F, Liu L (2021) Comparison of physicochemical properties and volatile flavor compounds of pea protein and mung bean protein-based yogurt. LWT 152:112390
Yang Y, Zheng Y, Ma W, Zhang Y, Sun C, Fang Y (2023) Meat and plant-based meat analogs: nutritional profile and in vitro digestion comparison. Food Hydrocoll 143:108886
Yépez A, Russo P, Spano G, Khomenko I, Biasioli F, Capozzi V, Aznar R (2019) In situ riboflavin fortification of different kefir-like cereal-based beverages using selected Andean lab strains. Food Microbiol 77:61–68

Yuliarti O, Kovis TJK, Yi NJ (2021) Structuring the meat analogue by using plant-based derived composites. J Food Eng 288:110138

Zaaboul F, Raza H, Cao C, Yuanfa L (2019) The impact of roasting, high pressure homogenization and sterilization on peanut milk and its oil bodies. Food Chem 280:270–277

Zahari I, Ferawati F, Helstad A, Ahlström C, Östbring K, Rayner M, Purhagen JK (2020) Development of high-moisture meat analogues with hemp and soy protein using extrusion cooking. Food Secur 9:772

Zander P, Amjath-Babu T, Preissel S, Reckling M, Bues A, Schläfke N, Kuhlman T, Bachinger J, Uthes S, Stoddard F (2016) Grain legume decline and potential recovery in European agriculture: a review. Agron Sustain Dev 36:1–20

Zhang P, Xu Z, Fan W, Ren J, Liu R, Dong X (2019) Structure dynamics and risk assessment of water-energy-food nexus: a water footprint approach. Sustain For 11:1187

Zhang J, Chen Q, Kaplan DL, Wang Q (2022) High-moisture extruded protein fiber formation toward plant-based meat substitutes applications: science, technology, and prospect. Trends Food Sci Technol 128:202–216

Zhao Y, Feng F, Yang Y, Xiong C, Xu M, Tu Y (2021) Gelation behavior of egg yolk under physical and chemical induction: a review. Food Chem 355:129569

Zheng Y, Sun M, Sun X, Sun C, Fang Y (2023) The role of protein concentration in heat-induced particulation of soy proteins at different pHs: structure and functional properties. Food Front 4:955–965

Zhou H, Hu Y, Tan Y, Zhang Z, Mcclements DJ (2021) Digestibility and gastrointestinal fate of meat versus plant-based meat analogs: an in vitro comparison. Food Chem 364:130439

Zhuang D, Abbas J, Al-Sulaiti K, Fahlevi M, Aljuaid M, Saniuk S (2022) Land-use and food security in energy transition: role of food supply. Front Sustain Food Syst 6

# Chapter 18
# The Future of Meat Analogues

**Arnout R. H. Fischer**

## 18.1 Introduction

Humans need food for energy and for building up and maintaining the body and brains. This requires food that provides relevant nutrients, some in large quantities, macronutrients, some in smaller quantities. Of the three macronutrients, carbohydrates, fat, and proteins, proteins have a pivotal role in the creation and maintenance of nerve and muscle tissue. A good complement of proteins and their composite amino acids is therefore essential for humans to thrive. While meat has a particularly good balance of amino acids, there are many problems with the sustainability of meat. Perhaps most prominently the inefficiency in converting feed into animal protein. But animal husbandry also can incur problems related to animal welfare, soil and surface water contamination with manure and ammonia, and greenhouse gas emissions. Given the attention to UN sustainability goals and the growing concern about environmental degradation and climate change, it is not surprising the market for meat alternatives has grown over the last decades (Dagevos and Verbeke 2022).

The current market and debate about alternatives to meat is characterised by a focus on meat analogues (Hoek et al. 2011). It is however unclear whether, and to what extent meat analogues will keep the centre stage. In the current chapter I will consider the future for meat analogues from the point of view of consumer sciences.

Before engaging in further analysis, it is important to set down some working definitions. Meat analogue and similar terms are not very strictly defined and are sometimes used interchangeably. In this chapter I take the approach to use meat analogue in a strict sense. Hence I use a working definition of *meat analogues* as

A. R. H. Fischer (✉)
Marketing and Consumer Behaviour Group, Wageningen University,
Wageningen, The Netherlands
e-mail: arnout.fischer@wur.nl

 
Ö. P. CAN et al. (eds.), *Food Analogues: Emerging Methods and Challenges*,
https://doi.org/10.1007/978-3-031-69872-9_18

products that (seamlessly) take over the role of a specific meat in a meal, are produced to mimic that meat in shape, colour, texture and preparation, have a similar nutritional contribution as the meat, and often have a name referring to the replaced meat. Typical examples are the Impossible Burger®, which tries to copy a beef patty including its bloody juice, and products marketed by Beyond Meat (e.g. Beyond Sausage), and Vegetarian Bucher (e.g. good Karma Shawarma). These are also the products that are sometimes labelled as "fake meat". I use the term *meat substitute* to refer to a broader group of products that take up the role of meat, but do not exactly mimic meat. This generally includes processed products like bean or lentil burgers, but depending on application can also refer to tofu used in curries to replace meat, falafel replacing doner or shawarma in a pita, or similar. I reserve the even broader term *meat alternative* for all products that replace meat even without explicitly claiming to substitute a meat. These include the use of beans in a chili or similar. Using distinct terms, suggests a clear delineation, however, in practice the boundaries are more blurred. Tofu in a curry may take the role of (e.g.,) chicken and hence be a meat substitute, while in a specific traditional tofu dish it should be classified as meat alternative. Jackfruit lacks protein, but if used as pulled pork substitute can be considered a meat substitute or even analogue. In line with the research on experience goods (Gilovich et al. 2015a, b), the proposed working definitions should be applied to how consumers ***use*** these products rather than being seen as an intrinsic property of the products themselves.

Given different approaches to meat alternatives, in this chapter I address the following question for the future of meat analogues.

How did the current popularity of meat analogues emerge?

The currently central position of meat analogues did not emerge in a vacuum, yet followed on from earlier societal and technological trends related to meat alternatives. To consider the future of meat analogues a sketch of the timeline in which they emerged is needed to provide the required context.

As meat analogues did not emerge in a vacuum they also do not exist in a retail environment without competition. This raises the question how this competition may influence the future of meat analogues. To answer that question I will provide a brief overview of that competition.

Building on these two introductory reflections I will then question why meat analogues are successful from a consumer point of view and what potential pitfalls they face. After which I will outline some potential futures of meat analogues and write a concluding section.

## 18.2 How We Got To Today

Given that humans are omnivores, we can access and utilise, but also need, a broad range of food stuffs to achieve a balanced diet. Meat, dairy and eggs are prime providers of high-quality proteins. Given the scarcity of food in most of human (pre) history, and the effort required to gain access to these high quality proteins it is not strange that humans tend to adopt diets with as large a proportion of animal protein as feasible. The historically high social status associated from especially meat rich diets further strengthens such drive towards meat consumption (Fischer et al. 2023).

Animal proteins have not always been dominant in human nutrition, and if fact were not dominant for most of the human population until fairly recently. Historically most proteins in human nutrition came from legumes and cereals (Grigg 1995a), and this is still the case in parts of the world (Grigg 1995b). Especially for the less affluent classes, who could not afford to consume meat more than once a week, if at all, legumes and cereals were the main source of proteins well into the twentieth century. The sources of staple proteins could be minimally processed lentils, chickpeas, beans or other protein rich plants (Fig. 18.1). In part such proteins remain important as they were and still are consumed in foods like bread (cereals) or spreads like hummus (chickpea) which consumers do not consume for their protein contents. In addition, for centuries, processed meat substitutes have been developed to take the place of meat in the diet. These include Indonesian tempeh. Tempeh is a fermented cake of soybeans, the history of which might trace back to the Middle Ages; although like the other discussed traditional protein sources, actual date of first production is not well documented. Tempeh has been a cheap source of nutrition and is still considered the poor men's beef in Indonesia. Other examples that probably stem from religious reasons to reduce the eating of meat, are the originally Chinese tofu, a curd of soy milk, China originating seitan, a structure from the wheat gluten after washing out starch, and falafel, chickpea-based balls or patties presumably originating in Egypt. All of these trace back to the Middle Ages or even antiquity. Many of these and other alternatives found their origin as a product to be used in festivals or as alternative to meat for religious reasons (be it permanently as in Hinduism or temporarily as in fasting for Ramadan).

In the twentieth century western industrialised countries were among the first where meat became the staple protein foods for the majority of the population. Until the mid-twentieth century meat had been a luxury for large proportions of the population, sometimes limited to a small portion once a week (generally on Sunday). What we would now call a flexitarian diet where most of the proteins were derived from legumes and grains was commonplace, but out of financial necessity rather than choice. It is no surprise that with the increasing welfare and industrialised agriculture in the mid-twentieth century, meat increasingly replaced legumes and cereals as the main source of protein in rich industrial countries. Especially after WWII, with increased power of unions, increasing wages, and due to agricultural advances in animal rearing and fodder production, the prevalence of meat in the daily nutrition increased in both the US and Western Europe (Grigg 1995a). This resulted in better nutrition for large groups of the population, in particular the lower middle classes and blue-collar workers. Nevertheless already from the mid-nineteenth century onwards vegetarianism as a lifestyle of moderation and a lifestyle exhibiting respect for animal welfare started to emerge among the higher classes. Especially in the United Kingdom and to a smaller extent elsewhere vegetarian initiatives as a

**Fig. 18.1** Top row: lentils, beans (unprocessed meat alternatives), and tofu (China second century?) (traditional processed meat substitute). Second row: falafel (Egypt/Levant fourth century?), seitan (China sixth century?) and tempeh (Indonesia between twelfth and eighteenth century?) (traditional processed meat substitute). Third row: textured vegetable protein (US, 1960s), Quorn® chunks (UK, 1985), Tofurky (US, 1995) (modern processed meat substitutes), Bottom row: Impossible burger® (US, 2016), Beyond Sausage® (US, 2018), Chick'N nuggets (US, 2020) (modern processed meat analogues). (All images from Wikipedia under creative commons licenses: 3 types of lentils by Justin Cormack (User:Justinc) accessed 7-nov-2023 https://commons.wikimedia.org/wiki/File:3_types_of_lentil.jpg; Ordinary Chickpeas in a ceramic bowl, User:AlixSaz accessed 7-nov-2023 https://commons.wikimedia.org/wiki/File:Ordinary_chickpeas_in_a_ceramic_bowl.jpg; BOSTempeh User:Snigdha4work accessed 7-nov-2023 https://commons.wikimedia.org/wiki/File:BOStempeh.jpg; Falafel-Jordanie.jpg, User:Ji-Elle Accessed 7 Nov 2023 from https://commons.wikimedia.org/wiki/File:Falafel-Jordanie.jpg: Japanese Tofu 001.jpg User:Ocdp Accessed 7 Nov 2023 from https://commons.wikimedia.org/wiki/File:Japanese_tofu_001.jpg: Seitan (140489989.jpeg) J doll Accessed 7 Nov 2023 from https://commons.wikimedia.org/wiki/File:Seitan_(140489989).jpeg: Textured vegetable protein.jpg User:NTK Accessed 7 Nov 2023 from https://commons.wikimedia.org/wiki/File:Textured_vegetable_protein.jpg: Quorn Stukjes (1) User:.Anja. Accessed 7 Nov 2023 from https://commons.wikimedia.org/wiki/File:Quorn_Stukjes_(1).jpg; Tofurky.jpg Aine Accessed 7 Nov 2023 from https://commons.wikimedia.org/wiki/File:Tofurky.jpg; Impossible_Burger_free_sample.jpg User:Dllu Accessed 7 Nov 2023 from https://commons.wikimedia.org/wiki/File:Impossible_Burger_free_sample.jpg; Beyond_Sausage_2.jpg User:Mx. Granger Accessed 7 Nov 2023 from https://commons.wikimedia.org/wiki/File:Beyond_Sausage_2.jpg; Vegan_ chicken _nugget.jpg incommunicado Accessed 7 Nov 2023 from https://commons.wikimedia.org/wiki/File:Vegan_%22chicken%22_nugget.jpg)

moral or even spiritual lifestyle choice were already present in the early twentieth century.

The agricultural revolution of the mid-twentieth century resulted in the first non-meat products developed in Western industrial agriculture being introduced into the market. Textured vegetable protein, lightweight, dried soy chunks with very long shelf life, were developed to maximise profits of soy production. These long lasting chunks were embraced by the food industry as intermediate product for many applications. As a mainstream product for end-users it did not take off. This may be due to the complexity of preparation (it needs to be soaked) and lack of flavour. In addition, at the time of its launch vegetarianism was an extremely small niche market. Nevertheless, it achieved and still has some use in specific consumer niches. For example, textured vegetable protein seems especially appealing to backpackers given its low weight and long shelf life without specific storage conditions. Next to introduction of textured vegetable protein, Small scale initiatives catering to the needs of vegetarians started to emerge more broadly as well in the 1960s; brought upon by the contemporary counterculture (Bloyd-Peshkin 2021). Asian alternatives such as seitan, tofu and tempeh became increasingly available across the Western world to cater for these needs. Yet the use of meat alternatives largely remained limited to the relatively small counterculture segment of the population.

Quorn® can be considered as one of the first modern, high tech meat substitutes. Quorn® is a mycoprotein based product, with mycoproteins grown in sugary solutions and brought together in meat like chunks. In the 1980s it was approved by the authorities and entered the market. Due to its meat like structure Quorn® became one of the first product lines that made a lasting impact. Although the mycoprotein is the main source of protein, the substrate in which it is grown requires high quality ingredient, and to structure the final product egg white is used. These processing steps make Quorn® a vegetarian but not vegan product. Nevertheless, the texture and flavour of Quorn® made it easier to use and to prepare a tasty meal, compared to many previous alternatives. Thus, it is sometimes claimed that Quorn® has prepared the market and consumers for the modern meat alternatives, culminating in the current meat analogues.

Following Quorn® more products were introduced. This resulted in the emergence of an initially small market segment from the early 1990s onwards where the emphasis increasingly was on ease of use and fitting in an established meal context. In retail, tofu shelves were expanded with Quorn chunks and increasingly with burger patty style alternatives, but also ready to eat preparations of traditional meat alternatives such as the tofu based Tofurkey roast. Branded vegetarian product lines like those of Vivera® (Nl, 1990) and Garden Gourmet® (Israel, 1986) were created. These vegetable, soy, or other products were easy to prepare and fit in the diet. However, shape and texture were far removed from that of meat. While taste and quality of these products gradually improved, they remained targeted at the small vegetarian market.

In 2005 Dutch dairy brand Campina (now FrieslandCampina) launched Valess® a dairy protein-based product structured with seaweed, and shaped like bread crusted escalopes. Although not resembling meat per se, Valess® had a firmer texture than most other products on the market at the time of its introduction. But its marketing was also different. One of the innovative features of the brand was that it was targeted at meat eaters who would like to try something else for once, rather than vegetarians. With that it became one of the first products that was targeted at a larger audience than vegetarians and these products may have contributed to the fledgling flexitarian lifestyle. In addition, the Valess® line was special as it is one of the first meat alternatives that was developed, produced and marketed by a food multinational, as Friesland Campina—with a multibillion Euro yearly turnover—is among the top ten global dairy companies.

By the end of the 2000s, several companies started to develop the meat analogues, meat alternatives that did not only provide proteins, taste and ease of preparation and fitting with the diet, but actually mimic meat as closely as possible on as many attributes as feasible. Beyond meat® was founded in the US in 2009, the Vegetarian Butcher™ opened shop in the Netherlands in 2010 and Impossible Burger® was launched in the US in 2011. These companies were among the first that fully focussed on creating products that resemble a wide range of different meat sources. These include blood like juices such as soy leghaemoglobin, which Impossible Burger gains from GM yeast lines. By copying meat closely, it became easy for consumers to substitute meat in the meal as they now could retain the same recipes and experiences. This strategy is however not without its risks, as consumers may have unrealistically high expectations of the new products. Actual meat may be relatively forgiving for the use of limited fat in frying, as the fat and moisture of the meat itself may support preparation. This facilitates consumers to prepare meat and in additions the time of heating meat is not that precise and the preparation of meat provides well-known cues as to how well done the meat is during cooking. These new products are sometimes less forgiving and hard to keep track off during preparation. They may get tough and dry more readily, may stick more easily to pans. When mimicking high quality meats like steak, where consumption preference for consumers may vary between blue (only seared on the outside) to well-done (cooked through and through), products are unlikely to be able to mimic the entire range of preparations suggested by the reference to the original meat product.

Nevertheless, the assortment of meat analogues rapidly expanded which created attention from major companies. In 2017, Nestle bought Garden Gourmet. In 2018, Unilever purchased the Vegetarian butcher. Morningstar Farms, a company producing frozen soy foods, was purchased by Kellogg's in 1999 and started to make refrigerated chicken analogues by 2020. Not only did major food companies buy promising meat analogue companies, but major catering outlets also started to adopt these products. Impossible Burger made a deal with the fast food chain Burger King. Burger King introduced the Impossible Whopper in the US in 2019. Given the use of genetically modified yeast to create the blood like leghaemoglobin juices, they could not use their patties in Europe. Unilever with its newly acquired Vegetarian Butcher brand was chosen to provide the patties for the Rebel Whopper in Europe. Although McDonalds, as major fast-food chain, already had some vegetarian options (McVeggie and McVegan), in 2021 it started to sell Beyond Meat products mimicking meat burgers more closely. These McPlant burgers achieved some success. Large retail organisations started to create store brands in substantial quantities by the late 2010s. Besides burgers, increasingly many products mimicking meat arrived in the market, often with a names referring to the original meat product. Examples are Beyond Sausage by Beyond meat, Chick'n Nuggets by Morningstar, or the great Escalope by the Vegetarian Butcher. In addition, other processed products that do not fully mimic meat, but take a similar function in the meal started to be more generally available. Such non-analogues include the more general availability of falafel, a range of burgers, stir fry pieces and other meal components. Among these products several explicitly mention their plant-based origin such as lentil burgers (where lentils as legume are the main protein source) and redbeetroot burger (where the protein contents can be doubted). This approach aligns with current trends of consumers that demand more transparency regarding the products they buy, as well as the emergence of plant-based diets, supported by the explicit mentioning of the plant origin of the burger.

Meat producers might have become worried about the success of meat analogues, as following a ban on the use of dairy names for plant based alternatives in 2017 (Verband Sozialer Wettbewerb eV v TofuTown.com GmbH (TofuTown), 2017), they started looking for similar bans on names for meat alternatives. Such restrictions were adopted in Italy and early 2024 in France[1] for some meat names. The case to ban use of meat names at EU level seems however considerably more complex than for dairy given the wide variety of terms used for meat products.

While so far the emergence of meat analogues appears to be a success story, there are also some concerns. In public perception these alternatives have long been seen as expensive compared to meat. While recently meat prices have been rising substantially, and the emergence of store brands of meat analogues resulted in prices for meat analogues to become equal to or lower then these of meat alternatives, this perception remains hard to shed (van der Meer et al. in preparation). In spite of initial price differences, meat analogues experienced a decade of rapid growth. However, growth has levelled off since 2022, at least in the Netherlands, which is among the countries with the strongest initial growth (Dagevos and Verbeke 2022). At the same time, the demand of meat has only marginally decreased since the introduction of meat analogues, which suggests that consumers may compensate flexitarianism with larger portions of meat on days they eat meat, or it may indicate a general trend towards eating more and larger portions of proteins. Whatever the cause, the combined development of stalled growth of analogues and a maintained high level of meat consumption limits the ambition of meat analogues to contribute to sustainable nutrition. Another reason for concern is that companies producing meat analogues target their products at a consumer segment that is likely to favour fresh, natural and unprocessed products. In contrast to the demands of these consumers, meat analogues are amongst the most heavily processed products in retail. Nevertheless, to date consumer purchase meat analogues with hardly any reference

[1] https://www.reuters.com/world/europe/france-issues-new-ban-meat-names-plant-protein-placate-farmers-2024-02-27/

to their ultra-processed nature. This makes for a paradoxical situation where those in favour of natural unprocessed food purchase ultra-processed products (Elzerman 2022; van der Meer et al. in preparation). This paradox may cause unpredictable changes in consumer opinion towards meat analogies if this latent ambivalence becomes widely experienced. The currently emerging trend (Innova Marketing insights 2023) to adopt plant based food in restaurants, catering and the creation of plant based brands in retail (Olthuis 2023) recognises the demand for a more natural portfolio of foods, and does not focus that much on mimicking meat, but more on capitalising on the plant based source. It may be that further growth in market share of meat alternatives will come from plant based diets or other emerging initiatives that complement the meat analogue portfolio. Nevertheless, the levelling of demand for meat analogues is a worrying observation, for all meat alternatives as it may suggest that the market for meat alternatives is saturated in which case competition between different meat analogues, substitutes and alternatives rather than against animal based products will be the main market challenge for meat analogues.

## 18.3 Competing Initiatives to Meat Analogues

Meat analogues are only one among a larger portfolio of protein sources. Consumers respond to some extent differently to these products (Onwezen et al. 2021; Siddiqui et al. 2022). If considering the potentially relevant consumer dimensions of how closely a product mimics meat, how processed it is, to what extent the source is fully plant based or whether there is some animal protein involved, and whether the original protein source is a new and unexpected, one might expect specific groupings of similar products on these dimensions that indicate successful groups products. The current range of initiatives does not show such clear patterns (Fischer et al. 2023), at least not yet. This may either indicate there is a lot of freedom in developing new alternatives, or that the market has not yet identified the most profitable solution, which may or may not be meat analogues. Nevertheless, given this plethora of alternatives, the future of meat analogues cannot be seen in isolation of other meat alternatives.

In the same 15 years between 2010 and 2024 that meat analogues became a dominant force among meat alternatives, much attention has also been paid to other sources of high-quality protein.

The alternative protein source that raised probably most interest was human consumption of insects (Van Huis et al. 2013). Originally classified as novel foods, the EU has recently started to approve insects on the market, grasshoppers and mealworms in 2021, and crickets and another species of mealworm in 2023.[2] Nevertheless, lasting feelings of disgust (Verneau et al. 2016; Wassmann et al. 2021), a willingness to try insects out of curiosity, without having the intention to adopt insect in the

[2] https://food.ec.europa.eu/safety/novel-food/authorisations/approval-insect-novel-food_en

regular diet (Tan et al. 2016), unclear marketing strategy where insect burgers were placed in the vegetarian shelf, with lower sensory quality and higher price than competing plant based products (House 2016), and the lack of product lines that brought superior properties to consumers caused by their insect origin (Fischer 2021) have resulted in failed introduction attempts. Together with strict food safety laws, and long-lasting regulatory approval timelines this has likely contributed to a general slowdown of the development of insects for human consumption. At the time of writing this chapter, most potential of insects for human consumption is seen in processed protein powders (Dagevos 2021), which would strongly reduce disgust response (Gmuer et al. 2016). However, given the high price of processed insect protein, use of purified insect protein may not be feasible in the competition with other alternative proteins. Much of the effort of insect breeders is now directed at the production of insects as animal feed, which has less strict food safety control, hardly any consumer issues (Verbeke et al. 2015b) and in the case of pet foods a promisingly high willingness to pay from pet owners without evoking the same levels of disgust (Villanueva et al. 2023).

Another alternative protein source is cultured meat. Hailed in the early 2010s with the claim it would be the product that would replace meat, it gained considerable attention. The first hamburger made of cultured meat was consumed in 2013 (Post 2012), but with an estimated cost of about €250,000 this was a proof of principle presentation rather than a production prototype. By the end of 2023, the development of cultured meat was still ongoing. The technological challenges to develop sufficiently large bioreactors and to produce large quantities of sustainable stem-cell growth-mediums at a reasonable price was not sufficiently solved for mass-market products to be sold at anywhere near reasonable prices for daily consumption. Transferring the required technologies from the medical field where most these technologies originated, where higher costs and small volumes are common, and where extremely high safety levels are required, to the cost competitive field of food production has been a barrier in this regard. In addition, by the end of 2023 cultured meat is still classified as novel food and can hence not be sold in the European market. Small scale samples are on the market in Singapore but given the quality of the product compared to the price, these seem more curiosity appealing products that one would try once, rather than products with the potential for a lasting market of repeated purchases. Besides the production and marketing challenges, consumer opinions also suffer from neophobia, a tendency to avoid new foods (Bryant and Barnett 2018; Verbeke et al. 2015a). In addition, evidence shows that consumer find it hard to classify cultured meat as meat, since no animal was killed, nor as a meat alternatives as it is made from animal cells (Bekker et al. 2017b, 2021). To allow the public to make sense of cultured meat, it has even been argued this may require partial societal redefinition of what is meant with meat (van der Weele and Driessen 2019). On the positive side, consumers readily relate cultured meat to sustainability (Bekker et al. 2017a). Overall, at the moment of writing cultured meat is technically, legally and societally not ready to be a competitor for meat analogues. Once these barriers are removed, the resulting product may be able to mimic meat even better than the analogues, and find a place in the markets, but that is for the future to show.

Next to insects and cultured meat, other sources of protein have also received attention. Algae where both micro algae (already commonplace in sports schools as supplements), and macro algae, also known as seaweeds, are seen as promising protein sources. While so far acceptance as functional supplement in for example protein shakes aimed at sport performance, seems to be high, consideration as food ingredient is met with less enthusiasm (Van der Stricht et al. 2024). Duckweed, another potential protein source, is more favourably considered as a complement or salad than as a meat alternative or protein source by consumers (de Beukelaar et al. 2019). Mycelium, "the root system of mushrooms", seems a promising way to turn difficult to use rest streams into protein, but this idea is only at the start of its development (Scholtmeijer et al. 2023; Fischer and Hilboesen 2025) with challenges in separating the protein from its substrates and which processing is needed to make it a valuable food ingredient. Pulses and legumes are also receiving some renewed attention, either processed into bean or lentil burgers or as ingredient, yet their market share seems to remain more or less stable.

Interestingly, plant-based trends are now more broadly promoting vegetable-based meals, with no clear protein ingredient. This trend relates to the use of jackfruit among vegans and vegetarians for its pulled pork like structure, although jackfruit has no relevant protein contents. Given the surplus of proteins in most Western diet this trend is not immediately worrisome, although for some strict adherents of plant-based diets, for example those who do not consume enough legumes and cereals, diets without clear protein ingredients may lead to some problems in the future.

## 18.4 Consumer Approaches on Meat Analogues

Meat analogues are targeted at vegetarian, vegan, but above all flexitarian consumers. Flexitarianism is a relatively new trend, where consumers deliberately refrain from eating meat a number of times each week (Dagevos 2015; Dagevos and Verbeke 2022). Compared to vegans and vegetarians who are generally more consciously considering food choices, flexitarians may be the consumer group best supported by products that replace existing meat products without the need to change any food practices or routines. To appease flexitarians, meat analogues have to conform to their needs, and we should therefore consider food choice motives of consumers.

### *18.4.1 Food Choice Motives*

Among the most used classification of consumer needs are the motives that are measured in the food choice questionnaire (Steptoe et al. 1995). The food choice questionnaire measures motivations of individual consumers on nine dimensions that are generally found to be important for consumers when they choose specific foods. The food choice questionnaire dimensions are found to be the same in many different countries (e.g. Markovina et al. 2015), suggesting they pick up on fundamental human motives. The nine dimensions comprise of: (1) Health of a food in terms of nutrients, (2) Mood in terms of cheering up consumers and helping to relax, (3) Convenience in terms of ease of preparation and purchase availability, (4) Sensory appeals in smell, taste, texture and looks, (5) Naturalness of contents in terms of ingredients, (6) Price not being too high, (7) Weight control as in being low in calories and fat, (8) Familiarity compared to regular meal, and (9) Ethical concern, both social and environmental. Reflecting the developing societal attention for sustainability, further developments of the food choice questionnaire suggested the subdivision of Ethical concerns into animal welfare, environmental sustainability and social justice (Onwezen et al. 2019), while it was later found that adding a local and seasonal dimension suffices to complement the original ethical dimension and made even better sense than splitting the ethical concerns dimension into three sub-dimensions (Verain et al. 2021).

### *18.4.2 Product Attributes*

For consumers to fulfil food choice motives, a product should help consumers to achieve their motives. To help consumers act, a product shows cues, observable properties that indicate attributes or underlying functionality that should align with consumer perceptions. In this light, consumer scientists distinguish between different attributes that make it easier, or harder for consumers to decide whether a product fulfils the function. The easiest to distinguish for consumer are search attributes. Search attributes can be observed without using the product and as such allow a consumer to find and choose a product in the store. Colour, shape and texture of products or their packaging are such search attributes. When using a product, consumers perceive experience attributes. Experience attributes are those consumers experience when handling, preparing and consuming a product (Nelson 1970). Taste, for example, is an experience attribute that cannot be assessed in the shop, but is obviously experienced when consuming the food. Some attributes are, however, neither searchable nor can be directly experienced. For example, the use of pesticides (or other environmental sustainability attributes) cannot be seen when choosing, nor experienced when using a product. Such attributes are called credence attributes as the consumer has to believe claims made by producers (Darby and Karni 1973; Grunert 1997). While, arguably, many credence attributes, such as pesticide use, can be observed through formal testing in a specialised laboratory, some attributes, such as labour conditions (social sustainability) may not even be established in a laboratory context. It has been argued that such extreme cases of credence attributes warrant an additional classification labelled Potemkin attribute (Jahn et al. 2005).

Another distinction is made is between extrinsic and intrinsic food attributes. Intrinsic attributes are those that are integral to a product and cannot be changed without changing the physical make-up of a product itself (Olson 1978; Steenkamp 1990; Zeithaml 1988). Typical intrinsic attributes are sensory aspects, but also naturalness and sustainability are intrinsic as they cannot be changed without changing the nature of the product. Extrinsic attributes on the other hand are attached to a product but can be changed without changing the nature of the product itself. Prices, brands, packaging are hence typical extrinsic attributes. In addition, claims attached to a brand or package should also be considered extrinsic as the cues conveying the message can be swapped without changing the contents of the product. This makes for a tricky situation where almost all credence attributes like health, sustainability etc., that in themselves are intrinsic properties can only be communicated by an extrinsic cue. Such extrinsic cues can be changed, falsified or otherwise tampered with, without the consumer having any control on the issue.

### *18.4.3 Relating Food Choice Motives and Attributes of Meat Analogues*

Health is among the most important motives. Healthiness of a product is an intrinsic attribute, which while experienced in the long run, can at the moment of consumption only be considered a credence attribute (see Table 18.1 for an overview). The extent to which consumers attribute healthiness to meat analogues differs broadly (Onwezen et al. 2021). Plant-based products are generally considered healthier than those stemming from other sources (Hartmann et al. 2022; Michel et al. 2021). Although Hartmann and colleagues also indicate that in many circumstances and countries perceived healthiness of meat analogues is at best equal to, and in many cases lower than that of meat. This suggests that consumer perception on healthiness of meat substitutes is not straightforward. Something that may have been induced by naturalness-health inferences (van der Meer et al. in preparation), or naturalness-acceptance relations (Fischer and Hilboesen 2025). In addition, the relatively high salt contents of many meat substitutes (Alessandrini et al. 2021) may suggest these products may indeed not be as healthy as sometimes implied.

The motive mood does not truly relate to an attribute that is intrinsic (embedded in the product) and can only be cued by extrinsic claims, which are part of the branding of products. This motive does however lead to immediate experience of being in a positive mood when consuming. Meat analogues in part fulfil this aim by

**Table 18.1** Relevance of different attributes to food choice motives of consumers

| Food choice motive | Search-experience-credence or potemkin attribute | Intrinsic or extrinsic attribute | Strategy for meat analogues |
|---|---|---|---|
| Health of a food in terms of nutrients | Credence (long term experience) | Intrinsic (but claims are extrinsic) | Implied claims through e.g., nutri-score optimization. But e.g., high salt contents and highly processed shedding doubt |
| Mood in terms of cheering up consumers and helping to relax | Experience | Extrinsic claims can be made | Claims through branding. Not strongly developed as strategy. Comparable with non-analogues |
| Sensory appeals in smell, taste, texture and looks | Experience | Intrinsic (but claims may in part be extrinsic) | Key focus of food technology for analogues to optimise and mimic meat as much as possible. |
| Convenience in terms of ease of preparation and purchase availability | Experience | Intrinsic (but claims may in part be extrinsic) | Key focus of food technology for analogues to harmonise with meat and simplify as much as possible. |
| Naturalness of contents in terms of ingredients | Credence | Intrinsic (but claims are extrinsic) | Claims through plant-based origin. Not strongly developed as strategy. Goes against ultra-processed properties of these products making for a paradox. |
| Price not being too high | Search | Extrinsic | Used to be at price premium, due to volumes and retail strategy. Price premium is reducing but perception remains negative. |
| Weight control being low in calories and fat | Credence | Intrinsic (but claims are extrinsic) | No specific claims made. |
| Familiarity compared to regular meal | Experience | Extrinsic claims can be made | Key focus of food technology for analogues to make it take the role and appearance as meat alternative as much as possible. |
| Ethical concern (general sustainability) | Credence | Intrinsic (but claims are extrinsic) | Implied claims aimed at environmental sustainability and animal welfare. |
| Ethical concern (local and seasonal) | Credence | Intrinsic (but claims are extrinsic) | Hardly claimed |

Note: This table was created by the author for the purpose of this chapter

providing the positive experience of the sensory properties of meat and being sustainably, but this may be less convincing compared to less processed alternatives.

Sensory appeal is an experience linking to intrinsic properties of a product. Such experience attributes can only be guessed at by consumers in retail, based on extrinsic cues such as serving suggestions. Meat analogues have invested heavily in creating products with similar look, feel and texture as the meats they mimic. Although they have achieved much, particularly with juicy burgers, other products have been harder to create and have been withdrawn from the market—probably after disappointment for consumers and subsequent disappointing (repeat) sales.

Convenience is another issue that is an experience attribute related to intrinsic product properties communicated through extrinsic cues. Convenience is one of the key approaches of meat analogues as meat analogues tap into existing knowledge about meal composition, allow replacement of meat with a close alternative, and the use of existing food preparation routines. Convenience reflects to ease of use, which is generally accepted as an important element for the adoption of new technologies (Davis 1989; Ronteltap et al. 2007, 2011; Venkatesh et al. 2012), and to compatibility, which is one of the predictors of diffusion of innovations (Rogers 1962/1995). Compatibility does, however, also require that the alternative behaves similar during preparation and has for example similar resilience to overcooking and similar sticking behaviour to grills as real meat, which is not always the case.

Naturalness of meat analogues is an intrinsic credence attribute. The usually plant based origin of meat analogues appeals to naturalness as a positive heuristic (Siegrist and Hartmann 2020). The production and texturing of meat analogues requires purification of ingredients, considerable amounts of flavouring and preservatives to be added, as well as mechanical structuring of the products. This makes the naturalness of these products rather ambiguous, with on the one hand these products being produced from natural, sustainable base materials, but on the other hand being ultra-processed to achieve their meat like substance. As the consumer segments that value naturalness and unprocessed product are also those that purchase meat analogues this makes for a paradox surrounding these products (Bastian and Loughnan 2017).

The price of meat analogues has long been substantially higher than that of conventional meat. This can be in part be understood from initially higher production costs, that lower volumes result in higher distribution costs per unit, and because of production by national brands based on higher advertising costs. In addition, retailers have added a price premium to these products as traditionally consumers purchasing these products are highly motivated to avoid meat and therefore show a low price-elasticity, i.e., low sensitivity to price for these products. Nevertheless for more widespread adoption beyond this niche group of committed consumers, lowering the price was considered an important, possibly unavoidable strategy (Denver et al. 2023; Realini et al. 2023). The emergence of store brands in the last few years, in combination with increased price of meats following the surge in energy and production prices of animal-based meat after the 2022 Russian invasion of Ukraine and subsequent boycotts of Russian gas and other products has considerably reduced the difference in price and several meat analogues are now cheaper than meat alternatives. Nevertheless, it appears that a lasting perception that meat analogues are expensive is holding consumers to some extent back.

Weight control is a motive that is generally not claimed by meat analogues, although the "A" nutri-score of most of these products in Europe suggests it caloric contents is not overly high.

The motive familiarity does not truly relate to an attribute that is either intrinsic (embedded in the product) or one that is cued by extrinsic claims, as familiarity is dependent on whatever the individual consumer if familiar with. The close mimicry of meat analogues capitalises on this experience, assuming almost all consumers are familiar with the meats that are mimicked. In addition, if products are perceived as familiar, consumer personality traits such as food neophobia, where consumer tend to shy away from new foods (Pliner and Hobden 1992) or even food technophobia, where consumer are reluctant to adopt new foods with technological origin (Cox and Evans 2008) may be less influential.

As with health, the perceptions of general sustainability related to environmental, societal and animal welfare differ between consumers, products, and countries (Hartmann et al. 2022; Michel et al. 2021; Onwezen et al. 2021). Most of the inferences consumers make appear to attribute relatively high environmental sustainability to meat analogues as being plant-based products, while there are few references to social sustainability. Animal welfare, albeit hardly mentioned explicitly, is probably important given the substitution of production animals with plants as source of protein.

### *18.4.4 Categorization*

Another take on how consumer consider meat analogues, is by considering how they categorise these products. Categorisation as a psychological process is a way in which humans create a knowledge structure that allows them to store new products in a consistent, relevant and easy to recall way (Loken 2006). In addition, by including new products in a category, consumers can make inferences about what to expect from the product (Kardes et al. 2004) even if they do not have personal experience with that specific product. The extent to which a product is analogous to meat, can for example influence sensory expectations, the convenience of preparation and other issues for a meat or meat alternative, even if these inferences are not directly cued by the product itself (van der Meer et al. in preparation). While this is highly efficient, it does come at the risk of negative experiences if the inferred quality is not achieved (Schifferstein et al. 1999), the consequent disappointment may have a long term negative effect on consumer perceptions of these products.

The way in which consumers classify a product determines to a large extent what properties are inferred. Products can be cross categorised into different categories depending on the way consumers consider them and the goal consumers envisage when considering the products. Nuts can thus be both a snack or a protein component for a meal. While this flexibility allows consumers to use products flexibly in different situations, the way they categorise the products becomes important. The role of supermarkets or other outlets in the way consumer categorise products cannot be ignored. When consumers were asked to freely classify a range of products containing animal derived and non-animal derived proteins, including meat analogues, as well as other non-protein products, consumers consistently followed the most common supermarket layout in grouping products. This seems largely independent of whether they followed omnivore, flexitarian, vegetarian or vegan diets (van der Meer et al. 2023). Participants grouped meat alternative separate from animal-based meat products, which follows the most common supermarket organisation, while they (with the exception of vegans) kept animal and non-animal dairy in one group. Nuts were consistently grouped with snacks, legumes and mushrooms with vegetables. Supermarket shelving effects are being investigated in field studies that investigate the placement of meat alternatives in the meat aisle of supermarkets. Preliminary findings suggest that this may have a small, positive effect on sales of meat alternatives but no discernible effect on meat sales (Piernas et al. 2021) suggesting that the impact of such interventions is on the one hand promising, but should also not be seen as a silver bullet to reduce meat consumption.

### 18.4.5 Skill Transfer, and Practice Theory

Much of the reasoning above follows (social) cognition approaches, where consumer decision making is considered to consist of sequence of function blocks connecting perceptions to decision making to behaviour. While this may inform how motives like convenience determine consumer choice, they are less relevant to inform how a product can be made convenient. One way to make a product convenient is by inducing analogy based learning, where knowledge from a well-known procedure is transferred to that of another product (Gentner and Holyoak 1997; Holyoak et al. 2010). This idea was expanded to consumer products, where attributes can help people who have given skills, to transfer entire schemata and routines to other products (Gregan-Paxton and John 1997). Such transfer of skill and knowledge thus allows consumers to make use of previous experiences gathered from similar situations.

From sociology, the importance of the context into which consumers acts is considered to be reflected in the (social) practices consumers engage in. Following a revised theory of practices (Schatzki 1996), this approach has gained considerable traction in sociological food science. Practice theory includes practical understanding of a product or situation in identifying the situation, to prompt it and to respond to it. Practices include knowledge of "rules" around a situation that allow or point consumers at undertaking specific action in specific situations; and provide a range of normative ends and emotions that can be pursued in a given situation, which ultimately depends on overall understanding of the situation. Practices are highly context and culture specific in outlining what can be done in a given situation. For example, Ethiopian consumers, did not consider kangaroo a source of meat as the meat of these animals is not eaten in their cultural context. But they could imagine this would be different for Australians (Bekker et al. 2017b).

Thus, analogies and transfer of knowledge can be used to design products and product communication that support how meat analogues as well as non-analogues can be used by consumers, and practice theory provides insight into how this would fit into larger (social) contexts. This allows manufacturers to create products that maximise use of existing consumer knowledge and hence create the highest possible convenience. This approach is the general approach taken by meat analogues. However, this approach also holds risks as it promises products to behave according to their analogue. Given the high context dependency of practices it may be impossible to create products that fit the practices of all consumers in a target group. As Fischer (2020) argued, this works well for products that for all practical purposes can be substituted for their originals in all situations. For example, soy and almond drinks allow for use in recipes requiring animal dairy such as pancakes and can be foamed for cappuccino or other coffee drinks. It can, however, also lead to disappointment when existing schemata and preferences do not work. For example preparing rare steak on a hot grill did not work for the (now withdrawn) first launched vegetarian steak. It turned out that such preparation practices were not foreseen as the plant-based steaks did not hold up well to a heated grill and could not mimic the structure, taste and bloodiness of rare steak. Similarly, meat alternatives that required strict timing did not allow established cooking practices where consumers worked on potatoes, vegetables and other meal components while meat is being cooked. This resulted in overcooked and tough meat replacements. Such disappointment likely occur if the meat analogue suggests too much similarity in preparation practice to existing products, while not being able to withstand the practice itself.

## 18.5 Here to Stay or Transition

At the time of writing, the increase in sales of meat analogues has decreased and seems to have arrived at a platform, while meat consumption remains omnipresent (Dagevos and Verbeke 2022). This raises the question what will happen next, and whether meat analogues will turn out to be a product group to stay or will prove to have been a temporary fashion and merely a catalyst to a shift to reduced meat eating. In other sustainability transitions, similar catalysts have been observed, for example hybrid electric-petrol cars have greatly supported the introduction of electric vehicles, in part allowing business to finetune the electrical technology and start developing the infrastructure, but also in part by allowing the public to get used to electric vehicles. Now that electric vehicles and charging stations are becoming more widespread (and policies to phase out petrol cars altogether are being developed) these hybrid cars are likely to disappear in the foreseeable future. Whether meat analogues will follow a similar path is not yet clear. I envisage several pathways into the future that seem promising.

1. The current meat analogues that mimic meat, are further developed towards even more undistinguishable copies of meat and remain the dominant approach in the

market of meat alternatives. This may be in a context where the current market of meat alternatives does not further grow for the long run and where meat analogues remain the dominant strategy. In this case the market of meat replacements will remain a niche market. But it may also be part of a scenario where increasing quality, availability and lower price of meat analogues together with a social dynamic turning against meat, allow substantial growth of the meat replacement market. In their retained dominant position, meat analogues will profit from this growth.

2. Meat analogues will be largely replaced by the next big trend. For example, less processed plant-based products. Given the concerns about the high level of processing of meat analogues this might seem an obvious scenario. However, preparing a nutritious meal from less processed plants requires skill and knowledge that not all consumers have, and the sensory properties and taste of these products may not appeal to all. The current trend of presenting less processed products, such as red-beetroot burgers, plays into this idea, but can only partially replace meat as such products only contain limited protein, making them fall short as full alternatives. Preparation of beans and lentils, and also of tofu and tempeh (being among the most ancient meat substitutes) requires cooking skills that go beyond a large group of the population.
3. Both animal meat and meat analogues will remain in the market as product group, but a portfolio of different protein alternatives will become common. While the market share of meat may decline, the replacement is partially taken over by products other than meat analogues. Intuitively this seems the most likely outcome. Current flexitarian diets, the range of different protein alternatives, and the discussion around the potential of other protein sources in consumer products such as mycelium grown in rest streams (Scholtmeijer et al. 2023), algae (Van der Stricht et al. 2024), duckweed (de Beukelaar et al. 2019), cultured meat (van der Weele and Driessen 2019), and insects (Fischer 2021). This will still require substantial changes in how society regards meat but may create a future where meat analogues retain a lasting, albeit less dominant position in the portfolio of protein products.

## 18.6 Conclusions

Claims that meat will have disappeared from the diet altogether by 2050 or even around the 2070s (50 years from now) are hard to validate. Given the current societal value that large groups of the population attribute to meat, I deem it unlikely meat will completely disappear from our diets for the foreseeable future. What we consume in our childhood and early adulthood determines our preferences and routines for large parts of our lives (e.g. Emmett and Jones 2015). While the younger generations, who will still be around in 50 years, generally report more vegetarian preferences compared to older consumers (e.g. Vergeer et al. 2020), vegetarian preferences even in these groups remains those of a small minority. They may even be indicative of a specific life stage where teenagers and adolescents adopting a vegetarian diet may revert to omnivore diets later in life (although the opposite, omnivore youngster turning vegetarian later in life is also conceivable). The current lack of increase in alternatives and lack of decrease in meat sales suggests that meat will remain an important element in our nutrition, unless dramatic interventions, such as extreme price increase of meat, occur, that go beyond marketing and consumer control.

It nevertheless remains likely there will be a demand for meat alternatives, and this demand may continue to grow. But whether these meat alternatives will be meat analogues, products based on alternative natural protein sources like duckweed, insects or mycelium, artificially created products like cultured meat or products from precision fermentation, basic and well known products such as unprocessed legumes, mushrooms, grains and nuts, yet other unforeseen products, or a portfolio of all of them is hard to foresee. Fifty year from now, meat analogues may have disappeared altogether, they may still be around as major alternative to meat, or may have become one among many types of products in a larger protein portfolio.

Whatever their role in the future, meat analogues played an important role in normalising meat reduction in the decade between 2010 and 2020, by reaching out to a large group of consumers that wanted to reduce meat intake without being fully vegetarian or vegan. In that capacity meat analogues have been a catalyser for the normalisation of meat free meals or even diets, in the consumer home, in fast food chains, and society in general. Whether their sales will pick up again after the current stagnation, whether they are here to stay at the current level, whether limited meat reduction will be all that can be achieved, whether other protein products take over the driving force in a protein transition is hard to predict. Not only product development and marketing but also larger societal dynamics towards meat eating, increasing impact of climate warming, and governmental interventions are likely to play large and unpredictable roles. Given the importance of dietary preference early in life, these dynamics may take a generation or more to be accepted, which given the current climate crisis may not be soon enough. It is not only meat consumption, but also broader use of resources such as energy, electronics, travel, and fabrics that may have met its limits and may require dramatic changes in consumer lifestyles, but how quickly and invasive these effects will force a change in lifestyle is hard to predict. To paraphrase the physicist Niels Bohr who claimed to paraphrase baseball legend Yogi Berra—it remains difficult to make predictions, especially about the future.

## References

Alessandrini R, Brown MK, Pombo-Rodrigues S, Bhageerutty S, He FJ, Macgregor GA (2021) Nutritional quality of plant-based meat products available in the UK: a cross-sectional survey [Article]. Nutrients 13(12):Article 4225. https://doi.org/10.3390/nu13124225

Bastian B, Loughnan S (2017) Resolving the meat-paradox: a motivational account of morally troublesome behavior and its maintenance. Personal Soc Psychol Rev 21(3):278–299

Bekker GA, Fischer ARH, Tobi H, van Trijp HCM (2017a) Explicit and implicit attitude toward an emerging food technology: the case of cultured meat. Appetite 108:245–254. https://doi.org/10.1016/j.appet.2016.10.002

Bekker GA, Tobi H, Fischer ARH (2017b) Meet meat: an explorative study on meat and cultured meat as seen by Chinese, Ethiopians and Dutch. Appetite 114:82–92. https://doi.org/10.1016/j.appet.2017.03.009

Bekker GA, Fischer ARH, Tobi H, van Trijp HCM (2021) Stuck in the middle with you: the role of similarity information on categorizing cultured meat. Food Qual Prefer:Article 104265. https://doi.org/10.1016/j.foodqual.2021.104265

Bloyd-Peshkin S (2021) From marginal to mainstream. In: Bloyd-Peshkin S, Whitaker C (eds) Curating culture: how twentieth-century magazines influenced America. Rowman & Littlefield, pp 121–133

Bryant C, Barnett J (2018) Consumer acceptance of cultured meat: a systematic review. Meat Sci 143:8–17. https://doi.org/10.1016/j.meatsci.2018.04.008

Cox DN, Evans G (2008) Construction and validation of a psychometric scale to measure consumers' fears of novel food technologies: the food technology neophobia scale. Food Qual Prefer 19(8):704–710. https://doi.org/10.1016/j.foodqual.2008.04.005

Dagevos H (2015) Exploring flexitarianism: meat reduction in a meat-centred food culture. In: Impact of meat consumption on health and environmental sustainability, pp 233–243. https://doi.org/10.4018/978-1-4666-9553-5.ch012

Dagevos H (2021) A literature review of consumer research on edible insects: recent evidence and new vistas from 2019 studies. J Insects Food Feed 7(3):249–259. https://doi.org/10.3920/JIFF2020.0052

Dagevos H, Verbeke W (2022) Meat consumption and flexitarianism in the Low Countries. Meat Sci 192:Article 108894. https://doi.org/10.1016/j.meatsci.2022.108894

Darby MR, Karni E (1973) Free competition and the optimal amount of fraud. J Law Econ 16(1):67–88

Davis FD (1989) Perceived usefulness, perceived ease of use, and user acceptance of information technology. MIS Q 13(3):319–340

de Beukelaar MFA, Zeinstra GG, Mes JJ, Fischer ARH (2019) Duckweed as human food. The influence of meal context and information on duckweed acceptability of Dutch consumers. Food Qual Prefer 71:76–86. https://doi.org/10.1016/j.foodqual.2018.06.005

Denver S, Nordström J, Christensen T (2023) Plant-based food—purchasing intentions, barriers and drivers among different organic consumer groups in Denmark [Article]. J Clean Prod 419:Article 138256. https://doi.org/10.1016/j.jclepro.2023.138256

Elzerman JE (2022) Consumer acceptance of meat substitutes: the roles of product, context and consumer characteristics. Wageningen University, Wageningen

Emmett PM, Jones LR (2015) Diet, growth, and obesity development throughout childhood in the Avon longitudinal study of parents and children. Nutr Rev 73(suppl_3):175–206

Fischer ARH (2020) Willekeurig consumentengedrag: Eigenlijk best logisch. In: de Zwarte IJJ, Candel JJL (eds) 10 Miljard monden. Hoe we de wereld gaan voeden in 2050. Prometheus, pp 245–251

Fischer ARH (2021) Eating insects—from acceptable to desirable consumer products [editorial]. J Insects Food Feed 7(7):1061–1063. https://doi.org/10.3920/JIFF2021.x008

Fischer ARH, Hilboesen O (2025) Consumer acceptance of mycelium as protein source. Food Q Prefer 122:105304. https://doi.org/10.1016/j.foodqual.2024.105304

Fischer ARH, Onwezen MC, van der Meer M (2023) Consumer perceptions of different protein alternatives. In: Meiselman HL, Lorenzo JM (eds) Meat and meat replacements: an interdisciplinary assessment of current status and future directions. Elsevier, Woodhead, pp 333–362. https://doi.org/10.1016/B978-0-323-85838-0.00005-5

Gentner D, Holyoak KJ (1997) Reasoning and learning by analogy – introduction. Am Psychol 52(1):32–34. https://doi.org/10.1037/0003-066x.52.1.32

Gilovich T, Kumar A, Jampol L (2015a) The beach, the bikini, and the best buy: replies to Dunn and Weidman, and to Schmitt, Brakus, and Zarantonello. J Consum Psychol 25(1):179–184. https://doi.org/10.1016/j.jcps.2014.09.002

Gilovich T, Kumar A, Jampol L (2015b) A wonderful life: experiential consumption and the pursuit of happiness. J Consum Psychol 25(1):152–165. https://doi.org/10.1016/j.jcps.2014.08.004

Gmuer A, Nuessli Guth J, Hartmann C, Siegrist M (2016) Effects of the degree of processing of insect ingredients in snacks on expected emotional experiences and willingness to eat. Food Qual Prefer 54:117–127. https://doi.org/10.1016/j.foodqual.2016.07.003

Gregan-Paxton J, John DR (1997) Consumer learning by analogy: a model of internal knowledge transfer. J Consum Res 24(3):266–284. https://doi.org/10.1086/209509

Grigg D (1995a) The nutritional transition in Western Europe. J Hist Geogr 21(3):247–261. https://doi.org/10.1006/jhge.1995.0018

Grigg D (1995b) The pattern of world protein consumption. Geoforum 26(1):1–17. https://doi.org/10.1016/0016-7185(94)00020-8

Grunert KG (1997) What's in a steak? A cross-cultural study on the quality perception of beef [Article]. Food Qual Prefer 8(3):157–174. https://doi.org/10.1016/s0950-3293(96)00038-9

Hartmann C, Furtwaengler P, Siegrist M (2022) Consumers' evaluation of the environmental friendliness, healthiness and naturalness of meat, meat substitutes, and other protein-rich foods. Food Qual Prefer 97:Article 104486. https://doi.org/10.1016/j.foodqual.2021.104486

Hoek AC, van Boekel MAJS, Voordouw J, Luning PA (2011) Identification of new food alternatives: how do consumers categorize meat and meat substitutes? [Article]. Food Qual Prefer 22(4):371–383. https://doi.org/10.1016/j.foodqual.2011.01.008

Holyoak KJ, Lee HS, Lu H (2010) Analogical and category-based inference: a theoretical integration with Bayesian causal models. J Exp Psychol Gen 139(4):702–727. https://doi.org/10.1037/a0020488

House J (2016) Consumer acceptance of insect-based foods in The Netherlands: academic and commercial implications. Appetite 107:47–58. https://doi.org/10.1016/j.appet.2016.07.023

Innova Marketing Insights (2023) Top ten trends for 2023. https://www.innovamarketinsights.com/trends/top-ten-trends-for-2023-redefining-value-in-a-volatile-world/

Jahn G, Schramm M, Spiller A (2005) The reliability of certification: quality labels as a consumer policy tool. J Consum Policy 28:53–73

Kardes FR, Posavac SS, Cronley ML (2004) Consumer inference: a review of processes, bases, and judgment contexts. J Consum Psychol 14(3):230–256. https://doi.org/10.1207/s15327663jcp1403_6

Loken B (2006) Consumer psychology: categorization, inferences, affect, and persuasion. Annu Rev Psychol 57(1):453–485. https://doi.org/10.1146/annurev.psych.57.102904.190136

Markovina J, Stewart-Knox BJ, Rankin A, Gibney M, de Almeida MDV, Fischer ARH, Kuznesof SA, Poínhos R, Panzone L, Frewer LJ (2015) Food4Me study: validity and reliability of food choice questionnaire in 9 European countries. Food Qual Prefer 45:26–32. https://doi.org/10.1016/j.foodqual.2015.05.002

Michel F, Knaapila A, Hartmann C, Siegrist M (2021) A multi-national comparison of meat eaters' attitudes and expectations for burgers containing beef, pea or algae protein. Food Qual Prefer 91:104195. https://doi.org/10.1016/j.foodqual.2021.104195

Nelson P (1970) Information and consumer behavior. J Polit Econ 78(2):311–329

Olson JC (1978) Inferential belief formation in the cue utilization process. ACR North American Advances

Olthuis L (2023, 14-11-2023). Hoe duurzaam en gezond is de volledig plantaardige 'Terra-lijn' van Albert Heijn? Volkskrant. https://www.volkskrant.nl/beter-leven/hoe-duurzaam-en-gezond-is-de-volledig-plantaardige-terra-lijn-van-albert-heijn~b97df6f7/

Onwezen MC, Reinders MJ, Verain MCD, Snoek HM (2019) The development of a single-item food choice questionnaire. Food Qual Prefer 71:34–45. https://doi.org/10.1016/j.foodqual.2018.05.005

Onwezen MC, Bouwman EP, Reinders MJ, Dagevos H (2021) A systematic review on consumer acceptance of alternative proteins: pulses, algae, insects, plant-based meat alternatives, and cultured meat. Appetite 159:105058. https://doi.org/10.1016/j.appet.2020.105058

Piernas C, Cook B, Stevens R, Stewart C, Hollowell J, Scarborough P, Jebb SA (2021) Estimating the effect of moving meat-free products to the meat aisle on sales of meat and meat-free products: a non-randomised controlled intervention study in a large UK supermarket chain. PLoS Med 18(7):e1003715. https://doi.org/10.1371/journal.pmed.1003715

Pliner P, Hobden K (1992) Development of a scale to measure the trait of food neophobia in humans. Appetite 19(2):105–120. https://doi.org/10.1016/0195-6663(92)90014-W

Post MJ (2012) Cultured meat from stem cells: challenges and prospects. Meat Sci 92(3):297–301

Realini CE, Driver T, Zhang R, Guenther M, Duff S, Craigie CR, Saunders C, Farouk MM (2023) Survey of New Zealand consumer attitudes to consumption of meat and meat alternatives [Article]. Meat Sci 203:Article 109232. https://doi.org/10.1016/j.meatsci.2023.109232

Rogers EM (1962/1995) Diffusion of innovations. Free Press

Ronteltap A, van Trijp JCM, Renes RJ, Frewer LJ (2007) Consumer acceptance of technology-based food innovations: lessons for the future of nutrigenomics. Appetite 49(1):1–17. https://doi.org/10.1016/j.appet.2007.02.002

Ronteltap A, Fischer ARH, Tobi H (2011) Societal response to nanotechnology: converging technologies-converging societal response research? J Nanopart Res 13(10):4399–4410. https://doi.org/10.1007/s11051-011-0473-1

Schatzki TR (1996) Social practices: a Wittgensteinian approach to human activity and the social. Cambridge University Press

Schifferstein HNJ, Kole APW, Mojet J (1999) Asymmetry in the disconfirmation of expectations for natural yogurt. Appetite 32(3):307–329. https://doi.org/10.1006/appe.1998.0208

Scholtmeijer K, van den Broek LAM, Fischer ARH, van Peer A (2023) Potential protein production from lignocellulosic materials using edible mushroom forming fungi [Review]. J Agric Food Chem 71(11):4450–4457. https://doi.org/10.1021/acs.jafc.2c08828

Siddiqui SA, Alvi T, Sameen A, Khan S, Blinov AV, Nagdalian AA, Mehdizadeh M, Adli DN, Onwezen M (2022) Consumer acceptance of alternative proteins: a systematic review of current alternative protein sources and interventions adapted to increase their acceptability [Review]. Sustainability (Switzerland) 14(22):Article 15370. https://doi.org/10.3390/su142215370

Siegrist M, Hartmann C (2020) Consumer acceptance of novel food technologies. Nat Food 1(6):343–350

Steenkamp JBEM (1990) Conceptual model of the quality perception process. J Bus Res 21(4):309–333. https://doi.org/10.1016/0148-2963(90)90019-A

Steptoe A, Pollard TM, Wardle J (1995) Development of a measure of the motives underlying the selection of food: the food choice questionnaire. Appetite 25(3):267–284. https://doi.org/10.1006/appe.1995.0061

Tan HSG, Fischer ARH, van Trijp HCM, Stieger M (2016) Tasty but nasty? Exploring the role of sensory-liking and food appropriateness in the willingness to eat unusual novel foods like insects. Food Qual Prefer 48(Part A):293–302. https://doi.org/10.1016/j.foodqual.2015.11.001

van der Meer M, Fischer ARH, Onwezen MC (2023) Same strategies—different categories: an explorative card-sort study of plant-based proteins comparing omnivores, flexitarians, vegetarians and vegans. Appetite 180:Article 106315. https://doi.org/10.1016/j.appet.2022.106315

van der Meer M, Onwezen MC, Fischer ARH (in preparation). (M)eat more plants: consumer's categorizations, inferences, and acceptance of plant-based proteins

Van der Stricht H, Hung Y, Fischer ARH, Verbeke W (2024) Consumer segments less or more willing to adopt foods with microalgae proteins. Food Qual Prefer 113:105047. https://doi.org/10.1016/j.foodqual.2023.105047

van der Weele C, Driessen C (2019) How normal meat becomes stranger as cultured meat becomes more normal; ambivalence and ambiguity below the surface of behavior. Front Sustain Food Syst 3:Article 69. https://doi.org/10.3389/fsufs.2019.00069

Van Huis A, Van Itterbeeck J, Klunder H, Mertens E, Halloran A, Muir G, Vantomme P (2013) Edible insects: future prospects for food and feed security, vol 171. Food and Agriculture Organization of the United Nations

Venkatesh V, Thong JYL, Xu X (2012) Consumer acceptance and use of information technology: extending the unified theory of acceptance and use of technology. MIS Q Manag Inf Syst 36(1):157–178. https://doi.org/10.2307/41410412

Verain MCD, Snoek HM, Onwezen MC, Reinders MJ, Bouwman EP (2021) Sustainable food choice motives: the development and cross-country validation of the sustainable food choice questionnaire (SUS-FCQ). Food Qual Prefer 93:104267. https://doi.org/10.1016/j.foodqual.2021.104267

Verband Sozialer Wettbewerb eV v TofuTown.com GmbH (TofuTown), Case C-422/16, (European Court of Justice 2017)

Verbeke W, Sans P, Van Loo EJ (2015a) Challenges and prospects for consumer acceptance of cultured meat [Review]. J Integr Agric 14(2):285–294. https://doi.org/10.1016/S2095-3119(14)60884-4

Verbeke W, Spranghers T, De Clercq P, De Smet S, Sas B, Eeckhout M (2015b) Insects in animal feed: acceptance and its determinants among farmers, agriculture sector stakeholders and citizens. Anim Feed Sci Technol 204:72–87. https://doi.org/10.1016/j.anifeedsci.2015.04.001

Vergeer L, Vanderlee L, White CM, Rynard VL, Hammond D (2020) Vegetarianism and other eating practices among youth and young adults in major Canadian cities. Public Health Nutr 23(4):609–619. https://doi.org/10.1017/S136898001900288X

Verneau F, La Barbera F, Kolle S, Amato M, Del Giudice T, Grunert K (2016) The effect of communication and implicit associations on consuming insects: an experiment in Denmark and Italy. Appetite 106:30–36. https://doi.org/10.1016/j.appet.2016.02.006

Villanueva V, Valdés F, Zavala S, Yáñez JM, Valenzuela C (2023) Perception of cat owners on the use of insects as feed ingredients for cats. J Insects Food Feed 9(11):1431–1443. https://doi.org/10.1163/23524588-20220106

Wassmann B, Siegrist M, Hartmann C (2021) Correlates of the willingness to consume insects: a meta-analysis. J Insects Food Feed 7(5):909–922. https://doi.org/10.3920/JIFF2020.0130

Zeithaml VA (1988) Consumer perceptions of price, quality, and value: a means-end model and synthesis of evidence. J Mark 52(3):2–22. https://doi.org/10.2307/1251446

# Index




Ö. P. CAN et al. (eds.), *Food Analogues: Emerging Methods and Challenges*, https://doi.org/10.1007/978-3-031-69872-9

The manufacturer's authorised representative in the EU is Springer Nature Customer Service Centre GmbH, Europaplatz 3, 69115 Heidelberg, Germany. If you have any concerns regarding our products, please contact ProductSafety@springernature.com

Printed and bound by CPI Group (UK) Ltd, Croydon, CR0 4YY
30/09/2026
02225598-0016